8th International Light Symposium: Re-thinking Lighting Design in a Sustainable Future (Light Symposium 2022)

IOP Conference Series: Earth and Environmental Science
Volume 1099

Copenhagen, Denmark
21-23 September 2022

ISBN: 978-1-7138-6443-1
ISSN: 1755-1307

Printed from e-media with permission by:

Curran Associates, Inc.
57 Morehouse Lane
Red Hook, NY 12571

Some format issues inherent in the e-media version may also appear in this print version.

Institute of Physics
Dirac House, Temple Back
Bristol BS1 6BE UK

Phone: 44 1 17 929 7481
Fax: 44 1 17 920 0979

techtracking@iop.org

Additional copies of this publication are available from:

Curran Associates, Inc.
57 Morehouse Lane
Red Hook, NY 12571 USA
Phone: 845-758-0400
Fax: 845-758-2633
Email: curran@proceedings.com
Web: www.proceedings.com

TABLE OF CONTENTS

Light Symposium 2022: Re-Thinking Lighting Design in a Sustainable Future Editorial

Peer Review Statement

Flames of Aegean: The Ec(h)o Screams of Nature: A 3d Printed Luminaire Series According to Sustainability Product and Lighting Design Principles 1
Effie Polydouli, Veroniki Korakidou

Calculation Model of Spaciousness in Rooms with windows-Experimental Procedure for Spaciousness Evaluation Using VR- 11
S Nishihara, M Okada, H Miyake, H Yamaguchi, N Yoshizawa

Reproduction of the Spatial Brightness on a High-Luminance Large Screen Display 22
Y Ishiwata, H Kage, K Harimoto, N Yoshizawa

Strategies for Achieving Circular Economy Goals in the Lighting Industry Through Design for Disassembly-Based Methodologies 32
K S Hickcox, A Smith

Bio-Mimetic Approaches to Kinetic Facades: A Design Proposal for a Light-Responsive Facade Module. 41
Anastasiia Scavée, Georgios Triantafyllidis, George Palamas

Lighting Inequality in an Urban Context: Design Approach and Case Studies 53
Luciana Martinez, Elettra Bordonaro

Ways to Study Changes in Pedestrians' Behaviour in the Artificially Lit Urban Outdoor Environment 65
Dmitrii Ingi, Pramod Bhusal, Paulo Pinho, Marketta Kyttä, Martin Parker

Social Interaction in Local Public Squares After Dark 77
Vkr Hennig, S Fotios, N Gentile, C Sternudd, M Johansson

Light Festivals: Entertainment Or Urban Renewal Tool? 88
M Bhatia

The Effect of Non-Uniform Urban Illumination Upon Pedestrians' Alertness and Sense of Safety 97
R Jedon, A Haans, Y De Kort

True Specs for Dimmed and Colour Tuned LED Luminaires 109
C Dam-Hansen, A Thorseth, D D Corell, A Bay, C Krause, C Kofod

Relationship Between the Gaze Point Movement and Spatial Brightness 120
H Kage, K Harimoto, Y Ishiwata, N Yoshizawa

Implementing a Digital Solution for Architectural Daylight Analysis in BIM Based Projects by Developing a New Plugin 130
Majid Miri, Elmira Ashtari

Describing the Characteristics of Light Field in Architectural Spaces Using Spherical Harmonic Function 141
Kazunori Yanagawa, Ayane Fujihira, Hideki Yamaguchi, Nozomu Yoshizawa

The Minimum Lux Paradox 149
Hilde Noreld Olaisen, Kristin Halldis Bredal

Urban Lighting: An Innovative Solution, Eco-Sustainable and Resistant to Corrosion 159
A Deodati, E Petrachi, G Vendramin, L Cosma, P Bene

Balanced Brightness Levels: Exploring How Lighting Affects Humans' Experiences of Architectural and Social Urban Contexts. 171
Mette Hvass, Ellen Kathrine Hansen

Drivers' Experiences of Presence Sensitive Roadway Lighting Match Experiences of Traditional Road Lighting – a Case Study in Finland 182
H Pihlajaniemi, E Juntunen, A Luusua

Window Geometry and Its Effect on the Experience of Illuminated Spaces – a Study of Three Daylit Architectural Cases 192
N Mathiasen, L Grønlund, A K Frandsen, M Harild

Lean Thinking into the Modular Construction of Industrial Buildings. Identifying the Role of Daylight 202
T Mavridou, L Doulos, N Nanos

The Appearances of Daylight – an Educational Method for Studying Daylight 210
L Grønlund

Creating an Architectural Experience in Terms of Spatial and Lighting Composition 220
Zuzana Tomková

In-Situ Evaluation of High-Performance Glazing Based on Illuminance and Glare 228
Caroline Karmann, Jan Wienold, André Kostro, Pietro Florio, Andreas Schüler, Jean-Louis Scartezzini, Marilyne Andersen

Physiological Effects of White and Coloured LED Lighting for Elderly People 238
Y Oe, Y Miura, Y Akuzawa

The Circadian Impact of Computer Monitors with Different Color Configurations 244
A Huguet-Ferran, D Kántor, S Hernández, B Garrido

Dynamic Lighting and Natural Ventilation in Patient Rooms at New Psychiatry Bispebjerg 255
C Volf, K Martiny, Pm Petersen, D D Corell, C Dam-Hansen

Identifying Nurses' Perception of a Lighting Installation in a Newly Built Hospital 267
Kathrine Schledermann, Thomas Bjørner, Michael Mullins, Torben Hansen

Evaluating an Integrative Lighting Design for Elderly Homes – a Mixed Methods Approach 278
Anna-Kristin Bochnia, Senja Maarit Ruohonen, Mihkel Pajuste, Ellen Kathrine Hansen

Designing Natural Atmosphere in Office Environment Through Daylight Responsive Dynamic Lighting 289
Verda Sigura, Ellen Kathrine Hansen, Henrik Clausen

Colour of Reflected Light: A Determinant of Sustainable Spatial Function 301
M Bhatia

The School Lighting Innovation Dilemma 309
Tove Karlsson, Jonas Kjellander, Åsa Machado, Olle Strandberg, Peter Götlind, Reine Karlsson

A Review of Lighting Research in Educational Spaces 319
S Angelaki, U Besenecker, C B Danielsson

Design Principles for Natural Lighting 330
P Thursfield, R Vd Ven

Analyzing Natural Lighting Conditions from the Perspective of Biophilic Design in Indoor Office Environments 340
Gulsah Dogan Karaman, Ayse Nihan Avci

Shining Light on Colors: Using a Puzzle Game to Teach Additive Color Theory 350
Dovile Perednyte, Katarzyna Ewa Flig, Kristinn Bragi Garðarsson, Malte Elkær Rasmussen, Sara Munk, Simonas Ceponis, Henrik Schønau Fog

Sociological Approach to the Territorialization of Night-Time Atmospheres 361
Houel Nicolas, Potard Saïg

An Examination of Reflected Glare Prediction Method Based on Luminance Distribution 373
M Sagawa, A Maeda, Y Mizokami, M Takashima, Y Oe, N Yoshizawa

Relationship Between Photon Flow and the Perception of Light Field Under Simple Lighting Environments 380
Otori Kaira, Ken Komazawa, Roland Schregle, Nozomu Yoshizawa

Artificial Light(ing) Or Electric Light(ing)? 390
Federico Favero, Ute Besenecker

Challenges in Lighting Education: A Recommended Practice 401
Paulo Sergio Scarazzato, Jéssica Cristine Da Silva Fonseca Matos, Ana Judite Galbiatti Limongi França, Taísa Dóccosse Pavani

Light4Health eLearning Course: Health Research for Interior Lighting Design. Re-Thinking Design Approaches Based on Science 413
Km Zielinska-Dabkowska, L Godley, F Kyriakidou, U C Besenecker, G Triantafyllidis

Envisioning Community Engagement in Smart Lighting Design 424
Leni Schwendinger, Philip R Ross

Urban-Centric Lighting Task Group: Tactical Lighting as an Innovation Strategy 432
A Barberá, B Teixidó

Dark Adaptation in Urban Environments: An Innovative Design Framework for Pedestrian Lighting 442
Chiara Ester Amoruso, Mette Hvam Larsen, Mette Hvass, Georgios Triantafyllidis, Ellen Kathrine Hansen

Amber is Not Yet the White Knight for Sustainable Outdoor Light 454
L C Radetsky

The Moon as a Light Source: New Sustainable Ways of Lighting Up Cities at Night 463
Camilla Monberg Rathsach, Mette Hvass

Optimizing Exterior Lighting Illuminance and Spectrum for Human, Environmental, and Economic Factors. 473
S M Simmons, S Baur, W Gillis, D Burns, H Pickerill

The Response of the Bonn Convention on Migratory Species to Light Pollution 484
Yana Yakushina

Energy-Efficient Lighting and Visually Impaired Users in Homelike Environments 493
P Mattsson

Dimming the Lights to Support the Overall Human Wellbeing in an Urban Context 505
Oktay Akanpinar

Light and Emotions. the Importance of Context........ 513
O Y Vara León

Universal Lighting Design: Towards Sustainable Lighting Solutions that Support Function, Experience and Human Diversity 522
Sl Nielsen, C Ryhl

Effects of Indoor Lighting Conditions and Window Views on Occupants' Well-Being and Behavior: A Systematic Review. 532
Natalia G Vasquez, Ricardo F Rupp, Rune K Andersen, Jørn Toftum

Lighting Design for Diversity: Learning from Low-Vision Rehabilitation 542
T B Øien, A K Frandsen

Light Sketching for Ecology: A Cooperative Design Tool for Balancing Human Experience and Ecological Impact 552
Philip R Ross, Niek Rutten

Author Index

IOP Conf. Series: Earth and Environmental Science 1099 (2022) 011001 doi:10.1088/1755-1315/1099/1/011001

Light Symposium 2022: Re-thinking Lighting Design in a Sustainable Future Editorial

With the greatest challenge of our time – the climate crisis – we begin to understand the need to rethink our ways of living and to maintain the ecosystems we are part of and depend upon. In that perspective, the role of lighting design takes on new aspects – we now understand light as an important tool in balancing the needs of both humans and the planet.

With available technological advances, possibilities are almost endless when working with light. Yet, some things never change - since the dawn of humanity, the sun has governed our bodies, rhythms and visual experience of the world - the more we are shielded in indoor environments, the more we seek a connectedness to nature and natural daylight.

At a time of rapid technological change, we can easily forget the importance of supporting sustainable development from a holistic perspective, considering not only energy consumption, but also sustainable behaviour, human well-being, better biodiversity for animals and plants, lifecycle assessments for products, connectedness, etc. However, we must solve the complex interplay between these aspects, if we are to meet the 2050 goal of reducing the CO^2 emission to net-zero and, at the same time, ensure a more sustainable approach to how we bring light into our environments.

Lighting design is a diverse field involving a wide range of disciplines, including lighting design, architecture and interior design, lighting engineering, virtual and media technology, cognitive and neuroscience, product development, amongst others. This diversity of approach, driven by collective desires for better lighting, certainly presents challenges, but through knowledge-sharing across different scientific fields, we can propose new ways of living with light: ways that consider the needs of people and the planet, and rethink the approach to lighting design in both research and practice.

In this context, Light Symposium 2022 (LS2022), held in Copenhagen Denmark in September 2022, is the 8th in a series of conferences, which have had the objective to bring together researchers, professionals and industry. The aim was to deliver a state-of-the-art outline of how to define sustainable approaches for lighting design to meet the UN sustainable development goals. Representatives from many educational programmes, the lighting industry, professional practice, the Danish Center for Light (DCL) and the International Association of Lighting Designers (IALD) participated in a program of presentations from researchers and practice, discussions, workshops, exhibitions and social activities, with the intention to exchange knowledge about how we together may rethink lighting design into a sustainable future.

The proceedings which follow here comprise all the papers presented at LS2022 and which had been reviewed by an international panel of peer-reviewers.

Georgios Triantafyllidis, Ellen Kathrine Hansen, Mihkel Pajuste and Michael Mullins

LS2022 Chairs

IOP Conf. Series: Earth and Environmental Science 1099 (2022) 011001 doi:10.1088/1755-1315/1099/1/011001

List of Reviewers:

#	first name	last name	country	affiliation
1	Marilyne	Andersen	Switzerland	Ecole Polytechnique Fédérale de Lausanne (EPFL)
2	Stavroula	Angelaki	Sweden	Royal Institute of Technology (KTH)
3	Thanos	Balafoutis	Greece	Hellenic Open University
4	Rigmor C.	Baraas	Norway	University of South-Eastern Norway
5	Pernille	Bech-Larsen	Denmark	Henning Larsen Architects
6	Ute	Besenecker	Sweden	Royal Institute of Technology (KTH)
7	Grega	Bizjak	Slovenia	University of Ljubljana
8	Luis Filipe	dos Santos	Denmark	Aalborg University
9	Lambros T.	Doulos	Greece	Hellenic Open University
10	Marie	Frier Hvejsel	Denmark	Aalborg University
11	Henrique	Galvan Debarba	Denmark	IT University of Copenhagen
12	Niko	Gentile	Sweden	Lund University
13	Natalia	Giraldo Vasquez	Denmark	Technical University of Denmark
14	Louise	Grønlund	Denmark	The Royal Danish Academy
15	Ellen Kathrine	Hansen	Denmark	Aalborg University
16	Lea	Holst Laursen	Denmark	Aalborg University
17	Mette	Hvass	Denmark	Aalborg University Copenhagen
18	Maria	Johansson	Sweden	Lund University
19	Reine	Karlsson	Sweden	InsideLight
20	Tove	Karlsson	Sweden	InsideLight
21	Panagiotis	Kontaxis	Greece	University of West Attica
22	Foteini	Kyriakidou	Sweden	Royal Institute of Technology (KTH)
23	Thorbjörn	Laike	Sweden	Lund University
24	Stine	Louring Nielsen	Denmark	Universal Design Hub - Bevica Foundation
25	Nanet	Mathiasen	Denmark	Aalborg University, Department of the Built Environment
26	Michael	Mullins	Denmark	Aalborg University
27	Rodrigo	Muro	Sweden	Royal Institute of Technology (KTH)
28	Mihkel	Pajuste	Denmark	Aalborg University
29	George	Palamas	Denmark	Aalborg University
30	Catherine	Perez Vega	Germany	Leibniz Institute of Freshwater Ecology and Inland Fisheries
31	Henrika	Pihlajaniemi	Finland	Oulu School of Architecture, University of Oulu
32	Bipin	Rao	Germany	Hochschule Wismar
33	Mimi	Ravn	Denmark	Aalborg University
34	Michael	Rohde	Germany	University of Wismar
35	Mandana	Sarey Khanie	Denmark	Technical University of Denmark
36	Kathrine	Schledermann	Denmark	Aalborg University
37	Georgios	Triantafyllidis	Denmark	Aalborg University
38	Carlo	Volf	Denmark	Region Hovestad
39	Emmanouil	Xylakis	Greece	University of Malta
40	Veronika	Zaikina	Norway	University of South-Eastern Norway
41	Dr. Karolina M.	Zielińska-Dabkowska	Poland	GUT LightLab, Gdansk University of Technology
42	Turid	Øien	Denmark	Aalborg University

Peer Review Statement

All papers published in this volume have been reviewed through processes administered by the Editors. Reviews were conducted by expert referees to the professional and scientific standards expected of a proceedings journal published by IOP Publishing Publishing.

- **Type of peer review:** Double Anonymous
- **Conference submission management system:** EasyChair
- **Number of submissions received:** 97
- **Number of submissions sent for review:** 93
- **Number of submissions accepted:** 55
- **Acceptance Rate (Submissions Accepted / Submissions Received × 100):** 56.7
- **Average number of reviews per paper:** 3
- **Total number of reviewers involved:** 279
- **Contact person for queries:**
  Name: Georgios Triantafyllidis
  Email: gt@create.aau.dk
  Affiliation: Aalborg University

IOP Conf. Series: Earth and Environmental Science 1099 (2022) 012001 doi:10.1088/1755-1315/1099/1/012001

Flames of Aegean: the ec(h)o screams of nature: A 3d printed luminaire series according to sustainability product and lighting design principles

Effie Polydouli, Designer & corresponding author. BA Interior Architect, MA Lighting Design, Vermiou 12, Athens, P.C 15235, Greece
effiepolydouli@gmail.com
Veroniki Korakidou, Adjunct Professor, Hellenic Open University, School of Applied Arts, Postgraduate Course in Lighting Design, Greece

Abstract. This paper documents the design process for the creation of a 3d printed luminaire series based on the principles of sustainable product and lighting design. In order to formulate the elements of what sustainable luminaire design should be, the lighting designer, made a recursion contemporary post-industrial design approaches, and concluded to state-of-the-art current trends in sustainable product and lighting design standards, technologies and materials. As a result of this interesting study of the principles of sustainable product and lighting design, this paper presents the design process step-by-step, as well as the sustainability design standards, which led to the final outcome, a series of original lighting fixtures entitled: *Flames of Aegean*.

1. Introduction

The purpose of this research is to first formulate the basic principles of sustainable product design methods and materials in relationship with the luminaire, as well as to track down contemporary trends and optimum technology uses for the production of green energy and sustainable lighting design solutions. The main research question involved is: what does a product and lighting designer need to know, in order to design and create sustainable luminaires?

The results of this research are offering insight into practical sustainable luminaire design and prototyping in an attempt to offer an example for the application of a set of sustainability principles and standards, as an outcome of the designer's research. These principles and standards are taking into consideration functionality and aesthetic issues, as well as practical design methodologies, reuse of materials, cost effectiveness of energy produced and the ecological impact of the recycling process of utilized lighting technologies. This paper will focus on a set of green and sustainable design principles, formulated as the result of a thesis dissertation for the postgraduate course of MA in Lighting Design at the Hellenic Open University, which concluded to the design and practical implementation of a series of luminaire prototypes entitled: *Flames of Aegean*.

2. What is Green design

In recent years the reference to sustainable design and green design is more and more intense. Climate change, global warming, resource depletion, waste generation are some of the serious effects that have been alarming environmental scientists warning us about the uncertain future of our planet. Although this concern may have intensified, no radical steps have yet been taken by governments to make vital changes and developments in this area [1]. However, there are quite a few examples of designers, such

as Heath Nash, Tomas Gabzdil Libertiny and Stuart Haygarth who are considering these warnings and have become the key proponents of a new set of green-oriented design methods. Heath Nash is an artist from Cape Town who is well known for the way he adopted the process of using garbage as raw material for his works, such as plastic bottles, white paper milk bottles or colourful tin cans which he mostly found in recycling centres. [2] Exceptional was the idea of the design studio, Freedom of creation, to use the method of 3D printing for the creation of Lilly lamp, on 2003. The method of 3d printing gave them the opportunity to produce three-dimensional shapes, which would be impossible to create using molds or any other factory technique, as Hobson mentioned at his article [3].
Green design can best be described as a set of goals for a designer in terms of reducing the use of non-renewable sources, both during the production process and in the final product. A product that follows the principles of green design must contribute to the reduction of environmental impacts during and after the use of products. Green designers aim to raise the consumers' awareness reducing the impact on the environment by introducing to the market, products made only from renewable resources. Innovation is the key characteristic of green designers, who discover themselves new design methodologies, materials and processes as part of their research.

3. Sustainability principles in product and lighting design

Although the main role of the designer has always been to combine aesthetic and functionality principles of product design, contemporary luminaire designers have already started to follow green design trends with regards to materials and methodologies, as well as, sustainable lighting design principles. So, let us have a look at the definition of sustainability principles. As mentioned in the ISO Committee: "*The principles of sustainable design is a set of criteria, which should be integrated for the protection of the environment and the life cycle of a product. The main goal of sustainable design is to anticipate and minimize the negative environmental impact in relation to the production, the use and the disposal of products, while it is important that the product does not lose its usefulness*" [4]. The European Commission also states: "*Sustainable design is part of a global approach consisting of a number of steps and criteria guided by the circular economy and the recycle of maximum natural resources*" [5]. More specifically, in the case of a luminaire design, the product should be based *"on energy saving and the avoidance of light pollution"*. Moreover, the designer should be aware of the product's life cycle, and specifically about the way it is *"absorbed or recycled in relation to the environment"*.
For the design of the series of luminaires, entitled: *Flames of Aegean*, two main principles of green design were followed. The first one was energy saving, by using design software (3D Studio max, Illustrator, Photoshop) leading to the reduction of the energy cost during the production of samples for the prototyping of the luminaires and the minimization of the non-desired prototypes. Second, prototyping was made using a 3d printer with sustainable and biodegradable materials. More specifically, a biodegradable and thermoplastic aliphatic polyester derived from renewable sources (such as corn starch, tapioca roots, starch, or cane) was used, which is a suitable biodegradable material according to the guidelines of sustainable design [6].

4. European standards for sustainability in product and lighting design

4.1 European standards for product design

The research about the basics steps of sustainable design based on the EU Commission, spontaneously created a connection point between sustainable design and the field of industrial design. As a result, the research was extended to the field of industrial design and so presenting the sustainability principles of both product and lighting design and how they can be combined for a "greener" future. Designers now, in order to design a product based on sustainability, must take into account the principles of sustainable design, without losing the uniqueness of their inspiration, following their restrictions. Contrariwise, the new materials and the sustainability principles, add inspiration and even, lead the new aesthetic trend in design. The main criteria for consideration by a designer, in order to integrate sustainable design principles, for the production of green products according to the EU are [6]:

- Consumption of raw materials
- Energy consumption

- Posts in the natural environment and other pollution, climate effects
- Impact on biodiversity

More specifically, some guidelines to be considered by designers during their design and prototyping process include some of the below objectives and principles related to [6]:

- Using less materials and resources to produce products
- Using materials and resources obtained with minimal environmental impact
- Produce as little waste and pollution as possible
- Reducing the ecological impact of the product's distribution
- Facilitating the reuse and recycling with intelligent design that makes disassembly easy

During the process of designing and prototyping the series of luminaires *Flames of Aegean*, the designer considered all the above mentioned sustainability criteria as well as lighting design sustainability guidelines as mentioned in the next section.

4.2 European standards for lighting design

A detailed overview in the individual areas that negatively affect and intensify the problem of climate change, certainly points to energy production and more specifically, to artificial lighting. Although the sun, has always offered light as energy to us, directly and naturally, during the last few decades, humankind has used artificial lighting recklessly. Usually, the first contact with the problem of lighting pollution comes from the realization that the starry sky has already disappeared from most urban areas. The effects of light pollution are evident, both in fauna and flora, as it is also a waste of valuable energy resources. Studies as mentioned by Babak Tafreshi, conclude that light pollution, in addition to wasting and destroying the image of the natural environment, has great consequences on ecosystems, so many birds, insects and other animals are disoriented by the artificial light in the evening sky with deadly consequences [7].

In relation to lighting design and sustainability, Professor Frangiskos Topalis states that *"the revolution in the use of reduced energy is considered as a primary change*" [8]. New technologies with much lower energy consumption replace existing ones with rapid succession over the recent years. Thus, the reduction of the heating rate increases dramatically, a positive fact, which helps a lot in the effort to tackle the environmental crisis. The replacement of incandescent bulbs, initially with CLF fluorescent bulbs and nowadays with LEDs turned out to be effective for the environment. However, all these technological changes in the field of lighting give rise, as Fairs reports" [9], to *"new paths and new combinations of technologies and materials for designers"*. As light bulb technologies evolve, the consumption of energy is becoming less and the intention of designers to come closer to a different design model is intensifying. Thus, the world of lighting design is adapting to a more sustainable mind-set. Therefore, specific standards for lighting designers apply, as stated by the EU guidelines.

More specifically, the EU has adopted many directives and regulations that apply to lighting, some of which are mandatory while some may be subject to modification. However, every luminaire must be CE certified by its manufacturer, which consists of EMC (electromagnetic compatibility) and LVD (low voltage directive). However, in the context of LVD (2014/35 / EU) are usually the standards: EN60598-1, EN60598-2-1, EN62471, EN62493, EN62031. Under the EMC (2014/30 / EU) are usually EN55015, EN61547, EN6100-3-2, EN6100-3-3 (TEE, 2021) Thus according to the T.O.T.E.E 20701-7 / 2021 and the European regulations as a result:

- Eco Design requirements of Energy Consumed Products (ERP), Lab Energy Labeling of Eco-Design Products
- Lab Energy Labeling of Eco-Design Products
- Low Voltage Directive (LDV) The RoHS, 2011/65 / EU
- Electromagnetic Compatibility (EMC)

- Waste Electrical and Electronic Equipment (WEEE)
- Restriction of hazardous Substances (RoHS)
- Energy Service Directive (ESD)
- Registration, Evaluation, Authorization and Restriction of Chemicals

According to the above, it appears that in order to make a holistic approach to the design of a lighting fixture, based on sustainability criteria, the lighting designer should take into account not only the optimum aesthetic and functional outcome but also calculate the ecological cycle of the luminaire. In order to achieve this, it is also necessary to study the suitable materials, the design methods, the waste losses, and the correct use of energy lamps. For the design of sustainable lighting fixtures, a set of specifications should always be followed, according to the design standards as mentioned above.

5. New materials and sustainable design methodology

In order to achieve her main goal and follow the sustainability design principles, the designer conducted a research comparing and analysing the sustainability of the available materials and the different methods of design and production for the prototyping of a modern lighting fixture. Emphasis was given to the guidelines mentioned in sections 4.1. and 4.2., as well as good practices of the use of new biodegradable materials in lighting design by pioneer designers. An inspiring approach is the one of the well-known designer Julia Lohmann. In her project, *Seaweed Lamp*, she used dried seaweed as the main material. The designer's intention was with her work to defend untapped resources, as presented by Patrick Sisson, in his article on the Dwell website [10]. Another inspiring approach is the Krill Design team, with their *Ohmie* table lamp, which is a 3D printed lamp made by orange peels, as presented in her article Alice Finney [11]. Another new material considered very promising is food waste, as not only is it recyclable but also gives a partial solution to the big problem of food waste. So, at the end of this overview, a designer today should ask him/herself: *What is the role of materials in the design and production of a sustainable product?*

5.1 Materials and design

We live in a world surrounded by materials. Materials are what give substance to what we see and touch in our daily lives. As humans, we stand out from the rest of the species, as we have the ability to think, imagine and design objects using materials. These items in combination with the appropriate materials can present an idea, an emotion or a hidden message of each designer or even evoke memories in each user. As Mike Ashby and Kara Johnson mentioned: *"The materials that will be used in a design model are not simple combinations or numbers. In the process of designing an object, it is important for the designer to find the best solutions, and those that are most important to humans. Materials that speak to peoples' soul and stimulate various emotions, while at the same time being a friendly example to the environment. After all, the main goal of most nowadays designers is sustainability and how to minimize the impact on the planet as much as possible"* [12]. Today, no specific material dominates the market, as there are thousands available, and many more can be produced by making combinations of existing ones with each other. There are known materials that rely on their sustainability in their basic ingredients and can either be recycled or reused. Some of these materials are cellulose, ceramics, glass, fibers, flax, bamboo, wood; these all share a big history within material culture, used since ancient times.
If we analyse the basic categorization of sustainable materials the emerging categories would be [13]:

- biodegradable materials
- recyclable materials
- materials that have already been recycled and
- materials that have the lowest rate of emission and generation of toxins and/or waste.

Some examples of new materials that designers use today for the production of various luminaires are recyclable paper and pulp, seaweed, food waste, clothing waste, mushroom mycelium. These examples and the new ways of designing and manufacturing products, is indicative of the research conducted by contemporary designers on the use of sustainable materials and methods which promotes low energy consumption and minimum environmental footprint. The need for all designers to connect their products with sustainability standards did not only foster the experimentation with new materials but also the implementation of alternative, innovative methods and techniques for prototyping and distributing products. The sustainability of materials depends also on their preparation. Specific procedures are followed in order to process any material and, at this point, it is important to mention that this process must follow the sustainability principles too.

The general processing of a material follows specific steps, such as the shaping process, the integration where needed, the refinement of surfaces and the final production of the product. Definitely, this process can affect the aesthetics, ergonomics and perception of the products in relation to their original design. However, they cannot be omitted nor separated from the sustainability standards as mentioned above. In the following paragraph we will analyze the more sustainable ways of processing materials.

5.2 Creating more efficient materials

The use and type of materials used for the production of products is something that directly affects its sustainability, as mentioned above. There are four main incentives for creating more efficient materials, which are in line with the European Directive [6]

- Reducing the depletion of non-renewable sources.
- Reduce the dependence of developed countries on imports.
- Restriction in demand for energy sources.
- Reduction of gas emissions due to mass production.

While the strategies for increasing more efficient materials are being added as:

- Advances in materials technology.
- Materials and more efficient design.
- Use of financial resources.
- Social adaptation - change of thinking and change of the way of life of the citizens themselves in each country.

All the new sustainable materials that the designers of green design discover or create need to go through a process in order to have their final form. This preparation could be simple and quick or on the other hand, time consuming and expensive. So, the designer should have a complete overview of the production of his product and specifically, of the design of a luminaire, as new energy-saving light bulbs should be procured.

6. Design process

6.1 Inspiration

The urgent need for a sustainable everyday life, the shift in the way a product is designed and produced, and the magic of lighting, spontaneously led to the desire to create a series of luminaires based on the principles of green design. Environmental global crisis, gave the impetus in terms of design to research artists and designer who use new ways of expression, new uses or even creation of new materials in order to produce sustainable creations. As mentioned in previous sections, the consumers now need to be introduced to a new, more sustainable way of life and such design products are means to promote and valorise concepts and ideas for a low energy consumption and zero waste future.

All this reckless use of raw materials, the exploitation and destruction of our natural resources, the unnecessary over-consumption, sparked a volcano of despair and cry for protest, which blew the central idea, the family of luminaires named, *Flames of Aegean*. Initially, the idea came out from the sound of the word "explosion". The emotional outburst, not only of those who try to understand and translate the violent abuse against mother nature, but also the anticipated explosive behaviour of nature, in response to her suffering all the consequences of the Anthropocene. In this approach, the metaphor of the screaming volcano as the echo of earth's suffering, represents the unexpected, the explosive despair, including in this scream all that nature cannot express in any other way.

6.2 Main idea-Design

In the initial sketches for the visual output of the luminaires, the sign of the volcano was connected with the geographical history of the Greek volcanoes, and more particularly, the Greek islands Milos, Santorini and Nisyros, which were created in prehistoric times by corresponding volcanic eruptions.

The specific islands are distinguished by their wonderful forms, their unique natural and geological elements and their textures due to volcanic soil, consisting of different layers of colourful rocks and minerals. The original idea was to transfer all these elements, even to the materials of the series of luminaires. Initially, a research was made to study the shapes of the Greek volcanic islands through satellite photos. Subsequently various isolated general forms were created based on these images. The family of the specific lighting fixtures is part of a category of interior, decorative and general lighting for home use.

All the designs for this lighting series were designed, step by step, with the Adobe Illustrator, Photoshop, 3D studio max and Sketch-up, in order to avoid mistakes in the production and follow the sustainability principles. A similar process was followed for the creation of the main material of the lighting fixtures, which was designed at the same programs. The main process started with sketching plastic forms on 3D max from a photo of the form of the natural material named "*slag*". The designer's idea was to simulate the texture of the certain material, which exists in volcanic areas. The texture and the peculiarity of the category of slags are interesting, as they have a porous form with small cavities and thicker walls. However, the desire in order to create the lighting fixtures with this specific material could not be materialised because their production and processing, would demand high temperature and a great deal of energy consumption. As a result, the designer followed a similar pattern and designed the texture on the 3D software, as illustrated on figure 1. The 3D model of the table/ wall lamp is presented on figure 2, 3. From the 3D model was created a 3D printed prototype and then a mold. The sustainable material, which decided for use, would be seaweed, paper pulp and shell concrete combined with chalk. Then, the materials would be placed by the injection process into the mold and thus, the luminaire would take its final shape.

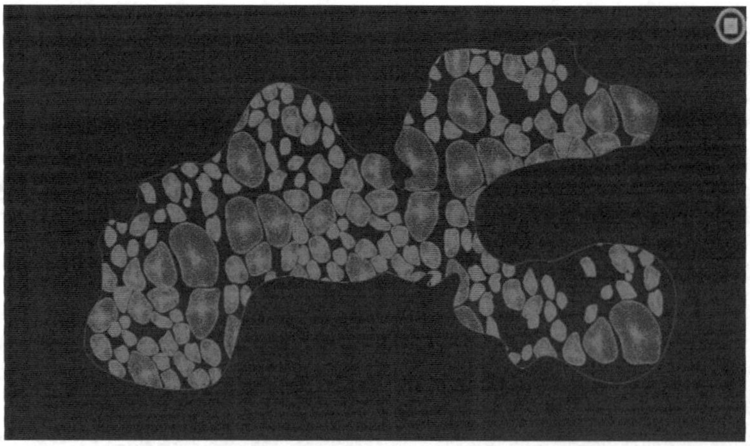

Figure 1. Pattern of the luminaire series Flames of Aegean created through 3D max. Image courtesy of the designer, Effie Polydouli

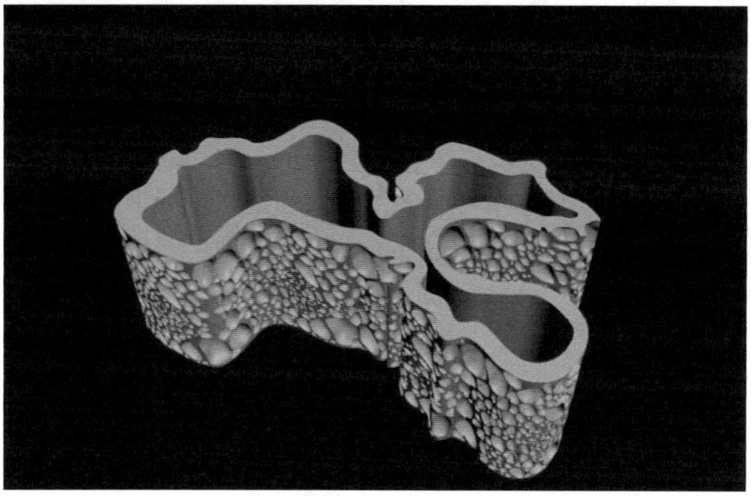

Figure 2. 3d model of the table/ wall lamp, Milos created through 3d max. Image courtesy of the designer, Effie Polydouli

Also, important was the selection of the lighting source that were chosen, as it had to follow the sustainability principles. So, after various efforts the best option was a LED tape with led power supply 3W with low consumption together with the transformer around the perimeter of the mold and just 4cm below the lid, so that it is not visible, figure 4.5.

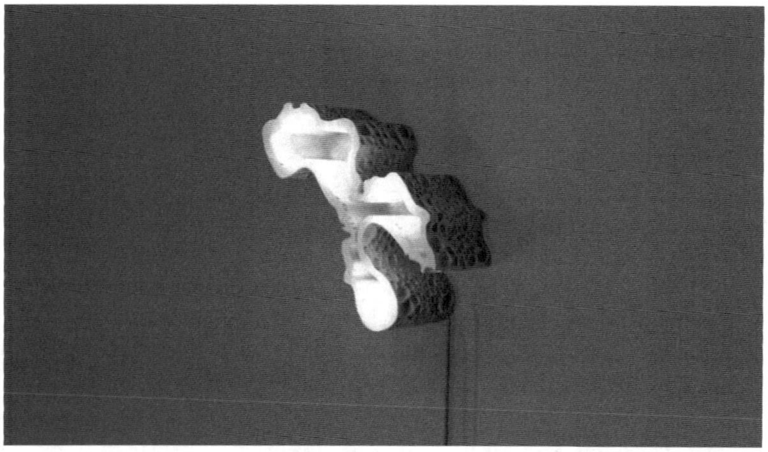

Figure 3. 3d model of the table/ wall lamp, Milos created through 3d max. Image courtesy of the designer, Effie Polydouli

Figure 4. Prototype of the table/ wall lamp, Milos of the luminaire series, Flames of Aegean, created on 3d printer. Image courtesy of the designer, Effie Polydouli

However, 3D printers which could be used for the printing of the luminaire directly on a new sustainable material, such as seaweed, paper pulp or shell cement mortar were not available in Greece at the time of production (July-August 2021). Technical restrictions, such as this, may complicate the work of the designer; however, obstacles create a fertile ground for further research, as sustainable design planning is the only solution to save the planet. There are printers that could print biodegradable materials, under certain conditions and in the future, there could be more models like this available for a researcher-designer who is interested in sustainable product design.

7. Conclusion

In our days, designers, architects and artists around the world are looking for inspiring ways to prevent the catastrophic consequences of artificial lighting and the whole industrial production related to the environment, so with this as a main goal nature is their inspiration in all our designs. Now, they give more importance and consider the countless use of plastics in everyday items and how they could be recycled. Also, they trying to discover and use new sustainable methods that could avoid further damage to the environment. However, the big bet for all designers is to sensitize the consumers and the ordinary art viewers to get closer to nature and moreover to respect her. All, the above concerns sparked the interest of artists and designers to find, create and generate new materials that could create products with the need for low energy and zero waste on the step of production. These new materials have one common goal, the sustainability of the product's life cycle and, subsequently, the protection of the environment.

Following the definition of sustainability and the urgent need for change in the sector of the production of luminaires an investigation was presented throughout this paper, on the basic principles, standards and guidelines for both product and lighting design. A truly sustainable product requires a combination of multiple factors: it could be made from recycled materials, and it has to also be recyclable. It could be made of biodegradable or locally managed materials that have sustainable characteristics. Another important stage is that that the production of these materials and end products has to be completed, in such a way as to reduce emissions of toxins and waste as much as possible.

IOP Conf. Series: Earth and Environmental Science 1099 (2022) 012001 doi:10.1088/1755-1315/1099/1/012001

The main criteria for sustainability in design are, the limited consumption of raw materials and energy, the waste generation rates and other pollutants, the climate impacts and the impact of the whole design and production process upon biodiversity. These are the basic points that a designer should keep in mind today when designing a sustainable lighting fixture. For a holistic approach to the design of a sustainable luminaire, in addition, the ecological cycle of the luminaire must have been calculated, the appropriate materials must have been studied, the way it is made, the waste losses and the use of proper light bulbs. It is crucial for a sustainable lighting fixture to have low energy consumption, the bulb life's cycle should also be considered, as it has to have a recyclability profile and, of course, it has to create the smaller environmental footprint.

In conclusion, the difficulty in accessible technologies, such as 3D printers who can apply biodegradable materials adhering to the principles of sustainable design was reflected in this paper, as shown through the different stages of creation of the family of the lighting fixtures *Flames of Aegean*. The limitations in relation to the new materials and the way of their use was at the same time an obstacle and a challenge for future creation, experimentation and research.

References

[1] M. Martty, "Sourceable," 6 October 2015. [Online]. Available: https://sourceable.net/difference-green-sustainable/. [Accessed 10 August 2021].

[2] H. Nash, "www.heathnash.com," [Online]. Available: http://www.heathnash.com/story.php. [Accessed 10 June 2021].

[3] H. Benedict, "Dezeen," 4 October 2016. [Online]. Available: https://www.dezeen.com/2016/10/04/video-interview-janne-kyttanen-lily-light-3d-printing-movie/.

[4] "Youmatter," 2021. [Online]. Available: https://youmatter.world/en/definition/definition-eco-design-examples-definition/.

[5] E. Commission, "Europian Commission," 2021. [Online]. Available: https://ec.europa.eu/growth/industry/sustainability/product-policy-and-ecodesign_en. [Accessed 24 July 2021].

[6] EE, 10 October 2009. [Online]. Available: https://eur-lex.europa.eu/legal-content/EL/TXT/PDF/?uri=CELEX:32009L0125&from=EN. [Accessed 12 July 2021].

[7] B. Tafreshi, "www.nationalgeographic.com," 3 April 2019. [Online]. Available: https://www.nationalgeographic.com/science/article/nights-are-getting-brighter-earth-paying-the-price-light-pollution-dark-skies. [Accessed 2 june 2021].

[8] L. O. S. K. Fragkiskos Topalis, Phototexnia, Athens: Tziola, 2018.

[9] M. Fairs, Green Design. Creative, sustainable designsfor the twenty-first century, Carlton Books, 2009.

[10] P. Sisson, "Dwell," 4 May 2016. [Online]. Available: https://www.dwell.com/article/laser-cut-lampshade-made-from-seaweed-fc2479ba.

[11] A. Finney, "Dezeen," 19 July 2021. [Online]. Available: https://www.dezeen.com/2021/07/19/ohmie-compostable-lamp-orange-peels/. [Accessed 29 July 2021].

[12] M. A. a. K. Johnson, "Materials profiles.Natural materials," in *Materials and Design*, UK, Butterworth-Heinemann Third edition, 2014, pp. 296-297.

[13] N. Hitti, "Dezeen," 28 August 2020. [Online]. Available: https://www.dezeen.com/2020/08/28/scarlett-yang-biomaterial-dress-central-saint-martins-fashion-design/. [Accessed 12 August 2021].

Calculation model of spaciousness in rooms with windows -Experimental procedure for spaciousness evaluation using VR-

S Nishihara[1], M Okada[1], H Miyake[2], H Yamaguchi[3] and N Yoshizawa[1]

[1] Department of Architecture, Tokyo University of Science, 2641, Yamazaki, Noda-shi, Chiba, 278-8510, Japan
[2]Arup, Iidabashi Grand bloom 8F, 2-10-2, Fujimi, Chiyoda-ku, Tokyo, 102-0071, Japan
[3] National Institute for Land and Infrastructure Management, 1, Tachihara, Tsukuba-shi, Ibaraki, 305-0802, Japan

* Correspondence: yoshizawa@rs.tus.ac.jp

Abstract. Spaciousness is an important psychophysical value of space which has impacts on building occupants' satisfaction and wellness. Previous research revealed through subjective experiments in spaces without windows that spaciousness is affected not only by the physical volume of space but also by lighting environment, and the effect can be estimated with a calculation model with average luminance and "dark-part-reduced visible volume". The purpose of this study is to expand the scope of the calculation model to spaces with windows through subjective experiment, and for that, to improve the experimental procedure for steady comparison of various spaces in size avoiding stimulus range bias. We determined "typical standard conditions" as three sizes of simple rectangular spaces with uniform general lighting to cover experiments which use comparative condition spaces in visible volume of $10m^3$-$2490m^3$. These are created with Virtual Reality to enable subjects to compare spaces apart from each other in reality, and to reduce the cost of preparing various spaces. We conducted a subjective experiment with this improved procedure in a room with windows and revealed that the accuracy of calculation model can be improved by additionally considering configuration factors of windows and applying different weighting factors to the volume corresponding to the windows.

1. Introduction

1.1. Background

Sufficient spaciousness according to the purpose of use and the number of users in the room is important for occupants' physical and mental health. However, considering the shortage of land especially in urban area, unnecessarily huge rooms with large openings must be avoided. Spaciousness can be affected not only by physical volume, but also the shape and proportion of a plan, the height of a ceiling, lighting environment, the size and number of windows, exterior view, interior furniture, and material finish such as colour and specularity. It is possible to achieve more spaciousness by optimizing these factors without increasing the physical volume.

In previous research, Inui et al.[1,2)] revealed from subjective experiments that lighting environment has significant effect on perceived spaciousness and the effect can be estimated from average work plane illuminance, the volume of a room and the configuration factor of windows. Miyake et al.[3,4)] improved the predictability of the model by using average luminance instead of average work plane illuminance and "dark-part-reduced visible volume" (hereinafter called "DRV", refer to figure 2 for details) instead of the volume of a room, but the scope of this model was limited to spaces without windows.

To further discuss spaciousness, it is necessary to clarify its definition. In the *Light. Res. Technol.* (1973), Inui et al.[1)] explained spaciousness as "the feeling that an interior provides of being open or enclosed", but in the *AIJ J. Technol. Des.* (1972), Inui et al.[2)] discussed the definition and characteristics of spaciousness in more detail in Japanese. Miyake et al.[3)] literally translated the definition by Inui et al.[2)] to English as "visual impression of volume of space". We basically follow this discussion by Inui et al.[2)], and here below is the translation of important characteristics described in it.

- "Spaciousness" is the impression of the size of a space that a person receives through visual perception. In short, "visual impression of the volume of a space".
- Spaciousness has the minimum and maximum value. The minimum value is 0 which corresponds to the situation that no space around the observer or complete darkness. The maximum value seems to converge to a certain value even in an open space without any walls.
- Spaciousness values do not describe comfort or discomfort even though there should be a range that is comfortable or acceptable for occupants.
- Spaciousness is a concept that is applicable both for spaces with and without windows. Even when there is no window, spaciousness could be sufficient with some physical volume and brightness.
- In this study, we consider spaciousness as a value that is specific to an observation position and direction. It is possible to define spaciousness of the whole space as an aggregated value of each position, whereas we do not consider it in this research.
- In this study, we limit the scope only to information obtained through visual perception. We do not consider other perceptual information on wind, sound, and so forth.

To investigate the effect of windows on spaciousness, it is necessary to define the difference between "spaciousness" which is a quality of space, and "openness" which is the quality of openings such as windows.

- Openness is the visual impression of openings such as windows to describe how the interior space is connected to the outside through it.
- Openness is the quality of openings and outside view, while spaciousness is the quality of space including both inside and outside. Openness is one of the factors which affects spaciousness.

Miyake et al.[3,4)] conducted subjective experiments in rooms without windows with the magnitude estimation (ME) method to quantify spaciousness of various spaces. They revealed that the effect of the lighting environment on spaciousness can be calculated with average luminance and "dark-part-reduced visible volume" (DRV)". They created the following calculation model using average luminance and DRV based on the experimental result.

$$\mathrm{Log}(Sp) = \mathrm{P} \cdot Sp_{\mathrm{model-1}} + \mathrm{Q} \tag{1}$$

$$Sp_{\mathrm{model-1}} = 0.07 \cdot \mathrm{Log}(L_{\mathrm{average}}) + 0.93\,\mathrm{Log}\,(\mathrm{DRV}) \tag{2}$$

where:

Sp	: Spaciousness (Geometric mean of evaluation values)
L_{average}	: Average luminance(cd/m^2)
DRV	: Dark-part-reduced visible volume(m^3)
P,Q	: Coefficient and constant for each experiment calculated from its result

"Visible volume" is the physical volume of the space which is directly visible from observers, excluding the space behind the observer and obscured by obstacles such as columns and furniture, etc. (figure 1 (a)). It is calculated as the sum of volumes of quadrangular pyramids whose bases are the surface of the room such as walls and columns and apexes are the viewpoint (figure 1, (b)). Visible volume is reduced by obstructions such as furniture and columns. The decrease of visible volume reduces spaciousness. DRV (figure 2 (a)) is calculated by assuming that the visible volume of low-luminance areas decreases in the same way as the presence of obstacles. The reduction rate is calculated from L/L_{average} and a sigmoid function.

There were some points to be improved in the previous study. The first point was that the evaluation values were relative to reference conditions for each experiment because the experiments were conducted with ME method. If the comparative condition was evaluated from different reference conditions for each experiment, it was difficult to compare evaluation values directly and analyze the factors of spaciousness. The second was that, since the experiments were conducted in multiple real spaces, preparing several experimental real spaces was very costly, and it was difficult to conduct experiments in distant spaces. The third was that the calculation model was able to be applied in a windowless space, however, it was not clear whether it could be used in spaces with windows.

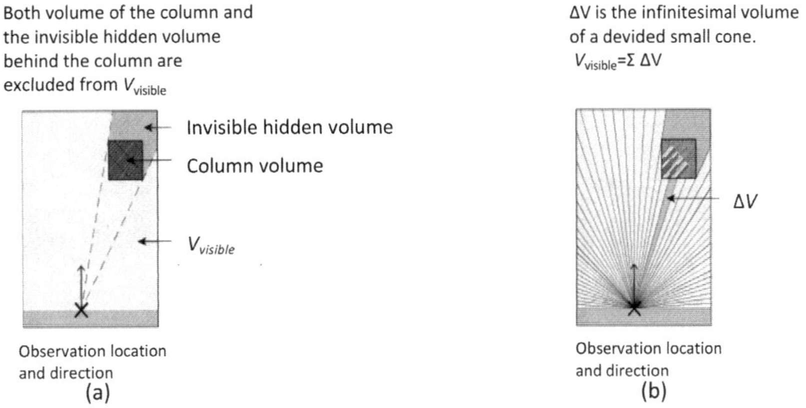

Figure 1. Visible volume.

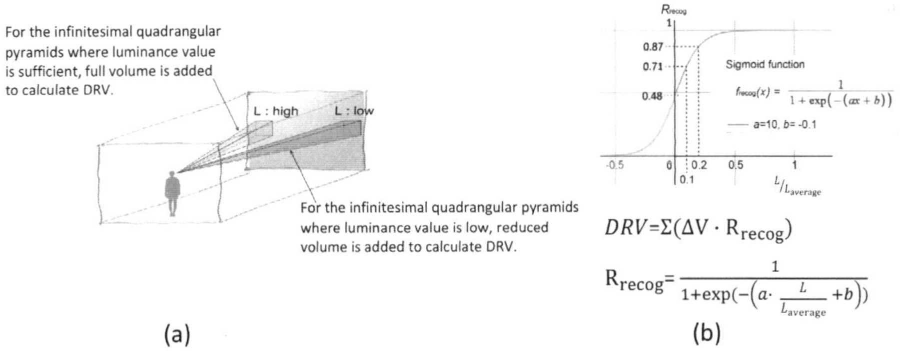

Figure 2. Dark-part-reduced visible volume.

1.2. Research purposes

The final purpose of this research is to improve the calculation model by taking into account the effect of the window, which will be described in section 3. In section 2, the improvement of the experiment procedure for getting the stable and practical data will be explained in advance.

2. Improvement of experimental procedures

In experiments conducted in previous studies[3,4], we found a tendency that the evaluation values vary widely among subjects when the reference condition was far different from the comparative condition in volume. Evaluating spaciousness of various volumes, it was necessary to establish the applicable range of volume for comparative conditions when compared with the reference condition. Preparing several reference conditions and knowing the differences in their spaciousness in advance, evaluation values of experiments using different reference conditions can be intercompared. We named reference conditions with such different volumes as typical standard conditions (hereinafter called "TSCs"). To conduct spaciousness evaluation experiments in spaces with various volumes, TSCs and experimental procedure were established through the following two experiments. The first was the experiment using seven spaces with different volumes to examine the applicable range of volume for comparative conditions when compared with one reference condition. The second was the experiment to clarify differences in spaciousness between TSCs to establish experimental procedure to intercompare spaciousness of spaces with various sizes.

2.1. Establishment of typical standard conditions

To quantify intersubject differences, we conducted subjective experiments to compare seven VR (Virtual Reality) spaces with different sizes. The experiment was conducted in 42 conditions in a round-robin fashion. The VR space was displayed using Oculus Quest, a head-mounted display. Subjects were 19 students in their 20s, and the experimental room was a shaded classroom where the vertical illuminance in front of the eye was kept constant by electric lighting. Specifications of 7 experiment spaces are shown in table 1.

Table 1. Specifications of 7 experiment spaces.

Space	a	b	c	d	e	f	g
Room type	Small space	Living room	Divisible space with divider curtain	Divisible space with no divider curtain	Medium space	Medium lecture room	Auditorium
Dimensions (mm)	W 2,910 D 4,500 H 2,770	W 3,640 D 6,370 H 2,400	W 5,070 D 5,380 H 2,600	W 5,070 D 11,450 H 2,600	W 6,500 D 9,730 H 2,670	W 9,600 D 15,000 H 3,950	W 17,500 D 21,815 H 5,525
$V_{visible}$(m^3)	28.1	39.3	68.1	147.2	160.5	412.7	1395.4
Reflectance (%): Ceiling/Wall/Floor	81/86/30	80/78/20	85/23/11	85/23/11	59/73/17	40/33/10	77/34/6
Reflectance (%): Desks/Chairs	-	18/13	-	-	-	84/13	-/2
Number of Desks / Chairs	-	2/4	-	-	-	46/138	-/203
Lighting Equipment	Dimming system, General Lighting	non-Dimmable General Lighting	Dimming system, General Lighting and Spotlight	Dimming system, General Lighting and Spotlight	non-Dimmable General Lighting	non-Dimmable General Lighting	Dimming system, Downlight
VR rendering							

The focal point of the experiment was asking the subjects to report the perceived spaciousness in ME method. The perceived spaciousness of the reference condition was set to be 100, then the subjects were asked to answer the perceived spaciousness of the comparative condition compared with the reference condition in a positive number. The evaluation flow of the subjective experiment is shown in figure 3.

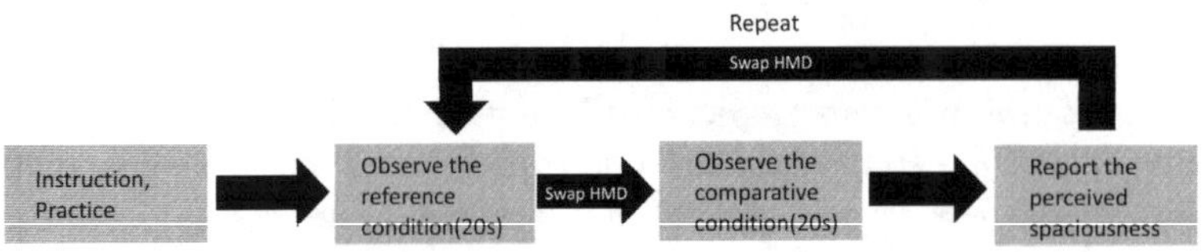

Figure 3. Evaluation flow of subjective experiment.

At the beginning of the experiment, the definition of spaciousness and the experimental procedure were explained to the subject by the experimenter. After that, the subject practiced the experimental procedure. These processes took about 10 minutes, during which time he/she adapted to the light environment of the experimental space. Next, the subjects observed the reference condition for 20 seconds, then he/she observed the comparative condition for 20 secomds and reported the perceived spaciousness.

I In figure 4, the horizontal axis is the volume ratio, and the vertical axis is the intersubject standard deviation of $\mathrm{Log}(Sp)$. It shows that the intersubject standard deviation increases as the volume ratio of the comparative condition to the reference condition increases, indicating that it is inevitable to prepare the reference condition whose volume is not significantly different from the comparative condition. We can see from figure 4 that, to keep the standard deviation below 0.15, the volume ratio should be between 4 and 0.25, therefore, TSC(S) corresponding to 10-155m^3, TSC(M) to 155-622 m^3, and TSC(L) to 622-2490 m^3 were prepared.

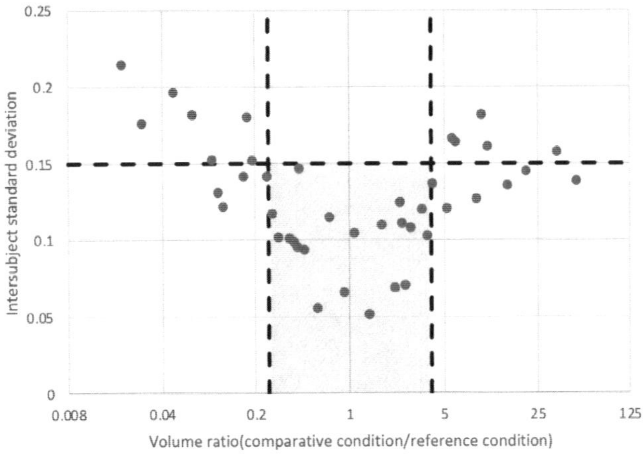

Figure 4. Relationship between volume ratio and intersubject standard deviation of evaluation values.

Table 2 shows the specifications of TSCs. TSCs are cubic spaces no furniture or columns. Vertical illuminance in front of the eye is all 200lx.

Figure 5 shows VR rendering and the plan of TSCs.

Table 2. Specifications of TSC (Typical Standard Conditions).

Typical standard conditions		TSC(S) : Small	TSC(M) : middle	TSC(L) : Large
Dimensions(mm)	Depth	3600	7200	14400
	Width	3600	7200	14400
	Height	3000	3000	3000
$V_{visible}$(m^3)		39	155	622
Vertical illuminance(lx) (at the observation point)		200	200	200
Range of applicable comparative space in $V_{visible}$(m^3)	Max	155	622	2490
	Min	10	39	155
	Recommended	10-78	78-311	310-2490

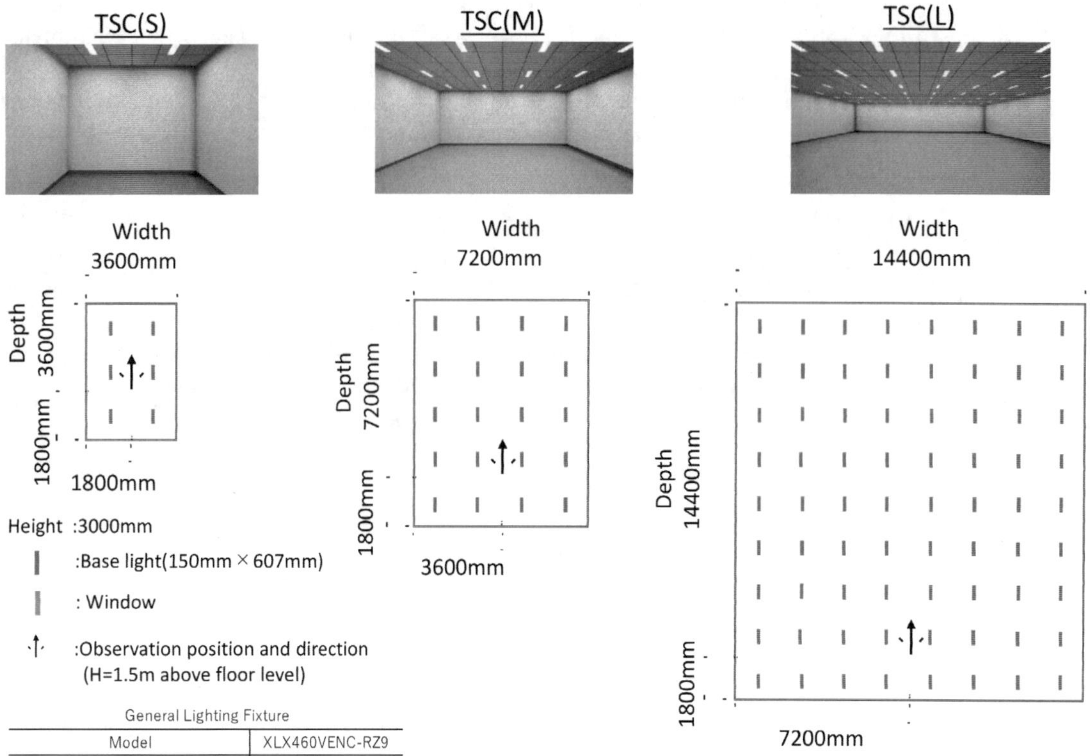

General Lighting Fixture

Model	XLX460VENC-RZ9
CCT	5000K
Luminous Flux per Luminance	3200lm

Figure 5. VR rendering and plans of TSCs (typical standard conditions).

2.2. Experiment to compare typical standard conditions

To evaluate the ratio of spaciousness between TSCs, we conducted experiments with TSC(M) as the reference condition and TSC(S), (L) as the comparative conditions. The TSC(L) was not used in the experiment described in section 3, but it was included in comparative conditions so that the evaluation values can be corrected in future experiments in which TSC(L) would be used as the reference condition. The subjects were 10 students in their 20s. The experiment was conducted in a classroom with all the blinds closed, where the illuminance in front of the eye was kept constant by electric lighting. As shown in table 3, the geometric mean of evaluated values was 57.0 for TSC(S) and 171 for TSC(L) when TSC(M) was the reference condition(=100). They were 1.75 (Log57.0=1.75) and 2.23 (Log171=2.23) in logarithm, respectively. TSC(S) and TSC(L) were evaluated 0.57 and 1.71 times larger than TSC(M). Therefore, the integer value of spaciousness evaluation should be multiplied by 0.57 or 1.71 when the reference condition is TSC(S) or TSC(L), respectively. If the evaluation value is logarithmized it should be plus 0.23 or minus 0.25 for TSC(S) or TSC(L), respectively.

Table 3. Spaciousness of TSC(S) and TSC(L) as TSC(M) is the reference condition.

Reference condition	Comparative condition	Integer evaluation value (Geometric mean)	Correction of integer evaluation value	Log evaluation value (Geometric mean)	Correction of log evaluation value
TSC(M)	TSC(L)	171	×1.71	2.23	+0.23
	TSC(S)	57.0	×0.57	1.75	-0.25

2.3. Comparison procedure using TSCs

The comparison procedure is shown in figure 6. As illustrated in blue, the first comparison is between

TSCs. The ratio was determined from the results of the experiment conducted in 2.2. The second comparison is between each TSC and the comparative conditions, which is done for each experiment. The visible volume of the comparative condition determines which TSCs are compared with. The range of visible volume between the reference condition and the comparative condition is 1/4 to 4 times and the space with the smallest difference between those volumes is used as the reference condition. Multiplying the ratio calculated by this comparison by the ratio between the TSCs, all evaluation values can be corrected to the value with TSC(M) as the reference conditions. This procedure makes it possible to compare indirectly and quantify spaces with volume that cannot be stably evaluated by direct comparison with TSC(M).

For example, in figure 6, the evaluation value of comparative condition A is 80 compared to TSC(S) as the reference. As the evaluation value of TSC(S) is 0.57 times that of TSC(M), the evaluation value of comparative condition A is 45.6(=100×0.57×0.8) relative to TSC(M).

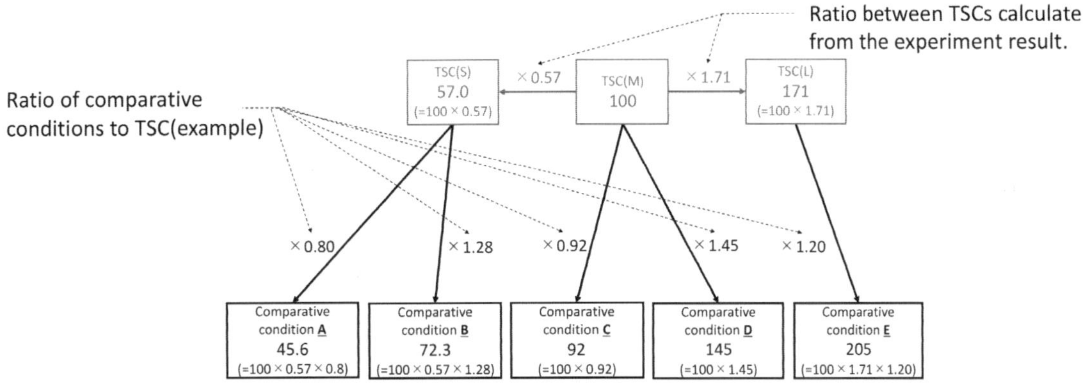

Figure 6. Example of direct and indirect comparison of spaces different in sizes.

3. Improvement of a calculation model for spaciousness in rooms with windows

3.1. Subjective experiment in rooms with windows

To expand the scope of the calculation model to rooms with windows, we conducted a subjective experiment in an experimental room in Building Research Institute, in Tsukuba, Japan. The room was rectangular, 173m^3 in volume, and separable at the center with a divider curtain. Condition A (figure 7(a)) is the whole room without the curtain, and condition B (figure 7(b)) is approximately half of the room with the curtain. Both spaces were 2.6 m in height.

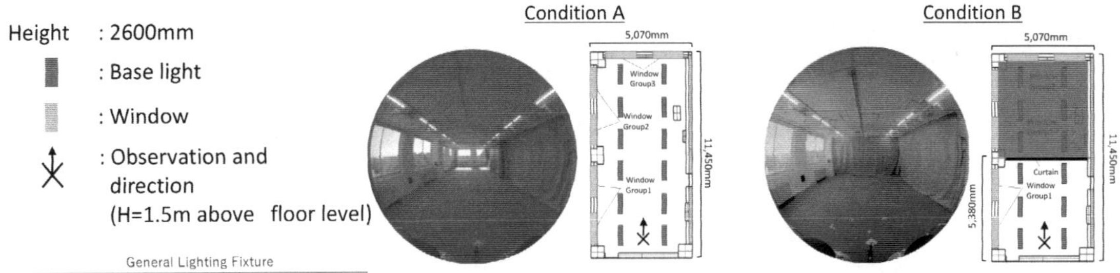

General Lighting Fixture

Model	XLX460VENC-RZ9
CCT	5000K
Luminous Flux per Luminance	6680lm

Figure 7. Specification of an experimental room.

Sky, trees, and buildings were visible through windows. Windows can be hidden with roll screens. We prepared eight patterns of opening/closing of the roll screen on windows and three levels for the intensity of the electric lighting. The subjects were 10 students in their 20s. They were asked to evaluate spaciousness in total of 25 conditions using ME method. If the illuminance in front of the eye fluctuated by more than 10% during a certain condition, this condition was conducted again after all the prepared conditions were evaluated, and if the illuminance still fluctuated by more than 10%, this condition was evaluated on another day. 25 conditions are shown in figure 8. TSC(M) was used as the reference condition for A1-A20 and TSC(S) for B1-B5. A1 and A9, B1 and B3 were similar conditions, whereas A1 and B1 had roll screens in front of the windows. They were included in this experiment for comparison with the results in the previous studies[3,4].

To calculate average luminance and DRV, we obtained luminance and distance information from the view position. A CCD camera was installed at the subject's position and fixed on a tripod. The center of the camera lens was at the subject's eye level, and the direction was the subject's line of sight. A measurement system for luminance and chromaticity distribution, L-CEPT (Luminance & Color Environment Photometric Tool) developed by Yamaguchi et al.[5], was used to obtain the luminance information. As to the distance information, we created a model of the space by using Rhinoceros, and the distance data from the view point to all the points in the scene were extracted by running the rtrace command on Radiance.

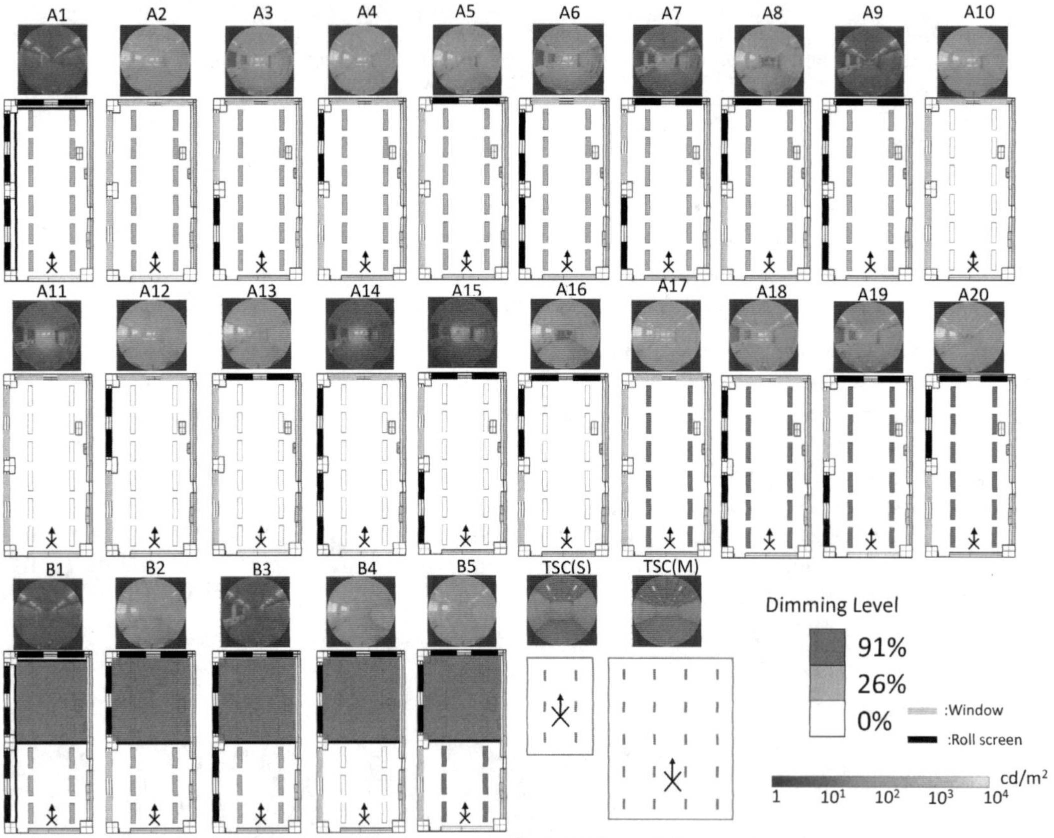

Figure 8. All experimental conditions (25 conditions).

3.2. Expanded calculation model

Figure 9 shows a scatter plot with the logarithmic values of spaciousness evaluation on the vertical axis and the calculated values of $Sp_{\text{model}-1}$ shown in section1.1 on the horizontal axis. The coefficient of determination of the linear regression R^2 is 0.25 and it shows a weak positive correlation. As shown on the left side of the figure, when only the windows near the subjects were open, the calculated value was low compared to the condition in which windows far from the subjects were open. In other words, the condition in which windows near the line of sight were open, the calculated values estimated by $Sp_{\text{model}-1}$ was strongly influenced by windows. Spaces with windows tended to appear more spacious than the ones without windows.

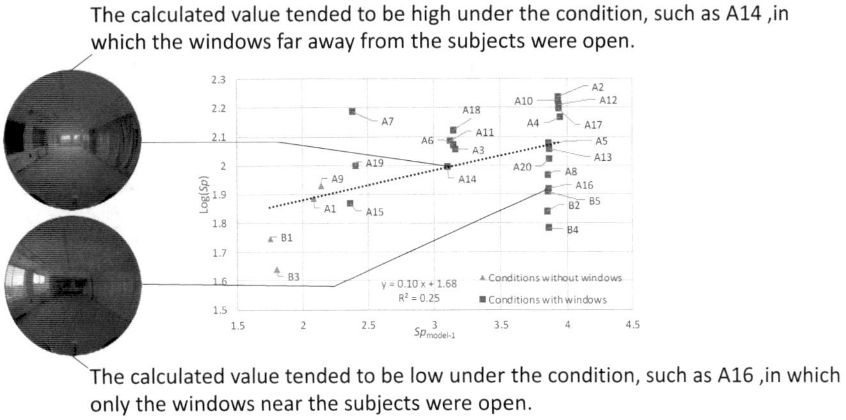

Figure 9. Relationship between $Sp_{\text{model-1}}$ and reported Log(Sp).

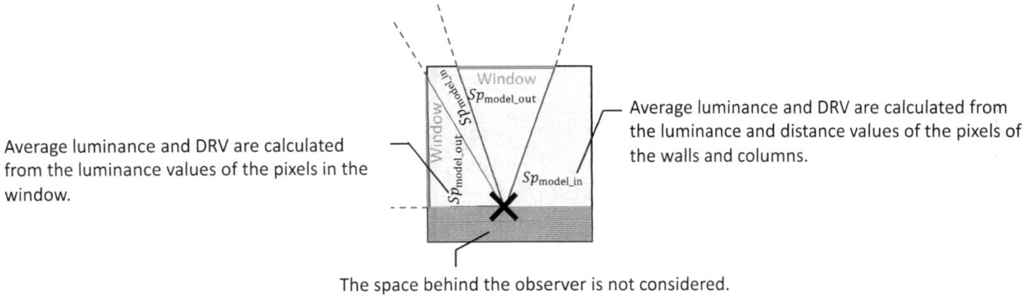

Figure 10. Separation border between $Sp_{\text{model\_in}}$ and $Sp_{\text{model\_out}}$.

As a result, $Sp_{\text{model}-1}$ was not applicable for rooms with windows, therefore we had to do some adjustments to the original calculation model. As shown in the figure 10, we made a border that divided the area. The area from the viewpoint to the window was calculated by $Sp_{\text{model\_in}}$. The rest of the area was calculated by $Sp_{\text{model\_out}}$. In each area, we obtained luminance and distance information, then estimated average luminance and DRV. We introduced the configuration factor to weaken the effect of windows near subjects. And we also used the weighting coefficients for both areas to balance the effects of $Sp_{\text{model\_in}}$ and $Sp_{\text{model\_out}}$. The coefficients were determined to optimize the coefficient of determination of the linear regression between evaluation values and the $Sp_{\text{model}-2}$.

$Sp_{\text{model-2}}$ is as follows:

.

$$Sp_{\text{model-2}} = \frac{Sp_{\text{model\_in}} \cdot U_{\text{in}} \cdot \mathrm{C}_{\text{in}} + Sp_{\text{model\_out}} \cdot U_{\text{out}} \cdot \mathrm{C}_{\text{out}}}{\mathrm{C}_{\text{in}}} \tag{3}$$

Where:

$Sp_{\text{model\_in}}$: Calculated value of spaciousness by $Sp_{\text{model-1}}$ for the area from viewpoint to wall or column, excluding windows.

$Sp_{\text{model\_out}}$: Calculated value of spaciousness by $Sp_{\text{model-1}}$ for the area from viewpoint to windows.

U_{in} : Configuration factor for the area where $Sp_{\text{model\_in}}$ is calculated (=1-U_{out})

U_{out} : Configuration factor for the area where $Sp_{\text{model\_out}}$ is calculated

C_{in} : Coefficient for the area where $Sp_{\text{model\_in}}$ is calculated ($= 1 - \mathrm{C}_{\text{out}}$)

C_{out} : Coefficient for the area where $Sp_{\text{model\_out}}$ is calculated (=0.83, in this case)

Figure 11 shows a scatter plot with the logarithmic values of spaciousness evaluation on the vertical axis and the calculated values of $Sp_{\text{model-2}}$ on the horizontal axis, which shows a strong positive correlation(R^2=0.89).

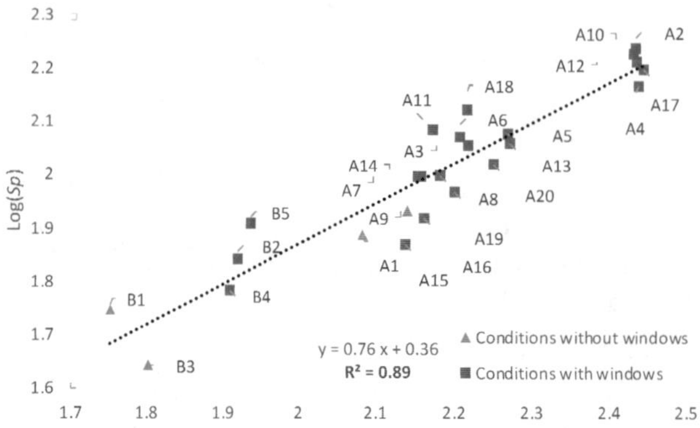

Figure11. Relationship between $Sp_{\text{model-2}}$ and Log(Sp).

4. Conclusion

As shown in section 2, we created typical standard conditions (TSCs) by using virtual reality (VR) and improved the stability of the experimental procedure for spaciousness evaluation. The evaluation values of all comparative conditions were compared directly with TSC(M), or indirectly even if the physical volumes were significantly different. We compared spaciousness of various spaces. The use of VR made it possible to compare spaciousness of various spaces conduct experiments in remote spaces. TSCs in VR also saved us a lot of works in preparing reference spaces for each experiment. With a model of the space and lighting data, average luminance, "dark-part-reduced visible volume (DRV)", and configuration factors can be calculated by simulation. Therefore, spaciousness can estimate quantitatively in the early design phase.

By the results of the subject experiments following the experimental procedure above, it was clarified that the calculated values of the original model tended to be high under the condition that only windows near the subject were open. Moreover evaluation values tended to be higher under the conditions with windows than the ones without windows. We improved the accuracy of the calculation model by adding the configuration factor of windows and multiplying the weighting coefficient for outside and inside effects.

TSCs created in this study can be used for experiments in spaces with a volume of 10-2490 m^3. By creating a set of TSCs with an even larger volume, it is possible to conduct spaciousness evaluation experiments in larger spaces. Because the border between outside and inside was set as shown in figure 10, $Sp_{\text{model}-2}$ ignored the distance information outside the window and took into account only the luminance information of the window. However, the perceived spaciousness would differ under the condition where a large building is seen just outside the window, or the horizon is seen in the distance even if the luminance on the window surface is the same. Further investigation would be necessary on the effects of factors outside on the spaciousness perception. Since the experiment was conducted on a sunny day, the luminance was higher outside than inside for all the conditions with windows in this study. Further experiments should be also conducted under the condition that luminance values inside the space are higher than the ones outside the window.

References

[1] Inui M and Miyata T 1972 Spaciousness in interiors. *Light. Res. Technol.* **5**(2) pp103-111
[2] Inui M Miyata T and Watanabe K 1973 Evaluation of spaciousness (1) *AIJ J. Technol. Des.* **192** pp 49-55
[3] Miyake H, Okada M, Yamamoto T, Yamaguchi H and Yoshizawa N 2021 QUANTIFICATION OF THE EFFECT OF LIGHTING ENVIRONMENT ON SPACIOUSNESS OF INTEIROR SPACE, *Proc. of the CIE MIDTERM MEETING 2021 (Kuala Lumpur),* Oral Presentation

OP48

[4] Miyake H, Yamamoto T, Yamaguchi H and Yoshizawa N 2022 QUANTIFICATION OF SPACIOUSNESS OF INTEIRORS CONSIDERING LIGHTING AND VISUAL ENVIRONMENT (PART 2): CALCULATION MODEL FOR SPACIOUSNESS IN ROOMS WITHOUT OPENINGS USING AVERAGE LUMINANCE AND "DARK-PART-REDUCED VISIBLE VOLUME, *J. Environ. Eng., AIJ,* Vol. **87**, No. 792, pp.62-73

[5] Yamaguchi H, Takasugi K, Sakurai M, Kato M and Hara, N 2019 DEVELOPMENT OF GENERIC COLORIMETRY SYSTEM FOR EVALUATION OF LIGHTING ENVIRONMENT, *Proc. of the CIE SESSION (Washington),* PO154

IOP Conf. Series: Earth and Environmental Science 1099 (2022) 012003 doi:10.1088/1755-1315/1099/1/012003

Reproduction of the spatial brightness on a high-luminance large screen display

Y Ishiwata[1], H Kage[2], K Harimoto[2] and N Yoshizawa[3]

[1]Department of Architecture, Graduate school of Science and Technology, Tokyo University of Science, 2641, Yamazaki, Noda-shi, Chiba Prefecture, 278-8510, Japan
[2]Taisei Corporation Taisei Technology Center, 344-1, Nase-cho, Totsuka-ku, Yokohama-shi, Kanagawa Prefecture 245-0051, Japan
[3]Department of Architecture, Faculty of Science and Technology, Tokyo University of Science, 2641, Yamazaki, Noda-shi, Chiba Prefecture, 278-8510, Japan

* Correspondence: yoshizawa@re.tus.ac.jp

Abstract. The purpose of this study is to verify whether the spatial brightness of the real space can be reproduced using the high-luminance large screen display. Firstly, we conducted the subjective experiment in the real space. Next, the linear scaling images to present on the high-luminance large screen display were made from luminance and chromaticity data measured in the real space. The maximum value presentable by the display was 1,200cd/m^2, therefore, luminance above 1,200cd/m^2 were blown out. Then, the subjective experiment was conducted using those images displayed on the large screen display. The spatial brightness was memorized and answered by adjusting the luminance inside the reference space, called "adjustment box", so that the brightness in the box became the same as the one in the real space. From comparing the logarithmic values of the brightness evaluation of the real space and the one of the linear scaling images on the display, we found that the spatial brightness evaluation on the display roughly matched to the one of the real space. It was also suggested that there is a possibility that the effects of windows on the spatial brightness evaluation may be different between the real space and the two-dimensional display.

1. Introduction

In recent years, high-performance displays have been developed which can reproduce the luminance distribution close to the real space. HDRI (High Dynamic Range Image) is appropriate for recording the luminance of the real-world scene which has a wide range from highlights to shadows. On the other hand, existing display devices have a narrower range of luminance that can be reproduced, and cannot directly display HDRI. A processing method called tone-mapping, which compresses HDRI tones to match the ones of display devices, is still being researched[1]. Pattanaik et al. proposed the tone-mapping operator (TMO) models of visual adaptation using a sigmoid function[2]. Kuang et al. reported the model that reproduces colour appearance considering colour adaptation[3]. The tone-mapping system proposed by Mikami et al. modulates the tonality of HDRI using gaze area information[4]. Choi et al. showed that the TMO yields better performance of the colour rendering by enhancing the contrast to compress the dynamic range and transforming chromatic adaptation that corresponds to the mismatch between the real scene and displayed images[5] Although some pieces of research have verified the performance of the TMOs by comparing tone mapped images displayed on a conventional LDR (Low Dynamic Range) monitor with HDR images on a HDR monitor[6][7][8], it is still necessary to verify whether they can reproduce appearance in the real scene by using HDRI displayed on a HDR monitor[1].

One of the important factors which determines the appearance in the real scene is the spatial brightness, which is the perceived brightness of the entire space. Its mechanism has been studied in many years, and it has become clear that the quality of the lighting environment cannot always be explained by an illuminance on a desk, which is one of the conventional elements for the brightness evaluation, and many studies have presented some models to predict the spatial brightness[9] [10] [11] [12] [13]. However, these models can be utilized mainly for the lighting environment with electric luminaires, and the prediction of spatial brightness under daylighting still needs to be discussed[12]. It is sometimes difficult to evaluate the spatial brightness of the lighting environment with daylighting which changes by the second[11]. Therefore, the evaluation method of spatial brightness using still images, which can always present the same lighting environment, may help to elucidate the mechanism of the spatial brightness.

The purpose of this study is to verify the reproduction of spatial brightness of the real space on the high-luminance large screen display through subjective experiments.

2. Methods: outline of the experiment in the real space

2.1. Experimental room

A subjective experiment in the real space was conducted in a room on the third floor of the Experimental Building in Chiba (35°55'N, 139°54'E), which had south-facing windows. The room was partitioned by a screen and was used as the experimental space shown as a red frame in figure 1. It was 6,400mm in wide, 8,600 mm in depth, 2,900mm in height, and its interior reflectance was 72% on the ceiling, 58% on the walls and the screen, and 7.6% on the floor. The experimental room had dimmable straight tube LED lamps (Panasonic: XLX450AENT RZ9, 5,200lm, 32.9W) on the ceiling.

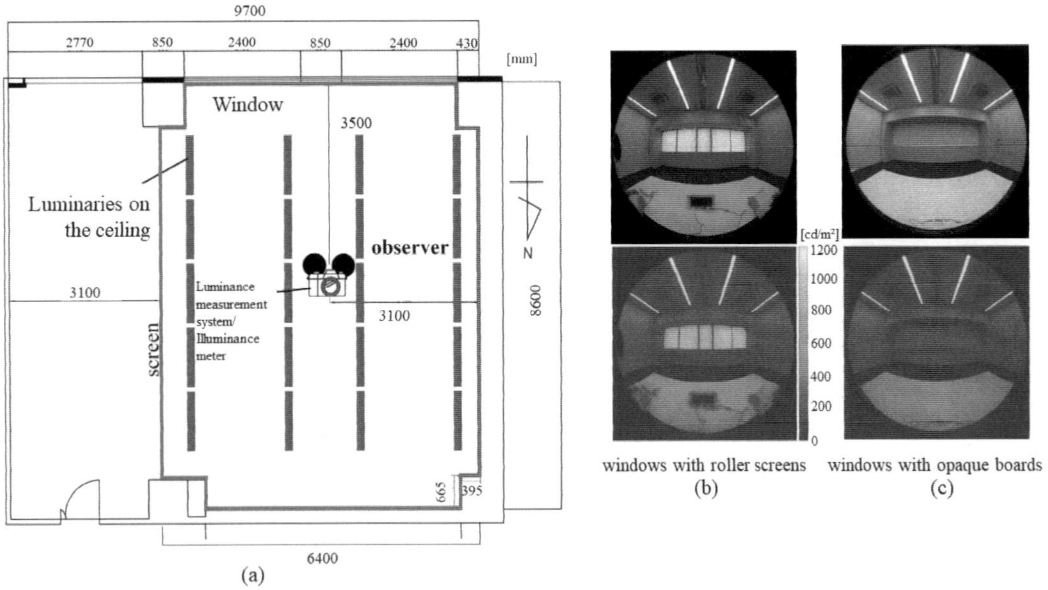

Figure 1. (a)Plan (b) Experimental condition with roller screens on the windows (upper: photograph, bottom: luminance distribution) (c) Experimental condition with opaque boards on the windows.

2.2. Experimental conditions.

Experimental conditions were determined by considering the maximum output luminance at the center position of the high-luminance display, that is 1,200cd/m² (= the performance limitation of this display).

6 conditions had a maximum luminance of less than 1,200cd/m^2 in the field of view and 8 conditions had a maximum luminance of 1,200cd/m^2 or more. 14 conditions in total are shown in table 1.

In conditions A/B/C under 1,200cd/m^2, only electric luminaires behind the viewpoint were turned on to avoid exposing excessively high luminance to observers, and their dimming levels were set to be 25%, 50%, and 100%. The windows were covered with opaque boards which had the same reflectance as the walls in the room. In conditions D/E/F, which had also the maximum luminance under 1,200cd/m^2, roller screens whose transmittance was 36% were placed on the windows for using daylight as the light source and the angle of venetian blinds behind the screens were adjusted so that the average luminance of window was set to the target values indicated in table 1.

In condition G/H, only daylight was used as the light source and the window luminance was set to be 1,200cd/m^2 or more. In condition I/J/K, the dimming level of electric luminaires was set to be 100% and daylighting was also used. In condition L/M/O, the windows were covered with opaque boards and all the electric luminaires were turned on with their dimming levels: 25%, 50%, and 100%.

Table 1. Experimental conditions.

	the maximum luminance	electric luminaires	dimming level	daylighting	window treatment	windows luminance
A	under 1,200cd/m^2	○[a] (luminaires behind the observation position were turned on)	25%	×[b]	opaque boards	-
B			50%			-
C			100%			-
D		×[a]		○[b]	roller screens	around 200~300cd/m^2
E						around 500~600cd/m^2
F						around 1,000cd/m^2
G	above 1,200cd/m^2					1,200cd/m^2
H						over 1,200cd/m^2
I		○[a]	100%	○[b]		around 200~300cd/m^2
J						around 1000cd/m^2
K						over 1,200cd/m^2
L		○[a]	25%	×[b]	opaque boards	-
M			50%			-
O			100%			-

[a] ○ means the electric luminaires were used. × means there were no electric luminaires.
[b] ○ means daylighting were utilized. × means daylighting was not utilized.

2.3. Experimental Procedure

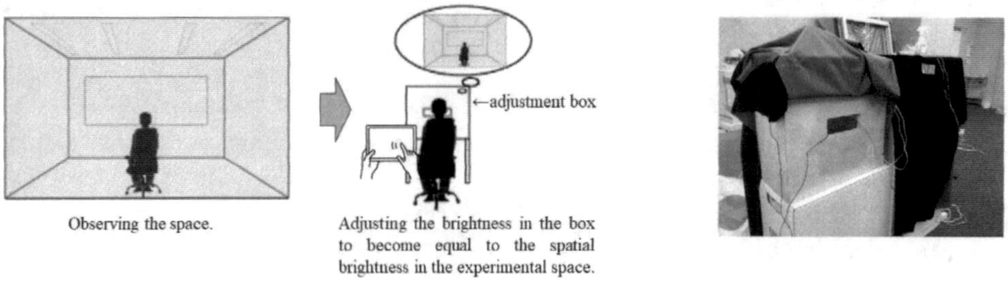

(a) (b)

Figure 2. (a) Experimental procedure and (b) adjustment box.

Since brightness evaluation in the real space and on the large screen display cannot be directly compared to each other, the adjustment method was used for evaluating the spatial brightness with the adjustment box as a reference. Kato et al. have shown that the adjustment method using the reference dome box (= the adjustment box) was determined to be acceptable for the spatial brightness evaluation since correlation was observed between the brightness in the dome box and the one in a rectangular scale model which mimics a room[9]. Firstly, a subject observed the experimental condition for 10 seconds, sitting 3,500 mm away from the windows and facing toward the window, then memorized the spatial brightness. Secondly, he/she moved to the adjustment box and looked inside through a peephole as shown in figure 2. The adjustment box was made of a white hemispherical dome with a diameter of 500mm, inside of which was uniformly illuminated with dimmable line tape LED light (PHILIPS: Hue light ribbon plus smart LED light). Then, the subject moved the slider on a tablet device to adjust the brightness inside the box to become equal to the memorized spatial brightness in the experimental space. Arithmetic average luminance in the box after adjustment by the subject was used as the brightness evaluation value. Average luminance was estimated from the illuminance value measured by an illuminance logger, which was installed under the peephole in the box. It took about 2 minutes to do this process for each condition.

Luminance and chromaticity in the experimental space were measured at the same time as evaluating the spatial brightness, using the luminance measurement system L-CEPT (developed by Building Research Institute in Japan) with DSLR camera (Canon EOS 5D Mark III) and a fish-eye lens (SIGMA 8mm F3.5 EX DG CIRCULAR FISHEYE).

2.4. *Subjects*

26 subjects (male: 12, female: 14) in their early 20s attended the experiment. In conditions A/B/C/L/M/O using electric luminaires, the experiment was conducted for each subject. In condition D/E/F/G/H/I/J/K using daylighting, it was necessary for subjects to experience the same lighting environment, avoiding significant fluctuations under daylight. Therefore, the subjects were divided into three groups and experiments were conducted for each group. Each subject evaluated the spatial brightness sitting at the evaluation position in turn. Vertical illuminances at the eye position toward the window were measured at 5 second intervals using an illuminance meter. If the fluctuation range of the illuminance during the evaluation (about 2 minutes) for each subject exceeded 10% of the average illuminance, the data were removed from the following analysis.

3. Methods: outline of the experiment with the high-luminance large screen display

3.1. *High-luminance large screen display*

Figure 3 shows the image of the high-luminance large screen display and table 2 shows its details. The maximum output luminance at the center of the display was 1,200cd/m^2. Gamut and white point of the display were measured and used when making images to present on the display.

Figure 3. High-luminance large screen display.

Table 2. Display specification.

Size	W:4,864mm/H:2,736mm
The maximum presentable luminance at the center of the display	1,200cd/m^2
Resolution	3,840×2,160
Bit depth	24bit
Gamma value	2.2

3.2. Luminance unevenness on the display.
The luminance at the corner of the display was about 1.2 times higher than the value at the center, even if the uniform image was displayed on it. Therefore, we made a correction filter to adjust the luminance unevenness of the display and applied it to the image files. Figure 4 shows the displayed white images (RGB (255, 255, 255)) with and without the correction filter. We can see that the unevenness of the luminance distribution on the display was reduced after applying the filter.

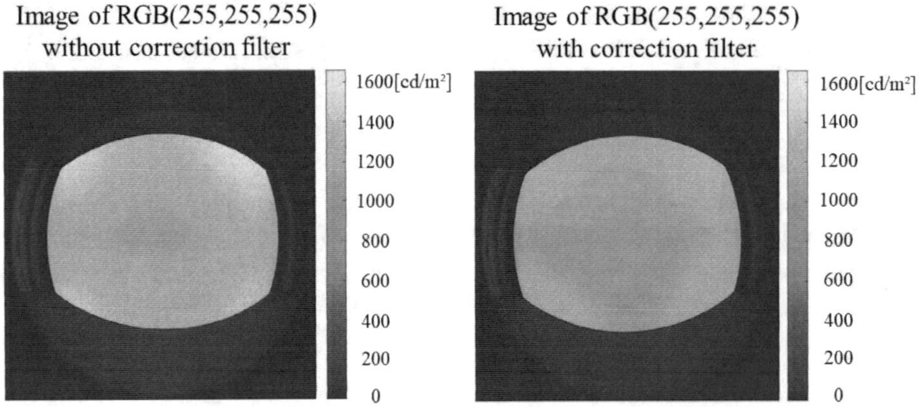

Figure 4. Adjustment of the luminance unevenness.

3.3. Images to present on the display
Figure 5 shows the flowchart of making the images to present on the display. Firstly, luminance and chromaticity images (XYZ data), which were measured in the experimental room with a fisheye lens and SLR camera, were converted from equisolid angle projection to central projection. Secondly, the correction filter made in 3.2 was applied to it. Thirdly, the HDR image (RGBE format) was made from XYZ data using Radiance command: pvalue, with a gamma value of 2.2. Fourthly, using the ra_xyze command in Radiance, pixels with luminance exceeding 1,200cd/m^2, that is the maximum luminance of the display, were blown out. Pixels which were less than 1,200cd/m^2 were tone-mapped by linear scaling with the luminance and chromaticity of the real space, which were based on the measured gamut and white point of the display. These images are hereinafter referred to as "the linear scaling images".

Figure 6 shows the relationship between the average luminance of linear scaling images on the display and that of the experimental room in the same area. The left graph is on the average luminance of the whole displayed area in the conditions of which maximum luminance was under 1,200cd/m^2, whereas the right graph shows the average luminance of the inside space excluding windows and ceiling lightings in all conditions. We can see that the average luminance was almost the same in all conditions, indicating that the linear scaling images were able to reproduce luminance of the real space.

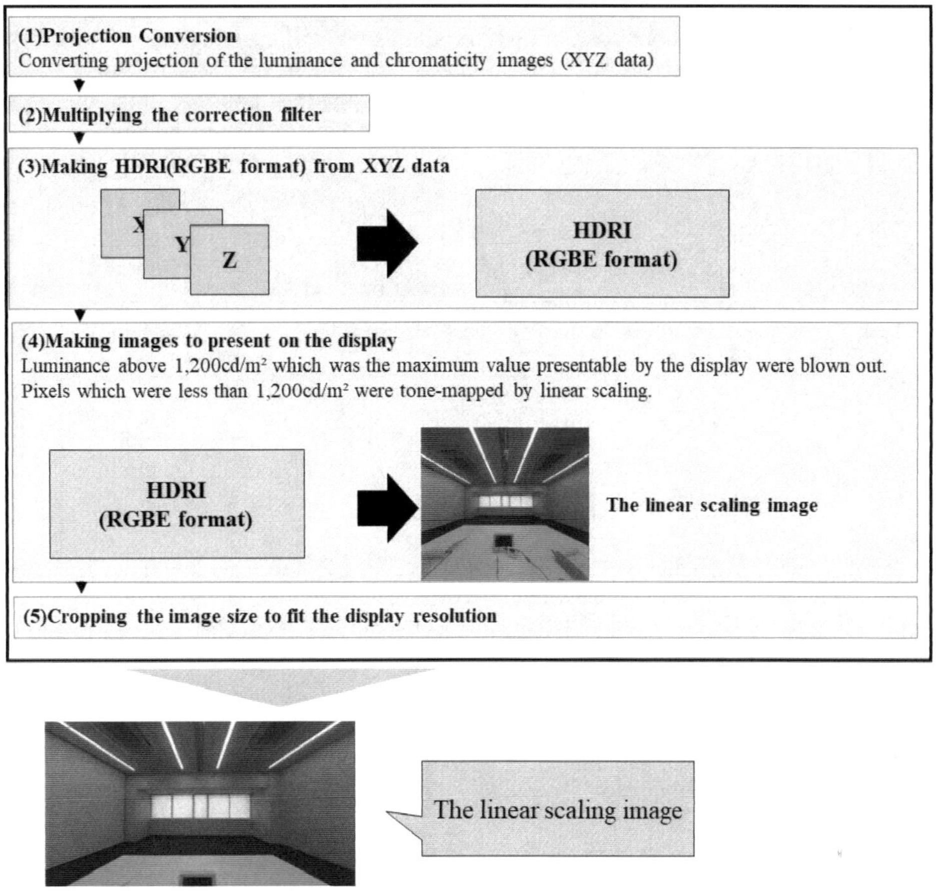

Figure 5. Flowchart for making the linear scaling images.

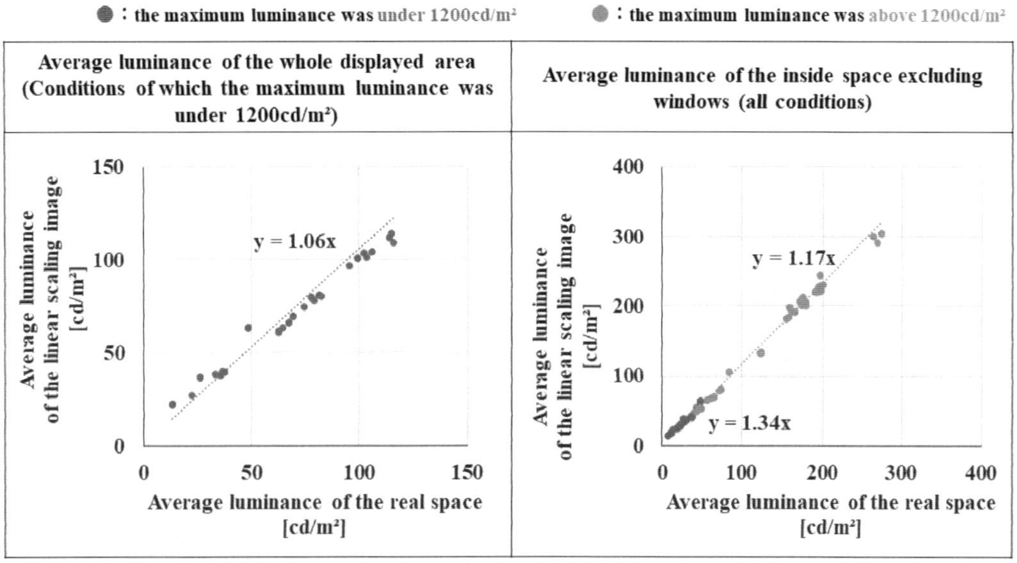

Figure 6. Comparing average luminance of the linear scaling images on the display and of the same area in the real space.

3.4. Outline of the experiment

The subjective experiment was conducted presenting the linear scaling images on the high-luminance large screen display. Figure 7 shows the evaluation position which was located from 1,000mm away from the screen so that the viewing angles of the objects and the room on the screen were identical to those when observing the real experimental room. The same subjects (26 students) also participated this experiment. 14 conditions were conducted with the evaluation method which was the same as in the real space.

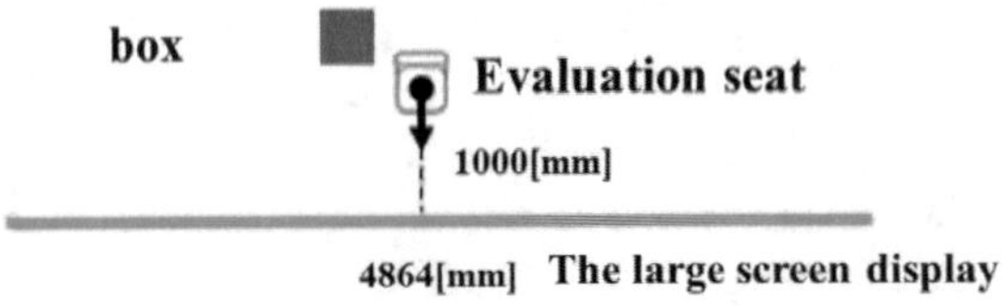

Figure 7. Evaluation position.

4. Result

4.1. Brightness evaluation of the experimental room and the linear scaling image

Average values of the brightness evaluation among subjects were calculated as follows: firstly, subjects were divided into groups based on the experimental conditions of light sources. Then, the brightness evaluation data in the groups, whose corresponding measured luminance were close, were divided again into sub-groups and calculated the average values of them.

Figure 8 shows the relationship between the average brightness evaluation values of the experimental room and those of the linear scaling images. Whether the maximum luminance of the condition exceeded 1,200cd/m² or not, the brightness evaluation on the high-luminance display was almost equal to that in the real space.

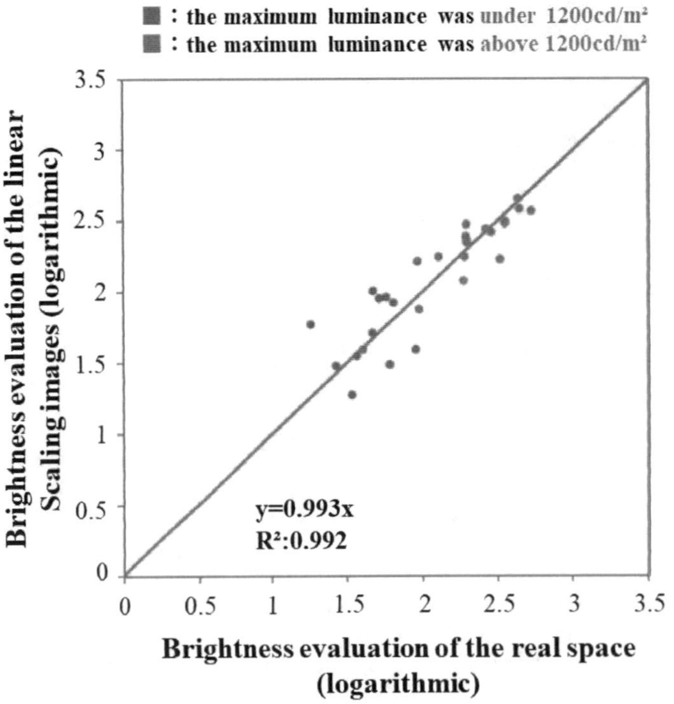

Figure 8. Brightness evaluation of the real space and the linear scaling images.

When evaluating the spatial brightness, some subjects might evaluate the brightness of the inside space excluding windows, not of all the space including the windows. Figure 9 shows the relationship between the brightness evaluation and the average luminance including/excluding the windows. When evaluating the brightness of the real space, the coefficient of determination R^2 was higher for the average luminance excluding the windows, while in the linear scaling images, the average luminance including the windows on the display explains the spatial brightness evaluation better. It indicates that the spatial regions affecting the brightness evaluation differ between the experimental room and the display.

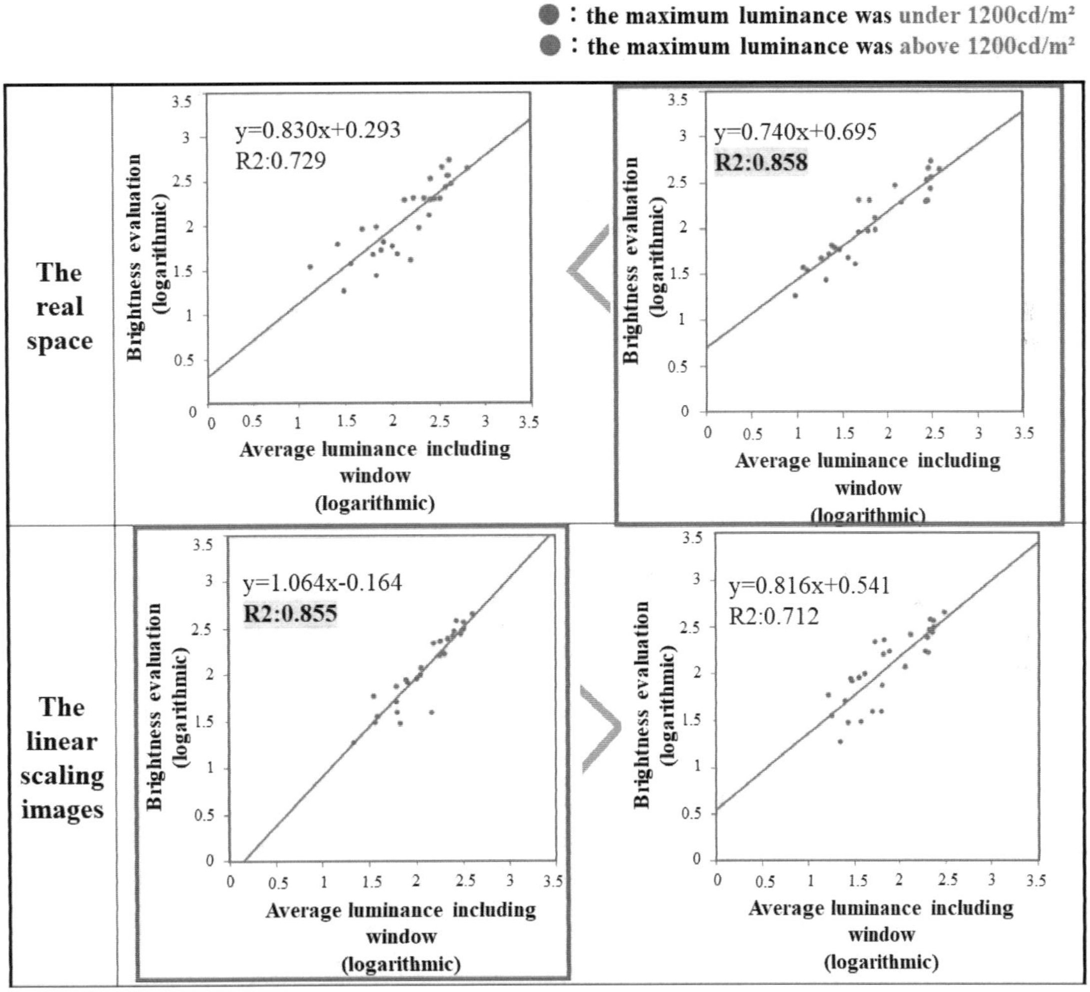

Figure 9. Coefficient of determination: the average luminance including windows (left) and the average luminance excluding windows (right).

4.2. Comparing the brightness evaluation by each subject

We verified the relationship between the spatial brightness and the average luminance including/excluding windows for each subject. The top graph in figure 101 shows the coefficient of determination of the linear regression between the brightness evaluation value in the real space and the average luminance including windows (blue bars) or excluding windows (orange bars). The bottom one is for the linear scaling images. We found that, in the real space evaluation, more than 2/3 of the total subjects had higher values of R^2 on the average luminance excluding windows, whereas in the evaluation using the linear scaling images on the display, more than 2/3 of the subjects' evaluation on spatial brightness could be better explained by the average luminance including windows.

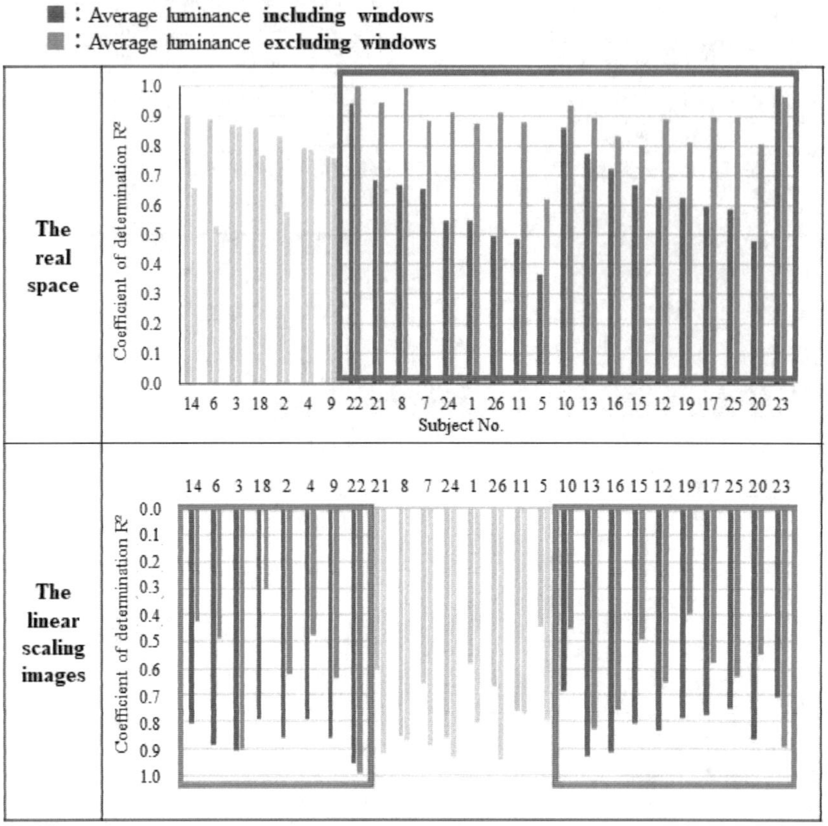

Figure 10. Coefficient of determination in the linear regression for each subject.

5. Conclusion

To verify the reproduction of spatial brightness of the real space on the high-luminance large screen display, subject experiments were conducted using the adjustment method. The experimental results show that brightness evaluation on the high-luminance display was almost equal to the one in the real space. On the other hand, when verifying the influence of the luminance distribution in the space on the brightness evaluation, there is an interesting possibility that the effects of windows on the spatial brightness evaluation may be different between the real space and the two-dimensional display. One reason might be that the experimental room had a simple configuration of lightness, making it difficult for subjects to perceive a three-dimensional effect especially when it is displayed on the two-dimensional monitor. In future work, we will further verify the reproduction of the spatial brightness in various rooms with more complicated lighting environment. In addition, it is necessary to verify the tone-mapping operator (TMO) for the high luminance levels outside the displayable range and for the low luminance levels which were sometimes difficult to reproduce on current displays.

6. References

[1] Reinhard E, Ward G, Pattanaik S, Debevec P, Heidrich W and Myszkowski K 2010 High Dynamic Range Imaging, 2nd Edition Acquisition, Display, and Image-Based Lighting pp 145-204 and pp 233-76
[2] Pattanaik S N, Tumblin J, Yee H and Greenberg D P 2000 Time-dependent visual adaptation for

fast realistic image display *Proc. Of the SIGGRAPH'00* p 47
[3] Kuang J, Johnson G M and Fairchilld M D 2007 iCAM06: A refined image appearance model for HDR image rendering *J. Vis. Commun. Image R.* **18** pp 406-414
[4] Yamauchi T, Mikami T, Miyata K, Nakaguchi T and Tsumura N 2012 Evaluation of Tone Mapping System using Gaze Area Information *J. Soc. Photogr. Sci. Technol. Jpn.* **75** No.1 pp 87-96
[5] Choi H, Kang H and Yun B July 2017 A Perceptual Tone Mapping HDR Images Using Tone Mapping Operator and Chromatic Adaptation Transform *J. Imaging Sci. Technol.* **61** No.4
[6] Newsham G R, Seetzen H, Veitch J A, Chaudhuri A and Whitehead L A May 22-25 2002 Lighting quality evaluations using images on a high dynamic range display *Proc. of the ARCC/EAAE Conf. on Architectural Research* Montreal pp 1-9
[7] Yoshida A 2008 December *Evaluation and Enhancement of HDR Image Appearance on Displays of Varying Dynamic Range* PhD Thesis Universität des Saarlandes(Saarbrücken)
[8] Yoshida A, Mantiuk R, Myszkowski K and Seidel H P Sep 2006 Analysis of Reproducing Real-World Appearance on Displays of Varying Dynamic Range *EUROGRAPHICS* **25** No.3 pp 415-426
[9] Kato M, Numajiri M, Yamaguchi H, Iwai W, Sakata K, Suzuki N, Hara N and Yoshizawa N Dec 2019 Study on applicability of mean luminance by image photometry as spatial brightness index *J. Environ. Eng. AIJ* **84** No.766 pp 1059-66
[10] Ko B, Koga T, Lu B, Hirata K, Mizuno M and Suzuki N 2013 Brightness of a space in terms of variation of luminance *J. Illum. Engng. Inst. Jpn* **97** No.8A pp 429-435
[11] Yonekura Y, Yoshizawa N, Tamura H, Kage H and Harimoto K Jun 14-22 2019 Validation of the spatial brightness estimation formula in offices with windows *Proc. of the 29th CIE session.* Washington D.C. pp 1561-8
[12] Yamaguchi H, Maruyama T and Shinoda H 2014 Space brightness evaluation for day-lit offices *J.Illum. Int. Jpn.* **98** No.11 pp 593-9
[13] Sakata K, Nakamura Y, Yoshizawa N and Takeda H Feb 2017 Spatial Brightness prediction model based on variance of contrast image *J.Environ. Eng. AIJ.* **82** No.732 pp 129-38

IOP Conf. Series: Earth and Environmental Science 1099 (2022) 012004 doi:10.1088/1755-1315/1099/1/012004

Strategies for achieving circular economy goals in the lighting industry through design for disassembly- based methodologies

K S Hickcox[1] **and A Smith**[2]

[1] Lighting Research Scientist, Pacific Northwest National Laboratory, Richland, WA 99354, USA

[2] Vice President of Technology and R&D, Finelite, Inc., Union City, CA, 94587, USA

E-mail: kathryn.hickcox@pnnl.gov [1]
E-mail: aaron.smith@finelite.com

Abstract. The significance of circular economy and its integration into core strategies for smart, sustainable, and inclusive growth, is increasingly being recognized by individuals, organisations, and nations in recent times. The term 'Design for Disassembly' may conjure up images about the end-of-life of a product or its disposal. However, design for disassembly connects to numerous circularity and sustainability goals over the lifetime of a product or project and can have a large impact on the carbon footprint of the product. 'Design for Disassembly' is a product development methodology which is in line with the vision of a circular economy, and supports increase in material efficiency, extends product lifetimes and improves recycling efficiency. Reduction of the imposed environmental risks and impact on the climate, by utilising circularity approaches such as remanufacturing, are tied to lowering of carbon footprint. This paper describes specific and actionable approaches that can be applied by luminaire manufacturers, specifiers, and other players in the lighting industry. The readers will learn about current tools and methodologies that can be used to improve iterative design, as well as measure, assess, and compare products or materials.

1. Introduction

In manufacturing and production, as in the specification of luminaires and lighting systems for architecture, there are numerous barriers which impede recovery, reuse, or remanufacture of parts and components. The major challenge for the lighting industry involves changing structures, practices, and economy related to lighting to transform a linear pathway to a circular one.

Currently, the practice within the lighting industry regarding the assembly and specification of lighting systems is unidirectional, with the end-goal being the installation and use of the lighting system, rather than the end-of-life of the system. This linear view of lighting production and use, limits innovation and implementation of strategies that could support a more circular approach to the built environment and the role of lighting therein.

This paper argues that it is critical and beneficial for manufacturers, specifiers, and all players within the lighting industry, to consider the disassembly process as well as the end-of-life (EoL) and potential reuse and recycling of components and materials, as early as the initial design stages.

One of the authors of this paper, has a sign on their desk reading "Start small, think big, get going", a great inspiration to remember every day! For the purposes of this paper, we might modify that phrase

[1] Author to whom any correspondence should be addressed

to read, “Start small, think circular, get going”. The greatest hurdle will be to change our way of thinking to design sustainable products with a cyclical project life cycle.

2. Challenge: How can lighting practitioners improve the sustainability of their designs?

2.1. Sustainable Design

Sustainability means different things to different people or industries. Within the lighting industry, sustainable lighting practices should ideally take into consideration the visual and non-visual needs of the people and should equitably benefit both society and the natural environment. In the past, sustainability approaches within the lighting industry have been tied to energy conservation, material transparency (communicating implicit impacts of components like embodied carbon, toxic ingredients or byproducts, and social responsibility in the supply chain), and light pollution (light trespass, sky glow, or impacts of lighting on wildlife and ecology). In response to more rigorous requirements, standards, and increased awareness in the lighting industry, these approaches have broadened to include other climate, or social and equitability impacts, such as reduction of raw materials extraction, circular design, and energy equity and justice [1].

In addition to the complex aesthetic, technical, and health-related lighting requirements in any project or product, lighting systems should encompass the core sustainability concept of the “triple bottom line”, which aims to address economic, environmental, and social impacts [2]. Lighting can have significant impact on economy, society, and the environment, at all stages of the life cycle. Lighting practitioners should commit to measuring their impact based on this triple bottom line in all stages of the lighting product or project.

The life cycle of a lighting system has many stages or phases (see Figure 1), starting with design and concept, extraction of raw materials, and moving all the way through the EoL and potential recycle, reuse, or remanufacture [1] [3] [4] . Past research and life cycle assessments (LCAs) have shown that the direct electricity consumed by the luminaire during the use stage of the luminaire life cycle has the largest environmental impact, reaching levels as high as 90% more impacts than other stages (in some life cycle categories) [5]. In order to reduce the large impact in the use stage, the lighting industry has responded by focusing sustainability efforts mainly on reduction of energy use. The transition from traditional technologies to light-emitting diodes (LEDs) over the past decades has had a tremendous impact on lighting efficacy and efficiency, lowering the operational carbon impact from lighting products and systems. This has created a somewhat unique situation in the lighting industry, as adoption of sustainable approaches related to other life cycle stages has had a slower rate. As reliance on fossil fuels has diminished, and as increased energy efficiency in lighting systems has decreased the overall operational carbon impact, the relative embodied carbon impacts are now an important focus.

Embodied carbon impacts can be seen at all life cycle stages of a luminaire. Based on previous research data and LCAs of luminaires and lighting systems [5] [3], we know that most of the total embodied carbon of a luminaire is released during the initial stages of its life cycle, such as production and manufacturing. Once this embodied carbon has been released during the initial stages, there are no further opportunities to recover it using modifications in later stages of the life cycle. Most of the environmental impact of a product is determined in its design stage [6], so any innovations or design decisions that help to limit the release of embodied carbon in the early stages of the life cycle are critical to reducing its environmental impact.

The worldwide increase in construction and development has contributed to an increased demand for luminaires and lighting systems. To avoid an irreversible and catastrophic climate change trajectory, average global temperatures need to be rigorously regulated to be maintained <2°C above pre-industrial levels [7]. Actions taken to reduce average global temperatures will support the reduction of risks and impacts from climate change. The lighting industry can act to limit global temperatures by reducing both operational and embodied carbon footprints. Design for Disassembly (DfD) is a framework to help lighting professionals address both economic and environmental concerns right from the initiation and designing through the entire lifespan of a lighting project or product. The DfD frameworks can help lighting practitioners optimize products and projects to reduce carbon footprints, improve efficiency and material health (by utilizing environmentally preferable materials), facilitate responsible EoL outcomes to transit into a circular economy for beneficial societal outcomes.

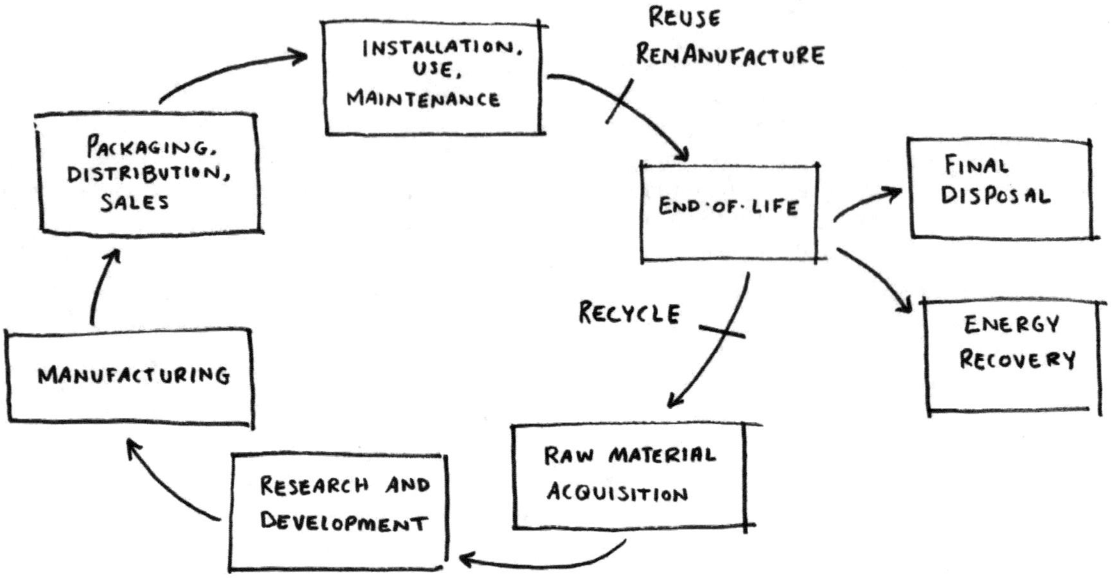

Figure 1. The different stages of the life cycle of a lighting product or system.

2.2. *Circular design*

The lighting industry traditionally follows a linear economic model, which is sometimes described as a "take-make-use-waste" model [8]. Materials travel in a linear fashion through the product and project life cycle, starting from resource extraction to use and eventual disposal. This common model assumes that the Earth has unlimited resources to extract; however, this is not the case. Increasing building stock imposes severe stress on the natural processes which renew some resources, while some resources that are non-renewable get permanently depleted. This model works in purely economic (not environmental) terms, as the use of scarce or hazardous resources is not always correlated to pricing and may provide an economical approach for producing luminaires and lighting systems [8]. In order to achieve a more advanced economic model that can simultaneously reduce embodied carbon components and reliance on finite raw materials, the industry will need to do more than simply recycle materials at the end of the product lifetime. The objective should be to adopt a circular economy model.

A circular economy model is a closed loop system that aims to reduce raw resource inputs, pollution, and carbon emissions. This type of model aims to eliminate waste and maximize the reuse, repair, and remanufacture of components. A circular economy model is designed to be restorative and regenerative. A transition from a linear economy to a circular economy can be seen as an opportunity to motivate economic competition by increasing sustainability and resource efficiency. The transition to a circular economy is an opportunity to generate new and sustainable competitive advantages. The European Union (EU) has recently recognized the finite nature of the Earth's resources by beginning the transition from a linear economy to a circular economy. The EU action plan for the Circular Economy claims, "The circular economy will boost the EU's competitiveness by protecting businesses against scarcity of resources and volatile prices, helping to create new business opportunities and innovative, more efficient ways of producing and consuming. It will create local jobs at all skills levels and opportunities for social integration and cohesion. At the same time, it will save energy and help avoid the irreversible damages caused by using up resources at a rate that exceeds the Earth's capacity to renew them in terms of climate and biodiversity, air, soil and water pollution." [9]

IOP Conf. Series: Earth and Environmental Science 1099 (2022) 012004 doi:10.1088/1755-1315/1099/1/012004

Modular designs and techniques that make products easier to disassemble and materials easier to separate, can help to save precious resources due to recovery and reuse of valuable materials and components. Many nations, inside and outside of the EU, have recognized that the interests of manufacturers, end-users and recyclers are not aligned and have responded by providing incentives to improve product design while preserving competition and enabling innovation [9] [10]. The DfD strategies and methodologies are aligned with sustainable product design approaches, which are necessary to support the circular economy.

3. Method: How to support circular economy with DfD

Circular product design requires a shift in thinking starting in the earliest stages of product development. The decisions made in the early stages of the design process will have significant effects through the rest of the product and project life cycle. Designing a product with recycled materials or unique design modalities that enables recycling at the EoL is not enough to support the circular economy. Designers need to incorporate a new set of design goals into their processes to include their effects on the triple bottom line, with the primary objective to improve circularity and modularity and prioritizing the reduction of both operational and embodied carbon.

At many stages in the product or project life cycle, a conflict between the environmental and economic goals can arise, causing a roadblock in the path to the circular economy. These conflicts often impede the journey to achieve sustainability goals and lead to an EoL outcome that results in waste. Lighting practitioners can positively impact EoL outcomes by incorporating DfD methodologies early in product development. Improved DfD strategies in early development stages improves the economic viability of EoL outcomes, such as increased homogeneity of recovered material at EoL, safer disposal of hazardous components and improved opportunity for reuse or remanufacturing of components or subassemblies [11]. DfD is a critical aspect of achieving a circular economy because it promotes modularity, which supports product life extension through repair or upgrades and improves recycling efficiency.

3.1. DfD background

Design for X (DfX) is a general strategy for improving a specific aspect of a product or project. DfX can be applied at a product scale or at a building scale. Some industries (e.g., automotive, and electronic) have adopted various DfX strategies; DfD, Design for Manufacturing (DfM), Design for Maintenance (DfM), and Design for Environment (DfE) [12].

DfD is a relatively new concept and was first defined in the 1990s. It has been known as a "critical twenty-first century discipline" [13]. Many papers have been written on the subject, and DfD has become an important topic in the design process for other industries, but it has not been incorporated into mainstream product or project design practices of the lighting industry [14]. Some lighting companies, mainly in Europe, are incorporating aspects of DfD into their practice to support the circular economy and corporate environmental goals [15] [16] [17] .

3.2. Quantifying DfD

Although disassembly activities are typically carried out in the use stage (i.e., during repair) or the EoL stage of a luminaire life cycle, DfD frameworks need to be applied as early as possible, potentially in the design and conceptualization stages of product development, when information such as materials and components, geometry, and disassembly/assembly tools required can be assessed and modified more easily. Most frameworks designed to measure or quantify improvements in disassembly of products focus on ease of disassembly (either time-based or motion-based) or the cost of disassembly [18] [19] [20] [11] [21] [22] [23]. As a part of the literature review for this paper, the authors found the existence of 22 different metrics and frameworks that have been proposed to quantify and analyze DfD in products, but none of them have been adopted as a primary accepted framework. The lack of an agreed-upon framework for quantifying DfD in lighting products and components presents a need and a potential opportunity to support the circular economy.

IOP Conf. Series: Earth and Environmental Science 1099 (2022) 012004 doi:10.1088/1755-1315/1099/1/012004

In 2021, the Chartered Institution of Building Services Engineers (CIBSE) and the Society of Light and Lighting (SLL) launched the Circular Economy Assessment Method "CEAM-Make", a digital tool which allows for the quantification of circularity in specific lighting products [8]. While the CEAM-Make tool is not a DfD framework and does not provide DfD guidance, it does measure some aspects that result from optimized DfD practices, including modularity of the product, techniques to minimize or simplify fasteners and screws, and minimizing component variation. If the lighting industry were to develop and adopt a similar DfD industry standard framework and measurement tool it would greatly support the move towards a transition to a circular economy.

4. Results: DfD rules and strategies

Early in the design stage of a lighting product, designers should apply DfD strategies to embed the potential for an increased product lifetime, aimed at increased serviceability and reusability. The disassembly activities are usually carried out either during the use stage (supports maintenance repair or replacing components), or during the EoL stage (reuse, recycling or remanufacture) of the product life cycle. Applying DfD strategies in early stages improves the homogeneity of recovered materials at the EoL for improved tangible recovery (when the components have monetary value) or non-tangible recovery (when there is no clear monetary value but there is an environmental value), and also supports safe disposal of hazardous parts such as electronic waste (e-waste) [11] [12] [24].

Five high-level DfD goals or design strategies that are important to consider in any lighting product development process have been presented in Figure 2. A list of actionable rules and practices associated with each of these high-level goals has been provided in Table 1. These actionable DfD rules and practices can be applied by lighting practitioners in early life cycle stages to optimize products and projects, reduce carbon footprints, improve efficiency, improve material health (utilizing environmentally preferable materials), facilitate responsible EoL outcomes, tie into the circular economy, and provide equitable benefit to the society [12] [25]. The DfD rules listed here are prevalent in DfD literature, but there is still room for innovation.

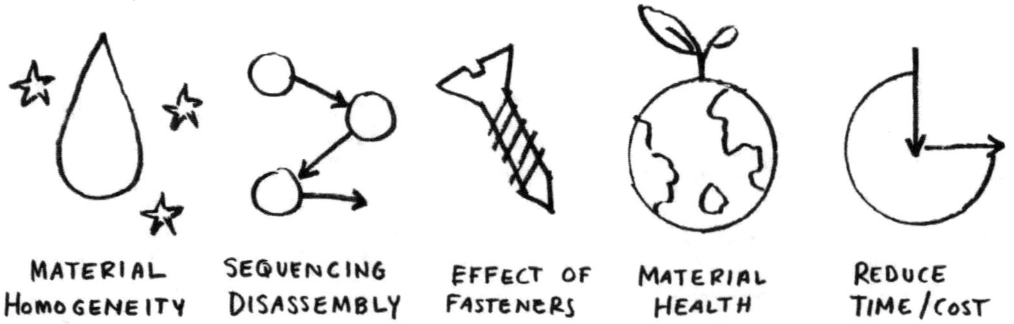

Figure 2. Five critical DfD goals for lighting product development.

4.1. Material Homogeneity

As part of any DfD implementation strategy, it is important to understand the concept of waste hierarchy. Waste hierarchy frameworks and directives have been developed by different nations and lay out a concept that preventing waste is the preferred option and sending waste to landfill should be the last resort [26] [27]. Within the waste hierarchy, some EoL options are preferred over others [28]. An important distinction related to DfD practice is between recycling and recovery. Recycling occurs when components can be separated into distinct material groups, so that each material can be treated independently. Recovery occurs when materials cannot be separated because they are part of a composite or they have been glued together, or when a coating or paint is applied to a material that cannot be removed. Components that cannot be separated are recovered by shredding, burning or other destructive techniques. Some energy can be recovered from burning these materials and small amounts may be recovered due to shredding, but it is preferrable to preserve material homogeneity by allowing for relatively straightforward material separation and recycling at EoL. A small amount of planning in the early stages can support these preferred outcomes at EoL.

Product designers should consider material homogeneity when planning a new product. A key strategy is to avoid adhesives or compounds that create new composite materials. Instead consider ways that two parts or components could be created out of one part and avoid an adhesive altogether. Product designers can improve material homogeneity, as well as environmental safety, by using molded-in colors or finishes, rather than wet paints, coatings, plating, or other hazardous finishes that are difficult to separate from the material at EoL [12].

In lighting product design, a common material homogeneity issue concerns the use of wire coatings or jackets. Coating on wiring is often made from Polyvinyl chloride (PVC), which is a chemical that has superior electricity insulating properties, but when disposed of, it releases toxic chlorine-based chemicals that build up in the environment [29]. Designers should be aware of common chemicals of concern in order to minimize the use of these in favor of safer alternatives. Wiring materials like Modified Polyphenylene Ether (mPPE) are safer to manufacture and are 100% recyclable. When adhesives are required, designers can look for water soluble adhesives, compatible adhesive materials, or adhesive materials that can be easily recycled when needed.

4.2. Disassembly sequencing

Disassembly sequencing techniques support the disassembly process by making it easier or by providing more information to the person performing the disassembly, repair, or maintenance. Markings can be embedded in parts to indicate how they should be removed, which can make disassembly easier in the field without requiring additional instructions. Labeling and instructions that are embedded in the fixture can indicate which key components in the fixture can be replaced, and how they should be replaced; for example, removal of common components such as LED drivers and LED arrays. A label with a matrix barcode or link could lead to the manufacturer's website for more in-depth instructions and guides to support servicing, disassembly, and recycling.

Another sequencing strategy involves grouping the product components by material type. Product designers may determine at the outset that all plastic components will be made from the same material (e.g., acrylic), all brackets will be steel, or all body components would be made from aluminum. These types of material grouping considerations make the components easier to separate at the EoL and require less effort for sorting parts.

4.3. Effect of fasteners

Fasteners play a key role in DfD, since thoughtful use of fasteners can make for a better end-user, maintenance, and production experience. Minimizing the use of fasteners wherever possible and use of fasteners that can be removed with common tools is highly recommended. One way to minimize fasteners is to integrate fastening features into other parts when possible. For example, instead of using a screw to attach parts or components, introducing a holding tab or a slot feature can help avoid the use of a fastener. Using a standardized set of fasteners will minimize the need for unique or multiple tools for repair, maintenance, or disassembly. These techniques can reduce assembly and disassembly time and will facilitate maintenance in the field. Product designers should also harmonize materials by using the same faster material type within a product. This will make the components easier to sort at EoL and promote recycling.

4.4. Material health

DfD embraces material health, which means that every new product design is an opportunity to choose materials that are safe for humans and the environment. There are well-recognized material health or material transparency reporting frameworks that support improved material choices. In practice, if a designer identifies a material they want to use but are unsure of the environmental or human health impacts, they can ask the vendor or manufacturer to disclose the materials used to create the product or ask for a report such as an Environmental Product Declaration (EPD), Health Product Declaration (HPD)

[30] or the International Future Living Institute's Declare Program [31]. Knowledge of the materials or ingredients that go into making a part or process, allows product designers to compare those materials against a list of known harmful materials and to make informed decisions about material choice. For example, although they appear in luminaires in small quantities, printed circuit boards (PCBs) include many rare and precious metals, such as tin, gold, nickel, and copper. It is possible to recover and recycle many of these precious coated plates and metals in the PCB, if it is properly disassembled and sent to a capable recycling facility [12]. Following-up on the material health of parts and components can help narrow down materials and vendors. In addition to identifying the environmental impacts from materials, designers can find ways to mitigate them; for example, those of PVC can be reduced by minimizing the number and lengths of wires and cables used.

4.5. Reduce time/cost
Reducing the time and cost required for repairs, modifications, and EoL disassembly supports the circular economy by making improved maintenance and EoL outcomes more economically viable. Many of the points listed above will support ease of assembly, production, and disassembly of parts or entire lighting systems, for example, product designers should think through both the assembly process and the disassembly or maintenance/repair process to ensure that snap-fits are in accessible locations and can be opened using standard tools.

Design for Disassembly Strategies	Design for Disassembly Rules and Practices for Lighting Practitioners
Material homogeneity	Use adhesives only if they are compatible with the parts which are to be joined
	Use only water-soluble adhesives
	Use joints, snap fits, or screws rather than welding, brazing, soldering or glue
Disassembly sequencing	Provide identification for separation points
	Place components in logical groups, for example according to recycling group
	Provide standard and permanent identification of material types on the product
Effect of fasteners	Reduce the number of fasteners and connectors
	Use easy to remove (i.e., accessible, easy to identify or common) fasteners
	Use fasteners of the same or compatible material for connecting components
	Part standardization such as use of same size screw throughout the product
Material health	Minimize the number and length of wires and cables used
	Avoid use of environmentally hazardous material
	Choose molded-in colors and finishes in lieu of using paints, coatings, and plating
Reduce time/cost	Locate snap fits so that they can be opened using standard tools
	Use joining elements that are easily separable or easy to remove or break free
	Use two-way snap fits or break points on snap fits
	Use techniques that minimize the number and variety of materials used
	Design fastening points so they are easy to access

Table 1. DfD Rules and Practices for Lighting Practitioners

5. Conclusion: DfD benefits across the life cycle
Even though DfD rules and strategies should ideally be applied in the early design processes, the benefits are evident at all stages of the lighting system life cycle. Successful application of DfD strategies in early design stages could allow for better recycling or recovery, even for difficult components, such as PCBs. As with any lighting product or project, the economic goals may not perfectly align with the environmental or circularity goals and can cause a conflict in the use or application of the lighting

system. In practical terms, this could mean that a lighting product with DfD strategies embedded in the early stages may still meet with destructive disassembly or may be tossed directly into the waste stream without the implementation of any disassembly-based processes. It is always true that a well-designed luminaire may suffer a destructive fate in a landfill; however, if the DfD-based thinking and modularity is not integrated in the designs from the onset, a destructive outcome is expected.

The best designed product that incorporates DfD practices will need to be applied and utilized within a circular economy in order to maximize the benefits. All lighting practitioners can support the circular economy by preferring lighting systems that incorporate DfD thinking. The best way for lighting specifiers (such as designers or lighting representatives) to support the circular lighting economy is to educate themselves about DfD and circular economy principles. Designers should incorporate DfD and circularity principles into design plans and inspire clients and manufacturers to do the same. Using the tools described in this paper, lighting practitioners can start small, think circular, and get going.

References

[1] Illuminating Engineering Society. 2020. Sustainable Lighting – An Introduction to the Environmental Impacts of Lighting. New York City, NY: IES/ANSI.

[2] Elkinton, J 2018. "25 Years Ago I Coined the Phrase "Triple Bottom Line." Here's Why It's Time to Rethink It." June 25. Accessed April 12, 2022. https://hbr.org/2018/06/25-years-ago-i-coined-the-phrase-triple-bottom-line-heres-why-im-giving-up-on-it.

[3] Tähkämö L, Bazzana M, Ravel P, Grannec F, Martinsons C and Zissis G 2013. "Life cycle assessment of light-emitting diode downlight luminaire - a case study." *The Int'l J. of Life Cycle Assessment*.

[4] Ellen MacArthur Foundation. 2019. "The butterfly diagram: visualising the circular economy." February. Accessed April 13, 2022. https://ellenmacarthurfoundation.org/circular-economy-diagram.

[5] Tähkämö L, Bazzana M, Ravel P, Grannec F, Martinsons C and Zissis G 2013. Life cycle assessment of light sources – Case studies and review of the analyses. Doctoral Dissertation, Helsinki: Aalto University School of Electrical Engineering Department of Electronics.

[6] Nations, United. 2015. Paris Agreement. United Nations.

[7] CIBSE, The Chartered Institution of Building Services Eng, and Society of Light and Lighting SSL. 2021. TM66: Creating a circular economy in the lighting industry. London: The Chartered Institution of Building Services Engineers/Society of Light and Lighting. https://www.cibse.org/knowledge/knowledge-items/detail?id=a0q3Y00000JruI1QAJ.

[8] Commission, European. 2015. Closing the loop - An EU action plan for the Circular Economy. Policy Document, Brussels: European Commission.

[9] The Guardian. 2014. Five countries moving ahead of the pack on circular economy legislation. The Guardian. October 29. Accessed April 05, 2022. https://www.theguardian.com/sustainable-business/2014/oct/29/countries-eu-circular-economy-legislation-denmark-sweden-scotland.

[10] Harivardhini S, Krishna K M, Chakrabarti A 2017. "An Integrated Framework for supporting decision making during early design stages on end-of-life disassembly." *J. of Cleaner Production* 558-74.

[11] Rayate V 2012. A tool for selection of design for manufacturing and assembly rules during product design stage while considering end of life conditions. Thesis Clemson University, Clemson University.

[12] Bogue R 2007. "Design for disassembly: A critical twenty-first century discipline." *Assembly Automation* 285-89.

[13] Abuzied H, Senbel H, Awad M and Abbas A 2020. A review of advances in design for disassembly with active disassembly applications. *Technical J., Engineering Science and Technology, an Int'l J.*.

[14] Components, Ingemann. n.d. How to design for later recycling. Ingemann Components. Accessed

April 05, 2022. https://www.ingemanncomponents.com/how-to-design-for-later-recycling/?lang=de.
[15] Stoane Lighting. 2021. disassembly information. Stoane Lighting. November 01. Accessed April 13, 2022. https://stoanelighting.com/disassembly-information.
[16] Signify. 2017. Signify Service tag: Making LED maintenance operations easier and faster. Signify. September 19. Accessed April 13, 2022. https://www.signify.com/en-us/blog/archive/led/signify-service-tag-making-led-maintenance-operations-easier-and-faster.
[17] Vanegas P, Peeters J R, Cattrysse D, Tecchio P, Ardente F, Mathieux F, Dewulf W and Duflou J R. 2018. "Ease of disassembly of products to support circular economy strategies." *Resources, Conservation & Recycling* **135** (Resources, Conservation & Recycling) 323-34.
[18] Das S K, Yedlarajiah P and Narendra R 2000. "An approach for estimating the end-of-life product disassembly effort and cost." *Int'l J. of Production Research* 657-73.
[19] Desai A and and Mital A 2003. "Evaluation of disassemblability to enable design for disassembly in mass production." *Int'l J. of Industrial Ergonomics* 265-81.
[20] Harivardhini S and Chakrabarti A 2016. "A new model for estimating End-of-Life disassembly effort during early stages of product design." *Clean Techn Environ Policy* 1585-98.
[21] Kroll E and Hanft T A 1998. "Quantitative Evaluation of Product Disassembly for Recycling." *Research in Engineering Design* 1-14.
[22] Soh S L , Ong S K and Nee A Y C 2014. "Design for Disassembly for Remanufacturing: Methodology and Technology." 21st CIRP Conf. on Life Cycle Engineering 407-12.
[23] Zygierewicz A. 2017. "The Ecodesign Directive (2009/125/EC)." European Parliament.
[24] Tuenge J, Hollomon B, Dillon H and Snowden-Swan L 2013. Life-Cycle Assessment of Energy and Environmental Impacts of LED Lighting Products Part 3: LED Environmental Testing. Richland, WA: DOE/ EERE.
[25] European Union. 2008. Directive 2008/98/EC of the European Parliament and of the Council. Brussels: EU.
[26] U.S. EPA, United States Environmental Protection Agency. n.d. Waste Management Hierarchy and Homeland Security Incidents. U.S. EPA. Accessed April 05, 2022. https://www.epa.gov/homeland-security-waste/waste-management-hierarchy-and-homeland-security-incidents.
[27] Lienig J and Bruemmer H 2017. *Fundamentals of Electronic Systems Design* (Cham, Switzerland: Springer)
[28] GreenPeace. 2003. PVC: The Poison Plastic. Greenpeace. August 18. Accessed April 05, 2002. https://www.greenpeace.org/usa/wp-content/uploads/legacy/Global/usa/report/2009/4/pvc-the-poison-plastic.html#:~:text=PVC%20is%20the%20most%20environmentally,water%2C%20air%20and%20food%20chain.
[29] HPD-Collaborative. 2018. HPD Version 2.1.1 - Health Product Declaration Open Standard. Wakefield, MA, USA: Health Product Standard Collaborative. https://www.hpd-collaborative.org/hpd-open-standard/#:~:text=The%20Health%20Product%20Declaration%C2%AE,product%2C%20and%20associated%20health%20information.
[30] Int'l Living Future Institute. 2012. Declare. The Nutrition Label for Products. ILFI. October 03. Accessed April 13, 2022. https://declare.living-future.org/.
[31] Crowther P 2005. *RAIA/BDP Environment Design Guide: Design for Disassembly - Themes and Principles*. (Brisbane: RAIA/BDP)

Bio-mimetic Approaches to Kinetic Facades: A Design Proposal for a Light-Responsive Facade Module.

Anastasiia Scavée, Georgios Triantafyllidis, George Palamas
Aalborg University, Copenhagen,
Department of Architecture, Design & Media Technology
E-mail: gpa@create.aau.dk

Abstract. Facades are important for controlling light entering a building and ensuring optimal conditions for occupants. But light can also cause discomfort, such as glare and overheating. Natural light changes with the time of day, the weather, and the season. Creating a responsive structure can enhance the visual quality and energy efficiency. Because of their properties, kinetic facades can react to various stimuli, such as heat and sunlight. In this paper, we discuss bio-mimetic approaches to kinetic facades and how they can be used to develop a design proposal for the light-responsive facade module. Motors drive most kinetic facades, which adds to the maintenance costs, thus,shape memory alloys (SMAs) are investigated as actuators. Facades should be visually appealing and reflect the bio-mimetic concept of the project while allowing natural light to pass through. The results show that the suggested solutions are both feasible and aesthetically pleasing, thus comparable to traditional kinetic facade workflows. In conclusion, if bio-mimicry is to be used in large-scale construction, it is imperative that it is examined for the effects of different environmental conditions on bio-mimetic patterns.

1. Introduction

Through the history of architecture, the role of a facade has evolved from being a defensive shell that protects from threats and instills a sense of safety, into a tool that allows us to stay connected with our surroundings. In spite of the need for sustainability, modern architecture is not always in harmony with natural environment. Even though numerous green zones are being integrated into roofs or atriums, the question is whether modern architecture appeals to us cognitively as a natural thing and what emotions it triggers. Numerous studies have shown how much space can influence our cognitive and affective states [8] [22] [11]. Moreover, light can affect us in various ways from boosting creativity and productivity to raising our mood and making us feel more optimistic by triggering both visual and non-visual pathways in our brain [25].

One of the main functions of a building's facade is to regulate the amount of light entering it and thus, to create optimal lighting conditions for the occupants. Accordingly, a good supply of light and daylight is essential for optimal conditions in offices and residences [2]. Although light is essential for optimal conditions in workplaces and residential buildings, it can also cause discomfort, such as glare and overheating. However, natural light changes depending on the time of day, the weather conditions, and the season [3]. By creating a responsive structure, the visual quality and energy efficiency can be enhanced. kinetic facades are able to respond to external stimuli, such as heat radiation and lighting, and due to their properties, they are becoming more prevalent in contemporary architecture. This paper focuses on a bio-inspired approach to kinetic facades and how it can be used to develop a design workflow for the light-responsive facade module.

The movement of a kinetic facade is usually achieved through motors, which adds to the cost of sustaining materials [19]. We explore the possibility of using SMA (shape memory alloy) and NiTinol (Nickel titanium alloy) wires in particular, as actuators. A "shape" criterion represents the characteristics of the design in terms of the visual appearance of the facade. The facade should be visually appealing and should reflect the bio-inspired concept of the project while passing sufficient natural light [23]. A bio-mimetic approach to solving design or engineering problems involves the imitation of elements and systems found in nature. Computational and generative design tools will be used in this project to achieve the idea of bio-philia. It is important that the design reflects irregularities, imperfections, but at the same time has regularity and harmony, like natural patterns. The adaptability of the responsive facade refers to the geometric properties of the generative design to adapt to constantly changing lighting conditions. The solution is expected to change its shape with a relatively fast speed and minimal energy consumption in order to meet the needs of the occupants [13]. Various bio-inspired facade patterns were tested and compared against regular geometry patterns in order to identify how people perceive and react to each. From the study, it was found that the bio-mimetic patterns elicited a significantly higher preference rating than the regular geometrical patterns. It was also found to be more interesting and appealing to people both in sunny and gloomy pictures. Moreover, participants enjoyed the indoor environment, which evoked good natural associations, such as a honeycomb. But some of the participants found the pattern to be quite disturbing. In conclusion, the study suggests that further investigation should be done to explore the effect of bio-mimetic patterns on different environmental conditions in order to determine the acceptability of bio-mimicry in large-scale construction.

Figure 1. University of South Denmark, responsive facade with moving mechanical parts.

2. Bio-inspired design

The term bio-mimicry, originated from Greek - bios meaning life, and mimesis meaning imitate - and has been widely used in industrial design, medicine, and engineering.The natural selection process is a driving force for the evolution of better suited species and traits based on pre-existing conditions [11]. As a result, more advantageous and best adapted organisms can survive [11]. Nature's most perfect systems can often be found in this iterative process, therefore, people have learned from nature by imitation while solving life's problems for thousands of years. Regardless of its popularity in other fields, bio-mimicry has recently gained popularity in architecture [17]. Based on the ability of nature to adapt to a variety of conditions, bio-mimicry is now widely used as a tool in the design of built spaces in order to meet certain challenges regarding thermal and visual comfort in buildings.

2.1. Bio-mimicry as a design practice

Bio-inspired design is usually divided into five main stages: planning, abstraction, searching a space of solutions, analyzing, and comparing the final solutions [9]. Data analysis is usually the next step after project planning. Since there are no direct links between biological systems and industrial needs, problems and possible natural analogies should be simplified. A popular method is to use bio-cards. Bio-cards describe a biological phenomenon's function and are used to generate design proposals and abstract descriptions based on conceptual ideas [16]. After the abstraction is complete, the next step is to decide which phenomena are relevant and applicable to the specific problem at hand. Finding a solution that could be applied requires to determine whether the phenomenon is transferable to an engineering domain.

Two main approaches are used in bio inspired design: (a) A problem driven or a technology pull approach [16] [10] which is based on identifying the problem in the biological domain that must be solved. (b) A biology push or solution driven approach where designers and engineers have to find a problem where an already known biological solution can be applied [16] [10] [24].

3. Shape Memory Alloys in Adaptive Facade Design

SMAs are often used as a substitute for muscles in robotics, which goes hand in hand with a bio-inspired approach. SMAs refers to alloys that are able to "remember" their initial shape. SMA is a family of materials that changes theirs physical shape when subjected to certain temperatures [6]. The shape of the material can be pre-programmed, or deformed, in low temperatures. When the material is heated again, it recovers its original shape (figure 6.1). Different types of SMA exist, such as Ni-Ti, CuAlNi, CuZnAl, AuCd, FeMn, etc [6]. NiTinol is one of the most popular SMA materials being used today [1]. Nitinol gained great popularity due to its unique properties combining ductility, bio-compatibility, corrosion resistance, and longevity (it can withstand up to 185000 cycles of deformations [20]).

Depending on the alloy composition and the temperature at which it is exposed, Nitinol will react differently. The activation temperature varies between 30C (86F) and 130C (266F). The material could be used in architecture for ventilation or facade structures, where it could change its shape depending on the temperature inside or outside of the building. Nitinol is activated through heating as it reacts to temperature changes. Electrical current is conducted through Nitinol, resulting in an increase in temperature due to the electrical resistance of the material. When the material increases its temperature, it will change its shape. Wires made of Nitinol have significant pulling force [1], which makes them a good choice for an adaptive facade as they can support heavy loads.

4. Challenges of Using Nitinol as a Facade Actuator

Fatigue is a common effect caused by cyclic loads and therefore deformations placed on the components. Deformation is difficult to estimate, since numerous factors affect that parameter, such as structural fatigue (strength declines under cyclic loads) and chemical fatigue [18].

It is important to consider how SMA can withstand a wide range of temperatures when designing facades. The composition and diameter of an alloy can affect how it deforms at different temperatures. Therefore, for temperate climates could be more suitable, where metals heat up during the day and become cold at night, but in extreme climates, SMA could be activated in an unpredictable way. Another challenge of using SMA as an actuator is the amount of power required to drive their shape change profile. SMAs require a lot of electric power, and this power could be difficult to provide if the facade is not connected to the grid or an alternative power source like Solar panels, and this power demand is likely to vary from season to season.

5. Designing Daylight Envelopes

One of the main functions of the facade is to regulate the amount of light that enters the building, and therefore to create an optimal lighting environment for the occupants. Although natural light, and daylight in particular, is essential for optimal conditions in workplaces and residential buildings, it is also associated with glare and overheating [2]. However, natural light is dynamic, the intensity varies according to the time of day, season, or the weather [7]. Creating a responsive structure would improve energy efficiency and visual appeal [13] [15].

The speed of light adaptation, as well as the frequency of lighting adjustments, must be taken into consideration when creating an autonomous dynamic facade. It is also important to address the question of whether people need to be shielded from direct sunlight, in order to avoid discomfort, or to have some kind of light patterns projected in the interior in order to maintain a connection with the outside and natural world. It is thus, extremely important to understand key factors that affect the perception of any design solution when exploring how people react to certain lighting conditions within an architectural environment.

Diversity, complexity, and intensity are aspects of designing disciplines that affect occupant stimulation [5]. In any interior, daylight is a constantly changing element. As time passes, we adapt to certain qualities of the interior (old stimuli) and stop responding to them. People can easily get bored when there is not enough stimulation or change, or they can get overstimulated if those changes are too drastic and frequent. This may result in either mental fatigue or sensory deprivation. It is therefore extremely important to keep medium levels of sensory stimulation [5]. Thus, in addition to regulating lighting in a space, a kinetic facade can also provide new stimuli, such as the "play of light" and light patterns in the room.

6. Design and Implementation

Three criteria were used to ensure that the design solution meets the project requirements in accordance with its vision.

- *Shape* is primarily a design criteria that represents visual requirements. The facade should reflect the bio-philic idea of the project, while allowing for sufficient light to pass through. Moreover, based the concept of bio-mimicry, the generated patterns should demonstrate irregularities and other imperfections, and yet still be harmonious, thus, computational and generative design methods will be employed to achieve this.
- *Bio-mimicry* can be used to solve engineering problems as a principle that imitates elements and systems in nature.
- *Adaptability* refers to the ability of a design to adapt to constantly changing lighting conditions depending on its geometry. The shape of the solution should be easily changed in a relatively short amount of time.

We aim to create an autonomous facade module that reacts only to external temperature changes by directly activating the actuators with electrical current. The next step was to find a pattern in nature that could be used as a visual reference for the facade design. Two different shapes were conceived and realized. The first shape was inspired by the patterns on the gills of mushrooms 5. Mushroom gills are thin structures, under the cap, which vary widely in pattern from species to species. Every one of them has a slightly different geometry, giving a nice pattern when they're layered together. We translated the pattern of the lines into a window

IOP Conf. Series: Earth and Environmental Science 1099 (2022) 012005 doi:10.1088/1755-1315/1099/1/012005

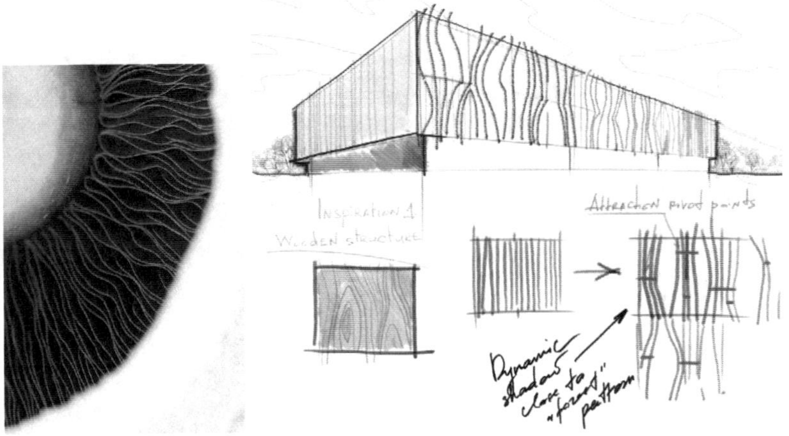

Figure 2. (Left) Mushroom gills as an inspiration for the facade pattern. (Right) Sketch of the actual facade, flexible strips and attached Nitinol wires in red.

shading module pattern with thin flexible sheets aligned vertically. Each of these modules will be reproduced throughout the facade. In selecting this pattern, we took into account the ease of fabrication, especially with digital fabrication technologies such as 3D printing and robotics assembly. The second shape was inspired by the lotus fruit and seeds. This pattern was recreated in 3D using Voronoi tessellation. Rather than a rigid material, the facade would be made of flexible tissue or material that would be fitted with a flexible SMA. As the wire contracts under certain temperatures, the opening gets smaller and less sunlight gets through the facade. It can also decrease in size, which causes the holes to widen (figure 6). In generative design methodologies, architects are only in control of variables, as the systems used are usually very complex, and cannot fully predict the outcome of the algorithm. Furthermore, when combined with software such as DIVA for illumination analysis, multiple iterations can be performed automatically to determine an appropriate opening ratio.

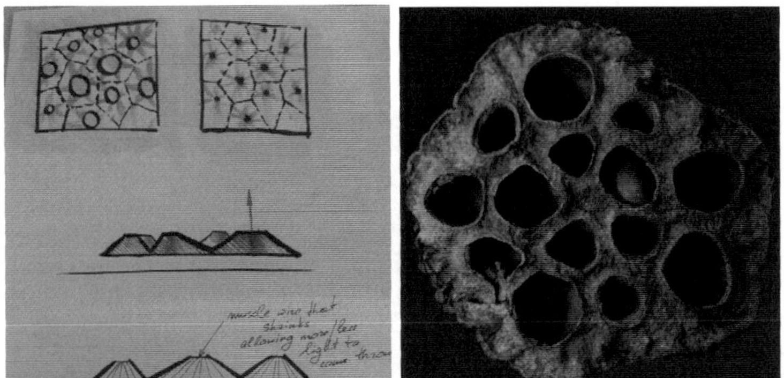

Figure 3. (Left) Sketch of the second concept (Right) Lotus seeds as an inspiration for the openings of the facade.

Figure 4. Generative Facades based on the concept of SMA controllable modules. (Left) A facade resembling mushroom gils, (Right) a Voronoi tessellation resembling lotus seeds sockets.

Figure 5. Perspective view of Voronoi tessellation facade

6.1. Prototype

A simulation was used to explore a range of possible solutions and to predict the behavior of the facade. Diva for Rhino was used for lighting simulations. The yearly daylight autonomy was simulated, including the mean daylight factor (figure 8). For the deformed vertical blinds, we used Kangaroo and the Weaverbird plugin to create a Voronoi network. It is possible to model nitinol actuators using Kangaroo, but we found it difficult to simulate the linear expansion of the SMA wire under different loads. Therefore, optimizing the form using a global optimizer such as an evolutionary algorithm might be a potential solution. In this case, the optimization process could take into account how much light passes through as well as the number of SMA wires attached to the modules and their hinge positions.

To test the behavior of Nitinol under a variety of scenarios and loads, a prototype was implemented. The module consists of a rectangular frame made of plywood with a series of equally spaced soft wooden stripes and 4 Nitinol wires attached to the stripes 6.1. The size of the unit was 45x45cm and was equipped with a single light sensor. By deforming the shape of the stripes more or less light was able to penetrate the unit, thereby providing a regulating effect. An Arduino microcontroller measures the amount of light in the environment through a light sensor, and based on the sensor readings, the Arduino activates the Nitinol wires 6.1. The prototype showed that Nitinol can responded to light stimuli, but it was challenging to control the light response. Further development of this method requires optimization of Nitinol wires, their arrangement, and control. Controlling the light entering the facade would be more precise with a closed loop controller such as a PID controller, but that was beyond the scope of this project. A PID controller is a common type of controller used in many industrial applications

IOP Conf. Series: Earth and Environmental Science 1099 (2022) 012005 doi:10.1088/1755-1315/1099/1/012005

such as robotics, automotive, aerospace, and manufacturing [4]. It is used to calculate an "error" value as the difference between a measured process variable and a desired set-point. The controller attempts to minimize the error by adjusting the process control inputs.

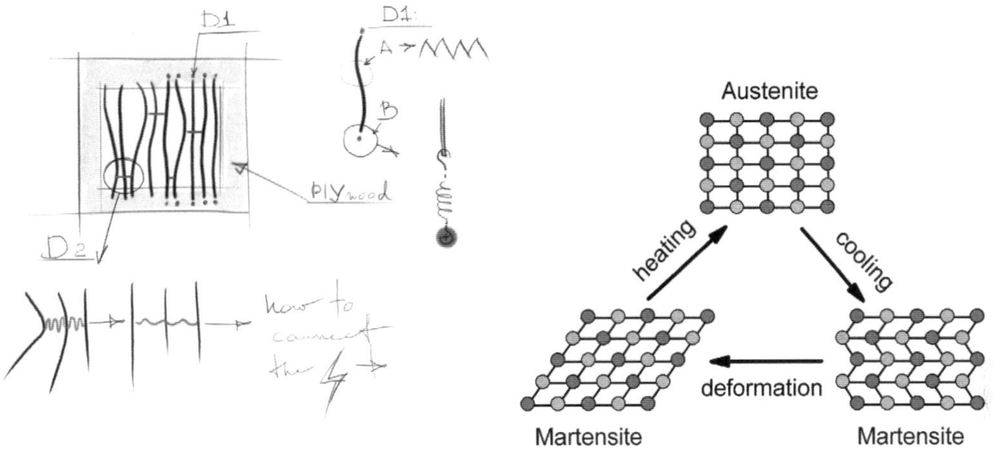

Figure 6. Sketch of the facade module prototype, and Illustration of the shape memory effect of Nitinol alloys. SMA alloys expand when heated and contract when cooled, (right)

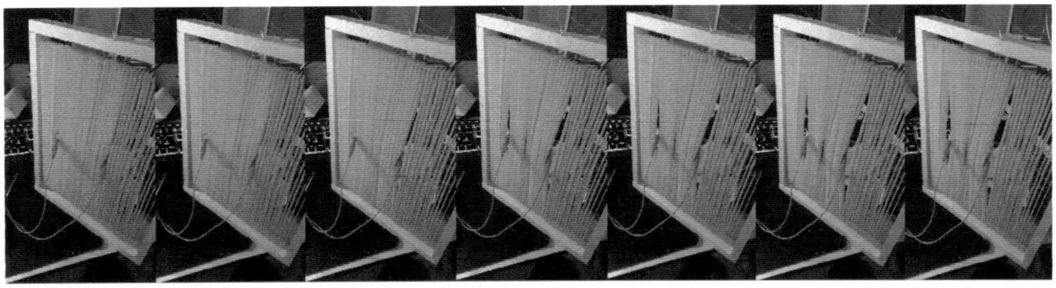

Figure 7. This time series shows the prototype module's horizontal strips being deformed as a result of Nitinol wire activation through electric current.

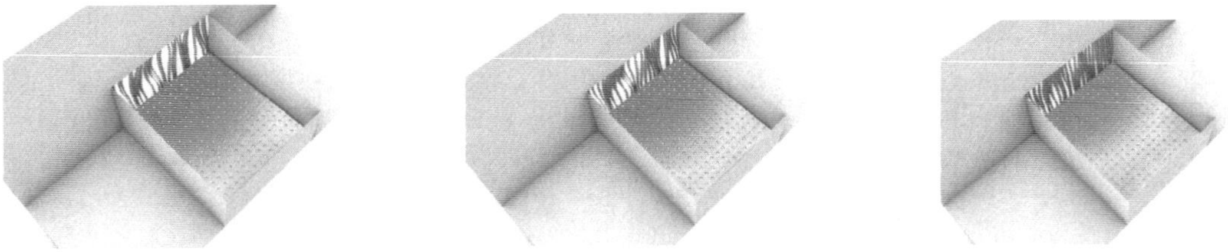

Figure 8. Daylight autonomy simulation. (Left) High open ration of the facade with Mean daylight factor - 2.1% (Center) Medium open ratio with Mean daylight factor - 1.9% (Right) Low open ration with Mean daylight factor - 1.6%

7. Evaluation

This study used Nitinol as a smart actuator to add flexibility to a facade design by copying natural processes. Therefore, we need to determine how people react to various facade designs in terms of aesthetics and other perceptive factors. Four rooms of the same geometry and position were created using Rhinoceros 3D, Grasshopper and Corona Render. Two lighting scenarios with direct sunlight and overcast sky were rendered for each room. Twenty-four participants were interviewed (15 males and 9 female). The age ranged from 22 to 54 years old, with an average age of 29 years. None of them had any design background. Participants were shown 8 different images of facades: two geometrical and two bio-inspired facades with two types of daylight 9. The testing process was divided into three phases. During the first phase participants asked to rate each image based on how attractive or pleasing they found it using a discrete analog scale of 1-10, with 1 being very unattractive, 5 being slightly unattractive, 6 being slightly attractive, and 10 being very attractive. The second phase of the experiment had participants rate the same pictures on how interesting/exciting they found the room using the same discrete analog scale of 10 points. In the third phase, participants had to evaluate 4 images with different facades but with only direct daylight. Participants were asked to select one or more words from the list of antonyms below to describe visual qualities of the room (table 1). More open-ended questions were posed at the end of testing to gather qualitative feedback.

Table 1. Visual qualities.

Organic	Geometrical
Beautiful	Ugly
Simple	Complex
Stimulated	Subdued
Pleasant	Annoying

8. Results

We use a modified version of the circumplex model of affect [21] to describe the different emotional states of our participants. This is a continuous two-dimensional model of affective states that describes the structure of emotions and their relationships in terms of two independent underlying dimensions: pleasure and arousal. Emotions are placed along the circumference of a circle and are defined by their level of pleasure and arousal. The circumplex model has been proven to be a very useful tool for representing and understanding emotions. It has also been used to investigate the emotional reactions of people to different stimuli, such as music [14] or to characterise four different emotion states (liveliness, relaxation, tense and gloom) for a workspace lit environment [12]. For facade C 9 most participants found the bee-shaped pattern and strong sunny pattern to be very appealing. The lines and sunny patterns on image B received the most mixed and evenly distributed grades in the first image. Image D received the least favorable ratings of all other images while "Very attractive" ratings were not given. The results show that people tend to rate rooms as less attractive when there is no direct sunlight present, and that rooms also appear less positive when there is no daylight. Most participants rated the facade on image A (~50%) and image C (~75%) as organic. The facade on image B was mostly described as Simple (95%) and Pleasant (40%). In contrast, the facade on image D was described as very annoying (25%) by a majority of respondents. The pattern

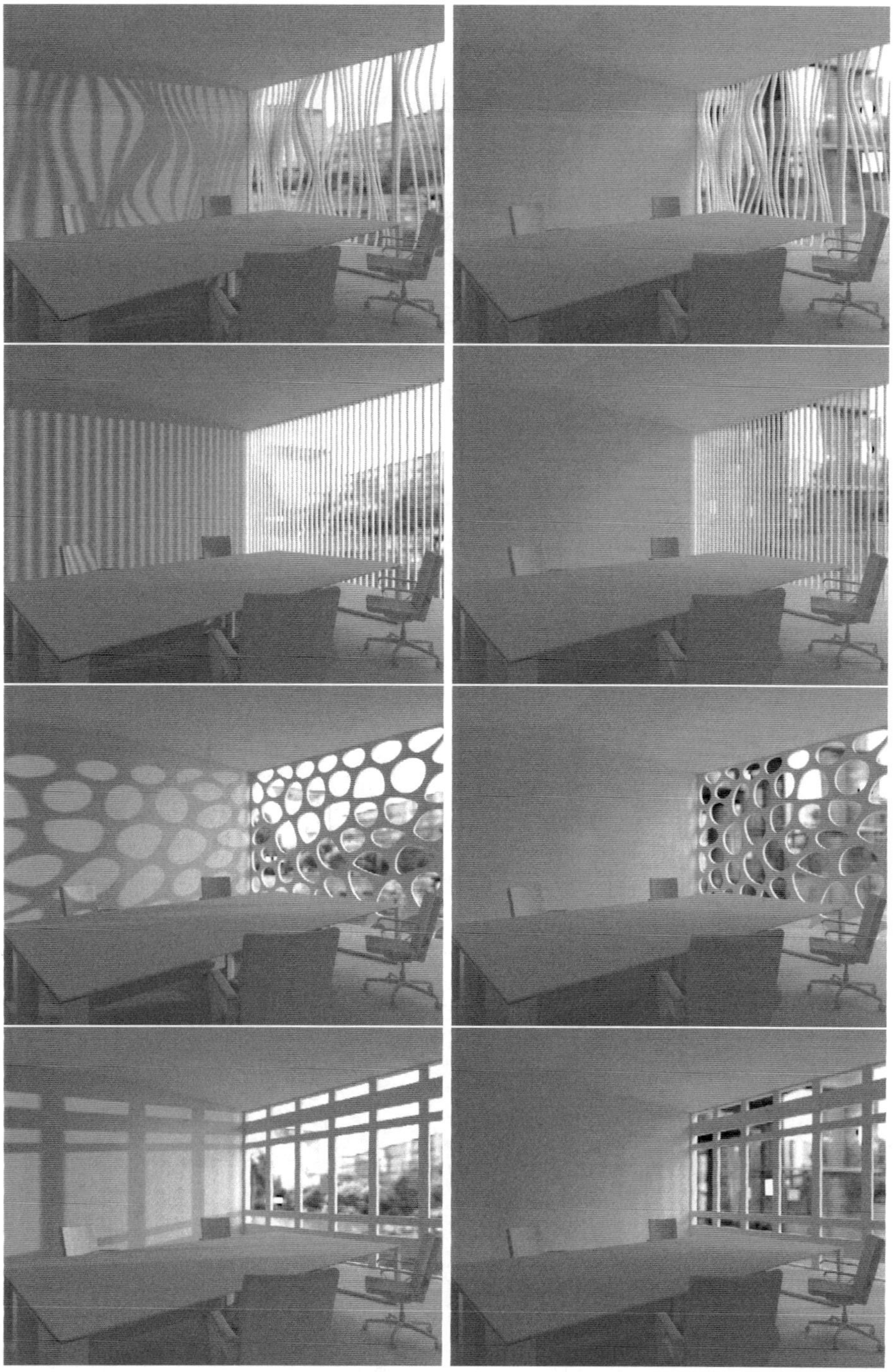

Figure 9. From top to bottom. A: Deformed Vertical Strips, B: Vertical Strips, C: Voronoi Tessellation, D: Regular Openings. All rooms were simulated under two different lighting conditions, direct sunlight and overcast.

in this image was difficult for the eye to discern, which resulted in it being rated annoying. Interestingly, the facade pattern in image C was rated at the same time as ugly and beautiful (25/0%).

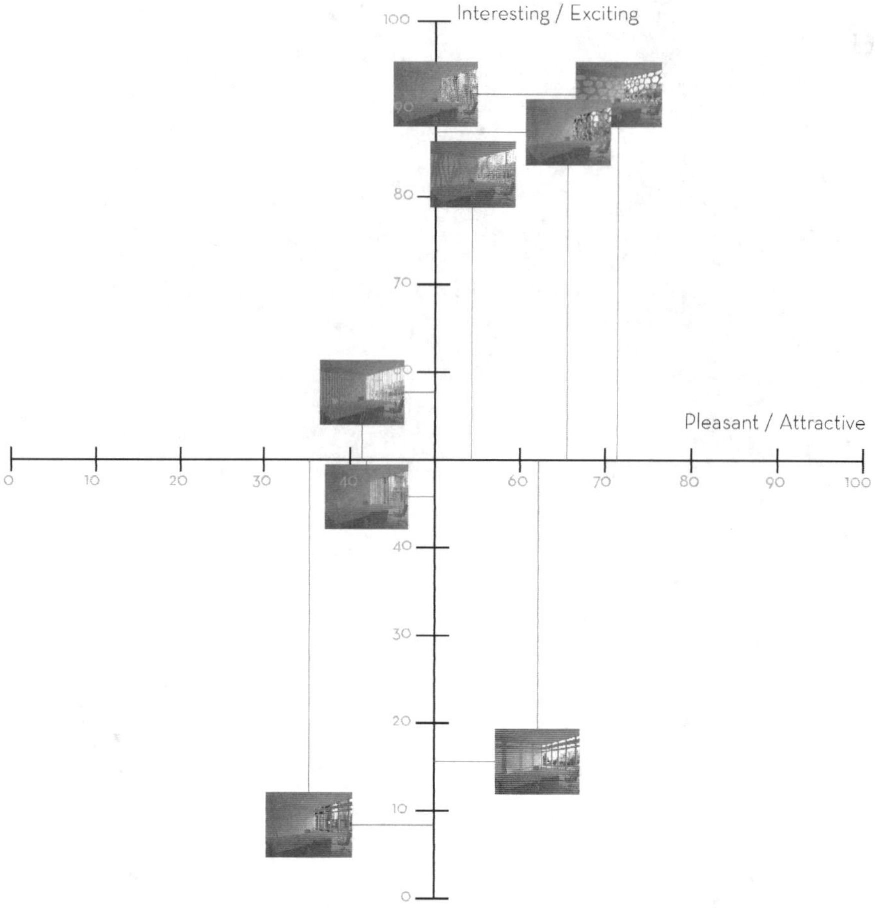

Figure 10. A circumplex model of affect was used to evaluate participants' responses to different facade designs under two different daylight scenarios.

Overall, organic patterns received more positive feedback than geometric patterns. Participants found them to be more interesting and pleasant looking in both sunny and gloomy daylight setups. People have given organic pictures generally positive feedback in an open-ended discussion after the test. An important concern is, however, the cognitive associations that people have with organic facades. In the eyes of some, they were very appealing and evoked positive associations, such as for example a "honeybee comb". However, the hole pattern turned some people off, and they found them frightening. A person referred to the facade with the twisted vertical blinds as "tentacles of a squid". Thus, it is extremely important to address the problem of triggering certain nature-based phobias of potential users while looking for natural inspired patterns.

9. Conclusions

A major concern when using electro-mechanical panels for adaptive facade design is its durability and power consumption. Nitinol has proven to be relatively durable in tests, and using a sheet instead of wire could be a more secure method. Additionally, the proposed design solutions meets all three success criteria, as it can adjust to the light, has a visual representation similar to the natural shape, and is based on bio-mimicry principles. As a result, Nitinol is a viable material for bio-mimetic facade design. Furthermore, it is easier to maintain and replace faulty parts, which may be useful for larger scale construction. It is however important to conduct additional testing in the future to verify Ninitol's durability, its financial benefits, and how it compares to other methods that use moving mechanical parts. Due to the uncertainty of the shape of the deformations or how they will behave under varying conditions, lots of experiments must be performed in order to optimize their use. Moreover, a controller is needed such as PID to make the adaptive facade more efficient and to reduce the amount of energy it consumes. Thus, it becomes clear why we have chosen to simulate the geometry and deformations of the facades. In addition, the human comfort in this space must also be considered. We will conduct more testing via VR to gain a better understanding of how people react to nature-inspired patterns. The majority of participants found the facade visually appealing, but some found it irritating, triggering unpleasant associations. The application of smart materials to the built environment can create unique structures that are constantly changing based on the environment. Thus, the aesthetic experience of the inhabitants should be taken into account in future work, as smart materials are a relatively new field of study and SMAs have tremendous potential for adaptive facades that can shape their appearance according to the time of day or season.

10. References

[1] *Shape Memory Alloy Valves.* 2015.

[2] Myriam BC Aries, Jennifer A Veitch, and Guy R Newsham. Windows, view, and office characteristics predict physical and psychological discomfort. *Journal of environmental psychology*, 30(4):533–541, 2010.

[3] Gregory N. Bratman, J. Paul Hamilton, and Gretchen C. Daily. The impacts of nature experience on human cognitive function and mental health. *Annals of the New York Academy of Sciences*, 1249(1):118–136, 2012.

[4] Christopher Schinkel Carlsen and George Palamas. Evolving balancing controllers for biped characters in games. In *International Work-Conference on Artificial Neural Networks*, pages 869–880. Springer, 2019.

[5] Gary W. Evans and Janetta Mitchell McCoy. When buildings don't work: The role of architecture in human health. *Journal of Environmental Psychology*, 18(1):85–94, 1998.

[6] Marco Formentini and Stefano Lenci. An innovative building envelope (kinetic façade) with Shape Memory Alloys used as actuators and sensors. *Automation in Construction*, 85(October 2017):220–231, jan 2018.

[7] Cristina Caramelo Gomes and Sandra Preto. Should the Light be Static or Dynamic? *Procedia Manufacturing*, 2015.

[8] Roddy M. Grieves and Kate J. Jeffery. The representation of space in the brain, 2017.

[9] Helena Hashemi Farzaneh and Udo Lindemann. *A Practical Guide to Bio-inspired Design.* 2019.

[10] Michael Helms, Swaroop S. Vattam, and Ashok K. Goel. Biologically inspired design: process and products. *Design Studies*, 2009.

[11] Emily Barbara Kennedy. Biomimicry: Design by Analogy to Biology. *Research Technology Management*, 2017.

[12] Dong Hyun Kim and Kevin Mansfield. Creating positive atmosphere and emotion in an office-like environment: a methodology for the lit environment. *Building and Environment*, 194:107686, 2021.

[13] Kyle Konis. *Validating performance from the perspective of end users.* Number 9783319394619. 2017.

[14] Gunter Kreutz, Ulrich Ott, Daniel Teichmann, Patrick Osawa, and Dieter Vaitl. Using music to induce emotions: Influences of musical preference and absorption. *Psychology of music*, 36(1):101–126, 2008.

[15] Maria Lorena Lehman. *Adaptive Sensory Environments.* 2016.

[16] Torben Anker Lenau, Sonal Keshwani, Amaresh Chakrabarti, and Saeema Ahmed-Kristensen. Biocards and level of abstraction. In *Proceedings of the International Conference on Engineering Design, ICED*, 2015.

[17] Marlén López, Ramón Rubio, Santiago Martín, and Ben Croxford. How plants inspire façades. From plants to architecture: Biomimetic principles for the development of adaptive architectural envelopes, 2017.

[18] M. J. Mahtabi, Nima Shamsaei, and M. R. Mitchell. Fatigue of Nitinol: The state-of-the-art and ongoing challenges, 2015.

[19] Marco Pesenti, Gabriele Masera, Francesco Fiorito, and Michele Sauchelli. Kinetic solar skin: a responsive folding technique. *Energy Procedia*, 70:661–672, 2015.

[20] S. W. Robertson, A. R. Pelton, and R. O. Ritchie. Mechanical fatigue and fracture of Nitinol. *International Materials Reviews*, 2012.

[21] James A Russell. A circumplex model of affect. *Journal of personality and social psychology*, 39(6):1161, 1980.

[22] Avishag Shemesh, Ronen Talmon, Ofer Karp, Idan Amir, Moshe Bar, and Yasha Jacob Grobman. Affective response to architecture–investigating human reaction to spaces with different geometry. *Architectural Science Review*, 2017.

[23] Mahdi Valitabar, Mahdi Moghimi, Mohammadjavad Mahdavinejad, and Peiman Pilechiha. Design optimum responsive façade based on visual comfort and energy performance. 2018.

[24] Carlos A M Versos and Denis A Coelho. Biologically inspired design: methods and validation. *Industrial Design-New Frontiers*, 2011.

[25] Andreas Wulff-Abramsson, Mads Deibjerg Lind, Stine Louring Nielsen, George Palamas, Luis Emilio Bruni, and Georgios Triantafyllidis. Experiencing the light through our skin-an eeg study of colored light on blindfolded subjects. In *2019 IEEE 19th International Conference on Bioinformatics and Bioengineering (BIBE)*, pages 609–616. Ieee, 2019.

IOP Conf. Series: Earth and Environmental Science 1099 (2022) 012006 doi:10.1088/1755-1315/1099/1/012006

Lighting inequality in an urban context: design approach and case studies

Luciana Martinez, Elettra Bordonaro

Light Follows Behaviour, Unit 7 411-413 Liverpool Road, London, N7 8PT, UK

luciana@lightfollowsbehaviour.com

Abstract. Over the past years, the focus on public spaces and the need for well-designed outdoor environments has been a key focus for those professions working in the sector. There is increasingly a push for more pedestrian-oriented spaces which support movement and activity and also have an extended use after dark. It is increasingly more apparent, however, where the divide exists between areas that are in well-off areas than those in more deprived areas. This is particularly apparent after dark where the lighting design more often than not, creates the most apparent social divides. Lighting becomes cooler, glare is abound, playgrounds are unlit and any sense of human scale is lost with a blanket lighting, or lack of lighting, approach becoming dominant. These spaces of course are not those featured in magazines, blogs or awards ceremonies. These are the spaces that continuously fly under the radar and hinder a bigger positive change. And although, in some cases, best practice guidelines are followed, the end result does not create inviting spaces after dark. We still regularly encounter pedestrian and residential spaces that are lit primarily with bulkheads or streetlights or where the only play areas are dark at 3pm in winter in Northern Europe. These spaces lack atmosphere and character, the very things pedestrian spaces require. With this in mind, we question how lighting design as a profession can evaluate a successfully lit outdoor space. Is it time to develop a new approach that takes into consideration the everyday user experience, not just the technical successes? And can this approach/methodology better inform best practice guidelines for the future and begin to close the gap of lighting inequality by raising the bar? Through our studio's work we have been exploring new methods to engage with the public on topics of lighting both in the UK and in Europe. A number of cities and municipalities are also increasingly working towards a more engagement-oriented approach that demonstrates lighting design in easy to understand terms and installations in situ which in turn helps us as designers better design outdoor spaces for the end users by creating a common language and understanding.

1. Introduction

Over the past decade, and most apparent in recent years as a result of the COVID-19 pandemic, the design of cities and the public realm has increasingly focused on creating spaces focused on improving the pedestrian experience and adapting to the need of evolving transport methods [1]. Creative design approaches to seating areas, investment in improved play spaces for children, and revised municipal and city design policies have resulted in an evolving public realm for pedestrians focused on usable, enjoyable outdoor space as opposed to maximising parking places and efficient traffic routes. Yet while spaces are increasingly welcoming by day, the after dark condition and design approach has not followed in tow.

2. Urban lighting and inequality

Lighting in public realm spaces plays a key role in strongly influencing how pedestrians move through and use spaces after dark, most notably in cities [2]. It is important to note that after dark is not solely related to late hours of the night given in many parts of the world darkness falls in early afternoon during winter months. Similarly, in many countries, use of outdoor spaces only happens after dark due to extreme weather conditions. This results in a very diverse approach to the use of cities and spaces after dark around the globe and throughout the year. Darkness does not signal a reduction or lack of activity outdoors in many cases which makes lighting a critical element in our experience of outdoor spaces. As a result, many cities have been upgrading and improving lighting to provide more welcoming city centres and welcoming residential developments [3][4]. Yet improvements are increasingly directly linked to socio-economic status of an area or assumptions stemming from outdated beliefs. Most crucial is the issue of safety and security in spaces and the resulting design approach in many cities results in a demarcation of space that is most visible after dark. It is increasingly visually apparent that there exists two separate set of rules and guidelines depending on where an area is located. Door lights become bulkheads, warm white light becomes cool white and typology of lighting resembles that typically found in industrial settings with acceptable applications significantly reduced.

This drastic shift is most evident in housing developments. It is important to note a distinction that exists to form the foundation of the issue where the word 'residential' is seen as more affluent while 'housing' (eg. housing estate, social housing, affordable housing, low income housing, etc.) downgraded to a separate tier of equipment as well as expectation as a simple internet search can confirm. Discussions of lighting design quickly turn towards crime prevention, deterring unwanted gathering of people, maximum efficiency and vandalism as the target criteria [5]. There is no mention of atmosphere, obtrusive lighting, character or function as priority. As a studio we have also observed this list of priorities is reflected in the topics residents of housing projects feel apply to them. Lighting is not an element of aesthetics but of prevention, and to the same effect residential lighting is not a category which applies to their situation, and more importantly lighting is not an element they feel they have control over or a say in. Artificial light shining into bedroom windows for example is often a consequence of living in housing developments and thick curtains are the solution adopted. Cool white light they were told results in better visibility and is associated with security lights and therefore implies improved safety [6]. Brighter is better for lighting and improves visibility. Of course, those working in the lighting design profession understand these statements are false and grossly over simplified, and in many cases outdated thinking from generations past, but we continue to see developments upgraded to LED technology as well as new developments which continue to light spaces without care to the after dark experience or wellbeing of residents. Notably, these projects are not those featured in lighting awards and publications relating to design but they are featured as technical and statistical successes to demonstrate efficient and cost effective examples of LED applications resulting in a reduction of a negative aspect of urban living.

How can we approach the design of such spaces in a different way which takes into consideration the needs of people living in the area, minimising the inequality gap and demonstrating a forward-thinking approach to lighting? Which methods could be utilized and how can they be implemented.

3. The "Users approach"

While, in principle, the existing approach is efficient design in terms of maintenance and energy consumption, it does little to support the needs of people. And while guidance for lighting is continuously being updated in an attempt to capture all requirements it is increasingly difficult if the evaluation of success is only set at a technical level. This idea of success is more in line with highways and motorways than homes and places for people. It also needs to be recognised that continuing to light housing developments and similar spaces intended for pedestrian use in a subpar manner hinders wider, holistic, improvements across our cities and only serves to exacerbate the existing social divide through increased lighting inequality in our cities[7][8]. Guidance documents relating to lighting are constantly in conflict with each other depending on the situation and particular design issues raised as well, with areas known or assumed to be less secure and safe falling into a separate category of guidance all together, as mentioned earlier, in terms of expectations.

Existing and evolving best practice guidelines, standards and regulations are extensive but it must be noted that updates often lag behind what is already being implemented by the lighting design community and are also formed and influenced by the profession. With that in mind, it is critical to understand where guidance is lacking or outdated and respond accordingly. It could also be argued that, in addition to existing guidance, a new approach, based on the understanding of local communities, should become commonplace to better understand the success of projects in relation to end-user response. This will help inform the profession as a whole to design more responsive spaces as well as help reduce lighting inequality by raising the bar and expectations for all projects equally.

It is becoming increasingly apparent that to deliver a successful project beyond a technical level it is critical to understand the needs of communities in which we work and respond to these needs in a creative and informed manner. This also helps to better understand the true issues associated with spaces directly from those who use or will use the space in the future. While the trend has been to impose design on communities in the past, this approach has seen the shortfalls inherent in such a design and in many cases fueled the negative aspects associated with challenging urban areas such as feelings of resentment and indifference. As a result, it is increasingly important to better understand who the end users of spaces are or will be, what their needs are and help them understand basic aspects of lighting to essentially break the cycle of poor lighting across our cities and help communities communicate their needs and aspirations in a more effective manner. This is an important step for the profession and municipalities to move away from referring to lighting only in terms of brightness and as a crime deterrent in relation to housing projects and to understand that poor lighting has a knock-on effect on the wider context of an area.

This new methodology should be based on a number of characteristics including engagement and user understanding, education and knowledge transfer to users, aesthetics of lighting, improved wellbeing for end users and reducing lighting inequality. This method can also serve to help document evolving trends and considerations and highlight ongoing challenges faced by designers due to constraints imposed outside of their control and, more importantly, move away from expectations within the profession of all projects of quality and merit resulting from high-end developments, bespoke installations and products and high budget allowances.

4. Background case studies

For a number of years our studio's work has focused on applying many of the elements an techniques listed above and developing innovative and forward thinking methods to communicate lighting design to the wider public and design professionals in easy to understand terms and help break preconceived notions related to lighting. A selection of case studies has been selected to demonstrate these principles and showcase an alternative approach to design and how this could be applied.

4.1. Brandon Estate, Southwark, London, UK

Brandon Estate was a project undertaken in collaboration with Configuring Light, an interdisciplinary research programme that explores the role lighting plays in everyday life to help build a better social knowledge basis for lighting design interventions. The brief focused on improving lighting on a social housing estate known for violence and gang activity in recent years following a number of incidents including a fatal stabbing witnessed by a number of residents. As a response the lighting was upgraded across the estate to LED drastically increasing light levels, glare and contrast.

Figure 1. Existing lighting condition was upgraded to LED to provide higher light levels

Figure 2. Existing lighting did not encourage the space to be used with many residents commenting they purposely avoid this route after dark due to security concerns.

The team spent multiple days both on site and on virtual platforms to better understand residents' current thoughts and concerns about the estate after dark. It quickly became apparent that many felt unease walking a specific route after dark due to security fears and poor lighting. The square, located at the estate's perimeter, functioned as a crucial entry point to the estate with large mature trees and homes adjacent. The cool, white flood lighting was mounted on residents' façade flooding light into windows and lighting the space to nearly 70lx. The premise of the project was to introduce a temporary lighting installation with drastically different lighting in the space to help dispel prior notions of appropriate lighting on housing estates and start to turn the conversations away from safety and security and focus on community aspirations for the site instead. Although simple in technical terms, the warm white festoon light introduced both functional considerations (low glare, high uniformity and required light levels) alongside aesthetic considerations (character, identity and joy) to the space immediately altering its appearance and creating enthusiasm and buzz among the community. Light levels were reduced by over 50%, colour rendering raised to above 80 and light spill through windows drastically improved while most importantly residents noted feeling safer walking through the space after dark as a result of the changes.

Figure 3. Warmer, more uniform light improved visibility and gave a welcoming atmosphere

Although only intended as a temporary installation to serve as a research tool, the overwhelmingly positive response led to residents convincing the local authority to make it a permanent feature. The installation and process has resulted in new dialogs beginning to emerge on what is possible in terms of lighting and design on housing estates across the area to deliver positive change as showcased with the lighting installation in the square and the realization that brighter and cool white light does not result in improved security after dark and the council acknowledging that a utilitarian approach to lighting can at times be part of the problem and that lighting to housing estates can take a variety of forms outside of bulkheads and floodlights.

Figure 4. The residents spoke with city officials about the lighting and the process, using the language and knowledge transferred to them by the team during the project meetings and interviews.

4.2. Moorings Underpass/Arnott Close, Thamesmead, London, UK

Serving as a main connection between two key areas of a large housing development known for crime, deprivation and vandalism, the Moorings Underpass was earmarked for improvement as part of an estate wide program aimed at working with local community groups to deliver the design for the small space. While the initial focus group was intended to be young primary school children, it became overwhelmingly apparent that teenagers were the key age group creating tension and vandalising property across the site. After consulting with the client, it was agreed a new approach would be undertaken to work with a youth centre on the estate and design the space in collaboration with them.

Once we began speaking with the group and brainstorming ideas, it came to light that many had never left the estate and therefore had limited references to refer to in terms of aspirations and design. A portion of the budget was therefore used to take the group on a research trip to central London and the vast array of opportunities lighting could provide. This was a small but important step to raise aspirations for the end result and expose the group to lighting installations and types they had not seen in the past.

Figure 5. A trip to central London helped the group expand aspirations for the new lighting

The new lighting to the underpass challenged the council to move away from utilitarian light fittings and install a simple lighting enclosure and luminaire detail which can be replicated in other similar areas. Technically, the lighting solution is more robust than what was installed initially and improved vertical lighting has resulted in a space whose perceived brightness drastically enhances the new tiled artwork and encourages pedestrians to use the route after dark.

Figure 6. Existing lighting condition in comparison to new lighting and artwork.

IOP Conf. Series: Earth and Environmental Science 1099 (2022) 012006 doi:10.1088/1755-1315/1099/1/012006

Although great fears were raised regarding longevity of both the tiled artwork and the lighting installation due to vandalism it is one of the first projects on the estate which has not been vandalised. More importantly the design process gave a forgotten space on the estate an improved appearance and the process of design resulted in a sense of ownership by a local community group. Although only a small intervention, it has resulted in growing enthusiasm to improve the adjacent area resident amenities nearby.

Figure 7. The work with the community we believe has ensured the artwork has not been vandalised and instead used as backdrop for many selfies and TikTok videos since its installation.

4.3. Shadwell Estate, Tower Hamlets, London, UK

As part of a larger housing estate improvement initiative, the communal spaces on the housing estate underwent a significant transformation with new landscape design and play features introduced to what was tarmac and parking spaces. The existing lighting was in dire need of improvement with legacy technology streetlights and bulkheads mounted onto building facades the only sources of light. Site constraints resulted in extremely limited instances where lighting could be introduced into the landscape and existing power connections on existing building facades were the sole viable option. While this would have illuminated pedestrian pathways to a sufficient level, lighting of play equipment was a key priority raised by residents, many of whom had small children.

Figure 8. Prior to the works the spaces were illuminated with legacy streetlights, which were not appropriate for the new landscape design proposed.

In order to light the play areas as well as reduce light spill through residential windows a simple catenary system was designed and installed. Alongside underbench lighting and tree uplights to key trees, the lighting design pushed what was expected and deemed acceptable for housing estates. It was the first catenary system installed by the client across it's housing stock. Residents comments of feeling as the estate was now a high end development pay testament to the importance of lighting and design together can influence public opinion of areas and instill a sense of pride. And although the deck access bulkhead lights were not included as part of the overall design as they were not part of the public realm improvement scope, a simple demonstration made apparent the importance of improving the lighting of these elements was key to reduce spill light and glare and more importantly, residents understood that alternate options were available to them.

Figure 9. The new lighting transformed the estate and drastically reduced spill light into windows while still lighting play areas

Figure 10. Building lighting was not changed (left image) on this occasion demonstrating the importance of reducing light spill and glare (right image).

Figure 11. Low glare door lighting installed on another area of the development demonstrates that an alternate, low glare option is possible.

5. Conclusion: A New Approach

As the lighting design profession continues to evolve and adapt to changing design trends and updated regulations, it is critical to acknowledge the importance of developing and adopting a design approach which includes qualitative analysis alongside quantitative data. An approach which is focused on local understanding and uniqueness of the place without relying only on preconceived notions should be driving design decisions. A number of actions would help propel a new approach forward including:

- Promoting alternate, qualitative methods of design and research to be used by design groups, offices and universities. This includes speaking about lighting in easy-to-understand terms which can allow non-lighting professionals to understand key and vital topics.
- Establishing post-implementation studies on existing projects. This can help better understand the impact of lighting in spaces but also promote alternate equipment and installations not previously considered.
- Understanding and acknowledgment of lighting inequality and the impact it is having on cities after dark. This includes more lighting professionals raising and tackling the issue in order to increase the number of city officers, municipalities and design professionals aware of the issues and downfalls if not addressed.
- Continuing to encourage cross-disciplinary discourse and design on the topic of lighting will ensure lighting designers are involved in projects from early stages through to completion. This is particularly important when designing public spaces in which gaining an understanding of a place and its communities requires time.

IOP Conf. Series: Earth and Environmental Science 1099 (2022) 012006 doi:10.1088/1755-1315/1099/1/012006

References:

[1] Gehl 2020 Public Space & Public Life during COVID 19
[2] Atkins S, Husain S and Storey A 1991 Influence of street lighting on crime and fear of crime
[3] 2014 Low energy street lighting: making the switch *Green investment bank report*
[4] 2015 UK Department of energy and climate change report energy consumption in the UK
[5] Kortava D, 2021 The controversial floodlights illuminating New York City's public-housing developments *The New Yorker*
[6] 2019 Secured by Design Homes 2019 Version 2
[7] Entwistle J, Slater D and Sloane M 2016 Tackling Inequalities in Public Lighting
[8] Nadybal S, Collins T and Grineski S 2020 Light pollution inequities in the continental United States: A distributive environmental justice analysis *Environmental Research* 189 (1-2)

IOP Conf. Series: Earth and Environmental Science 1099 (2022) 012007 doi:10.1088/1755-1315/1099/1/012007

Ways to study changes in pedestrians' behaviour in the artificially lit urban outdoor environment

Dmitrii Ingi, Pramod Bhusal, Paulo Pinho, Marketta Kyttä, Martin Parker

dmitrii.ingi@aalto.fi, pramod.bhusal@aalto.fi, paulo.pinho@aalto.fi, marketta.kytta@aalto.fi, martin.parker@taltech.ee

dmitrii.ingi@aalto.fi

Abstract. Lighting technologies have been rapidly developing in recent years together with the advancement in digital and connected technologies. Nevertheless, while achieving quantitative goals with the help of technical developments, there is always a need to correlate them with their impacts on users' experience. This paper focuses on the changes in human behaviour whilst walking under an artificially lit outdoor environment. A systematic literature search was conducted from which three groups of a pedestrians' behaviour were identified, namely walking (including activity levels and walking patterns), microscopic mobility (including walking speed, gait characteristics and use of walkway space) and gaze behaviour. An additional round of a literature search was done for each of the group mentioned above. Final set of the articles consisted of 41, 13 and 12 papers for walking, microscopic mobility behaviour and gaze behaviour subgroups respectively. The existing knowledge on possible changes in pedestrian behaviour caused by various artificial lighting settings was summarized. In addition, the methods to gather data about these changes applied in the selected papers were reviewed, and the perspectives for video-based analysis and online public participation geographic information system methodologies were discussed.

1. Introduction

The role which outdoor lighting plays in our current, 24/7-world cannot be underestimated. Due to different layers of artificial outdoor lighting, people can continue their activities after dark, feel safer and enjoy the urban environment. It goes without saying that leisure time for most people, including their necessary, optional, and social activities, is held after dark when they have finally completed a workday. In addition, people who need to work night shifts also use outdoor public spaces intensively even during night hours. The lighting aspect is particularly relevant in the autumn and winter periods in the northern countries when daylight hours are significantly reduced.

Lighting technologies have been rapidly developing in recent years together with the advancement in digital and connected technologies. Nevertheless, while achieving quantitative goals with the help of technical developments, there is always a need to correlate them with their impacts on users' experience. The evaluation of pedestrian experience in the lit environment can be approached based on three types of human responses: perception, evaluation, and behaviour [1]. These responses to lighting were systematically reviewed by J. Rahm in his doctoral thesis [2]. While methods to study pedestrians' perception and their evaluations are sufficiently developed to implement them in real-world studies, methods to study changes in pedestrians' behaviour are less known and usually limited with direct observations. However, to fully evaluate pedestrians' experience in the lit environment lighting professionals need to analyse and correlate the data of all three human responses mentioned above. This paper focuses on behavioural aspects of the pedestrian experience in the artificially lit urban outdoor environment, as well as on the review of different data collection methods used to study changes in a pedestrian behaviour.

The first objective of this paper is to summarise and analyse existing knowledge on possible changes in pedestrian behaviour caused by various artificial lighting settings (e.g., characteristics and targeted lighting design). Secondly, it reviews and compares methods to gather data about changes in pedestrian behaviour applied in the selected papers, identifying and discussing perspectives for authors' upcoming experiments.

2. Method

A systematic literature search was conducted between December 2021 and February 2022 by the author of the paper. The flowchart of the process is shown in Fig.1. The initial keywords in this work were pedestrian* AND light* AND behav*. The terms were searched in titles, abstracts, and keywords of the papers in two digital databases: Scopus and Web of Science. After the abstracts' screening and filtering process, a set of articles (21) was identified using key inclusion and exclusion criteria listed below. Then selected papers were fully read and topically grouped into three main subgroups, namely walking (including activity levels and walking patterns), microscopic mobility behaviour (including walking speed, gait characteristics and use of walkway space) and gaze behaviour. Relevant articles used as references in the selected set were also included in the final set. Afterwards, building upon reading material, additional search terms mentioned in the flowchart were used for the second round of a literature search with the same key inclusion and exclusion criteria. The second round was conducted in three digital databases: Scopus, Web of Science and Google Scholar. Final set of the articles consisted of 41, 13 and 12 papers for walking, microscopic mobility behaviour and gaze behaviour subgroups respectively. It is important to declare that some articles simultaneously covered several topics (e.g., speed of walk and gaze behaviour). Selected papers were fully read and analysed to achieve previously stated aims.

In addition, it is worth noting that there were 7 papers which also described behaviour although could not be linked with the three identified groups that were harvested during the second round. They will be mentioned in the 4th chapter.

Key inclusion criteria:

- Studies are dedicated to any type of pedestrian behaviour after dark.
- Both laboratory and field studies which focus on outdoor or public spaces experiences.
- All scientific publications available in English including reviews, conference papers, book's chapters.
- Availability of the study's results and a method description.

Key exclusion criteria:

- Articles which study traffic lights and street crossing behaviour (e.g., signalised crosswalks), lighting-warning systems as well as focus on driver's experience.
- Studies where lighting was considered in combination with other aspects. For example, "challenging environments" (i.e., irregular surface and low lighting).
- Studies which were focused on subjects with disabilities (e.g., visually impaired subjects and glaucoma).

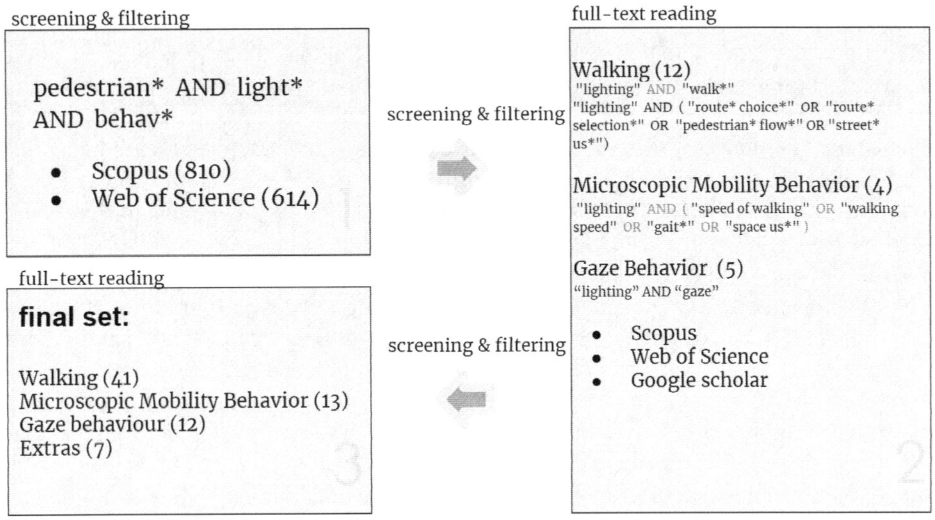

Fig. 1 – The flowchart of the literature search process

3. Results

3.1. *Impacts of artificial lighting on pedestrians' behaviour*

3.1.1. Walking. The perception of the built urban environment after dark dramatically differs from the perception under daylight. This fact, in its turn, leads to shifts in various urban experiences, activity levels and walking patterns of people. It was repeatedly found that outdoor lighting, being one of the components of the built environment, has a significant impact on different forms of physical activity after dark [3-20]. Firstly, it was statistically substantiated that ambient lighting conditions affect the number of people who walk or cycle [3-6]. It was found that conversely to daylight, darkness leads to notable reductions in pedestrians and cyclists, and that more people cycle and walk on cycle trails and footpaths after dark when they are artificially illuminated. In addition, the findings of study [6] demonstrated a significant reduction in footfalls after the clock change (i.e., when it dark outside) on 5 out of 6 sites on weekdays and half the sites at weekends. Secondly, the presence of good quality street lighting can increase the use of pedestrian streets and public spaces after dark by different user groups promoting walking and active lifestyle [7-11, 13-16, 18-20]. It is important to note that inadequate street lighting can also be considered as a walking and cycling barrier [12, 17, 21, 22]. For instance, the finding extracted from an interview: "For example, at a certain time of the day, a poor street lighting makes me take the bus instead of walking alone in the dark" depicted the relevance of lighting conditions in the selection of travel mode [21].

An artificial lighting environment has been found to affect the pattern of movement in terms of route choices. The study [23] explored the impact of altered visual perception after dark (i.e., in relation to the spatial configuration) on most frequently used routes and found the correlation between selected or most frequently used routes and illuminance values. In addition, according to the results of the study [24], people tended to avoid dark spots at night-time residential streets. The authors assumed that dark spaces, turning themselves into visual boundaries, can be largely attributed to decreased perceived safety and reassurance. Indeed, the results of studies [25-32] indicated that the reassurance of people (i.e., the feeling of confidence a pedestrian might gain from road lighting [33]) can be put in relation with outdoor lighting, and, moreover, can be considered as a key contributor to enhancing or limiting (e.g., compelled detours) walking at night-time. It is worth noting that blind spots, such as dark spaces between streetlights, under tree branches, and dark winding street corners, had a significant input to perceived safety, especially among female students [31]. In addition, the study [28] showed that urban greenery and street lighting must be explored and designed together, since their interaction influences perceived safety and impacts the walkability of the neighbourhood. Nevertheless, outdoor lighting can have an influence on walking also without a direct impact on perceived safety and reassurance. For instance, the walkability of an area can be related to visibility [34], perceived restorativeness [35, 36] or personal evaluations and preferences [31, 37, 38]. It is notable that the time spent in the environment can be altered by environmental features, including lighting [19].

It is important to highlight that when people are choosing a route, they also consider the quality of the lighting environment. The study [7] showed that the improved lighting schemes have a markable impact on the number of people using the streets. Moreover, there were several studies attempted to investigate how certain lighting variables influence walking patterns. It was found that among other factors average illuminance on the pavement affects pedestrian route taking decisions (i.e., pedestrians avoid routes with average illuminance of less than 5 lux and prefer routes with the average illuminance of 8 lux and above) [39]. Besides, the study [31] concluded that students' preference of the brightest lit route resulted in more foot traffic relative to other routes. Additionally, it was found that a decrease in the colour temperature of ambient lighting from 6500K to 3200K reduces the comfort level of pedestrians, as well as that colour temperature had the strongest influence on the choice of a route by pedestrians, compared with dimming [40]. Lastly, the use of targeted lighting design and visual cues such as projected light or chasing light-emitting diode (LED) lighting, had a potential to influence people's movement and behaviour [41, 42].

3.1.2. Microscopic mobility behaviour. Besides walking in a broad sense, lighting conditions can also impact on more precise parameters of human mobility, namely walking speed [42-48], gait characteristics [46, 47, 49, 50-53, 55] and use of the walkway space [42, 45, 54].

Suggesting that accelerated walking speed might indicate pedestrian discomfort, the authors of the study [43] showed that in the selected urban street a decrease in lighting does quicken pedestrian steps in comparison with the same street when well-lit or at daytime. In contrast, several laboratory experiments concluded that during walking a decrease in illumination leads to fall in a walking speed [44, 47] or a step rate [48]. The study [44] examined possible effects of dynamic lighting on walking. It was found that the participants walked significantly slower under dimmed than static lighting conditions, even after the illuminance had increased, and the larger were differences between the dimmed lighting levels and the full lighting, the longer was walking time. When comparing two different lighting conditions (high pressure sodium (HPS) and ceramic metal halide (CMH)), the authors of the study [45] reported that the mean speed was significantly slower under HPS lamps along the straight part of the path. Finally, the impact of targeted lighting design and visual cues on walking speed was also discussed [42, 44, 55]. It is worth noting that two studies [1, 54] did not find any walking speed changes with the change in lighting conditions.

There were several laboratory studies which found that the reduced illumination has an influence on various gait characteristics during walking which, for instance, can be indicators for slips, trips and fall accidents [46, 47, 49, 50-53]. For example, the lighting environment can affect such gait characteristics as stride length, stride time, double support time gait variability of stride length and stance time [49], stance phase ratio and toe clearance [47]. Moreover, changes in gait patterns under low illumination can be enhanced during obstacle avoidance tasks [51, 52] or when walking on stairs [41], as well as can be differently impacted among various user groups [46, 47, 53]. It was also found that reduced lighting conditions are significant on perceived risk of tripping [51, 52].

Outdoor lighting has also been found to affect the way people use a walkway space [42, 45, 54]. The authors of the study [54] observed pedestrian placement on pedestrian paths. Participants tended to vary their placement and change their path more with a fluorescent tube luminaire (FTL) light source than with a LED light source, which was explained by improved visual accessibility with the LED light sources. Moreover, the observations of pedestrian walking behaviour revealed a more centric walking path along the sidewalk under LED illumination, particularly with the optimised LED luminaire. Changes in passengers' behaviour such as an increase of the use of the left-hand side for descending, and a shift from the centre to the opposite side of the staircase after introduction of the targeted lights (i.e., handrail lighting on the stairway) were reported in the study [42]. In addition, it was noticed that the average path for lighting application with CMH was systematically closer to the centre line compared to application with HPS, although the difference, in absolute terms, was quite small [45].

3.1.3. Gaze behaviour. Lighting conditions can also affect a pedestrian's gaze behaviour while walking in terms of critical distances, viewing directions, the fixation number and fixation durations. It consequently impacts the way people execute their visual tasks such as object detection, visual identification, visual orientation and on the perception of the urban environments.

Pedestrians spent between 40% and 50% of their time looking at the footpath, but not all that time was spent on critical visual tasks [56]. Identification of other pedestrians and their emotions also appeared to be significant during walking [56-58]. It is notable that after dark the path was more likely to be fixated and other people less likely to be fixated compared with daylight [57]. The apparent importance of fixation on other pedestrians was also highlighted in the study [58]. Their results showed that the proportion of time that fixations were on pedestrians was 14%, the proportion of fixations at critical moments that were on pedestrians was 23%, and the probability of an approaching pedestrian being fixated at least once was 86%. In addition, the study [48] concluded that in the conditions of mesopic vision (i.e., 10 lux), older adults more often looked at the floor or the edge, especially without edge emphasis, which meant that higher levels of illuminance were associated with a decreased need for directing one's gaze towards the ground.

There were few studies analysing the distance and duration of fixations on other pedestrians and found a tendency to fixate at 15 m and for a duration of 500 ms [59, 60]. The judgement of other people's emotions from facial expressions can also be influenced by lighting conditions. For instance, to be assured in accurate identification of the facial expressions, the study [61] suggested requirements of a minimum luminance of the face of 0.1–1.0 cd/m^2 at 4 m distance, and a luminance above 1.0 cd/m^2 for identification at 10 m. Additionally, it was found that human's sense of where another person is looking can be biased by the lighting direction and eye glint, which also affects people's sense of gaze direction, varies with the lighting conditions [62].

Eye tracking records collected and discussed in the studies [57, 58, 63] suggested a tendency to peripherally detect 10 mm obstacles at approximately 3.4 m ahead under horizontal photopic illuminances of up to 0.9 lux. It is important to note that lighting requirements and a critical distance can vary, according to the scotopic/photopic ratio of the lighting and the age of the observer. In addition, it was reported that both the trial time and detection distance of older participants during a detection of an upcoming step were affected by light level, hence, their vulnerability may be reduced if a better lighting environment is provided [64].

Ultimately, relative luminance and saliency of a various scene areas may attract the visual attention of pedestrians. The study [65] found that fixations within 2° and 10° viewing angle were predominantly distributed in bright and significantly salient regions of the outdoor visual environment.

It is worth noting that background lighting conditions may impact spontaneous eyeblink activity, as well as that spontaneous eyeblink activity rate changes can occur when lighting levels suddenly increase [66].

3.2. Methods used to gather data about changes in pedestrian behaviour

3.2.1. Walking. There were both quantitative and qualitative approaches to study pedestrian behaviour which can relate to walking. Firstly, the total number of pedestrians who use the site can be recorded automatically by traffic counters [3, 4, 5] or by direct observation (e.g., a gait-counting method) [6, 7, 23, 40, 42, 67] for specific periods of time. Afterwards, count data collected for after dark conditions can be compared with daylight values by using the biannual daylight-saving clock changes approach

[3, 4, 6] or the whole-year approach [5]. To compare the usage of various artificially lit spaces or to evaluate changes after the lighting installation improvements, it is possible to implement real-time field observations [7, 23, 40, 67] or to analyse footage from fixed cameras [42]. These counts can be further used for the creation of the generative natural movement patterns [23]. Moreover, structured and semi-structured observations can be utilised to investigate stationary and transitory activities and measure their duration time [18], as well as people's movement and occupancy patterns [23, 42]. It is important to note that observed data should be carefully filled in protocols. A behavioural observation method might be combined with a lighting survey to allow researchers access to the context and meaning surrounding people's behaviour [39]. In addition, physical activity levels of the experiment participants can be assessed by accelerometery [11].

Another approach was to collect and analyse self-reported data through on-site and online pedestrian surveys [19, 32], interviews [8, 15, 22, 41] and environmental walks [28, 68]. Surveys were used for measuring physical activity [9, 10, 13] (e.g., the Youth Physical Activity Questionnaire [13]), walking and its characteristics (i.e., walking times, frequency, and duration) [16, 19, 26, 27] (e.g., the International Physical Activity Questionnaire [16]), and evaluations of behavioural intents and space perceptions in order to understand likely behaviour in particular sites [24, 29, 34, 38]. It is notable that a couple of studies used a real-time approach, according to which, participants filled in the surveys while walking along the sites with the help of a paper questionnaire [32] and a mobile phone application [30]. Interviews were used to gather qualitative data about perceived supports and barriers of physical activity [8], utilitarian walking [15] or walking routes [22], as well as unique knowledge from various groups of stakeholders or staff [41, 42]. Finally, there were several approaches of guided structured [28] and random [68] environmental walks. These approaches allowed researchers to gather subjective environmental evaluations, behavioural preferences, and ideas how to improve the neighbourhood in relation to walking.

3.2.2. Microscopic Mobility Behaviour. Walking speed was examined in both full-scale laboratory [1, 44] and field [43, 45, 54, 58] studies. It was measured as walking time a pedestrian took to cross an established distance [43] or the whole experimental site [1, 44]. The study [45] described a novel method based on Video Analysis of Pedestrian Movement (VAPM). The method includes adaptation of existing video technologies to measure actual microscopic movement (i.e., placement and speed). It was claimed that VAPM can be successfully combined with observer-based assessments. Pedestrian observations, in their turn, can also be performed to study walking speed (e.g., unexpected slowdown) [54]. Gait characteristics were studied in laboratory conditions with the help of infrared and reflective markers placed on anatomical landmarks of each participant and a tracking system such as an eight-camera motion analysis system [46, 47, 49, 50, 53]. Perceived risk of tripping was evaluated through self-reported protocols [51, 52]. Finally, the use of a walkway space can be assessed with such tools as direct observation in field studies [42, 54], analysis of recorded footage [42] or using VAMP [45].

3.2.3. Gaze behaviour. Studies related to gaze behaviour were predominantly implemented with the help of wearable eye-tracking systems. Among them, there were experiments carried out in controlled laboratory settings [48, 64], as well as studies which analysed the data obtained in real-world situations [56-60, 63, 65]. In these studies, participants were asked to wear eye-tracking devices (different manufacturer, but typical gaze position accuracy was between 0.5° and 1.0° and sampling frequency was around 25-30 Hz) and to walk along the defined routes, both freely and under a researcher's control. After that, measured eye movements were superimposed on a simultaneously recorded videotape to analyse gathered data and correlate it with aspects of the built environment or other people's presence. The critical distances within scenes can be estimated by relative size of reference objects in the field of view [60]. It is important to note that in most naturalistic studies, experiments were conducted during day and night [56-58, 65]. In addition, research questions connected with recognition of another person' intentions, emotional states or gaze directions can be done in laboratory settings using target images of actors [61] or faces produced with 3D graphical rendering software [62] showed on the screen under various lighting conditions or in digitally simulated lighting scenes.

4. Discussion and conclusion

This paper focuses on the changes in human behaviour whilst walking under an artificially lit outdoor environment. A systematic literature search was conducted using key words. During the search process three groups of pedestrians' behaviour (i.e., walking (including activity levels and walking patterns), microscopic mobility (including walking speed, gait characteristics and use of walkway space) and gaze behaviour) were identified. This division is a product of the authors' subjective view, and it was undertaken to ease the report process in describing existing knowledge on possible changes in pedestrian behaviour caused by various artificial lighting settings.

Several studies, which also can be associated with people' behaviour, were not included in the main body of the paper because they cannot be clearly assigned to the three identified groups. These studies were predominantly focused on various forms of interaction with other people and the environment. They covered such topics as impacts of light on personal space requirements [69], critical distances between pedestrians [70], approach and avoidance behaviour [71], de-escalation of aggression [72] and the issues connected with interaction between a human and urban environment through lighting means [73-75].

The current literature demonstrates a decent potential of lighting in influencing on various types of behaviour, as well as a great variety of research approaches and methods. Numerous findings described in the Chapter 3 show that the lighting environment has an impact on pedestrians' behaviour, depending on the ambient lighting condition (e.g., daytime or night-time), the presence of lighting posts, as well as the quantitative and qualitative aspects of particular lighting solutions. The absence of outdoor lighting or its poor quality seems to be a significant barrier to active lifestyle in a general city scale and can alter the pattern of pedestrian movement because of route and travel mode choices. To the date, articles which investigate how certain lighting variables influence the walking patterns are limited, although there were several findings suggesting that average illuminance on the pavement and the colour temperature of a light source tend to influence route choices [39, 40]. The reasons behind the changes in walking behaviour can be largely attributed to decreased perceived safety and reassurance, but also to perceived restorativeness, better visibility and subjective preferences. Moreover, the potential of targeted lighting design in issues of supporting pedestrians' behaviour was noticed [41, 42] and need to be further studied. An important aspect in relation to outdoor lighting along pedestrian paths could also be the possibility to walk smoothly and maintain a comfortable speed. It seems that reduced lighting conditions can influence the walking speed and the placement of a pedestrian on a walk path, decrease his or her gait's quality and increase the perceived risk of tripping. Besides the quantity of light, spectrum, dynamic routines, and targeted lighting design also have a potential to impact microscopic mobility behaviour. Lastly, it is important to note that lighting conditions may affect the way people execute their visual tasks such as object detection, visual identification, visual orientation and on the perception of the urban environments.

There were many different approaches in relation to gathering data about changes in pedestrian behaviour, including both quantitative and qualitative methods. Nevertheless, one of the current disadvantages for these methods is their limitation to a certain urban environment and difficulties in the collection of data in a city scale. In addition, it goes without saying that the development of a multi-tool, allowing gathering and correlating data on various types of changes in pedestrians' behaviour will reveal new opportunities in studying the impacts of lighting conditions.

It is worth discussing that the video-based analysis has a potential as a promising multi-tool for studying pedestrian behaviour in urban environments. First, this is due to the wide range of features currently offered by image-processing algorithms, computer vision and machine learning fields. There is a possibility to improve and further develop the approach suggested by M. Johansson et. al., where the authors adapted existing video technologies to measure microscopic movement, i.e., placement and speed [45]. For instance, in terms of analysing data, the VAPM method can be enriched with the analysis related to gaze behaviour and microscopic mobility behaviour, namely by capturing gaze directions with head posture estimation [76] and by measuring gait features with extracting body parts from the image [77]. Other possible improvements can relate to capturing facial features and/or expressions [78], action recognition [79] and crowd behaviour [55] analysis. Secondly, the method has a decent prospect because of steady integration of video cameras in our daily lives through video security surveillance and smart city applications. This fact gives the opportunity for researchers to work with a great amount of video data, supporting clear and stable conclusions about impacts of lighting environments. Finally, when the impacts and correlations between outdoor lighting and human behaviour will be better studied, it would be possible to control lighting installations based on real-world images captured from video cameras. Nevertheless, to date, the ethical issues corresponded with an operation of surveillance data represent a major barrier in a detailed assessment of pedestrian behaviour.

Another promising approach to study walking patterns of pedestrians is applying an online public participation geographic information system (PPGIS) methodology [80]. GIS-based measures were already used in previous research studies related to built environment and lighting [10, 15, 23], digitising the positions of street lighting posts. Hypothetically it is also possible to correlate the physical locations of lighting posts with related lighting sources' characteristics or even show the lighting spatial distribution of the area. Public participation (PP) helps to map the user's knowledge and experience to different urban places, making city-scale participation possible. The correlation between HardGIS and SoftGIS knowledge layers reveals the possibility to study human behaviour and experiences in relation to the physical environment. The approach has a means to potentially reach a larger number of participants than traditional methods and to include various groups of people and languages. The data collected through PPGIS may be subjected to scientific standards of data quality [81], but generally can be served for both research and planning/design objectives. For instance, N. Davoudian and A. Mansouri conducted an interview where respondents were asked to indicate a walking route they usually take during the day on a map of the area [22, 39]. Afterwards they were asked whether they would take the same route at night for the same purpose as daytime or describe the reasons why an alternative route will be chosen. It goes without saying that this methodology can be successfully digitised and applied on a bigger scale.

References

[1] Rahm J and Johansson M 2018 Assessing the pedestrian response to urban outdoor lighting: a full-scale laboratory study PLoS ONE 13(10):e0204638

[2] Rahm, J., 2019. Urban outdoor lighting: Pedestrian perception, evaluation and behaviour in the lit environment. Ph. D. Lund University

[3] Fotios S and Robins C J 2022 Effect of ambient light on the number of motorized vehicles, cyclists, and pedestrians Transport Res. Record: J. of the Transportation Res. Board 2676 issue: 2 593-605

[4] Uttley J and Fotios S 2017 Using the daylight savings clock change to show ambient light conditions significantly influence active travel J. of Env. Psych. 53 1-10

[5] Fotios S, Uttley J and Fox S 2017 A whole-year approach showing that ambient light level influences walking and cycling Lighting Res. & Technology 51 issue: 1 55-64

[6] Unwin J, Symonds P and Laike T 2017 Proc. of the Lux European Lighting Conf. - Lighting for Modern Society, ed M B Kobav et al (Ljubljana) pp 245-250

[7] Painter K 1996 The influence of street lighting improvements on crime, fear and pedestrian street use, after dark Landscape and Urban Planning 35 issue 2-3 193-201

[8] Addy C L, Wilson D K, Kirtland K A, Ainsworth B E, Sharpe P and Kimsey D 2004 The American J. of Public Health 94(3) 440-443

[9] Lee C and A V Moudon 2008 Neighbourhood design and physical activity Building Res. & Information 36 issue 5 395-411

[10] Edwards N, Hooper P, Knuiman M, Foster S and Giles-Corti B 2015 Association between park

features and adolescent park use for physical activity Int. J. of Behavioral Nutrition and Phys. Activity 12, 21
[11] Jago R, Baranowski T, Zakeri I and Harris M 2005 Observed environmental features and the physical activity of adolescent males American J. of Preventive Med. 29, issue 2 98-104
[12] Bonaccorsi G, Manzi F, Del Riccio M, Setola N, Naldi E, Milani C, Giorgetti D, Dellisanti C and Lorini C 2020 Impact of the built Environment and the neighborhood in promoting the physical activity and the healty aging in older people: an umbrella review Int. J. Env. Res. Public Health 17(17)
[13] Hermosillo-Gallardo M E, Sebire S J and Jago R 2020 Perception of safety and its associaion with physical activity in adolescents in Mexico American J. of Preventive Med. 58, issue 5, 748-755
[14] Nabipour M, Nasseri S H and Saber E T 2021 Comparison and evaluation of built environment factors for developing pedestrian urban travels Fuzzy Information and Eng. 13, issue 4 505-521
[15] Lee C, Lee C, Stewart O T, Carlos H A, Adachi-Mejia A, Berke E M and Doescher M P 2021 Neighborhood environments and utilitarian walking among older vs. younger rural adults Frontiers in Public Health 9:634751
[16] Corseuil Giehl M E, Hallal P C, Brownson R C and d'Orsi E 2017 Exploring associations between percieved measures of the environment and walking among older adults J. of Aging and Health 29 issue: 1 45-67
[17] Rydenstam T, Fell T, Geleta Buli B, King A C and Bälter K 2020 Using citizen science to understand the prerequisites for physical activity among adolescents in low socioeconomic status neighborhoods - the NESLA study Health & Place 65 (2020) 102387
[18] Al Odat S M and Al Kurdi N 2020 Lively streets: the role of streetscape elements in improving the experience of commercial street users in Amman, Jordan J. of Settlements and Spatial Planning 12, no. 1 1-12
[19] Akpinar A 2016 Factors influencing the use of urban greenways: a case study of Aydin, Turkey Urban Forestry & Urban Greening 16 123-131
[20] Mateo-Babiano I 2016 Pedestrian's needs matter: examining Manila's walking environment Transport Policy 45 107-115
[21] Ferrer S and Ruiz T 2018 The impact of the built environment on the decision to walk for short trips: evidence from two Spanish cities Transport Policy 67 110-120
[22] Davoudian N and Mansouri A A 2017 Int. Conf. on Env. Psych. Book of Abstracts (A Coruna) p 63-64
[23] Dwimirnani P, Karimi K and Palaiologou F 2017 Proc. of 11th Space Syntax Symp. (Lisbon) #129
[24] Komatsu Y, Egoshi M, Ikeda T and Tamada T 2011 The distribution and effects of dark spots at night-time residental streets J. of Env. Engineering (Transactions of AIJ) 76 Issue 666 641-647
[25] Ferrer S, Ruiz T and Mars L 2015 A qualitive study on therole of the built environment for short walking trips Transport Res. Part F: Traffic Psych. and Behaviour 33 141-160
[26] Park Y and Garcia M 2020 Pedestrian safety perception and urban street settings Int. J. of Sustainable Transportation 14, No. 11 860-871
[27] Lee J S, Park S and Jung S 2016 Effect of crime prevention through environmental design (CPTED) measures on active living and fear of crime Sustainability 2016,8(9), 872
[28] Rahm J, Sternudd C and Johanson M 2021 "In the evening , I don't walk in the park": the interplay between street lighting and greenery in perceived safety Urban Design Int. 26 42-52
[29] Kobayashi S, Nakamura Y and Hosoya S 2017 A low illuminance lighting arrangement to enchance the natural experience in a resort area J. of Illuminating Engineering Institute of Japan 101 issue 2 82-87
[30] Hughes N, Ryan B, Hallewell M, Coad N, Grant A, Parrot N, Roberts S and Thompson K 2020

Identifying new concepts for innovative lighting-based interventions to influence movement and behaviours in train station Lighting Res. & Technology 52 issue: 8 976-990
[31] An H, Chung H and Chang J 2020 Gendered perception of night-time walking safety in urban-rural campus town - focusing on the CPTED project and street lighting J. of the Architectural Institute of Korea 36 Issue 12 77-87
[32] Mattoni B, Burattini C, Bisegna F and Fotios S 2017 Conf. Proc. 2017 IEEE Int. Conf. on Env. and Electrical Eng. and 2017 IEEE Industrial and Commercial Power Systems Europe (EEEIC / I&CPS Europe) (Milan) p 1-5
[33] Fotios S, Unwin J and Farrall S 2015 Road lighting and pedestrian reassurance after dark: A review Lighting Research and Technology 47 issue: 4 449-469
[34] Prihatiningrum A, Ramawangsa P A and Seftyarizki D 2021 Artificial lithtin strategies on campus pedestrian path in increasing the safety of walking in night (case study: university of Bengkulu) IOP Conf. Series: Earth and Env. Sci. 738 012042
[35] Nikunen H, Puolakka M, Rantakallio A, Korpela K and Halonen L 2013 Perceived restorativenes and walkway lighting in near-home environments Lighting Res. & Technology 36 issue: 3 308-328
[36] Kobayashi S, Nakamura Y and Hosoya S 2017 A low illuminance lighting arrangement to enchance the natural experience in a resort area J. of Illuminating Engineering Institute of Japan 101 issue 2 82-87
[37] Nasar J L and Bokharaei S 2017 Lighting modes and their effects on impressions of public squares J.of Env. Psych. 49 96-105
[38] Casciani D 2020 A virtual experience of the human-scale urban lightscape The Human and Social Dimensions of Urban Lightscapes (Springer Cham) pp 79-97
[39] Davoudian N and Mansouri A A 2016 Proc. of CIE 2016 Lighting Quality and Energy Efficiency (Melbourne) pp 588-595
[40] Bystryantseva N V, Antonov R A and Balaev A F 2020 Assessment of the impact of street lighting on usage activity of pedestrian traffic spaces on the example of the square in Saint Petersburg IOP Conf. Series: Materials Sci. and Engineering 962 042001
[41] Hughes N, Ryan B, Hallewell M, Coad N, Grant A, Parrot N, Roberts S and Thompson K 2020 Identifying new concepts for innovative lighting-based interventions to influence movement and behaviours in train station Lighting Res. & Technology 52 issue: 8 976-990
[42] Ryan B, Hallewell M, Hughes N, Coad N, Eliseu J, Lira M, Grant A, Parrot N and Thompson K 2021 Field testing two innovative lighting interventions to influence waitning behaviours and movements on stairways in train stations Preprint doi.org/10.1177/14771535211016811
[43] Donker S F, Kruisheer J and Kooi F 2011 Env. 2.0: the 9th Biennial Conf. on Env. Psych. ed A Haans et al (Eindhoven) pp 125-126
[44] Pedersen E and Johansson M 2018 Dynamic pedestrian lighting: effects on walking speed, legibility and environmental perception Lighting Res. & Technology 50 issue: 4 522-536
[45] Johansson M, Laureshyn A and Nilsson M 2020 Video analysis of pedestrian movement (VAPM) under different lighting conditions - method exploration Energies 2020,13,4141
[46] Berard J R and Ann Vallis L 2006 Characteristics of single and double obstacle avoidance strategies: a comparison between adults and children Experimental Brain Res. 175 21-31
[47] Choi J-S, Kang D-W, Shin Y-H and Tack G-R 2014 Differences in gait pattern between the eldery and the young during level walking under low illumination Acta Bioengineering and Biomechanics 16, nr 1 3-9
[48] Itoh N 2006 Visual guidance of walking: effects of illumination level and edge emphasis Gerontechnology 5, No. 4 246-252
[49] Huang C-K, Chien J-H and Siu K-C 2017 The reduced lighting environment impacts gait characteristics during walking Int. J. of Industrial Ergonomics 61 126-130
[50] Cho S-Y, Ryu Y-U, Dong Je H, Jeong J H, Ma S-Y and Kim H-D 2013 Effects of illumination on toe clearence and gait parameters of older adults when stepping over an obstacle: a pilot

study J. of Phys. Therapy Sci. 25 Issue 3 229-232
[51] Li K W, Chen Y, Li N, Duan T and Zou F 2020 Assessment of risk of tripping before and after crossing obstacles under dimmed lighting conditions Work 66, no 3 551-559
[52] Li K W, Chen Y, Zou F, Li N and Duan T 2019 Perception of risk of tripping under lighting and obstacle conditions Human Factors and Ergonomics in Manufacturing & Service Industries 29, Issue 6 529-536
[53] Hamel K A, Okita N, Higginson J S and Cavanagh P R 2005 Foot clearance during stair descent : effects of age and illumination Gait & Posture 21, Issue 2 135-140
[54] Markvica K, Richter G and Lenz G 2019 Impact of urban street lighting on road users' perception of public space and mobility behavior Building and Env. 154 32-43
[55] Haans A, Corbetta A, Kumar P and Toschi F 2017 Int. Conf. on Env. Psych. Book of Abstracts (A Coruna) p 219
[56] Davoudian N and Raynham P 2012 What do pedestrians look at night? Lighting Res. & Technology 44 issue: 4 438-448
[57] Fotios S, Uttley J, Cheal C and Hara N 2015 Using eye-tracking to identify pedestrians' critical visual tasks, part 1. Dual task approach Lighting Res. & Technology 47 issue: 2 133-148
[58] Fotios S, Uttley J and Yang B 2015 Using eye-tracking to identify pedestrians' critical visual tasks. Part 2. Fixation on pedestrians Lighting Res. & Technology 47 issue: 2 149-160
[59] Fotios S, Uttley J and Fox S 2018 Exploring the nature of visual fixations on other pedestrians Lighting Res. & Technology 50 issue: 4 511-521
[60] Fotios S, Yang B and Uttley J 2015 Observing other pedestrians: investigating the typical distance and duration of fixation Lighting Res. & Technology 47 issue: 5 548-564
[61] Fotios S, Yang B and Cheal C 2015 Effects of outdoor lighting on judgements of emotion and gaze direction Lighting Res. & Technology 47 issue: 3 301-315
[62] Palmer C J, Otsuka Y and Clifford C W G 2020 A sparkle in the eye: illumination cues and lightness constancy in the perception of eye contact Cognition 205(2020)104419
[63] Fotios S and Uttley J 2018 Illuminance required to detect a pavement obstacle of critical size Light Res. & Technology 50 issue: 3 390-404
[64] Cheng T-J, Yang B, Holloway C and Tyler N 2018 Effect of environmental factors on how older pedestrians detect an upcoming step Lighting Res. & Technology 50 issue: 3 405-415
[65] Jiand X, Li J and Yang B 2021 Luminance and saliency have impact on pedestrians' fixation distribution during natural walking: evidence from mobile eye-tracker Lighting Res. & Technology 53 issue: 4 359-372
[66] Doughty M J 2013 Effects of background lighting and retinal illuminance on spontaneous eyeblink activity of human subjects in primary eye gaze Eye & Contact Lens: Sci. & Clinical Practice 39 - Issue 2 138-146
[67] Rahm J and Johansson M 2021 Assessment of outdoor lighting: methods for capturing the pedestrian experience in the field Energies 2021,14(13),4005
[68] Patching G R, Rahm J, Jansson M and Johansson M 2017 A new method of random environmental walking for assessing behavioral preferences for different lighting applications Frontiers in Psych. 8:345
[69] Adams L and Zuckerman D 1991 The effect of lighting conditions on personal space requirements The J. of General Psych. 118 - Issue 4 335-340
[70] Berghuis W J H, den Hartog I, Romijn A R and Haans A 2014 Proc. Experiencing Light 2014 ed Y A W de Kort et al (Eindhoven) p 133
[71] Veenstra L and Koole S L 2018 Disarming darkness: effects of ambient lighting on approach motivation and state anger among people with varying trait anger J.of Env. Psych. 60 34-40
[72] de Kort Y A W, Ijsselsteijn W A, Haans A, Lakens D, Kalinauskaite I and Schietecat A 2014 Proc. Experiencing Light 2014 ed Y A W de Kort et al (Eindhoven) pp 94-97
[73] Pihlajaniemi H and Luusua A 2020 Playful and playable lighting in smart cities: towards a holistic framework of design Making Smart Cities More Playable ed A Nijholt (Springer) pp 149-177

[74] Poulsen E, Morrison A, Jorgen Andersen H and Jensen O B 2013 Mobile HCI '13: 15th Int. Conf. on Human-Computer Interaction with Mobile Devices and Services (Munich) pp 217-226
[75] Casciani D 2014 Urban social lighting. Exploring the social dimension of urban lighting for more sustainable urba nightscapes (Milano: Politecnico di Milano)
[76] Murphy-Chutorian E and Trivedi M 2009 Head Pose Estimation in Computer Vision: A Survey body parts. IEEE Transactions on Pattern Analysis and Machine Intelligence. 31 pp 607-626
[77] Kim W, Kim Y 2018 Human Body Model using Multiple Depth Camera for Gait Analysis Action. International Conference on Software Engineering, Artificial Intelligence, Networking and Parallel/Distributed Computing (SNPD). pp 70-75
[78] Zadobrischi E et al 2020 Detection of Emotional States Through the Facial Expressions of Drivers Embedded in a Portable System Dedicated to Vehicles. 28th Telecommunications Forum (TELFOR). pp 1-4
[79] Dawar N and Kehtarnavaz N 2017 Continuous detection and recognition of actions of interest among actions of non-interest using a depth camera EEE International Conference on Image Processing (ICIP). pp 4227-4231
[80] Brown G and Kyttä M 2014 Key issues and research priorities for public participation GIS (PPGIS): a synthesis based on empirical research Applied Geography 46 122-136
[81] Fagerholm N, Raymond C M, Stahl Olafsson A, Brown G, Rinne T, Hasanzadeh K, Broberg A and Kyttä M 2021 A methodological framework for analysis of participatory mapping data in research, planning and management Int. J. of Geographical Information Sci. 35 -Issue 9 1848-75

IOP Conf. Series: Earth and Environmental Science 1099 (2022) 012008 doi:10.1088/1755-1315/1099/1/012008

Social interaction in local public squares after dark

VKR Hennig[1,4], S Fotios[3], N Gentile[2], C Sternudd[1] and M Johansson[1]
[1] Environmental Psychology, Dept. of Architecture & Built Environment, Lund University, Sweden
[2] Division of Energy and Building Design, Dept. of Building & Environmental Technology, Lund University, Sweden.
[3] School of Architecture, The University of Sheffield, Sheffield, UK
[4] Author to whom any correspondence should be addressed. E-mail: katarina.hennig@abm.lth.se

Abstract. This paper explores social interaction in local public squares under different lighting conditions. At its best public squares are social spaces that engender a sense of belonging, increase the quality of life and wellbeing of individuals. It is proposed that outdoor lighting would be essential to the use of the public realm after dark, but empirical results regarding lighting conditions and social aspects of life in public squares are limited. Based on a socio-physical conceptual model of the transactional relationship of the user, the lit environment and the behavioural outcome, this study investigated active social interaction in daylight compared to after dark. A field study was conducted in two local public squares in Malmö, Sweden. The occurrences of which visitors were being alone, in pairs, or in groups of three or more (N=2522), and verbal or non-verbal interaction amongst those in company of another person were recorded. The lit appearance of the two squares after dark, was assessed with HDR-photography and photometric measurements; portraying dissimilar spatial, spectral and intensity characteristics. The results of social interaction show dissimilar patterns of the two squares; an increase in social interaction in EL after dark was observed in one of the squares, while a decrease in the afternoon and no significant difference was displayed in the evening after dark in the other square. It is suggested that lighting may sustain patterns of social interaction after dark, however it might be, that the company of another is especially important after dark.

1. Introduction

UN Sustainable Development Goal (SDG) 11 targets the provision by 2030 of universal access to safe, inclusive and accessible public spaces in cities and human settlements.[1] Consequently, UN Habitat III policy paper *The Right to the City and Cities for All* promotes quality public spaces that are participatory, enhance social interaction, and engender sense of belonging.[2] The implementation of SDG 11 requires focus at a community level to make tangible differences to the everyday lives of people.[3] Urban theorists argue that the social and psychological health of modern communities requires quality public spaces, here defined as 'publicly accessible spaces' which generate public life, and support active and passive social interaction.[4-7] Social life in public spaces increase the quality of life and wellbeing of individuals.[4, 8] Measures of social interaction in public space may therefore be used to assess the level of health and vitality of communities.[9] Urban public spaces include a variety of urban settings such as streets, parks and squares. Outdoor lighting is essential to promote the use of public squares after dark because it supports movements and stationary activities.[10, 11] However, most research on the qualities of urban design and its role in facilitating social life in public space concern daylit conditions. Frequented public spaces are meaningful, protective of rights of different user groups, and responsive; that is, designed to serve the needs of users.[9, 12, 13] Users are here defined as those who frequent public spaces and rely on them for active and passive engagement.[13] Perceived environmental qualities such as accessibility, safety, comfort, pleasure and sociability encourage use of public spaces by day.[5, 14] While lighting research on pedestrians suggests that electric lighting may support these needs,[15] field-studies on the association between the lit environment and social life of public squares are still limited. Ethnographic approaches have revealed affective capacities of light and darkness by 'light walks', and thus accentuated the on-going flow of shifting impressions and feelings which are embodied in the experience of moving through an urban lightscape.[16] Lighting design research with a social oriented scope have stressed the need to study people's activities, pattern of use, and appraisals of lit environmental settings, to understand how lighting might encourage sociability and support users' daily life in public space.[9]

This paper reports on lighting conditions and social interaction in local public squares, which constitute 'everyday life spaces' set within the context of a neighbourhood.[17] While a local public square is a publicly accessible space, the sense of 'publicness' may vary along a continuum from 'public' to 'private', and so may the type of interactions vary from strangers to friends.[17] It is a space which offers an opportunity to be co-present with people, who might be either unknown (unacquainted), categorically known (the woman from the flower-shop), acquainted, or even a friend.[17, 18] Social interaction in a local public square might be expressed as a brief glance of recognition (face engagement), a greeting, a small chat or even spending time socializing with friends.[17, 18] Though merely observing others, to be co-present with strangers, to see and be seen, may induce a sense of belonging. A local public square may therefore serve as a social space *where interrelatedness with other human beings is affirmed*.[19] Public solitude (passive social interaction) may be a lone pleasure derived from people-watching. Public sociability though, by definition involves, spoken interaction between people in dyads, triads or larger groups.[17]

1.1. Aim
The intent of this study was to investigate active social interaction in daylight (DL) and in electric lighting (EL) after dark, in two local public squares in Malmö, Sweden. Three objectives were targeted. (1) To compare occurrences of people visiting the squares being alone, in pairs or in larger groups. (2) To assess the share of verbal and non-verbal interaction amongst those in company of others. (3) To investigate if there is any differences in visitors presence in the squares after dark between age groups.

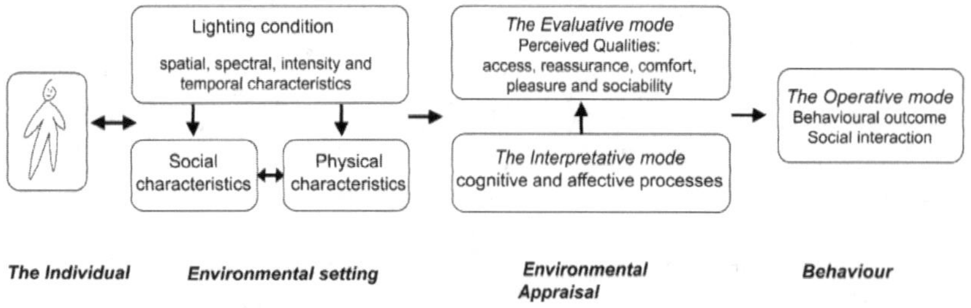

Figure 1. A socio-physical conceptual model of the transactional relationship between the Individual, the Environmental setting and the Behavioural outcome.

1.2. *A transactional-contextual framework*
To interpret the behavioural relationship between the individual (user), the lit environment of a public square and the behavioural outcome in terms of social interaction we employ a framework, that suggests a dynamic interplay between people and their every-day environmental settings.[20] With this perspective a behaviour is viewed in its socio-physical and temporal context. In any given setting physical aspects are closely linked to social ones, also giving spatiotemporal patterns to the occurrences of behaviours.[20] Furthermore physical aspects (limits, spatial arrangement and characteristics including lighting condition) may either facilitate or impede behaviours depending on the individual's appraisal of the setting.[20] We propose a socio-physical conceptual model (Figure 1) which shows the transactional relationship between the individual (with personal traits, abilities and needs), the environmental setting (with social and physical characteristics including lighting condition), the environmental appraisal

(including cognitive interpretative and evaluative processes), perceived qualities (access, reassurance, comfort, pleasure and sociability), and the behavioural outcome (social interaction). The configuration of the model departs from Stokols' modes of human environment transactions; the interpretive, the evaluative, the responsive and the operative mode.[21] It is stipulated that the individual's appraisal of a setting and therefore her behaviour, is conditioned by the lighting condition in terms of spatial, spectral, intensity and temporal characteristics.[22, 23]

2. Method

2.1. Direct observation of users' active social interaction

Direct structured observations of users' active social interaction was conducted at two local neighbourhood squares in Malmö, Sweden. In this method the observer adopts, as far as feasible, a non-participant pure observer role to avoid reactivity of those being observed.[24] Direct observation here involved watching, listening and recording types and frequencies of individuals' active social interaction, in DL and after dark in EL. A scheme for coding active social interaction was developed and tested in a pilot study. Amongst those involved in active social interaction, defined as being in the company of others, the occurrences of verbal or non-verbal interaction was recorded. A scan-sampling technique was employed for the recording of events of active social interaction in each setting.[25] Each event involving a social interaction was given an ID-number, used for field notes to provide qualitative information of social interaction and to discern any spatiotemporal pattern.

2.2. Settings

The two squares, Kirseberg square and Lindeborg square (Figure 2), were selected for having several features in common; function (as a local centre with services and amenities), size (in terms of surface area), physical setting and spatial arrangement (with design features such as benches, trees and planting). However, the lighting installations and lighting conditions were dissimilar in terms of spatial light distribution, intensity, uniformity, level of contrasts in the visual field, spectral power distribution (SPD), correlated colour temperature (CCT), CIE general colour rendering index (CRI) and differences in scotopic/photopic (S/P) luminance ratios. Therefore the after dark appearances and ambience of the squares are different, which enabled a comparison. Both neighbourhoods have equivalent number of inhabitants; approximately 5300 in Kirsebergsstaden in northern Malmö and approximately 5000 in Lindeborg in southern Malmö.

2.2.1. Kirseberg square has a surface area of approximately 3100 m^2. To the north there is an area (zone A) for stationary occupancy with design features including benches in each corner, a sculpture and a boule court. Cherry trees and rose bushes flank each side. To the east there is a one-way street mixed with vehicle, cyclists and pedestrians (zone B). To the south there is a pedestrian route with access to a frequented grocery store and to the west a pedestrian route along a residential building (zone C). The surrounding buildings are low rise (between one and four stories high) and were constructed in the 1960's. A technical description of the lighting installation is shown in Table 1 and Table 2. The square has 12 lampposts (3.7 m high) with two asymmetric reflector luminaires, each containing a 70W metal halide lamp (MH). The low height of the lamppost combined with a high luminous output of the lamp results in large contrasts between bright and dark areas (see the technical assessment of the luminous appearance, section 3.1).

2.2.2. Lindeborg square has a surface area of approximately 3100 m^2. The area for stationary occupancy (zone A) subjectively suits its purpose well; with a soft-scape of lime trees and cherry trees, cut hedges of beech and well composed flowerbeds with perennials, a water feature and a little sculpture. The surrounding buildings are low rise (between one and two stories high), constructed in the 1970s and host commercial services, gym, church and an elementary school. A pedestrian and cyclist's path (zone B) to the north gives access to the commercial service building and another path (zone C) along the western side provides a linkage to the surrounding neighbourhood. A technical description of the lighting installation is provided in Tables 1 and 2. There are 11 lamp posts (4.2 m high) placed at intervals of 20

IOP Conf. Series: Earth and Environmental Science 1099 (2022) 012008 doi:10.1088/1755-1315/1099/1/012008

m along paths and arranged to accompany flowerbeds and seating. The lanterns, with opal diffusers, have an omnidirectional distribution and 70 W high pressure sodium (HPS) lamps (see the technical assessment of the luminous appearance, section 3.1).

Table 1. Specification of Luminaire types

	Luminaire types						
Square	Type	Name	Light distribution	Optics	Shield	Height	Qty.
Kirseberg	road-luminaire	Philips, Copenhagen	asymmetric	reflector	upwards	3.7 m	2x12
	spotlight	SILL, Plane projector	rotational symmetric	reflector	lamels		
Lindeborg	park lantern	DEFA, Helena	omnidirectional	opal diffuser	glare rings	4.2 m	11

Table 2. Specification of Lamp types

	Lamp types						
Square	Type	Name	Luminous flux (lm)	CCT(K)	CRI (Ra)	S/P ratio	Qty.
Kirseberg	MH	CDO-ET 70W/828	7030	2800	84	1.3	2x12
	HPS2	HST-DE 150W	1500	2000	25	0.5	1
Lindeborg	HPS	SON Pia Plus 70W	6000	2000	25	0.5	9
	MH	CDO-ET 70W/828	7030	2800	84	1.3	2

2.3. Data collection

Observations were conducted for the same times of day in the two weeks before and two weeks after the autumn 2020 daylight savings clock change. This enables a comparison of users' active social interaction between the two lighting conditions and which offsets other confounding factors; i.e. the time of day factor is held constant and seasonal factors are sufficiently constant during the sampling periods. The spatiotemporal patterns of users' behaviour in a given time frame are supposedly similar, e.g. inhabitants of each neighbourhood would hypothetically perform the same activities and social interaction as part of their daily life rhythm.[20] Sampling in each square was performed on six days, two week-days and one weekend-day. Sampling sessions were scheduled in the afternoon at 16.15 to 17.00 and in the early evening at 17.30 to 18.15. In total, 18 hours of sampling was conducted and 2522 events were recorded. Prior to each sampling session, conditions such as precipitation and air-temperature, the present sky-condition (e.g. clear, semi-overcast, over-cast), the state of vegetation and the lighting condition (DL) or (EL) were recorded. Field notes regarding the ambient feeling, pace of movements and interactions were also taken prior to sampling sessions, in combination with notes on social interaction IDs these provided qualitative data used for a written narrative.

2.3.1. Sample. Individuals present at each of the square were classified visually by their apparent gender and by their apparent age into four age groups: children (0 - 12 years old approximately), teens (13 - 19 years old approximately), adults (20 - 64 years old approximately) and elderly (over 65 years old). Due to the observers non-participant role a visual classification of the individuals was employed. Table 3 shows the demographics of the visually classified sample.

Table 3. Demographics of sample based on a visual classification during observations.

	Age-group									Gender				
	0 – 12 yr.		13–19 yr.		20 – 64 yr.		> 65 years		Total	Female		Male		Unidentified
	N	%	N	%	N	%	N	%	N	N	%	N	%	%
Kirseberg	147	10.7	140	10.2	867	63.0	222	16.1	1376	743	54.0	627	45.6	0.4
Lindeborg	94	8.2	294	25.7	558	48.7	200	17.5	1146	606	52.9	538	46.9	0.2

2.3.2. Sampling units. To enable observations of individuals' social interaction from a visible and audible distance each square was divided into three sampling units (zones A, B and C), as shown in Figure 2. These zones were chosen due to their spatial arrangements and specific function; A is designed

IOP Conf. Series: Earth and Environmental Science 1099 (2022) 012008 doi:10.1088/1755-1315/1099/1/012008

as a 'social' area with seating, B is a pedestrian and cyclist path with access to commercial shop(s) and other services(s) and C is a path linking the surrounding residential areas of the neighbourhood to its local square. A sampling session of 45 min enabled 3 rotations between the sampling units, with one rotation every 15 minutes always starting in sampling unit A, continuing to B and hereafter to C.

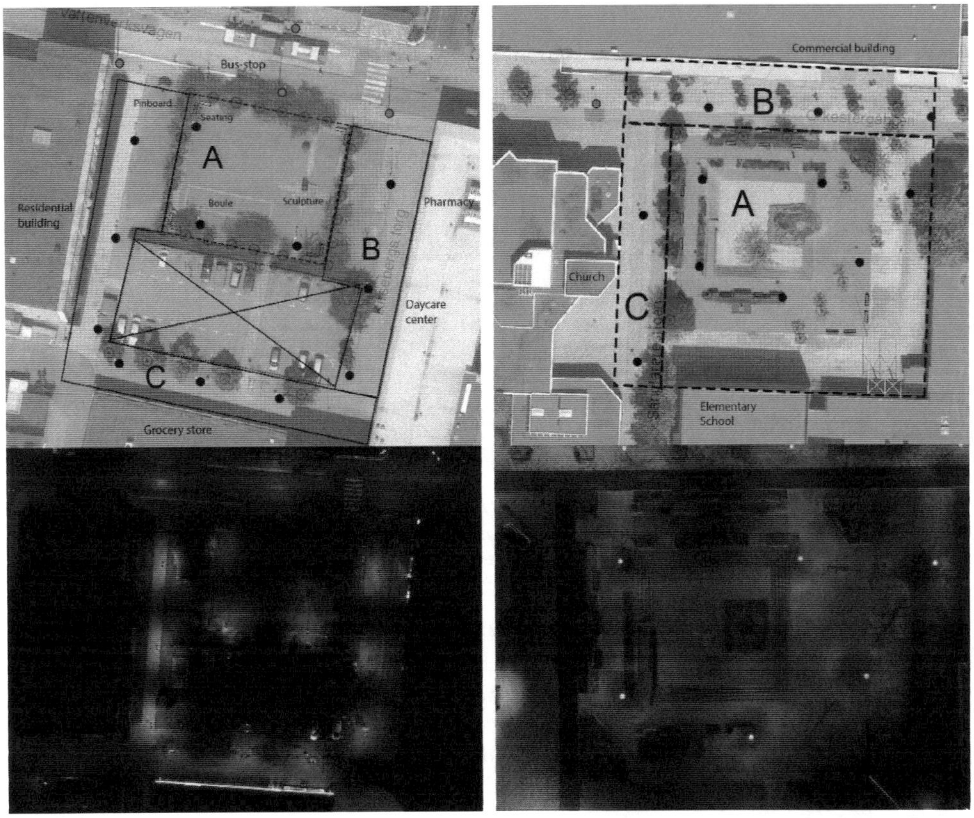

Figure 2. Plans of the two squares with sampling units (zone A, B, C) and after dark lighting layouts. From left Kirseberg Square. and to the right Lindeborg Square.

2.4. Technical environmental assessment

EL conditions were assessed in terms of spatial, spectral, intensity and temporal characteristics. The assessment was carried out as follows:

- HDR-images were captured for vital viewpoints in each sampling unit (zones A, B, C). These images were converted into luminance maps, calibrated against luminance spot measurements on a board placed in each image frame.
- HDR-images of spheres were taken in each sampling unit to convey facial recognition.
- Horizontal illuminance was measured at ground level across the whole square in a grid with a grid size of 3 m x 3 m.
- Horizontal illuminance was measured on paths (in zones B and C) in compliance with SS-EN 13201-3:2016.
- SPD, intensity and S/P-ratios of light sources were retrieved from the manufacturer.

2.5. Data analysis

Data analysis was carried out in IBM SPSS in two steps to establish the occurrences of active social behaviour in daylight compared to after dark in electric lighting in each of the squares:

Step 1: A frequency analysis for each of the selected behavioural categories:

- active social interaction operationalized as being alone, in pairs or in a group of three or larger,
- verbal versus non-verbal interaction amongst those individuals engaged in active social interaction.

Step 2: A Pearson's Chi-square test to test for any differences of behaviour between:

- lighting condition in DL and after dark in EL,
- age groups in DL and after dark in EL.

The level of significance was set to $p \leq 0.05$ with Bonferroni correction for multiple comparisons.

3. Results

3.1. Technical environmental assessment

HDR-images with corresponding luminance maps are shown in Figures 3 and 4. An overview of the illuminance measurements is shown in Table 4. The illuminances in Kirseberg square have a high average and a low uniformity, resulting in large contrasts between bright and dark areas, especially in zone A, the stationary area for occupancy. Lit windows in the residential building (zone C) and the shop windows (zone B and C) contribute to a lit impression of the square. Lindeborg square on the other hand, has a low average illuminance but higher uniformity. The lit scene has a warm ambiance, diffuse and soft, with poor modelling characteristics of objects and spatial elements and poor colour rendering due to the low CRI of the HPS lamp. Along the commercial building linear fluorescent tubes cause a high contrast in the visual scene.

Figure 3. HDR-images with corresponding luminance measurements in Kirseberg square, from the top:

- Zone A - There are big contrasts between dark and bright areas on the horizontal level with luminance levels in the range of 0.1 – 15 cd/m^2.
- Zone A (bright corner) - The illuminated sphere with luminance levels in the range of 1 – 15 cd/m^2 depicts good facial recognition.
- Zone A (dark corner) - The dark sphere with low luminance levels in the range of 0.1 – 1 cd/m^2 depicts poor modelling and poor facial recognition.
- Zone B (path) - The façade is dark while the path is bright with luminance levels in the range of 1 – 15 cd/m^2.
- Zone C (path) - The path is bright with luminance levels in the range of 1 – 15 cd/m^2.

Table 4. Measurements of horizontal illuminances and uniformity of illuminance.

Square	Zone	$\bar{E}$ (lx)	E min (lx)	U_o
Kirseberg	All	23.4	0.6	0.03
	B (path)	35.9	8.9	0.25
	C (path)	25.5	6.5	0.26
Lindeborg	All	5.8	0.4	0.07
	B (path)	7.1	3.9	0.55
	C (path)	11.1	4.7	0.42

Figure 4. HDR-images with corresponding luminance measurements in Lindeborg square, from the top:

- Zone A - This area is characterized by a uniform and diffuse light distribution with low luminance levels in the range of 0.1 – 1 cd/m^2 and a poor colour rendering due to the low CRI of the HPS lamp. This also results in poor modelling of steps, benches and hedges.
- Zone B (path) - The luminance levels along the path are low in the range of 0.1 – 1 cd/m^2. There is a huge contrast to the light levels along the shops where luminance levels exceed 10 cd/m^2.
- Zone B (path) - The light distribution on the sphere depicts a fair modelling with luminance levels in the range of 1 – 10 cd/m^2 which give a fair facial recognition.
- Zone C (path) - The average illuminance levels are 11 lx with a uniformity of 0.42. Luminance levels are in the range of 1 - 10 cd/m^2.
- Zone C (path) - The light distribution on the sphere depicts poor modelling due to the diffuse distribution of light. Luminance levels are in the range of 1 – 3 cd/m^2.

3.2. Social interaction in Daylight and in Electric lighting after dark

3.2.1. Alone or accompanied? Results of the Chi-Square test, shown in Table 5, for Kirseberg square. in the afternoon session display a significant difference between the level of social interaction in DL compared to after dark in EL (χ^2 (2, N=603) = 6.58, p=0.038). Social interaction of people being in pairs or in group of three or larger decreases after dark. Results for the early evening session display no significant difference between the level of social interaction in DL and after dark in EL (χ^2(2, N=772) = 2.74, *n.s.*). The results of Chi-Square test, shown in Table 5, for Lindeborg square display a significant difference between the level of social interaction in DL and after dark in EL for both the afternoon session (χ^2(2, N=540) = 11.64, p=0.003) and for the evening session (χ^2(2, N=605) = 12.45, p=0.002). After dark in EL less people are present alone, more people instead are present in pairs both in the afternoon and in the early evening session. The presence of people in groups of three or larger also increases in EL.

3.2.2. Narrative of the spatiotemporal patterns and social interaction
Narrative 1 – Kirseberg square, Afternoon in DL on the 7th of October. The sky is clear. Long afternoon shadows stretch across zone A. A pair, a girl and a woman, is sitting on a bench enjoying the late afternoon sun, which still lingers in the north eastern corner of zone A. An elderly woman rests on her walker, she is smoking. As the bus stops a few people get off, they head in different directions, some of them cross zone A and head for the grocery store. A man and two boys halt at the sculpture and the boys start climbing. Later a man on his bike rush towards the children's day-care centre in zone B. He picks up his son and chats to the lady at the centre, while the little boy patiently waits on his side on the curb-stone. Two chatting teenage girls walk by, heading for the grocery store. An elderly woman, a beggar, sits on the pavement outside the store. There is a gentle flow of people entering and exiting the grocery

store. Occasionally people stop and exchange a few words with the begging woman. In zone C an elderly man walks his dog in a slow pace. A man in his fifties walks along the path, he stops to chat with a woman on her balcony. Kirseberg square, Early evening in EL on the 13th of October. There is an evening chill. Four men in zone A are playing boule, the game is not intense though. They are chatting and drinking rather than playing, it seems. The pace of people walking to and from the grocery store is intensified.

Narrative 2 – Lindeborg square, Afternoon in DL on the 2nd of November. The sky is partly overcast with clouds passing quickly, autumn leaves whirl in the wind. Four boys in their teens are hanging out by the benches in the north east yet sunny corner of zone A. A boy on scooter roams around. Every now and then someone enters and exits the grocery store. Two teen girls walk by, they receive attention and comments from the boys. Lindeborg square, Early evening in EL on the 2nd of November. The sky is dark, it is still warm for the season, 16 degree Celsius. A teen boy is sitting on a bench texting on his mobile. He is anxiously looking out for his friends. He makes a call. A few minutes later another boy arrives and soon they are an enclave of six. A man walks by he is heading towards the fitness centre. The flow of people to and from the grocery store is steady at this hour.

Table 5. Frequencies for visitors being alone, in pairs or in groups of three or larger in DL and in after dark in EL conditions, measured in the afternoon and in early evening in the period with (DST) and in the period after daylight savings clock change.

Square	Social interaction	DL		EL		Total	Pearson's Chi-Square Tests		
		N	exp.	N	exp.	N	χ^2	*p*	df
Afternoon at 16.15-17.00									
Kirseberg	Alone	211	226	151	136	362			
	Pairs	122	110	54	66	176			
	Group ≥3	43	40	22	25	65			
	Total	376		227		603	6.58	0.038	2
Lindeborg	Alone	188	173	120	135	308			
	Pairs	86	88	70	68	156			
	Group ≥3	30	43	46	33	76			
	Total	304		236		540	11.64	0.003	2
Early evening at 17.30-18.15									
Kirseberg	Alone	212	203	229	238	441			
	Pairs	94	105	134	123	228			
	Group ≥3	49	47	54	56	103			
	Total	355		417		772	2.74	0.255	2
Lindeborg	Alone	168	146	124	146	292			
	Pairs	88	100	112	100	200			
	Group ≥3	47	57	66	56	113			
	Total	303		302		605	12.45	0.002	2

3.2.3. Verbal or non-verbal after dark? The results of Chi-Square test, Table 6, suggests a significant difference between the level of active social interaction in terms of non-verbal or verbal amongst those people being present in pairs or in groups of three or larger, in DL compared to after dark in EL for both squares. In Kirseberg square results suggest that people being present in pairs or groups of three or larger are engaged in verbal interaction in EL ($\chi^2(1, N=573) = 6.010$, $p=0.016$), while in Lindeborg square people are less verbal after dark with ($\chi^2(1, N=546) = 23.221$, $p<0.001$).

3.2.4. Home turf of the teens? Which age groups visit the squares after dark? The frequencies for visiting people per age-group in DL and EL after dark are shown in Table 7. In both squares children and elderly are more likely present in DL than after dark in EL. In Lindeborg square, teens have a significantly higher presence in the square after dark, which suggests a spatiotemporal pattern of the teens in this neighbourhood.

Table 6. Frequencies for verbal and non-verbal interaction amongst those involved in social interaction in DL and after dark in EL, measured in both timeslots in the period with (DST) and in the period after daylight savings clock change.

		DL		EL		Total	Pearson's Chi-Square Tests		
Square	Active social interaction	N	exp.	N	exp.	N	χ^2	p	df
Kirseberg	Non-verbal	114	100	72	86	186			
	Verbal	195	209	192	178	387			
	Total	309		264		573	6.010	0.016	1
Lindeborg	Non-verbal	45	70	107	82	152			
	Verbal	207	182	187	212	394			
	Total	252		294		546	23.211	<0.001	1

Table 7. Frequencies for visiting people per age group in DL and after dark in EL, measured in both timeslots in the period with (DST) and in the period after daylight savings clock change.

		DL		EL		Total	Pearson's Chi-Square Tests		
Square	Age group	N	exp.	N	exp.	N	χ^2	p	df
Kirseberg	Child	95	78	52	69	147			
	Teen	87	74	53	66	140			
	Adult	417	461	450	406	867			
	Elderly	133	118	89	104	222			
	Total	732		644		1376	25.28	<0.001	3
Lindeborg	Child	58	50	35	44	94			
	Teen	132	156	162	138	294			
	Adult	295	296	263	262	558			
	Elderly	122	106	78	94	200			
	Total	608		538		1146	16.49	<0.001	3

4. Discussion

This study investigates active social interaction in DL compared to EL after dark in two local public squares in Malmö, Sweden. To interpret the relationship of the individual (user), the lit environment of a public square and the behavioural outcome in terms of social interaction, we proposed a socio-physical model, Figure 1, stipulating that the individual's appraisal after dark and therefore her behaviour is conditioned by the characteristics of lighting. A technical environmental assessment of the two squares, section 3.1, depicted two dissimilar visual scenes after dark; Kirseberg square has a high average illuminance level but very low uniformity resulting in large contrasts between dark and bright areas and particularly so in the social area (zone A). Lindeborg square has a low average illuminance but a higher level of uniformity across the whole square. Observations of social interaction at both squares also reveal different patterns; In Kirseberg square the rate of social interaction decreases in EL in the afternoon, but there are no significant differences in the evening. Lindeborg square displays the same pattern in both timeslots, with less people being alone and more people being in pairs and in groups of three or more in EL. This suggests that being in company of another person is important during darkness. However the pattern of active social interaction in groups of three or more is increased in the evening only in Lindeborg square. It implies that spatial light distribution is important to sustain patterns of social interaction.

Results suggests that teens have a higher presence in EL than in DL in Lindeborg square, while this pattern is not evident in Kirseberg square. Although this study is limited to two squares, it suggests that lighting may sustain social interaction and spatiotemporal pattern after dark, as was displayed for teens in Lindeborg square. However, the results also indicate that for small children and elderly the after dark conditions do not sustain their presence at any of the squares. Results for verbal and non-verbal interaction are contradictive. Research on pedestrians indicate that lighting may impede or facilitate walking through the support of accessibility, reassurance, comfort and pleasure.[15] Investigations on movements and stationary activities in local public squares also imply that lighting may sustain use of public squares.[11] However further attention and focus on how lighting may support social life in public space is needed, as the provision of inclusive and participatory public spaces are considered a universal right. Future research should strive to investigate user's appraisal in relation to the spatial, spectral, intensity and temporal characteristics of lit environments in relation to the afforded social and physical qualities of the settings.

5. Conclusion

This investigation indicates that EL may sustain spatiotemporal patterns of social interaction in local public squares after dark. Spatial distribution might be an important lighting characteristic for social interaction. However further research is required to confirm how different lighting characteristics might impede, facilitate or vitalize social interaction in public spaces.

References

[1] UN. Transforming Our World: The 2030 Agenda for Sustainable Development A/RES/70/1;. New York, USA: UN General Assembly, 2015.
[2] UN. *UN Habitat III Policy Paper 1: The Right to the City and Cities for All* 2017. New York: United Nations.
[3] Vaidya H and Chatterji T. SDG 11 Sustainable Cities and Communities. *Actioning the Global Goals for Local Impact*. 2020, pp.173-185.
[4] Cattell V, Dines N, Gesler W, et al. Mingling, observing, and lingering: everyday public spaces and their implications for well-being and social relations. *Health Place* 2008; 14: 544-561.
[5] Mehta V. Evaluating Public Space. *Journal of Urban Design* 2014; 19: 53-88.
[6] Amin A. The Good City. *Urban Studies* 2006; 43: 1009. research-article.
[7] Madanipour A. Why are the design and development of public spaces significant for cities? *Environment & Planning B: Planning & Design* 1999; 26: 879. Article.
[8] Francis J, Giles-Corti B, Wood L, et al. Creating sense of community: The role of public space. *Journal of Environmental Psychology* 2012; 32: 401-409.
[9] Casciani D. *The Human and Social Dimension of Urban Lightscapes*. Cham, Switzerland: SpringerBriefs in Applied Sciences and Technology, 2020.
[10] Boyce PR. The benefits of light at night. *Building and Environment* 2019; 151: 356-367. Review Article.
[11] Hennig VKR, Gentile N, Fotios S, et al. User behaviour in public squares after dark. *Under peer review* 2022.
[12] Carr S, Francis M, Rivlin LG, et al. *Public Space*. New York: Cambridge University Press, 1992.
[13] Altman I and Zube E, H. *Public places and spaces. Human behavior and environment : Advances in theory and research*. New York: Plenum Press, 1989.
[14] Whyte WH and Madden K. *How to turn a place around: A Handbook for creating Successful Public Spaces* New York: Project for Public Spaces 2001.
[15] Fotios S and Johansson M. Appraising the intention of other people: Ecological validity and procedures for investigating effects of lighting for pedestrians. *Lighting Research & Technology* 2017; 0: 1-20.
[16] Sumartojo S, Edensor T and Pink S. Atmospheres in Urban Light. *Ambiances* 2019. DOI: 10.4000/ambiances.2586.
[17] Lofland L. *The Public Realm: Exploring the City's Quintessential Social Territory*. New York: Aldine De Gruyter, 1998.
[18] Goffman E. *Behavior in public places : notes on the social organization of gatherings*. 1st Free Press paperback ed.: Free Press, 1966.
[19] Crowhurst Lennard SH and Lennard HL. *Livable cities: People and Places: Social and Design Principles for the future of the city*. New York: Gondolier Press, 1987, p.9.
[20] Bonnes M and Secchiaroli G. *Environmental psychology: a psycho-social introduction*. London: Sage, 1995.

[21] Stokols D. Environmental psychology. *Annual review of psychology* 1978; 29: 253-295.
[22] Veitch JA, Fotios SA and Houser KW. Judging the Scientific Quality of Applied Lighting Research. *Leukos* 2019; 15: 97-114.
[23] Veitch JA. Psychological processes influencing lighting quality. *Journal of the Illuminating Engineering Society* 2001; 30: 124-140. Article.
[24] Sussman R. Observational Methods: The first Step in Science. In: Gifford R (ed) *Research Methods for Environmental Psychology*. 1st ed.: John Wiley & Sons, 2016, pp.9-27.
[25] Altmann J. Observational study of behavior: sampling methods. *Behaviour* 1974; 49: 227-267.

Light festivals: Entertainment or urban renewal tool?

M Bhatia

Msc in Architectural Lighting Design from KTH Royal Institute of Technology (2019); Bachelors in Architecture from Chandigarh College of Architecture (2014), India

Email: mbhatia.ld@gmail.com

Abstract. This paper is part of the author's thesis wherein she attempted to investigate the hypothesis that a light festival is not only an entertainment based light show but could also be a sustainable renewal strategy for dead urban spaces. 'Public space is our open-air living room' [1], where people meet, interact and replenish their body and mind away from work and home. However, due to various reasons, the current pandemic being one of them, has led to the present generation being seen as an indoor generation [2]. As a result, while a lot of urban spaces are tending to be 'abandoned and left areas that have lost their function' [3]. The author's thesis is a qualitative study to check the feasibility of light festivals as a sustainable strategy to tackle the lifelessness of such dead urban spaces. This paper in particular, presents a qualitative case study of 4 light festivals. This was achieved through both primary sources in the form of personal interviews with organisers and systematic online desk study of the secondary sources of information available as reports by various national and international agencies, journals and peer reviewed papers published by researchers in various online journals of lighting and architecture. These festivals were chosen on the basis of their initial intentions of inception to match the impetus for this study. The success of the festivals is studied with respect to achievement of initial objectives. These are then qualitatively analysed under the categories of image regeneration, revenue generation and job generation to assess the holistic impact of light festivals at various locations in the world. The results of the study suggest that light festivals add playfulness and a sense of identity to public urban scape. The interventions, being temporary in nature, may disappear, but the memory of the event remains in collective memory. Through this, light festivals could potentially transform the views of inhabitants, so that they in turn, can transform their public spaces. The study also suggests that light festivals can be be socially and economically profitable. In conclusion, a light festival could be a successful contender as a sustainable urban renewal strategy.

1. Introduction

'Public space is all around us, a vital part of everyday urban life: the streets we pass through on the way to school or work, the places where children play, the local parks in which we enjoy sports and sit at lunchtime. Public space is our open-air living room' [1], a place where people meet, interact and replenish their body and mind away from work and home. It is a welcoming space that caters to people of all ages, gender, caste, creed and economic status. 'Public spaces are the primary site of public culture. They are a window into the city's soul' [4]. While they form the first impression of a city, reflecting its social and ideological spirit, it is also true that in the last decade, a large percentage of the population across the globe has pushed itself indoors due to various reasons, the current pandemic being only one of those. The present generation is being seen as an indoor generation. [2] Naturally, with most of us staying indoors, a lot of open urban spaces are tending to be comatose, 'abandoned and left areas that have lost their function' [3]. 'Lack of surveillance and usage' of 'these lost spaces', unperceived by the public eye, accentuates the 'possibility of criminal activities, therein [3]. During day, the city is perceived by the combination of light distribution and contrasts between set ups. During the night the scenario flips completely. The city is hidden and 'all orders are erased' [5]. The city transforms into a dark blank canvas, revealing by means of highlights or illuminated. Artificial light, thus, becomes an important tool in the place making aspect of the nocturnal landscape of a city, creating 'a canvas for people to explore and fill with life. Illumination acts as a backdrop for the spectrum of human activity after dark. Its immense power includes the ability to alter spaces; as a medium, it allows designers to play with shades, colour and intensities.' [6].

In an academic lecture by Henrik Gidlund of Trafikverket Stockholm, he talked about a park that was not being used due to the poor image it had in the local residents' minds. But this park did not receive visitors even after putting on general outdoor lights as the people had a negative image of the place. So, the Trafikverket came up with the solution of putting bright coloured and visually loud lights in this park to attract attention. The result was that people started visiting the park, even though it was loud and garish. Over a period of time, the park started receiving a decent number of visitors and the lighting was made more visually comfortable. As mentioned in Sherouk Badr Taha's thesis on Light memory, as a design tool, light memory is an episodic memory that only takes visual inputs into account [7]. As per Atkinson, R.C. and Shiffrin, R.M, a long-term memory is formed when there is repetitive bold impact on the short-term memory formation [8]. This led to the hypothesis that light festivals have the potential to alter the long-term collective memory and association with a space, reinstating the public space with potential for multiple activities and a positive image. But for an urban renewal tool to be tested for being feasible and sustainable in the long run, economic and social aspects too could be considered to have a holistic approach.

2. Methodology

This research employs a qualitative approach 'that seeks to understand phenomenon in context specific settings' [9] Hence, for this paper, the phenomena of light festivals is analysed through case studies of four such festivals. This was achieved through interviews of primary sources and systematic online desk study of the secondary sources of information available in the form of reports by various international agencies, journals and academic thesis work.

2.1. Criteria for case study selection

The festivals were chosen on the following initial intentions of inception, as presented in LUCI's 2019 publication of Cities & Lighting. [10]

- Change the image of the city- Aim of using light festivals 'to bring new dimension to the image of the city at a local, regional, national or international level. Light festivals intended to be an opportunity to show the city in a new light, either by highlighting its positive aspects or by revealing a new facet of its image' [10]
- Foster citizen participation and social cohesion – 'The interactive, playful and approachable nature of light means that light festivals are often seen by cities as an opportunity to engage citizens. Some cities organise these festivals for 'mobilising inhabitants around urban renewal initiatives' [10]
- Showcase local creative talent – Light festivals aimed at 'boosting local artists and designers by showcasing their work' [10]
- Bring in new meaning to urban landscape – Aim of highlighting existing architecture or portraying the everyday spaces with a new character.

IOP Conf. Series: Earth and Environmental Science 1099 (2022) 012009 doi:10.1088/1755-1315/1099/1/012009

2.2. Light festivals selected

- Lights in Alingsas, Sweden
- Light City Baltimore, USA
- Urban Light celebration, Medellin, Colombia
- GLOW Eindhoven, Netherlands

2.3. Data collection

2.3.1. Lights in Alingsas, Sweden. The data was collected from a personal interview with the organisers of the event- Alingsas Energi and the festival website [11].

Figure 1. Installation at Lights in Alingsas

Figure 2. Installation at Lights in Alingsas

2.3.2. Light city Baltimore, USA. The data was collected from Economic impact report of Light City 2018 [12] and Architectural blogs [13] [14] [15].

Figure 3. Installation at Light City, Baltimore

Figure 4. Harbour at Light City, Baltimore

2.3.3. Urban light celebration, Medellin, Colombia. The data was collected from LUCI publication The Social Dimensions of Light 2011 [16].

Figure 5. Installation on the Medellin river

Figure 6. Installation on the Medellin river

2.3.4. GLOW, Eindhoven, Netherlands. The data was collected from previous master thesis works and the official website of the festival [17] [18].

Figure 7. Projection mapping installation at GLOW

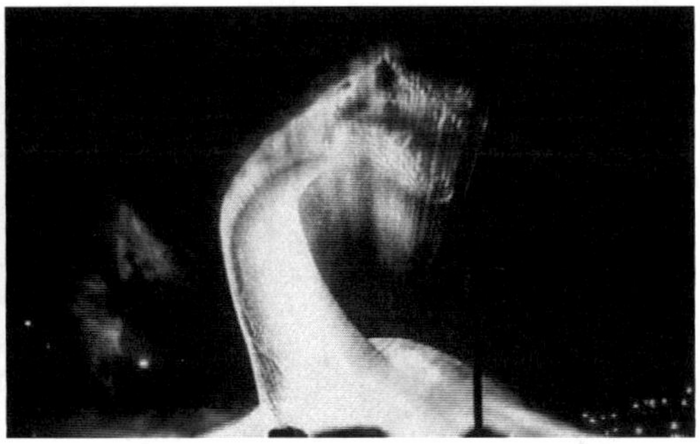

Figure 8. Dynamic projection mapping installation at GLOW

2.4. Criteria for evaluation

Figure 9 shows the broad components of sustainability. Since the phenomenon of light festivals is a temporary one and does not involve any permanent construction or infrastructural intervention, it could be suggested as less environmentally invasive as opposed to other permanent additions to a public space, followed by their upkeep and maintenance. To consider the social and economic aspects, the following criteria is arrived at.

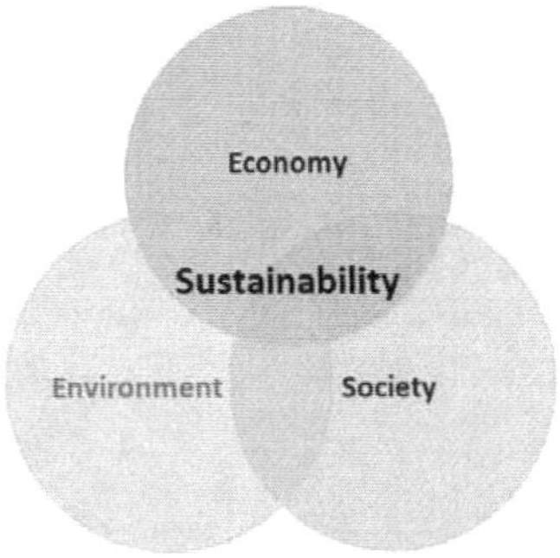

Figure 9. Components of sustainability

2.4.1. Image regeneration. To assess the effectiveness in change of perception that the population has of a public space and consequently of the city. This also includes the feeling of safety/lack of it, felt in a public space. This criterion takes root in the long-term memory formation process aided by rehearsal of short-term memory inputs as mentioned in section 1.

2.4.2. Job generation. To assess the inclusion of the local population in the program of a light festival. This includes permanent and seasonal job creation.

2.4.3. Revenue generation. To assess the financial return on investment for light festivals to become a viable solution for any city municipality and stakeholders to invest in.

3. Analysis of collected data

The data collected through methods mentioned in section 2.3 was categorised under the three heads that are used as parameters for assessing the success of each light festival.

Table 1. Comparative analysis of the four case studies.

Evaluation criteria	Lights in Alingsas	Light City Baltimore	Urban light celebration, Medellin	GLOW Eindhoven	Category
Visitor count	80,000	442,500	4 million	750,000	Image + Revenue
Types of sites chosen	Popular places +dead spaces	Central tourist spots with safety concerns + residential areas.	City centre + river corridor + 80 other locations	City centre +industrial area (Strejp-S)	Image
Change in perception	Some installations are made permanent. Inhabitants feel proud.	63% were present only for the festival. 87.4% felt it changed perception of the city.	Top 10 places in the world to see Christmas lights from most violent city in the world in 1616	Strejp-S seen as a hub for local lighting design. Synonymous with new ideas and technologies.	Image
Visitor/ Inhabitants review	Demand for more such events.	92.3% felt it added value. Festival became new tradition. 82% satisfied visitors	Not available	59% visitors had 'more fun than expected'	Image
Successes	Lighting integrated in planning stages.	73% repeat visitors expected. Changes in safety concerns (82% felt safe).	Included in social planning policies.	75% expected repeat visitors +43% repeat visitors. Rated in top 5 best light festivals.	Image
Jobs created	No new jobs. Provides platform for students to build contacts.	36 temporary jobs Platform for local designers in Neighbourhood lights	530 temporary jobs +4580 temporary businesses +Seasonal employment to single mothers	5 Permanent jobs	Job
Young designer/ student involvement	50 students work in teams lead by professional designers.	Local students participate in workshops. Local artists create light-art in communities	No involvement as all the designing is done by EPM	Dedicated space and event (New Light) for local designers to participate in the festival	Job
Community involvement	Local school, professional schools participate	330 volunteers. 60+ building facades lit up as part of neighbourhoods light to show civic pride.	Involved in major decisions of the festival. Provides jobs for single mothers.	Volunteers help with the organisation. GLOW Run – a marathon brings the city together	Job
Budget	470,000 Euros	2.6 million Euros	2.6 million Euros	1.6 million Euros	Revenue
Revenue-Hotels	1/3rd annual revenue	2 million euros	60% hotel occupancy	Not documented	Revenue
Revenue – Local Businesses	Not documented	17.67 million euros	Not documented	Not documented	Revenue
Revenue – Cafes/ restaurants	1/3rd annual revenue	9.4 million euros	Not documented	Not documented	Revenue
Total revenue	Not documented	37.9 million euros	Not documented	12.8 million	Revenue

4. Discussion

Although all four festivals have been organized with similar objectives the organizational outcomes are quite different. By bringing forth the commonalities and comparing the differences, we can understand the success of light festivals under the wider categories of image regeneration, revenue generation and job generation.

4.1. Success in Image regeneration

All four festivals attract large number of visitors annually. This reflects positive image of the host cities that these festivals can project onto the world. Baltimore and Medellin, that were previously infamous for nefarious activities, have managed to turn this around by introducing the vitality into their urban spaces with these festivals, as shown in figures 10 and 13. GLOW has become a premier light festival comparable to Fete de Lumieres, as represented in figure 11. Lights in Allingsas on the other hand has gained name among design students as best lighting workshop, as seen in figure 12. These festivals manufactured these dazzling images of their host cities. The visitor reviews of safety felt during the festival in Baltimore, as shown through the survey results in figure 14 and the enormous number of people visiting Medellin are clear indicator of the potential and success of light festivals as a solution to fight the feeling of being unsafe.

Figure 10. Installation on Medellin river, changin the way people view and interact with the space

Figure 11. Installation transforming the experience at Philips stadion at GLOW

Figure 12. Design students making an installation at Lights in Alingsas

Figure 13. Inner Harbour at Light City, Baltimore

IOP Conf. Series: Earth and Environmental Science 1099 (2022) 012009 doi:10.1088/1755-1315/1099/1/012009

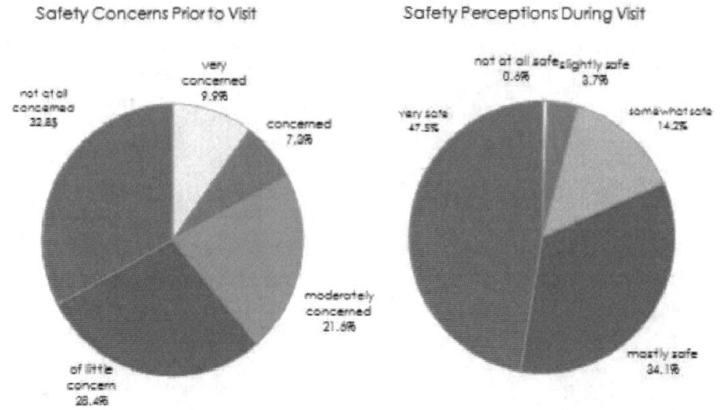

Figure 14. Survey results of safety concerns in public spaces before and during the Light City Baltimore

4.2. Success in Job generation

As seen in the examples of Alingsas, Medellin and Eindhoven, light festivals inspire and initiate better lighting practices and integration of lighting in urban policies, in the long term. This reflects upon informed and sustainable lighting practices, while giving a boost to the lighting industry of the host city. Light festivals in these case studies vary in scale and format and hence the number of people required to organize them also vary a lot. Alingsas, being focused in the educational aspect of lighting, does not require any extra job posts to be formed for the organization of the festival. The work force consists of international and local students, as shown in figure 12, so need for volunteering is eliminated. On the other hand, Baltimore, Medellin (figures 15) and Eindhoven are large scale festivals that require people to work in the organization all year round. They create few permanent jobs and a large number of temporary jobs. Consequently, these festivals also depend heavily on volunteering for the execution and smooth running of the festivals. This instils a feeling of involvement and civic pride among the inhabitants, inspiring social cohesion. Baltimore takes a step further in the community involvement by launching new branches of the festival, in the form of Neighborhood lights, as shown in figure 16, Brilliant Baltimore, etc. that gives a sense of ownership and pride to the inhabitants. When it comes to job opportunities for designers, the festivals under study, do not contribute to the volume of job opportunities, but act as platforms for building valuable contacts (Alingsas), gain more hands-on knowledge (Alingsas, Baltimore) and marketing young talent (Eindhoven Strejp-S).

Figure 15. Community involvement in making art work for Urban lights in Medellin

Figure 16. Neighbourhood lights at Light city, Baltimore

IOP Conf. Series: Earth and Environmental Science 1099 (2022) 012009 doi:10.1088/1755-1315/1099/1/012009

4.3. Success in Revenue generation

The large number also has massive ramifications in revenue generation. Light festivals work as successful tourism policies for their host cities especially in time of low tourism flows. This can also be seen by the overall satisfaction of visitors of a light festival, as shown in figure 17 for Light City, Baltimore. All four festivals succeed in generating revenue for local establishments that cater to food, beverage and transport. The hotels too record increased bookings in each festival. The numbers available on investment and revenue generated, confirm LUCI's (Lighting urban community international) research on the return on investment of light festivals. Though the cost of organizing a light festival is not meagre (Alingsas is the smallest in scale among the four case studies and its organization requires 470,000EUR), but the traditional urban renewal methods adopted by cities across the world require heavy investments that take years to shape up. Historically, some efforts have succeeded [19] and some are known to have failed and invited criticism [20]. Light festivals, because of their temporary nature, offer 'a flexible solution' to some urban apathies. All four festivals choose a combination of sites – popular and dead spaces, for their festivals. By different means and in varying measures, these festivals revitalize the dead spaces with light art, while attracting the numbers based on the appeal of popular spaces in a new light.

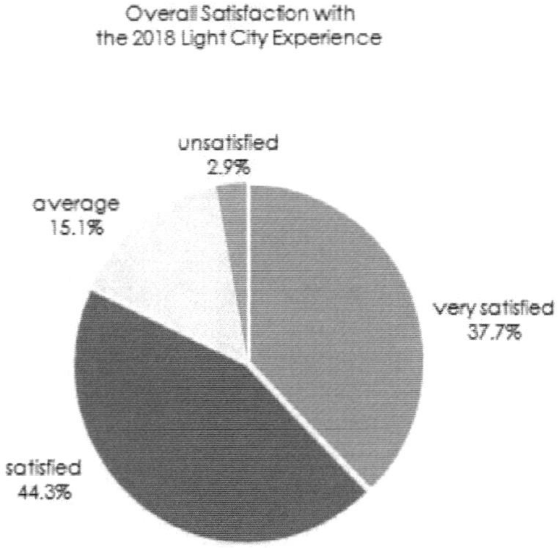

Figure 17. Survey results for overall satisfaction of visitors at Light city, Baltimore

5. Conclusion

Light festivals as a phenomenon have increased its foothold in many big and small cities across the globe. Through the process of light memory formation and the repetitive nature of such festivals, they make a significant impact on the perception of public spaces in a city and on the overall idea of that city. This proves the functionality of light festivals as an urban renewal tool. As presented through section 3, light festivals generate high revenue giving high returns on investment, while also generating seasonal jobs for the local population. As a pilot study of light festivals as a sustainable step towards urban renewal, light festivals seem to satisfy the evaluation criteria. The study leaves scope for further research through assessment of the sustainability markers for a light festival, by comparing them with standard urban renewal tools and their impact on the public spaces of the city.

References

[1] UN-Habitat_01. (2015) *Global Public Space Toolkit: From Global Principles to Local Policies and Practice. United Nations Human Settlements Programme* p 124. Retrieved from unhabitat.org: https://unhabitat.org/global-public-space-toolkit-from-global-principles-to-local-policies-and-practice
[2] Sharma S (Apr 2020) The Indoor Generation *The Times of India*, https://timesofindia.indiatimes.com/life-style/health-fitness/de-stress/the-indoor-generation/articleshow/75451919.cms
[3] Khalid, N. S. (2020) Exploring the Prescriptive and Descriptive Lost Space in the Setting of Urban Fabrics of Kuala Lumpur in A. Almusaed, A. Almssad, L. T. Hong (eds.), *Sustainability in Urban Planning and Design, IntechOpen, London.* 10.5772/intechopen.89907
[4] Rupa C K (2015) *Importance of public spaces in cities* 10.13140/RG.2.1.1656.1125. https://www.researchgate.net/publication/281749329_Importance_of_public_spaces_in_cities
[5] http://olssonlinder.se/projects/bling-bling/
[6] Arup (2015) *Cities Alive: Rethinking the Shades of Night* p 23. https://www.arup.com/perspectives/publications/research/section/cities-alive-rethinking-the-shades-of-night
[7] Badr S T (2018). *Light Memory, as a Design Tool* (Dissertation). Retrieved from http://urn.kb.se/resolve?urn=urn:nbn:se:kth:diva-229744
[8] Atkinson R.C. and Shiffrin R.M. Storage and retreival processes in long-term memory 1969 *Psychological Review* **Vol. 76**, No. 2, pp 179-193 1969 Storage and Retrieval LTM Shiffrin, Atkinson.pdf (iu.edu)
[9] Kaul, L. (2009) *Methodology of Education Research.* (Noida : Vikas Publishing)
[10] *Cities and Lighting*, 2019, ed M B Page and N Junagade, (Lyon: LUCI association) pp 16-17
[11] https://www.lightsinalingsas.se/en/
[12] *Economic impact study and audience research 2018 Light city* (2018) Baltimore office of promotion and the arts
[13] https://www.baltimoresun.com/entertainment/bs-fe-lightcity-impact-20180627-story.html
[14] https://brilliantbaltimore.com/light-city/
[15] https://www.promotionandarts.org/events-festivals/light-city
[16] The social dimensions of light. 2011, ed A Colombani, N Junagade, et al (Imprimerie Rochelaise) pp 81-87
[17] Annual report of GLOW 2017, https://issuu.com/jetvetter/docs/def_-_jaarverslag_glow_eindhoven_20
[18] Brown C F L (2015) on *Trip the light fantastic: The role of lighting festivals and events in urban design* (Dissertation). Retrived from https://issuu.com/cashelflbrown/docs/150822_final_dissertation
[19] Fontenot A *30 Inspiring Urban Renewal Projects* https://www.socialworkdegreeguide.com/30-inspiring-urban-renewal-projects/
[20] https://en.wikipedia.org/wiki/Urban_renewal

The effect of non-uniform urban illumination upon pedestrians' alertness and sense of safety

R Jedon[1], A Haans[1] and Y de Kort[1]

[1] Human-Technology Interaction, Industrial Engineering and Innovation Sciences, Eindhoven University of Technology, Eindhoven, The Netherlands

R.Jedon@tue.nl

Abstract. The influence of urban lighting on the visual tasks of pedestrians are well known and, likewise, studies have explored the relationship between urban lighting and pedestrians' feelings of safety. But there are other, less investigated, possible effects of light in mesopic situations. Research on street lighting for pedestrians has traditionally focused on minimal required illuminance for various visual performance tasks (e.g., obstacle detection and face recognition). However, it is not only pedestrians' visual performance that is important for safety, but also their general attentiveness to the environment. Despite this, the psychological concepts related to the individuals' attention, such as alertness, arousal and anxiety, have not been considered in street lighting research to date even though they too can play an important role in pedestrian's attention and safety. Since proper metrics for these concepts are needed, we conducted a pilot study to test their sensitivity. We decided for a conceptual replication of a field experiment from 1916 and managed to partly replicate the findings. More importantly, we learned meaningful lessons from this endeavour. Implications for future research are discussed.

1. Introduction

Urban lighting plays an essential role in ensuring people's safety after dark. Yet, given the current environmental issues and threat of climate change, there is a growing demand for more sustainable urban lighting. Such sustainable lighting should be more efficient in energy consumption and minimize the negative outcomes of light pollution. Research on street lighting for pedestrians has therefore focused on minimal required illuminance for various visual performance tasks, such as the effect of lighting on obstacle detection or face recognition.[1-4] We argue that it is not only the visual task performance that should be of interest in research on lighting for pedestrians but also the general state of alertness of the individual.

We understand alertness as a cognitive state of enhanced readiness and openness to incoming information.[5-9] This is reflected in faster and more accurate reactions to incoming stimuli.[5,7] Highly alert person is also able to discern even small changes in the environment.[9] The state of alertness is linked with both arousal (i.e., general state of both cognitive and physiological activation)[5,8] and anxiety.[10,11] Anxiety, as a state of readiness to potential future threats, is negatively related to feelings of safety.[12,13] From an evolutionary perspective though, anxiety can be considered a benefit to cognition as it facilitates physical readiness and faster detection of potential threatening information.[11,14-16] We can see that increased levels of alertness, arousal and anxiety could lead to improved objective safety (i.e., obstacle detection, prevention of traffic collision) but could also influence perceived safety (for further discussion see Jedon, Haans & de Kort, 2022; accepted for publication in Lighting Research & Technology). Despite this, these psychological concepts have not been considered in urban lighting research to date.

In 1916, Burtt claimed that mere range of individuals' sight is insufficient for their safety if they are also not attentive to their surroundings and able to react to possible dangers.[17] Therefore, to compare pedestrians' performance under different lighting conditions, he employed an auditory two choice task (ATCT), a short memory task, and a motor coordination task in a field experiment. In general, his findings suggest better performance on all three tasks under non-uniform lighting conditions (see Table 1). In Burtt's explanation, these positive performance outcomes were caused by two factors. First, by changes in the intensity of illumination through which the pedestrians walked. Second, by the alternations between the illuminated areas under the lamps and the dark, unlit, regions in between them. These effects, however, may result from different mechanisms. The dark spots and reduced ability to see ahead might have caused higher arousal and/or anxiety, resulting in heightened alertness and attention, and hence better performance on the ATCT. However, such changes in alertness and attention may also be associated with more favorable increases in arousal, such as when the non-uniform lighting conditions would be perceived by the participants as novel and stimulating.

Table 1. Results from Burtt's study[17] (adapted from Burtt, 1916; public domain).

	Number of trials	% of trials in which performance under non-uniform was superior	Average superiority of non-uniform, in %	% of trials in which uniform had greater M.V.	Average M.V. uniform / M.V. non-uniform
ATCT	18	100	17.0	50	0.98
Short-term memory	31	84	8.1	81	1.32
Motor coordination	22	77	3.6	82	1.48

Notes: The uniformity of illumination is defined as the mean percent variance to the average. Burtt considered both horizontal and vertical illumination. The greater the mean variation (M.V.) in performance across trials the lower the degree of attention.

Burtt's study, despite asking relevant research questions, received little attention. In particular, no replication of Burtt's study, or its parts, has been done to our knowledge. Therefore, we decided that a good starting point for disentangling the relations between street lighting, alertness, arousal and anxiety (i.e., the sense of safety) is to perform a conceptual replication of Burtt's experiment using modern day reaction time equipment and self-report measures for assessing alertness, arousal and perceived safety. To investigate the applicability of these measurements and the sensitivity of these metrics to changes in the urban environment (including lighting) we conducted a pilot study.

We designed a quasi-experimental field study to test two hypotheses: 1) Responses will be faster and more accurate on the streets with lower uniformity of illuminance than on the streets with high uniformity of illuminance; 2) The perceived safety will be lower on the streets with lower uniformity of illuminance than on the streets with high uniformity of illuminance.

2. Method

2.1. Participants

Sample size was determined first most with feasibility in mind. Given that the experiment could only start in the last week of April 2021, we had a short window of opportunity before the late sunset would make it too troublesome for participants to come to the test site in the late hours. The late sunset and COVID19 restrictions, including a partial lockdown and evening curfew, prevented us from inviting more than 8 participants per night. Based on the availability of the researchers and considering possible days with poor weather conditions, we expected to run about 40 participants at best. A sensitivity analysis conducting in G*Power[18] for a within-subject ANOVA with 4 repeated measures (i.e., 4 street sections; see *2.2 Design*) and assuming a .50 correlation between repeated measures showed that with 40 participants we had .90% power to detect a small to moderate effect size of $f = .21$ at $\alpha = .05$.

Unfortunately, technical issues with one of the responses boxes forced us to have 3 rather than 4 participants simultaneously per session. As a result, only a total of $n = 29$ people participated in the experiment. Participants were recruited from the J.F. Schouten database of Eindhoven University of Technology (TU/e), Eindhoven, The Netherlands. All participant that took part in the experiment before April 29, the day the evening curfew was dropped, received a curfew employer statement from the TU/e to be exempted from the curfew. Out of the 29 participants, 13 (44.8%) were women and 16 (55.2%) were men. Average age of the participants was $M = 23.1$ ($SD = 3.3$; range 19 to 30). Of the 29 participant, 12 (41.4%) and 10 (34.5%) indicated to be "very familiar" and "somewhat familiar" with the TU/e campus respectively. Of the remaining participants, 3 (10.3%) indicated to be "a bit familiar" with the campus and 4 (13.8%) to be "not familiar at all". Participants received € 10,- euro as a compensation for their participation.

2.2. Design

We conducted a one condition (4 different street sections) within-subject experiment with response time and accuracy in an auditory choice task as objective markers of alertness and self-reports of sleepiness, arousal, safety and lighting appraisal as subjective outcomes. The four streets differed, amongst other things, in the uniformity of lighting (see Table 2). The experiment was conducted in multiple sessions, each with three participants partaking at the same time, but with each starting on a different street section. To reduce the time needed to move from one section to the other, and to keep the experiment manageable, we did not fully randomize the order of the street sections. Instead, Section 2 always followed Section 1, Section 3 followed Section 2, and Section 4 followed Section 3. In other words, a participant starting on Section 3, would subsequently walk Section 4, then Section 1, and ending with Section 2. During each 5 min walk participants engaged in two simultaneous tasks. We selected a dual task approach to increase perceptual load and reduce the chance that the slower and/or inaccurate responses are caused by lack of concentration.[19] The primary task was walking along the street, following coloured ribbons that marked the path. An auditory two choice task (ATCT), which was adapted from the earlier study by Burtt[17] was the secondary task.

Reaction times and accuracy to the ATCT were recorded. At the end of each walk participants completed a short survey. The experiment was approved by the Ethical Review Board of TU/e.

2.3. Setting

The experiment took place on the TU/e campus during six days between 22nd of April and 5th of May 2021, at least 15 minutes after sundown. Participants started between 21:00 and 22:45hrs. Weather conditions were rather similar between days, except for May 5 where it started to rain. The average temperature was 9ºC (range 3 to 15).

The four test streets on the university campus were reasonably free of traffic at the time of the experiment. The four street sections had an average length of 277.5 meters (range 215 to 390 meters). Each street section was marked with coloured ribbons (corresponding to the colours in Figure 2) that the participant had to follow. The street sections differed in various aspects beside the lighting. On Section 2, for example, participants walked on the pedestrian pathway that was separated from the main road by a several meters wide strip of grass and trees. On Section 3, in contrast, the participants were asked to walk on the middle of road as the pedestrian pathway directly next to the street was partly inaccessible due to construction work on an adjacent building.

To quantify the lighting conditions in each of the four street sections, several lighting measurements were made. On the preceding evening to the first experiment day, extensive measurements were made to obtain estimates of each street section's average horizontal (E_h) and vertical illuminance (E_v) as well as uniformity (E_{min}/E_h) within a measurement area that was representative for that particular street section. The length of the measurement area was determined by the distance between two consecutive lamp posts and varied between 30.0 to 40.6 meters (see Figure 1). To keep measurements comparable across street sections, the width of the measurement area was the same for all street sections: 5 m. This was based on the width of the road of Section 3. Horizontal and vertical illuminance was measured in each point of—depending on the distance between the poles—a 3 by 10 or a 3 by 14 measurement grid as recommended in European road lighting standards (see Figure 1).[20] Measurements were made with LMT pocket-lux meters. In addition, we measured the spectral power distributions of the luminaires with a JETI specbos 1211-2 spectroradiometer. For each street section, the Correlated Colour Temperature (CCT) and Dominant Wavelength (Wl) are reported in Table 2. In addition to these extensive measurements, average horizontal and vertical illuminance of each street section's measurement area were obtained daily using three easily identifiable measurement points (see Figure 1). These daily measurements reveal that inter-street difference in vertical and horizontal illuminance remained similar across the six days of experimentation (see Table 2 for details).

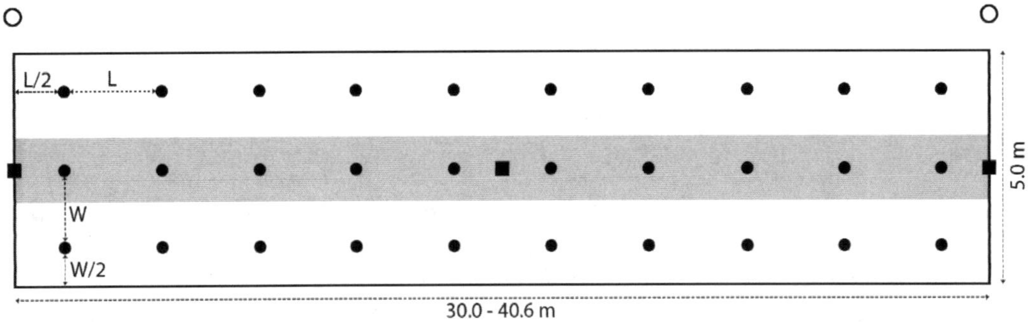

Figure 1: Illuminance measurement area (black box) here displayed around a footpath (grey). Black dots represent the measurement grid for the extensive illuminance measurements. For each street section there were always 3 measurement points perpendicular to the walking direction, with distance W = 1.67 m. Distance L depended on the distance between the two lampposts (open circles). There were at least 10 measurement points parallel to the footpath with max. L = 1.0 m. Black squares show the three measurement points for the daily measurements.

Table 2: Light measurements per street section: Average horizontal illuminance (E_h), average vertical illuminance (E_v), uniformity (U), Correlated Colour Temperature (CCT), and dominant wavelength (Wl)

	Extensive					Daily			
	Illuminance			SPD		Horizontal		Vertical	
	E_h	E_v	U	CCT	Wl	E_h (SD)	Range	E_v (SD)	Range
Section 1*	2.72	1.19	0.18	1940	589	3.50 (0.44)	2.93 – 3.85	0.93 (0.31)	0.38 – 1.20
Section 2	1.20	0.60	0.17	1951	588	1.11 (0.12)	0.90 – 1.25	0.41 (0.07)	0.33 – 0.55
Section 3	13.60	6.32	0.46	1926	589	16.3 (1.99)	13.4 – 17.83	5.75 (1.25)	3.48 – 7.25
Section 4	4.15	2.30	0.07	1918	588	5.29 (0.66)	4.45 – 6.00	2.02 (0.33)	1.70 – 2.55

Notes: SPD stands for Spectral Power Distribution. All measurements are in lux, except CCT (in Kelvin) and WI (in nm). The extensive measurements were obtained on the day before the experiment using a 3 by 10 to 3 by 14 measurement grid (see Figure 1). Vertical illuminance (E_v) was measured at a height of 1.5 meter in walking direction (from the start point). Uniformity is U = minimum E_h / average E_h. Daily measurements were taken every day of the experiment using the three measurement points in Figure 1. The daily measurements for E_h and E_V were calculated by taking the weighted average of the three measurements giving 0.5 weight to the two outermost measurements and a weight of 1 to the middle one.[21] For each street section, averages across all six days, their standard deviation (SD) and range are provided. * A small part of street Section 1 was lit by different luminaires. These had a CCT of 2812 K and a WI of 584 nm.

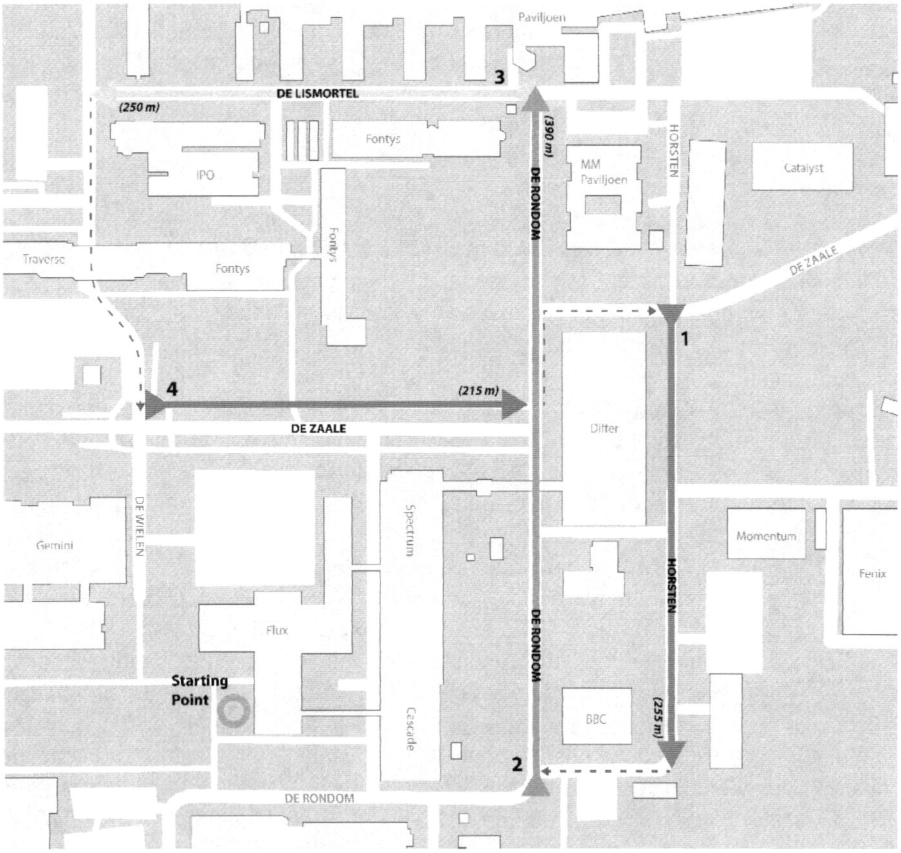

Figure 2: Map of the experiment setting

2.4. Procedure

After the participants arrived on the designated meeting location (see Figure 2), they read and signed the informed consent form. After a short introduction to the experimental procedure, each participant was guided to the starting point of their respective first street section. Here, they received a survey and pen attached to a clipboard worn around the neck. They were asked to complete the first page of the questionnaire about demographics (age, gender and familiarity with the test site). When they completed the survey, participants were asked to put on a backpack carrying a laptop to which were attached a two-button response box and a pair of headphones. Next, the participant received further instructions, part of which were during the walk also provided by pre-recorded voice instructions through the headphones.

After these instructions, the participant was asked to press any button on the response box to start the experiment. In the first session, participants were asked to complete a practice session for the

ATCT, consisting of 10 trials. They were instructed via a pre-recorded voice messages to press as fast and as accurate as possible the leftmost green button in response to a high pitch sound, and the right most red button in response to a low pitch sound. During these trials they would receive feedback as to their response accuracy and whether they had responded fast enough. No such feedback was given during the experimental sessions. If everything was clear to the participant, then they were asked to press a button to start the experiment. When the participant started walking, the experimenter moved out of sight as to not interfere with the measurements. For 5 minutes the participant walked the street section back and forth while following the coloured ribbons and performing the ATCT. After five minutes, the participant was asked, using a pre-recorded message, to walk to the middle of the street section and to complete the remaining pages of the questionnaire.

When they completed the survey, they were escorted to the next street section by another experimenter. When arriving at the start of their second street section, the participant started walking without practicing the auditory choice tasks. After having walked all four street sections, the experimenter escorted the participant back to the meeting location, where they were thanked and reimbursed for their participation.

2.5. Measures

2.5.1. Alertness and attention. Alertness was measured with an auditory two choice task (ATCT). Although established methods such as the auditory Psychomotor Vigilance Task (PVT)[22] are typically used for this purpose, we wanted to use a response task that resembled most closely the method employed by Burtt.[17] In our ATCT, participants would hear at random intervals one of two randomly selected sounds; a low and a high pitch sound. Each time, the participant was asked to give the correct response to each stimulus as fast and as accurately as possible. For the low and high pitch sound, we used a brief 500ms stimulus of 220 and 880 Hz respectively. The two stimuli were matched in loudness using equal loudness curves. The interstimulus interval was randomly varied between 2000 to 10000 ms, always in full thousands of ms. Participants had to respond within 2000 ms. Non-responses were treated as missing values. The total duration of the ATCT was 5 min. The ATCT was programmed in E-prime 3[23] running on a HP ZBook Studio G5 Mobile Workstations, and the auditory stimuli were presented to the participants using Sennheiser HD 265 linear headphones. Responses were recorded using Millikey SV-2 response boxes,[24] where pressing the left most green button was the correct response for the high pitch sound, and the right most red button for the low pitch sound. Each participant was presented with 40 to 52 auditory stimuli per street section. Total number of responses per street across all participants ranged from 1325 to 1331. Participants failed to respond to the auditory stimulus in 50 (0.9%) out of 5361 trials, leaving 5311 responses for further analysis. Three metrics were derived from the ATCT: the reaction time (RT), the proportion of error responses (PE), and the coefficient of variation (CV_{rt}). CV_{rt} is calculated by dividing the standard deviation of the reaction time (SD_{rt}) by the mean reaction time across each person and street section and reflects the participant capacity for sustained attention on the task.[25]

2.5.2. Self-reported sleepiness. Sleepiness was measured with the Stanford Sleepiness Scale (SSS).[26] The SSS is a single item instrument asking people to report on their degree of sleepiness. Participants can answer using a 7-point response format ranging from "Feeling active and vital; alert; wide awake" (coded with a 1) to "Almost in reverie; sleep onset soon; lost struggle to remain awake" (coded with a 7). There were no missing responses.

2.5.3. Self-reported arousal. Self-reported energetic and tense arousal were measured using the Activation Deactivation Adjective Checklist (ADACL).[27] Both types of arousal are each measured with 5 items, in each of which a term is presented to the participant, such as "Active" (for energetic arousal) or "Fearful" (for tense arousal). We added Dutch translations to the English terms (i.e., "actief", "nerveus", "vitaal", "gejaagd", "krachtig", "angstig", "levending", "verward", "pittig" and "gespannen"). Participants are asked to indicate the extent to which they are feeling those terms. They can respond using a 4-point response format ranging from "Definitely feel" (coded with a 1) to "Definitely not feel" (coded with a 4). All responses were reverse coded to ensure that a higher score means more energetic and tense arousal. We averaged the responses to each set of 5 items to calculate four scores per participant (i.e., one per street) for both energetic and tense arousal. These aggregated scores were used in the analysis. Reliability (Cronbach's α) ranged from $\alpha = .74$ to .88 for energetic arousal measures obtained in different streets, and from $\alpha = 87$ to .93 for tense arousal. There were no missing responses on either the energetic or the tense arousal questions.

2.5.4. Perceived environmental safety. Perceived environmental safety was measured using three items from van Rijswijk, Rooks and Haans.[28,29] Example items include "How safe or unsafe do you judge this environment?" and "How uneasy or comfortable do you feel with the idea of having to walk into this environment?". Participant could respond using a 5-point response format, ranging, for example, from "very unsafe" (coded with 1) through "neither unsafe nor safe" (coded with 3) to "very safe" (coded with 5). Across all items, streets, and participants, there were 3 (0.9%) missing responses. By averaging across the tree items four scores were calculated for each participant, each corresponding to a street. These aggregated scores were used in the analysis. For one person and on two streets no reliable perceived environmental safety score could be calculated as the participant responded to less than the required three items. Reliability ranged from $\alpha = .79$ to .93.

2.5.5. Perceived lighting quality. Perceived quality of lighting was measured using the 6 items from van Rijswijk and Haans.[29] These items were originally intended to measure the perceived darkness of the environment as well as the perceived quality of the lighting, but as in the research by van Rijswijk and Haans, the two measures were found to correlate very highly with $r = .82$ and $p < .001$ (using within-subject correlations).[30] Hence we decided to calculate a single perceived quality of lighting score across all eight items. Example items include "How good or poor do you think the quality of the lighting in this nocturnal environment is?" and "How dark or light is this environment in comparison to other urban environments at night?". Participant could respond using a 5-point response format, ranging, for example, from "very poor" (coded with 1) through "neither poor nor good" (coded with 3) to "very good" (coded with 5). Across all items, streets, and participants, there were 7 (1.0%) missing responses. By averaging across the six items four scores were calculated for each participant, each corresponding to a street. These aggregated scores were used in the analysis. Due to missing responses, no reliable perceived lighting quality score could be calculated for one person on one street. Reliability ranged from $\alpha = .87$ to .91.

3. Results

3.1. Differences in response accuracy, reaction time and the coefficient of variation

Overall, participant gave the correct response in 5154 (i.e., 97.0%) of the 5311 trials. Based on a series of Shapiro-Wilk tests—one for each street section—with $\omega \geq .97$ as the criterion for sufficient normality, the PE was not found to be normally distributed in any of the street sections with $\omega \leq .88$. Using the Skillings-Mack test, we found no statistically significant differences between streets sections on PE, with SM = 0.4 and p (with ties) = .929 (for descriptive statistics, see Table 3). There were no outliers (i.e., with standardized scores of $|Z| \geq 4$)

Reaction times, also, did not meet our normality criterion, with $\omega \leq .84$ (for descriptive statistics, see Table 3). Therefore, we used the inverse of the reaction times (1/RT) or speed in the remainder of the analysis. After this transformation, the Shapiro-Wilk tests showed sufficient normality for all street sections with $\omega \geq .98$. To test for differences in speed, we used a Mixed Linear Model with the inverse of the reaction times (1/RT) as dependent variable, street as fixed, and person as random factor. We found a statistically significant effect of street section with Wald $\chi 2(3) = 23.8$ and $p < .001$. Post-hoc paired comparisons revealed that people responded, on average, faster in Street Section 4 (471 ms) as compared to the Sections 1 (487 ms), 2 (498 ms) and 3 (489 ms), with $Z \geq 2.95$ and $p \leq .003$. Removal of 25 outliers (i.e., with standardized scores of $|Z| > 3$) did not alter the interpretation of these findings. As Street Section 4 was the most non-uniform of the four sections, this finding is consistent with the findings of Burtt's experiments on ATCT performance,[17] and thus in support of Hypothesis 1.

With $\omega \leq .95$, CV_{rt} scores did not meet our normality criterion, and hence the Skillings-Mack test was used to test for differences between street sections on sustained attention. No statistically significant differences were found with SM = 0.9 and p = .818 (for descriptive statistics, see Table 3). There were no outliers (i.e., with standardized scores of $|Z| \geq 4$)

3.2. Differences in sleepiness and energetic and tense arousal

Responses to the sleepiness item were found to have sufficient normality for all street sections, with $\omega \geq .98$. There were no outliers (i.e., with $|Z| > 3$). A repeated measures ANOVA with sleepiness as dependent variable and street section as factor, showed no differences in sleepiness between the four streets, with $F(3,84) = 0.38$, and $p = .750$ (Huynh-Feldt corrected). Descriptive statistics per street section are reported in Table 3.

Energetic arousal was not found to be normally distributed, with $\omega \geq .95$. There were no outliers with (i.e., with $|Z| \geq 4$). Using the Skillings-Mack test, we found no differences between streets sections in self-reported energetic arousal, with SM = 1.7 and p (with ties) = .580. Descriptive statistics per street section are reported in Table 3.

Tense arousal also was not found to be normally distributed, with $\omega \leq .83$. There were no outliers (i.e., with $|Z| \geq 4$). Using the Skillings-Mack test, we found no differences between streets sections in self-reported tense arousal, with SM = 2.7 and p (with ties) = .205. Descriptive statistics per street section are reported in Table 3.

Table 3: Means (M), medians (Mdn), and standard deviations (SD) per street section on the reaction time (in ms), proportion of errors (PE), coefficient of variation (CV_{rt}), sleepiness, energetic and tense arousal, perceived environmental safety (Env. safety), and perceived lighting quality.

	RT (ms)	PE (%)	CV_{rt}	Sleepiness	Energetic arousal	Tense arousal	Env. safety	Lighting quality
Street 1								
M	541	2.9	23.7	2.7	2.4	1.6	3.7	3.1
Mdn	489	2.2	22.8	3.0	2.6	1.4	4.0	3.2
SD	141	3.2	5.4	1.0	0.8	0.9	1.1	0.8
Street 2								
M	547	2.6	24.4	2.6	2.4	1.6	3.7	2.9
Mdn	516	2.2	21.9	3.0	2.4	1.2	4.0	2.7
SD	130	2.9	6.8	0.9	0.6	0.8	0.9	0.8
Street 3								
M	550	3.3	23.9	2.6	2.5	1.5	4.0	3.8
Mdn	526	2.3	24.0	3.0	2.6	1.4	4.0	4.0
SD	156	3.7	6.4	1.0	0.8	0.7	0.9	0.7
Street 4								
M	527	3.0	24.9	2.8	2.4	1.5	3.7	2.8
Mdn	499	2.2	24.2	3.0	2.6	1.4	3.7	2.8
SD	139	3.5	7.3	1.1	0.8	0.7	1.0	0.8

Notes: Sleepiness was measured on a scale from 1 to 7; Energetic and tense arousal were measured on a scale from 1 to 4.; Perceived environmental safety and lighting quality were measured on a scale from 1 to 5.

3.3. Differences perceived environmental safety and quality of the lighting

Perceived environmental safety was not found to have sufficient normality, with $\omega \geq .94$. There were no outliers (i.e., with $|Z| \geq 4$). Using the Skillings-Mack test, we found streets sections to differ in perceived environmental safety to a statistically significant extent, with SM = 63 and p (with ties) = .033. Post-hoc paired comparison revealed that Street Section 3 was perceived as safer (M = 4.02) as compared Sections 1 (M = 3.70), 2 (M = 3.70) and 4 (M = 3.75), with SM ≥ 3.4 and p ≤ .042 (in support of Hypothesis 2). No other statistically significant differences in perceived safety were found between the other street sections. Descriptive statistics per street section are reported in Table 3.

Perceived lighting quality was not found to have sufficient normality, with ω ≥ .94. There were no outliers (i.e., with |Z| ≥ 4). Using the Skillings-Mack test, we found street sections to differ in lighting quality, with SM = 31.3 and p < .001. Post-hoc paired comparison revealed that Street Section 3 had better lighting quality evaluations (M = 3.79) as compared Sections 1 (M = 3.07), 2 (M = 2.89) and 4 (M = 2.77), with SM ≥ 15.2 and p < .001. No other statistically significant differences in perceived lighting quality were found between the other street sections. Descriptive statistics per street section are reported in Table 3.

3.4. Correlations between variables

To investigate the relationships between the various variables including in our theoretical model, we estimated all pairwise correlations using the within-subject method by Bland and Altman.[30] For this purpose we calculated an average speed on the ATCT for each participant and each street section. We found a moderate negative correlation between energetic arousal and sleepiness of r = -.58 (see Table 4). As expected, higher energetic arousal was associated with less sleepiness. Tense arousal also had a negative association with sleepiness albeit to a lesser extent than energetic arousal with r = -.25. As expected, lighting quality was positively correlated with perceived environmental safety with a moderate correlation of r = .53: Better quality lighting was associated with higher appraisals of the safety of the environment. In line with our expectations, we found a differential association between perceived environment safety on one hand, and energetic and tense arousal on the other hand. Whereas, as expected, tense arousal was associated with lower appraisals of environmental safety, with r = -.43, energetic arousal was found to be positive related to environmental safety, with r = .34. Given the strong relation between lighting and safety, energetic arousal showed a similar differential association with lighting quality with r = .29 and -.27 respectively. We found a positive correlation between PE and ATCT speed (1/RT) which suggests the presence of a speed-accuracy trade-off. However, we found only very small and statistically non-significant correlations between ATCT speed and any of the self-report measures. In this sense, contrary to ATCT speed, the PE and CV metrics showed more promising correlations with the self-reports. Exclusion of outliers did not affect the interpretation of these correlations.

Table 4: Correlations (within subject) between reaction time (1/RT), proportion of error responses (PE), coefficient of variation (CV_{rt}), tense arousal, energetic arousal, environmental safety and lighting quality

	PE	CV_{rt}	Sleepiness	Energetic arousal	Tense arousal	Environmental safety	Lighting quality
1/RT	.29**	.03	-.07	.06	.12	.07	-.01
PE		-.08	.18	-.23*	-.1	-.01	-.02
CV_{rt}			.14	-.06	.16	.06	.01
Sleepiness				-.58***	-.25*	-.21	-.15
Energetic arousal					.13	.33**	.29**
Tense arousal						-.40***	-.27*
Environmental safety							.52***

*Notes: *p < .05. **p < .01, ***p < .001.*

4. Discussion and Conclusion

4.1. Pilot findings

We managed to partially replicate Burtt's[17] findings about the positive effect of non-uniform lighting on pedestrian's attention, indicated by their reaction speed. People responded, on average, faster on Street 4, which had the lowest uniformity. Self-reported sleepiness and energetic arousal did not vary between streets. Differences on self-reports were in line with previous studies showing positive effects of illuminance levels and uniformity on lighting appraisal and perceived safety.[31-36]

The correlation between energetic arousal and SSS is in line with our expectations as the energetic dimension is presumed to be linked with the sleep-wake cycle.[16,27] Next, we found correlations of both dimensions of arousal with environmental safety but in different directions. This demonstrates the importance of differentiating between different types of arousal in research on pedestrian safety perceptions. Third, we found a correlation between environmental safety and lighting quality. Such correlation is in line with the concept of spatial brightness; more perceived light implies to people that their range of sight increases and therefore they should be able to detect and react to incoming dangers faster.[33]

Last, we found no correlations between the ATCT and any of the other metrics. An explanation might be that although response time tasks are commonly used as a measure of alertness (e.g., PVT, Go / No-Go) they might not be sensitive enough, or confounded by effort invested. This needs to be further investigated, for example by having more extreme conditions (e.g., very dark, very bright), a more demanding dual task (e.g., scanning the environment rather than walking), or the addition of effort measures. Using other metrics, namely the PE and CV, showed more promising correlations with the self-reports and might be therefore of further interest in future research.

4.2. General reflections

Due to the lack of experimental control and other confounds typical for field experiments we should not draw firm conclusions and should not attribute any effects found solely to the uniformity of the street lighting. It is impossible to make claims on the causal relationship between lighting characteristics, including uniformity and RTs, in a quasi-experimental field study. Nonetheless, we have learned various important lessons for future research and did find indications of the relation between lighting, arousal, alertness and anxiety in mesopic outdoor conditions of pedestrians. The discussion of how this relates to spectral power distributions, and visual and non-visual pathways is, currently not our prime concern. Our main aim now is at disentangling the relationships between safety perceptions on one hand and measures of alertness and arousal on the other. There is a demand for greater experimental control to establish the sensitivity of the metrics and test the links and correlations between the constructs and their measures properly.

In this paper we pointed out several unresolved points in the research on urban street lighting. Psychological constructs such as alertness, vigilance and/or anxiety are not considered, at the moment, in pedestrian lighting research. However, our results show that they are relevant even when studying subjective safety. We will need a better understanding of the influence of different dimensions of arousal on individual's attention and safety and whether these results were mediated through increases in alertness, anxiety or both. This would also have implications for both social and traffic safety of pedestrians.

References

[1] Mao Y, Fotios S. Lighting for pedestrians: Does multi-tasking affect the performance of typical pedestrian tasks? Lighting Research and Technology. Epub ahead of print 26 March 2021. DOI: 10.1177/14771535211002617

[2] Fujiyama T, Childs C, Boampong D and Tyler N. Investigation of lighting levels for pedestrians: some questions about lighting levels of current lighting standards. In: Walk21-VI, Everyday Walking Culture 6th International Conference of Walking in the 21st Century.

Zurich, Switzerland, 22-23 September 2005: 1-14.
[3] Knight C. Field surveys of the effect of lamp spectrum on the perception of safety and comfort at night. Lighting Research & Technology 2010; 42(3): 313-329.
[4] Yang B and Fotios S. Lighting and recognition of emotion conveyed by facial expressions. Lighting Research & Technology 2015; 47(8): 964–975.
[5] Brown VJ, Bowman EM. Alertness. In: Ramachandran V.S. (ed) *Encyclopedia of the Human Brain*. University of Michigan: Academic Press, 2002: 99–110.
[6] Chandrakumar D, Keage HA, Gutteridge D, Dorrian J, Banks S, Loetscher T. Interactions between spatial attention and alertness in healthy adults: A meta-analysis. *Cortex* 2019; 119: 61–73.
[7] Matthias E, Bublak P, Müller HJ, Schneider WX, Krummenacher J, Finke K. The influence of alertness on spatial and nonspatial components of visual attention. *Journal of Experimental Psychology: Human Perception and Performance* 2010; 36(1): 38-56.
[8] Petersen A, Petersen AH, Bundesen C, Vangkilde S, Habekost T. The effect of phasic auditory alerting on visual perception. *Cognition* 2017; 165: 73–81.
[9] Shapiro CM, Auch C, Reimer M, Kayumov L, Heslegrave R, Huterer N, Driver H, Devins GM. A new approach to the construct of alertness. *Journal of Psychosomatic Research* 2006; 60(6): 595–603.
[10] Cornwell BR, Mueller SC, Kaplan R, Grillon C, Ernst M. Anxiety, a benefit and detriment to cognition: Behavioral and magnetoencephalographic evidence from a mixed-saccade task. *Brain and cognition* 2012; 78(3): 257–267.
[11] Kastner-Dorn AK, Andreatta M, Pauli P, Wieser MJ. Hypervigilance during anxiety and selective attention during fear: Using steady-state visual evoked potentials (ssVEPs) to disentangle attention mechanisms during predictable and unpredictable threat. *Cortex* 2018; 106: 120–131.
[12] Öhman A. Fear and anxiety: Overlaps and dissociations. In M Lewis, JM Haviland-Jones and LF Barrett (eds.) *Handbook of emotions.* 3rd ed. New York: The Guilford Press, 2008: 709–728.
[13] Adolphs R. The biology of fear. *Current Biology* 2013; 23(2): 79–93.
[14] Bateson M, Brilot B, Nettle D. Anxiety: an evolutionary approach. *The Canadian Journal of Psychiatry* 2011; 56(12): 707–715.
[15] Eysenck MW, Derakshan N, Santos R, Calvo MG. Anxiety and cognitive performance: Attentional control theory. *Emotion* 2007; 7(2): 336–353.
[16] Thayer RE. Toward a psychological theory of multidimensional activation (arousal). *Motivation and Emotion* 1978; 2(1): 1–34.
[17] Burtt HE. The effect of uniform and non-uniform illumination upon attention and reaction-times, with especial reference to street illumination. Journal of Experimental Psychology 1916; 1(2): 155-182.
[18] Faul F, Erdfelder E, Buchner A and Lang AG. Statistical power analyses using G*Power 3.1: Tests for correlation and regression analyses. Behavior Research Methods 2009, 41: 1149-1160.
[19] Fotios S, Uttley J, Cheal C and Hara N. Using eye-tracking to identify pedestrians' critical visual tasks, Part 1. Dual task approach. Lighting Research & Technology 2015; 47(2): 133–148.
[20] CEN (2016). Road lighting - Part 3: Calculation of Performance. EN 13201-3.
[21] van Bommel W. Road Lighting: Fundamentals, Technology and Application. Cham (CH): Springer; 2015.
[22] Loh S, Lamond N, Dorrian J, Roach G and Dawson D. The validity of psychomotor vigilance tasks of less than 10-minute duration. Behavior Research Methods Instruments & Computers 2004, 36: 339-346.
[23] Psychology Software Tools (2016). E-prime 3.0 [computer software].

https://support.pstnet.com/.
[24] Labhackers (nd). https://labhackers.com/
[25] Flehmig HC, Steinborn M, Langner R, Scholz A and Westhoff K. Assessing intraindividual variability in sustained attention: Reliability, relation to speed and accuracy, and practice effects. Psychological Science 2007; 49: 132-149.
[26] Shahid A, Wilkinson K, Marcu S and Shapiro, CM. Stanford Sleepiness Scale (SSS). In: A Shahid, K Wilkinson, S Marcu and CM. Shapiro (eds.) STOP, THAT and One Hundred Other Sleep Scales. New York: Springer Publishing, 2012: 369-370.
[27] Thayer RE. Activation-deactivation adjective check list: Current overview and structural analysis. Psychological Reports 1986; 58(2): 607–614.
[28] Van Rijswijk L, Rooks G and Haans A. Safety in the eye of the beholder: Individual susceptibility to safety-related characteristics of nocturnal urban scenes. Journal of Environmental Psychology 2016; 45: 103–115.
[29] Van Rijswijk L and Haans A. Illuminating for safety: Investigating the role of lighting appraisals on the perception of safety in the urban environment. Environment and Behavior 2018; 50(8): 889–912.
[30] Bland JM and Altman, DG. Calculating Correlation Coefficients with Repeated Observations: Part 1—Correlation within Subjects. British Medical Journal 1995, 310, 446. doi:10.1136/bmj.310.6977.446
[31] Peña-García A, Hurtado A and Aguilar-Luzón M. Impact of public lighting on pedestrians' perception of safety and well-being. Safety science 2015; 78: 142–148.
[32] Portnov BA, Saad R, Trop T, Kliger D and Svechkina A. Linking nighttime outdoor lighting attributes to pedestrians' feeling of safety: An interactive survey approach. PloS one 2020; 15(11): 1-21.
[33] Boyce PR. Human factors in lighting. 3rd ed. Boca Raton: Crc Press, 2014, pp.1-690.
[34] Fotios S, Monteiro AL and Uttley J. Evaluation of pedestrian reassurance gained by higher illuminances in residential streets using the day-dark approach. Lighting Research & Technology 2019; 51(4): 557–575.
[35] Narendran N, Freyssinier J and Zhu Y. Energy and user acceptability benefits of improved illuminance uniformity in parking lot illumination. Lighting Research & Technology 2016; 48(7): 789–809.
[36] Bullough J, Snyder J and Kiefer K. Impacts of average illuminance, spectral distribution, and uniformity on brightness and safety perceptions under parking lot lighting. Lighting Research & Technology 2020; 52(5): 626–640.

IOP Conf. Series: Earth and Environmental Science 1099 (2022) 012011 doi:10.1088/1755-1315/1099/1/012011

True specs for dimmed and colour tuned LED luminaires

C Dam-Hansen[1,4], A Thorseth[1], D D Corell[1], A Bay[2], C Krause[2], and C Kofod[3]

[1] Department of Electrical and Photonics Engineering, Technical University of Denmark, Frederiksborgvej 399, 4000 Roskilde, Denmark
[2] Viso Systems, Vesterbrogade 12, 1., 1620 Copenhagen V, Denmark
[3] Energy Piano, L. F. Cortzensvej 3, 2830 Virum, Denmark

[4] Corresponding author e-mail: cadh@dtu.dk

Abstract. Colour tunable and dimmable LED lamps and LED luminaires are used increasingly in both domestic and professional lighting sector. The documentation of energy and light quality performance are often very limited for only for a single setting of multiple possible settings. Fifteen LED lamps and eight LED luminaires that are dimmable and tuneable white have been selected for demonstration of the variations in these parameters and possible errors in lighting design simulations. These have been tested in over twenty modes of operation including dimming from 100% to 25%, and correlated colour temperature (CCT) settings from 2200K to 6500K. The inherent standby power of these colour tuneable devices is measured, and the overall efficacy is evaluated in normal usage situations. In average for all tested LED lamps and luminaires 49% and 33% of the energy consumption is used in standby mode for 1h and 2h on time pr. day, respectively. The extensive set of test results show that there can be very large variations in e.g. the luminous flux, luminous efficacy and stroboscopic visibility measure (SVM), while others are very consistent over the modes of operation. The luminous flux shows variations of down to 22% of the maximum and rated value, and these lowest values occurs at the lowest CCT. The best performing has over 90% at all CCT settings. For the luminous efficacy most test results show a maximum value at 4000K or at the maximum CCT, and variations down to 20-50% of the maximum efficacy. The SVM is shown to increase to values over 1 and up to 2.4 at low dimming levels or low light output, demonstrating that stroboscopic effect will be visible for the majority of people and hence a problem in such settings. This is seen for approx. half the tested while the others have SVM < 0.4 for all settings and will not be visible over all settings. These results demonstrate the need for testing over the possible modes of operation to be able to correctly design and simulate a lighting installation using dimmable and colour tunable LED lamps and luminaires. The needed new dataformats and adoption of these from test laboratories to lighting software, in order for lighting designers and energy engineers to be able to make use of the extensive test results, are discussed.

1. Introduction

Over the last decade, LED lighting products have proven to be reliable and energy-saving and are now the preferred choice over all traditional light sources. This is due to the high energy efficiency, long lifetime and high light quality achievable, but LED lighting technology also provides an unprecedented potential for adjustment and control of the light output. This is in terms of dimming i.e., control of the luminous intensity, but also in terms of control of the spectral power distribution (SPD) of the emitted light. This is based on color mixing systems using both various colored LEDs and/or white LEDs at different correlated colour temperature (CCT) This is used for tunable white light sources, where the white light emission can be controlled in CCT, but also for color tunable light sources emitting light over a large gamut also covering saturated, colored light. It is the former that is interesting for general lighting installations.

There is a rapid increase in the use of dynamic LED products and lighting controls from both household smart lamps with app-control and dynamic luminaires in office environments and health care institutions. They are e.g., used to make the light follow the rhythm of daylight and for stimulation of nonvisual effects of light, typically to support alertness and circadian rhythm [1].

Modern lighting installations in both new and renovated commercial buildings often consist of such controllable LED solutions. Where possible, installers prefer to use the same luminaire model throughout, and to meet local minimum illumination requirements by permanently dimming specific luminaires. Some installations also control the CCT during the day or for special purposes. However, dimming and colour/CCT controls affect both energy efficiency, colour quality and temporal light modulation (TLM) and the accompanying temporal light artifacts (TLA)[2].

The standby power consumption of these dynamic LED products will affect the overall energy consumption and efficacy of a light source or lighting system dependent on the usage pattern over time. A study of office lighting and of lighting control systems[3] has found that the standby energy consumption accounts for around 30% of the total energy consumption but it may rise to 55% in extreme cases. This was based on actual occupancy data and simulation-scenarios of the total yearly lighting consumption for different combinations of lighting control systems (LCSs) and lighting poser density including standby consumption. It was concluded that a complete switch off LCSs during unoccupied hours is needed to preserve savings from LCSs.

IEA 4E Solid State Lighting (SSL) Annex [4][5] has measured the energy consumption associated with the new features that are being incorporated into SSL products and specified a new measurement method, including settings and parameters to test. They have made indicative test measurements following the development over the last seven years with analysis including more than 200 LED lamps and luminaires. In this period some manufactures have lowered their standby consumption by better components and/or control strategy while others have not focused on this issue. The IEA 4E SSL analysis has revealed that the standby consumption varies substantially from 0.08 W to 2.71 W. The EU regulation[6] in force from September 2021 sets a maximum standby power of 0.5 W. In the rest of the world there is typically no standby power limitations except for California, USA, that already in 2019 demanded the standby power limited to a maximum of 0.2 W [7] .

Since the EU regulation[6] only sets performance requirements for light sources at full load, manufacturers only report standby power and energy consumption at full load together with the photometric and colorimetric parameters, describing the efficiency and light quality. For tunable white lamps/luminaires the luminous flux may be very different depending on the choice of CCT. Anyhow the manufacturers typically only provide one luminous flux value without specification of the CCT.

The effects of dimming and CCT setting are typically not taken into account in light source documentation. Therefore, the actual performance parameters in dimmed and CCT settings are not known. Luminaire data is usually stored in IES and/or EULUMDAT files, and these typically contain information about luminous intensity distribution (LID), luminous flux, active power, CCT, and colour rendering index (CRI) for one setting only. These are then used in lighting design and simulation software, which can lead to large miscalculations of energy consumption and undesirable lighting quality at the used settings. As an example, temporal light modulation (TLM) issues typically increase when dimming is applied. If the energy consumption of lamps and luminaries are assumed to be linear with light output, control setting, or constant with colour temperature, the energy calculated energy consumption for lighting installations can be erroneous.

LED lamps and LED luminaires should generally be measured using the globally agreed methods contained in the international test standard CIE S 025/E:2015 [8]. The standard includes only limited information on the treatment of tunable and dimmable luminaries in terms of measurement methods. ANSI/IES TM-38-21[9] specifies a recommendation for measuring "tunable white" lamps, luminaires, and light engines. In 2022 CIE is beginning the process of revising CIE S 025/E:2015 [8] in technical committee TC 2-97, work that may include methods for more lighting devices with more advanced controls. The question is how these dynamic light sources should be characterised, by which and how many settings, and how to share the data to enable efficient and accurate lighting design and simulation of systems. And the next question is how should the data be exchanged from laboratories to lighting designers and energy engineers. ANSI/IES TM-33-18 [10] describes a new standard xml-format that allow for data at different settings, and adds new characteristics e.g., SPD, different color rendering parameters, and TLA metrics, like the stroboscopic visibility measure, SVM, to the normally used LID, active power and luminous flux. As this American standard has not been widely adopted, efforts are made in this work to support and influence Global Lighting Data Format (GLDF) [11] e.g., through organisations like IES, ISO and CIE.

In this work a selection of commercially available LED lamps and LED luminaires that are dimmable and tunable white have been selected for demonstration of the variations in performance parameters and possible errors in lighting design simulations. These are tested in over twenty modes of operation including dimming from 100% to 25%, and CCT settings from 2200K to 6500K.

2. Measurements of lamps and luminaires

A range of dynamically controllable LED lamps and LED luminaires was selected for extensive testing to demonstrate the problems mentioned. Models were chosen from different manufactures, and single artefacts of each were bought on the market in the period 2020-2022. The anonymized 23 artefacts covered a range of light sources primarily sold for the domestic sector and includes

- 15 LED lamps with nominal power ranging from 4.5 W – 12 W. The nominal luminous flux range covered 250 lm to 1000 lm with E27 or GU10 type sockets
- 8 LED luminaires including ceiling fixture (panel, cylinder, ring and build-in spot), wall fixtures, wall frameless panel and table lamps. The nominal power range covered 4.5 W – 24 W and the nominal luminous flux range was 300 lm to 2800 lm

All lamps/luminaires were controllable via a manufacturer specific apps communicating over a Bluetooth, Zigbee, or Wi-Fi connection protocol. All artefacts were dimmable and except for one they were all also tuneable white and/or colour-tunable. The apps for controlling the dimming and CCT settings used sliders set by touch control. Some of the apps had a numerical scale on these controls indicating dimming level and/or CCT values and others did not. On the former setting values had to be estimated.

2.1. Measurement methods

All artefacts or devices under test (DUT) were tested by measurements with an integrating spherespectroradiometer according to the international test standard CIE S 025/E:2015 [8]. All DUTs were powered at 230 V AC and 50 Hz using a stabilised programmable AC power supply (Elgar, CW1251P). The electrical parameters were measured using 4 wire connection at the DUT with a power analyser (WT3000, Yokogawa) and included RMS voltage, RMS current, active power and power factor.

The photometric quantities were measured in a Ø1.9 m integrating sphere (Instrument Systems, ISP2000) fibre connected to a spectroradiometer (Instrument Systems, CAS 140CT). The system was calibrated to total radiant flux and corrected for the change in sphere transmission caused by changing the DUT. The parameters of main interest here were: the luminous flux measured in lm [12, Secs. 17-21–039], the correlated colour temperature (CCT) measured in K [12, Secs. 17-23–068], and the luminous efficacy defined as the ratio of luminous flux and active power, measured in lm/W [12, Secs. 17-21–089]. In addition, the colour coordinates and colour rendering index[13] were calculated from the measured spectral power distribution.

In relation to TLM and TLAs, the stroboscopic visibility measure (SVM) M_{VS} [14] and Short-term flicker index P_{st}^{LM} [15] were measured. This was done according to the guidelines in CIE TN 012 [16].

A custom built TLM measurement system was used. It was based on a Ø10mm Si-photodiode (United detector Technology, SN-19907) and the signal was amplified using a variable gain transimpedance amplifier (Femto, DLPCA-200) and digitized using an analog to digital converter (National Instruments, NI-USB-9162). Calculations are done using a custom-built LabVIEW routine.

One of the issues of measuring LED light sources is the stabilization of the light output and power consumption of the DUT. According to CIE S 025/E:2015 [8], the DUT must be on for at least 30 min. and the variation of measured parameters must be less than 0.5% over a 15 min time window. When measuring many different settings of a specific LED lamp or luminaire, stability should normally be achieved before measurement of each setting, however, a 15 min verification was not practically possible in this study, due to the need for manual control through an app. Consequently, stability was checked manually over 1-3 minutes before measurements. The measurement procedure used to test the DUTs was as follows:

- stabilizing the DUT at the nominal CCT setting at 100% dimming level by monitoring the luminous flux and active power as a function of time, in accordance with CIE S 025/E:2015 [8].
- Setting the dimming level to 100, 75, 50 and 25% consecutively and doing a measurement at each setting.
- Changing the CCT and repeating the measurement at each dimming level.

The CCT settings used were 2700 K, 4000 K, 5000 K and the minimum (typically 2200 K) and maximum CCT (typically 6500 K) possible for the specific DUT. This protocol was used as accurately as practically possible. Some DUTs could not be set to all the specified CCTs. By default, the CCT given in the apps was used to set the CCT during measurements, however in case CCT values were not given in the app the CCT setting was done through monitoring of the measured CCT. By default, the dimming level given in the apps by percentages was used for dimming the DUT during measurements. Some apps did not show any percentage indication on the dimming slider and in these cases the dimming setting was done by using the active power consumption relative to the full load active power.

Photometric and TLM parameters were measured in different setups and not at the same time. It was a difficult task to reproduce the same CCT and dimming level for these two measurements due to the resolution and sensitivity of the app and controls used. Effort was made to ensure this by using a spectral irradiance measurement along with the TLM measurement and monitoring the CCT and consumed electrical power to match earlier settings for photometric measurements.

2.2. Standby power

The investigated LED lamps and luminaires were all dynamically controllable via Bluetooth, Zigbee or Wi-Fi and had a power consumption even when not providing light, i.e. a standby power consumption. This affects the energy consumption of an installation over time, and the effective efficiency of a light source. Therefore, the IEA SSL Annex[4][5] has defined a new key term "overall efficacy":

$$\mu_{\text{overall}} = \frac{\phi_v t_{\text{on}}}{P_{\text{on}} t_{\text{on}} + P_{\text{standby}} t_{\text{standby}}} = \frac{\phi_v}{P_{\text{on}}} \left(\frac{1}{1 + \frac{P_{\text{standby}}}{P_{\text{on}}} \frac{t_{\text{standby}}}{t_{\text{on}}}} \right) \text{ [lm/W]}, \quad (1)$$

which describes the effective efficiency over an evaluation time $t_{on} + t_{standby}$ where the light source is only on for a time period of t_{on} while being in off-state for t_{standby}. Here, ϕ_{v} is the luminous flux emitted. P_{on} and $P_{standby}$ are the powers in the on and off states. When there is no standby power or no standby time period the overall efficacy equation reduces to the luminous efficacy, i.e. ϕ_{v} divided by P_{on},. It is worth noticing that the overall efficacy can become much lower than the luminous efficiency in case t_{on} is much lower lower than t_{standby}, and especially if $P_{\text{standby}}/P_{\text{on}}$ is high. That means that this is of higher importance for low power LED lamps than for high power LED luminaires. The standby power consumption P_{standby} was measured using a power analyser (Yokogawa WT3000) when the DUT was powered on at 230 V AC 50 Hz using a stabilised power supply (Elgar, CW 1251P), but set to the off-state in the app and emitting no light. The procedure for measuring standby power consumption is described in IEC 62301 [17].

3. Results

In this section the measurement results on the LED lamps and luminaires are presented with an emphasis on the parameters that shows a large and significant variation as a function of the CCT and dimming settings. Here, the luminous flux and luminous efficiency are most important to lighting design and energy calculations. In regard to temporal light modulation, the measured SVM values are of highest importance and shows the largest variations.

3.1. Luminous flux and efficacy

The measurements of luminous flux as a function of CCT setting showed large variations for the majority of the DUTs. In Figure 1, this variation is illustrated by the measured luminous flux at the five CCT settings, relative to the maximum luminous flux for the DUT which may be at any of the CCT settings. These values are measured for each DUT at full load, 100% dimming level.

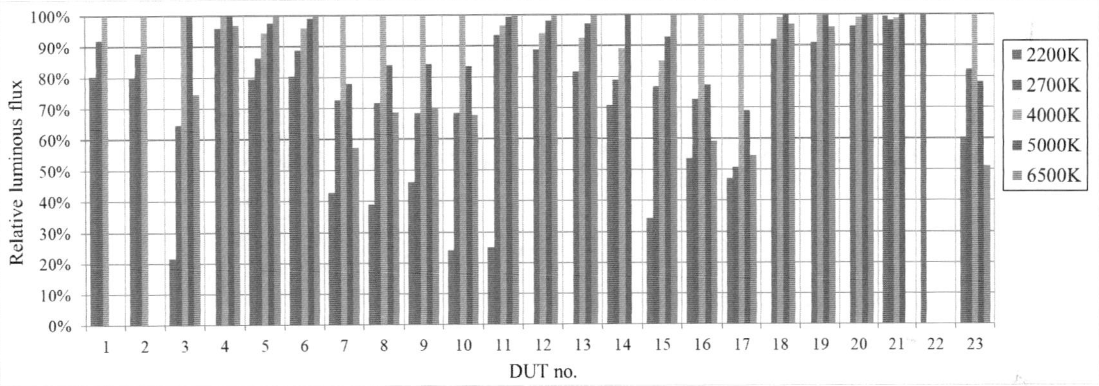

Figure 1. Luminous flux, Measured luminous flux for the five CCT settings relative to the maximum luminous flux for each of the 23 DUTs.

It is observed that the majority of the DUTs can be changed in CCT over 4 or 5 settings, while one cannot be changed in CCT. The maximum luminous flux is often at the rated CCT, informed by the manufacturer, and is seen to be at 4000 K for the majority of DUTs, and the relative luminous flux for the other settings are below 85%. The minimum relative luminous flux is found at 2200 K for all DUTs except for one. For one third of the DUTs, the luminous flux at 2200 K is very low at 22 – 53% of the maximum luminous flux. Four LED luminaires and one LED lamp shows a much higher consistency and has more than 90% relative luminous flux at all CCT settings.

In Figure 2, the measured luminous efficacy at full load or 100% dimming level as a function of the measured CCT is shown for the 14 LED lamps and the 8 LED luminaires. It is observed that the measured CCT are not exactly at the set CCT values. There are three types of variation of the measured luminous efficacy as a function of CCT. The first is characterised by having the highest efficacy at the highest CCT and it decreases for lower CCTs. These are shown with solid lines in Figure 2. The second type exhibit the highest efficacy at 4000 K, corresponding to the results in Figure 1 where the luminous flux decreases for both lower and higher CCT around peak at 4000 K. These are shown with dashed lines in Figure 2. A relative change in luminous efficacy from the maximum value 20-50% is observed for both the first and second type of variation.

The third type shows almost no variation in the efficacy, these are shown as dotted lines and corresponds to the results seen in Figure 1 that showed almost no variation in the luminous flux. It is observed that the luminous efficacy of the LED lamps is generally higher than for the LED luminaires.

The maximum efficacy of 132 lm/W for LED lamps is found at a high CCT of 6500 K and the average efficacy is around 90 lm/W for the tested DUTs. For the LED luminaires the maximum efficacy is 100 lm/W, while the average efficacy is around 70 lm/W.

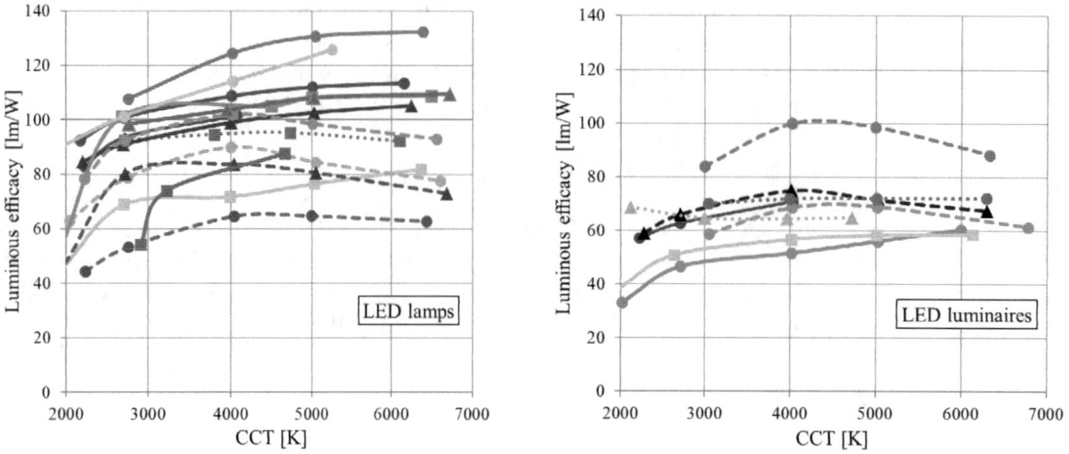

Figure 2. Efficacy, Measured luminous efficacy at full load as a function of measured CCT for 14 LED lamps (left) and 8 LED luminaries (right).

3.2. Stroboscopic visibility measure

The TLM has been measured for a representative selection of the DUTs, and the measured SVM are presented for five LED lamps and three LED luminaires, which demonstrates the large differences in variation pattern of the SVM values as a function of dimming level.

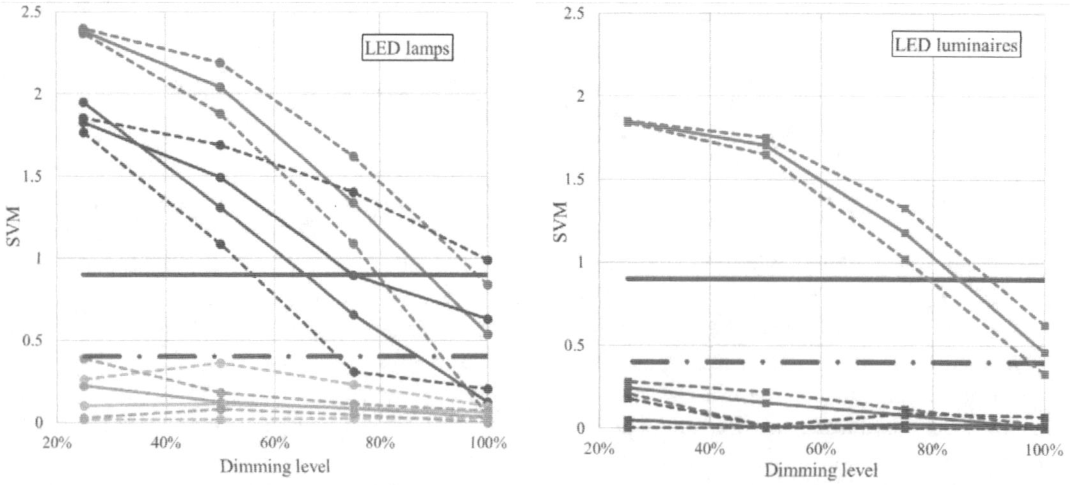

Figure 3. SVM, Measured average SVM (solid line and markers) for the different CCT settings of the DUT as a function of the dimming level, maximum and minimum SVM values are shown with same colour dashed lines and markers. Left: 5 LED lamps, right: three LED luminaires. The EU requirement for SVM is shown with grey solid and dashed-dotted line.

The SVM was measured at up to five CCT settings for each DUT and for the four dimming levels. The variation observed as a function of dimming level does not change much for each CCT setting. Therefore, the variations for CCT settings are not explicitly showed. In Figure 3, the average, maximum and minimum measured SVMs at each dimming level are shown. To the left this is done for five LED lamps and to the right this is shown for 3 LED luminaires.

The EU regulation[6] requirement for SVM is that it must be below 0.9 at full-load and this requirement will be lowered to 0.4 from September 2024. These two values are shown in Figure 3, as grey solid and dash-dotted lines. Requirements only apply at full load, i.e. at 100% dimming level.

It is observed that there are two types of markedly different behaviours of the measured SVM as a function of the dimming level. The one is characterised by low SVM values at full load or 100% ranging from under 0.4 to close to 1 for the different CCT settings, i.e., a large variation. It is observed that the SVM values increase for each lower dimming level and ends at SVM values from 1.8 to 2.4 at 25% dimming level. This behaviour is observed for three of the LED lamps, of which two were of filament type and one an LED luminaire. The stroboscopic effect is pr. definition visible for 50% of the observers at a value of SVM = 1. This means that for these devices SVM will be visible at dimming levels of 50% or lower. This is not the case for the other type of behaviour. It is characterised by a very low SVM value of 0.1 or less at full load or 100% and only a small increase in the SVM value at lower dimming levels. This is observed for two of the LED lamps, where the maximum SVM values measured are below 0.4, and for two of the LED luminaires where the maximum SVM values measured are below 0.3. That means that a majority of observers will not be able to observe a stroboscopic effect in the light from these LED lamps and luminaires. For these, the stricter EU 2024 requirement is fulfilled at all settings. And these will also be able to comply with the recommendation of the European workplace lighting design standard EN 12464[18] stating that "lighting systems should be designed to avoid the negative effects of flicker and stroboscopic effect throughout the full dimming range", but without establishing maximum limits. It is the power supply and the electronic driver that determines the behaviour of the TLM and hence the SVM values. Luminaries have room for larger and more complex drivers which can deliver this low SVM behaviour at all settings. But it is seen in a E27 A60 bulb where the available space for electronic drivers is limited. The TLM waveforms measured for these are characterised by low modulation at high frequencies, at 2000 Hz and 4000 Hz for the LED luminaires tested. For the other type of behaviour, the TLM waveforms are pulse width modulation (PWM) waveforms with 100% modulation depth, at frequencies lower than 2000 Hz up to which people are sensitive to the stroboscopic effect. This gives rise to high SVM values.

3.3. Standby power

The standby power was measured for 23 of the artefacts, covering 15 LED lamps and 8 LED luminaires. The majority shows a standby power in the range from 0.19 to 0.47 W, which comply with the EU requirement. However, a single LED lamp has a standby power of 0.61 W and four of the LED luminaires have standby powers in the range from 0.76 to 0.96 W, which are not compliant.

In Figure 4 the rated efficacy is shown together with the calculated overall efficacy for two on time situations of 1 h and 2 h over a day of 24 h. These are typical ON time for lamps/luminaires used in the domestic sector [19]. This is shown for each of the 23 LED lamps and luminaires The DUTs have been sorted to show the DUTs with the highest rated efficacy to the left.

It is observed that the overall efficacy is much lower for both 1h and 2h on time than the rated efficacies. The lowest overall efficacy is found for is a 4.5 W LED luminaire (DUT no. 20) with a high standby power of 0.76W. In this case the overall efficacy is only 14 lm/W and 23 lm/W for 1h and 2h on time, respectively. This is close to the luminous efficacy of incandescent lamps which have been banned from the market due to low efficacy. Best performance is observed for a linear LED luminaire (DUT no. 6) with a low standby power of 0.19W giving overall efficacy of 82 lm/W and 91 lm/W for 1h and 2h on time, respectively. On average for all tested LED lamps and LED luminaires 49% and 33% of the energy consumption is used in standby mode for 1h and 2h on time pr. day, respectively.

Non-compliance does not seem to adhere from technical obstacles, as interestingly the California Energy Commission's so-called Modernized Appliance Efficiency Database System [20] [21]_shows that more than 300 smart lamps/luminaires on the USA market have standby power lower than 0.2 W and a large part is also below 0.1 W.

A way to obtain substantial reduction of the standby power was described by IEEE [22], which was first developed for IoT connected devices on a battery. In this concept, a low power radio receiver is added to the lighting product and is the only thing that stays connected but only for 2 milliseconds out of every 100 milliseconds. Use of this concept for a product with standby power 0.2 W thus reduce the standby power to 0.004 W.

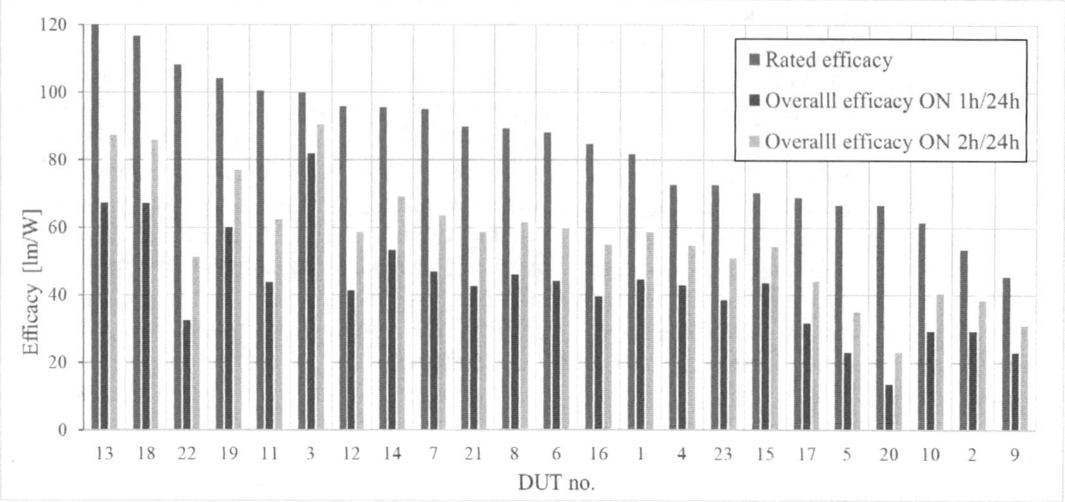

Figure 4. Overall efficacy, rated and calculated overall efficacy corresponding to 1 h and 2 h on time over 24 h for 23 of the LED lamps and luminaires, ordered according to rated efficacy.

In relation to the stroboscopic visibility measure, an analysis of the measured standby power and SVM, does not reveal any correlation between the two for the LED lamps and luminaires tested in this work.

4. Discussion

The results of this study show that dimming and colour tuning affect efficiency and light quality (in terms of SVM) of all artefacts – but also with considerable differences. This study is primarily based on household light sources. Assuming that conclusions may also apply to professional light sources the results underline that there is a real risk that power consumption in full-scale professional lighting installations is miscalculated and TLA effects underestimated. As building codes in many countries set more and more strict limitations on building energy consumption, it is evident that data like the ones provided by this study is critical to avoid excessive building energy consumption and unacceptable TLA. More data is also important to make true comparisons between alternative lighting products and solutions. This study also points to some of the practical difficulties of producing data and making such data available in the whole value chain:

- Performing an extended suite of measurements across several dimensions is highly time-consuming work. Rigorously studying combinations of all parameters cause the workload to and testing costs increase exponentially. Further, regulating both CCT and dimming levels in the lab proved to be very difficult to perform systematically because of inconsistent control systems. Thus, improved light measurement control systems and software automated measurement protocols are essential components in more detailed characterisation
- Existing light measurement standards such as CIE S 025/E:2015 [7] do not specify how to interpret the term dimming and do not advise on light measurement protocols for dimming/colour tuning during measurements, and which steps are necessary. IES TM-38 [8] lays out a possible framework for specifically characterising white tunable light sources.

- Once measured, the detailed data needs to be stored in formats that can be shared with downstream users such as consumers, lighting desingers and specifiers and energy engineers. The well-known text formats EULUMDAT and IES LM-63-19[18] are not suited. Major standardization bodies have already worked for some time on new XML-based formats: IES [10], CIE TC-2.83, ISO 274/JWG 1/TF BIM, CEN/TC 169/WG 1/TF BIM). XML-formats are machine-readable and in general much more flexible and allows for addition of more data.

- Design and energy calculation software such as DIALux and Relux must be able to use the more detailed tested performance data as a function of colour tuning, dimming and use pattern. Consequently, collaboration has been started with the GLDF initiative [10] aiming at a much more comprehensive dataformat for lighting design software.

- Conscious product selection by both consumers and professional lighting specifiers relies on awareness of the possible negative effects of dimming, colour tuning, and standby consumption. The purpose for bringing more details into the planning stages is obviously to make energy calculations of whole buildings as accurate as possible, while maintaining awareness of light quality.

5. Conclusion

A wide selection of dimmable and tunable white LED lamps and LED luminaires have been selected for demonstration of the variations in performance parameters. These have been tested in over twenty modes of operation including dimming from 100% to 25%, and correlated colour temperature (CCT) settings from 2200K to 6500K. The inherent standby power of these colour tunable devices have been measured, and the overall efficacy have been evaluated in normal usage situations. In average for all tested LED lamps and luminaires 49% and 33% of the energy consumption is used in standby mode for 1h and 2h on time pr. day, respectively.

The extensive set of test results show that there can be very large variations in the luminous flux, luminous efficacy and SVM, while others are very consistent over all modes of operation. The luminous flux shows variations of down to 22% of the maximum and rated value, while the best performing has over 90% at all CCT settings. For the luminous efficacy most test results show a maximum value at 4000K or at the maximum CCT, and variations down to 20-50% of the maximum efficacy. The SVM is shown for some devices to increase to values over 1 and up to 2.4 when applying dimming levels, demonstrating that stroboscopic effect will be a problem in such settings. Some had SVM < 0.4 for all settings and will not pose a problem with stroboscopic effect.

These results demonstrate the need for testing over the possible modes of operation to be able to know the performance and correctly design and simulate a lighting installation using dimmable and colour tunable LED lamps and luminaires. The needed new data formats and adoption of these from test laboratories to lighting software, in order for lighting designers and energy engineers to be able to make use of the extensive test results, have been discussed.

Acknowledgement

This work was mainly supported by ELFORSK through the project 353-014 International standards for lighting, global energy savings. This project runs until 2023 and includes various activities of standardisation to support energy savings in lighting. For more information see the project website [23].

This work was partly supported by EMPIR 19NRM02 RevStdLED, Revision and extension of standards for test methods for LED lamps, luminaires and modules and EMPIR 20NRM01 and MetTLM, Metrology of Temporal Light Modulation. The projects (19NRM02 RevStdLED and 20NRM01 MetTLM) has received funding from the EMPIR programme co-financed by the Participating States and from the European Union's Horizon 2020 research and innovation programme."

References

[1] CIE, "CIE statement on non-visual effects of light recommending proper light at the proper time," CIE - International Commission on Illumination, 2019.
[2] CIE, "CIE 249:2022 Visual Aspects of Time-Modulated Lighting Systems," Vienna, Austria,

2022.
[3] N. Gentile and M. C. Dubois, “Field data and simulations to estimate the role of standby energy use of lighting control systems in individual offices,” *Energy Build.*, vol. 155, pp. 390–403, Nov. 2017.
[4] IEA, “SSL Annex Task 7: Smart Lighting – New Features Impacting Energy Consumption,” IEA 4E SSL Annex, 2016.
[5] C. Kofod, “Solid State Lighting Annex: Task 7: Smart Lighting – New Features Impacting Energy Consumption, Second Status Report (Unpublished),” 2022.
[6] European Commission, “Commission Regulation (EU) 2019/2020 of 1 October 2019 laying down ecodesign requirements for light sources and separate control gears - Document 02019R2020-20210901,” 2019. [Online]. Available: https://eur-lex.europa.eu/eli/reg/2019/2020/2021-09-01. [Accessed: 02-May-2022].
[7] CALIFORNIA and ENERGY COMMISSION, “CALIFORNIA CODE OF REGULATIONS Title 20. Public Utilities and Energy Division 2. State Energy Resources Conservation and Development Commission CALIFORNIA ENERGY COMMISSION,” 2021.
[8] CIE, *CIE S 025/E:2015: Test Method for LED Lamps, LED Luminaires and LED Modules*. CIE - International Commission on Illumination, 2015.
[9] IES, “ANSI/IES TM-38-21, Photometric and Electrical Measurements of Tunable-White Solid-State Lighting Products,” IES - Illuminating Engineering Society, 2021.
[10] IES, “ANSI/IES TM-33-18 Standard Format for the Electronic Transfer of Luminaire Optical Data,” 2018.
[11] DIAL GmbH and RELUX Informatik AG, “GLDF | Global Lighting Data Format.” [Online]. Available: https://gldf.io/. [Accessed: 01-Apr-2022].
[12] CIE, *CIE S 017:2020 ILV: International Lighting Vocabulary*, 2nd editio. Vienna, Austria: CIE - International Commission on Illumination, 2020.
[13] CIE, “CIE 13.3-1995, Method of measuring and Specifying Colour Rendering Properties of Light Sources,” 13.3, 1995.
[14] CIE, “CIE TN 006:2016 Visual Aspects of Time-Modulated Lighting Systems – Definitions and Measurement Models,” Vienna, Austria, 2016.
[15] IEC, “IEC TR 61547-1:2020 Equipment for general lighting purposes - EMC immunity requirements - Part 1: Objective light flickermeter and voltage fluctuation immunity test method,” International Electrotechnical Commission, 2020.
[16] CIE, “CIE TN012:2021 Guidance on the Measurement of Temporal Light Modulation of Light Sources and Lighting Systems,” CIE (Commission International d’Eclairage), Vienna, Austria, Jan. 2021.
[17] IEC, “IEC 62301:2011 Household electrical appliances - Measurement of standby power,” IEC, 2011.
[18] CEN/TC 169, “EN 12464-1:2021 Light and lighting - Lighting of work places - Part 1: Indoor work places,” 2021.
[19] C. Kofod, “PremiumLight IEE/11/941/SI2.615944 Assessment of the initial situation in the participating countries, Final version,” 2013. [Online]. Available: https://www.visosystems.com/wp-content/uploads/2022/05/Assessment-of-the-initial-situation-in-the-participating-countries-PremiumLight-IEE11941SI2.615944-Energy-piano-30-06-2013.pdf. [Accessed: 12-May-2022].
[20] California Energy Commission, “Modernized Appliance Efficiency Database System,” 2016. [Online]. Available: https://cacertappliances.energy.ca.gov/Pages/Search/AdvancedSearch.aspx. [Accessed: 02-May-2022].
[21] CALIFORNIA and ENERGY COMMISSION, “CALIFORNIA CODE OF REGULATIONS Title 20. Public Utilities and Energy Division 2. State Energy Resources Conservation and Development Commission CALIFORNIA ENERGY COMMISSION,” 2021.

IOP Conf. Series: Earth and Environmental Science 1099 (2022) 012011 doi:10.1088/1755-1315/1099/1/012011

[22] D. K. McCormick, "IEEE technology report on wake-up radio : an application, market, and technology impact analysis of low-power/low-latency 802.11 wireless LAN interfaces," 2017.
[23] A. Thorseth and C. Dam-Hansen, "ELFORSK project 353-014 International standards for lighting, global energy savings — DTU Research Database," *DTU Orbit*. [Online]. Available: https://orbit.dtu.dk/en/projects/elforsk-project-353-014-international-standards-for-lighting-glob. [Accessed: 12-Apr-2022].

IOP Conf. Series: Earth and Environmental Science 1099 (2022) 012012 doi:10.1088/1755-1315/1099/1/012012

Relationship between the gaze point movement and spatial brightness

H Kage[1], K Harimoto[1], Y Ishiwata[2] and N Yoshizawa[2],

[1] Taisei Corporation, Technology Center, 344-1 Nase-cho, Totsuka-ku, Yokohama-shi, Kanagawa Prefecture, 245-0051, Japan,
[2] Department of Architecture, Graduate School of Science and Technology, Tokyo University of Science, 2641 Yamazaki, Noda-shi, Chiba Prefecture 278-8510, Japan

mthhnk00@pub.taisei.co.jp

Abstract. This study aims to verify the relationship between the spatial brightness evaluation and average luminance on a moving gaze point tracking by targeting offices with different luminance distributions. Subjects wearing eye-tracking devices evaluate the spatial brightness of the various light environments displayed on a large screen display. In the results, the tendency for gazing at the window is found to be high as compared to the other parts. The average luminance is calculated by integrating the time on the path of the moving gaze point. Furthermore, we analyse the relationship between the average luminance of each part of the light environment on the display and the brightness evaluation value. The average luminance considering the movement of the gaze point at the time of brightness evaluation has higher explanatory power,0.305, for the spatial brightness than the average luminance,0.259, in the entire visual field area. In contrast, the explanatory power of brightness evaluation was not improved drastically, even considering the gazing time and luminance. In future, the relationship between the evaluation of the spatial brightness and the gaze point/gaze time can be analysed by considering the absolute value of the luminance of the window surface and the luminance ratio with the surroundings.

1. Introduction

Generally, the conventional light environments in-office spaces are uniform considering the workability at a desk[12]. However, in recent years, the technology associated with the subdivision of spatial applications, daylight utilization, and LED lighting has substantially progressed. A light environment with different luminance for each part of the space is being provided[3]. The workers who work in the office space perceive the brightness not only from the task area but also from the luminance of the entire space, including daylight. Light environments with various luminance of each part, such as the ceiling, wall, window, desk, and floor of the space, have a different way of perceiving the spatial brightness from the conventional light environment where the luminance of each part of the space is uniform. In a space where the luminance of each part of the space varies, the worker sees different luminance each time he or she follows his or her eyes. For this reason, it may feel a little brighter / darker than the spatial brightness appropriate for performing work. Several studies have been reported on predictive models of physical quantities that explain spatial brightness[45678]. The evaluation formula, proposed by Taka et al, uses the standard deviation of the logarithm of the brightness value and the average brightness[9]. It shows that the brightness evaluation of the space decreases as the variation in brightness increases.

To design a light environment that considers the worker's spatial brightness, it is necessary to understand the relationship between the perception of space brightness and the space observation method. In the previous research focusing on the gaze point when evaluating the spatial brightness, Ishida et al. reported the relationship between the subject's attention and the spatial brightness evaluation for residential lighting[12]. However, the subject's gaze point was declared by the subject, and the actual gaze point was not tracked. Therefore, the relationship between the spatial brightness and the area that was actually gazed at was not considered.

This study aims to understand the relationship between spatial brightness and average luminance on a moving gaze point path. We targeted various offices with different luminance distributions in the space. Furthermore, the gaze point of the occupant was detected for each part of the space using the eye-tracking device. The outline of the research is shown in figure 1.

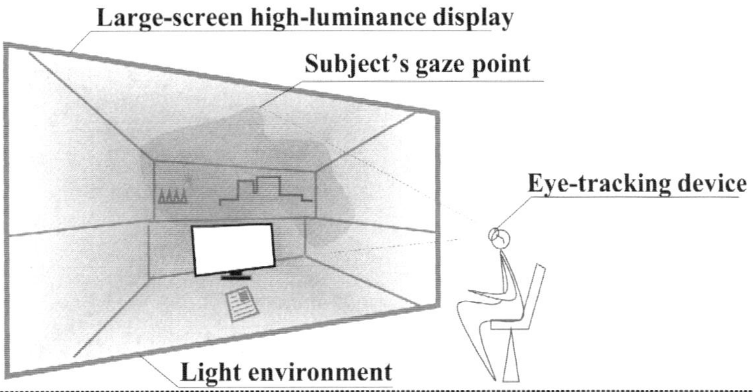

Figure 1. Research outline

2. Method

2.1. Experimental outline

Generally, a person unconsciously moves his/her eyes to observe space. Furthermore, it is known that the person is affected by the brightness of the moving gaze point and the gaze time. Therefore, an eye-tracking device was used in this study to accurately grasp the gaze point of the occupant. The experiments in a light environment are generally evaluated in a full-scale space. However, it is difficult to conduct sufficient repetitions due to the limitations, like the experimental location, daylight conditions, and securing the subjects.

Therefore, in this study, we used a display to produce the same luminance and spatial brightness as the actual space. This made it possible to conduct the experiments targeting multiple offices and daylight conditions in one experimental space with a single display. Table 1 summarizes an overview of the display (Silicon Core Technology). Figure 2 shows the image of the display, and Figure 3 shows the experimental layout.

In the experiment, we selected seven offices with different lighting methods, the evaluation seat orientation and the presence or absence of a daylighting device, targeting the conventional uniform light environment and the light environment where there is a difference in luminance of each part of the space using a daylighting device, A total of 22 conditions were prepared. Table 2 lists the details of the experimental cases. The eye-tracking device (Tobii Pro Glasses3) extracted the position of the subject's gaze 50 times per second. Figure 4 shows the eye-tracking device used in the experiment, and the heat map image of the subject's gaze point obtained by the eye-tracking device is given in figure 5. With this method, the brightness of the position where the subject is looking at, can be grasped from the time-series data of the position where the subject is looking and the image of the captured brightness.

IOP Conf. Series: Earth and Environmental Science 1099 (2022) 012012 doi:10.1088/1755-1315/1099/1/012012

Table 1. Display screen information.

Size of the screen	Width: 4,864 mm and height: 2,736 mm
Maximum value presentable by the display	1,200 cd/m²
Resolution	3,840 × 2,160
Bit depth	24 bit
Gamma value	2.2

Figure 2. Image of the high-luminance large screen display,

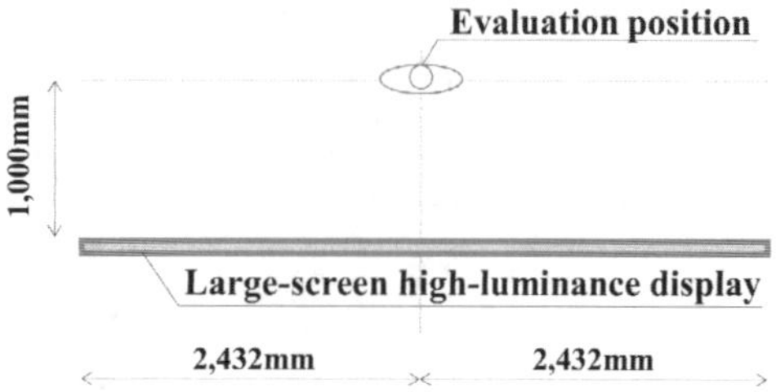

Figure 3. Experimental layout

Table 2. Experimental cases

Case	Office	Evaluation seat direction	Light environment conditions	Average luminance [cd/m²]	Office with window
1	Experimental room	window	Window luminance 500 cd/m²	76.1	○
2	Experimental room	window	Window luminance 1200 cd/m²	80.9	○
3	Experimental room	window	100 %	300	○
4	Office with up/downward lighting and daylighting device	window	Daylight and artificial lighting 300 lx	114	○
5	Office with up/downward lighting and daylighting device	wall	Daylight and artificial lighting 300 lx	118	○
6	Office with up/downward lighting and daylighting device	window	Only artificial lighting 300 lx	78.0	○
7	Office with up/downward lighting and daylighting device	wall	Only artificial lighting 300 lx	81.1	○
8	Office with up/downward lighting and daylighting device	wall	Only artificial lighting 300 lx at night	71.7	○
9	Office without ceiling	window	Low reflectance of the ceiling	71.7	○
10	Office without ceiling	wall	Low reflectance of the ceiling	59.1	○
11	Office without ceiling	window	High reflectance of the ceiling	75.4	○
12	General office	window	Desktop illuminance 700 lx	128	○
13	Office with daylighting device	window	With daylight	288	○
14	Office with daylighting device	wall	With daylight	176	-
15	Office with daylighting device	window	Without daylight	132	○
16	Office with individual lighting control system	wall	Uniform 700 lx	66.8	-
17	Office with individual lighting control system	wall	Non-uniform 700–500 lx	54.7	-
18	Office with individual lighting control system	wall	Uniform 500 lx	58.1	-
19	Office with individual lighting control system	wall	Non-uniform 500–300 lx	52.3	-
20	Office with individual lighting control system	wall	Non-uniform 500–200 lx	44.9	-
21	Office with individual lighting control system	wall	Uniform 300 lx	41.4	-
22	General office	window	Desktop illuminance 700 lx	93.2	○

Figure 4. the Eye-tracking device

Figure 5. Heat map image of the subject's gaze point obtained by the eye-tracking device, Case3 subject1

2.2. Experimental procedure/evaluation method

In the experiment, a subject wearing an eye-tracking device evaluated the spatial brightness of the actual space image reproduced on the display. At that instant, the trajectory of the subject's gaze point was measured. The presented light environment was converted into an image that could be displayed on the screen. The image was produced using Radiance by measuring the luminance and chromaticity distribution. The luminance and chromatic distribution were measured using a single-lens reflex camera equipped with a fisheye lens and L-CEPT system (Luminance and Colour Environment Photometric Tool: developed by Building Research Institute in Japan). The measurements were recorded from the viewpoint of the subject, which was at a distance 1.2 m from the floor. Figure 6 shows a part of the light environment image used in the experiment. The experiment was conducted on 22 subjects belonging to the age group of 20–30 years.

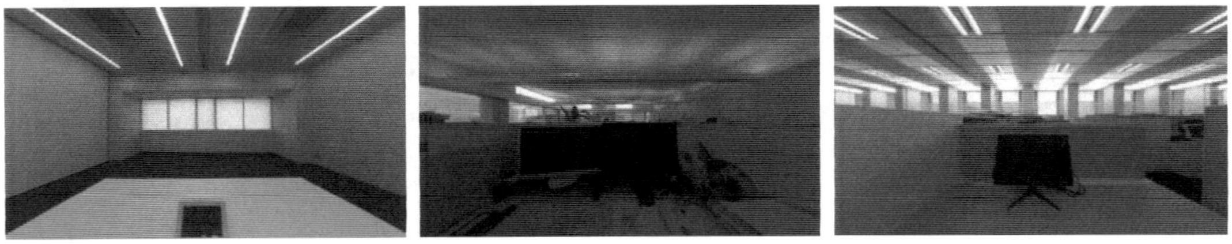

Figure 6. The light environments presented on the display in experiment (left) is Case3 Experimental room, (center) is Case4 Office with up/downward lighting and daylighting device, and (right) is case13 Office with daylighting device

2.2.1. Adaptation / observation method

The subjects were allowed to get adapted to the experimental space for 10 min. The experimental space is 9m × 14 m in size, and a display is installed. The indoor reflectance is composed of the ceiling 0.1, walls 0.6, and floor 0.2. During the experiment, in order to eliminate the influence of light from sources other than the display, the indoor lighting was turned off, and a blackout curtain was installed to block the daylight from the window. Subsequently, the subjects were observed and asked to evaluate the image of the space displayed on the screen. Figure 7 shows the method of adaptation and observation.

First, the subject was instructed to observe his/her own light environment for 20 s. During this time, the subject was not allowed to evaluate brightness. This was followed by observing the gray image consisting of the questions for the following 10 s. The questions were displayed in a monochromatic gray image with the same average luminance as the presented light environment. Therefore, it could avoid the effects associated with the subject's adaptability. Finally, the subject was instructed to observe the light environment for 10 s and evaluate the brightness of the entire space on a 7-point scale. A total time of 40 s was considered for the measurement target to capture the movement of the gazing point. Figure 8 shows the experimental situation.

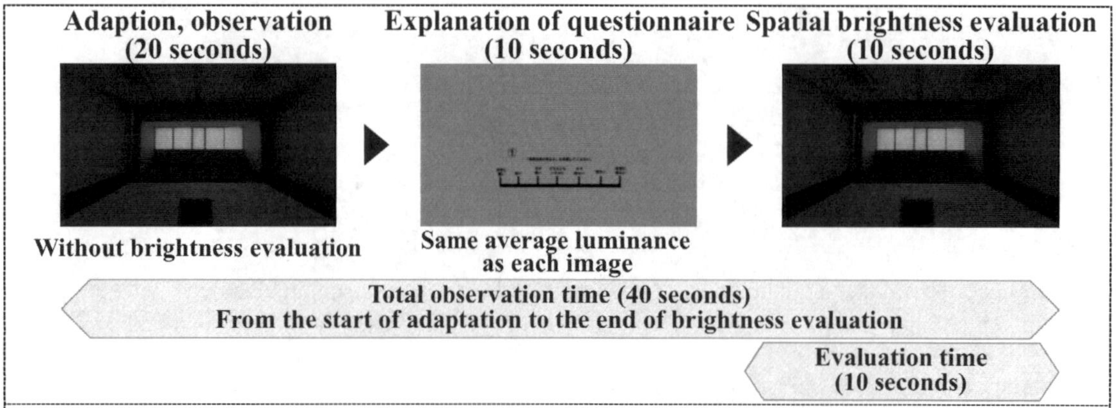

Figure 7. Adaptation and observation, evaluation procedure, observation time zone

(a) (b)

Figure 8. Experimental situation, (a) displaying questions, (b) evaluating spatial brightness

2.2.2. Correspondence between gazing point position and luminance data

The coordinates of the brightness data measured by a single-lens reflex camera equipped with a fisheye lens are different from the coordinates of the gazing point position measured by the eye tracking device. Therefore, the calibration of the camera image was calibrated, and the coordinates were accordingly adjusted to match those of the eye-tracking device. The luminance value of the gazing point was the average luminance value of 3 × 3 pixels centered on the coordinate data.

2.2.3. Analytical method

The gaze time and the relationship between average gaze luminance and spatial brightness evaluation value were analysed using the following different methods.

a) Gaze time

To understand the subject's tendency toward the gaze point, the gaze time ratio was analysed for each part of the space. Figure 9 shows the Field of view in calculating gaze time. For the target parts, all parts of the space and all the parts without workspaces or windows were considered. The workspace included the partitions and displays. Furthermore, the analysis targeted 15 conditions in total with windows.

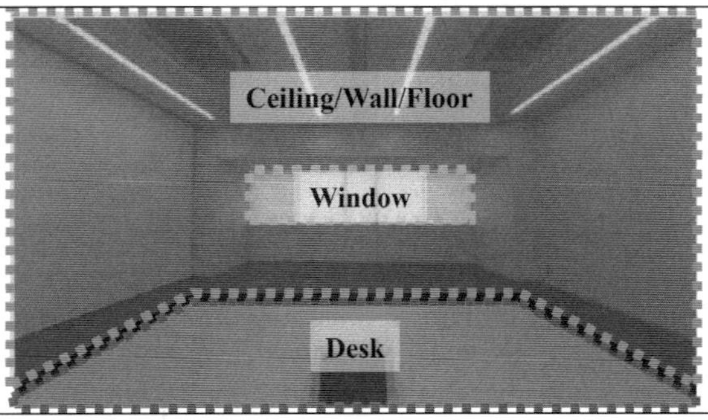

Figure 9. Field of view in calculating average gaze time

b) Types of visual field range (with or without the movement of the gazing point)

Figure 10 shows the different types of visual fields for calculating the average luminance. The "average luminance in the entire visual field area" is a general calculation method that does not consider the movement of the gazing point. This method calculates the average luminance in the area of the presentation image only. In contrast, the "average luminance of the gazing point" method calculates the average luminance by integrating the time zone on the locus of the moving gazing point. By comparing the values obtained from these two methods, the need for considering the movement of the gazing point in the brightness calculation and the brightness evaluation was analysed.

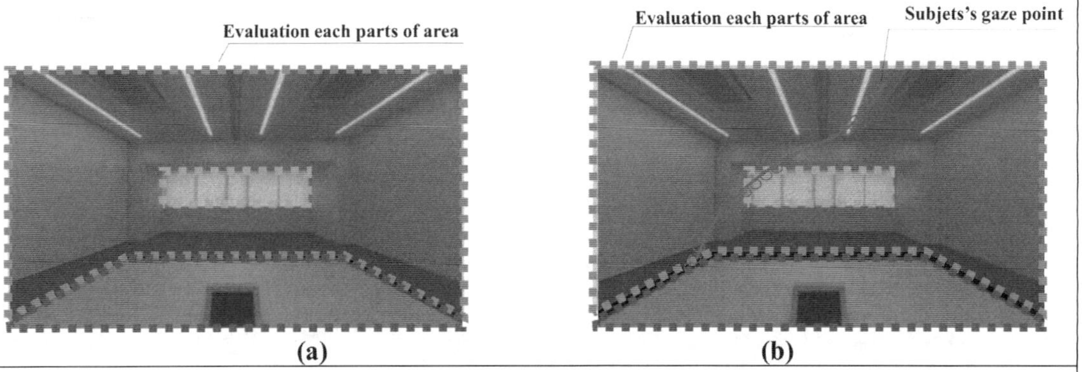

Figure 10. Field of view for calculating average gaze brightness, (a) is average luminance in the visual field area without considering the gazing point movement, and (b) is average luminance of the gazing point with considering the gazing point movement

IOP Conf. Series: Earth and Environmental Science 1099 (2022) 012012 doi:10.1088/1755-1315/1099/1/012012

c) Relationship between the average luminance of the gazing point and spatial brightness evaluation

Understanding the contribution of each part of the space toward the average luminance and brightness evaluation of the gazing point is very important. Figure 11 shows the four different types of the space parts. The spaces, namely all the parts, other than workspace (ceiling/wall/floor/window), other than window, and other than workspace / window (ceiling/wall/floor), were analysed.

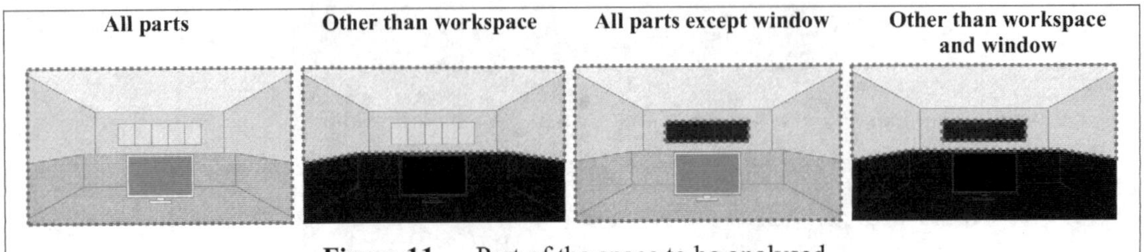

Figure 11. Part of the space to be analysed

3. Results

3.1. *Gaze time for each part of the space*

To understand the tendency of the subject's spatial observation, we analysed the gaze time in each part of the space. The left, center, and right subfigures in Figure 12 display the ratio of the solid angle of each part of the space to the screen, the gaze time ratio in each part, and the gaze time ratio per unit solid angle, respectively. For the target office, the ratio of the solid angle was 90 % or more of the whole by the desk, ceiling, wall, and floor, whereas the for the window, it was 10 % or less. The gaze time ratio of each part was 80 % or more for the ceiling/wall/floor and workspace, whereas for the windows, it was short. Conversely, the gaze time ratio per unit solid angle was increasing in the order of windows, ceiling/floor/wall, and desk.

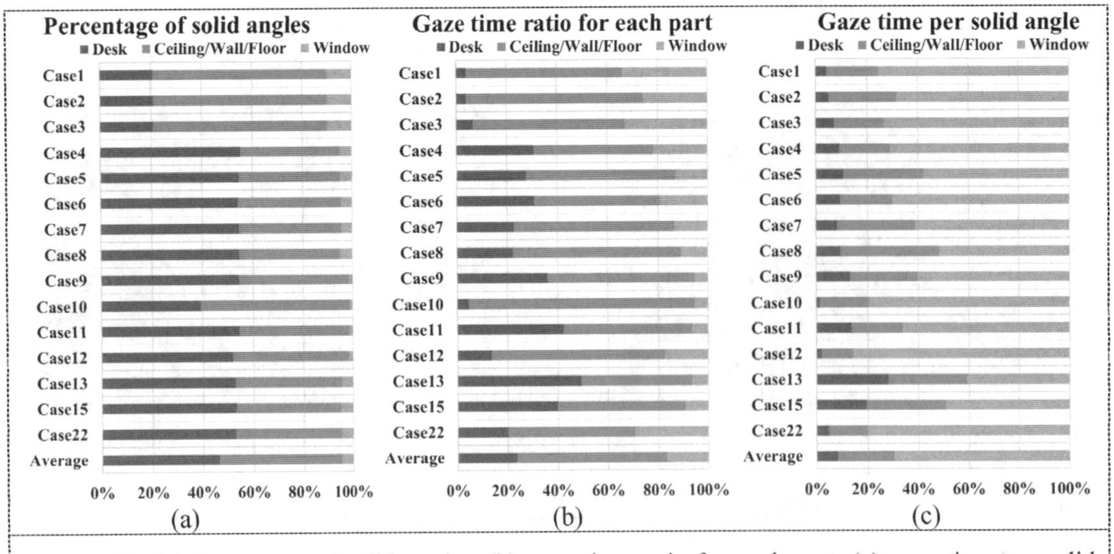

Figure 12. (a) Percentage of solid angles, (b) gaze time ratio for each part, (c) gaze time per solid angle

3.2. *Types of visual field range (whether or not the movement of the gazing point is considered)*

Figure 13 provides a comparison of the "average luminance of the entire visual field area" and the "average luminance of the gazing point" with the brightness evaluation. The target parts were "all the parts" and "parts other than the workspace." For "all the parts," the "average luminance of the gazing point" exhibited a higher explanatory power (coefficient of determination). Contrastingly, even when the "average luminance of the entire visual field area" was targeted at "a part other than the workspace," the explanatory power increased to the same extent.

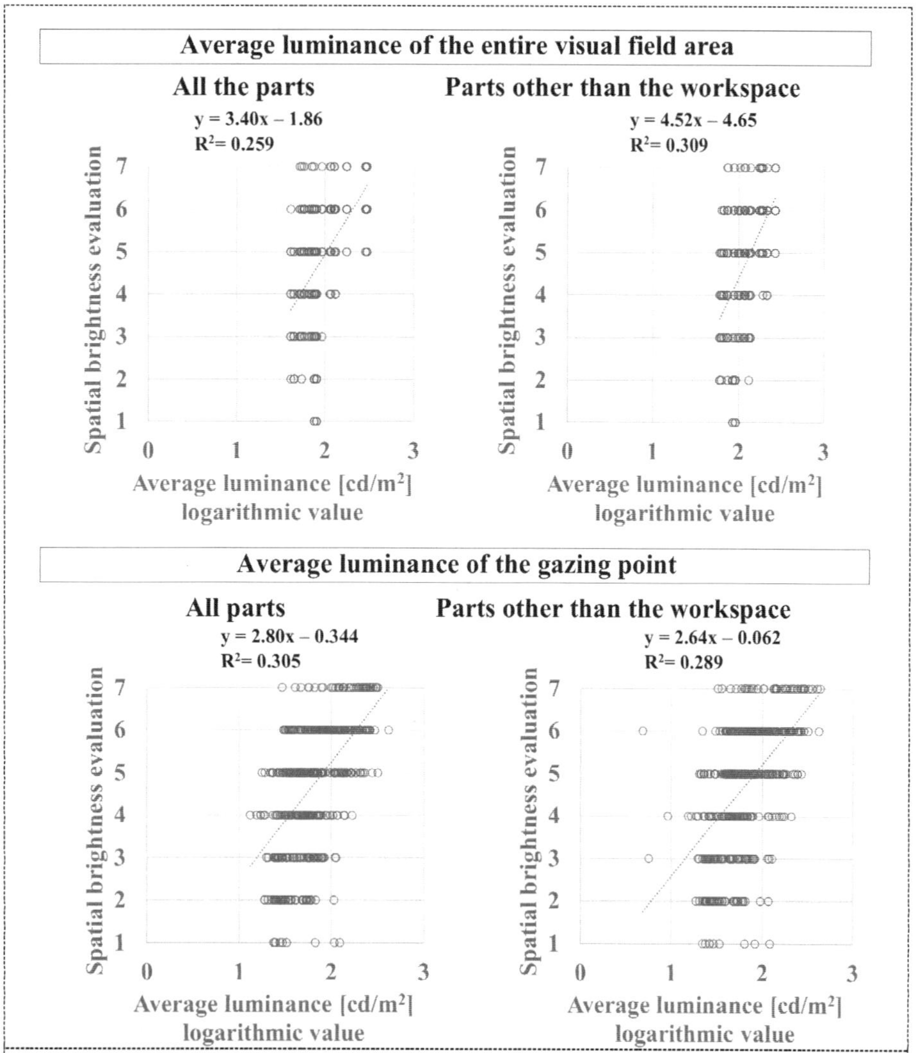

Figure 13. Comparison of the "average luminance of the entire visual field area" and the "average luminance of the gazing point."

3.3. *The relationship between spatial parts and spatial brightness*

Figure 14 shows the results of classifying the parts of the space (see Figure 11) and helps in determining the relationship between the average luminance of the gazing point and the spatial brightness evaluation during the evaluation time zone. A comparison between the spaces with or without workspace can be seen on the left and right sides of Figure 14. Whereas a similar comparison between the spaces with or without the window can be found on the upper and lower side of Figure 14. The best results are obtained when all the parts (including the workspace and windows) were targeted (as seen in the upper left corner of Figure 14).

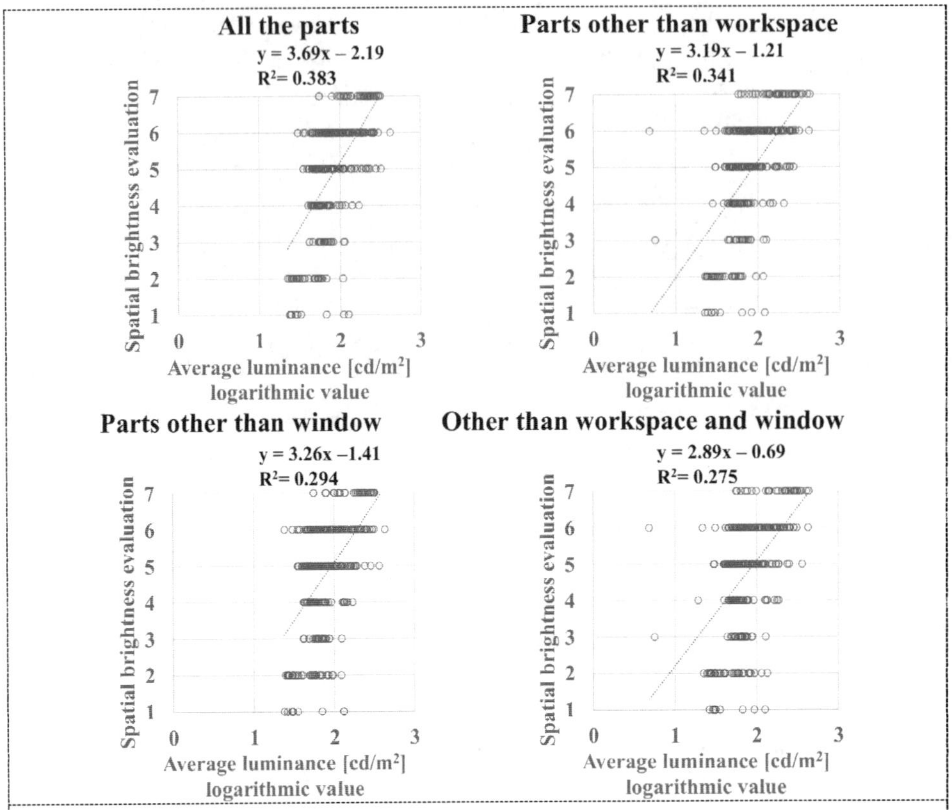

Figure 14. Explanatory power of spatial brightness evaluation using average luminance as an explanatory variable considering the subject's gazing point.

4. Discussion

In this study, we analyzed the relationship between the subject's gaze and spatial brightness evaluation of the luminance of the space by three methods.

In the analysis of gaze time for each part of the space, by comparing the gaze time for each area at the time of the subject's brightness evaluation, it was observed that the windows tend to have a small solid angle but a long gaze time per solid angle, whereas the workspaces tend to have a short gaze time per solid angle. A comparison of the gaze time per solid angle showed that the window was more attractive than the other parts of the space. In the future, by analyzing the effects of the window's position, the absolute value of the luminance, and the luminance ratio with the surroundings on the gaze time, it will be possible to clarify the factors that make it easier for the worker to gaze.

In previous studies, it was common to use the average luminance of each part as the variable that explains the spatial brightness. However, this method does not take into account the luminance that the worker is watching, and there is a possibility that the brightness felt by the worker cannot be accurately grasped. Therefore, in this study, we compared the explanatory power based on the time average of the luminance that the worker sees with the eye-tracking device and the average brightness of the area. When the entire space was targeted, the explanatory power was higher when the gaze point was considered. Still, the numerical value of the explanatory power was around 0.3 as a whole. In this study, only the central vision of the subject was considered, not the peripheral vision. In the future, it is necessary to proceed with the analysis, including the influence of peripheral vision. Considering the gazing point, the average luminance did not exhibit an exceptionally high explanatory power.

By classifying the parts of the space (see Figure 11) and helps in determining the relationship between the average luminance of the gazing point and the spatial brightness evaluation during the evaluation time zone. In the brightness evaluation, "workspace/other than windows (= ceiling/wall/floor)" was considered to be the most dominant. However, "workspace" and "window" were also observed as parts of the space, which affects the brightness evaluation.

5. Conclusion

To design a light environment that considers the gaze point of the worker, the relationship between the tendency of the gaze point to move and the spatial brightness evaluation by subject experiments using eye-tracking were analysed.

The gaze time per solid angle of the subject was higher in the window than in other parts, and the window tended to be more attractive. The calculation method for average luminance when considering the movement of the gazing point at the time of the brightness evaluation was presented. It was found that the explanatory power of the brightness evaluation was higher than the average luminance in the entire visual field area. The workspace and window also affect the brightness evaluation. This shows that spatial brightness can be improved by increasing the luminance of the gazing part of the space or by arranging the hi-luminance part of the space at the gaze position.

In contrast, the explanatory power of spatial brightness evaluation was not improved drastically, even considering the luminance and time of gazing. Future work can consider and include the following. The effects of the central vision and peripheral vision on the spatial brightness evaluation; the contrasting effect of the gaze point. Additionally, individual analysis, calculation of brightness perception at the gazing point, and analysis based on the time average can also be performed.

References

[1] JIS Z 9110:2010 General rules of recommended lighting levels, *Japanese Standards Association*
[2] JIS Z 9125:2007 Lighting of indoor work places, *Japanese Standards Association*
[3] CEN (European Committee for Standardization): EN 12464-1:2011 Light and lighting - Lighting of workplaces - *Part 1: Indoor workplaces*
[4] Hopkinson R G 1957 Assessment of brightness: What we see *Illum Eng.* **52** pp.211-22
[5] Loe D L et. al. 1994 Appearance of lit environment and its relevance in lighting design: Experimental study *Lighting Res. Technol.* **26**(3) 119-133
[6] Kato M et. al. 2019 Study on applicability of mean luminance by image photometry as spatial brightness index *J, Env. Eng. AIJ.* **84** No 766 p 1059-66 (in Japanese)
[7] Yonekura Y et. al. 2019 Validation of the Spatial Brightness Estimation Formula in Offices with Windows *Proceedings of CIE Centenary Conference*
[8] Yoshizawa N et al. 2019 Examination of the application range of the average luminance *Proceedings of CIE Centenary Conference*
[9] Ko B and Koga T: Brightness of a Space in Terms of Variation of Luminance (*Journal of the illuminating Engineering Institute of Japan* Vol **97** No.8A pp 429-35, 2013.8(in Japanese)
[10] Kage H et al. Experimental Study on Illuminance Distribution in the Task and Background Area of Open-office Lighting, *Proceedings of CIE Centenary Conference* 2019
[11] M SareyKhanie, B M 't Hart, J Stoll, M Andersenand W Einhäuser Integration of eye-tracking methods in visual comfort assessments
[12] Ishida K and Inoue Y 2009 Influence of lighting equipment position and attentive point on suitable illuminance for relaxation environment *J, Env. Eng. AIJ.* **74** No 644 p 1123-29 (in Japanese)

Acknowledgments
We are grateful to Mr. Toshiki Torimitsu of Tokyo University of Science, for helpful discussions.

IOP Conf. Series: Earth and Environmental Science 1099 (2022) 012013 doi:10.1088/1755-1315/1099/1/012013

Implementing a Digital Solution for Architectural Daylight Analysis in BIM Based Projects by developing a new plugin

Majid Miri, Elmira Ashtari

Sweco AB, Stockholm, Sweden
AftabSoft, Huddinge, Sweden

majid.miri@sweco.se, eashtari@kth.se

Abstract. To achieve a successful architectural project, it is essential for architects to truly understand and collaborate with other designers and engineers during the whole project. Regarding daylight analysis, what architects decide about the fenestration types, sizes, and their placements, interior layout design, interior or exterior materials, and finally the outer shape of the buildings may change the quantity and quality of natural light in the interiors. Moreover, such decisions about room configurations, fenestration, and materials can also directly or indirectly affect thermal comfort which is needed to be checked and confirmed by energy experts in the project. In this paper, a newly developed digital solution, and a new plugin called AftabRad is introduced. Its main goal is to export the 3D model from the BIM Modeling software to Radiance, as its daylight simulation core, and present the results back to the BIM software. This plugin can help daylight specialists to be fully or partially involved in any projects to assist architects or urban designers to easily calculate different types of static or dynamic daylight metrics by themselves. However, daylight specialists can do daylight calculations in the final stage or the project if it is needed. In such a project, since most of the daylight analyses for different types of alternatives can be done by architects, and regarding the nature of BIM-based project, by having direct dialogs with energy experts, the probability of achieving a well-daylit space, being thermally and visually comfortable, and fulfilling the daylight requirements can be much higher in comparison with doing daylight simulation just at the end of a project.

1. Introduction

Including daylighting strategies from early to the end in the building and urban design process is crucial when it comes to the compliance of environmental certification systems (like LEED or BREEAM) or when the aim is to simply create an interior or exterior space with sufficient access to natural light while still thermally comfortable. Moreover, since the end of 2018 when the first European Standard (EN 17037) dealing with daylight provision, exposure to sunlight, view out, and glare in buildings is released, doing daylight analyses in all stages of the architectural design process is more and more needed. However, generally, until now, daylight specialists start to be involved in a project when the other designers and engineers like architects, structural engineers, mechanical engineers, etc., finish their jobs. Therefore, in such a project, there are several limitations and restrictions to improving the quality and quantity of natural light in and around the buildings. The below diagram together with inspiration from the MacLeamy curve shows different possibilities of making changes in a project and its cost during an architectural design process. As shown, by taking into account enough daylight access for interior spaces as one of the design decision-making factors from an early stage in a project, it makes it possible to create a better-daylit space while still complying with other disciplines' needs such as energy requirements.

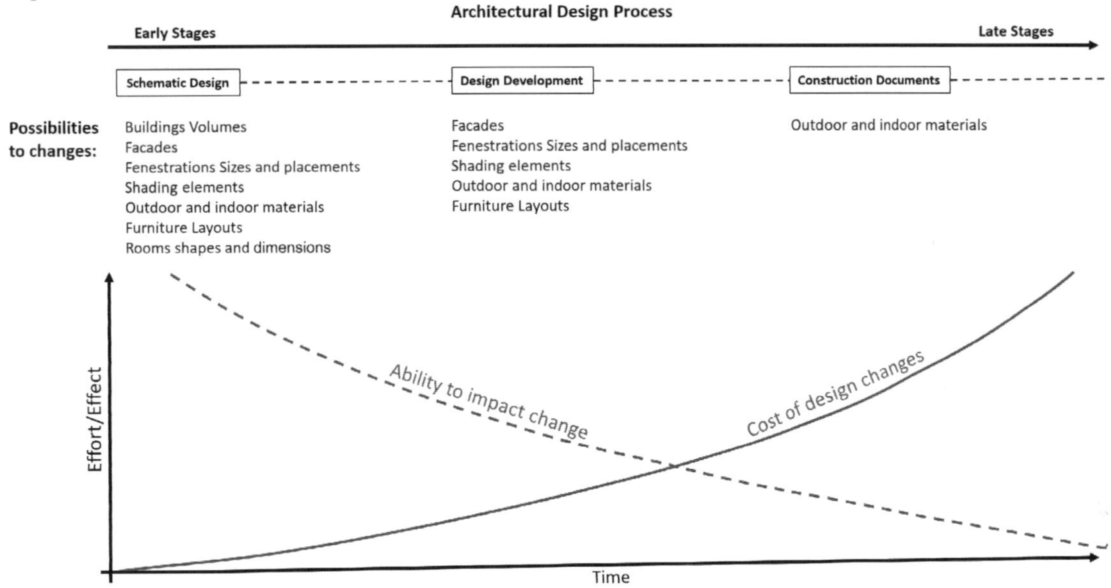

Figure 1. The modified MacLeamy Curve, "The effort curves in the design and construction industry" (Marcus, Kirby & Krygiel, 2017)

Moreover, to achieve a successful project, it is usually needed to satisfy all other disciplines such as mechanical, acoustic, and structural engineers. To efficiently collaborate between different disciplines, we need to have a core platform where each group can easily upload their works to the model and get updated with the latest changes on the model. Nowadays, working with BIM-based 3D modeling software is very common in architectural firms. Although architectural firms have adopted using BIM-based 3D modeling programs like Revit, Archicad, or Microstations for more than a decade around the world, they usually use other computer programs to do any type of simulation. Such programs mostly have one-way connections with such BIM-based 3D modeling software. So, in the conventional method, to do analyses like electrical and natural lighting, wind, and thermal comfort, the model is usually exported to an auxiliary simulation program from the 3D modeling software, and then all the preparations, calculations, and presenting the results are done in that simulation software. However, if we can use a computer program that can directly communicate with the 3D modeling software, it will be possible to save lots of time and reduce mistakes, and problems which could happen when exporting and importing to and from other software.

Revit has had an accessible API (Application Programming Interface) for more than 2 decades, which can be used to access almost any of its commands in the software. Thus, among all main BIM 3D modeling software that has been used by architectural consultants, scripting and creating new tools is much more convenient in Revit. Therefore, in this paper, we are going to introduce and discuss a new add-in that has been developed by the authors under the Autodesk Revit platform. This tool can save time and minimize the mistakes and associated problems when exporting the analysis model directly from Revit to the simulation core and importing the results to Revit. This new add-in can work as a supportive decision-making tool for all architects and urban designers/planners in their everyday works.

2. Building Information Model (BIM)

In ISO/TS12911, Building (Construction) Information Model (BIM) is defined as a "shared digital representation of physical and functional characteristics of any built object, including buildings, bridges, roads, process plant". In opposite to traditional CAD tools that help their user to create stand-alone 2D or 3D drawings and objects, in a BIM-based 3D modeling software, not only can we create and modify

3D models and represent them in 2D or 3D, but also each element of the model includes some non-geometrical information relevant to that element.

In a conventional CAD-based design workflow, each group in the design team usually works in separate files. Thus, the collaboration in such a working style can be quite low. However, in a BIM-based model on which all elements and data designed or needed by different professionals should be included, the design team like architects, landscape architects, mechanical, civil, and structural engineers, etc., should be able to work on a single model to be able to update and be updated with the latest changes live during the design process. Thus, working on a local server-based or cloud-based central model file is very common for every BIM-based model file. Consequently, in such a project, the level of collaboration between different disciplines is much higher.

3. BIM-Based Design Workflow

During working on a BIM-based project, a 3D model with an acceptable level of detail including all elements like architectural, structural, mechanical, electrical, etc., should be prepared before producing any 2D drawings. Therefore, creating a single 2D drawing can take much more time compared with a CAD-based workflow. However, once such a model is ready, many drawing sheets like plans, sections, or elevations can be automatically produced separately (Marcus, Kirby & Krygiel, 2017). Moreover, in addition to all geometrical information that should be included for each object, different types of none geometrical data should be assigned to each element. For example. information like the thickness, layers, and materials of each surface of elements like windows, floors, walls, ceilings, furniture, architectural or structural columns, beams, etc., should be defined in a BIM-based model. Thus, although working on a BIM-based project doesn't necessarily mean reaching a higher level of expertise or quality than a CAD-based project, extracting data from the model, quantifying each type of element, and creating a variety of 2d and 3D drawings are much more convenient in a BIM-based model rather than a CAD-based model.

4. Daylight Analysis Inside a BIM-Based Model

Data like geographical information, building orientation, rooms and space types, and materials of each element, etc., that are needed to do any daylight analysis can already exist in the BIM-based 3D model file. Moreover, views that are needed to do any photo-realistic rendering simulation should be included in the Model too. Therefore, doing any type of natural light simulation inside the model needs much less additional information rather than a conventional method that uses a third-party program. Anyway, based on the type of daylight analysis that is needed for each project, sometimes it is necessary to add some information to the project which is not included in the model. Some of the most important ones are as follows:

- Selecting the rooms, spaces, or surfaces on which we should do our analysis
- A relevant weather data file of the place
- The required illuminance level that is usually needed to do most of the dynamic daylight metrics like daylight autonomy
- Surrounding obstructions and topography if they do not exist in the model.

All in all, for a BIM-based project that needs to do daylight simulations, as most of the information that is required to do a daylight simulation should be already included in that model, doing a daylight analysis can be much more convenient and efficient with a lower risk of mistakes than before. Creating drawing results including a false-color background in plans can be mostly done automatically in a BIM-based model while in a CAD-based model every single drawing has to be made separately and mostly in another third-party simulation program. Moreover, in such projects, the interaction between daylight specialists, architects, energy experts, etc., should be much easier and faster than before.

However, working with a BIM model is not always easy. One of the most important challenges that we as daylight specialists probably face is the size of the project which can directly affect the calculation accuracy and speed. As including all objects from all disciplines involved in a BIM-based project makes the model quite heavy, to be able to do an accurate simulation within a reasonable time, we have two main choices to reduce the size of the simulation file that could differ in each project. The first choice is to exclude some individual elements or all elements of a category from exporting to the simulation program by using a section box or hide/isolate tools. The second method is to simplify some elements that cannot be skipped when exporting to the simulation program and save them as a design option or in another layer/workset in the main model or directly save them in a detached model without affecting the main central model that is used by all other disciplines. Thus, the simulation program can work with a simpler model when doing its calculation.

5. Existing Daylight Analysis Tools for BIM Based 3D Modelling Software

To do any type of daylight simulation for a BIM-based 3D model, they are two common ways. The most usual method that most of the daylight specialists do is to export the model to another daylight/light or energy simulation software such as Climate Studio/Diva4Rhino, LadyBug/Honey Bee, Ecotect, IES-VE, IDA ICE, Dialux Evo, Relux, etc., by saving them as GBXML, IFC, DWG, DXF, OBJ files. Information like surface materials, room names, types and dimensions, and geographical information that usually exist in any BIM-based 3d model are usually missed when exporting to another file type except IFC and GBXML formats. Regarding the IFC and GBXML files, as those computer programs that can handle such files are not BIM-based software, what is imported is usually different from the original BIM-based model and needs to validate the imported geometry, and materials of each surface. Moreover, as most of the software that can handle such files are energy/HVAC analysis software like IES-VE or IDA ICE, to be able to make it works it is quite often that their users need to simplify or remodel such imported 3D models. (Miri, Ashtari, 2019)

Table 1. Comparison between different daylight analysis Software having a Revit add-in

Software	When in the Design Process	Results Visualizer Software	Simulation Engine	Metrics	Access to Radiance Files	Materials
Climate Studio	Early to the End	Rhinoceros	Radiance	DF, Lux, Cd/m2, UDI, sDA, ASE, aDGP, ...	Yes*	by Revit Category
Sefaira	Early	Sefaira	Cloud-Based Radiance	DF, Lux, sDA, ASE, SunHours	No	by Sefaira Category
Lightstanza	Early	Lightstanza	Cloud-Based Radiance	DF, Lux, DGP, aDGP, UDI, sDA, ASE, cDA	No	by Lightstanza Category
Naviate Daylight	Early to the End	Naviate / Revit	Radiance	Only DF	No	by Revit Category
Elumtools + Licaso®	Early to the End	Elumtools / Revit	Own Engine	DF, Lux, Cd/m2, UDI, sDA, ASE, aDGP, ...	...	by Revit Materials
Autodesk Insight 360	Early to the End	Revit	Own Cloud-Based Engine	DF, Lux, DA, sDA, ASE	...	by Revit Materials
AftabRad	Early to the End	Revit	Radiance	DF, Lux, Cd/m2, UDI, sDA, ASE, aDGP, ...	Yes	by Revit Materials

*In Climate Studio all Radiance files are saved in AppData's Temp folder

The other method to do such analyses is doing it directly inside the BIM-based software and applying all the information needed for simulation and existing in the model directly from the model in the calculation. Programs like Sefaira, Lightstanza, Climate Studio, Naviate Daylight, Elumtools, and Autodesk Insight 360 are add-ins that can help to do daylight simulation directly from Revit. As you can see in table 1, among those five programs, Sefaira and Lightstanza are web/cloud-based programs, so they do not present the results in Revit and they are more suitable for the early stages of the design process. Climate Studio shows the results in Rhinoceros, while Elumtools can show the points values or false-color calculation surfaces in Revit, or its own window. The Autodesk Insight 360 is the only program that can fully present the results in Revit. Moreover, regarding the material specifications, except for the Climate Studio, assigning specific types of materials rather than common opaque or transparent materials is not usually possible in these computer programs. In Climate Studio, Sefaira and Lightstanza, the material of each face is not exported, so their materials are directly or indirectly based on the Revit category. And about the Naviate Daylight, it can just do Daylight Factor calculation until now, so, it has not yet ready to do other types of daylight metrics. Finally, except for the Climate Studio, the rest do not provide their users with access to all calculation files. Therefore, it makes it more difficult for their users to validate the exported model to the simulation core.

Considering what is said above, that is why we thought that the market is lacking a tool directly communicating between Revit and simulation core together with providing all the calculation files accessible for their users to provide them the ability to validate and do their simulation as they wish. To make the simulation workflow more efficient, such software should also use all the information that can be used in any daylight simulation and are available in a BIM-based model. And finally, using a validated and open-source simulation program like Radiance (Ward and Shakespeare, 1998), it should assure its users to reach accurate results if they follow the simulation program requirement.

6. Development of a New Daylight Analysis Tool for a BIM Based 3D Modeling Software

Among all the daylighting simulation software that are available in the market, Radiance (Ward and Shakespeare, 1998) as an open-source physically based, backward raytracing lighting simulation tool has been used as a daylight calculation software directly or through other simulation programs for more than 30 years. Its calculation results have been validated for different types of buildings and shading elements (Mardaljevic, 1995; Reinhart and Walkenhorst, 2001; McNeil and Lee, 2013). Since it is a Dos-command-based program, it is not very user-friendly software. That is why a few graphical user interfaces (GUI) has been developed during the last 20 years. Desktop Radiance, Rayfront, and Ecotect (Roberts and March, 2001) were some of the pioneers. Open Studio (Guglielmetti, Macumber and Long, 2011), Diva4Rhino(Jakubiec and Reinhart, 2011)/Climate Studio, and LadyBug (Roudsari, Pak, and Smith, 2013) are some other tools that use Radiance as their daylight calculation core. However, none of them are designed to be used directly in a BIM-based program.

Regarding the need to do a quite fast but still reliable simulation for a variety of alternatives during the design process (Østergård, Jensen, and Maagaard, 2016) and the importance of making the right decision in the early stages to make it more possible to reach a successful passive design (Kanters, Horvat and Dubois, 2014), the needs for an analysis tool which can help architects to easily evaluate their designs from early stages until the end when the complexity level of the model becomes much higher can be quite high. Such a tool should not only facilitate the communication among the BIM software and calculation core but also have a user-friendly interface to be used by architects and engineers with a basic knowledge of simulation. Thus, in the following, our newly developed tool and the way of using it by architects and engineers for their daily applications will be briefly discussed.

The AftabRad add-in that has been released in its more stable version recently by the authors is developed to estimate the daylight and sunlight availability in and around the buildings from early to the end of the project. In addition, to directly export the 3D model from the BIM 3D modelling software to Radiance, most of the information like geographical specifications, building orientation, room, space types, or material specifications usually needed by Radiance and existed in the BIM-based model can be exported to Radiance and the calculation results can be imported from Radiance without using any third-party programs. However, based on the project needs and the selected daylight metric, there is some other information that does not usually exist in a BIM model, so it is needed to be added manually through its interface. For example, the height of surface analysis grids, spacing distances between the grids, their offset from the surrounding walls, the weather data file of the place, and occupancy schedules are some of them. Thus, we need to import or add such information directly into the calculation process. All in all, such a program can both save the 3D model preparation and minimize the mistakes and errors that are related to exporting and importing to and from other applications.

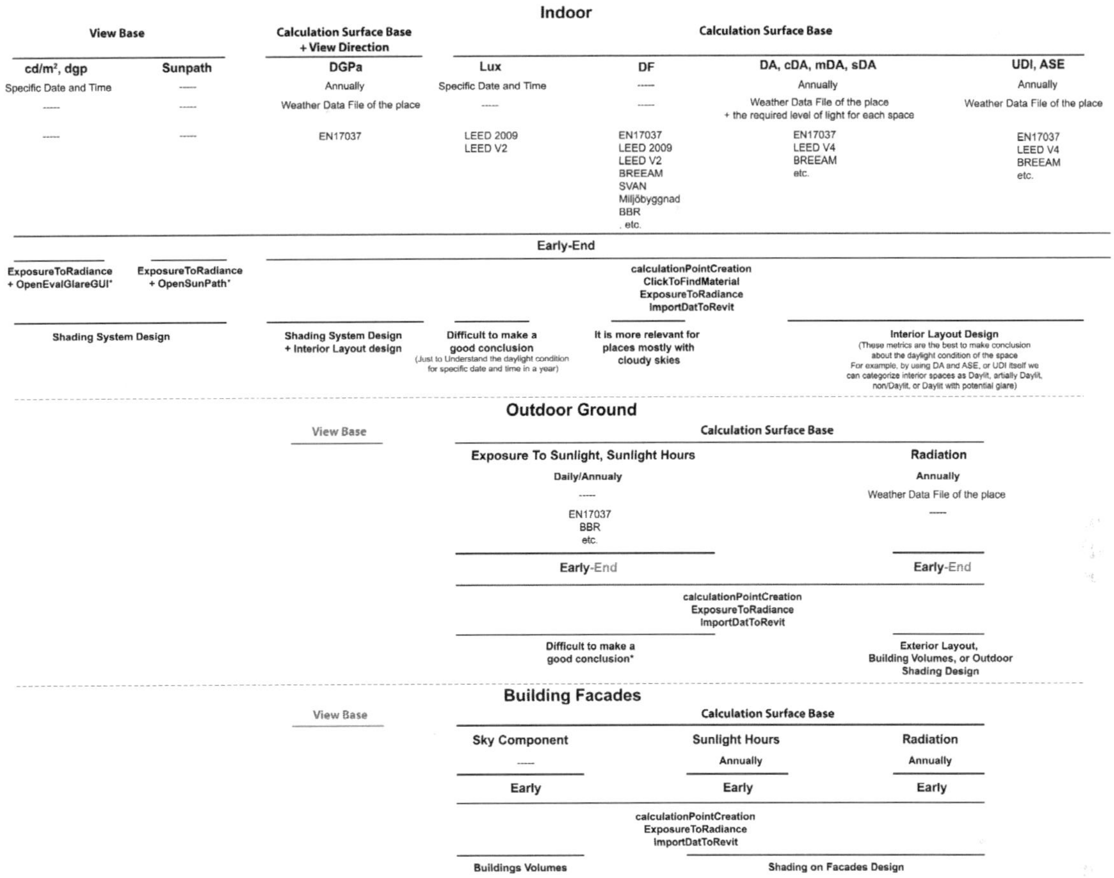

Figure 2. The above image shows the diagram on which the toolbox of this add-in is based. It represents when and for what kind of project AftabRad tools should be applied.

This Add-in is made to be suited for both users with advanced and basic daylight simulation knowledge in their everyday work or research. For advanced users, it can help them by providing all files, like rad, mat, sky, and pts files that are needed by Radiance to do any type of daylight simulation. Then they can continue their research by using any other already developed computer programs or using some other programming languages like Python, Perl, C++, C#, etc., to do any custom-made analyses. However, ordinary users generally need to evaluate their design against the most common metrics that are usually asked by most of the building design standards, rules, or environmental certifications that are available in the Add-in. The possibility to present the results in Revit, as pdf files, or even through web links can help them to directly communicate with other designers, engineers, or clients.

Its main simulation engine to do any static or dynamic daylight analyses is Radiance. Regarding dynamic daylight metrics, it uses the two-phase method or the DDS (more accurate solar component) version of the two-phase method. However, to do a sunlight hours study, this add-in uses its own algorithm that is based on precomputing the sun positions in the selected period and saving them in a couple of sky files. Such a method can save lots of time when doing the sunlight hours study. Moreover, to reduce the simulation time, it offers its users to use Accelerad binary files instead of ordinary Radiance files to make the calculation time much faster by using Optix^{TM}, a graphics processing unit (GPU) based ray tracing engine (Reinhart and Jones, 2015). Additionally, it can also cut the simulation time by dividing the calculation points into smaller pieces and letting the computer calculates them parallelly.

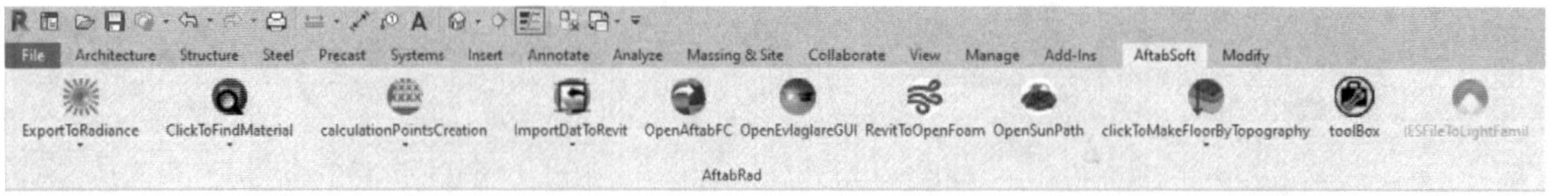

Figure 3. In this image, you can see the AftabRad tab and most of the tools in this add-in.

Below you can see a chronological order of the work that we should usually follow to simulate any types of daylight metric in different stages of the project by using some Revit or AftabRad tools.

6.1. 3D Model Preparation Works

6.1.1. Filtering

Filtering refers to only exporting the needed objects to the simulation program. Generally, to achieve the results with quite a good accuracy and in an acceptable time, each energy simulation software suggests an acceptable range of complexity to accurately and efficiently work with. In this add-in, there are four different ways for manipulating the level of details and preparing the model when exporting the 3D model from Revit to Radiance. These four methods are as follows:

- Using the Revit 3D View Section Box
- Isolating or hiding the selected elements or categories in the current view. This can be done with the permanent or temporary Hide, Isolate or Show tools, or by using the Design Options or Workset tools.
- By selecting exporting categories in the Add-in
- By making an area limitation to exclude objects in some categories in the Add-in

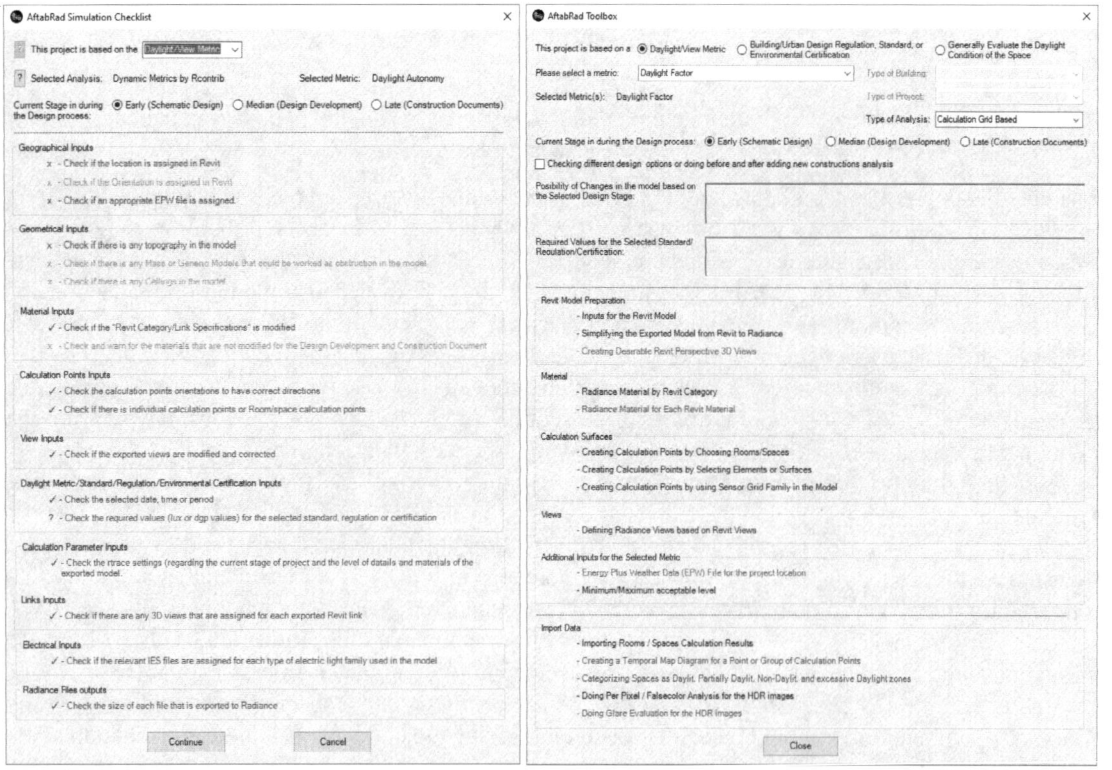

Figure 4. The right image shows the toolbox page, while the left one shows the simulation checklist page in the AftabRad add-in

6.2. Daylight Simulation Toolbox for Architects

Regarding the nature of BIM models, and what they can offer to their users, here in this add-in, there is a toolbox that can help architects with a basic knowledge of daylight simulation to start and finish their daylight analysis within acceptable accuracy.

The toolbox is a step-by-step interactive guidance tool that can give some hints to the users about how to prepare the model, what kinds of Revit or AftabRad tools, and which commands inside those tools should be applied during the design process to fulfill the project needs, and finally how to present and interpret the results. All this guidance is based on the current design stage of the project or follows the relevant standards, regulations, or environmental certifications of the project. It can also help its users if they just want to do an analysis based on a specific daylight metric too. Therefore, this tool provides more assurance for architects as the main user of this add-in when they start and finish any required daylight analysis during the project.

6.3. Daylight Analysis Preparation Works

6.3.1. Modifying Material Definition

To assign a Radiance material definition for each surface of every element in the model, there are three different ways in this add-in that are briefly explained as follows:

1. Using the Revit material definition and translating its relevant information into an appropriate Radiance material.
2. Applying a default reflection or transmission value for each Revit category
3. Redefine the Radiance material definition for each Revit material that is used in the model
4. Modifying the Radiance materials when the whole 3D model is translated and exported for daylight calculation in Radiance.

6.3.2. Selecting Area of Interest/Calculation Surfaces

To create the calculation points, this add-in provides five different ways that we can apply one or some of them together. Those four methods are shortly explained below:

1. By placing calculation points manually wherever is needed in the model
2. By selecting any surfaces in the model as the calculation surfaces
3. By selecting the room or space names for each selected level to create calculation grids based on their boundaries
4. By selecting floors of the selected levels from which the surface grids are offset

In all these methods, we can define the offset distance of the calculation points from the selected surfaces.

6.3.3. Creating Views from Revit 3D Views

Regarding the project needs, sometimes we need to do some view-based daylight analysis such as Glare analysis. Therefore, this add-in can choose and translate any 3D views defined in Revit as Radiance views. Moreover, in this add-in, there is another page to double-check the exported views parameters and preview them before starting to render in Radiance that sometimes can be time-consuming.

6.4. Doing Daylight Analysis

6.4.1. Various Types of Analyses in AftabRad Revit Add-in

Here in below, you can find all types of daylight analyses that are possible to do in this add-in.

- Daylight analysis
    - Static analysis for a specific date and time, or one type of skies like Illuminance, Luminance, and Daylight Factor
    - Dynamic analysis for a specific period or a full year like Radiation (kWh/m2), Kilo Lux Hour/m2, Shadow Range for a specific date, Periodic or Annually Sunlight

Hours, One Day Sunlight Hours, UTCI based Sunlight Hours, Exposure2sunlight (based on EN 17037), Annual DGP, Daylight Autonomy, Maximum Daylight Autonomy, Continues Daylight Autonomy, Useful Daylight Illuminance, Annual Sunlight Exposure

- View analysis
    - View Analysis (based on EN 17037)
    - Visibility Analysis

6.4.2. Simulation Checklist

However, before starting any analysis there is a simulation checklist tool that can help its users to validate all parameters, exported elements based on the current stage of the project, and also selected standards, rules, or certifications. Therefore, by using such a simulation checklist the mistakes that could happen even for professional daylight specialists will be reduced.

6.4.3. Validating the exported model to Radiance

As there are still some bugs in Revit, although it is quite rare, if you see any strange results that are against your expectation as a daylight specialist, in addition to exporting the 3D model to Radiance, you can export it as another 3D file like, obj, vrml, stl, or glb. Then you can open such a file in any other 3D modeling software and try to find and solve the issue in Revit.

6.5. Extracting the Results

6.5.1. Per Pixel Analysis Page

The main goal of developing this part of the add-in is to present and analyze the HDR images which are created by Radiance or some other simulation software. This tool makes it possible the evaluation of an HDR image corresponding to its relevant Revit 3D view through a variety of options.

6.5.2. Evalglare Analysis Page

Here, on this part of the AftabRad plugin, glare analyses can be done for each HDR image that are made by Radiance. The main engines for this kind of analysis are "evalglare" which is a Radiance-based tool being developed by Jan Wienold at the Fraunhofer Institute in Germany and "findglare" which is a Radiance command. Therefore, this page is an interface to detect glare sources and calculate glare indexes such as DGP, DGI, UGI, UGR, etc., for each Radiance-based HDR image.

6.5.3. Importing the calculation data files

With this tool, all types of calculations that are done with the help of this add-in can be presented inside Revit. The statistic results for each room can be extracted and shown here too.

6.5.4. Sharing the Results

Other than presenting the results inside Revit, the AftabRad add-in gives a possibility to share the results with other designers or clients by providing a web link or 3D pdf file including the 3D model and false-color results. This way can facilitate the collaboration between architects themselves, architects and clients, and architects and other disciplines that do not use Revit as the main 3D modeling software in their daily work.

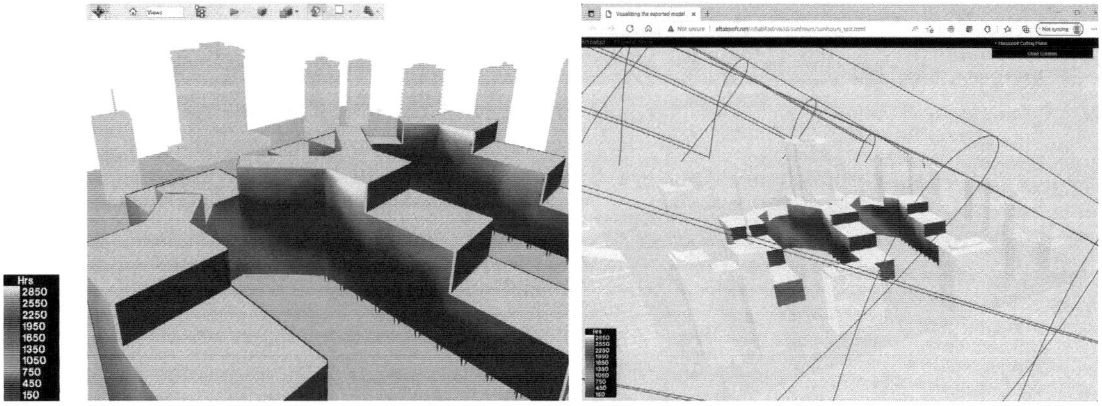

Figure 5. The left image shows an example of a 3D pdf file while the right one shows another way of facilitating communication with other designers or clients through a web link by including the 3D model, calculation results, and its relevant sun-path diagram

7. Conclusion

Since BIM-based 3D modeling software like Revit are computer programs usually used by designers or engineers with different skills, one of the main goals of this add-in is to make the daylight simulation possible for anyone involved in a BIM-based project during the design process of a building or urban area even if they have a basic knowledge about daylighting and daylight simulation. Providing some user-friendly and interactive graphical user interfaces to facilitate the connection between Revit and Radiance can make the daylight analysis much more convenient for all those that are involved in the project. Moreover, a toolbox is designed to help users to choose the right tool and commands to reach the requirement of evaluating the daylight condition of any interior or exterior spaces from early to the end of the project. By using a simulation checklist tool, the mistakes that could happen for even professional daylight specialists can also be reduced and assure more accurate results for everyone that is using this add-in. And finally, although this developed solution is designed to be used just inside Revit, it will be possible to import a BIM-based model that is created by other BIM 3D modeling software through some other extension file like IFC in near future. Moreover, as Revit is not the only BIM 3D modeling program in the market, in addition to developing this add-in toward making it more user-friendly, our other target for the future will be to provide this add-in for other BIM 3D modeling software too.

References

[1] CEN - European Committee for Standardization (2018). Daylight in Buildings (EN 17037:2018).

[2] Guglielmetti, Rob, Dan Macumber, and Nicholas Long (2011). OpenStudio: an open source integrated analysis platform. *Proceedings of the 12th conference of international building performance simulation association.*

[3] http://aftabsoft.net/aftab-rad.php

[4] Jakubiec, J. Alstan, and Christoph F. Reinhart (2011). DIVA 2.0: Integrating daylight and thermal simulations using Rhinoceros 3D, Daysim and EnergyPlus. *Proceedings of building simulation. Vol. 20. No. 11*, 2202-2209

[5] Jones, Nathaniel L., and Christoph F. Reinhart. "Validation of GPU lighting simulation in naturally and artificially lit spaces." Proceedings of the IBPSA Building Simulation Conference. 2015.

[6] Kanters, Jouri, Miljana Horvat, and Marie-Claude Dubois (2014). Tools and methods used by architects for solar design. Energy and Buildings 68, 721-731.

[7] Mardaljevic, John (1995). Validation of a lighting simulation program under real sky conditions. *International Journal of Lighting Research and Technology 27.4*, 181-188.

[8] McNeil, Andrew, and E. S. Lee (2013). A validation of the Radiance three-phase simulation

method for modelling annual daylight performance of optically complex fenestration systems. *Journal of Building Performance Simulation 6.1*, 24-37.
[9] Miri, M. and Ashtari, E., 2019. Daylight Study from Early to the End of Architectural/Urban Design process by Developing an Add-in for Revit. In 16th International Conference of the International Building Performance Simulation Association, Building Simulation 2019, 2 September 2019 through 4 September 2019 (pp. 2745-2753). International Building Performance Simulation Association.
[10] Østergård, Torben, Rasmus L. Jensen, and Steffen E. Maagaard (2016). Building simulations supporting decision making in early design–a review. Renewable and Sustainable Energy Reviews 61, 187-201.
[11] Reinhart, Christoph F., and Oliver Walkenhorst (2001). Validation of dynamic RADIANCE-based daylight simulations for a test office with external blinds. *Energy and buildings 33.7*, 683-697.
[12] Roberts, Andrew, and Andrew Marsh (2001). ECOTECT: environmental prediction in architectural education.
[13] Roudsari, Mostapha Sadeghipour, Michelle Pak, and Adrian Smith (2013). Ladybug: a parametric environmental plugin for grasshopper to help designers create an environmentally-conscious design. Proceedings of the 13th international IBPSA conference, August 2013.

Describing the characteristics of light field in architectural spaces using spherical harmonic function

Kazunori Yanagawa[1], Ayane Fujihira[1], Hideki Yamaguchi[2] and Nozomu Yoshizawa[1]

[1] Department of Architecture, Faculty of Science and Technology, Tokyo University of Science, 2641, Yamazaki, Nodashi, Chiba, 278-8510, Japan
[2] National Institute for Land and Infrastructure Management, 1, Tachihara, Tukubashi, Ibaraki, 305-0802, Japan

* Correspondence: yoshizawa@rs.tus.ac.jp

Abstract. The direction, density, and diffusivity of light are important indicators of spatial characteristics in describing a three-dimensional light environment. Mury presented a method for describing, measuring, and visualizing the structure of light fields using spherical harmonics in terms of changes in the density and direction of light in three-dimensional space. We extended this study by using higher-order spherical harmonics, which would represent more diverse characteristics of the light environment. We also quantitatively described the light environment as numerical values and investigated the correspondence between these numerical values and human perceptual quantities. As a result, we confirmed that there is a certain degree of correspondence between the "complexity" quantified by the spherical harmonic and the "complexity" perceived by people when observing real space.

1. Background and purpose

Current lighting design is usually conducted by using the information on surface-bound and two-dimensional distribution of illuminance and luminance. Koenderink [1] confirmed that human observers have a mental representation of the light field by experimenting with how an object would look like to a human observer when it was introduced at an arbitrary location in a scene. In other words, it is very important to understand the three-dimensional light field in lighting and architectural design to realize the appropriate appearance of objects in spaces.

The term "light field" appeared in the publication by Gershun[2] in the 1930's. By introducing the concept of radiant light density and light vector, Gershun described the quantity and transfer direction of the radiant power through space. In the 1990's Adelson and Bergen[3] proposed the Plenoptic Function to describe the structure of ambient light in luminous environment by mathematically representing the optic array. The concept of light field have been further developed in the fields of computer graphics and perceptual psychology. Mury[4] used spherical harmonic decompositions to represent natural light as the combination of different order components.

Lynes and Cuttle[5] introduced the indexes of scalar illuminance and illumination vector to the field of lighting design and research to explain the "flow of light" in architecture. Xia used first- and second-order spherical harmonics to describe the diffuseness of light[6], and discussed its relation to the vector/scalar illumination ratio. However, the potentiality of such kinds of indexes, related to the light field, has not been yet fully developed in the architectural and lighting design.

The final purpose of this research is to describe the characteristics of light field by using spherical harmonics for evaluating the light environment in architecture. Current methods to quantify the flow of light in architecture are limited to aggregated information such as cubic illumination and low-order spherical harmonics. In this study, higher-order spherical harmonics were used to describe more complex properties of the light environment. We also investigated the validity of this method by comparing it with the perceived complexity of the light field in real spaces evaluated by some observers.

2. Method

2.1. Light field measurement method

A 360-degree camera, Theta Z1, was used to capture the light field. The camera was properly calibrated in advance and set at a height of 1500 mm in four different buildings (Buildings A~D) to capture photographs and obtain 360-degree luminance images by using L-CEPT (Luminance & Color Environment Photometric Tool), which was developed in the National Institute for Land and Infrastructure Management[7]. Figure 1 shows the actual images taken by the above process, each having different light environment characteristics. Building A is a space with 360-degree continuous horizontal openings located in Osaka, Japan. Building B is in Kagawa prefecture and its entrance lobby has huge windows on its sidewall and some fluorescent tube lights on the ceiling. Building C is located in Kanagawa and has a space with small circular openings in all directions. Building D is a space with a skylight and a continuous series of small windows in the horizontal direction, located at Naoshima island in Kagawa.

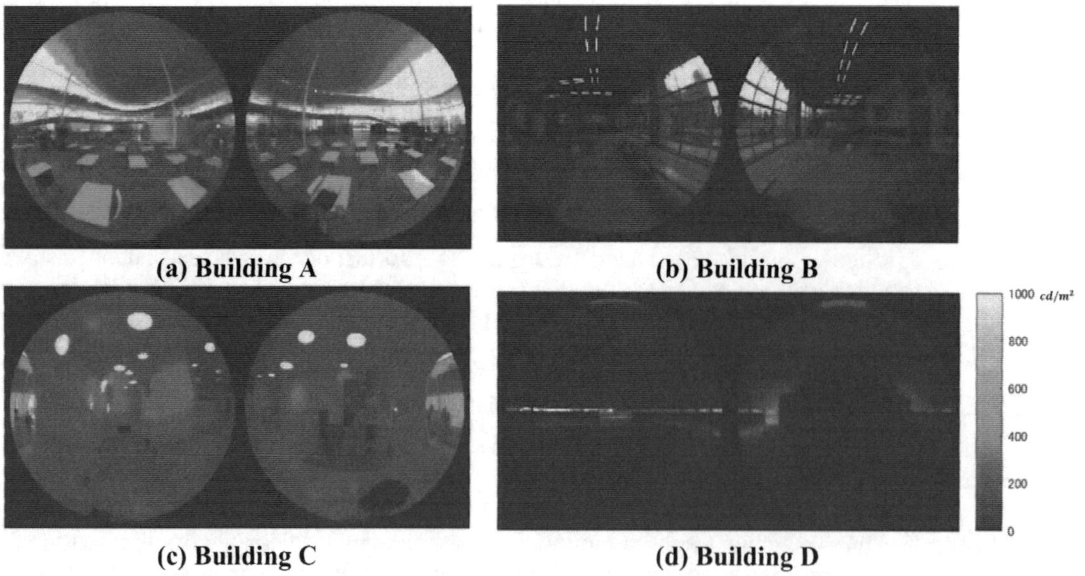

Figure 1. 360-degree luminance images measured in four buildings (Buildings A~D). These images in Building A/B/C/D were captured at 12:00 pm on October 23, at 11:00 am on December 10, at 8:40 am on Nonember 12, and at 9:20 am on December 11, respectively.

2.2. Perceptual evaluation of light field by subjects

After photographing the light field, subjects were asked to evaluate their impression of it. First, they adapted to the lighting environment for approximately 10 minutes in the architectural space. Then, they stood in the position where the luminance image was taken, rotated 360°, and observed the light field. Finally, they rated the "complexity" of the light environment on a 7-point scale: 1 is the least complex (simple light field), 4 is neutral, and 7 is the most complex. The number of subjects in Building A and C were 5 each, and in Building B and D were 7 each. Not the same subjects attended the evaluation among Building A~D.

2.3. *Spherical Harmonic function*

Spherical harmonic (SH) functions are used in physics as wave functions to draw the ideal orbit of an electron. The properties of this function have been applied to a variety of fields. Especially in the field of computer graphics, its properties are applied in describing "ambient light". In this paper, we use the properties of the SH function to capture the characteristics of the light field in architectural spaces.

2.3.1 Definition of Spherical Harmonic function. Locally, the light field is a function of all direction and thus a spherical function. Any spherical function $f(\theta,\varphi)$ can be reconstructed by the sum of its spherical harmonics.

$$f(\theta,\varphi)=\sum_{l=0}^{\infty}\sum_{m=-l}^{l} C_l^m\, y_l^m(\theta,\varphi) \tag{1}$$

The measured light field can be described by the spherical harmonic function y_l^m and the scaling factor C_l^m[8]. The SH functions in general are defined on imaginary numbers but we are only interested in approximating real functions over the sphere, therefore, in this paper we will be working only with the Real Spherical Harmonics, which we denote as SH functions. The SH function is traditionally represented by the symbol *y*:

$$y_l^m(\theta,\varphi)=\begin{cases}\sqrt{2}K_l^m\cos(m\varphi)\,P_l^m(\cos\theta) & m>0\\ \sqrt{2}K_l^m\sin(-m\varphi)\,P_l^{-m}(\cos\theta) & m<0\\ K_l^0P_l^0(\cos\theta), & m=0\end{cases} \tag{2}$$

$$K_l^m=\left[\frac{(2l+1)}{4\pi}\frac{(l-|m|)!}{(l+|m|)!}\right]^{1/2} \tag{3}$$

Where *P* is associated Legendre function and K_l^m is just a factor to normalize the functions. l is a positive integer starting from 0, but m takes signed integer values from –l to l. And the scaling factor C_l^m[8], which describes how much of the feature is in the three-dimensional direction, is calculated by the formula below:

$$C_l^m=\int_{\omega} f(\theta,\varphi)y_l^m(\theta,\varphi)d\omega \tag{4}$$

Here, $f(\theta,\varphi)$ is the luminance value of the captured data and ω is the solid angle. By examining the magnitude of the scaling factor (C_l^m), we can identify the SH function that characterizes the light field of the scene.

2.3.2 SH function visualization when plotted as spherical function. SH functions at a certain order have a fixed direction and can be visualized in a couple of ways[8] [9]. One standard way is to distort a unit sphere, by scaling each point radially by the absolute value of the function and colouring it based on the sign (red for positive, blue for negative.). A three-dimensional visualization of the SH function is shown in figure 2.

When the spherical harmonic order m is zero (in the central column of figure 2), the SH functions do not depend upon longitude, and are referred to as zonal harmonics (ZH). When $\ell = |m|$ (bottom-right in the figure), there are no zero crossings in latitude, and the functions are referred to as sectorial harmonics. For the other cases, the functions checker the sphere, and they are referred to as tesseral harmonics.

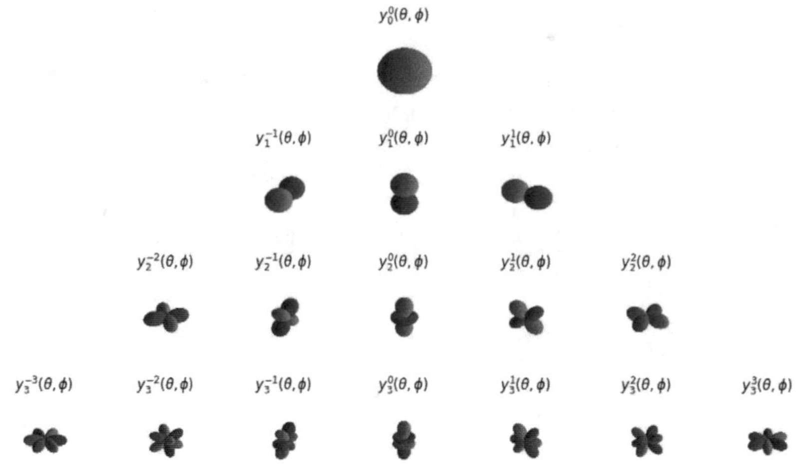

Figure 2. The first five orders l plotted by distance from the origin and by colour on a unit sphere. Red are positive values and blue are negative.

2.3.3 Drawing Approximate Images. To reconstruct the approximated function (notated by f capped with a tilde), we just take the reverse process and sum scaled copies of the corresponding SH functions[8]:

$$\tilde{f}(\theta,\varphi)=\sum_{l=0}^{n-1}\sum_{m=-l}^{l} C_l^m \, y_l^m(\theta,\varphi) \tag{5}$$

Using this formula, an approximate image was drawn for the 360-degree luminance image (light field). In this case, l=20, i.e., 441 scaling factors (C_l^m) were used to create the approximate image. Figure 3 shows the approximate image created by the above procedure, and we can confirm that the features of the light environment are well described.

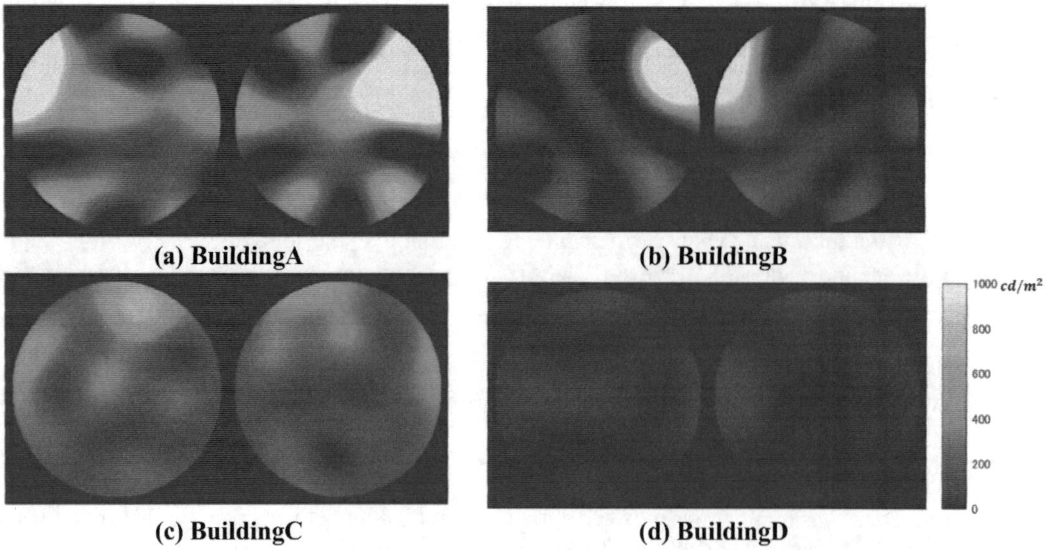

Figure 3. Approximate images drawn using SH functions in four buildings (Buildings A~D). The order was used up to l=20 (i.e., 441 scaling factors) and an approximate image was created.

3. Analysis and Result

3.1. Describing light field using Low-order-scaling factor

When an approximate image is drawn using zeroth-order scaling factor, all pixels have the same luminance value. It is the coarsest approximation, and the image is marked on the first line of figure 2. The luminance value (Li) is obtained by the multiplication of the zeroth-order scaling factor and the zeroth-order SH function (constant).

$$Li = \left(\frac{1}{\sqrt{4\pi}}\right)C_0^0 \tag{6}$$

Comparing the average value of the captured luminance data with *Li* calculated by equation (6), we see that the zeroth-order scaling factor accurately describes the average luminance per pixel of the captured data. That is, the zeroth-order scaling factor describes the brightness of the space.

Table 1. Average luminance of each pixel(cd/m^2).

	Building A	Building B	Building C	Building D
Photograph data	440.82	127.74	209.40	24.06
Approximate image	442.03	128.58	210.91	23.95

According to Mury[4][10][11], zeroth-order scaling factors correspond to scalar illuminance, and first-order scaling factors correspond to illumination vectors. By using these properties, Xia[6] says that diffuseness can be obtained from the lower-order scaling factors. The D_{Xia} calculated here varies from 0 for fully collimated light to 1 for fully diffuse light. The strength of each order l can be calculated using equation (8) [12].

$$(D_{Xia})_{Nomalized} = 1 - d(L_1)/d(L_0)/\sqrt{3} \tag{7}$$

$$d(L_l) = \left[\sum_{m=-l}^{l} (C_l^m)^2\right]^{1/2} \tag{8}$$

D_{Xia} was obtained for the four buildings photographed, with Building D, C, A, and B showing the highest light diffusion to lowest in that order.

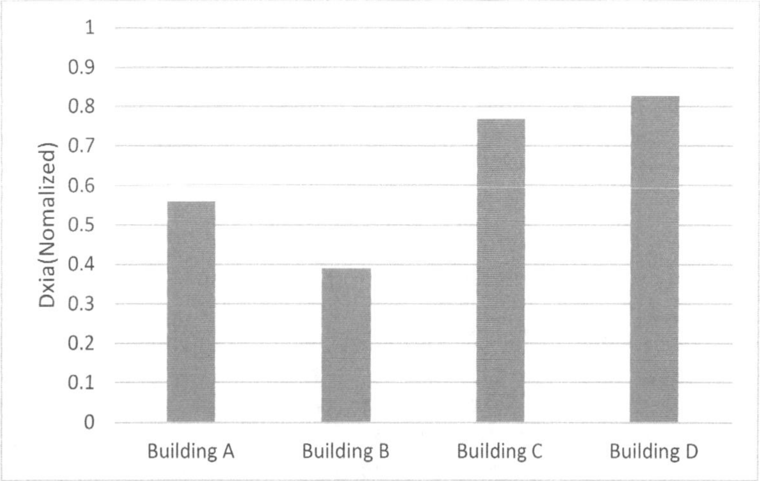

Figure 4. Diffuseness of light calculated using Dxia. Normalized to 0~1, where 0 represents perfectly parallel light and 1 represents perfectly diffuse light.

IOP Conf. Series: Earth and Environmental Science 1099 (2022) 012014 doi:10.1088/1755-1315/1099/1/012014

3.2. Describing light field using high-order-scaling factor

The more scaling factors used, the more complex the light environment can be described. The distribution of scaling factors for Buildings A~B is shown in figure 5. The values of the scaling coefficients are different for each light environment, describing features according to the space. In other words, by observing the scaling factors with large values and the SH functions corresponding to their orders, we can read the characteristic quantities of the light field. Then, in order to detect at which order (m, l) the feature quantity appears, the value of the scaling factor was normalized to 0 to 1. Normalization was done by dividing the value of each scaling factor by the value of the zeroth-order scaling factor. After that, the threshold value was set to 0.1, and the order having the features and the number of detected features were examined.

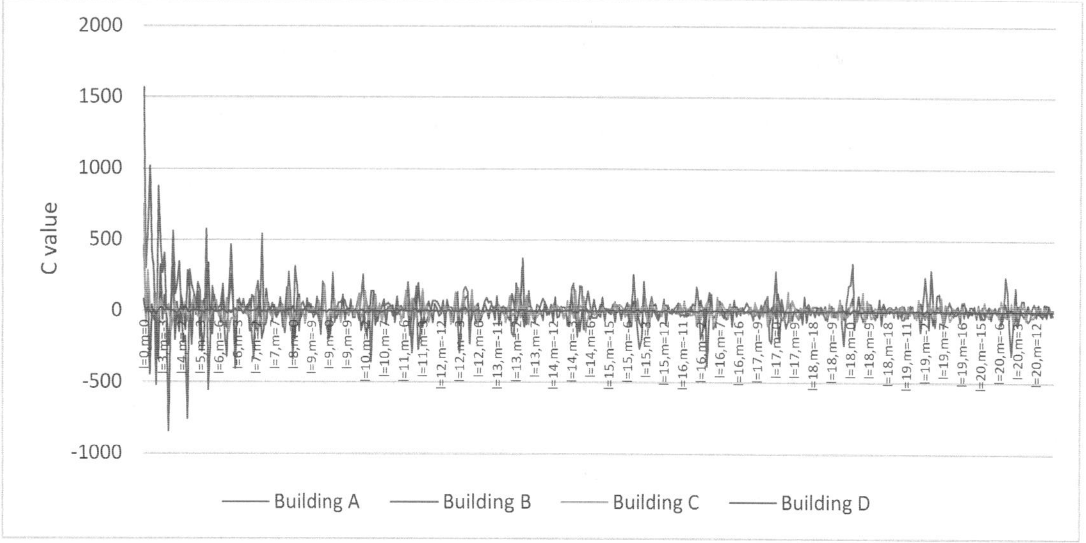

Figure 5. Scaling factor for a certain order l, m (calculated by Equation 4).

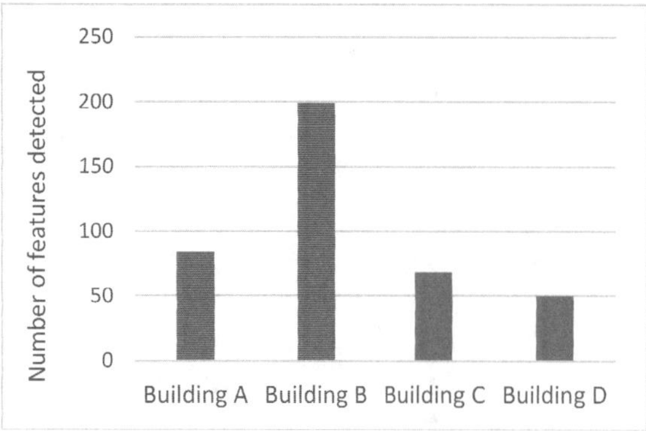

Figure 6. Complexity of light fields quantified by scaling factors.

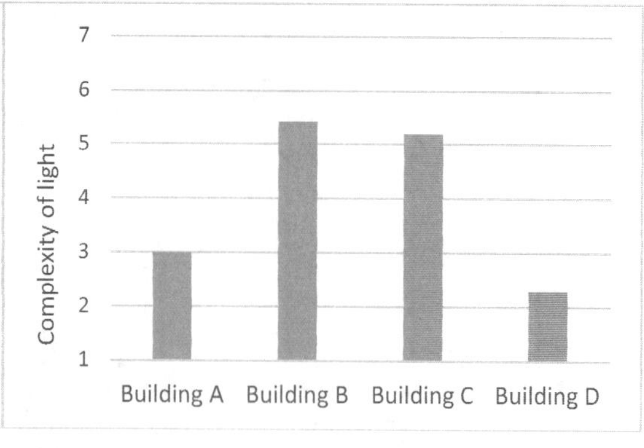

Figure 7. Complexity of the light field read by the subject.

The number of features detected in Buildings A~D is shown in figure 6. The higher the number of features, the more complex the light field. In other words, Building B has the most complex light field, and Building D has the simplest light field, which could be quantitatively described.

Next, by comparing figure 6 and figure 7, the correspondence between the quantitatively determined "complexity" and the perceived "complexity" was investigated. In both cases, the most complex space is Building B and the simplest space is Building D. In other words, there was some correspondence between the two, and there is a possibility that the "complexity" of the light field as perceived by people could be explained by the scaling factor features computed by SH function.

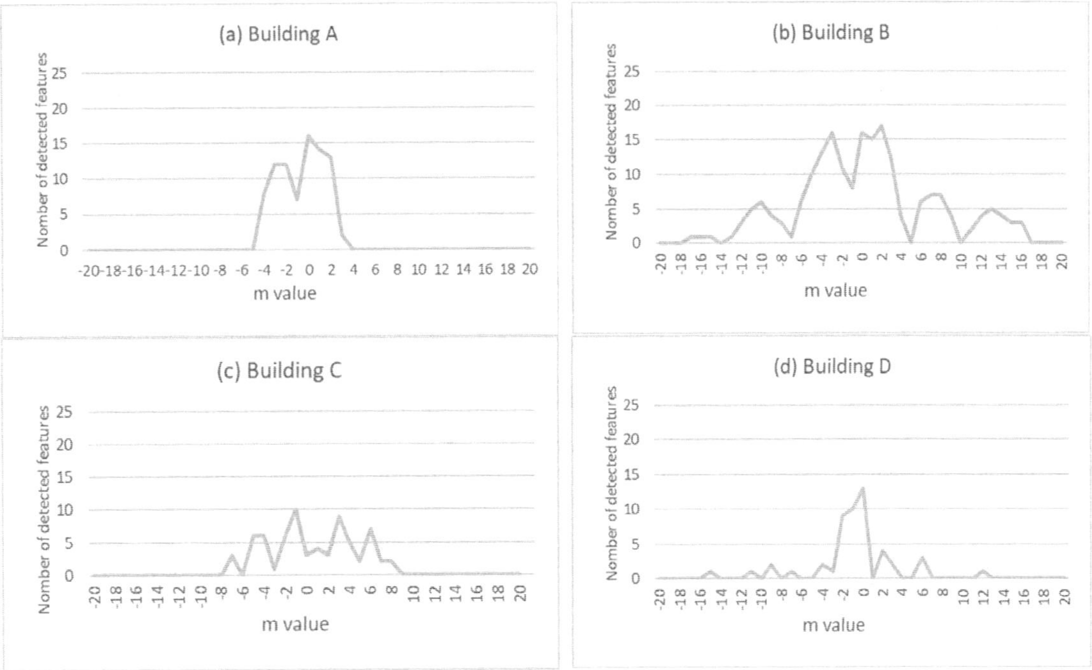

Figure 8. Number of features detected in each order m (threshold 0.1).

We will examine this in more detail. For each order m, the number of features detected was determined for each Building A~D. This allowed us to read what kind of features the light field of each building has. By observing the value of the scaling factor and the shape of the SH function corresponding to that order (figure 2), we were able to read the characteristics of their light fields. Figure 8 shows that in Building A and D strong features are detected on the order m=0, indicating that there are large light sources in the vertical direction. Building B has many features in various directions, including higher orders. In other words, it can be read as a space with a large light source in the vertical direction and bright light sources in various directions. Building C has features at m=3,6,-1,-2, indicating that it is a space with light sources in all directions.

By observing the shape of the SH function at the order corresponding to the feature values of the detected scaling factors, the feature of the light field can be read to some extent.

4. Conclusion

As the order of the scaling factor increases, it becomes available to grasp the detailed characteristics of the light field. From the value of the scaling factor with each order and the shape of the SH function corresponding to it, we can read the three-dimensional light distribution, and it is likely possible to capture the brightness of space, the complexity of the light field, and the characteristics of the luminance distribution. In this study, we introduced a new formula and method to describe the light field using SH function. In order to detect finer features of light filed, we quantified the light field using higher-order spherical harmonics and investigated the correspondence with human perception. As a result, it was found that the perceived complexity of the light field may be described by scaling factors using SH functions. In the future, the methods presented in this paper will be applied to the further analysis of light environments and the discussion of psychometric quantities. Specifically, luminance images created in advance using light simulations could be available to architects and lighting designers allowing them to quantitatively evaluate the light environment during the design phase.

References

[1] Koenderink JJ, Pont SC, van Doorn AJ, Kappers AML and Todd JT 2007 The visual light field *Percept.* **36** pp 1595–1610
[2] Gershun A 1939 The light field *J. of Math. and Phys.* **18** pp 51–151
[3] Adelson EH and Bergen JR 1991 The plenoptic function and the elements of early vision *Comput. models of visual proc.* pp 3–20
[4] Mury AA 2009 The light field in natural scenes *Delft: Faculty of Industrial Design Engineering, Delft University of Technology*
[5] Cuttle C 1971 Lighting patterns and the flow of light *Light. Res. Technol.* **3** pp 171–189
[6] Xia L, Pont SC and Heynderickx I 2017 Light diffuseness metric Part 1: Theory *Light. Res. Technol.* **49** pp 411–27
[7] Yamaguchi H, Takasugi H, Sakurai M, Kato M and Hara N 2019 DEVELOPMENT OF GENERIC COLORIMETRY SYSTEM FOR EVALUATION OF LIGHTING ENVIRONMENT *Proc. of the 29th CIE SETTION* PO154
[8] Green R 2003 Spherical Harmonic Lighting: The Gritty Details *GDC2003 Proc*
[9] Sloan PP 2008 Stupid Spherical Harmonics (SH) Tricks: *GDC2008*
[10] Mury AA, Pont SC, Koenderink JJ 2007 Light field constancy within natural scenes *Appl. Opt.* **46** pp 7308–16
[11] Mury AA, Pont SC, Koenderink JJ 2009 The structure of light fields in natural scenes *Appl. Opt.* **48** pp 5386–95
[12] Stock RD, Siegel MW 1996 Orientation invariant light source parameters *Opt. Eng.* **35** pp 2651–60

Acknowledgments

We are grateful to Osaka University of Arts for their cooperation in taking data for the analysis.

IOP Conf. Series: Earth and Environmental Science 1099 (2022) 012015 doi:10.1088/1755-1315/1099/1/012015

The minimum lux paradox

Hilde Noreld Olaisen[1], Kristin Halldis Bredal[1]
Zenisk Lighting Design, Oslo, Norway

Hilde@zenisk.no

Abstract. Through our work with urban masterplans, studying lit places and spaces in Norway for more than 20 years, and observing procedures of procurement and installations, we see some clear trends. There is a general tendency to light with the scope of augmenting visibility. Meaning that most effort is geared towards making spaces more visible with light, rather than creating good lighting environments. Lighting standards that set minimum lux level requirements should be a subject for thorough debate. They result in a never-ending increase in lighting installations and light levels. As lighting designers, we feel responsibility for the impact our work has on nature and society. Today we know that artificial light at night has both advantages and disadvantages to humans, in addition to negative impact on ecology and use of resources (3-11). Through a series of examples from the authors practice we will show how our studio is approaching sustainable lighting design. Working with lighting strategies, the focus is to give our clients tools to reduce light levels and maintain the best visual presentation. For this purpose, light distribution needs to be thoroughly planned and tested. When installations are completed, strategies to evaluate results, and discover possibilities to reduce more must be implemented.

1. Why focus on lighting that increases visibility when we can do so much more

The scope of the lighting design studio Zenisk is, and has always been, better quality of life through stimulating visual experience and environments. Through our work with urban masterplans, studying lit places and spaces for more than 20 years, and observing procedures of procurement and installations, we see some clear trends. We claim that there is a general tendency to light with the scope of increasing visibility. Meaning that most effort is geared towards making spaces more visible by making them brighter, rather than creating good lighting environments. The unfortunate effect is observed by citizens, and reported by the local press, that is buzzing with complaints about new LED light installations. We have seen headings like: “Ingrid and Bjoern felt their hometown suddenly looked like an airport or gas-station", and “Our cities are bathed in a horror-movie-like white light” (1, 2).

Research on the development of lighting worldwide confirms that light pollution is a rapidly increasing environmental hazard (3-7). Its visibility has been growing much recently, compared to natural nocturnal light from stars and the moon. We do not know the long-term consequences and impact it has. The First Atlas of Artificial Night Sky Brightness showed that more than 60% of the world population (99% of the population of USA and Europe), and almost one-fifth of the world’s terrain is under light polluted skies (8). There is reliable evidence that this artificial extension of the day produces serious adverse consequences to both human health and the environment (9-11).

1.1. Lighting procurement decisions have the potential to reduce light pollution and cut costs, while also delivering high quality lighting to customers. *However, most lighting is planned based on* regulations and guidelines that recommend **minimum** lux level and advise about **minimum** evenness. The regulations focus on safe circulation of car traffic but are lacking specific tools to analyze the surrounding light situation, and to evaluate negative effects of over-lighting, such as light pollution, waste of energy and negative consequences to living creatures. Public authorities also regard lighting as a safety measure to prevent crime, even though studies struggle to prove its relation to crime rates (12). Lighting is planned with software where it is measured in lux levels and evenness. Then held against the recommended levels in the regulations. This exercise does not consider the complexity and considerations needed to design lighting that stimulates the visual mechanism, rather the emphasis is on quantitative and measurable values. Lighting standards that set minimum lux level requirements should be a subject for thorough debate. Apparent brightness is only loosely correlated with luminance and, more importantly, the relationship between luminance and apparent brightness will change depending on context (13). This is why, lux levels, for the most part are irrelevant, and we should consider light distribution, and place-specific factors when planning lighting. When quantitative values are the only requirement to light a space, they unfortunately also remain the only tool to enhance it with light. Good-willed intentions of creating brighter and safer public spaces for people seems result in a never-ending increase in light levels, where one area or building, needs to "outshine" the adjacent to appear bright and appealing.

The authors collaboration started years back with a discussion about how lighting designers work can contribute to the sustainable development of cities. After having researched sustainable lighting design through a master thesis, held against experience drawn from two decades of Zenisk practical work with master lighting plans, Zenisk set out to further developed company goals and tools to create 0 CO2 designs. **The minimum lux paradox remains a central part of this discussion.** It grows more relevant as climate changes demand drastic changes in every aspect of our lives. A study done in 2021 estimated the midnight power emissions of the world's continents and nation states to have a cost of as much as $81.46 billion (14), The ongoing energy crisis is now sending cost through the roof, this must be a good moment to address the issue.

Through examples from studio Zenisk's practice we will show our approach to sustainable lighting design. In specific we will discuss the following questions:

A) How can we communicate lighting quality amongst stake holders in an urban lighting project?
B) 1) Which criteria can replace "minimum lux", to ensure a lit environment that stimulates vision, health, and social behavior, while limiting light pollution.
   2) How does place-specific factors influence the experience of a lit space.

2. How can we create awareness about lighting quality among stake holders in urban lighting projects?

Working with lighting strategies for city areas, the intent is to give overall control and direction for the light planning. This includes giving tools to be able to reduce light levels and maintain good visual presentation. The knowledge of the city needs to come from the people living there. We encourage the city to invite all stakeholders to a workshop where they are invited to take part in the analyses of how existing lighting is functioning for life in the city. What lighting is/is not successful and why. Stakeholders are everybody from planners to social workers and specific interest groups (special needs, environmental, heritage etc). The first action is to give a basic understanding about how light works and what it does. To put everyone on the same track we have found that we need to explain the act of seeing/perceiving. How the human eye reacts with light and the fact that we all have our own automatic dimming function in the iris. We do it through a quick lecture called "10 things you need to know about seeing". But before this lecture we inspire the workshop attendees by showing examples of well-lit spaces and share experience from other places that successfully have changed glare to glow. This warms up the crowd before the actual workshop begins. Participants are divided into mixed groups with a big map on the table. We are often met with the best response when addressing the potential in the project. Addressing how money and resources can be saved by investing in thorough planning and user-oriented design is imperative. When stakeholders are on board, they get a basic understanding of what light can do, they give us as designers understanding of the specific qualities and challenges of the place. Successful user participation has been made and can continue. For this purpose, there are several tools that can be used, and adapted, depending on project and client type.

2.1. Tools for user participation in lighting design and how they can be used to advocate for lowering light levels.

A collaborative workshop **where stakeholders are divided in groups** can help the designer understand the place in a social context: Use, habits, history, and local pride. For the stakeholders it raises the awareness of use of the city during dark hours. Positive and negative sides is addressed with a constructive and schematic approach. When simple question is asked, the discussion brings out new information and different views. The designer can pinpoint unique qualities, thus boosting the owner and user's pride, and seeks solutions to target their challenges. We learned that asking too complicated questions is counterproductive, because people get scared to answer the "wrong" thing and often rather stay silent. Simple exercises like:

1. Where do people feel safe/unsafe at day/night? Why?
2. What are night-time highlights? How does it look now? And how would we like it to look? does the evening presentation influence use?
3. How does circulation flow? Which factors influence circulation? Does lighting play a role. Does it work as intended?

This type of questions lays the basis for stakeholders to start elaborating history and vision for the place in a social and practical context. We merge in to discussing "soft and hard qualities". Soft qualities being the identity through narrative and the content that creates history, status, experience and so on. The hard qualities are the physical structures such as buildings, streets, bridges, plazas, and other structural infrastructure. We refer to these qualities when discussing how lighting and light levels can influence the appearance, the atmosphere, and the use of a place from a broader perspective, that includes lowering levels and tidying up. **How** you see is what you get.

The **Night walk with site analysis** is the conclusion of the workshop and a must for the designer to see the existing lighting of the place and compare with the impression of the place at day. At the same time, you get an overview of the relationship stakeholders have with the place and with existing lighting installations. For the stakeholders it is always a real eye opener. The goal is to have them really looking at the space - to explain and help them understand how light reveals the space: what do we see and why. Which areas are perceived as bright and pleasant? Which are perceived as dark. How does lux level, brightness, contrast, and glare influence the perception? Visual exercise include:

1. Looking at long and near views
2. Light distribution, brightness, and visibility
3. Glare – experienced glare
4. Volumes and framing
5. Colours and physical contrast
6. Layers of light and light distribution (road, facade, commercial, active first floors…)

Once the night walk is finalized **Night Photography** is an effective way of documenting and remembering what we saw. The photographs can also be used to make a comparison of the **Night experience** with the d**ay experience.** When seeing back-to-back which character and impression a space has during the day versus the night it is easy to identify what can be improved in the latter.

Image 1 and 2. Day and night photography of Vågen in Stavanger

Illustrative **Visualizations (paired with text)** are great tools to show how a place "could look" at night. They are used to test and show lighting design concept and ideas, and they help to inspire and encourage stakeholders to invest in them. The visualization is used as a representation of how the space can be perceived with a specific lighting. The emphasis is on representation of light rather than creating an image that is photo-realistic.

Image 3. Visualisation of Vågen in Stavanger with controlled and dimmed façade light.

The **Test lighting** is often the final verification of the lighting design. It can also be used in an earlier stage to test an idea, show a concept in real life, and to test suited lighting equipment. This is the perfect tool for testing light distribution, discover potential for lowering levels and verifying how they influence the impression of a space. Glare control is an important part of this exercise. We push to test low levels; to show that with a well-planned light distribution the space is nonetheless easy to orientate and read. When it possible to control and dim down fixtures in a larger area, an interesting exercise is to dim down or turn of light in adjacent areas, and show how this influences the impression of a selected area.

Very few studies address architectural and urban projects as contexts to study end-user participation (15). *In general, Active user participation in the design process with direct involvement of the user has been seen as the best level of user participation* (16). Seeing is believing, says the proverb, and in our experience seeing the lighting environment as a whole is the only effective way to verify how different design-concepts can influence it. Though, "seeing is also deceiving", as of all our senses, the eyes are the most easily deceived" (17). "A guided verification process" with an experienced lighting designer that explains how we see, gives a broader understanding about additional qualities of light, not only the lux requirement.

3. Which criteria can replace "minimum lux", to ensure a lit environment with positive impact on health and social behaviour, while limiting light pollution?

With sustainability as a scope, we can say that we are designing to find the "best lit effect" and while using less resources. To understand which design-tools support which effort towards sustainable lighting we have categorized our approach in four main categories: 1) Impact on health and social behaviour, 2) Light Pollution 3) Circularity of lighting equipment, 4) Energy consumption. Below the schematic overview of design-tools and sustainability impact in the key design stages for points 1 and 2 are shown.

Table 1. Design approach and sustainability impact in different design stages. 1) Impact on health and social behaviour

Design stage:	Analysis	Design	Verification	Installation	Use & Post occupancy
Sustainability impact					
Visual system	1)Mapping of light levels in zones, related to use, and visual importance 2)Evaluation of surfaces and reflectance in area.	Creating hierarchy of light levels related to use, and visual importance. Determining optimal light distribution, for visual impact.	Evaluation of fixtures: type, quantity, output and glare (lighting calc and test lighting). Verification of visual output.	Focusing installed lights and adjusting control system	Post occupancy study: evaluating project specific sucess criteria with clients. Reduce light levels – evaluate – repeat.
Circadian system	1)Mapping of brightness and cct in zones, related to circadian effects.	Determining optimal light intensity and wavelength for circdian entrainment.	Calculate circadian entrainment. Melonipic EDI	Focusing installed lights and adjusting control system	Post occupancy study: Questionnaire on users alertness, concentration and sleep. Reduce light levels – evaluate – repeat.
Mood and motivation	Identifying place specific behaviour, culture, history and use.	Creating lighting designs that enhance place specific qualities.	User+stakeholder involvement. Presentation and evaluation of lighting concept(s).	User+stakeholder involvement. Focusing installed lights. Look at several options and decide together.	Post occupancy study: evaluating project specific sucess criteria with clients. Reduce light levels – evaluate – repeat.

Table 2. Design approach and sustainability impact in different design stages. 2) Impact on Light pollution

Design stage:	Analysis	Design	Verification	Installation	Use & Post occupancy
Sustainability impact					
Providing darkness (for all living beings that need it to rest)	1)Mapping of light zones, related to use, and time of use. 2)Evaluation of possible "dark zones"	Creating hierarchy of light zones related to use. Minimize light according to need. Advocate for surfaces with high reflectance	Test lighting with evaluation of needed light levels for visual orientation and feeling of safety. Try lighting different surfaces and reducing levels.	Focusing installed lights and adjusting control system.	Post occupancy study: evaluate hours of darkness, and % of dimming achieved. Reduce light levels – evaluate – repeat.
Minimize spill light and skyglow	1)Evaluation of spill light, glare and possible preventive measures.	Shield Light sources. Direct light only where it is needed. Reduce light levels to a minimum.	Lighting calc and test lighting with evaluation of spill light, glare and light distribution. Try lighting different surfaces and reducing levels.	Focusing installed lights, and adjusting control system	Post occupancy study: Evaluate % of reflected light that has been reduced. Reduce light levels again – evaluate – repeat.
Minimize impact on species in area	1)Mapping of vulnerable species in zones. Measuring current brightness and cct.	Adapting and reducing, light hours, light levels, light distribution, and spectral distribution. Specific	Check spectral distribution with spectrometre.	Focusing installed lights, and adjusting control system	Collaboration with biologists to follow development of vulnerable species. Reduce light levels again – evaluate – repeat.

		preventive measures for vulnerable species.			

In the design stage in each category, there are basic questions about the character of the lighting that needs to be answered to meet the sustainability goal.

1) What to light, and with which light distribution?
2) What should the minimum and maximum luminance be?
3) Which extent in space should the lighting have?
4) When should the light be turned on, and off?
5) Which light-intensity is proper at the different times the light is turned on?
6) Which CCT and spectral power distribution should be used, and which should be avoided?

Providing answers to these questions is needed to ensure "the best lit effect" and adhering to the sustainability criteria. Another point could be added to make it explicit:

7) What, where and when to NOT light.

This is described indirectly in points 1-6 but should be made crystal clear to insure the limitation of light pollution.

Image 4. Example of how reductive measures can be set for lighting characteristics of urban architecture in a city centre.

Tabell 1. Eksempel på begrensninger som kan settes til belysning for en plass i et bysentrum

Kilde: Zenisk 2021

SONE: Bysentrum ELEMENT	REDUSERE LYS			REDUSERE TID FOR BRUK			TILPASSE SPEKTRALFORDELING	
	MAX TILLAT: LYSNIVÅ (cd/m2) LYSFLUX (lm)	TILLAT OMFANG	TILATT LYS MOT HIMMELEN UWLR MAX %	NED DIMMING	NAT SENKING	SLUKKING	FARGE TEMP	KRITISKE BØLGELENDER
HISTORISKE BYGG	Fasade luminans (Lb)* XXX **cd/m2**	Fasadebelysning: Kun 1. og 2. etg	UWLR X %	**Min 30%** Hverdag: 18.00 - 22.00 Helg/ Høytid: 22.00 -24.00	Hverdag min 90% 22.00 - 06.00 Helg/ Høytid min 70% 00.00 - 08.00	Hekking gås 15 mars - 15 april 19.00 - 07.00	2200K	365 - 565 nm 15 mars - 15 april 19.00 - 07.00 **NB: Skjerme interiørlys fra vinduer**
MODERNE BYGG	Fasade luminans (Lb)* XXX **cd/m2**	Fasadebelysning: Kun 1. etg	UWLR X %	**Min 30%** Hverdag: 18.00 - 22.00 Helg/ Høytid: 22.00 -24.00	Hverdag min 90% 22.00 - 06.00 Helg/ Høytid min 70% 00.00 - 08.00	Hekking gås 15 mars - 15 april 19.00 - 07.00	2500K -2700K	365 - 565 nm 15 mars - 15 april 19.00 - 07.00 **NB: Skjerme interiørlys fra vinduer**
LANDEMERKER	Fasade luminans (Lb)* XXX **cd/m2**	Fasadebelysning: Opp til 50 m	UWLR X %	Tidsperiode % dim xx klokkeslett	Tidsperiode % dim xx klokkeslett	Sårbar art og Tid for slukking xx	XX K	XX - XX nm XX dato - xx dato xx tid - xx tid
TRÆR	Lysflux fra armatur XXX **lm**	Ikke permanent belysning. Sesongbelysning: Desember - Mars	UWLR X %	Tidsperiode % dim xx klokkeslett	Tidsperiode % dim xx klokkeslett	Sårbar art og Tid for slukking xx	XX K	XX - XX nm XX dato - xx dato xx tid - xx tid
LYSSKILT	Skilt luminans (Ls)* XXX **cd/m2**	Kun 1. etg	UWLR X %	Tidsperiode % dim xx klokkeslett	Tidsperiode % dim xx klokkeslett	Sårbar art og Tid for slukking xx	XX K	XX - XX nm XX dato - xx dato xx tid - xx tid

If we attempt to design to limit a sustainability impact while using only the minimum lux requirement as criteria, it becomes clear how insufficient it is. Try asking: "How can we minimize the impact of light on species in the area by installing minimum 15 lux?" or "How can we support mood and motivation with lighting by installing minimum 30 lux? The answer is simple, you can`t! Questions about the character of lighting should be considered as criteria, rather than the minimum lux requirement. As research leads us to discover ever new connections between the properties of light and how human and nature responds to these. The criteria and design approach must be updated.

4. How does place-specific factors influence the experience of a lit space.

To determine which character light shall have to enhance a space, a thorough analysis of all relevant place-specific variables should be performed. As before mentioned the relationship between luminance and apparent brightness will change depending on context (13). Light levels are measured using the photopic luminous efficiency function, V(λ). It converts energy measured in irradiance/radiance into light level measurements in illuminance/luminance. Brightness perception however is nonlinear. Brightness responses might be influenced by a wide variety of parameters, including the presence of sparkle, the colour gamut of the illuminating source, and the reflectance characteristics of objects in the scene being evaluated (18). This means that in order to control the brightness response in a scene, a complete overview of the situation to light is important. The lighting designer should consider the qualities of urban architecture: surfaces, colour schemes and related reflectance, as well as the existing lighting in a space. In addition, urban spaces are dynamic, they change, almost continuously, during the day (and dark hours), with seasons, with the weather and with time. The uses of a space changes with these variables and will require different lighting scenarios. It is suggested to take these factors into consideration to provide "the best lit effect", that abides for different uses and users, at a specific place and time. For example, lighting a vertical façade that provides spatial relation to the surrounding environment, can be a better approach in creating a well-lit environment than lighting a dark street. Light pointed down on the street have less luminance at the eye than a vertical surface, in addition the street is often dark. In the Norwegian context horizontal lighting differs much from a dry evening to a wet evening. It will need a significant amount of light to appear bright on a dark rainy evening.

Image 5. Example of a visualization showing low façade lighting and reduced street lighting.

In the conference presentation, 5-10 case studies will be shown, where user involvement and analysis of place specific factors have been central to design with low light levels while enhancing local identity and pride.

Conclusion

To verify how light can influence a space, seeing the lighting environment as a whole, is the only effective way. A guided verification process through user-participation, with a lighting designer that explains how and what we see creates understanding for the complexity of lighting. This is fundamental to comprehend why the minimum lux requirement is insufficient as a lighting criteria.

To control the brightness perception of a space, and reduce light pollution, a description of the character of light must be addressed in the lighting design. An analysis of place specific factors is necessary to determine the best suited character of light for each space, time, and use. This impacts social sustainability.

An interesting input in the debate about light levels is made by Steve Fotios and his research. He argues light levels generally are too high, because of the design of lighting research. In a lecture he explains that when respondents to light tests are asked about their preferred light level, the majority tend to choose the level in the middle of the scale. Regardless of where the scale starts or ends (19). This supports the authors hypothesis that evaluating results before and after installations are completed can unveil possibilities to reduce more. Evaluation tools must be implemented to find the maximum level of light that is really needed to stimulate vision, health, and social behaviour: **See, evaluate, inform about changes, set adjusted levels, collect feedback, evaluate feedback, adjust levels – repeat!**

References

[1] (https://www.morgenbladet.no/aktuelt/reportasje/2020/11/20/vindmoller-skaper-rabalder-men-hva-skjer-nar-kalde-hvite-ledlys-invaderer-norge/)
[2] https://www.morgenbladet.no/ideer/debatt/2020/02/07/led-er-kommet-for-a-bli/
[3] Garrett, J.K.; Donald, P.F.; Gaston, K.J. (2019) Skyglow extends into the world's Key Biodiversity Areas. Anim. Conserv. 23, 153–159.
[4] Grubisic, M.; Haim, A.; Bhusal, P.; Dominoni, D.M.; Gabriel, K.M.A.; Jechow, A.; Kupprat, F.; Lerner, A.; Marchant, P.; Riley, W.; et al.(2019) Light Pollution, Circadian Photoreception, and Melatonin in Vertebrates. Sustainability, 11, 6400.
[5] Owens, A.C.S.; Cochard, P.; Durrant, J.; Farnworth, B.; Perkin, E.K.; Seymoure, B. (2019) Light Pollution is a Driver of Insect Declines. Biol. Conserv. 241, 108259.
[6] Supronowicz, R.; Fryc, (2019) Urban park lighting as a source of botanical light pollution. Photon-Lett. Pol. 11, 90–92.
[7] Falcón, J.; Torriglia, A.; Attia, D.; Viénot, F.; Gronfier, C.; Behar-Cohen, F.; Martinsons, C.; Hicks, D. (2020) Exposure to Artificial Light at Night and the Consequences for Flora, Fauna, and Ecosystems. Front. Neurosci. 14, 1183.

[8] CINZANO, P., FALCHI, F. & ELVIDGE, C. D. 2001a. The first World Atlas of the artificial night sky brightness. Monthly Notices of the Royal Astronomical Society,328,689-707.
[9] KRAUS, L. J. (2016) Human and Environmental Effects of Light Emitting Diode (LED)Community Lighting.
[10] Cho, Y.; Ryu, S.-H.; Lee, B.R.; Kim, K.H.; Lee, E.; Choi, J.(2015) Effects of artificial light at night on human health: A literature review of observational and experimental studies applied to exposure assessment. Chronobiol. Int. 32, 1294–1310.
[11] NAVARA, K. J. & NELSON, R. J. 2007. The dark side of light at night: physiological, epidemiological, and ecological consequences. J Pineal Res, 43, 215-24.
[12] MARCHANT, P. 2004. A Demonstration That the Claim That Brighter Lighting Reduces Crime Is Unfounded. British Journal of Criminology,44.
[13] Toward a model of outdoor lighting scene brightness MS Rea PhD, LC Radetsky BS and JD Bullough PhD Lighting Research Center, Rensselaer Polytechnic Institute, 21 Union St., Troy, NY 12180, USA)
[14] First Estimation of Global Trends in Nocturnal PowerEmissions Reveals Acceleration of Light PollutionAlejandro Sánchez de Miguel 1,2,3,*,† , Jonathan Bennie 4,† , Emma Rosenfeld 1, Simon Dzurjak 1and Kevin J. Gaston 1
[15] Between organisation and architecture: end-user participation in design, Marianne Stang Våland
[16] Wulz,1986;Yanki, 2008; Jusan,2010a
[17] Hare, Guesses at Truth, ca.1848
[18] Spectrum and Scene Brightness Perception U. C. Besenecker1, & J. D. Bullough1
[19] Fotios, Lecture Oslo, 20 nov. 2019

IOP Conf. Series: Earth and Environmental Science 1099 (2022) 012016 doi:10.1088/1755-1315/1099/1/012016

Urban lighting: an innovative solution, eco-sustainable and resistant to corrosion

A Deodati[1], E Petrachi[1], G Vendramin[1], L Cosma[2], P Bene[2]

[1]Niteko Illuminazione, Montemesola, Italy
[2]CETMA (Technologies Design and Materials European Research Centre), Brindisi, Italy

alessandro.deodati@niteko.com

Abstract. Salt fog and atmospheric agents are two of the main causes of the corrosion affecting urban lighting luminaires. This research starts right from the study of this phenomenon. After a short analysis of the corrosion process occurring in traditional aluminum lighting bodies, it highlights the advantages of using an eco-sustainable polymeric material to produce highly resistant luminaires, a perfect example of the application of the circular economy model on products. The study focuses on the choice of green materials and low environmental impact production techniques, which lead to the production of efficient and durable urban lighting luminaires. Besides being eco-sustainable and recyclable, these materials make it possible to obtain highly customized designs, which can be molded in any geometric shape thanks to the production technique. Moreover, remote control devices can be fully integrated within the lighting fixtures, so they do not affect products' aesthetics adversely. In support of this thesis, the research shows examples of installations where the design of these innovative luminaires met both aesthetic and functional needs of Municipalities and it focuses on the benefits deriving from the combination of green materials and LED technology, which guarantees innovative luminaires in the public lighting sector, completely environment-friendly and future-oriented.

1. Introduction

Technological development has changed the way of experiencing everyday life over the years. It detached men from nature, which has now become a simple observer of its work. However, the approach to technology is changing, and market needs are increasingly oriented towards green and sustainable solutions in any field of application.

The use of eco-sustainable materials, i.e., the bio-based or recovered polyethylene, is at the heart of this research, whose innovation lies in a patent pending manufacturing technique. These materials are resistant to both corrosive and polluting agents, so they make it possible to produce sustainable and circular urban lighting luminaires, totally immune to corrosion.

The aim of this research is to produce durable and resistant lighting bodies – with a highly customizable design – by combining the virtuosity of circular economy with efficient and made in Italy luminaires.

2. Methods

The corrosion process of metals is slow, yet unstoppable. The extent of corrosion depends on the physical properties of the liquid film formed on the surface of those metal structures exposed to atmosphere, and on factors such as temperature, relative humidity and composition, which frequently affect the corrosion process in a complex way and with antithetical effects.

Metals exposed to marine and industrial atmospheres do not corrode to negligible speeds like those exposed to moisture-free atmospheres. The strong contamination of chlorides and anthropogenic pollutants is the reason why metal bodies installed in these areas corrode rapidly.

Therefore, the ISO 9223 standard divides the environments into different corrosion categories, according to the following parameters: temperature, relative humidity, deposition of sulfur dioxide and sodium chloride deposition.

When exposed to aggressive environmental conditions – such as coastal or pollutants-rich environments – luminaires' frames are subject to corrosion under organic coating phenomena (delamination, blistering). This causes structural embrittlement and a change in their aspect, involving replacement costs and worsening of the related environmental footprint.

The study was conducted in cooperation with an external research organization. It validated and optimized the methods of use of a specific eco-sustainable polymeric material to produce urban lighting luminaires, thus immune to the phenomenon of corrosion.

This is a plant 100% recyclable polymer with both the same molecular structure and technological characteristics as the fossil origin one. It is obtained by polyolefin blends containing recyclable bio-based polyethylene, which is made up of ethylene from renewable origin and obtained from the processing of sugar cane. However, the production cycle has advantages in terms of environmental impact, since it has lower carbon footprint.

According to the carbon footprint index – which is expressed in equivalent tons of carbon dioxide (CO_2) – it generates lower amounts of greenhouse gases during all its manufacturing processes.

Therefore, bio-based PE can substitute its petrochemical equivalent without problems – which is a great advantage – since it is produced from a renewable, non-fossil, and eco-sustainable resource.

2.1 Corrosion problems in public lighting: comparison between metals and polymers

Lampposts, luminaires and lighting fixtures are clearly very exposed to atmosphere, and they need to keep unchanged their aspect for decades to resist the attack caused by corrosion mechanism.

Corrosion is a natural and irreversible process, which is present everywhere. It implies a slow and continuous consumption of the material initially involved, where – as a consequence – there is a worsening of its physical properties and characteristics. To remedy this situation, it is important to use at source a correct and effective corrosion protection, as it helps to save both money and resources in the long term, even if it requires a higher advance expense.

2.1.1. When and why metal-corrosion occurs. It is possible to synthetize the corrosion phenomenon of metallic materials in the following reaction: metal + aggressive agents = corrosion products.

Therefore, corrosion phenomenon occurs when there is the simultaneous presence of a conductive metal, an electrolyte (you only need a slight humidity of the surface) and oxygen to generate the chemical reaction.

The metallic material intervenes in the corrosion process through its chemical composition, that is:

- By its nature (in case of pure metals) or by both its nature and relation with its other constituents (in case of alloys)
- By its structure, where this term is to be understood in a very broad sense, because it includes different properties such as geometric characteristics, nature, morphology and distribution (and hence position, shape and extension) of the phases that constitute it; but also type and consistency of the existing reticular defects, surface conditions, microstructural characteristics of the grain boundary, effects of mechanical stresses on the material itself, temperature, etc.

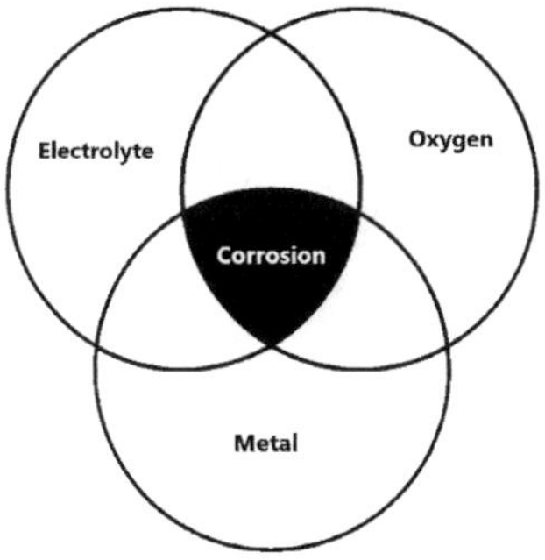

Figure 1. Conditions for the atmospheric corrosive reaction.

To solve the problem from its root, many industrial sectors are choosing more and more often to use polymers which, by their nature, are not subject to the corrosion phenomenon.

Polymer materials were introduced with the coming of nanotechnologies. They are significantly less expensive than metals processed with machine tools, but equally reliable, rigid and resistant in many applications. Recently, they were even used instead of stainless steel to produce heart surgery instruments.

2.1.2. Final considerations. When analyzing painted aluminum lighting fixtures for urban lighting, the corrosion phenomenon of main interest is represented by corrosion under organic coating (delamination, blistering) – of which filiform corrosion can be considered its particular case. It is evident that, in such circumstances, maximum resistance to corrosion can be guaranteed by those organic coatings able to both develop the highest barrier protection against aggressive agents (water, oxygen, specific ions) and guarantee an efficient adherence to metallic substrates even in strong humidity conditions.

In fact, in case of organic coatings (especially paints and varnishes) the main problem is related to their intrinsic porosity: in the long term, both atmosphere oxygen and humidity will penetrate by diffusion and will affect the metal/coating interface of the metallic mass. The resulting effect is the typical blistering on the surface of the painted layer: oxygen and atmospheric humidity cause the deterioration of the metal alloy, which means development of corrosion products with high specific volume at the interface, such as to create the swelling of the paint layer. Therefore, by using a same type of paint or varnish, these coatings will give an adequate protection in relation to the thickness used for protecting the material: the greater the thickness of the paint, the longer it takes for the atmospheric aggressive agents to spread within the metal.

In addition, close attention must be paid to the assembly phases of the different metallic components, in order to minimize the possible formation of cracks in correspondence of the joints. This is in order to avoid crevice corrosion phenomena.

It is thus possible to slow down the corrosion process. However, with the use of polymers, such as biopolyethylene and recovered polyethylene, it is possible to eliminate the problem at the root, because these materials are naturally immune to corrosion.

2.2. Recovered materials

According to the study, it is thus possible to produce urban lighting luminaires by using biopolyethylene, and also polyethylene from clean and homogeneous production waste and selected and post-consumer polyethylene.

The resulting luminaires are completely circular, since they are sustainable during all the manufacturing stage. Moreover, at the end of their service life, they can be recycled through mechanical shredding and be reused to produce new luminaires.

The use of a "secondary raw material" – so called to underline the fact that technical and chemical characteristics of the recycled material are very similar to the initial ones – allows minimizing waste, so respecting the sustainability concept of circular economy. In addition, it reduces the use – and therefore the consumption – of fossil resources.

In this way, it contributes to the reduction of greenhouse gas emission – particularly CO_2 emissions – and offers a further possibility to dispose of the product, thus facilitating the reduction of the volume of waste.

Figure 2. Example of a luminaire made of biopolyethylene.

Figure 3. Example of a luminaire made of post-consumer shredded polyethylene.

Figure 4. Example of luminaires made of polyethylene deriving from clean and homogeneous shredded production waste.

Both luminaires made of plant based and recovered polyethylene keep the same mechanical properties. The only difference is their final look. In case of recovered materials, its surface is "floral". On the contrary, it is homogeneous in case of biopolyethylene materials.

Luminaires made of the selected materials are immune to corrosion, since they have no metal inside, which is one of the main causes for the growth of the corrosion process; in addition, they are impact resistant, because the geometry of the lighting fixtures – combined with thicknesses, mechanical properties and a low-pressure production process – give the luminaires the highest degree of protection against impacts (called IK), according to CEI EN 62262 (CEI 70-4) standard.

2.3. Sustainability of plastics as raw materials

In Europe 99% of virgin plastics are produced from the raw materials oil and natural gas and fossil fuels are also used to generate the heat needed during the production process.

This results is the emission of approximately 1.2 kg of CO_2 into the atmosphere for every kg of plastic produced during the production stage alone. If we also include CO_2 emissions generated by the extraction and refining of the fossil fuels, then the production of 1 kg of plastic results in total direct emissions of CO_2 of 1.7 kg (figure 5).

Mechanical recycling of plastics generates no direct emissions, while the indirect emissions are around 0.5 kg of CO_2 per kg of plastic. These indirect emissions can be reduced by using renewable sources for the power generation. The emissions generated by bioplastics depend on the raw materials used in their production. If raw materials of fossil origin are used then the emissions from bioplastics are the same as those for ordinary plastics. If they are of plant origin, then a negative emission factor is estimated of about 1 kg of CO_2 per kg of bioplastics removed from the atmosphere during the growth of the biomass.

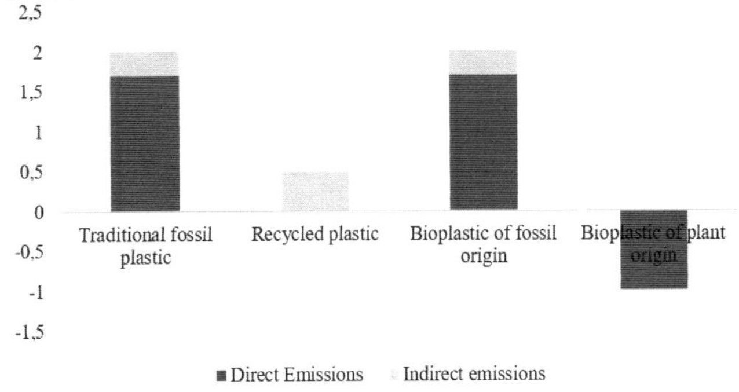

Figure 5. Emission factors of virgin fossil plastics, recycled plastics, fossil origin bioplastics and bio-based plastics. In literature, there are no data about indirect CO_2 emissions related to bio-based plastics.

2.4. *The adopted calculation hypothesis*

The aforementioned study made it possible to develop a parametric model, allowing the optimization and validation of thicknesses, geometries, and thermo-mechanical performances of these polymer luminaires by means of both thermo-fluid dynamic and structural numerical analysis.

In particular, in order to analyze dissipative behavior of both luminaire and its subcomponents, a thermo-fluid dynamic calculation model based on the finite volume method was developed.

The thermo-fluid dynamic calculation model developed by this study allows solving the Navier-Stokes equations expressing mass, momentum and energy conservation for fluids in differential form.

Concerning the luminaire system, the fluid has been represented by the air in its inside – which has a key role in heat dissipation – as it was also confirmed by the analysis performed. Instead, only the energy conservation equation was solved within the volume occupied by the solid components.

Moreover, thermo-fluid dynamic equations were coupled with a model for the calculation of the heat transmitted by radiation between the internal surfaces. A turbulence model was not included, because the analysis confirmed a low Reynolds number, and therefore the existence of an exclusively laminar fluid motion (figures 6 and 7).

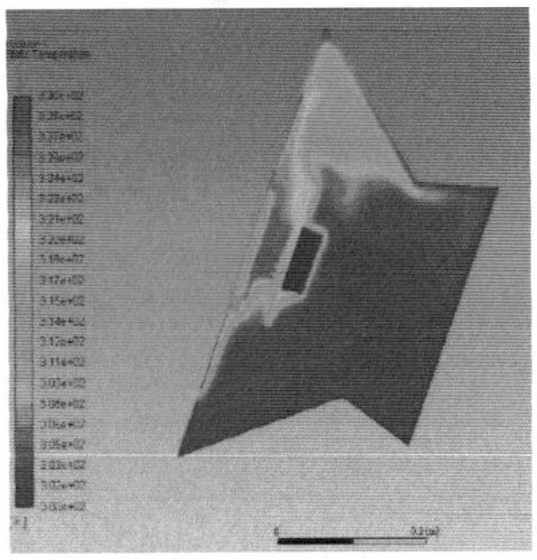

Figure 6. Calibration model: temperature contour (K).

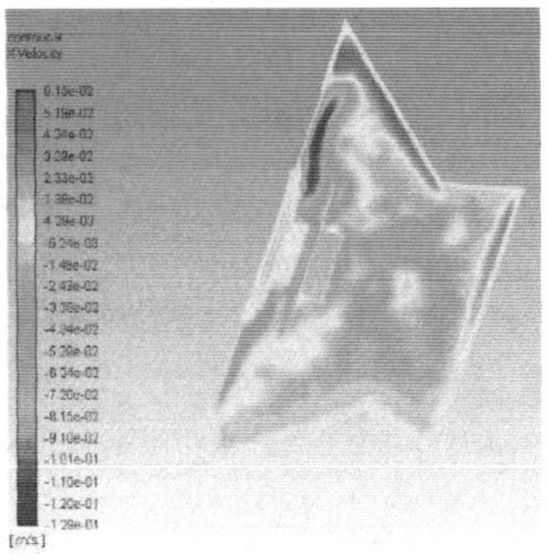

Figure 7. Calibration model: vertical speed contour (m/s).

Luminaire's thermal behavior was studied starting from the calibrated thermo-fluid dynamic model. Then, the curves representing the minimum volume of air that a PE lighting body should enclose were drawn. These minimum volume of air is based on the power rating of the luminaire, so that the temperature at LED's solder-point could be lower than the acceptable limit.

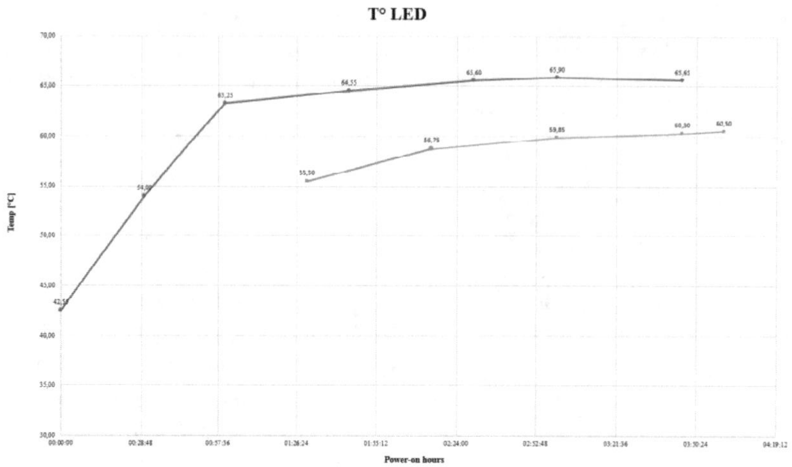

Figure 8. Ts thermal test on LEDs (top line 70W, lower line 60W).

These thermal tests defined the LED's maximum power usable within the polymeric lighting bodies, in order to guarantee a LED's (L90B10) service life longer than 100.000 hours even in environments with temperatures up to 45 °C.

Table 1. Projected L90.

	If = 50mA	If = 100mA
Ts = 55°C	-1.494.348	253.444
Ts = 85°C	**200.452**	**115.531**
Ts = 105° C	65.454	60.573

2.5. The rotational molding technique

Luminaire's geometry was optimized from a structural point of view according to the results of thermo-fluid dynamic analysis. The static analysis made it possible to develop luminaires models meeting the resistance checks with a wide margin of safety, and it allowed the optimization of frames' thickness while respecting their geometries.

The selected materials are suitable for rotational molding – also known as rotomolding – a low pressure and high temperature plastics processing technology, based on the adhesion of plastic materials to rotating hollow molds.

This technological process allows realizing hollow objects with constant thicknesses and without welding spots. It makes also possible to produce large-dimensions products with very complicated contours, which are difficult to obtain with other automated techniques such as injection molding or thermoforming.

This technique is particularly advantageous compared to the other techniques, not only because its process is extremely ductile, but also because its molds have limited costs.

Thanks to rotomolding, it is possible to produce recessed objects with both complicated outlines and well-defined surfaces. It is also possible to mold identical or similar products simultaneously, or different sections of a part in different colors on a single arm. It is often possible to put in plastic or metal inserts as integral parts of the products, and to produce double-layered objects, thus providing the designer with the maximum freedom when designing the workpiece.

The machines consist of one to four carrier molds, rotating as planets do. When heated, they rotate at low speed along two axes, thus allowing the material to adhere homogeneously to the walls of the mold.

Rotational molding is a manufacturing technique with a low environmental impact, as the melting of the powders occurs without direct contact with flames or heat sources. The machine heats the environment in which the mold is located, and the mold in turn heats and shapes the raw material contained in it by induction.

The research has thus shown that it is possible to produce luminaires with different geometric shapes, by combining the potential of rotomolding technique with the flexibility of the biopolyethylene material. In this way, luminaires result batch-dyed, UV resistant and thus free from the discoloration or delamination problem of the paint affecting the analogue aluminum lighting fixtures.

3. Results

3.1. Summary of the comparative LCA project of urban lighting products whose bodies are made of different materials. Drafted according ISO 14040 and 14044 standards.

From the comparison of a same type of luminaire made of five different technologies and compositions, and assuming a durability of 10 years for an aluminum lighting body and a durability of 40 years for an HDPE body in an environment rich in aggressive agents, it appears that:

- Life cycle of HDPE products has a lower environmental impact than aluminum products
- The result comes mainly from the fact that 4 life cycles were considered for the aluminum products – due to shorter durability of the products – and 1 life cycle was considered for the HDPE products
- There are also some small differences between the HDPE products, where those products containing 90% of recycled materials are more eco-sustainable than the fossil and bio ones.
- Concerning the most affected impact categories and excluding the use phase, the most affected categories are those about reduction of mineral resources, climate change, reduction of fossil resources and ecotoxicity of the water.

Table 2 shows the difference of eco-indicators of the five examined systems (excluding the use phase), where 1 point (pt) stands for the eco-indicator measuring the total impact of a person in Europe in one year. Consequently, 1 mPt is a millipoint of the above-mentioned eco-indicator.

The report was made by the University of Bari (Italy) through its Industrial Ecology Solutions S.r.l. (IES S.r.l.) spin-off. It describes the results and the technical-scientific aspects of the Life Cycle Assessment study – drafted according ISO 14040 and 14044 (ISO 2021 a, b) standards – of a same urban lighting product with an aluminum lighting body or a polyethylene lighting body made by rotational molding.

Table 2. Eco-indicators of the five systems (excluding the use phase) with transport by ship.

	Transport by ship
	mPt
LCA die-cast aluminum body	15.09
LCA turned aluminum body	24.39
LCA HDPE body (90-10)	7.99
LCA bio-based HDPE body	9.30
LCA fossil-based HDPE body	8.74

IOP Conf. Series: Earth and Environmental Science 1099 (2022) 012016 doi:10.1088/1755-1315/1099/1/012016

3.2. The potential of design

The high resistance and variety of the forms – guaranteed by the combination of rotational molding and material's flexibility – make it possible to realize uniquely designed luminaires that meet the needs of Municipalities and professionals. Aesthetics is essential in case of urban lighting luminaires. This is way design needs to meet aesthetical needs of the place where the luminaires will be installed, and practical needs of Municipalities that require energy saving, durable lighting bodies and the least possible maintenance.

Figure 9. Example of a polyethylene luminaire characterized by the overlapping of truncated conical geometries.

The result is a luminaire with slender volumes, ideal to be installed along tree-lined boulevards (figures 10 and 11), because it merges with branches and makes light appear more scenic.

To produce a similar die-cast aluminum luminaire, it would have been necessary to use more molds, and to assemble the different final parts – connected through bolted joints – where, consequently, crevice corrosion phenomena may occur, because they represent critical points.

Instead, with rotational molding, the molded object results in a single body produced with a single material, thus simplifying and speeding up the final assembly of the lighting body.

Figure 10. Installation of polyethylene luminaires along a tree-lined boulevard in the lake city of Sirmione, northern Italy.

Figure 11. Detail of a polyethylene luminaire installed in the lake city of Sirmione, northern Italy.

IOP Conf. Series: Earth and Environmental Science 1099 (2022) 012016 doi:10.1088/1755-1315/1099/1/012016

For large objects with complex geometries, rotational molding technique allows significantly lower production costs than the die-cast aluminum ones.

It is thus possible to produce geometries characterized by volumes' subtraction (figure 13), so avoiding the problems related to other molding technologies: for example to undercuts, which are typical of those parts that cannot be molded in die-cast aluminum or in plastic injection molding with a simple two-parts mold, because the material would get stuck during the opening of the mold or during the expulsion of the molded piece.

Hence the high freedom of expression in design, and the possibility to produce lighting bodies made of batch-dyed biopolyethylene or recovered materials, which are thus free from UV discoloration.

Figure 12. Polyethylene luminaire with full and soft volumes.

Figure 13. Polyethylene luminaire characterized by volumes' subtraction.

3.3. Lighting results

Biopolyethylene lighting bodies keep the advantages of the latest LED technology in terms of electric energy consumption and optic efficiency.

In table 3, it is possible to compare the data concerning the installation in figures 10 and 11 of 24 LED luminaires with biopolyethylene lighting bodies of 80 W to illuminate a motor traffic road and 24 more luminaires of 50W to illuminate the sidewalks, which contributed to an annual savings of electricity costs amounting to €14.651 and to annual savings in carbon dioxide emissions of 38 t. The previous lighting system had 24 poles with 125 W mercury vapor lamps – 3 for each pole – and a 250 W metal halide floodlight.

It should be noted that, apart from the huge energy savings, particular attention was focused on adapting lighting to standards and on achieving the best visual comfort by using 3.000 K color temperature LEDs with a minimum color rendering index of 80.

Table 3. Consumptions and annual savings of an installation with polyethylene LED luminaires.

Compared parameters	Units of measurement	Traditional luminaires	LED luminaires
Luminaire's average lifetime (L90B10)	H	9.000	100.000
Number of luminaires	pcs	96	48
Total effective power	kW	16,50	3.12
Hour use per year	dd/yy	365	365
Electricity price	€/kWh	0.25	0.25
Average useful lifetime	yy	2.1	22.8
Annual electricity consumption	kWh/yy	72.270	13.665
Annual emissions of carbon dioxide	t/yy	46.97	8.88
Annual consumption tonnes of oil equivalent	toe/yy	6.21	1.17
Annual savings	**%**		**81.0**
Annual savings in electricity consumption	**kWh/yy**		**58.605**
Annual savings in electricity costs	**€/yy**		**14.651**
Annual savings in carbon dioxide emissions	**t/yy**		**38.09**
Annual savings of tonnes of oil equivalent	**toe/yy**		**5.04**

The result of using the selected polymer are efficient LED luminaires whose resistant and eco-sustainable lighting bodies have a service life longer than 50 years, and which are ideal to be installed in marine environments or in those environments exposed to polluted agents.

Mechanical properties remain unchanged from -60° C to +80° C and there are no gases, annoying smells and toxic vapors during the rotomolding process.

These 100% recyclable biopolyethylene luminaires are currently installed in two Italian cities close to watercourses (figures 10, 11 and 16) that benefited from design customization, lighting body's resistance and the economic saving guaranteed by the LED lighting system.

Not only an energy but also a qualitative redevelopment, with a significant increase of lighting values achieved by meeting M3 category standards for the road and P1 standards for sidewalks. Plus, the achievement of both overall and longitudinal uniformity values greater than 0.7 and a TI index value of glare lower than 7% (figure 15 and table 4) by meeting current regulations against light pollution with zero emissions from 90° up (figure 14).

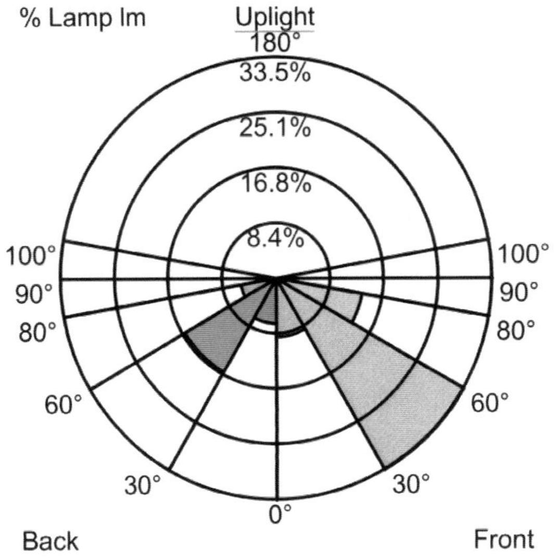

Figure 14. LCS diagram.

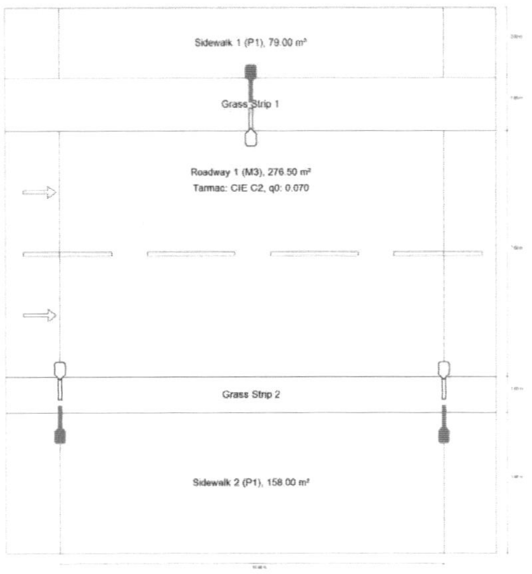

Figure 15. Project parameters.

Table 4. Results of the simulation.

	Symbol	Calculated	Target
Sidewalk 1 (P1)	E_{av}	16.93 lx	[15.00-22.50 lx]
	E_{min}	6.72 lx	≥ 3.00 lx
Roadway 1 (M3)	L_{av}	1.33 cd/m^2	≥ 1.00 cd/m^2
	U_o	0.75	≥ 0.40
	U_l	0.70	≥ 0.60
	TI	7%	≤ 15%
	R_{EI}	0.87	≥ 0.30
Sidewalk 2 (P1)	E_{av}	15.51 lx	[15.00-22.50 lx]
	E_{min}	4.97 lx	≥ 3.00 lx

Since this green material does not shield radio signals – as it occurs in aluminum lighting bodies – the design of these new luminaires can be protected by inserting antennas and wireless technology devices – which are essential for the communication between people and connected devices in a future-oriented Smart City – directly within the lighting body. For example, the latest 5G network.

The remote-control device is thus incorporated into the lighting body through a mechanism that allows a safe opening of the luminaire – also at height – and guarantees a fast and easy maintenance.

Figure 16. Installation of polyethylene luminaires in the coastal city of Vieste, southern Italy.

4. Conclusions

Author's aim was to group the best of research in eco-sustainable materials and LED technology for a new way of conceiving lighting, particularly in urban contexts.

The wide choice of shapes and finishes allowed by rotational molding, makes it possible for architects and lighting designers to express their creativity in luminaire's design. In addition, service life of those luminaires whose covers are made of these selected polymers has considerably lengthened, especially in aggressive environments. Therefore, they have a significantly lower environmental impact if compared to the aluminum covers.

In this way, the goal was reached: a 100% eco-sustainable production that made it possible to produce resistant, durable, and uniquely designed and Made in Italy luminaires.

The use of both these particular polyolefin blends and production technologies with low environmental impact, led to high performance and eco-sustainable products, with the stated aim of having a positive impact on the planet and of producing circular lighting bodies which combine luminous efficiency, environmental respect and the possibility to choose very creative designs.

5. References

[1] Deodati A and Vendramin G 2022 *Soluzioni contro la corrosione negli apparecchi di illuminazione urbana* Maggioli Editore (Santarcangelo di Romagna, Italy) pp. 8-26

[2] Novati G 2022 *Plastics in Italy a vice or a virtue? Technical Report* Ecco. The Italian Climate Change

[3] Spekreijse J, Lammens T, Parisi C, Ronzon T, Vis M 2019 *Insights into the European market for bio-based chemicals* JRC

[4] Witecka W K, Hauser P D, Sartor O 2021 *Breakthrough Strategies for Climate-Neutral Industry in Europe* Agora Energiewende

[5] University of Cambridge Institute for Sustainability Leadership (CISL) 2019 *Industrial Transformation 2050* Material Economics

[6] Tassielli G, Notarnicola B, Renzulli P A 2022 *Comparative Life Cycle Assessment (LCA) project of urban lighting products whose bodies are made of different materials. Drafted according ISO 14040 and 14044 standards* IES (Industrial Ecology Solutions S.r.l.)

Balanced Brightness Levels

Exploring how lighting affects humans' experiences of architectural and social urban contexts.

Mette Hvass[1] and Ellen Kathrine Hansen[1]

[1]Aalborg University, Copenhagen, Denmark. Department of Architecture, Design and Media Technology

Corresponding author: mh@create.aau.dk

Abstract.
This paper explores how lighting and darkness influence human experiences of architectural and social public urban contexts in dark hours. Tram stations in Aarhus, Denmark, are used as cases to investigate how brightness levels influence human sensory experience of the local space, the surroundings, co-presence with other people, and the activities performed in a specific context. Furthermore, this paper describes a transdisciplinary process, where methods from natural science, social science and arts/humanities are combined in three pre-analyses and a main field experiment. Lighting is often related to safety while darkness is related to fear. However, the findings from the main field experiment of this project indicate that lower and balanced brightness levels can sharpen our senses and create a relaxed atmosphere. Additionally, lower brightness levels enrich perceptions of and connections with spatial and social surroundings, thereby increasing perceived safety. We argue for context specific field experiments based on pre-analyses and the use of a transdisciplinary process, for drawing nearer to people's immediate senses when exploring experiences of urban lighting. The ultimate goal of the studies is to inspire to solutions that exploit the architectural and social potential of lighting by lowering and balancing brightness levels, all while reducing energy use and light pollution.

1. Introduction

1.1. Balanced brightness levels

People generally believe that bright light is necessary to be able to see well, but less light can also increase visibility and a sense of connectedness between humans and their spatial and social surroundings. This connectedness is expressed in the following quote from a field experiment at a tram station, n Aarhus, Denmark in the dark hours. During the field experiment test participants experienced first a bright and then a dimmed lighting setting:

'I actually start looking around in a different way because I have a feeling that I can see the surroundings. I'm in harmony with the surroundings now. It feels different from when I was in the very bright light…it is no longer the case that I am a target, exposed, and they can only look at me. Now, we can just look at each other.' (Test participant 10, female, 56, dimmed lighting)

We highlight this quotation because it represents an example of the human sensory experience of a balanced brightness level in an urban context. Immediate senses are formulated in response to the dimmed lighting, and both the architectural and the social aspects of lighting are mentioned. The feelings of both the atmosphere in the dimmed lighting setting (being in harmony), and perceived safety in the bright lighting setting (being a target and exposed), can be interpreted due to the choice of words of the test participant.

The field experiment constitutes one out of four studies in a PhD thesis about lighting in the urban context. The four studies were performed over a period of three years. They consists of three pre-analyses (study 1, 2 and 3) and a main field experiment (study 4):

Study 1 was performed as a literature study of how lighting and darkness are used in scenography to enhance atmosphere and the dramaturgy of a play. The purpose of the study was to compare the use of lighting in scenography and in urban public spaces to collect inspiration for how lighting and darkness can enhance the architectural and social qualities of an urban space [1].

Study 2 was performed as a field study, a qualitative explorative study about the role of lighting at two tram stations in two different urban contexts. The purpose of the study was to investigate if the lighting supported the architectural and social qualities at the two tram stations and to deside a direction for the research in study 3 and 4 [2].

Study 3 was performed as a lab study, a quantitative study of human experiences of six lighting settings. The purpose of the study was to examine human experience of different ratios of lighting between two lit spaces. The two lit spaces were a) one lit local space where the test participant was placed and b) a surrounding lit space [3].

Study 4 was performed as a field experiment and the experiment was based on the results from the three pre-analyses (study 1, 2 and 3). The purpose of the field experiment was to study human sensory experience of two lighting settings, with different ratios of lighting between a local space (a tram station) and the surroundings (the urban context around the station), in a real-life context [4].

This paper describes how human experiences of lighting were collected and how the theme of balanced brightness levels (hereinafter BBL) was formulated in the process of the four studies. The focus of this paper evolves around the main field experiment and how this experiment contributes to the field of lighting design research. The pre-analysis in the field (study 2) and the pre-analysis in the lab (study 3) are briefly described to explain how these studies lead to the main field experiment (study 4). Furthermore, the paper describes how research methods from natural science, social science and arts/humanities were combined in a transdisciplinary process of conducting the four studies. 'The Architectural Experiment' – model [5] was used to structure this process. The four studies are further described in two conference papers, a journal paper, as well as a book chapter [1,2,3,4].

1.2. Urban lighting – understanding the effects of lighting

Research on the experience of urban lighting has a long tradition of utilizing an engineering approach with a focus on visibility and perceived safety [6,7,8,9]. In recent years, interest in how lighting affects human behavior and social interaction has increased in social science [10,11]. Likewise, researchers in architectural lighting design have received inspiration from the notion of social lighting [12]. However, knowledge about the role of lighting in urban contexts as seen from an architectural point of view seems to be lacking. There is a need for a holistic architectural approach to lighting research which combines knowledge about the ability lighting has to; create atmosphere, enhance the visual identity of a space, support design intentions, and to ensure that both the technical and the social properties of lighting are incorporated.

Lighting designer Roger Narboni states that a holistic thinking is needed in outdoor lighting design. He sees the need to change from light urbanism, where lighting is designed to produce visibility. According to Roger Narboni, lighting should be adapted to nocturnal urbanism in the future. The lighting should adapt to '*citizens' needs, new infrastructures, and new ways of using the city after dark*'[13]. To support this change from light urbanism to nocturnal urbanism, we need to study humans' experiences of lighting in different urban contexts and to combine research disciplines. Peter Boyce states: '*The point is which lighting conditions are most suitable depends on context. Until the importance of context is acknowledged, there is little likelihood of achieving a finer understanding of the effect of lighting in all its complexity'* [14]. Likewise, Yvonne de Kort emphasizes that tests should be done in field studies after the effect of light has been confirmed in the laboratory [15]. Furthermore, she describes that she seeks to gain knowledge about lighting in relation to, for example, vision, comfort, health, performance, the atmosphere, and energy savings, and this requires collaboration between various research fields and combining methods [15]. In line with these statements, this project is aimed at fostering a deeper understanding of how lighting can be adapted to nocturnal urbanism by combining methods from various research fields and conducting a field experiment based on pre-analyses in a specific context.

2. Aim

The aim is to increase understanding of how brightness levels affect the human sensory experience of architectural and social urban contexts, and how these experiences can be explored in a transdisciplinary process.

3. Background

3.1. Sustainable urban infrastructure

Globally, the majority of the earth's population lives in cities, and it is estimated that this urbanization will be further developed in the future. An energy-efficient infrastructure is necessary to support the development of sustainable cities [16]. A political decision in 2013 led to an agreement about a national green transport policy in Denmark, and tram infrastructures in the three largest cities became a part of this initiative. When it comes to increasing the use of public transportation, the pedestrian's experience of using public transportation plays an important role [17]. In Denmark, it is dark for half of the year when people commute to and from work. Therefore, to ensure attractive public transportation, we need to investigate not only how lighting can create visibility but also how lighting and darkness can enhance the experience of architectural and social urban contexts.

Tram stations were chosen as a case for studies of the architectural and social potential of lighting in outdoor spaces for several reasons. The tram represents a sustainable infrastructure and at the time this project started the tram had just been taken into operation. Furthermore, the tram stations are placed in different urban, suburban, and rural contexts and as the project developed it was possible to focus the studies on a specific context. The focus became urban stations, and one specific context were chosen for the final experiment to concentrate on how the lighting and darkness could enhance the architectural and social qualities of this specific context. Additionally, one of the industrial partners of the PhD project, Holscher Design, was in charge for the design of the tram stations. This collaboration gave access to information about design intentions and how the lighting design had been implemented during design and construction phases of the stations.

3.2. Lighting influencing architectural and social urban contexts

The architectural urban context, refers to lighting and darkness in relation to the atmosphere of the space [18,19,20] and the physical context, such as the buildings, objects, space, surroundings, materials, surfaces, etc. With architectural lighting design, one should ensure that the shapes of objects are visible, shadows are present, colors and textures are visible in a local space, and details in the physical surroundings are visible. While, the social urban context refers to people in the urban built environment, the co-presence of people, and their behavior in a public space. The social values of lighting can be described as, being able to see people in a pleasant light in a local space, as well as the ability to see people in the surrounding space. These values also include creating a safe and relaxed atmosphere that provides social enhancement, accessibility, and hospitability [12], as well as creating lighting that supports human behavior and the activities performed in the space [11].

3.3. Brightness, illuminance, and luminance levels

To describe levels of lighting, both unmeasurable and measurable terms have been used, due to the transdisciplinary nature of this project. Brightness is unmeasurable, as brightness describes the subjective experience of light. Brightness describes the state of the light–dark pattern in the eye and the actual pattern of brightness in the field of vision [21]. Illuminance and luminance are measurable. Illuminance describes the amount of light that falls on a surface, measured in lux [21]. While luminance describes the amount of light reflected from a surface, measured in candela per square meter (cd / m2) [21]. As luminance measures the light leaving a surface in a specific direction, luminance better describes how the human eye perceives a surface's brightness [22]. For many years, existing lighting standards and regulations have relied mainly on the illuminance levels on horizontal surfaces, likely because they are easy to measure (with a lux meter). Current standards also include luminance, although illuminance is still the more commonly used measure in practice [22].

3.4. Human sensory experience

During studies, emphasis is placed on the fact that experiencing outdoor lighting is a human sensory experience, which should be experienced by being bodily present in the urban environment [23]. In a test situation, human night vision should be adapted to the experience of fine nuances between light and dark areas in an urban space [14]. Sound, smell, and touch are also human senses that have an influence on how a place is experienced [24]. Pallasmaa touches upon the complexity of evaluating the quality of a space by stating: *'The quality of a space is not merely a visual perceptual quality as it is usually assumed. The judgement of environmental character is a complex multi-sensory fusion of countless factors that are immediately and synthetically grasped as an overall atmosphere, ambience, feeling, or mood'* [19].

This project seeks to get an understanding of how human sensory experiences of the lighting and darkness can be collected in the urban complex context, where countless factors can have an influence on how the architectural and social context is experienced. Presence of lighting in the urban space can often be taken for granted by people who are not used to talk about experiences of lighting [10]. Therefore, a combination of research methods is used to seek to understand test participants' vocabulary about experiences of lighting and darkness.

3.5. Balance between lighting and darkness – atmosphere and perceived safety

Lighting is often associated with safety and darkness with fear. However, too much light can *'wipe away the sense of a place'* as Pallasmaa describes it [23]. Therefore, brightness should always be explored in relation to darkness as Gernot Böhme observes: *'Brightness is what turns sight into a real capability in the first place and enables visible things to be seen in reality...Light is not the only precondition of visibility. Darkness is another. True, light and darkness are asymmetrical. Light is a precondition for seeing at all, whereas darkness (interacting with light) is a precondition for our seeing something'* [25]. Nick Dunn and Tim Edensor explore the multiple meanings and uses of darkness across time and space. Specifically, they look at how darkness has been laden with negative attributes throughout history, forgetting the positive aesthetic and sensory experiences that darkness can create. They draw attention to this historical tradition of relating darkness to danger and lighting to safety in urban spaces without considering the aesthetical values of darkness [26]. Lighting and darkness can thus both be linked to the understanding and feeling of a pleasant/unpleasant atmosphere or a safe/unsafe space. A particular lighting design can never match all users of a public space, but a more detailed understanding of the experience of lighting and darkness is needed to challenge biased assumptions about the link between a high brightness level and perceived safety [26]. Throughout time, phenomenologists have described the connection among light, darkness, and the atmosphere. In his work about atmospheres, Böhme uses the stage as a metaphor to highlight a situation in which light and darkness are tools for producing atmospheres [27]. Similarly, Pallasmaa states: *'Deep shadows and darkness are essential, because when the sharpness of vision is dimmed the unconscious peripheral vision and tactile fantasy is invited'* [23].

Nonetheless, we also need lighting at night. The benefits of light at night are described by Peter Boyce as greater safety for pedestrians and reduced fear of crime [6]. Regulations and standards for lighting provide general rules to ensure that the illuminance level is sufficient in relation to safety and visibility in urban contexts. At the tram stations in Aarhus, Denmark, the illuminance level was based on two norms: 'Bane Norm BN2-81-1' [28] and 'Danske Vejregler, standsningssteder for letbaner' [29].

This project focus on perceived safety defined as subjective experiences of safety [30]. The investigations focus on perceived safety in relation to the lighting in the local space, where a person is situated, combined with, the importance of being able to perform long-range detection of possible threats, in the surroundings [30,31,32]. Leon van Rijswijk and Antal Haans highlight that there may exist an intuitive or learned association between lighting and safety, such as the mere presence of lighting may directly affect people's perception of safety of an environment [30]. This intuitive or learned association is challenging for studies of perceived safety and therefore different approaches should be used to gain understandings of this. Lighting research conducted primarily in lab studies shows that test participants feel safer if the illuminance level is higher [7,8,9,33]. This project seeks to combine research methods and investigate the balance between atmosphere and perceived safety both in a lab study and in a field experiment in relation to a specific urban context.

To structure the research and to frame the four studies, the following two research questions was asked:

1. How do brightness levels affect human experience of architectural and social urban context?

2. How can research methods be combined to explore experience of brightness levels in the urban context?

4. Methods

4.1. Research methods – research disciplines

Methods for exploring outdoor lighting vary within research disciplines. Within natural science, research on outdoor lighting is primarily focused on measurable facts and reproducible quantitative tests. Research mainly involves lab studies in controlled environments to eliminate biased findings. The results must be quantifiable through measurements or questionnaires, which can be statistically analyzed. In social science, humans are studied in urban environments after dark with a focus on how artificial lighting is experienced in different situations and how these experiences affect behavior. Research is primarily done through field studies, interviews, and observations. Finally, when it comes to research within the arts/ humanities, the architectural investigation of scale, spatial hierarchy, and the atmosphere is rooted in the architectural design tradition of experimentation in a real context. Methods are combined, and a problem is often examined through field studies; observing spaces and people, and registering through photos, sketches, models, or mock-ups. The three disciplines represent three approaches to research. To understand the complexity of lighting and particularly the experience of brightness levels, we study how to combine the research methods.

4.2. The architectural experiment – a transdisciplinary process model

The combination of methods and knowledge across research disciplines in the BBL project is based on the transdisciplinary process model 'The Architectural Experiment', hereinafter the AE model (figure 1) [5]. The AE model illustrates a transdisciplinary and knowledge-based design process in horizontal steps 1-5. The three vertical criteria represent knowledge and methods from three research disciplines, natural science, social science, and the arts/ humanities, which become part of the design process in each step. The AE model is aimed at crossing research boundaries and refers to theories on innovative processes such as 'knowledge boundaries' [34] and 'knowledge brokering' [35], combined with theories on design research [36]. The AE model has been used throughout the BBL project to structure the process and the implementation research methods for each study.

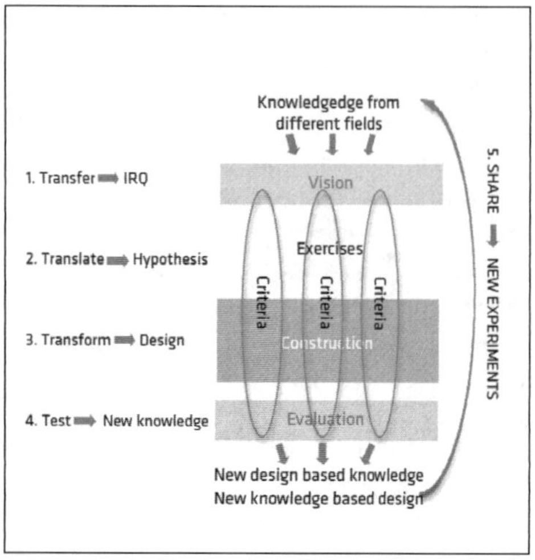

Figure 1. The AE model [5]

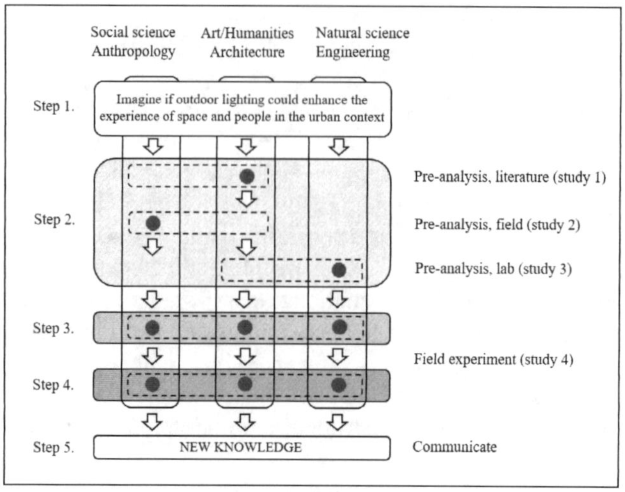

Figure 2. Process model for the BBL project

4.3. Combining disciplines in the BBL project

The horizontal bands in figure 2 illustrate how the four studies of the BBL project are related to the steps of the AE model. The vertical bands illustrate research disciplines. The black circles illustrate the main research approach of each study, and the dotted lines illustrate the fact that methods from other research disciplines have been used during the study. The arrows illustrate how knowledge from one study has been transferred to the next. During the field experiment (study 4), knowledge from the pre-analyses (study 1,2 and 3) was implemented for formulating the research design (step 3) as well as for performing and evaluating the results (step 4). Meanwhile, research methods from natural science, social science, and the arts/ humanities were combined and incorporated into the process. Overall, figure 2 illustrates how the BBL project is aimed at crossing knowledge boundaries by combining knowledge and methods in all stages of the project. In the following two sections, the methods used in the three pre-analyses and the field experiment are described briefly further information can be found in papers about the studies [1,2,3,4].

.

4.4. Pre-analyses in the field and lab

Study 2, the pre-analysis in the field [2], was based on ethnographic methods, anchored in social science. This was combined with an architectural approach to analyse space and atmospheres. During three days in January 2019, a total of 20 people participated in short interviews while they were waiting for the tram at two tram stations. James P. Spradley's descriptive question matrix was chosen for action-based data collection [37,38]. The interviews was supplemented with photo registrations and observation of the spaces, people, and activities at and around the stations [39].

Study 3, the pre-analysis in the lab [3], was primarily based on methods from natural science. The study was performed in a period of three weeks in January 2020 in the lighting lab at Aalborg University, Copenhagen. The experiment included 30 participants who experienced six lighting settings with different ratios of brightness levels in a lit zone where the test participant and the interviewer were placed and a lit zone representing the surrounding space. Information where gathered through a questionnaire with semantic differential scale. Through twelve questions test participants were asked about their experiences of the atmosphere in the local space, co-presence with other people, and perceptions of the surrounding spaces during the six lighting settings.

4.5. Field Experiment, Study 4

The field experiment took place in November 2020 at the Nørreport tram station during two weeks. One week with the existing bright lighting setting and one week with a dimmed lighting setting. The existing illuminance level was measured to approx. 165 lux in the sheltered waiting area and in the surroundings, a level of approx. 1- 6 lux was measured along the facades. The illuminance level was dimmed approx. 80% in the sheltered area in the dimmed lighting setting.

During the main field experiment both methods from arts/humanities, social science and natural science were combined. The research was conducted in collaboration with anthropologist Karen Waltorp from Copenhagen University. Her anthropological methods were combined with the architectural methods of the authors and to measure illuminance and luminance levels, engineering methods were used.

Ethnographic interviews were conducted as comparative go-along interviews with participant-produced pictures. The go-along interviews included 10 participants. The interviewer met each participant at the Nørreport tram station twice, in the existing bright lighting setting and in a dimmed lighting setting. The interview included a tram ride to the nearby station and back again. Each recorded interview lasted approx. 30 minutes. At each interview, the participants were asked to comment on the atmosphere in the waiting area , the experience of the surrounding urban context, the experience of co-presence with other people, and the experience of lighting in relation to activities being performed [24]. Participant-produced pictures were taken before the first meeting, during the first interview, and then during the second interview [40]. During the interview, the test participants were asked to take pictures with an Ipad in relations to the topics discussed.
Architectural methods used on site were observations, sketching, photo registrations and time lapse videos which registered experiences of rhythms of daylight, urban light, traffic, and people [41,42].
Engineering methods were used to gather quantitative data about illuminance and luminance levels, such as lux measurements on horizontal surfaces at ground level at the station and on the pavement in the surroundings. Likewise, luminance measurements were performed on vertical surfaces at the station and on the surrounding buildings (at selected points at eye height), and through luminance maps based on high dynamic range (HDR) photos [43].

5. Results

5.1. Pre-analyses in the field and lab

Study 2, the pre-analysis in the field. Field notes were transcribed and analyzed through qualitative traditional coding [44]. It was found that the brightness level in the sheltered area at the station was out of balance in relation to the brightness level in the surrounding context. The high differences in the brightness levels made the architectural and social qualities of the space invisible, as the contrast could not be compensated for with the human-sensitive night vision [2].

Study 3, the pre-analysis in the lab. Answers from the questionnaires were analyzed separately using Friedman analyses of variance (ANOVAs). For quantifying differences in luminance levels among the six settings, luminance measurements and luminance maps calculations were performed. When the balance between brightness levels was examined in a lab study (study 3), the results showed that the low brightness level in the local space increased the perception of the atmosphere as being relaxed and private. Furthermore, the low contrast in the brightness level ratio between the local space and the surrounding context increased the perception of the area as being relaxed and harmonious [3].

5.2. Field Experiment, Study 4

The recorded interviews were transcribed and coded in Nvivo12. First, a deductive sorting of the data was performed to detect and recognize themes. Subsequently, an inductive coding of the data was performed to introduce new angles and perspectives to the project. The findings from the go-along interviews in the field experiment (study 4) demonstrated that the sensory experiences of space and people in the urban context was affected in the dimmed lighting. The participant-produced pictures (study 4) were effective for understanding the participants' immediate thoughts about the experience of lighting in the urban context. The pictures led to sudden realizations and triggered memories for the participants [40]. It was discovered that issues that were difficult to describe in words was easier to describe when the test participant explained the content in a picture. During this process, a vocabulary was developed. Dimming and adjusting the brightness level according to the surrounding brightness level created a relaxed atmosphere at the station for those waiting for the tram. A connectedness to the surroundings and the people there was felt, thus increasing the feeling of perceived safety. The adjustments of the brightness levels also helped to ensure the visibility of the architectural and social qualities of the urban context. While waiting at the station, the majority of the people appreciated the dimmed lighting. But when test participants arrived at the station and got off the tram, they perceived the brightness level as too low due to the high brightness level in the tram.

Luminance maps (figure 3) provided measurable facts about the luminance levels in the urban scene as well as a visual hierarchy of the luminance levels. The luminance maps also revealed how the sheltered area at the station became transparent, as intended by the designers, when the brightness level was lowered. By lowering the brightness level, in the sheltered area, the lighting was not reflected the glass panels which divide the platform into two parts and the surroundings became visible again.

Figure 3. Photos and luminance maps illustrating how brightness levels are balanced when the lighting is lowered in the sheltered area at the station; the surroundings became visible when the contrast was lowered (study 4).

6. Discussion

The findings from the BBL project illustrate the fact that by combining methods in a transdisciplinary process, we can obtain a nuanced understanding of how lighting can influence architectural and social qualities in the urban context. In the lab test, general knowledge was produced and proved to be valuable for the design of the context specific field experiments. This revealed that lowering and balancing brightness levels resulted in increased visibility, a relaxed atmosphere, and a higher level of perceived safety.

The American philosopher and pragmatist John Dewey argue that problems and situations in practice should dictate which methods should be used to investigate a problem [45]. One cannot have knowledge about the world without being an actor in it, as bodily experiences related to a topic, or a problem should be gained in the actual context to acquire access to participants' immediate senses [46]. While the specialist in lighting research, Yvonne de Kort, points out that when it comes to lighting research tests should be done in the field after the effect of light has been confirmed in the laboratory [15].

In existing research on urban lighting, opinions about lowering brightness levels are diverse within the research disciplines. When illuminance levels are lowered, some would investigate changes in perceived safety whereas others would investigate changes in experience of atmospheres alone. But can this balance between light/darkness and safety/atmosphere be established? Humans have different needs, and each urban context has different circumstances that should be taken into consideration. Regulations and standards give general recommendations about urban illuminance levels, but the findings about visibility, the atmosphere, and the perceived safety in these studies advocate for the idea that general recommendations should be supplied with context specific recommendations.

7. Limitations

The findings of this study must be seen in light of some limitations. Lowering illuminance levels at a tram station caused security concerns for Aarhus Tram and a permission was difficult to obtain which caused delays. In addition, several limitations were related to the COVID-19 pandemic e.g., recruiting test participants and investigating the social potential of lighting when people had to cover their faces with masks. A week before the field experiment at Nørreport tram station began, restrictions regarding the COVID-19 pandemic were announced. Due to warnings concerning a second lockdown in Denmark, several already recruited test participants withdrew, and it was difficult to recruit new test participants. Therefore, the results of the experiment should be interpreted considering the relatively small number of test participants.

8. Conclusion

The conclusion should be seen in relation to the above-mentioned limitations which caused that the amount of test participants was reduced to ten people. Further studies should be performed to support the conclusions presented below.

The conclusion can be divided into collected knowledge about how lowered and balanced brightness levels affect the human experience of the architectural and social urban contexts and furthermore, knowledge about methods for exploring lighting in the urban context.

First, the BBL project contributes information on how an urban context is experienced when brightness levels are lowered and balanced. Three important findings are presented:

- In the lab (study 3) and the field experiment (study 4), atmosphere was perceived as relaxed when brightness was lowered and balanced.
- Through lowering and balancing brightness levels, an increased visual connection to the people in the surroundings was established, and this led to an increased feeling of perceived safety.
- Spatial connectedness and social connectedness between the space and the surroundings, as well as the humans there, were established when the brightness level was balanced.

Second, the BBL project provides an example of how to explore the complexity of lighting in the complex urban context. Three important approaches are pointed out:

- Through conducting research in a context specific field experiment, it was possible to focus on brightness in relation to spatial characteristics, the humans present, and specific activities being performed.
- Performing ethnographic go-along interviews with participant-produced images gave access to participants' immediate senses. Thoughts about light, space, co-presence with other people, and activities, were described in the situation, in the own words of the test participants, while walking, talking, and taking pictures.
- Luminance maps showed both quantitative and qualitative details about the luminance levels and the ratio of luminance levels in a local space (the station) and the surrounding urban context.

The BBL project indicates that outdoor lighting can facilitate more than being able to see in the dark. It explores how the balance between lighting and darkness, as well as the balance between the brightness level in a local space and its surroundings influence the human sensory experience of the atmosphere, perceived safety, and spatial and social connectedness. The project provides an example of how a lower brightness level can lead to more visibility when the lit urban context is taken into consideration, and when the nature of nocturnal urbanism in a specific context is taken into consideration.

9. Acknowledgement
We would like to thank our two industrial partners of the PhD project Schréder and Holscher Design, Also, a thanks to Aarhus Light Rail and AFA JCDecaux to make the field experiment at Nørreport tram station possible. We would also like to thank co-authors Kevin Van Den Wymelenberg (Study 3) and Karen Waltorp (study 4) and test participants in lab and field studies. A final thanks goes to the Lighting Design Lab at Aalborg University Copenhagen.

References
[1] Hvass M & Hansen E K 2019. Potentials of light in urban spaces defined through scenographic principles. *NAF Symposium*, June 2019. Proceedings Series 2021-1
[2] Hvass M & Hansen E K 2020 Architectural and social potential of urban lighting, a field study of how brightness can affect the experience of waiting for public transportation. *Proceedings PLEA Conference,* September 2020
[3] Hvass M, Wymelenberg K V D, Boring S & Hansen E K 2021 Intensity and ratios of light affecting perception of space, co-presence and surrounding context, a lab experiment. *Building and Environment*, Vol **194**, 107680
[4] Hvass M, Waltorp K & Hansen E K 2022 Lights out? Lowering Urban Lighting Levels and Increasing Atmosphere at a Danish Tram Station. In *Lighting Design in Shared Public Spaces* edited by Shanti Sumartojo pp. 151-172 (New York: Routledge)
[5] Hansen E K & Mullins, M 2014 Lighting Design: Towards a Synthesis of Science, Media Technology and Architecture. *Smart and Responsive Design*, vol **2** pp 613-620
[6] Boyce P R 2019 The benefits of light at night. *Building and Environment* vol **151** pp 356-367
[7] Boyce P R, Eklund N H, Hamilton B J & Bruno L D 2000 Perceptions of safety at night in different lighting conditions, *Lighting Research and Technology* vol **32** pp 79–91
[8] Nasar J L & Bokharaei S 2017 Impressions of lighting in public squares after dark, *Environment and Behavior* vol **49** pp 227–254
[9] Fotios S & Gibbons R 2018 Road lighting research for drivers and pedestrians: the basis of luminance and illuminance recommendations, *Light. Res. Technol.* vol **50** pp 154–186
[10] Sumartojo S and Pink S 2020 *Atmospheres and the Experiential World - Theory and Methods* (New York: Routledge)
[11] Slater D, Sloane M & Entwistle J 2015 *Configuring light: Staging the Social.* http://www.configuringlight.org/. Accessed 10. October 2020
[12] Casciani D 2020 *The Human and Social Dimension of Urban Lightscapes* (Cham: Springer)
[13] Narboni R 2016 From light urbanism to nocturnal urbanism. *Light & Engineering* vol **24** pp 19-24
[14] Boyce P R 2014 *Human Factors in Lighting*, Third Edition. (CRC Press; Taylor & Francis Ltd.)
[15] de Kort Y 2021 Opinion: On becoming smart. *Lighting Research and Technology* vol **53** p 4
[16] United Nations. *World Urbanizations Prospects 2018: Highlights*. New York, USA
[17] Rahm J & Johansson M 2016 Walking after dark - A systematic literature review of pedestrians' response to outdoor lighting. *Environmental Psychology Monographs* vol **26**
[18] Rasmussen, S E 1966 *Om at Opleve Arkitektur* (København: Gads Forlag)
[19] Pallasmaa J 2014 Space, Place and Atmosphere: Emotional and Peripheral Perception in Architectural Experience. *Lebenswelt* vol **4** pp 230-245
[20] Zumthor P 2006 *Atmospheres. Architectural Environments. Surrounding Objects (*Basel: Birkhäuser)
[21] Tregenza P & Loe D 2014 *The Design of Lighting* (New York: Taylor & Francis)
[22] Dubois M C, Gentile N, Laike T, Bournas I & Alenius M 2019 *Daylighting and Lighting under a Nordic Sky* (Lund: Studentlitteratur AB)
[23] Pallasmaa J 2012 *The Eyes of the Skin: Architecture and the Senses* (Chichester: Wiley-Academy)
[24] Pink S 2015 *Doing Sensory Ethnography* (London: Sage)
[25] Böhme G 2017 *The Aesthetics of Atmospheres* ed J P Thibaud (London: Routledge)

[26] Dunn N & Edensor T 2020 *Rethinking Darkness: Cultures, Histories, Practices* (London: Routledge)
[27] Böhme G 2013 The art of the stage set as a paradigm for an aesthetics of atmospheres. *Ambiances* vol **315** pp 2 -8
[28] Banedanmark 2015 Banenorm BN2-81-1. https://www.bane.dk/da/Leverandoer/Krav/Tekniske normer og regler/Banenormer.
[29] Vejregler, standsningssteder for letbaner http://vejregler.lovportaler.dk
[30] van Rijswijk L & Haans A 2018. Illuminating for safety: Investigating the role of lighting appraisals on the perception of safety in the urban environment. *Environment and behavior*, vol **50** (8) pp 889-912
[31] Boyce P R & Gutkowski J M 1995. The if, why and what of street lighting and street crime: A review. *International Journal of Lighting Research and Technology*, vol **27**(2) pp 103-112.
[32] Caminada J F & Van Bommel, W J M 1984. New lighting criteria for residential areas. *Journal of the Illuminating Engineering Society*, vol **13**(4) pp 350-358
[33] Johansson M, Pedersen E, Maleetipwan-Mattsson P, Kuhn L & Laike, T 2014. Perceived outdoor lighting quality (POLQ): A lighting assessment tool. *Journal of Environmental Psychology*, vol **39** pp 14-21
[34] Carlile P R 2002 A Pragmatic View of Knowledge and Boundaries: Boundary Objects in New Product Development *Organization Science* vol **13** pp 442-455
[35] Hargedon A B 2002 Knowledge Brokering: Linking Learning and Innovation. *Research in Organizational Behavior* vol **24** pp 41-85
[36] Koskinen I, Zimmerman J, Thomas B, Redström J & Wensveen S 2011 *Design Research Through Practice. From the Lab, Field, and Showroom* (Croydon: Elsevier)
[37] Spradley JP 1980 *Participant observation* (New York: Rinehart and Winston)
[38] Spradley J P 1979 *The ethnographic interview* (New York: Rinehart and Winston)
[39] Cullen G 1997 *The Concise Townscape* (London: Architectural Press)
[40] Pink S 2021 *Doing Visual Ethnography* (London: Sage)
[41] Edensor T 2012 *Geographies of rhythm: Nature, place, mobilities and bodies.* (Ashgate: England)
[42] Lefebvre H 2013 *Rhythm Analysis: Space, Time and Everyday Life* (London: Bloomsbury)
[43] Inanici M 2006 Evaluation of high dynamic range photography as a luminance data acquisition system, *Lighting Research and Technology*, vol **38** pp 123-136.
[44] Bjørner T 2015 *Qualitative methods for consumer Research.* (Copenhagen: Hans Reitzels)
[45] Brinkmann S 2013 *John Dewey, science for a changing world* (New York: Routledge)
[46] Dewey J 1934/1980 *Art as Experience* (New York: Perigee Books)

Drivers' experiences of presence sensitive roadway lighting match experiences of traditional road lighting – a case study in Finland

H Pihlajaniemi[1], E Juntunen[2] and A Luusua[1]

[1]Oulu School of Architecture, University of Oulu, P.O.Box 4100, FI-90014 University of Oulu, Finland

[2]VTT Technical Research Centre of Finland Ltd., Kaitoväylä 1, 90570 Oulu, Finland

henrika.pihlajaniemi@oulu.fi

Abstract. This research aimed to test and study presence sensitive roadway lighting in a housing area in Finland, and to evaluate drivers' experiences and attitudes of it. The lighting adapted both to motor vehicles using the road and to the measured traffic density along it. The case study was conducted on a collector road in Salo, a town in southern Finland. New, controllable LED lighting with PIR (passive infrared) presence sensors was built along the road, and test scenarios were designed, programmed, and tested. Drivers' experiences and attitudes of the lighting were collected in a three-phase evaluation with questionnaires from the community of about 1000 households using the road as part of their daily mobility. The results indicate that as an experience, presence sensitive lighting in a road environment was at least as positive as traditional, uncontrolled lighting. The experiences of presence sensitive lighting did not differ from the experiences of uncontrolled lighting regarding pleasantness, uniformity, glare, and road visibility. Most of the drivers (86 %) did not notice any dynamic change in lighting. When informed about the tested lighting strategies, most of the participants (72 %) would prefer either one of the intelligent lighting modes to be the permanent lighting solution. The results encourage the use of intelligent lighting in striving towards more sustainable lighting solutions while maintaining user comfort and safety.

1. Introduction

There is a growing need to reduce energy consumption of road lighting while maintaining user comfort and safety. Applications of adaptive and intelligent lighting technologies offer solutions for this demand. The concept of adaptive lighting refers to an intelligent lighting system, which is equipped with sensors and actuators that allow the system to dynamically adapt to the current conditions or to the needs of individual users [1]. Adaptation in lighting systems can be achieved through sensor networks that gather information from the environment, such as the flow of traffic, and feed it to a closed-loop adaptive system. Through adaptive behaviour, the system can offer energy savings and cost efficiency. [2,3] One example of adaptive systems is *presence sensitive roadway lighting*, i.e. lighting that is programmed to adapt to the presence of vehicles or people. The energy saving potential has been demonstrated both in simulations and in real world case studies [4,5,6]. When designed wisely, adaptive lighting can offer, besides energy savings, added experiential value for illuminated environments [7].

There is still a lack of real-world research scrutinizing the experiences of drivers in adaptive and intelligent roadway lighting situations which hinders the introduction of applications by cities and municipalities. A study from 2011 by Atıcı et al. [8] focused on a pedestrian setting, noting that "other forms of mobility cases, such as cyclist and vehicular, should also be analyzed." Recent research from Finland concerning dimmed road lighting in an urban road suggests that it is feasible to reduce "road lighting intensity when car headlights are available" from the drivers' point of view [9]. To address this gap in knowledge, our research designed and studied presence sensitive roadway lighting in a housing area. The lighting that was tested in our study adapted both to motor vehicles using the road and to the measured traffic density along it, enabling us to evaluate drivers' experiences and attitudes concerning it. A traditional lighting scenario was also implemented, as we wanted to compare the experiences of traditional and presence-sensitive lighting. Our research questions were the following: How do drivers' experiences of presence sensitive roadway lighting compare to experiences of uncontrolled, traditional lighting? What are the experiences, opinions, and attitudes of informed participants regarding a real-world implementation of two differing intelligently controlled lighting scenarios: presence sensitive lighting, and presence sensitive lighting based on the measurement of traffic density? The case study was conducted along a collector road in Salo, a town in southern Finland. The study was part of the Sencity – Intelligent Lighting as a Service Platform for Innovative Cities project [10].

2. Methods and materials

2.1. Research setting and participation process

Our case study consisted of a real-world test of presence sensitive roadway lighting, which adapts both to motor vehicles and to the measured traffic density along a collector road. A new, controllable LED lighting with PIR (passive infrared) presence sensors was built along the road, 41 light poles altogether, and test scenarios were designed, programmed, and tested. The luminaire type was Easy LED PRO Flow S 30-850 UP-R 740, with luminaire wattage of 76.0 W, lumen output of 8470 lm, and correlated colour temperature of 4000 K. The mounting height was 10 meters. The manufacturer of the presence sensor and intelligent lighting control system was Lumine. Users' experiences of the lighting were collected with questionnaires from the surrounding community of about 1000 households that use the road for daily mobility, and from other interested inhabitants of the city. The roadway was used only by cars; the light traffic route for pedestrians and cyclists was located separately in a surrounding forested area.

The study consisted of three scenarios, which tested three different lighting control methods (Figure 1). *The first scenario ('Traditional' lighting control method)* ran from 23 Jan – 5 Feb 2017. In this phase, the new lighting was controlled in a basic manner (daylight level based control): during the bright period of the day, the lights were turned off, and during the dark period, they were on at a 100 % control level. The sensor detected the threshold lux level and turned the lights on and off automatically. During *the second scenario ('Presence sensitive' lighting control method*; 6 Feb – 26 Feb 2017), a presence-based dynamic control was added to the previous basic daylight level based control. In this phase, the lighting was controlled dynamically so that it was always brightened around a moving car to the maximum control level of 100 %, and in those parts of the road where there was no traffic, it was dimmed down to a 20 % control level. The dimming and brightening were done softly using a 3 second ramp up and down. The bright area around a detected vehicle consisted of five streetlights: the one which detected the car with a passive infra-red (PIR) sensor as well as two forward and two backward lights. In *the third scenario ('Presence and traffic density sensitive' lighting control method;* from 6 March 2017 onward, questionnaire deployed 19 Jun – 2 July 2017), a third control method was introduced to complement the two former ones. The introduced method was based on the measurement of traffic density: the lower control level of lighting adapted to the amount of traffic detected along the route. When traffic was dense, for example during the commute-heavy periods in the morning and in the evening, the control level of the lighting was dropped to 70 % on those parts of the road where no traffic was detected. During times of moderate traffic, the level was 40 %, and with the lowest traffic density during the night, the level was dropped to 20 %. The times of sunset and sunrise in Salo varied along the test period being on 23 Jan 16:16 and 09:03; on 6 Feb 16:53 and 08:30; on 6 March 18:07 and 07:11; and on 19 Jun 22:59 and 03:58, respectively.

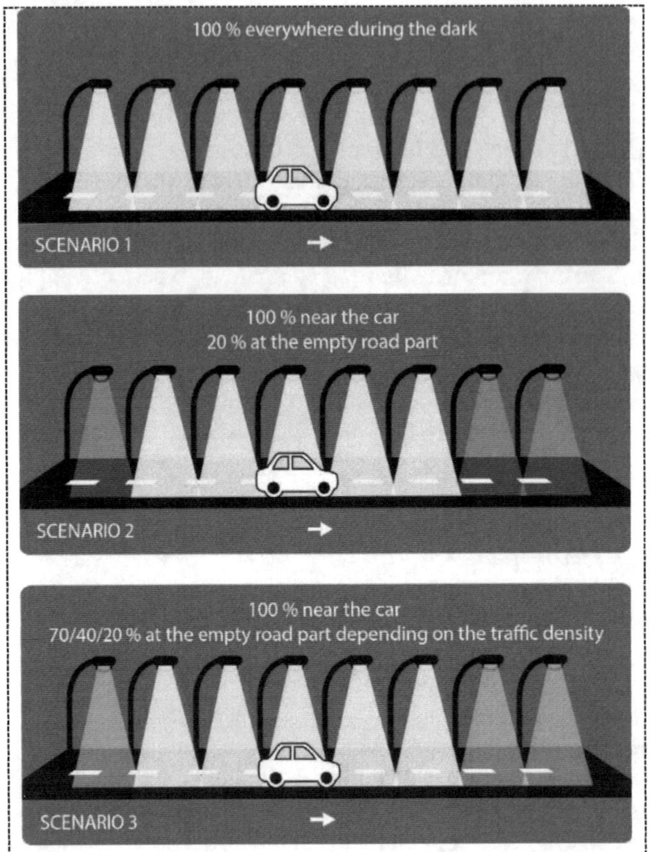

Figure 1. The lighting control principles of the scenarios. Scenario 1: Daylight level based control, Scenario 2: Presence sensitive lighting control, Scenario 3: Presence and traffic density based lighting control.

To study road users' experiences of the test site, questionnaires were deployed. During the first scenario, questionnaires were sent to all inhabitants of the area. While testing for the second scenario, the questionnaire was only sent to those who had explicitly agreed to participate again in the first questionnaire, but the questionnaire was also advertised in local media; this process was repeated for the third questionnaire. In the first and second phase of the testing period, the questionnaires used were almost identical. Prior to answering the questionnaire, participants were asked to drive along the road with the test lighting during the dark period of the day. Since there was no sidewalk or cycle lane on the side of the collector road, feedback from walkers and cyclists could not be obtained. Table 1 contains the number of participants answering to the questionnaires and some demographic data.

Table 1. Information about participants answering the first questionnaire (Q1), the second questionnaire (Q2), both the first and second questionnaire (Q1&Q2) and the third questionnaire (Q3).

	participants	female	male	age under 20-59	age 60+	not local
Q1	124	54	71	97	28	1
Q2	100	41	59	82	18	1
Q1&2	94	39	55	76	18	0
Q3	52	17	35	41	11	1

To gain background information regarding our participants, we asked about their usage of the road and its prevailing conditions during the time when they evaluated the lighting. Other questions concerned their overall impressions of the lighting; the experienced color of the lighting; their experiences regarding the amount of lighting on the road surface and on the surrounding environment; the experienced evenness (uniformity) of lighting; and experiences of glare. The participants were also queried on road visibility – how well they could see the road in front of them and other people moving on it or in the surrounding environment. Participants were also allowed to freely give personal feedback about the lighting. Moreover, participants were asked whether they had noticed any changes in the lighting during different times of the day or during their test drive. Those who answered to the second questionnaire were queried on whether they had noticed any change in the lighting after the first questionnaire. Most of the questions were formulated as scales with ratings on a scale of 0 to 5, with an added possibility for open-ended qualitative commentary.

Table 2. Results about the energy consumption simulation based on PIR sensor data of the date 8.2.2017.

The energy consumption at the test collector road		
SCENARIO 1	**SCENARIO 2**	**SCENARIO 3**
1.1 kWh / lighting fixture	0.5 kWh / lighting fixture	0.6 kWh / lighting fixture
Energy saving as compared to Scenario 1	**55 %**	**45 %**

For the evaluation of the first two scenarios, the participants did not receive any information about the new lighting other than it was accomplished with LED lights and that the lighting control had been developed during that same winter and spring. This was because the objective was to gain feedback from participant experiences that were as genuine as possible, unaffected by previous knowledge. During the testing of the third scenario, the approach was the opposite. Participants received detailed information of the three different lighting control methods that had been tested in each of the scenarios. We were interested in participants' attitudes towards lighting in general, especially those toward intelligent lighting as well as the three tested control methods. Furthermore, our objective was to gain more reflective viewpoints rather than mere reports of raw experiences. At this stage, the questionnaire did not require participants to visit and observe the lighting on site due to seasonal reasons. The third questionnaire was deployed in the summer, and outdoor lighting everywhere in the country was almost completely turned off at the time because of the long daylit periods during the summer months in northern latitudes. In Salo, outdoor lighting was used only for a couple of hours in the dead of night. Thus, a site visit component at this stage did not make sense.

IOP Conf. Series: Earth and Environmental Science 1099 (2022) 012018 doi:10.1088/1755-1315/1099/1/012018

To disseminate information about the lighting, our team had developed an urban dashboard (The City Monitor) [11] in the form of a website. In the dashboard, scalable charts, dynamic visualizations of the lighting behavior, and verbal descriptions of each lighting control method were presented. The scalable charts illustrated the average lighting and energy consumption levels. Figure 2 illustrates the information provided in the website simulation of the scenario 2. In the evaluation process, the participants could choose to answer questionnaires either in an electronic or printed form. With the third questionnaire, only the electronic one had interactive and dynamic simulations, but the printed one contained the same information in picture and textual mode. The PIR sensor data of a single date (8.2.) was used for simulating lighting behavior with the three different control methods, allowing comparison of the energy consumption [11]. The results of the comparison (presented in table 2) indicate great energy saving potential for both scenarios two and three.

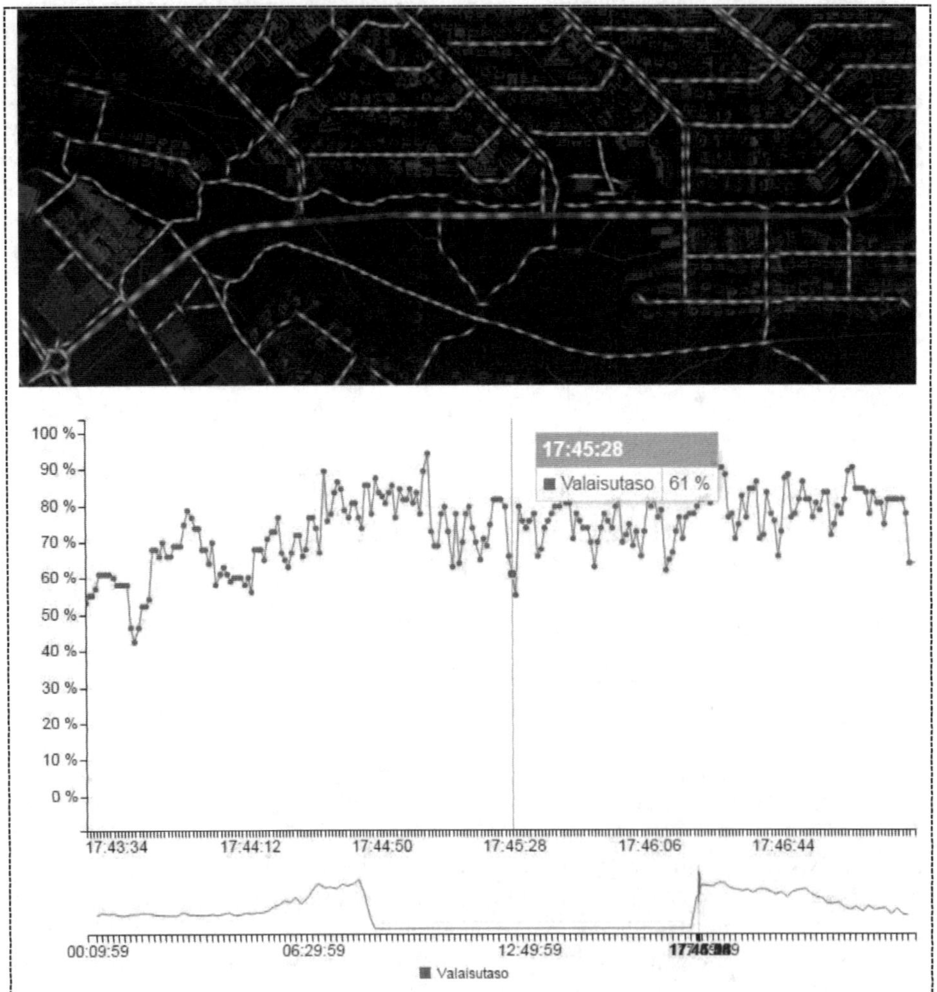

Figure 2. A screenshot of a simulation of scenario 2 as visualized on the dynamic interactive web page.

2.2. Data analysis methods

The participant answers to all three questionnaires were organized into Excel sheets and checked for validity. A few participants for the first and second questionnaire were rejected based on information indicating that the participant had not been driving a car or been a passenger in a car along the test road. The answers of all accepted participants were included in the analysis and description of experiences concerning the scenarios 1 and 2, the number of participants being 124 and 100 respectively, but in the statistical comparison of the two scenarios, only those participants were included who had answered both the first and the second questionnaire (n = 94). The answers for the open-ended questions were thematised into important themes and aspects, and explanatory factors for experiences were collected.

The objective was to show that the experience of presence sensitive lighting would be as positive as traditional lighting. A result of having no differences between the experiences of the two scenarios would indicate that. To test whether there were statistically significant differences between the experiences of the scenario 1 and 2, the dependent samples t-test for paired samples was selected as the statistical analysis method. The data included a good sample of 94 road users evaluating both lighting scenarios, the traditional and presence sensor controlled. In the test, the difference scores (D) were used as variables, so for statistical analysis, the difference between the values of each question were calculated for every participant. As this distribution of the differences was fairly symmetrical, testing could be performed as a dependent samples t-test and corresponding 95% confidence interval calculation. If there was no difference between the sample pairs, then the mean of the difference scores would be equal to zero. Therefore, the null hypothesis to be tested was $\mu_D \neq 0$.

3. Results and analysis

3.1. *Traditional and Presence-sensitive scenarios*

Figure 3 illustrates the results concerning the experiences of all participants evaluating scenarios 1 (traditional lighting, n = 124) and 2 (presence sensitive lighting, n = 100). Generally, the participants gave both lighting scenarios high scores in all the aspects which were queried, and the scores were similar regarding both traditional lighting and presence sensitive lighting. Concerning the general appearance of lighting (0 = very unpleasant, 5 = very pleasant), the mean of the scores was 3.98 with traditional lighting and 3.94 with presence sensitive lighting. When participants were asked whether anything in the lighting disturbed them, 81 % answered "no" with traditional lighting and 82 % with presence-sensitive lighting. While evaluating the uniformity of the lighting on the road (0 = not uniform at all, 5 very uniform) the mean was 3.64 with traditional lighting and 3.70 with presence sensitive lighting. Lighting did not seem to cause much glare while driving, as the mean of the scores (0 = very much, 5 = not at all) was 4.14 with traditional lighting and 4.16 with presence sensitive lighting. Participants were also asked separately how well they were able to see the roadway and others moving on the road and its surroundings (0 = very badly, 5 = very well). Here the mean concerning roadway visibility was 4.16 with traditional lighting and 4.17 with presence sensitive lighting. The visibility of others moving on the road and in the surroundings was experienced as being slightly worse: the mean of the scores was 3.53 with traditional lighting and 3.77 with presence sensitive lighting. This relates also to the evaluation of the amount of light separately on the road and in the surroundings: more participants experienced that there was a sufficient amount of light on the road surface (89 % with traditional lighting, 92 % with presence sensitive lighting) than in the surroundings (74 % with traditional lighting, 79 % with presence sensitive lighting).

The participants were given the possibility to verbally comment on their numerical evaluations, which helped us further understand the experience beyond the scores. Positive feedback was given to both scenarios regarding a sufficient amount of light (*"not too bright", "a sufficient amount of light", "bright"*), the tone of light (*"soft, good tone"*) and the absence of glare. However, some of the same aspects were also criticised in a few comments (*"too weak", "bright LED-lights where lumen values are raised by raising the colour temperature are not sufficient for human eyes"*). Obviously, critical comments were given to the question of whether anything disturbed the participants in the lighting. These negative notions included experiences of glare (several comments), unevenness of lighting, inefficiency of lighting, costs, and the perceived redundancy of lighting. Also, a concern regarding not having enough light by the pedestrian crossings and cross-roads in general was raised.

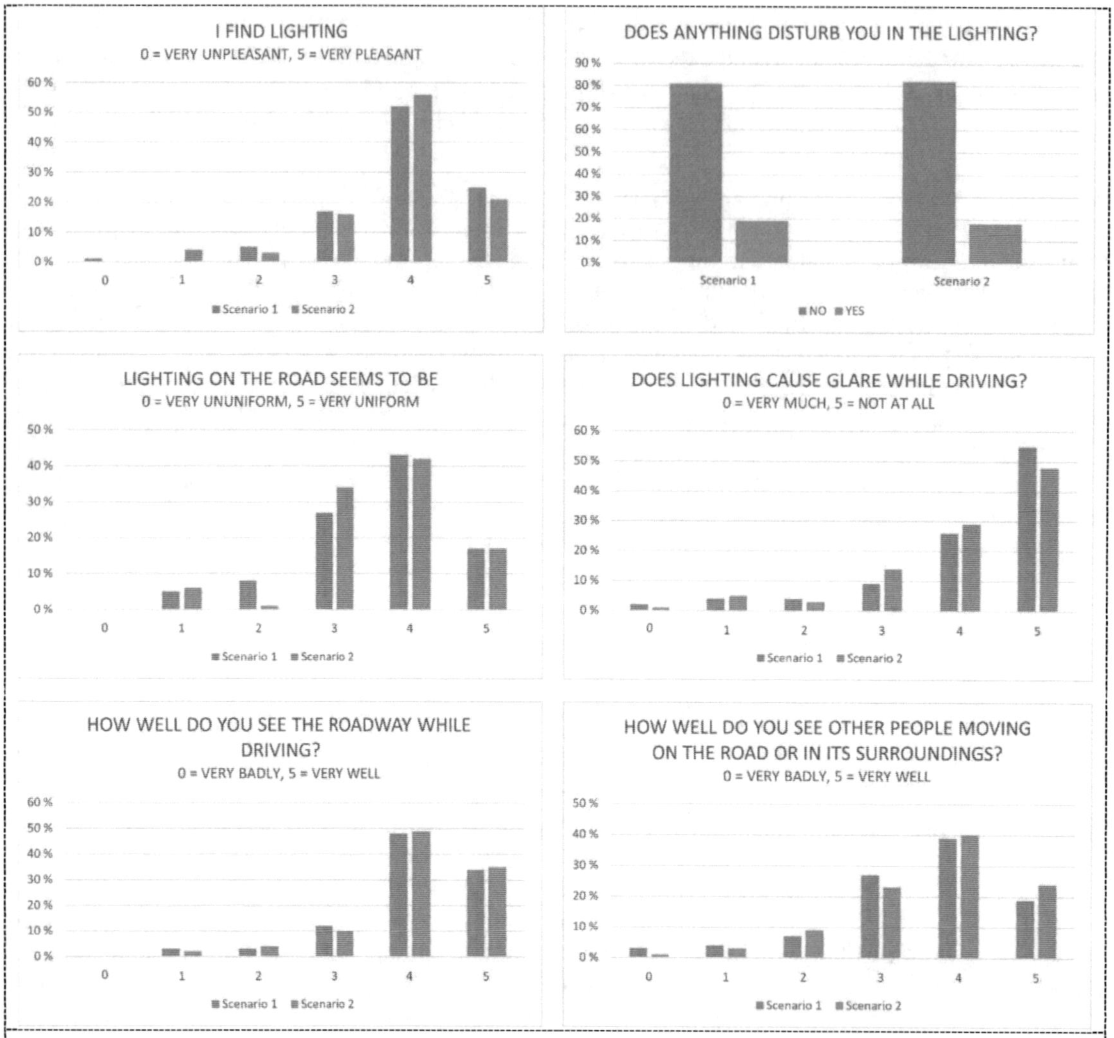

Figure 3. The comparison of participants' scores of Scenario 1 (Traditional lighting) and Scenario 2 (Presence sensitive lighting)

Table 3 presents the statistics from the dependent samples t-tests conducted. For all the questions concerning 1) the appearance of the road lighting in general; 2) the evenness of light distribution on the road surface (uniformity); 3) experiencing glare when driving and; 4) experience of seeing the roadway when driving, the statistical analysis showed no difference between the compared conditions. The p-value in these was between 0.48 – 0.89 indicating that the null hypothesis had to be abandoned and the detected variation between the results is coincidental. The only exception was the question *"When driving, how well do you see the other users of the road and the environment?"* Here the results indicated a slight statistical significance in favor for the presence sensor controlled lighting scenario: $t(94)=-2.296$, $p=0.024$ (dependent samples t-test). Thus, it could be stated that the experiences of presence sensitive road lighting are at least as positive as experiences of traditional lighting, in light of the aforementioned questions.

Respondents to both the first and the second questionnaire were questioned on whether they had noticed any differences in lighting during different times of the day. With the traditional lighting, 97 % had not noticed any differences, and with the presence-sensitive lighting, 93 % had not noticed any differences. There were no differences in the control method during different times of the dark time with the first two scenarios. However, depending on the time of day, the amount of traffic is altered and due to this, also the time when lighting is on at the 100 % control level changes. Additionally, there were two extra questions in the second questionnaire. Firstly, participants were asked whether they had noticed any change in lighting after the first questionnaire. Out of those respondents who had answered also to the first questionnaire, 76 % had not noticed any change. The comments included approximately the same number of remarks about brighter and better lighting as about dimmer lighting; however, most of the evaluations indicating dimmer lighting had added a question mark or also doubted that the reason may have been snow conditions or their own imagination. A couple of participants had noticed that a few broken luminaires had been replaced with new ones, and that lights had been on during a bright day. This was probably due to the testing of the replaced luminaires. Two participants commented about noticing the dynamic dimming and brightening of lighting. One of them had been walking while making the observation. The other commented that the lighting brightens when a car is approaching. Secondly, the participants were asked whether they had noticed any change happening in lighting while they were moving along the test road with a car. 86 % of the respondents answered "no". Of those participants who reported noticing something, only four described noticing dynamic dimming and brightening. The others commented on different types of experiences, for example, they described the lighting to be better, brighter in the dark, having a more pleasant tone, or being dimmer.

Table 3. The results of dependent samples t-tests using difference scores as variables (n=94): mean, standard error, 95 % confidence interval and p-value are presented.

	Mean	SD	95 % CI	p-value
Appearance of the road lighting	0.043	0.775	[-0.116, 0.201]	0.596
Light distribution on the road surface	-0.064	0.878	[-0.244, 0.116]	0.482
Glare when driving	-0.021	1.136	[-0.254, 0.211]	0.856
Visibility of the road when driving	-0.011	0.796	[-0.174, 0.152]	0.897
Visibility of road users and environment when driving	-0.234	0.988	[-0.436, -0.032]	0,024*

3.2. Informed opinions of all scenarios

Here, we report on answers to the open-ended questions of questionnaire 3 that involved all scenarios.

When queried about the participants' opinions regarding *scenario 1* (*"What do you think about this type of a lighting control method where the lighting is kept at the same level?"*), most participants considered this method to be inadequate or not sensible in some way; two participants were of the opinion that it was *"old-fashioned"*, and three thought that it was simply *"bad"*. Most didn't consider this to be a sensible way to light since *"technology enables better control without added costs"*, or they had concerns regarding light pollution or wasteful energy consumption. Curiously, one participant noted that they had thought this was the only possible way, but having perused our questionnaire, the participant concluded that this was *"old-fashioned"*. Thus, it must be taken into consideration that this might be true for some other participants as well. Even so, some of our respondents had a positive opinion of the traditional lighting scene, thinking it was *"sensible and 'nice' (safe)"*, simply *"ok"* or that it *"improves visibility for the elderly"*. However, these opinions were in the minority, with most stating that they would prefer a different type of a lighting control method. Of all the participants, 28 % wanted to have this scenario as a permanent lighting solution.

Concerning *scenario 2*, (*"What do you think about this type of a lighting control method where the lighting is dimmed where there are no passers-by?"*) the responses were decidedly more positive. Several participants described the lighting control method in the scenario as *"good"*, *"very good"* or *"sensible"*, and many others still gave more in-depth positive answers: *"I think it's sensible to control lighting according to need"*; *"Good. It doesn't glare and it's better for the surrounding environment as well"*; *"Good idea, as long as it brightens when there's passers-by"*; *"Sounds logical"*. Several noted that it would save energy in their opinion. Some also had requirements for the system: Firstly, that the control method must function reliably. Secondly, that it would need to take into account all types of traffic, especially at a cross-roads or pedestrian crossing. However, some clearly dissenting voices were also expressed. One notable opinion was that scenario 2 was *"unnecessarily fancy"*. One other was concerned that it might add to feelings of unsafety. Others simply expressed negative opinions, stating that the scenario was *"not good"*, *"not nice"*, *"suspicious"* or *"bad"*. However, these were in the minority, and unfortunately, these negative opinions were also left unexplained. Of all the respondents, 45 % wanted to have the scenario 2 as a permanent lighting solution.

For our *scenario 3* (*"What do you think of a lighting control method like this where the lighting is dimmed where there are no passers-by, and the lighting is also controlled by the density of traffic?"*). Once again, the answers proved to be mostly positive. *"Best option"*, *"I support this [idea]"*, *"Good"* and *"Sounds good"* were common answers. Many participants also gave more detailed reasons: *"This is good too, why to light up empty roads." "Amounts of traffic do indeed vary a lot. It's more important to see further ahead during rush hour. Usually, light is needed more when there are more other road users."* One stated that it was a *"Good compromise between safety and energy consumption"*. Again, there was an important minority of negative opinions as well. Some expressed that it seemed difficult to understand, or that it was suspicious. Accurate sensing of passers-by was doubted by one participant. In general, fear of faulty operation seemed to be the most prevalent negative opinion. This is understandable from a commonsense point of view, as a more complex system can be seen as riskier. One also pointed out that more "moving parts" means more maintenance which might cost more. One laconically stated, *"As long as it works"*. 55 % of the participants accepted scenario 3 as a permanent lighting solution along the road, and 72 % either one of the scenarios 2 or 3.

At the end of the questionnaire, participants were asked how gaining more information about the lighting affected them. The overwhelming majority of opinions stated that it had a positive effect on them and that intelligent lighting control had even surprised them positively. One participant was frustrated that the information on the costs of devices was not available. We were surprised at how interested and motivated our participants were to learn more about street lighting control methods; however, it must be noted that participants had self-selected to participate. Nevertheless, there are many individuals with a desire for more knowledge on urban lighting among the general public.

4. Discussion and conclusion

The results of the real-world case study in Salo indicate that drivers' experiences of presence sensitive roadway lighting are at least as positive as the experiences of traditional, uncontrolled lighting concerning many factors which are essential from the perspective of traffic safety and visual comfort. These included the experienced uniformity of lighting on road surface, the experience of glare, and the visibility of road while driving. In addition, the presence sensitive lighting was experienced as being as pleasant as traditional lighting. In the statistical testing of the results with dependent samples t-tests no statistically significant difference between the two tested lighting scenarios was detected. However, concerning the visibility of other people moving on the road and in the surroundings, the test results indicated with slight statistical significance better results with presence sensitive lighting. One explanation for this surprising result might be the experienced focus effect as the illuminances were higher near the driver than in the far distance along the road. The results of the recent study of dimmed, static road lighting and car headlights [9] interestingly relate to this and open up new research questions and applications for presence sensitive and intelligent lighting. When the participants were asked specifically about negative experiences, almost the same percentage of them (19 % with the traditional lighting, 18 % with presence sensitive lighting) reported that something disturbed them. Interestingly, most of the drivers (at least 86 %) did not notice any dynamic change in lighting while driving along the test road. When informed about the tested lighting strategies, most of the participants (72 %) would prefer either one of the intelligent lighting modes to be the permanent lighting solution. The attitudes towards intelligent lighting were generally positive: it was deemed to be wise and modern solution – saving energy and costs, providing lighting when needed. While these results are extremely encouraging, they also warrant more research into the experienced safety and comfort of presence sensitive roadway lighting from different road users' perspective.

Acknowledgements
SenCity Project was funded by TEKES The Finnish Funding Agency for Innovation and by participating cities (Helsinki, Lahti, Oulu, Raahe, Salo, and Tampere), companies (C2 Smartlight, Elisa, GreenLED, Misal, Nokia, and Valopaa), and research institutions (Univ. of Oulu and VTT. The authors gratefully acknowledge S. Kourunen, T. Pukinkorva and Jari Päkkilä from the University of Oulu, E.M. Sarjanoja, R. Vääräniemi and J. Eskeli from VTT, and all the participants for their contributions to the case study.

References

[1] Hughes R F and Dhannu S S 2008 Substantial Energy Savings through Adaptive Lighting 2008 *IEEE Canada Electric Power Conf.* IEEE 2008 pp 1–4

[2] Fostervold K I, Larsen P J, Lillelien E, Mjøs T and Berg M O 2010 Energy efficient lighting systems consequences for environment, health and human factors *Proc. of CIE 2010 Lighting Quality and Energy Efficiency*

[3] Ceriotti, M et al. 2011 Is There Light at the End of the Tunnel? Wireless Sensor Networks for Adaptive Lighting in Rad Tunnels *Proc. of the 10th ACM/IEEE Int. Conf. on Information Processing in Sensor Networks* IEEE 2011 pp 187–198

[4] Lau, S P, Merrett G V and White N M 2013 Energy-efficient street lighting through embedded adaptive intelligence *Int. Conf. on Advanced Logistics and Transport (ICALT 2013)* Sousse Tunisia 29-31 May 2013 pp 53–58

[5] Ouerhani N, Pazos N, Aeberli M and Muller M 2016 IoT-based dynamic street light control for smart cities use cases, *IEEE Conf. Public. in 2016 Int. Symp. on Networks, Computers and Communications (ISNCC)* Yasmine Hammamet Tunisia 11-13 May 2016 pp 1–5

[6] Boyce P R, Fotios S, Richards M 2009 Road lighting and energy saving, *Light. Res. Technol.* 41 (3) (2009) pp 245–260

[7] Pihlajaniemi H 2016. *Designing and experiencing adaptive lighting. Case studies with adaptation, interaction and participation.* Doctoral dissertation. University of Oulu Graduate School; University of Oulu, Oulu School of Architecture, Acta Universitatis Ouluensis H 3.

[8] Atıcı C, Ozcelebi T and Lukkien J J, Exploring user-centered intelligent road lighting design: a road map and future research directions, *IEEE Transactions on Consumer Electronics* vol. 57 no. 2 May 2011 pp 788-793

[9] Bozorg S, Tetri E, Kosonen I and Luttinen T 2018 The Effect of Dimmed Road Lighting and Car Headlights on Visibility in Varying Road Surface Conditions *LEUKOS* 14:4 pp 259–273

[10] Pihlajaniemi H, Juntunen E, Luusua A, Tarkka-Salin M and Juntunen J 2016 SenCity – Piloting Intelligent Lighting and User-Oriented Services in Complex Smart City Environments, *Proc. of eCAADe 2016 Conf.* pp 669–680

[11] Pihlajaniemi H, Luusua A, Sarjanoja E-M, Vääräniemi R, Juntunen E and Kourunen S 2017 SenCity City Monitor as a platform for user involvement, innovation and service development *Proc.* of *eCAADe 2017 Conf.* pp 561–570

IOP Conf. Series: Earth and Environmental Science 1099 (2022) 012019 doi:10.1088/1755-1315/1099/1/012019

Window geometry and its effect on the experience of illuminated spaces – a study of three daylit architectural cases

N Mathiasen (1), L Grønlund (2), A K Frandsen (1), and M Harild (3)

1 AAU, Department of the Built Environment, Denmark
2 Royal Danish Academy – Architecture, Design, Conservation
3 AAU, Create, Lighting Design

corresponding author: nam@build.aau.dk

Abstract Looking at the architecture of modern housing in Denmark, it seems to express an understanding that equals more daylight with good daylight. With coated energy glass, the windows have increased in size, making glass into a façade material. This means that windows are no longer just holes in the façade but rather make up the entire façade itself. This changes the spatial relationship between the window design and distribution of daylight within the interior space, though we seldom address this lacking a vocabulary and methods. This paper sets out to investigate and experiment with how we can describe and document this change in daylight conditions and how these influence our visual environment, using photographs to record this. The methodology focusses on observations of daylit spaces in three different housing examples in Copenhagen from the 1800s to today. By observing the three daylit spaces, it becomes clear that differences in the size, shape, and position of window apertures influence the qualities of light in a space significantly. Considering that when it comes to window apertures different designs affect the experience of light in a space significantly, it is important to take this into account when designing with daylight. It is also important from a sustainability point of view to include this awareness in future design approaches.

1. Introduction: experiencing a daylit room

The geometry of a window determines the character of the daylight in a room. Size, shape and position of a window have a significant influence on the illumination of a room with daylight [1]. This can be explained through the physics of light, where light is defined as radiation which travels in straight lines and interacts with the surfaces it meets, thereby allowing light to penetrate a room based on the design of the aperture. The illumination of a space is therefore directly related to the *geometry of the window.*

Light in a space is traditionally measured on a horizontal plane 0.85m above the floor, described through the daylight factor (DF). The DF indicates the relationship between the exterior and interior light as a measure of the illuminance within a room relative to the outdoor illuminance incident on a horizontal plane from an unobstructed overcast sky. The DF is expressed as a percentage for various points in a room, indicating the amount of light and thereby also the variation of light throughout the space. [2]. This is one of the main properties of illuminated spaces, a property which is relatively easy to register if these spaces can be simulated and measured using well-known units of light. The DF is one of the most often used measures of light in a space even though it has its limits, as it does not provide any information regarding the climate, latitude or orientation of the room in question.

Apart from being measurable, light-filled spaces are also perceived by humans, the inhabitants of the space. Engineer R.G. Hopkinson (1913–1994) also stated this in his book *Daylighting* [3]: *Light is the visual manifestation of radiant energy and, consequently, is intimately related to human sensations.* When daylight openings vary in size, shape and position, illuminated spaces vary in appearance. This paper is investigating the impact of different aperture designs on the illumination of spaces and how they are perceived. This paper is thereby centred on the experience of visual environments, more specifically, the qualitative aspects of daylit spaces. In order to provide a focussed study on the qualitative aspects of light, the quantitative aspects are intentionally left out. A future study could benefit greatly by looking at both the quantities and qualities of light.

At present daylight design is often described through quantitative aspects, such as efficiency, energy consumption and climate conditions, for good and unignorable reasons. Advanced computer-aided design techniques allow daylight situations of a building to be simulated in 3D programs, providing information on energy consumption and indoor climate. Simulations carried out in relation to specific climate conditions [4] can predict the daylight levels and distribution in a particular space [5]. Information which relies on quantitative data to ensure that standards, guidelines, or certifications are met, all important to support sustainable development and to decrease energy consumption of our build environment [6]. However, these simulations often leave out the aspect of how the illuminated space may be perceived. Aspects that support our vision and could thereby potentially diminish the use of energy. These qualitative aspects are often highlighted for different user groups, e.g. visually impaired people, for whom they are essential [7].

Qualitative aspects of daylight design are difficult to measure, but they are undoubtedly experienced and of importance to people occupying the space [8]. Acknowledging the importance of the qualitative aspects of daylight, this paper investigates systematic methods to register them. Through three cases, the visual impact of the design of daylight apertures in relation to illuminated spaces is studied.

2. Objectives and research question

The focus of the paper is on how daylight, window aperture and space interact and create the perceived light *in the space* and *on the objects*. Through the interplay of light penetrating the geometry of a window aperture, surface reflectance and space boundaries, the visual environment of the space is created. The three cases represent three different designs of daylight aperture that are commonly known and differ significantly.

The objective is to describe the relationship between the geometry of the aperture and the perceived effect of the light in the interior space. The experience of light is studied in relation to both the overall illumination of the space and how a three-dimensional object is perceived in the light.

The aim of systematic registration is to gather knowledge on the perceived light, which potentially can qualify architects' daylight design process and give building clients a more specific vocabulary for the spatial qualities of daylight.

Focusing on the visual environment of the daylit space, the main question is: *How do various aperture designs affect the way light is perceived in the space?*

To answer the question, the following sub-questions have been posed: Firstly, *how do the different designs of daylight aperture determine the overall light distribution, and how is this perceived?* Second, *how does the framing of the window affect the way daylight is perceived in the space?* And thirdly, *how is a three-dimensional object perceived when exposed to light emerging from different daylight apertures?*

IOP Conf. Series: Earth and Environmental Science 1099 (2022) 012019 doi:10.1088/1755-1315/1099/1/012019

Figure 01. Daylight penetrating a traditional cross-window on an overcast day.

3. Theory

This project draws on the Nordic tradition of both studying and acknowledging the impact of daylight qualities in architectural design. The Norwegian architect and Professor of Architectural Theory and History Christian Norberg-Schultz (1926–2000) described how the local environment has an impact on traditional architecture [9] and argued that daylight is one of the most significant elements of space and place [10]. In Sweden, architect Sven Hesselgren (1907–1993) focused on how people perceive light [11].

This is pursued by architect Professor Anders Liljefors, who establishes a precise vocabulary on light with a clear distinction between measurable and perceived light [12]. In Denmark, lighting design research has focused on the *"geometry of light"* [13] both as a tool of design and of understanding. Architect Poul Henningsen (1894–1967) designed the PH-lamp with the ambition of making lamps that reflected rays of light and created a lighting design without glare and with a balanced shadow pattern based on this geometric understanding of light [14]. This approach was developed further at the School of Architecture at the Royal Danish Academy of Fine Arts and taught by architect Mogens Voltelen (1908–1995), who worked with Poul Henningsen. The colleague of Mogens Voltelen architect Sophus Frandsen (1927–2013) transformed the geometric understanding of light from artificial light in the field of daylight design. He studied how to describe daylight that penetrates windows systematically, including the spatial and qualitative effects. His studies were done in spaces/labs designed for the purpose. The advantages of studying light in a lab is that it is possible to control and change the surface reflectance and size and position of the daylight apertures. His theory and results were published in the article: "*The Scale of Light*" [15], which has been the inspiration for the study presented in this paper regarding the overall impression of the spaces and the details of the shadow patterns.

With the point of departure within architecture and the subsequent attention on spaces and being in spaces, this paper focuses on how light is perceived in a space, in other words, it focuses on spatial light distribution, surface texture, the form of three-dimensional objects and shadow patterns. Computer simulations and calculations can be extremely useful, but the subtle and delicate surface texture and shadow pattern on three-dimensional objects are still difficult to reproduce in computer simulations. By referring to the Nordic theory on daylight and architects' attention to the way light is perceived, the aim is to (re)introduce useful tools to describe qualitative properties of daylight.

4. Methodology

The study is based on case studies performed as onsite observations [16] of the daylit spaces in three housing examples in Copenhagen from the 1800s to today. Each case involves a sidelight, but the designs of the daylight apertures differ significantly, from the traditional cross-windows to a bay-balcony window and a curtain wall, presenting different relationships between the design of the aperture and the daylight in the interior space.

All three cases are located within a building complex in an urban setting in Copenhagen, Denmark. All three have an unobstructed view of pleasant surroundings and open sky. The oldest case, with classical cross-windows, is oriented towards the east and has a view of a historical square in the centre of Copenhagen; the case with a bay-balcony window is in a modern movement apartment building oriented towards the west and facing the lakes of Copenhagen; and the newest building with triangular balconies is oriented towards the south and overlooks a preserved national park. Except for the case with the cross-window, the cases have balconies. Though the balconies influence the daylight penetration, they are not discussed in this study. The interiors of all three cases have been painted with white matte paint with approximately 80% reflectance. They have light-coloured wooden floors.

The observations have been made onsite by the authors, focusing on three themes: 1) the overall light distribution, 2) the mediation of light from the exterior to the interior, focusing on the design of the window framing and 3) the form-enhancing effects of the light on a three-dimensional object in the space.

As the three cases are oriented differently, the observations have been made without direct sunlight penetrating the spaces. The inhabitants of the apartments have kindly given access to their private homes, but it was not possible to visit Case 2 on a day where there was an overcast sky, hence the study of the three-dimensional objects was only conducted for Cases 1 and 3.

The three cases have been documented through a series of systematic photo registrations. Recorded with the following camera: Canon EOS 5D Full-frame DSLR / Objective: EF 24-105mm f/4L IS USM. Each photo series relates to one of the three themes mentioned above, described respectively in subchapters 6.1., 6.2. and 6.3. (See fig. 03, 04, 05).

Figure 02. Three cases, exterior view. Left to right: *Gråbrødre Torv*, *Vestersøhus* and *VM house.*

5. Cases: three window typologies

The window designs included in the study vary in size, shape and position. The windows have been selected due to their design, though they also represent three different periods of architectural history. The window typologies are the cross-window, the bay-balcony window and the curtain wall, representing an increase in window size. Each of the window typologies are well known and common in a Danish context, though the details and execution can vary greatly.

5.1. Cross-window

The cross-window has a four-light casement with a mullion and a transom forming a cross. The casement window opens on hinges like a door with a closing mechanism (fastener). The two lower casements are about twice the height of the top sashes. The mullion and transom form a cross, and the window is therefore usually called a “cross-window”. This window has secondary glazing. This type of window is very common in Danish residential housing from the 18th and 19th centuries. [17].

Case 1 *Ildebrandshuset* on Gråbrødretorv was built in 1729 and is part of some of the very old residential buildings in Copenhagen. They are named Ildebrandshusene – the Firehouses – as they were constructed after the first Great Fire of Copenhagen in 1728. The existing window is not the original window, as the cross-window was not introduced until the late 18th century due limitations in the production of glass. Developments in the production of glass in the late 18th century meant that after the second Great Fire of Copenhagen in 1795 cross-windows became widespread. In 1729 it was not possible to produce as large panes of glass as needed for the four-light casement window [18]. Therefore, the original window would probably have been a multiple-light casement window, and the glassed area might also have been smaller [19].

5.2. Bay-balcony window

The bay-balcony window is a window that integrates a balcony and has corner windows letting light in from two sides. The window area is larger than the traditional cross-window, meaning there is often only one large bay-balcony window in a living room compared to cross-windows, where there are typically several.

Case 2 *Vestersøhus* by the lakes of Copenhagen was designed by architects Kay Fisker and C.F. Møller and was built in 1936 as one of the first and most excellent examples of the bay-balcony houses [20]. This type of residential housing was very popular during the heyday of Danish Functionalism. The focus on light and fresh air promoted building structures that included balconies with easy access to the exterior [21].

5.3. Curtain wall

The curtain wall is a façade concept where the façade does not carry any structural load, as opposed to traditional façade systems, where the façade is load bearing [22]. This allows freedom with regard to designing the façade and opens up the possibility of constructing the whole façade entirely from glass. However, in a Danish context it must shield from various weather conditions, including intense sunlight during the summer and cold weather during the winter. Hence to provide the spaces with a comfortable indoor climate, special glass that can create a balance between heating and cooling must be used. This glass technology often lowers the light transmission.

Case 3 *VM Houses* designed by Plot was built in 2005. The building was one of the first housing complexes in Ørestaden, a newly developed environment on the outskirts of Copenhagen. The playful building was constructed using concrete, steel and glass. The housing complex consist of 80 different apartment types, most of them with more than one level. Each apartment has been designed as one open space, allowing the inhabitants to subdivide it into more spaces as they wish. The façade is entirely made of glass subdivided into smaller areas, allowing for doors to the balcony. The glass façade allows light to penetrate the apartments unhindered and offers the inhabitants an undisturbed view of the outside.

6. Findings: the interrelation between window and the experience of a daylit space

The following three subchapters present the findings from the study, focusing on 1) how the different designs of the daylight apertures determine the overall light distribution, 2) how the framing of the windows affects the perceived daylight in the space and 3) how a three-dimensional object is perceived in light coming from different daylight apertures.

Figure 03. Three cases, interior view of the overall light distribution. The photos represent different experiences of the spaces according to different positions in the space. Top to bottom: cross-window, bay-balcony window, and curtain wall.

6.1. *Illuminated space*

The experience of the overall light distribution of the three spaces varies (Fig03). In Case 1, the light level is experienced as generally high and evenly distributed even though the façade wall is darker than the sidewalls, as it only receives reflected light from the space. Light from the window on an overcast day is scattered softly onto the sidewalls, where a significant pattern of brightness reflects the design of the window. These variations in light level are stimulating and apply a character to the space. The areas of brightness are also experienced on the floor and ceiling, which entails a pattern of bright and less bright areas. The brightness fades further away from the window, creating an experience of a soft shaded environment with a myriad of grey tones, which are especially visible on the walls. Furthermore, the semi-gloss surfaces on the wooden panels around the daylight apertures create reflections which are especially visible at a short distance.

In Case 2, the light level is experienced as generally high and reaching rather far into the space. A larger bright area is experienced on the floor, ceiling and on the sidewall fading into the room but with less nuances than in Case 1. The visual environment is experienced as bright with few contrasts and easy to comprehend.

Case 3 on the contrary is experienced with large contrasts. The level of light is experienced as very high close to the window but decreases rather quickly further into the space, giving rise to contrasts that reduce the ability to see the space properly. The window occupies the whole façade, which further creates an intense backlight. Therefore, the experience of the space differs significantly depending on which way you are facing the curtain wall. Furthermore, the semi-gloss floor creates reflections and thereby glare that can be disturbing for vision.

Figure 04. How the design of the aperture modulates the daylight of the space. From left to right – Case 1: cross-window, Case 2: bay-balcony window, and Case 3: curtain wall.

To conclude, the three spaces differ regarding the variation of light in the space. In Case 1, there are areas of pure reflected light over the window and in the corners of the façade wall, creating a rhythm between precise direct daylight and soft reflected daylight and thereby producing different options depending on preferences. In Case 2, there is the same type of variation between the direct and the reflected light in the space, with direct light in front of the window and reflected light in the corners, though not as a rhythm as in Case 1. In Case 3, the direct light is evenly distributed along the curtain wall, with precise light that accentuates details on the ceiling and the walls, without variation.

6.2. Light transition

The framing of the window affects the perceived daylight in various ways (Fig. 04). In Case 1, the wooden framing of the window mediates the transition between the daylight level of the exterior and the interior. The framing modulated by panels with profiles softens and allows a gradation of light, diminishing the potential glare from the exterior light and thereby affecting the experience of the interior light positively. Though the framing in Case 2 has less detailed profiles, it still creates an effective gradation of light from the exterior to the interior and softens the experience of the light coming through the window. As Case 3 has no framing the transition between the light level of the exterior and the interior is very contrasted. Intense light settles on the sidewalls, ceiling and floor, creating a very bright area close to the window. Whereas Case 1 and 2 had a window framing with details mediating light, Case 3 has only the framing of the industrial window product without further architectural mediation between the exterior and interior light. The few but very functional elements of the window – the plastic joint and the ventilation opening – get all the attention, as the light accentuates them.

6.3. Form-enhancing shadow pattern

Three-dimensional objects are perceived differently depending on differences in daylight apertures (Fig.05). Whereas the two previous observations focused on light *in the space*, the last focuses on light *on an object* and thereby on the smaller shadow pattern that enhances form and details. To see light on a three-dimensional object, a sphere has been used. The sphere can also represent a human body or face. In Case 1 and 3, a sphere has been photographed and observed in 5 positions with 0.5 m in between on a straight line perpendicular to the middle of the window 0.85 m above the floor, starting as close to the window as possible. Observations up to a distance of 2.5 m inside the room from the window have therefore been made.

In Case 1, there is great contrast between the top and bottom of the sphere. In the position close to the window, the directionality of the light on the sphere differs depending on the position. Close to the window, the directionality of the light is very steep, but it lowers the further away from the window the sphere is positioned. The direct light from the sky is present close to the window and creates a very precise shadow pattern with a clear bright and dark side, enhancing the shape of the sphere. The reflected light takes over further away from the window, the sphere gets a softer shadow pattern where the roundness of the sphere disappears, and it becomes difficult to perceive the shape. These variations are

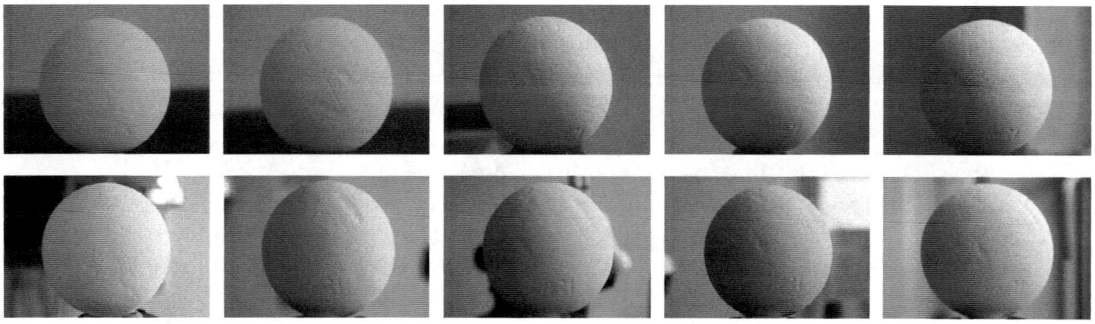

Figure 05. How a three-dimensional object receives light from a cross-window, Case 1 (top) and a curtain wall, Case 3 (bottom). The window is positioned to the right of the spheres.

not only present from the façade and inwards but also parallel to the façade. As there are two windows, there will be a rhythm of direct light and reflected light depending on whether the sphere is in front of the window or between the windows.

In Case 3 the sphere is more evenly lit in the various positions. The light from the whole façade illuminates the whole sphere and makes it difficult to distinguish whether it is close to the window or further away from the window. Even though a vague highlight is noticed on the sphere closest to the window, there is very little variation in brightness and the shadow pattern on the sphere in general. Furthermore, varying the position parallel to the façade would not make the appearance of the sphere differ, as the curtain wall will light the sphere evenly. Whereas the appearance of the sphere in Case 1 varies from looking clear and round close to the window to looking more like a sponge deeper in the space, the appearance of the sphere in Case 3 is very much the same wherever it is placed in the room. That makes a large difference in experience of a three-dimensions object according to the design of the daylight aperture.

7. Discussion and conclusions: balancing the qualities of light in daylight design

Three different cases have been observed onsite in relation to the themes and theory developed by Nordic architectural lighting research primarily discussing and unfolding the qualities of light and the experience of the visual environment. Observing the three daylit spaces, it becomes evident that the size, shape and position of a window aperture influences the qualities of light in a space significantly.

When comparing the three different cases, we find a milder transition in Case 1 with the cross-window, which is both smaller and is a hole in the wall. The modulation of the daylight apertures and the surrounding wall soften the light and not least lessen the experience of glare when looking towards the daylight apertures, especially compared to Case 3 with the curtain wall, but also to some extent Case 2 with the bay-balcony window. The size of the daylight apertures in relation to the façade has significance: the larger the light aperture is, the greater the perceived contrast is in the space.

Likewise, the three cases are very different when it comes to variation in light. In Case 1, there are both areas of very direct and precise light close and parallel to the window but not in the areas opposite the window, between the windows and inwards into the space with soft reflected light. Though the space is small, there is variation in the light. In Case 2 there is less variation, but there are some areas with either precise and direct light and areas with soft reflected light. In Case 3 the wall-to-wall window results in very uniform light that accentuates details both at the ceiling and the walls.

How the details of the framing of the windows are done influences the experiences of glare. The perceived light becomes visible in the modulated transitions between outside and inside when these are designed and shaped by frames, profiles, and ornaments. These shaped transitions transmit light and reduce glare. Thus, details are highlighted, e.g. the profile in Case 1 & 2, and when there are not any transmitting details, as in Case 3, then the plastic joint is accentuated by the light.

Combining the observations of the sphere from Case 1 and 3, it can be concluded that the two different daylight apertures produce two very different light settings that produce very different visual appearances of objects. The light from the curtain wall creates an evenly lit environment without many nuances in the shadow pattern. On the contrary the light from the cross-window provides a space with very dynamic light. As there is variation in shadow pattern across different areas of the space, a variety of visual appearances of three-dimensional objects are produced.

In the study accounted for in this paper, the observations were made in real spaces. In real spaces, all minor and subtle details are visible and can therefore influence the evaluation of the experience of a daylit space. This is a preliminary study attempting to develop alternative ways to systematically register and observe aspects of the perceived light in a space. It is emphasised that this study has not yet developed a thorough vocabulary of the qualities of daylight. As most studies are carried out on computers, it is of course the future challenge to convert onsite experiences such as these into computer programs.

Having found that the various designs of window aperture affected the experience of light in a space significantly, it can be concluded that it is important to take such knowledge into account when designing with daylight, balancing window size, shape and position in relation to the intention of the space both when it comes to functionality and visual appearance. Unfortunately, these aspects are not in focus much, as present agendas mean that focus is on energy consumption. In a Danish context this is reflected in the guideline on daylight design in the Building Regulation which has gradually moved from an awareness of the importance of a comfortable visual environment to primarily supporting energy consumption [23]. By leaving out the experience of light, daylight design might end up disregarding the fact that people need to occupy spaces with a comfortable visual environment.

References

[1] Johnsen, K & Christoffersen, J (2008) *Dagslys i rum og bygninger. SBi-anvisning 219* (København: Statens Byggeforskningsinstitut, Aalborg Universitet)

[2] Christoffersen, J; Johnsen, K & Petersen, E (2002) *Beregning af dagslys i bygninger. SBi-anvisning 203* (København: Statens Byggeforskningsinstitut, Aalborg Universite)

[3] Hopkinson, R G; Petherbridge, P & Longmore, J (1966) *Daylighting* (London: University College, Garston, Watford, London, Building Research Station,, England, William Heinemann Ltd.) p.2

[4] Mardaljevic, J (2011) Daylight science: A brief survey and suggestions for inclusion in the architectural curriculum. In: B.S. Matusiak & K.F. Ante eds., 2012. *Nordic Light and Colour.* Trondheim: NTNU – The Faculty of Architecture and fine Arts. pp.73-94.

[5] Mardaljevic, J; Anderson, M; Roy, N and Christoffersen, J (2011) Daylighting metrics for residential buildings. *CIE 27th Session*, Sun City, South Africa.

[6] Mardaljevic J; Heschong, L; Lee, E S (2009) Daylight metrics and energy savings, *Lighting Research + Technology*, 0: 1-23

[7] Mathiasen N & Frandsen A K (2018) Lighting Design as a Universal Design Strategy to Support Functional Visual Environments, *Transforming our World Through Design, Diversity and Education: Proceedings of Universal Design and Higher Education in Transformation Congress 2018.* Craddock, G, Doran, C, McNutt, L & Rice, D (red.). Amsterdam: IOS Press, Bind 256. s. 752-759 8 s. (Studies in Health Technology and Informatics, Bind 256)

[8] Boyce, R B (2003) *Human Factors in Lighting*. London: Taylor & Francis.

[9] Norberg-Schultz, C (1980) *Genius Loci. Towards a Phenomenology of Architecture* (London: Academy Edition)

[10] Norberg-Schultz, C (1993) *Nattlandene. Om byggekunst i Norden* (Oslo: Gyldendal Norsk Forlag)

[11] Hesselgren, S (1967) *The Language of Architecture* (Lund: Studenterlitteratur)

[12] Liljefors, A (2000) *Seende och Ljusstråling* (Stockholm: Kungliga Tekniska Högskolan)
[13] Voltelen, M (1976) *Belysningslære* (København: Kunstakademiets Arkitektskole, Institut for byggeteknik, Belysningslaboratoriet)
[14] Jørstian, T & Munk Nielsen, P E (2000) *Tænd. PH lampes historie* (København: Gyldendal)
[15] Frandsen, S (1989) *The Scale of Light - A New Concept and It's Application* (Paris: 2nd European Conference on Architecture)
[16] Denzin, N & Lincon, Y (1998) *Strategies for Qualitative Inquiry* (Thousand Oaks, Californien: Sage Publication) p 3
[17] Svendsen, S P and Hammer, C (2010) *Windows words* (Vallensbæk: Knutzon Graphic)
[18] Elgstrøm, K & Svendsen, S P (1999) *Vinduer. Bevaring eller udskiftning* (København: Arkitektur Forum)
[19] Engquist, H H (1973) *Strandgade 30. En Christianhavnsk Gaards Historie gennem 300 Aar* (København: Henry L W Jensen)
[20] Millech, K & Fisker, K (1951) *Danske arkitekturstrømninger 1850-1950* (København: Østifternes Kreditforening)
[21] Fisker, K (1999) *Formprincipper. Strejftog i den nyere arkitekturs historie* (København: Arkitektens Forlag)
[22] Svendsen, S P and Hammer, C (2010) *Windows words* (Vallensbæk: Knutzon Graphic)
[23] Mathiasen, N; Frandsen, A K & Grønlund, L (2022) Dayligth conditions in housing – Its role and priority in Danish building regulation. *Architecture, Structures and Constuction* (Springer, Volume 2, issue 1) pp 23-37

Lean Thinking into the modular construction of industrial buildings. Identifying the role of daylight

T Mavridou[1,*], L Doulos[2] and N Nanos[3]
[1] School of Mechanical and Design Engineering, University of Portsmouth, UK
[2] School of Applied Arts and Sustainable design, Hellenic Open University, Greece
[3] School of Civil Engineering and Surveying, University of Portsmouth, UK

*Corresponding author: dora.mavridou@port.ac.uk

Abstract. The necessity of energy equilibrium in industrial building stock must be addressed at design level in order to optimise use of materials without jeopardising the functional outcome. To do so effectively, the designer should incorporate a light use vs availability balance that will harness as much natural light resources as possible without creating unfavourable conditions due to over-exposure and thermal gains/losses. This work is an exploration between the relationship of sustainability with Lean thinking in the backdrop of sustainable and Lean construction. Exploring the benefits of lean and sustainable construction is about exploring how one can bring about great environmental, economic and social gains with the minimization of energy use, natural waste, environmental impacts as along with the creation of a healthy and productive work environment are the focus of sustainability in the construction industry. In effect answering how Daylight and Lean Thinking Principles can lead to optimizing Industrial buildings in terms of design and execution.

1. Introduction

Currently, one can observe a significant change in corporate environments and decision making. The new dawn of corporate social responsibility demand from end users and stakeholders alike as a result of pressing socioeconomic [1] and environmental issues [2]. In this climate environmental concerns attract more and more attention and become central parts in the global development agenda. Work presented in the past from [3] has irrefutably presented the great contribution of the construction industry in terms of environmental impact demonstrating its insatiable demand for earth extracted materials that reaches 60% while being responsible for 50% of the man-made CO_2 emissions in the atmosphere. Despite the fact that a considerable amount of research has taken place in the past in terms of a building's environmental and pollution footprint, as this is determined in terms of air, noise, energy and water consumption amongst others, less attention was paid to the effect the building design and construction process has in that regard.

Additionally, work in the field has shown that the industrial sector is one of the major consumers of energy with data from 2011 showing that the sector was responsible for 26% of the total energy consumption according to [4] while in the US it represents an even higher proportion of 31% [5] with only 7.5% of that energy coming from renewable sources of energy in the same year [6]. Research in different sectors in the US identified a "waste" of energy on non-productive activities of 15% with the majority of that (80%) being directly or indirectly attributed to lighting and air conditioning [7].

Tsangrassoulis et al. [8] has undertaken research in the past that aimed at examining a hybrid daylighting system while Doulos et al. [9–11] investigating the relationship that hybrid dimming and daylighting systems have while Kontadakis et al. [12] focused on the impact of active sunlight redirection systems for daylight. On top of the above, a variety of studies have been undertaken on the study of daylight systems and devices inside different building types and the effects those have [13, 14] providing ample of evidence and scope for the incorporation of those elements in a holistic approach to the design paradigm. In that effect daylight can be an effective sustainable construction strategy based on Lean thinking principles. This can provide important environmental and socioeconomic benefits with the reduction of the overall environmental impact due to reductions in emissions. Furthermore this strategy requires no significant maintenance resources for and is a passive form that requires no additional energy.

The above all signified the importance of such work in the field of conceptual design and construction where the synergy between those different aspects of modular and lean construction principles along with the importance of maximizing the utilization of available natural resources such as lighting and how that can lead to the identification of optimal solutions in terms of resource waste. This paper aims at addressing the environmental, social and economic progress by ensuring a continuous increase of embodied added value in the Architecture, Engineering and Construction (AEC) industry by providing a framework for the synergy between lean construction and sustainability in the industrial buildings in terms of modularity and daylight requirements. As a result, delivering not only reduced energy consumption, environmental impact and waste for the design and construction process but also improve the quality, safety and health aspects for the end users.

Due to the particular needs of an industrial envelope and estate in terms of big, unobstructed areas and adequate work light conditions without jeopardizing sustainable construction and ethical footprint it is becoming important for the stakeholders to explore industrial construction, sustainability and their relationship. With the implementation of Lean Thinking principles in design and construction one can attain benefits in environmental, economic and social terms with the elimination of non-value adding activities and processed in terms of design and implementation. Additionally, the apparent economic and environmental benefits of Lean Thinking compound to the potential social improvement aspects that such an approach can have with its objective being the efficient use of resources in the design, construction and use of buildings putting a major emphasis on the resources necessary to improve the overall environment and user health while minimizing waste. A series of studies have investigated the relationship between lean construction and Lean Thinking (LT) [15] investigating the review of the bibliography about the Sustainability through Lean Construction (LC) while others [16] dealt with a more systematic review of the state-of-the-art literature in lean and green.

2. Methodology

The project will use qualitative research methods. The qualitative approach is considered an effective method for conducting extensive review and identifying research gaps. A systematic approach will be used to determine the characteristics and requirements of industrial buildings. The aim is to analyse the characteristics of industrial buildings and the relationship between LC and Sustainability for optimisation of the corresponding building type. A qualitative research to identify the key components that link modular construction to the above will be conducted. This approach aims to identify similarities between these issues. This method will drive to recognition of a tool that will follow these common values and optimize industrial buildings in relation to Lean thinking.

The qualitative approach is an exploratory research that focuses on the collection of secondary data collected from sources such as published articles, books, magazines and web portals. After the collection of the data will follow summary, synthesis and comparison of the information obtained in order to develop critical knowledge about the investigated area and address the research questions.

3. Industrial Buildings

An industrial building is designed according to its functional needs despite morphological choices. Their main feature is the simple geometry. Their design is a "shell" with purely functional purposes, reasons of security and flexibility, follows the prevailing one construction technology, available building materials and project the image of the business. As a point to note, the buildings consist mainly of one-dimensional volumes with rectangular construction.

Table 1. Characteristics of Industrial Buildings and Requirements during the design Stage.

Characteristics of industrial buildings	Requirements during the design stage of industrial buildings
Construction of metal or concrete	Aesthetics and customer's corporate identity
Continuous changes	Flexibility in current and future use
Land	Maintenance Requirements
Large roof	Needs of the production line
One dimensional volume	Specific material
Shape (Rectangular / simple geometry)	Speed of construction
Short life cycle	Sustainability / Environmental performance
Sufficient penetration of Daylight	

The concept of LC is based on the application of LT and Philosophy in the construction industry. The construction sector uses intensively resources and produces waste with a significant impact on the environment [17]. In the same vein, Grohmann [18] showed that the amount of material and workforce wasting ranges anywhere up to 30% of the actual construction itself in a world that suffers from the lack and cost of clean water. Furthermore, [19] has shown the reflection of such waste in labour costs, which can cause a 6% increase in total costs. Therefore, LC positions itself at the right place to effect change in the production management system of the construction sector eliminating all types of wastes.

Modular Construction buildings are created and controlled in a factory environment, which allows the high quality of finishing and the use of different materials such as wood, steel or concrete [20]. Some of the key features of Modular Construction are the safety of construction, the reduction of construction time, the reduction of waste, the flexibility enabling the creation of many types and the reuse for various types of buildings of any size. Historically, the design of these structures was done with simple lines and simplicity. In recent years, Modular Construction has evolved into a much more complex and aesthetically acceptable set which not only meets the basic needs of the user but can also allow constant change depending on the needs that arise. Yu et al. [21] identified Modular construction systems as an agglomeration of 3 discreet ideas, those of prefabrication, standardization, and dimensional coordination. One of the most important aspects of modular products and their implementation is the potential for effective flexibility and reduced process complexity [22].

One of the most important benefits of Modular Construction is the rapid construction and the short time between the beginning of the construction project and the delivery of the final project to the customer [23] In addition, the risk of delays due to extreme weather conditions is minimized [24]. Finally, Modular Construction can and does save about 40% of construction time compared to traditional techniques factory-built in a controlled environment [24, 25]. Such a method of construction offers various advantages compared to the conventional way of on-site construction such as the reduction of project time, but also the increased quality characteristics. Therefore, standardization can be considered as a process with multiple advantages throughout the manufacturing process. Modularity offers improvements to the construction of a building throughout its life cycle, from development to resale [26]. The use of Modular philosophy in construction is a method of value management, where internal (production) and external (customer) needs are balanced [27]. It enables construction flexibility but also reduced complexity in the construction process [22].

Concluding, Modular Construction is the most effective way to reduce waste in the construction industry. The significant reduction in materials but also time is what plays such an important role in designating this way of construction as Sustainable Building. The above in combination with the Modular design form allows, under appropriate forecasting conditions, the deconstruction of the structure in the future and the reuse of these elements in new constructions outside the space [20]. There are multiple benefits stemming from the use of modular construction. Nikmehr et al. [28]

identified those in terms of waste minimization while Boyd et al. [29] focused in the improvement recorded in the overall quality. However, the role of daylight and how it can be adapted in the Modular Construction with a beneficial way hasn't examined further.

4. The role of Daylight

Natural light has many benefits for humans and the environment. Increases productivity, reduces errors while reducing electricity consumption for lighting. At the same time, it creates a pleasant environment that cannot be easily or economically achieved with artificial lighting [30]. Research says that productivity can be increased by 10% and errors can be reduced by 30% [31]. Studies correlate lighting with performance and health of users. Misuse of natural light can lead to poor quality of work, low productivity and increase of errors as a result the creation of a waste. Low lighting in the working plane can lead to employee fatigue, headaches, stress and accidents. In addition, excessive lighting can also cause health problems. Light regulates our circadian rhythm, our alertness, concentration and cognitive performance amongst many other non-visual functions [32].

Research shows that 20% of world electricity consumption comes from electricity consumption for artificial lighting [33], a number that exceeds the percentage of total annual nuclear energy production worldwide [34]. In addition, electricity consumption in the construction sector for artificial light represents about 11% of energy use in domestic and 18% in commercial buildings [35], generating billions of tons of carbon each year. Global electricity consumption for lighting will continue to expand and is expected to rise in the future, with CO_2 emissions to be doubled and even more.
Although sustainability is promoted to save energy and reduce environmental impact, the use of artificial lighting cannot be easily reduced [36]. Finding a solution of a suitable natural lighting system could reduce the consumption of lighting, while it could reduce the cooling loads of the internal equipment as much as possible. By achieving a reduction in the consumption of artificial lighting, an additional 10-20% of the energy use of refrigeration loads can be saved [37].

By applying natural lighting strategies, the reduction of the use of luminaires can be achieved, with a direct reduction of greenhouse emissions, as well as with consumables and with the maintenance of lamps. Incorporating lighting strategies can reduce the total energy cost in buildings such as commercial or social welfare by one third (1/3) [37]. Furthermore, daylight provides many benefits, but this requires an application study that includes elements of visual comfort and thermal gain. Proper study of the daylight characteristics and its integration into the building design will lead to the achievement of appropriate, comfortable and energy efficient systems [38]. Proper use of natural light can reduce the two main consumers of a building, electric lighting and cooling as the misuse of luminaires excessive amounts of heat in buildings, which create an increased need for cooling. The use of daylight can reduce by up to 30% the amount of electricity consumed for lighting [30]. The daylight harvesting in order to be efficient needs to be designed in a way that meets the required lighting levels using photosensors [39].

While the part of the daylight in industrial building is crucial, its impact in modular building hasn't yet been correlated (Figure 1). As an example, the opening dimensions chosen to be implemented should conform to multiparametric criteria and ensure light uniformity as described in EN 12464 while maintaining an economic profile in terms of construction while making optimal use of available daylight for the examined area where a modular building can be placed (Figure 2).

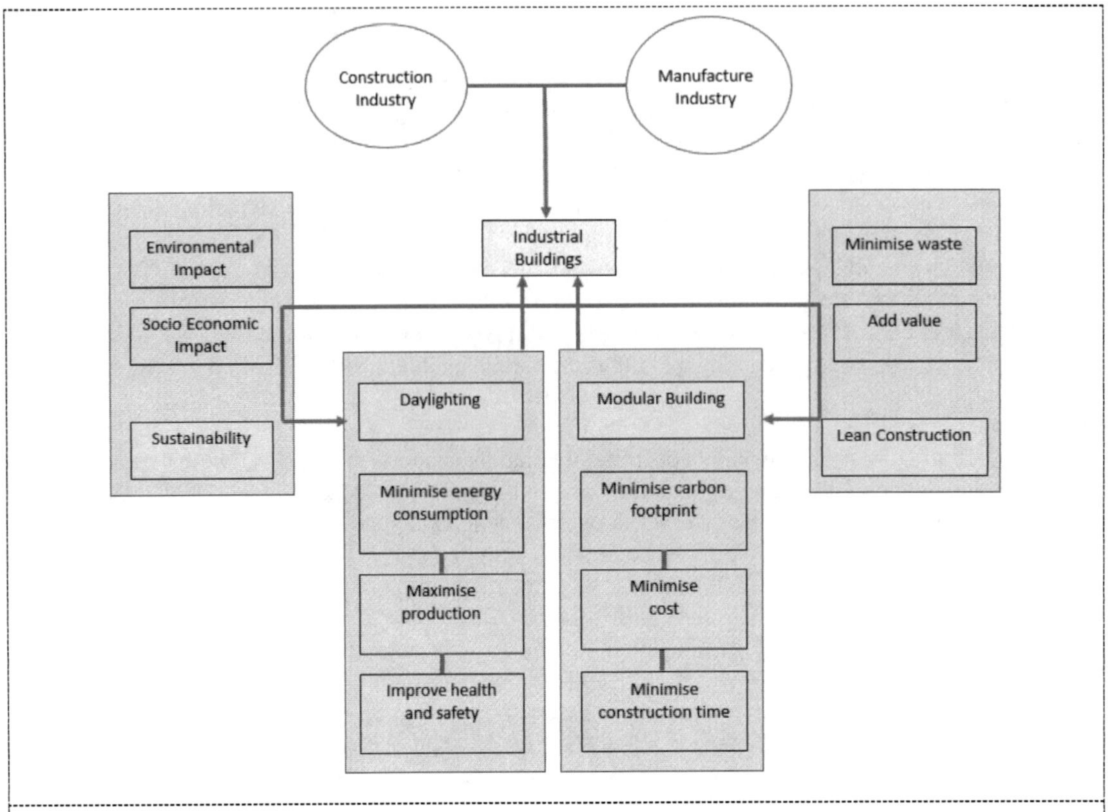

Figure 1. Path of interdependency between Sustainability, Lean Construction and Daylight.

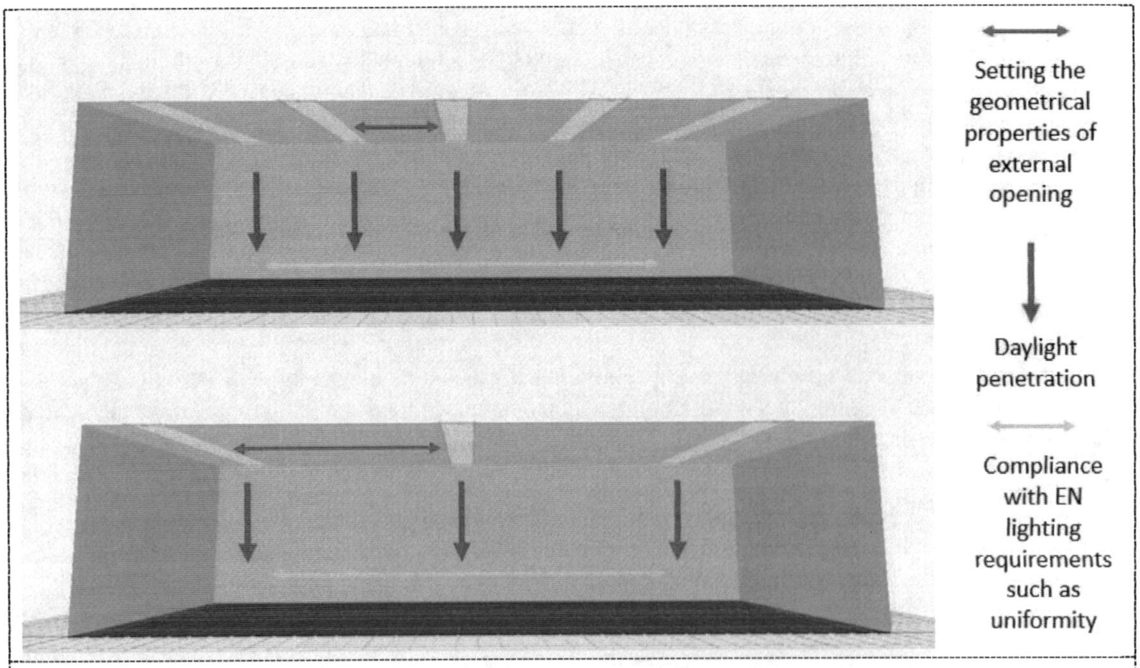

Figure 2. Setting the geometrical properties of external opening for a modular construction a) concerning the most beneficial solution having compliance with European Norm lighting requirements and b) considering the area of the modular building for the corresponding daylight data.

5. Discussion and Conclusions

Lean Thinking provides multiple benefits in terms of the environment, the economy and social aspects of a project while sustainability has as an objective the efficient use of resources in the design, construction and use of a building. Therefore, the emphasis on resources related to the environment and user health becomes the apparent focus of sustainability in the construction industry. In that regard, as the principles of Lean Thinking in Construction are those of minimizing waste of material use and time allotted to maximize added value the similarities between the concepts of Lean Construction and those of sustainable and modular construction become apparent. Nowadays industry produces more waste than any other sector of human activity worldwide [40]. The above in combination with the Modular design form allows, under appropriate forecasting conditions, the deconstruction of the structure in the future and the reuse of these elements in new constructions outside the space [20].

For the purpose of this work the authors approached the concept of lean, modular and sustainable construction in a manner that allows meaningful conclusions to be drawn comparing and analysing them for intersection points to be identified. In that respect, multiple environmental, economic and social benefits of daylight have been identified and the ability of Lean thinking to aid in harnessing those benefits has been explored. Therefore, the ability of Natural light to influence visual comfort and improving users' health have been highlighted with the relevant increases in productivity, and improvement in efficiency being directly affected. Furthermore, the overall reduction in energy consumption, directly contributes to the reduction of pollutants which in turn both directly and indirectly affect the environmental footprint and affects the whole life cycle cost of the building in question. It has therefore been shown that the use of daylight is an effective strategy for sustainability with savings in the order of up to 20-30% of the total energy demand of a building in addition to the real economic cost and benefits. As such it presents a holistic approach that provides multiple benefits to the stakeholders and end-users alike with reductions in the overall thermal equilibrium and specially the cooling load while reducing the initial costs of HVAC and reducing overall waste.

As the cornerstone of Lean Thinking philosophy revolves around identifying ways to achieve more with less, not taking advantage of a resource like the sun and the Natural daylight that it provides in abundance would not play strongly in its favour. Therefore, utilising the sun as a source of illuminating a building's interior illuminate or for collecting, storing and distributing heat load through the use of technologies such as passive solar heating are typical examples of Lean Thinking and Philosophy and must be exploited to achieve our sustainable future goals in construction. Natural Lighting has been the point of application of Lean Construction in the present research with the aim of this study of a methodology that optimizes industrial buildings through Natural Lighting in various areas.

6. References

[1] Pacala S, Sokolow R. 2004 Stabilization Wedges: Solving the Climate Problem for the Next 50 Years with Current Technologies Science 305(5686):968-972

[2] Smalley R 2004. Future global energy prosperity: the Terawatt Challenge Materials Research Society Fall Meeting, Boston, MA.

[3] World Watch Institute 2005 Vital signs 2005 In: Mastny L (ed.) The Trends That Are Shaping Our Future, W.W. Norton & Company New York.

[4] Eurostat 2013 Final energy consumption by sector Eurostat, European Union http://epp.eurostat.ec.europa.eu/tgm/table.do?tab= table&init=1&plugin=1&language=en&pcode=tsdpc320, accessed Oct 21, 2013.

[5] EIA 2012 Manufacturing Energy Consumption Survey U.S. Energy Information Administration.

[6] EIA 2012 The Annual Energy Outlook 2012 – with projections to 2035 U.S. Energy Information Administration, U.S. Department of Energy, Washington, DC 20585 (Retrieved from https://www.eia.gov/outlooks/aeo/pdf/0383(2012).pdf)

[7] ORNL 2012. ORNL/TM-2012/504. U.S. Manufacturing energy use and greenhouse gas emissions analysis. https://www.energy.gov/sites/prod/files/2013/11/f4/energy_use_and_loss_and_emissions.pdf, (accessed 10 October 2017)
[8] Tsangrassoulis A, Doulos L, Santamouris M, Fontoynont M, Maamari F, Wilson M, Jacobs A, Solomon J, Zimmerman A and Pohl W 2005 On the energy efficiency of a prototype hybrid daylighting system Sol. Energy 2005, 79, 56–64
[9] Doulos L, Tsangrassoulis A, Kontaxis P, Kontadakis A and Topalis F 2017 Harvesting daylight with LED or T5 fluirescent lamps? The role of dimming. Energy Build. 2017, 140, 336–347
[10] Doulos L, Tsangrassoulis A and Topalis F 2008 The role of spectral response of photosensors in daylight responsive systems. Energy Build. 2008, 40, 588–599
[11] Doulos L, Tsangrassoulis A and Topalis, F 2008 Quantifying energy savings in daylight responsive systems: The role of dimming electronic ballasts. Energy Build. 2008, 40, 36–50
[12] Kontadakis A, Tsangrassoulis A, Doulos L and Topalis F 2017 An active sunlight redirection system for daylight enhancement beyond the perimeter zone. Build. Environ. 2017, 113, 267–279
[13] Kontadakis A, Tsangrassoulis A, Doulos L and Zerefos S A 2018 A Review of Light Shelf Designs for Daylit Environments. Sustainability 2018, 10, 71
[14] Mushtaha E, Kana'an B A, Al-Jawazneh R A and Hammad R S 2016 Effect of using different light pipe parameters on the daylight quality in buildings: The case of Jordan. Int. J. Green Energy 2016, 13, 1590–1598
[15] Marhani MA, Jaapar A, Bari NA, Zawawi M 2013 Sustainability through lean construction approach: A literature review. Procedia-Social and Behavioral Sciences. 2013 Nov 8;101:90-9.
[16] Garza-Reyes J A 2015 Lean and green–a systematic review of the state-of-the-art literature. Journal of Cleaner Production. 2015 Sep 1;102:18-29.
[17] Pinheiro M D 2003 Construção Sustentável – Mito ou realidade VII Congress National of Environment Engineering, Lisbon
[18] Grohmann M Z 1998 Redução do Desperdício na Construção Civil: Levantamento das Medidas Utilizadas pelas Empresas de Santa Maria VI Congress Internacional of Industrial Engineering. Univ. Federal FluminenseUFF, Brazil.
[19] Pinto T P 1995 De volta à questão do desperdício Construction, Vol. 271, pp. 34-35.
[20] MBI 2011 Using off-site construction to eliminate waste in design phase Modular Building Institute MBI and BD+C Modular Advantage, source: http://www.modular.org/htmlPage.aspx?name=Eliminate_Waste, (10 October 2011).
[21] Yu H, Al-Hussein M, Al-Jibouri S, Telyas A. Lean transformation in a modular building company: A case for implementation. Journal of management in engineering. 2013 Jan 1;29(1):103-11.
[22] Marshall R and Leaney P 1999 A system Engineering Approach to Product Modularity Proceedings of the Institution of Mechanical Engineers Part B.
[23] Kawecki L R 2010 Environmental performance of modular fabrication: calculating the carbon footprint of energy used in the construction of a modular home (PhD thesis) USA: Arizona State University; 2010
[24] NAHB 2006 Lean Construction. Report Marlboro, MD: National Association of Home Builders;
[25] Smith R E 2011 Prefab Architecture: a Guide to Modular Design and Construction John Wiley & Sons, New Jersey
[26] Rogan A L, Lawson R M and Bates-Brkljac N 2000 Value and benefits assessment of modular construction Report. Berkshire, UK: Steel Construction Institute: 2000
[27] Björnfot A and Stehn L 2004 Industrialization of Construction – A Lean Modular Approach Proceedings of the 12th Annual Conference of the International Group for Lean Construction, Elsinore, Denmark.

[28] Nikmehr B, Hosseini M R, Rameezdeen R, Chileshe N, Ghoddousi P and Arashpour M 2017 An integrated model for factors affecting construction and demolition waste management in Iran. Engineering, Construction and Architectural Management. 2017 Nov 20.
[29] Boyd G, Zhang G 2013 Measuring improvement in energy efficiency of the US cement industry with the ENERGY STAR Energy Performance Indicator. Energy Efficiency. 2013 Feb;6(1):105-16.
[30] IESNA 2000 Illuminating Engineering Society of North America. The IESNA lighting handbook, 9th edition. New York: IESNA
[31] ILO 2014 Physical Hazards, Indoor Workplace Lighting. International Labour Organization OSH Brief No. 3c. (Retrieved from http://www.ilo.org/wcmsp5/groups/public/---americas/---ro-lima/---sro- port_of_spain/documents/presentation/wcms_250198.pdf).
[32] Vandewalle G, Maquet P and Dijk D J 2009 Light as a modulator of cognitive brain function. Trends in cognitive sciences 2009 Oct 1;13(10):429-38.
[33] IEA 2012 World Energy Outlook IEA Publications, 9 rue de la Fed eration, 75739 Paris Cedex 15, November 2012.
[34] McSmith A 2006 A bright idea: How changing light bulbs helps beat global warming (and cut bills). Retrieved from http://www.independent.co.uk/environment/a-bright-idea-how-changing-light-bulbs- helps-beat-global-warming-and-cut-bills-406494.html, (accessed 05 October 2017).
[35] EERE 2011 Building Energy Data Book Office of Energy Efficiency & Renewable Energy. (Retrieved from https://openei.org/doe-opendata/dataset/buildings-energy-data-book)
[36] Leslie R 2003 Capturing the daylight dividend in buildings: why and how? Building and Environment 38(2):381-385.
[37] Ander G 2016 Daylighting. Retrieved from https://www.wbdg.org/resources/daylighting, (accessed 05 October 2017)
[38] Levy B, Wegman D, Baron S and Sokas R 2011 Occupational Health—Recognizing and Preventing Work-Related Disease and Injury 6th Ed. Oxford University Press, England (Retrieved from http://sgh.org.sa/Portals/0/Articles/Occupational%20and%20Environmental%20Health%20-%20Recognizing%20and%20Preventing%20Disease%20and%20Injury.pdf).
[39] Bell J and Burt W 1996 BRE Designing buildings for daylight (BR288) Publisher BRE, ISBN:9781860810268
[40] Meadows D 2011 Designing out waste. Environmental Design & Construction. http://www.edcmag.com/articles/87434-designing-out-waste#comments (accessed July 29, 2011).

The Appearances of Daylight – an educational method for studying daylight

L Grønlund

Royal Danish Academy – Architecture, Design, Conservation
anch@kglakademi.dk

Abstract.
The central theme is daylight. In other words, seeing and sensing it, registering it through photographs and drawing the tangible conditions that form the framework for the appearances of daylight. Basically, it is about practising the ability to see differences – to see the differences in the daylight and thus heighten an awareness of how to choose to design with daylight in architecture. In educating architects, we are keenly aware of encouraging students to develop an awareness of the importance and effects of daylight. This paper will examine how the experience of daylight through a specific qualitative method can provide students with new knowledge through a phenological approach: gaining knowledge by studying the daylight in the home of each student where they live their daily lives, perhaps seeing the daylight or overlooking it. Their home becomes a laboratory for studying and documenting daylight through different formats and using different tools. There is a need to develop new teaching methods that can give students insight and tools for seeing daylight, founded on a belief that through this experience and knowledge they can design architecture with nuanced, diverse experiences of daylight. In a sustainable perspective, robust aesthetic sustainability must begin with the qualities of space, of which daylight is one of the most significant elements.

1. Introduction: *the study of daylight* [1]

The study of light is perhaps the most important thing for an architect, for a sculptor and their friend, the photographer. It is a matter for everyone whose work is aimed at people, at understanding the conditions of life and exposing the nature of the immediate surroundings [2]. This how professor and architect Erik Christian Sørensen clarifies the importance of studying and using daylight in architecture.

At present, there is a tendency to focus primarily on the quantity of daylight rather than its quality. This could cause the potential to create interplay between daylight and architecture to be easily overlooked. The intensified focus on the quantity of light is apparent in the requirements stipulated in the building regulations for documenting energy consumption, including the use of electricity for lighting, heating and, possibly, cooling in connection with overheating. This means that the documentation of lighting often plays a bigger part in a construction process than when daylight is actually experienced and perceived [3]. As a result, considerations of how daylight occupies space and how it is experienced as a human being can become secondary. [4]

Therefore, it is important to focus on the qualitative aspects of daylight and on its significance for how a space is experiences, and perhaps rediscover the wide variety of effects and appearances of daylight in terms of forming different spaces with diverse daylight atmospheres.

Light can be described in many ways: from the scientific approach with quantitative units of measurement such as lux and lumen to the humanities [5] approach focused on experiencing and perceiving light, described by its qualitative aspects. In the study programme for architects at the Royal Danish Academy – Architecture, Design, Conservation, we have a strong tradition of teaching students a qualitative approach [6] [7] to understanding daylight, and this article builds on this, but with a limited framework concerning how daylight is experienced.

This method is developed on the basis of previous research and as research-based teaching.

2. Research questions

There is a need to develop new teaching methods that can give students the tools to see and experience daylight according to a belief that experience, and knowledge can enable them to design architecture with nuanced and multifaceted experiences of daylight.

The aim of this article is to provide insight into a method of study from a qualitative and experience-based approach. Therefore, the overall research question is formulated as follows: *How can we as educators and researchers impart methods to our students that will enable them to examine and study the appearances of daylight in their home?*

This is followed by a sub-question that more specifically asks for special requirements for the framework in which the method is tested: *What role does the choice of space play as a framework for studying daylight?*

To clarify and discuss these questions, the article includes the following chapters: *Theory*, where the phenomenological approach is unfolded; *Method*, where the framework and tools are explained, *Findings*, where various examples of the assignment are displayed and described; and finally, *Summary and discussion*.

3. Theory: *Being in a space – sensing and perceiving the light*

Scientific terms can be used to describe light and in so doing render its quantitative qualities comprehensible. Nevertheless, light has an intangible aspect that is described through its qualities. As humans, we can detect light with our eyes, feel it's warm rays, be in it, be enclosed by it, or be on the edge of it. This intangible aspect of light is the focus of phenomenological optics, and which can be scientifically equated with the quantitative facts.

When you move in light and sense it in a room, you register and understand it intuitively. You can do this without an explanation of why and how the light flows illuminates a space and how it is dispersed: the light is in the room, and you sense it. This immediate sensing and intuitive registration of the surroundings, where it is the phenomenon that is registered and not the understanding of what is defined, is the phenomenological approach to the world.

Phenomenology deals with the matter itself [8]. In a phenomenological approach, the starting point is a sensory examination of the light, where the appearance of the light is understood through a bodily experience. The body as the sensory device is the primary optic, i.e., how light and space are experienced by the body. In order for the body to experience of space and light, architect Steen Eiler Rasmussen argues in the following, *It is not enough to see the architecture. It must be experienced in order to feel how it is designed to perform certain tasks and attuned to a period's whole perception and rhythm of life. You must live in the space of the building, feel how it closes around you, how you are actually led from one space to another. You must sense the material effects, discover what the colours mean in their particular contexts, how they depend on the orientation of the rooms in relation to sunlight and windows.* [9]. The quote is from his book *On Experiencing Architecture* and describes how an encounter with space must involve all senses and not just the prevailing sense of sight.

A similar phenomenological approach is taken by Finnish architect and theorist Juhani Pallasmaa, who, like Rasmussen, describes the encounter between body and space as something that must not be reduced to the domain of the sense of sight. But even seeing can possibly be done in many ways, and as Carsten Thau writes in the preface to the Danish edition of Pallasma *Arkitekturen og sanserne* (original title: *The Eyes of the Skin*), the book draws attention to the joy of stopping to observe [10].

The phenomenological approach is based on the thinking of the German philosopher Edmund Husserl (1859–1938). Husserl is considered the founder of phenomenology. The philosophy is further developed by his student Martin Heidegger, and subsequently the French philosopher Maurice Merleau-Ponty has built on the philosophical direction through a description of a focus on the bodily presence and being. In the article *How the world emerges*, the Danish philosopher Zahavi introduces what phenomenology is, through Husserl: *To Husserl, the phenomenon is the object's own way of appearing. The phenomenon is that which manifests itself by itself, that which manifests itself, that which reveals itself. (…) Phenomenology draws our attention to the appearance of the object. But it not only makes us aware of the objectivity of the object, it also makes us aware of the subjective correlation of appearance. When we examine prominent objects, we also reveal ourselves as those to whom the object appears* [11]. (…) *It is precisely the subject, and more specifically, the bodily subject, which takes the perspective through which the object emerges* [12].

Phenomenology is thus less concerned with how an object is described from a scientific point of view, and more focused on the way in which an object appears, comes into view and is experienced by the bodily subject, as Zahavi describes it. He further describes in the article how Husserl's philosophy inherently has a critique of the objectivism of science and its attempt to provide an absolute description of reality. Here, Zahavi claims that Husserl wants to point out the importance of the world of life, *which is the world we take for granted in our daily lives* (…) *and which we do not question. It is the world we sense, move around in and are born, live, and die in* [13]. Husserl argued that science, in its search for the absolute, has supplanted this world of life, forgetting that it itself is grounded in this world of life, *since science is originally based on the phenomena or rather on a desire to explain the phenomena. Therefore, phenomenology is also often referred to as a pre-scientific description of the world* [14]. Husserl's point of view of *the world of life as the world we take for granted in everyday life* [15] is directly reflected in the basic intention of the approach to light, is described in this article. Namely an immediate experience and bodily sensation of light in a space – the light that we take for granted and that we are surrounded by in our daily life, and which permeates our day from sunrise to sunset.

The phenomenological experience of light will stand in contrast to the measurable scientific description of light solely by making use of experience and bodily sensing as its tool. Neither the body nor the senses have fixed units of measurement at their disposal but rely exclusively on subjective experiences. It is the conviction of the article that a description of the phenomenological approach to the phenomenon of light can make it possible to get closer to the actual experience of light.

4. Methodology: *The framework, the depictions of daylight and the assignment*

4.1 The framework

This chapter reviews the framework for the task, as well as the tools with which light is to be studied. The framework for the assignment is the student's own home, typically an apartment, because it is a familiar space that provides a sound foundation for studying light. Students are familiar with their own spaces and they may have already had some realisations and experiences of the light – both subconsciously and consciously. The experience of light is particularly associated with memories [16] which can be activated by using the students' own home as the framework for the assignment. Another parameter for the choice of place is that it must be a space to which they have access over a period of time, which is another reason their own home is chosen.

Before students are presented with the assignment, they read the text *The Light in the Shadow* by Sophus Frandsen [17]. Then they attended a double lecture. In the first part, they were introduced to the phenomenological method for studying daylight and daylight theory, including Anders Liljefors' [18] seven points for determining the visual properties of light. In the second part, they were introduced to working with photography, especially becoming aware of different photographic strategies, such as the difference between documenting and registering, contrasted with interpreting or portraying daylight in a space. Based on this knowledge, they were given an assignment that is divided into four parts: part 1, drawings, and three other parts comprising photographic series.
Initially, the choice of tools and forms of depiction are elaborated in 4.2–4.4.

4.2 Depicting daylight in photographs
In the assignments, the camera is chosen as a tool and the photograph is chosen as the form of representation, due to its ability to depict what is seen in a representative, similarity-related manner. Photography theorist Susan Sontag defines photographs as follows: *Photographs are captured experiences, and the camera is the ideal weapon of consciousness to capture them with* [19]. Her quote precisely describes the way in which the assignments make use of the photograph: as a framework for maintaining an experience of the effect of light.

A photograph can capture a moment, which is why a specific lighting situation is selected in that moment. Since the photograph captures only a moment of the light in the room, it can visualise other shades and facets of the light. Unlike many other practices today, which focus on the medium of film and moving images, this tool has been deselected. It is notably the freezing of a moment or moments of light that the assignment seeks to focus on. The individual photograph points to an awareness and presence within the photographed space that is directed towards the slowness and lingering at this very moment of light. Or in Christian Norberg-Schulz's words, *taking the time to see and take the pictures* [20]. The experience of light in a space is a sensuous three-dimensional experience, and the photograph as a depiction of the sensed space is a two-dimensional piece of paper of *a thin slice of both time and space* [21]. Overall, a basic prerequisite for all forms of representation is that they cannot present a bodily experience of the world in time, but rather depict *one* experience of *one* world. This basic prerequisite is also embodied in this assignment.

4.3 Geometric representations in drawings
As a secondary form of representation, the photographs are supplemented by drawings, plans and sections and a street section. The drawings are prepared as pure drawings without shadows, etc.
As drawings constitute a well-known format, they also serve as a scale within which to understand the assignment.

4.4. The assignment
The assignments range from large scale, i.e. a street view, to a micro scale of the interaction of materials with light. The following is an overview of the various sub-assignments:

Drawings: comprises a layout drawing, 1:50, and a street view, 1:200, with a drawing of a no-sky line. This drawing of the no-sky line is essential for understanding the significance of the exterior space to the brightness of the light inside the home. This is followed by the three photographic series: 1) *Over time*; 2) *Daylight outside, the daylight aperture and daylight in the room;* and 3) *Close-up*.

Over time: This study basically involves the students recording daylight changes over time in a room of their home. The changes can occur over short or prolonged intervals of time. For each photograph taken, the time is registered and supplemented, if necessary, by a short note. For each series, the camera is placed in the same position, and artificial light must not be switched on during the study period, as the focus is solely on daylight. The photographs are taken in daylight, which can start when the sun begins to rise and end when it has completely set.

Daylight outside, the daylight aperture and *daylight in the room:* The second task focuses on looking at the specific geometric conditions that create light in a room, i.e., the relation between daylight outside, the aperture through which the light penetrates the room and the daylight in the room. In this study, daylight is examined and documented from three different positions in an indoor space: the photographer directly faces the daylight aperture and photographs it; the daylight aperture is photographed from the side, and daylight aperture is photographed with the photographer facing away from the aperture. The daylight is then photographed from two outdoor positions: the daylight aperture seen from street level and the sky.

Close-up: this involves the close-up photographing of details, objects in the room, surfaces and materiality illuminated by daylight, specifically in relation to the character of the various materials.

5. Findings: selected examples

Examples from each sub-assignment were selected based on 60 students' work on the assignment (three classes of BA students, from 2020, 2021 and 2022). All the assignments contain answers to all four sub-assignments. The selection is representative of the general type of answers, and this will be elaborated on through descriptions of the individual examples.

The works shown below are from different students' work, as this provides a more nuanced overview. The students have submitted their assignments in a pre-designed layout so that everyone has submitted images of the same size, images that are set up in the same way, etc. This was deliberately done to create a clear comparative set of materials.

The selected examples of works are displayed and described below:

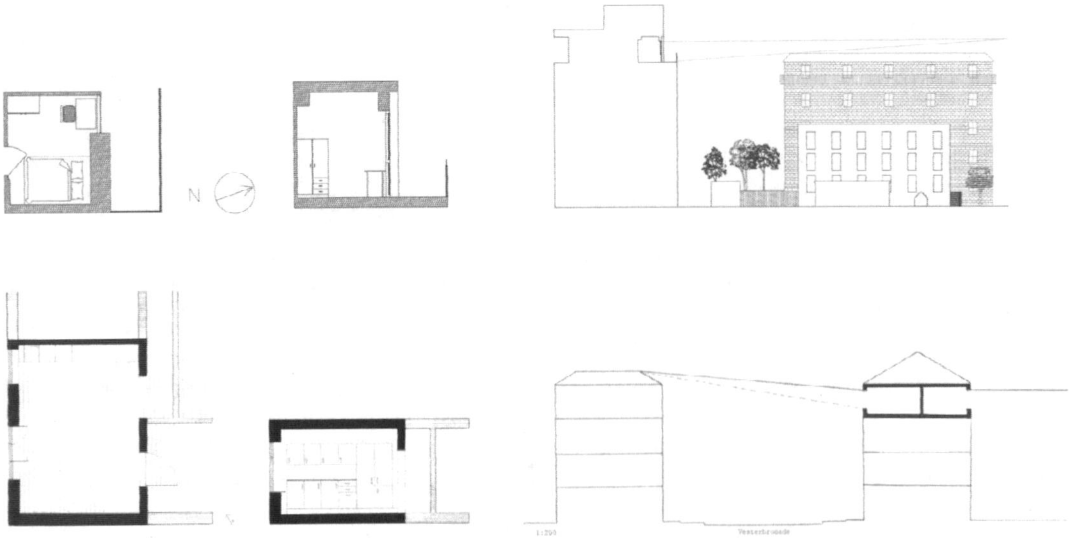

Figure 01. The drawings: geometry

5.1 The drawings: geometry

By drawing their own home, the students gain an objective view of the spaces in which they live on a daily basis. Their first awareness of the location of the home in relation to orientation is achieved by drawing a north arrow on the plan. The chosen room is measured on site and in some instances, furniture is drawn in. The furniture helps scale the room and render visible how the space around the window is inhabited. For instance, the top drawing of Figure 01 depicts a worktable placed in the daylight aperture. In addition to the layout drawing of the room's geometric dimensions, the section also specifies the height of the room and the location of the window in the facade. This gives rise to a focus on the connection between the size of the window and the experience of the light in the room, especially in relation to its height. By drawing the close context with and drawing the no-sky line, the significance of the surrounding obstructions to skylight in the home becomes clear. The drawings thus form a basic understanding of daylight in the space, both on an urban scale and within the chosen space.

The experiences from the drawings are passed on to the photographic studies.

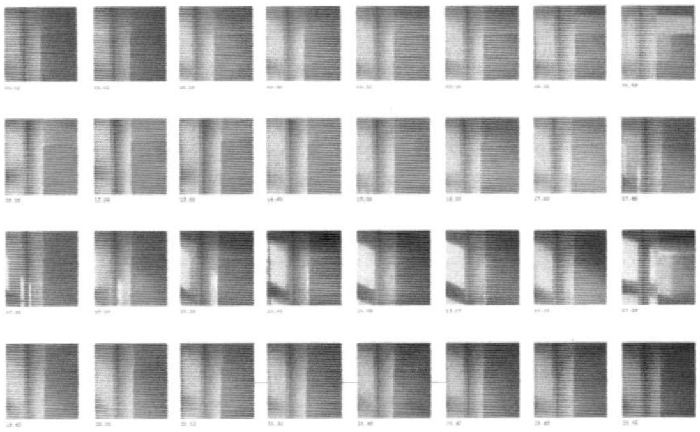

Figure 02.1 Over time

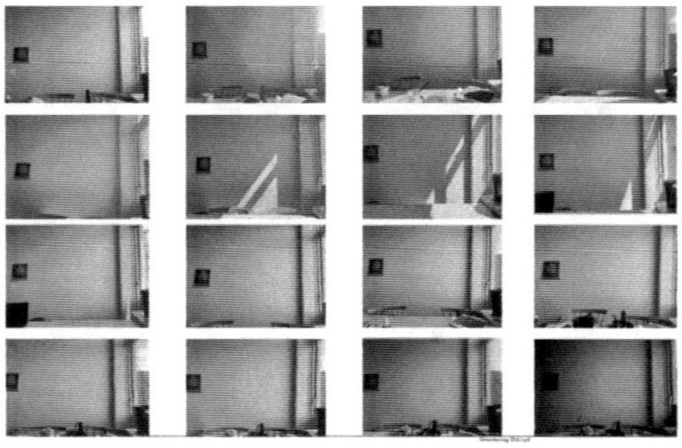

Figure 02.2 Over time

5.2 Over time

In the photographic series over time, pictures are taken from early morning, during the day and finally when the sun has set. The size of the section of the room that students have chosen to photograph differs from one student to another. Some have focused on the details of a transition, while others have focused on larger segments of the space. The most successful series of daylight changes over time is when a student chooses a surface in a room that enables changes both big and small to become visible. In Figure 02.1, an edge is selected that forms the transition from a window frame to a cabinet. During the series, the shadows from the window bars become noticeable and appear with differing intensity. The warm tones of sunlight, especially in the early morning hours, also appear very clear. Choosing a smaller segment of the space, without scale and crucial details revealing the scale, makes the series more abstract. The abstract series, of course, is bound to the space, but it also has another more lyrical character that evokes many other associations. Figure 02.02 shows a larger part of the room, so that both the scale of the room and the function can be seen, and an exterior is also visible. In the series, the shades of light become the most obvious – but the markings of sunlight on the wall are also visible. In particular, the shadow from the frame around the window appears with differing intensity. The no-sky line can also be seen. In this series, another aspect is seen: how the space in front of the window is used, with dining table, chairs, etc. Throughout the day, various everyday objects are also visible, showing the life that is lived in this particular space.

The two series depict in different ways the importance of selecting the motifs for the series of photographs in terms of how clearly the change of daylight is rendered visible. This is true both when choosing a part of a space (Figure 02.1) but also in Figure 02.2, where more of the space is framed.

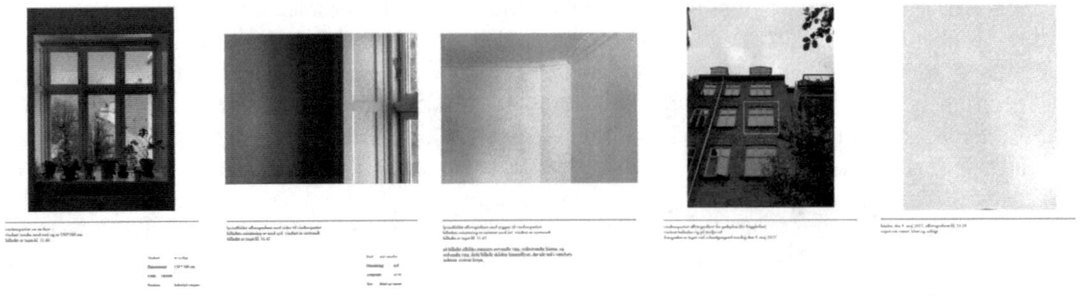

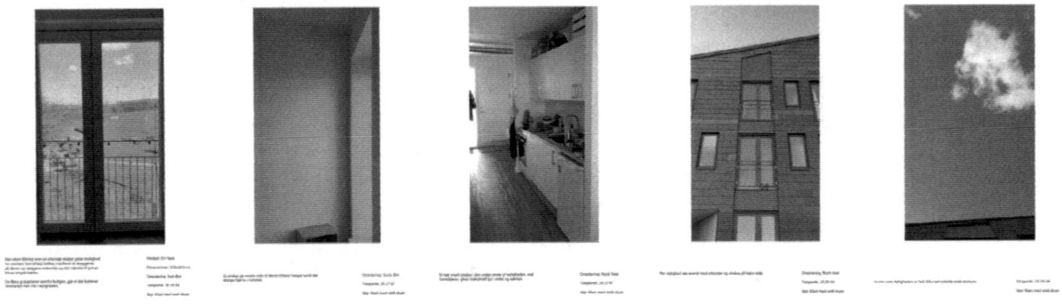

Figure 03.1 *Daylight outside, the daylight aperture and daylight in the room* (upper row)
Figure 03.2 *Daylight outside, the daylight aperture and daylight in the room*

5.3 Daylight outside, the daylight aperture and daylight in the room

In this sub-assignment entitled: *Daylight outside, the daylight aperture and daylight in the room*, the focus is on a single daylight aperture. Two different series from two homes are chosen in the above. Several factual conditions are stated below the pictures, including orientation, size of the window, time of day, etc.

One important insight focused on by this sub-assignment is an understanding of the different experiences of daylight depending on the position of the body in the space. Therefore, it was decided that the photographs should be taken from specific positions in the space.

In the series, the photographer's body is in the following positions: 1. facing the window; 2. looking at the window from the side; and 3. turned with his/her back to the window. The last two photographs are of the window seen from street level and the sky.

In the first vertical row of photographs, the window is photographed from the front, documenting the size, shape and position of the window and a view out to the context. The window is photographed in backlight. In the next position, 2., *from the side of daylight aperture* it becomes clear how daylight is experienced differently in the room from that position. Here the focus in on the transition from the window to the first few metres into the room. The importance of focusing on this part of the room is also developed in the article *'The light, the facade and the room'*, which precisely specifies that it is within the first 2.5 metres from the window that daylight intensity is greatest, measured in DF (Daylight Factor) [22]. The two photographs of 2., *from the side of the daylight aperture, reveal* the significance of the window frame's design. How the form of the window and the profiles mediate the transition from outside to inside. In the last position, *3., with the daylight aperture behind*, seeing where the daylight is in the room becomes more obscure. In the top series, the daylight is clearly seen on the wall. This position is also the most difficult, as there is typically no wall that daylight can 'illuminate and thus become visible. But compared to photograph no. 2, one can get a sense of the light in the room. The last two photographs in the series, *the daylight aperture seen from street level* and the sky should give a view of the exterior design of the window. The image of the sky is essential for clarifying the relationship between the daylight inside the room and the daylight outside, in this case the light from the sky and the sun.

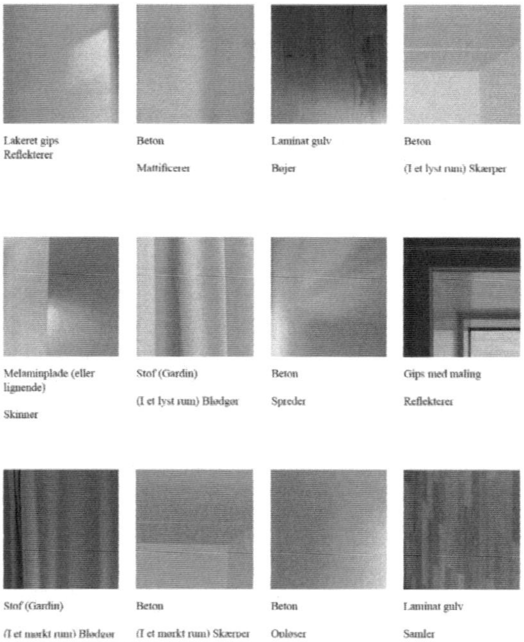

Figure 04.01 *The close-up*

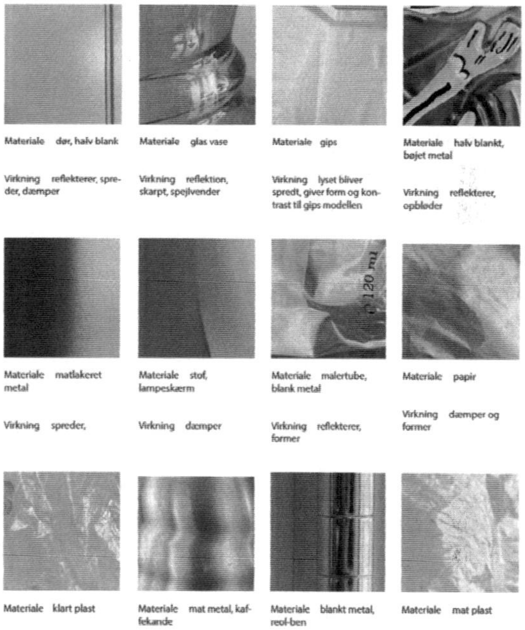

Figure 04.02 *The close-up*

5.4 Close-up

The focus of this series, *Close-up*, is to depict on a small scale how light combines with different materials to form different light effects. This relationship is unfolded in a series of photographs of various surfaces and objects. Two different answers to this assignment are shown above. Below each image is a description of the material and the nature of the surface. The particular aim of the *Close-up* sub-assignment is to create an understanding that light is scaleless, making it visible on all scales, even if it is easily overlooked, and especially on a small scale.

The series depicted in Figures 04.01 and 04.02 create multifaceted studies into different types of surfaces and materials, all of which are found in the home. The focus of Figure 04.01 is on the different surfaces in a room whereas Figure 04.02 focuses more on the surfaces of various objects. Together, the two figures constitute a nuanced insight into what this Close-up format has to contribute.

6. Summary and discussion: using a qualitative method to study daylight

The introduction to this article refers to the tradition of teaching daylight at the Royal Danish Academy – Architecture, Design, Conservation, of which this article is a continuation. This article has a limited format for framing how to develop a teaching method to study the experience of daylight. The chapter *Theory: Being in space* unfolds the phenomenological approach forming the theoretical approach to the experience and perception of the bodily encounter with light. The method chapter describes the framework for the assignment and how different parts of the assignment, lectures and literature form an introduction to the method. The type of assignments and their composition are also reflected upon. In the findings, the various sub-assignments are reviewed with selected examples of works from students. Based on these parts of the article, the primary research question unfolds: "*How can we as educators and researchers impart methods to our students which enable them to examine and study how daylight illuminates an interior space?*" There are multiple ways to answer this question, as the article shows. The first is about a clearly defined and delimited study, which in this case comprises a phenomenological approach as the starting point for how light is experienced. Next, as is also described in the method section, the formats for how the surveys are to be conducted are crucial. The main emphasis is on the photographic approach to the study of daylight, combined with drawings. These formats have been developed through own and earlier research and subsequently transformed into this series of assignments as research-based teaching. The link to previous research and testing of the formats is important. The method section specifies the assignments that are given, and they are concrete and briefly described in relation to scope and submission. In the examples, it is clear that students must 'spot' the light and become aware of how to use the photograph to document the experience. Both in the series *Over time,* by being aware of where the changes of light can be seen in space, but also how the choice of motif reveals an awareness of the light. In the series *Daylight outside, the daylight aperture and daylight in the room*, the relation between exterior, interior and the scale-lessness of light is emphasised through the five different photographs. Especially the series, *From the side of the daylight aperture,* the character of the window as a light modulator is revealed. The window creates a transition between inside and outside and the awareness of this is greatest in the part of the room with the highest intensity of daylight – which is also visible in the photographs. In the last part, *Close-up*, a wide variety of different encounters between various types of surfaces, materials and light are captured. The basis is given in the drawings where scale, orientation and the no-sky line form the geometric foundation.

Working on a specific assignment in a pre-defined layout creates simplicity and clarity, both for the individual student, but in the overall material. The works of all students are gathered in a common archive which forms the basis for cataloguing different types of daylight in dwellings. Because the students themselves have been through the assignment and contributed their own surveys, the overall work will also be relevant and useful for them later on in their studies. This leads us to the sub-question regarding the assignment framework: *What part does the choice of space play as a framework for studying daylight*? By examining a living space, which their own home is, students gain an objective view of the known as well as an opportunity to study light over a prolonged period of time. This also creates a phenomenological attention to daylight so that it becomes visible in its manifestation and not 'merely' as part of the world they take for granted and where its meaning is easily overlooked and forgotten. Another aspect is also that by using their own home as a setting or laboratory for the study, the students' view of daylight does not stop when the assignment is over but will continue because they will continue living in the 'laboratory' in the future. This aspect is not depicted in the examples, but it was expressed by the students in the post-assignment evaluation.

In a sustainable perspective, robust aesthetic sustainability must begin with an understanding of the qualities of space, of which daylight is one of the most significant effects for creating the atmosphere of the space. Using this method, the students have demonstrated their ability to capture an experience of daylight in their homes, and hopefully these phenomenological insights will lead to a future architecture with a more nuanced awareness of the importance and appearances of daylight.

References

[1] Introductory parts of this article are based on the dissertation: Grønlund, Louise. Lysvirkninger : fremkaldelser af rummets arkitektur, PhD dissertation. Aarhus School of Architecture, 2015
[2] *Lys er godt for øjnene*. Erik Christian Sørensen, 5, http://lysnet.com/media/LYSnET%20Norden%20moder/20100525131832776.pdf
[3] Osterhaus, Werner, and Merete Madsen. "Dagslyset som arkitekturens alfa og omega." Lys, pp. 40–41.
[4] Grønlund, Louise. Lysvirkninger : fremkaldelser af rummets arkitektur
[5] Collin, Finn, and Finn Collin. *Videnskabsteori for humanistiske fag*. 1st edition, Copenhagen: Hans Reitzel, 2017.
The humanities, deal with man as a subject, i.e., as a thinking, feeling, acting, and communicating being, as well as with the products of human thoughts, actions, and communications. These include writings, works of art, tools, buildings, and clothing.
[6] Frandsen, Sophus *"Lyset i skyggen,"* in *Til Poul Kjærgaard*, 1982, pp.177–185.
[7] Voltelen, Mogens, *Belysningslære : kompendium fra forelæsninger*, 7th edition (Kbh.: Kunstakademiets Arkitektskole, Instituttet for byggeteknik, Belysningslaboratoriet, 1978).
[8] Pallasmaa, Juhani, *Arkitekturen og sanserne*, ed. Carsten Thau and Martin A4 – Colding Keiding
Cornelius Holck A4 – Fogh, Annemette (Copenhagen: Arkitektens Forlag, 2014). p.11
[9] Rasmussen, Steen Eiler, *Om at opleve arkitektur* (Aarhus: Archi Tegn, 2012)., p.33
[10] Pallasmaa, *Arkitekturen og sanserne*. p.9
[11] *Filosofi & arkitektur i det 20. Århundrede* (Copenhagen: Kunstakademiets Arkitektskoles Forlag, 2006). Dan Zahavi, Edmund Husserl – Hvordan verden træder frem, p.9.
[12] ibid., p.11
[13] ibid., p.13
[14] Mathiasen, Nanet, *Nordisk lys og dets relation til dagslysåbninger i nordisk arkitektur* : PhD dissertation (Copenhagen: Det Kongelige Danske Kunstakademis Skoler for Arkitektur, Design og Konservering, 2015), p. 64
[15] *Filosofi & arkitektur i det 20. Århundrede* (Copenhagen: Kunstakademiets Arkitektskoles Forlag, 2006). Dan Zahavi, Edmund Husserl – Hvordan verden træder frem, p.13
[16] Corrodi, Michelle, Klaus Spechtenhauser, and Gerhard Auer, *Illuminating: Natural Light in Residential Architecture*, Living Concepts, vol. 3 (Zürich: Birkhäuser, 2008). pp.208–210
[17] Frandsen, Sophus *"Lyset i skyggen,"* in *Til Poul Kjærgaard*, 1982, pp.177–185.
[18] Liljefors, Anders. Bättre belysning : om metoder för belysningsplanering. Statens råd för byggnadsforskning, T17:1990. Stockholm: Byggforskningsrådet, 1990
[19] Sontag, Susan. Fotografi : *Essays Om Billede Og Virkelighed.* Fremad,
[20] Norberg-schulz, Christian. "Arkitektur Gjennom Kameraøyet." *Byggekunst*, no. 6 (195AD).
[21] Sontag, Susan. Fotografi : *Essays Om Billede Og Virkelighed.* Fremad, p.30
[22] Christoffersen, Jens, and Merete Madsen. *"Lyset, facaden og rummet."* Arkitekten, 106, no. 13 (2004): pp. 33–34.

IOP Conf. Series: Earth and Environmental Science 1099 (2022) 012022 doi:10.1088/1755-1315/1099/1/012022

Creating an architectural experience in terms of spatial and lighting composition

Zuzana Tomková

PhD student, Faculty of Architecture, Brno University of Technology, Czech Republic

E-mail: tomkozuzana@gmail.com

Abstract. The light plays crucial role in forming the architectural experience and my interest is to examine how architects create bridges between the poetic and practical potential of light. A critical discussion about this topic is relevant especially in relation to the perceived lighting environment and role of light in today's architecture. How was the role of light changing in our society during historical periods? What is the role of light in architecture nowadays? Which architectural experience supports this role and what are the tools to capture it? Different ideas about light, linked to the aesthetic sense of successive historical periods lead to differently configured spaces. Understanding the enjoyment of the experience of spaces in history, but also today and determining its aesthetic value requires attention to details. Architecture can shape our daily experiences in material and detail and together with light form the basis of complete perception. I believe that consistent work with daylight in architecture is a catalyst for change and can significantly help us rediscover the relationship with nature. I believe, that by raising awareness about experiential approaches of working with a daylight will help us to find our way back to nature, origins and rise quality of architectural objects.

1. Introduction

There are different analysis methods today how to work with light – represented by standards. Relying too much on the scientific methods, there is a risk that lighting design will be reduced to a quantitative exercise, ignoring the qualitative, aesthetic and experiential aspect. The quantity of light is decisive for functions of the space, but lighting in architecture is not only a sophisticated technical system and the quality of the light is also very important.

Light play crucial role in forming the ultimate enjoyment of the architectural experience. The intention of my work is to examine how architects manage to create bridges between the poetic and practical potential of light, when creating beautiful lit spaces. That is why I focus on research of extensive use of technical tools to predict and design visual environment. A critical discussion of the use of these tools are very relevant especially when they are in relation to the perceived lighting environment and role of light in today's architecture. This analysis can be beneficial, as an inspiration for lighting designers and architects, spreading the education about the light. As professor Juhani Pallasmaa says: „In order to understand human experience, we must shift from the scientific processes of measuring to the courage and desire to live and encounter architecture directly through our very act of living." [1]

2. How the role of light in architecture was changing in our society during historical periods

A work of architecture is the result of many factors – technical, economic, social conditions and particular historical period. Knowledge should lead us to respect the values of space and light.

Building is an autonomous organism that can resonate far beyond the period in which it originated and the style to which it belongs. Light, one of the fundamental materials for the construction of space, never ages.

2.1. Harmony with nature

There was a strong symbolism of light in classical architecture. Greek architecture, thanks to the favorable local climate and sunlight was in total harmony with nature. We can also observe this continuity of interior light with that of the exterior in Roman basilica.

2.2. Symbolical lighting

Medieval architecture deepens symbolical lighting, in meaning of connection to spiritual world. In Gothic has light strong transcendental meaning and sacred character. Very typical was filtered light – chromatic system of windows, which separated interior of cathedrals from any other natural space and creating world inside the world. Intention was to create uplifting effect.

2.3. Light – a physical element that provided humanity with knowledge of space

Renaissance light returns to the idea of classical, when rediscover spatial effects of light to create more universal immaterial space. Architects were exploring scientific instruments to understand the reality, which is approach closely connected with later contemporary studies of space. Da Vinci created light and shadow theory, which then developed during Baroque era.

2.4. Dramatic lighting

Baroque era is typical for filtered lateral light and dramatic optical effects, which were achieved due creating the light box, or side lighting. Daylight was not good for a shadow play, that is why connection with outdoor was filtered and regulated again.

Industrial revolution in 18th century brings back functional view of space and light. We can see the light as a spiritual substance, where illusion of the weightlessness and immateriality, bring out light as a poetic element of architecture. No deep shadows, but softened surface, represents infinity of continuous, universal space. [2]

2.5. Universal light

Plummer [3] talks about the change in the light quality at the beginning of the twentieth century. He connects much broader understanding of the perceptual, physiological and psychic experience of different spaces associated with the way light behaved in a space. Until this point, light was having metaphysical connotations.

This changed with the early works of Corbusier and Gropius, who had a vision to build a new open society. As they were blurring borders between architecture and figurative art, the theme of light and transparency of space was important topic. Gropius' Bauhaus, the Villa Savoy by Le Corbusier - two of the most celebrated models of functionalism, share the same conception of light.

They are using term 'universal light', which supports the geometric consistency of the space. Universal light is an ideal midday light, inclined at 45°, which penetrates the interior without changing quality, since all divisions between the interior and exterior are temporary and occasional, and do not 'separate' different worlds, but at most distinguish 'zones,' separate a 'microclimate' within the same environment" [4]

3. Role of the light in architecture nowadays

German author and poet Paul Scheerbart, notable herald of expressionist architecture in Europe wrote about ideal behind the image of a liberated, open society, which developed in 20th century. He said: „It would be as though the earth robed itself in diadems of diamonds and enamels. We will possess an earthly paradise and will no longer have any need to yearn for the paradise of heaven. "

The exaltation of technology and progressive optimism of the last decades of 20th century led to a new anthropocentrism. Humans seemed able to take control of the world by their own efforts and believed they were the center of the universe.

This believe fluently continues to begin of the 21st century, when climate crisis began to be debated more publicly. Nowadays, we can observe how our behavior is driving our planet to ruin and that humanity's relationship with nature is in a crisis. That is why, remembering our dependence on natural phenomena is core topic of the coming decades.

I believe that consistent work with daylight in architecture can significantly help us rediscover the relationship with nature. In my work I am not focusing on technological aspects such as reducing energy or efficient using of sources. I want to point human and psychological aspects, showing atmospheres of the light in spaces.

4. Architectural experience

During the past few decades, an experiential approach of architecture, has progressed. This thinking is based on the philosophies of Martin Heidegger, Edmund Husserl, Maurice Merleau-Ponty, and many other contemporary philosophical writers.

In the book Art as Experience, John Dewey links a role of architecture in relation to the nature of experience itself as well as to our understanding the passing of time and history. This visionary pragmatist philosopher writes: "Art is always the product in experience of an interaction of human beings with their environment. Architecture is a notable instance of the reciprocity of the results in this interaction [...] they not only influence the future, but they record and convey the past". [5] Unlike many aspects in construction, in its performance, structural essence, formal and dimensional properties can be studied scientifically, experiential and mental meaning of the entity can only be experienced.

Our architectural experience probably involves some combination of cognitive, emotional and sensory experience. Sauchelli [6] proposes to use cognition to capture pleasure as a central feature of the architectural experience. Understanding the enjoyment of the experience and thus determining its aesthetic value requires knowledge in the form with attention to details such as depth, edge detection, colour and light, ...

Steven Holl defines space as the essence of architecture. The architectural synthesis of foreground, middle ground, and distant view, together with all the subjective qualities of material and light, form the basis of complete perception. As a catalyst for change, architecture's ability to shape our daily experiences in material and detail is subtle yet powerful. When sensory experience is intensified, psychological dimensions are engaged.

Steven Holl, together with Frank Lloyd Wright, Alvar Aalto, Louis Kahn, Peter Zumthor and many others, approached architecture in a poetic, philosophical and metaphorical manner, usually arise from personal experiences, observations and beliefs. In their essays, we can find a lot of the experiential and phenomenological attitude to their architectural objects. I will look closer to some of them.

5. Functional harmony

We can see the internal complexity of his architectural projects of Alvar Aalto. As he said on his inaugural lecture as member of the Academy of Finland in 1955. "Whatever our task, whether large or small, in every case, opposites must be reconciled. Almost every formal assignment involves dozens, often hundreds, sometimes thousands of conflicting elements that can be forced into functional harmony only by an act of will. This harmony cannot be achieved by any other means than art. The final value of

IOP Conf. Series: Earth and Environmental Science 1099 (2022) 012022 doi:10.1088/1755-1315/1099/1/012022

individual technical and mechanical elements can only be assessed afterwards. A harmonious result cannot be achieved with mathematics, statistics, or probability calculus" [7] [8].

Alvar Aalto´s Villa Mairea is frequently listed as one of the most important one-family houses of the 20th century. In the 1920s, modernist architects used lighting as a practical means to divide small open-plan flats into dining, living and work zones. At Villa Mairea, its role was very different – shadow mattered as much as light, and atmosphere as much as utility. Villa Mairea, surrounded by natural woodlands stands out as Aalto's gift of poetic inspiration and the art of play. In its interior, and its lighting, Villa Mairea is offering intimacy and homeliness and in time of the rigid modernism, brings a more human built environment.

Aalto designed Villa Mairea with the intention of blending the interior and exterior environments. This is mainly achieved using specific types of architectural elements and materials, merging the natural and artificial.

View of the library from the living room - Photo by Lindman Photography

6. Light, the giver of all presences

Louis I. Kahn was one of the leading architects in America during the twentieth century. His stay at the American Academy in Rome in 1950 and his associated studies and analysis of ancient buildings in Italy, Greece and Egypt had a lasting effect on Kahn's understanding of architecture.

On February 12, 1969, Kahn gave a lecture at the School of Architecture at Swiss Federal Institute of Technology in Zürich, entitled Silence and Light. This lecture explains Kahn's spiritual understanding of architecture, which goes deeper than simply constructing buildings. He also gives a vision of a belief in sustainable architecture that prefigures the twenty-first century's focus on green technology. In this lecture, Louis Kahan emphasized the psychological functionality of buildings. For him, there are spaces inhabited by our consciousness (hence light and silence).

„You can say the light, the giver of all presences, is the maker of a material, and the material was made to cast a shadow, and the shadow belongs to the light" [9].

IOP Conf. Series: Earth and Environmental Science 1099 (2022) 012022 doi:10.1088/1755-1315/1099/1/012022

The first significant commission of Louis Kahn was Yale University Art Gallery in New Haven, Connecticut (1951–1953). In his projects, Kahn explored numerous strategies and construction options for reinforced concrete and perceptual effects. Part of these strategies is Kahn's use of natural light. In the Yale Art Gallery, Kahn surrounded the public stairs with a concrete drum, in which protruding concrete triangles set in the ceiling hang over our heads. The drama, amplified by the addition of natural light, is an architectural chiaroscuro that creates an incredible contrast. “It is a world of elements in which one imposes one's will on uncompromising material and where even the simple act of climbing the stairs becomes a heroic expedition." [10].

In his later works, Kahn rose above function and heroism. At the time he began working at the Yale Center for British Art (1969–1974), Kahn progressed from dogmatically promoting the masses in the sharp relief of daylight to a more comprehensive understanding of steel-reinforced concrete and natural light as his sparkling partner. Kahn no longer imagined silence competing with light, but as two brothers. According to one of Kahn's remarkable ideas, silence itself comes from the depths of the natural world. Kahn no longer responded to the paradoxes of concrete and the resistance of natural light by enforcing will, he approached it humbly.

As Michael Cadwell mentioned in his book Strange Details, we can compare it perfectly on the staircase in the drum, which reappears here, as well as at Yale University Art Gallery. There's something unreal about this drum. Concrete surface atmospherically circles. The great will is suppressed by it and is subject to natural light phenomena. It is no longer a struggle of matter and emptiness, light and shadow, man with nature, but participation. This mood carries us up the stairs, the concrete cover wraps us, and above us the light falling through the glass block of the drum ceiling breaks. Unlike its predecessor at the Yale Art Gallery, this drum does not imprison us inside with a feeling of anxiety, but it lifts us.

Left: Yale University Art Gallery Stairs - Photo Jeffrey Scott, 2009Right: Yale Center for British Art · New Haven, Connecticut- Photo by Lindman Photography

The concrete of the Center for British Art will stop us with its lightness. Kahn revived the concrete mix by adding volcanic ash, creating perfect forms for the formwork, and mixing the concrete so that its cement slurry flowed to the perimeter, creating a shiny surface. The surface of the concrete looks atmospheric - clouds of ash tones float along its silky surface. If we compare it with the ceilings of Yale University Art Gallery, which are 0.9 meters thick, consist of a grid of triangular openings that draw the eye upward into dimly-lit, three-sided pyramidal spaces, we see the great shift towards the relation of mass and light. Concrete is not a supplement to light, but its partner. Light takes on a material value.

Left: Yale University Art Gallery - New Haven - Photo by Lionel Freedman - 1953
Right: Yale Center for British Art by Louis Kahn, New Haven CT (1974) – Photo by Elizabeth Felicella

7. Rooms orchestrated with light

In November 2016, Steven Holl received The Daylight Award for Daylight in Architecture. His ability to orchestrate rooms with light is unparalleled; the emotional intensity of his rooms and buildings is impressive. "Space is oblivion without light. A building speaks through the silence of perception orchestrated by light," wrote Holl in 2006 at the start of his book 'Luminosity/Porosity' [11].

This sentence is, in many ways, the complementary counterpart of Louis Kahn's famous verdict, "the sun never knew how great it was until it hit the side of a building." The spiritual similarity that reverberates in both quotations is not a coincidence. Steven Holl refers to Louis Kahn as one of the architects who had an early influence on his own work with daylight.

To Steven Holl, daylight is not merely a means of shaping the atmosphere of his buildings but enriching the sensual perception of their users. Great example how Holl is working with a daylight, connecting the time and natural principles of the sun is the entrance hall at the Cranbrook Institute of Science (1998). Holl uses effects of light distorted through different types of glass. Here the constantly changing effects are of sunlight and shade, of streaks, informal patterns and the occasional spectral projection through prisms built into the facade.

These work in concert to animate the space as part of a highly experiential series of internal and external spaces. Wall is modified in appearance only by the instrument of light and the passing of time.

Cranbrook Institute of Science, Bloomfield Hills, United States. 1998 – Photo by Steven Holl Architects

8. Conclusion

As light never changes, we have all history to learn about successful tools how to consciously work with daylight. Observing different approaches during historical periods made me think about the role of the light in architecture today.

As was previously stated, I see one of the biggest topics of today as current climate change and separation of humans from the nature. I believe that consistent work with daylight in architecture can significantly help us to rediscover our relationship to nature and led us to pursuit of love to natural beauty against manmade existential crisis.

With this intention, I am exploring the experience of an interaction of human beings with their environment, where daylight plays an important role. Through the works of architects with experiential and phenomenological approach, I am reminding unscientific nature of architecture as a fusion of technology and intuition, knowledge and emotion, dreams and reality. Unlike many aspects in architectural objects can be studied scientifically, experiential and mental meaning of the space can only be experienced. That is why the analysis of the different architectural cases were made through the observation.

This work tent to be an inspiration for the architects, underlying the internal complexity of architectural object, where atmosphere matter as much as utility "We tend to underestimate the psychological power and the health aspects of light, particularly in our society, which is often driven more towards commercial ends. As architects, it is our task to bring these aspects to the table, and to remind our clients and all the experts involved in building design today of the psychological and biological importance of light." [12]. I believe, that by raising awareness about different approaches how to work with a light in our spaces will help us to find our way back to nature, origins and rise quality of architectural objects.

I will close with the sentence of Luis Kahn: „ I only wish that the first really worthwhile discovery of science would be recognition that unmeasurable you see, is what they are really fighting to understand, and the measurable is only a servant of the immeasurable, that everything that man makes must be fundamentally immeasurable." [9].

References

[1] Pallasmaa, J. 2005. The Eyes of the Skin: Architecture and the Senses. Chichester, England: John Wiley & Sons.

[2] Valero Ramos, Elisa, 2015, Light in Architecture The Intangible Material, Newcastle upon Tyne: RIBA Publishing,

[3] Plummer, H. 2003. Masters of Light, Architecture and Urbanism, Tokyo

[4] Portoghesi, P. 1994. Light and modern architecture. In Light & Space: Modern Architecture, Futagawa, Y. ed., Ga International Co Ltd, Tokyo.,

[5] Dewey, J. 1934. Art as Experience. New York: Putnam

[6] Sauchelli, Andrea, 2011 Functional beauty, architecture and morality: A beautiful Konzentrationslager, The Philisophical Quarterly, Published by Blackwell Publishing, 9600 Garsington Road, Oxford ox4 2DQ , UK, and 350 Main Street, Malden, MA, 02148, USA,

[7] Aalto, A. 1997, Art and technology. In: G. Schildt ed. Alvar Aalto in His Own Words, Helsinki: Otava.

[8] Aalto, A. 1997, The trout and the stream. In: G. Schildt ed. Alvar Aalto in His Own Words, Helsinki: Otava.

[9] Kahn Louis I., February 12, 1969 - Silence and Light: The Lecture at ETH Zurich,

[10] Cadwell Michael, 2012, Strange Details, MIT Press, Camridge, USA

[11] Holl Steven, January 6, 2006, Luminosity / Porosity, Publisher:Toto

[12] Q&A 20 March 2013, Steven Holl – Point of View. Interview by Martin C. Pedersen, Metropolis Magazine

In-situ evaluation of high-performance glazing based on illuminance and glare

Caroline Karmann[1], Jan Wienold[1], André Kostro[2], Pietro Florio[2], Andreas Schüler[2], Jean-Louis Scartezzini[2], Marilyne Andersen[1]

[1] Laboratory of Integrated Performance in Design (LIPID), École Polytechnique Fédérale de Lausanne (EPFL), Switzerland
[2] Solar Energy and Building Physics Laboratory (LESO-PB), École Polytechnique Fédérale de Lausanne (EPFL), Switzerland

Corresponding author: caroline.karmann@kit.edu

Abstract. High-technology glazing panes are designed to meet the need for natural light, while optimizing key functions such as reducing solar heat gain in spaces and/or redirecting daylight to the rear of the spaces. However, the actual performance in terms of both illumination and glare protection of these glazing systems is often unknown. This study aims to compare the visual comfort performance of four complex fenestration systems by means of point-in-time measurements under real sky conditions and for different seasons. The tested systems included: two glazing types with embedded seasonal solar protection, namely, MicroShade^{TM} and Coolshade^{TM}, one prototype of daylight redirecting glazing with embedded micro-mirrors (GEMM), and one clear glazing. We conducted our tests in side-by-side nearly identical spaces in the NEST building in Dübendorf, Switzerland. Results showed a higher daylight provision for clear and daylight redirecting glazing. Discomfort glare assessment from daylight showed that MicroShade^{TM} was substantially more efficient at addressing glare due to sunlight for high sun angles when compared to Coolshade^{TM}. The performance of the GEMM prototype was highly dependent on the glazing pane used and one pane provided a promising performance when it comes to mitigating glare risks for high sun positions. However, none of the systems were able to provide an adequate glare protection for lower sun positions. The performed measurements provide a basis for the calibration of simulation models.

1. Background

The extensive use of glazing as a building façade has become a common feature in commercial buildings. These building types require careful design decision-making that accounts for user's needs. Among these, access to daylight appears to be of primary importance, both in terms of the appreciation of occupants for natural light – approximately 80% of occupants claim to prefer daylight to electric lighting [1–3] – as well as in terms of the non-image forming effects of light, with notably the crucial role of (day)light for synchronizing circadian rhythms [4,5] and its effects on other bodily processes necessary for our well-being and health [6–8]. In parallel, the access to views out through windows has shown its positive impacts on human workplace satisfaction, increasing productivity, focus, life satisfaction, stress modulation, and patient recovery time [9–15], while views also satisfies fundamental human needs for visual information about location, time, weather and activities outside the building [16,17]. Thus, the extensive use of glazing in commercial buildings seems to be a sound approach, and high-performance glazing can offer promising combinations of properties when it comes to ensuring protection against excessive solar gains and/or glare, or bringing natural light deeper into the space.

The International Commission on Illumination (CIE) defines glare as the "condition of vision in which there is discomfort or a reduction in the ability to see significant objects, or both, due to an unsuitable distribution or range of luminances or to extreme contrasts in space or time" [18]. Glare is common in workspaces and is a source of disturbance to building occupants [9] that can affect their productivity [19]. Multiple metrics have been established to characterize discomfort glare from daylight from objective measurements. Of these, hybrid metrics are based on both saturation and contrast effects, amongst which the Daylight Glare Probability (DGP) [20] – adopted for daylight glare prediction in the European standard EN 17037 [21] – was shown to be one of the most robust and reliable metrics [22,23].

High-technology glazing systems are designed to meet the need for natural light, while optimizing certain key functions like managing solar heat gains. However, this usually requires a trade-off. Reducing heat loads generated by solar gains for instance can also indirectly affect indoor illumination. Similarly, light-redirecting glazings, which aim at increasing daylight penetration through embedded reflective or refractive components, can cause discomfort due to glare from daylight and do not prevent per se from overheating. In general, it is therefore difficult to optimize glazing for multiple objectives and complex fenestrations systems are often only tested for the objective for which they have been optimized.

2. Objectives

This study aims to compare the visual comfort performance of four complex fenestration systems by means of measurements under real sky conditions, with a focus on the depth of penetration of daylight in the space (illuminance distribution) and the effectiveness of the offered protection against glare from the sun. Evaluations were conducted for both low (winter) and high (summer) sun positions and all the systems allowed to maintain at least a partial view to the outside. This study does not cover the topics of daylight spectral power distribution and outdoor view clarity.

3. Methods

3.1. Experimental protocol

We conducted our tests in nearly identical side-by-side spaces located within the SolAce unit in the NEST building in Dübendorf (CH) [24]. In these spaces, the upper part of the façade was designed to test complex fenestration systems. We tested a total of four systems including two glazing types with embedded seasonal solar protection, namely, MicroShade™ [25] and Coolshade™ [26], one daylight redirecting glazing prototype with embedded micro-mirrors (GEMM) [27,28] and one clear glazing. The glazings were mounted successively in the upper part of each space (see Figure 1(a) and 2). We carried out the on-site monitoring during the summer and winter seasons on sunny days. We focused on point-in-time measurements and conducted seven pairwise comparisons spread over the two seasons (see Table 1).

3.2. Experimental space

SolAce is living lab that can be used for a variety of in-situ experiments, including energy performance and carbon neutral living assessments [29]. Our side-by-side testing spaces are South-West oriented and can be separated from the rest of the module by means of curtains, as represented in Figure 1(a). The facade was designed with operable windows and external shading fabric in the lower part, and with fixed shading panes in the upper part. We performed our tests with the curtain of the lower part down and without artifical lighting to better focus on the upper part. Prior to testing, we measured the reflectance of the indoor surfaces, and conducted measurements of indoor lighting conditions with the upper windows blacked out to ensure enough similarity between the side-by-side spaces.

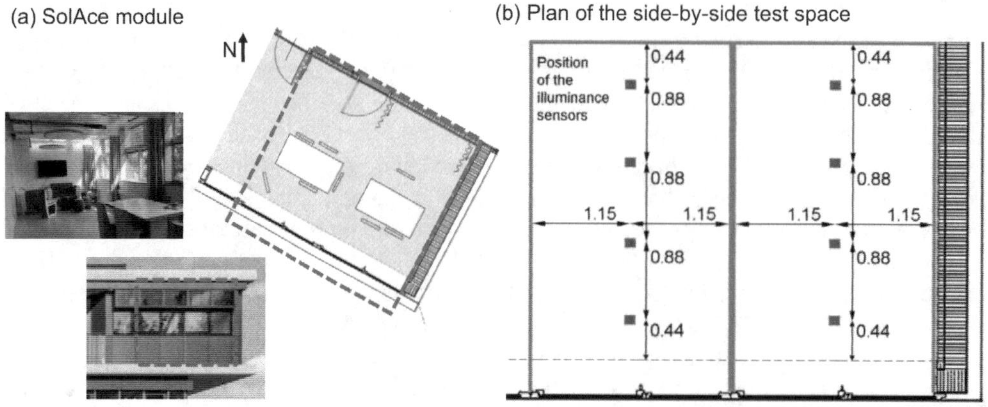

Figure 1. (a) Layout and views of the indoor space in the SolAce unit with in red the side-by-side test space; (b) Plan of the side-by-side test space (numbers are reported in meters) with the location of the illuminance sensors

Figure 2. Views of the two rooms with the GEMM (left), MicroShade™ (center) and Coolshade™ (right) glazings fitted on the upper part of the façade. The textile blinds of the lower windows were kept down during the testing

Table 1. List of glazing comparisons and test dates for each season

Comparisons	Upper part of the left space	Upper part of the right space	Summer[a]	Winter
1	Coolshade™	GEMM[b] (Summer: C+E) GEMM[b] (Winter: D+E)	Jun. 03, 2020	Feb. 23, 2021
2	Coolshade™	MicroShade™	Jul. 13, 2020	Feb. 22, 2021
3	GEMM[b] (Summer: B+C)	MicroShade™	Jul. 14, 2020	-
4	Clear glazing	GEMM[b] (Winter: D+E)	-	Feb. 23, 2021
5	Clear glazing	MicroShade™	-	Feb. 24, 2021

[a] The clear glazing was not available in Summer

[b] GEMM is a prototype and different glazing panes were used. The panes used involved: prototypes "C&E" for the comparison to Coolshade™ in Summer, prototypes "B&C" for the comparison to MicroShade™ in Summer, prototypes "D&E" for the comparison to Coolshade™ and to the clear glazing in Winter in Summer. This alternation of glazing pane was the result of glazing damage and incompatibility of window pane size between the left and right spaces, and not an experimental design choice.

3.3. Fenestration systems

Four fenestration systems were compared to one another in different pairwise configurations:

- Coolshade™ (aka., Swisslamex Coolshade™) is a laminated safety glass with a mesh of solar protection lamellae (1.27 mm wide, spaced 1.27 mm apart and inclined at 30°) inserted by lamination. The mesh can be customized and integrated into different types of glazing. Coolshade™ was a prototype from Glaströsch and was mounted within the original NEST SolAce unit.

- MicroShade™ is a micro-structured perforated foil that can be integrated into double or triple insulating glazing units. The perforations are slanted through the thickness of the sheet to provide additional shading from the sun.
- GEMM includes a polymer film which contains small reflective parabolic micro-mirrors comparable to the horizontal slats of venetian blinds with micrometer dimensions (50-150 micrometer). The concept behind this glazing is to redirect the incoming solar radiation to the rear of a room in winter. For the summer, the original design included an additional layer of horizontal micromirrors aimed at blocking the sun, but this extra layer was not integrated in the prototype.
- Finally, and in order to have a baseline, we added a clear glass to our selection of fenestration systems. In order to remain consistent with the performance of current systems, we have chosen a triple insulated glazing with a high solar heat gain coefficient. This glazing however was not delivered on time for the summer studies and could only be included in the winter comparisons.

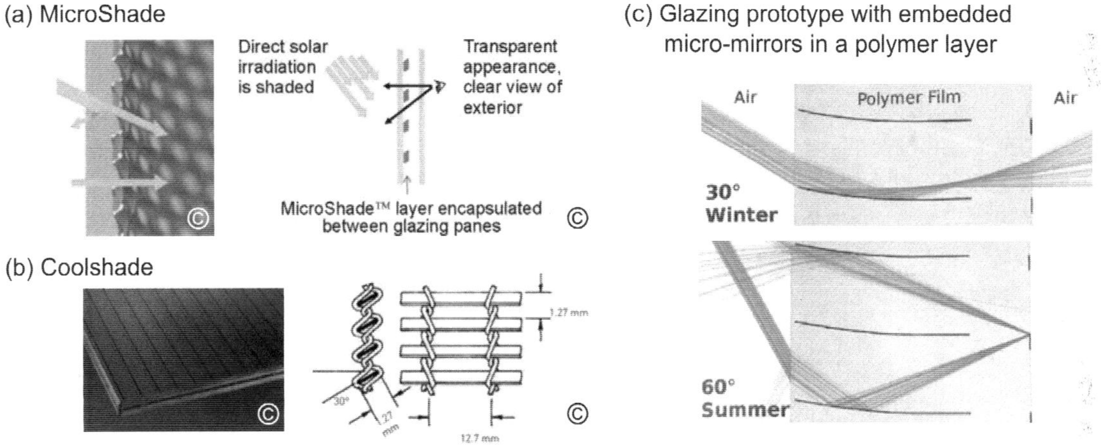

Figure 3. Conceptual diagrams and drawings from the fenestrations systems: (a) MicroShade™ (image source: [30]) ; (b) Coolshade™ (image source: [26]); (c) GEMM (adapted from Gong et al., 2016 [31]).
Note that the GEMM prototype used in this study **only** included to horizontal redirecting micro-mirrors.

Table 2. Glazing properties

Glazing	Nr. of panes	U-value	SHGC	t_{vis}
Coolshade™	3	0.9	0.47	NA
MicroShade™ (MS-A)	3	NA	0.1-0.33(a)	0.49
GEMM (prototype)	2	NA	NA	NA
Clear glazing (Silverstar EN2plus 3-fach)	3	0.8	0.53	0.74

(a) The Solar Heat Gains Coefficient (SHGC) of MicroShade™ depends on the angle of incidence (0.1 for high sun position and 0.33 for low sun position)

3.4. Testing equipment and analysis methods

We performed a series of point-in-time measurements simultaneously in both spaces. For the illuminance measurements, we relied on eight horizontally mounted Delta Ohm HD2021 TBA photodiode sensors (accuracy: ±10% of reading (lux)) connected to an Onset HOBO data logger. Four sensors were distributed in the center of the room as presented in Figure 1(b). For glare assessment, we used two digital single-lens reflex cameras Canon EOS D70 equipped with a fish eye lens and Kodak neutral density (ND) filters, which were calibrated using the procedure described in Pierson et al (2020)

[32]. As the images were taken with the sun in the field of view, we relied on aperture sizes (f) of 18 or 22. While capturing the images, we also measured the vertical illuminance right next to the camera with the illuminance meter LMT Pocketlux 2 (accuracy: ±7% of reading (lux)), and the luminance of a reference sample in the space using the luminance meter Minolta LS100 (accuracy: ±2% ±1 digit of reading (cd/m^2)). The measurements enabled us to calibrate the images according to the luminance ranges measured in the space. We derived the DGP by running Evalglare version 3.02 [33,34] on the calibrated HDR images. For some images we encountered pixel-overflow for the sun-disk even using the ND filters. Those images were corrected by adjusting the sun-pixels while matching the vertical illuminance for the full image using a correction procedure in Evalglare. We relied on DGP glare thresholds and their corresponding predictions (which were established based on the feedback from study participants) for the interpretation of the glare results (see Table 3).

Table 3. DGP thresholds and their corresponding predictions

DGP thresholds	Prediction
<0.35	Imperceptible
0.35-0.40	Noticeable
0.40-0.45	Disturbing
>0.45	Intolerable

We report the results in terms of absolute values – for both illuminance and glare – and compare the glazings in pairs according to the experimental design shown in Table 1. The results are grouped into (1) comparisons involving daylight redirecting glazing (GEMM), and (2) comparisons between the solar control glazing.

4. Results

4.1. Discomfort glare from daylight with sun in the field of view

The glare measurements were taken under clear sky conditions with the sun in the field of view while the viewing direction was horizontal and perpendicular to the façade (directly facing outside). These represent critical scenarios, which has the advantage of informing us about the possible protection by the different complex fenestration systems. The results are reported in Table 4. Winter solar elevation was in the range of 30-33° and summer solar elevation in the range of 50-53°.

The performance of each system remains steady during each season for both Coolshade™ and MicroShade™. For GEMM, the performance depends strongly on the window pane prototype in place. Although the choice of the glazing pane studied was not part of the objectives of this research (but a question of availability of panes), we realized while analyzing the results that fabrication quality could significantly affect their properties for incident angles in the range 45-55°, in particular in regard to the direct-direct transmittance of the pane and therefore to the induced glare.

Table 4. List of glazing comparisons and test dates for each season

Season	Date	Time	Solar azimuth(a)	Solar elevation(b)	Left	Right	DGP left	DGP right
Summer	2020-06-03	15:40	239°	53°	Coolshade™	GEMM (pane E)	0.68	0.91
	2020-07-13	15:49	239°	52°	Coolshade™	MicroShade™	0.72	0.36
	2020-07-14	16:03	243°	50°	GEMM (pane C)	MicroShade™	0.29	0.41
Winter	2021-02-22	14:38	197°	31°	Coolshade™	MicroShade™	0.93	0.79
	2021-02-23	14:15	190°	32°	Coolshade™	GEMM (pane E)	0.87	1
	2021-02-23	15:07	205°	30°	Clear glass	GEMM (pane E)	1	0.97
	2021-02-24	14:00	186°	33°	Clear glass	MicroShade™	1	0.86

(a) Azimuth is measured in degrees clockwise from north
(b) Elevation is measured in degrees up from the horizon

IOP Conf. Series: Earth and Environmental Science 1099 (2022) 012023 doi:10.1088/1755-1315/1099/1/012023

4.1.1. Daylight redirecting glazing

The GEMM prototypes are in part 'hand-made' from different fabrication batches and we observed fairly large variations between the different prototypes made available. In the winter, we used the GEMM pane E in comparison to the clear glazing and to Coolshade™ (measurements performed at different moments so with slightly different conditions). DGP values for the GEMM pane E reached 0.97 and 1 while it reached 1 for the clear glazing and 0.87 for Coolshade™ (see **Figure 5**). All these values are associated with an intolerable glare perception. For the summer measurements, the GEMM pane E led to a similarly high DGP value of 0.91 while Coolshade™ showed a DGP of 0.68 (still intolerable glare). We used the GEMM glazing pane C for the comparison to MicroShade™ (see **Figure 4**), which led to a DGP of 0.29 for the GEMM pane C (imperceptible glare), and 0.41 for MicroShade™ (disturbing glare). Given that not all systems could be measured for all conditions (cf. Table 1), these results show us that, beyond the volatility in glare mitigation for the GEMM panels due to their still prototype-level production process, all systems generally tend to have a poor glare protection performance for the low sun angles found in the winter months.

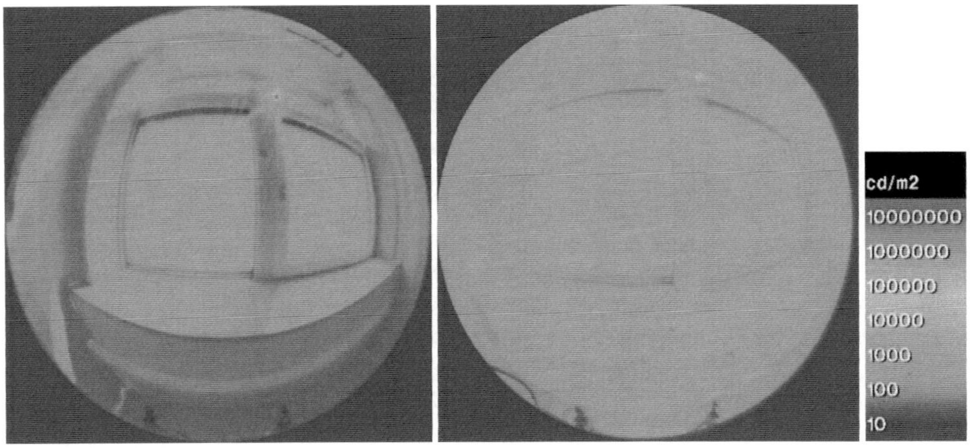

Figure 4. High Dynamic Range (HDR) images of GEMM (left) and MicroShade™ (MS-A) (right) glazings with sun in the field of view taken on July 14th 2020 at 16:03

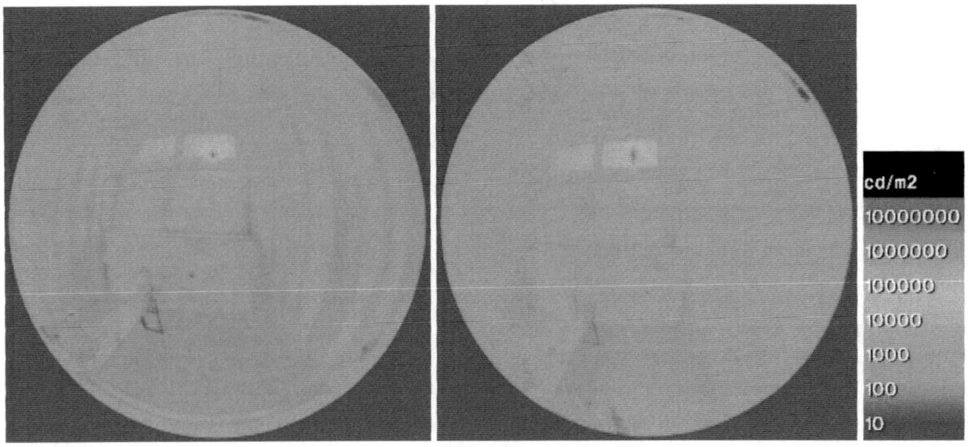

Figure 5. High Dynamic Range (HDR) images of Coolshade™ (left) and GEMM (right) glazings with sun in the field of view taken on February 23rd 2021 at 14:15

4.1.2. Seasonal solar control glazing

Concerning the seasonal solar control glazing, predictions showed that MicroShade^{TM} was able to provide higher glare protection compared to Coolshade^{TM} given the summer sun angles, with a DGP of 0.36 (which corresponds to a 'noticeable glare' rating), while Coolshade^{TM} reached 0.72, rated as 'intolerable' (see Figure 6). However, both fenestration systems performed very poorly for winter sun angles and all measurements were consistent with an intolerable glare (see **Figure 7**). We note that these glazings are designed to offer a seasonal solar control and are in this sense permissible to the winter sun for the solar heat gains. The unprotective behavior towards glare in winter is therefore expected.

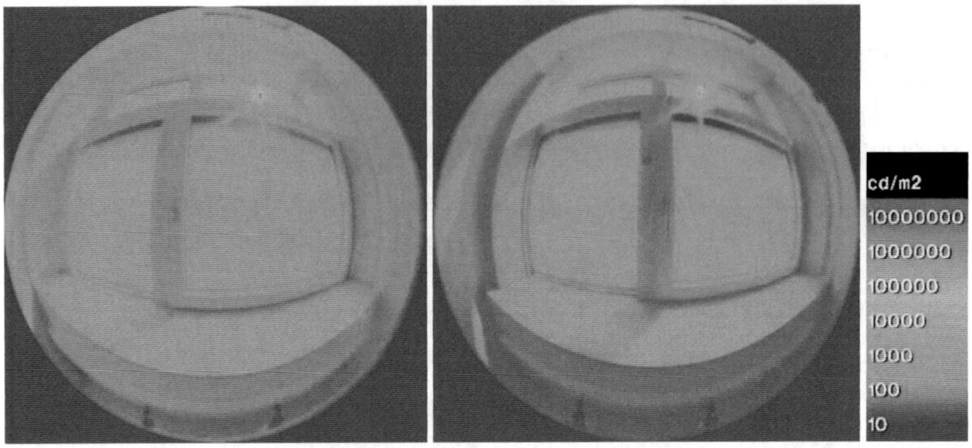

Figure 6. High Dynamic Range (HDR) images of Coolshade^{TM} (left) and MicroShade^{TM} (MS-A) (right) glazings with sun in the field of view taken on July 13th 2020 at 15:49

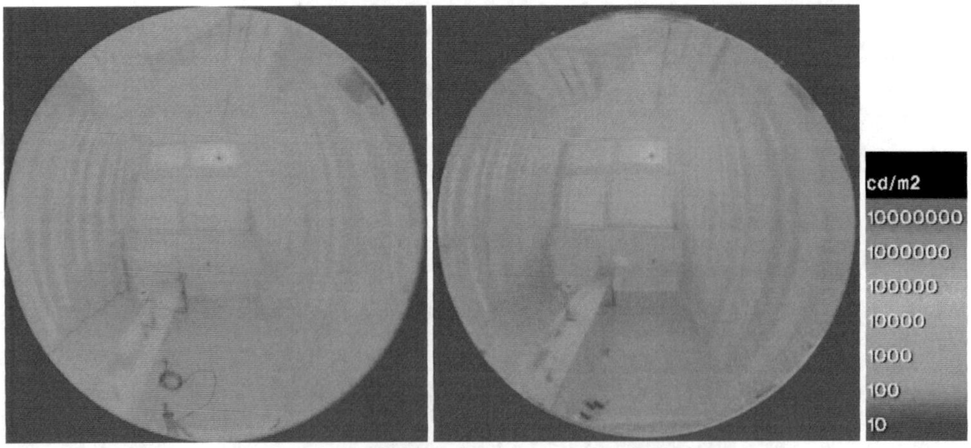

Figure 7. High Dynamic Range (HDR) images of Coolshade^{TM} (left) and MicroShade^{TM} (MS-A) (right) glazings with sun in the field of view taken on February 22nd 2021 at 14:38

4.2. Illuminance

4.2.1. Daylight redirecting glazing

Illuminance measurements focused on the daylight redirecting glazing (**Figure 8**) showed that GEMM consistently provided higher illuminance levels in the space than the glazing with seasonal solar control. This difference is larger for MicroShade^{TM} (75% reduction) than for Coolshade^{TM} (63% reduction) under summer sunny conditions. We also found that clear glazing provides higher daylight illuminance than GEMM in winter, both under direct (35% increase) and indirect sunlight (that is, when the direct sun is not facing the façade) (55% increase).

4.2.2. *Seasonal solar control glazing*

Illuminance measurements were focused on glazing with seasonal solar control (see **Figure 9**) and confirmed the advantage of light provision for Coolshade^{TM} compared to MicroShade^{TM} (14% higher illuminance level for a sunny summer day, and 77% for a sunny winter day). Coolshade^{TM} being worse for glare than MicroShade^{TM}, it was therefore expected that the former would be better at letting in daylight.

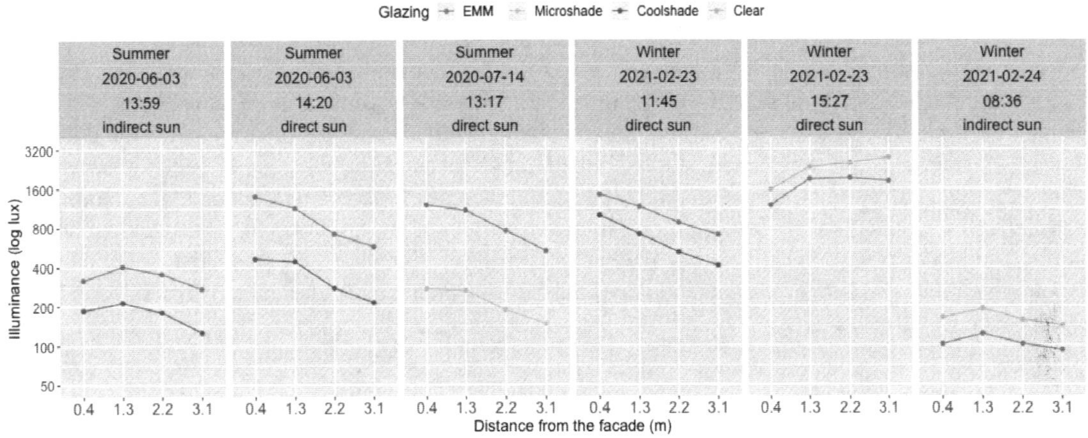

Figure 8. Sample of illuminance measurements (log scale) as a function of room depth (in meters) for the daylight redirecting glazing. Reference glazing: GEMM

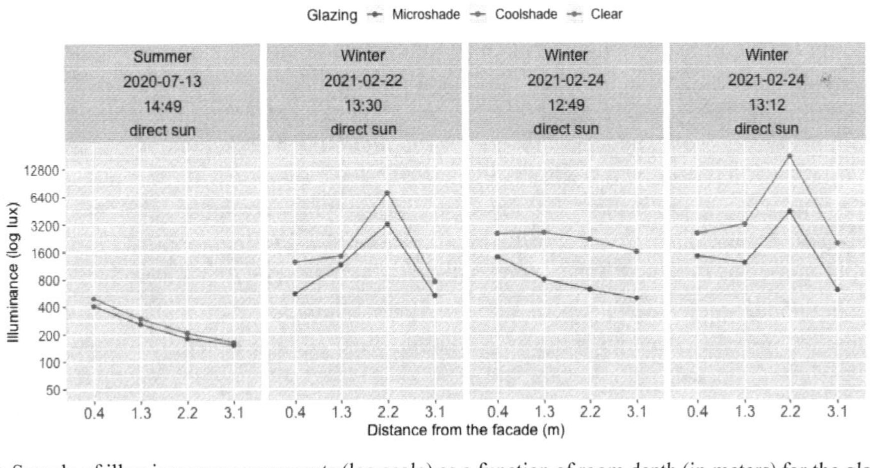

Figure 9. Sample of illuminance measurements (log scale) as a function of room depth (in meters) for the glazing with seasonal solar protection. Reference glazing: MicroShade^{TM}

5. Discussion

In this experimental study, we tested the first prototypes of the GEMM micro-structured glazing, which are still far from being optimized. Variations in the manufacturing processes induced a variation in the actual geometries which impacted the visible transmittance and its angular dependence. In the original and theoretical design of GEMM, the direct transmittance was supposed to be less than 1% for a solar elevation of 48° and above. The fabricated prototypes did not reach this target. Further, only the first set of nearly horizontal micromirrors are implemented, while the second set of vertical ones for backwards redirection are still missing. Therefore, the effect of the blocking angle for overheating protection (and glare) is not yet achieved. However, for the prototype pane C, the glare protection seemed promising for the summer sun position. Glare protection provided by this pane outperformed the solar control glazing while allowing more daylight into the back of the space. But given the extremely large differences between the windows tested, it would be necessary to finalize the prototypes and conduct further studies to make a conclusive statement about the performance of this new fenestration system.

We conducted this study to compare four glazings using a side-by-side configuration but the specifics of the set up, issues with system delivery / quality and capricious weather made it impossible to ensure a complete and clean dataset for the two seasons. The resulting point-in-time measurements thus only represent a snapshot of the actual performance of the systems and only for critical sun positions (i.e., when the sun is in the field of view and viewing direction is perpendicular to the façade) in both summer and winter. We should note that in real office settings, an occupant's desk would more typically be positioned perpandicularly to the façade and the viewing direction would thus be parallel to the window instead. In that case, the sun is typically in the peripheral field of view and situations as critical as those investigated should thus only occur for a limited time. In addition, clear skies are not necessarily representative of all conditions encountered throughout the year. Under these circumstances, continuous measurements would have been more representative of the systems' performance but would also have required continuous access to additional testing facilities. Simulation-based analyses could nicely complement this gap by providing annual evaluations, which could also be run for different desk positions.

6. Conclusion

Complex fenestration systems are becoming increasingly popular and it is necessary to adequately characterize their performance in terms of daylight provision and glare protection from direct sunlight in order to anticipate the visual comfort conditions provided to occupants. In this study, we compared the illuminance distribution and discomfort glare risks associated to four complex fenestration systems by means of measurements under real conditions for both winter and summer sun positions. The results showed, as expected, a higher daylight provision for clear and daylight redirecting systems. Daylight glare assessment showed that MicroShade™ was substantially more efficient at addressing glare from sunlight for summer sun angle compared to Coolshade™ (glare prediction of 'noticeable' to 'disturbing' compared to 'intolerable'). However, the two systems equally failed at providing a glare protection from sunlight in winter, which confirms a strong impact of the sun position on the effectiveness of these fenestrations systems. The performance of the GEMM prototype was highly dependent on the glazing pane used but one pane provided a promising performance for the summer sun position with a 'imperceptible' glare prediction. All the systems allowed to maintain the view to the outside, which is a very laudable benefit in comparison to the more traditionally used opaque shading systems.

7. Acknowledgments

This study was funded by the ETH Board as part of the research project SolAce | Recomfort 'Perception based Human Comfort through a Multi-Functional Solar Façade and Renewable Energy Integration' submitted to EMPA NEST by EPFL Researchers. We would like to thank M. Enrico F. Marchesi, Innovation Manager NEST for his help during our stay at the NEST building infrastructure. We would also like to thank MicroShade™ for their in-kind donation of the glazing.

8. References

[1] Heerwagen JH. The psychological effects of windows and window design. Proceedings of the 21 st Annual Conference of the Environmental Design Research Association. Champaign-Urbana, Illinois, 1990, p. 6–9.
[2] Veitch JA, Hine DW, Gifford R. End users 'knowledge, beliefs, and preferences for lighting. Journal of Interior Design 1993;19:15–26.
[3] Osterhaus WK. Discomfort glare assessment and prevention for daylight applications in office environments. Solar Energy 2005;79:140–58.
[4] Brainard GC, Hanifin JP, Greeson JM, Byrne B, Glickman G, Gerner E, et al. Action spectrum for melatonin regulation in humans: evidence for a novel circadian photoreceptor. Journal of Neuroscience 2001;21:6405–12.

[5] Berson DM, Dunn FA, Takao M. Phototransduction by retinal ganglion cells that set the circadian clock. Science 2002;295:1070–3.
[6] Smolensky MH, Hermida RC, Reinberg A, Sackett-Lundeen L, Portaluppi F. Circadian disruption: New clinical perspective of disease pathology and basis for chronotherapeutic intervention. Chronobiology International 2016;33:1101–19.
[7] Roenneberg T, Merrow M. The circadian clock and human health. Current Biology 2016;26:R432–43.
[8] Lockley SW, Evans EE, Scheer FA, Brainard GC, Czeisler CA, Aeschbach D. Short-wavelength sensitivity for the direct effects of light on alertness, vigilance, and the waking electroencephalogram in humans. Sleep 2006;29:161–8.
[9] Aries MB, Veitch JA, Newsham GR. Windows, view, and office characteristics predict physical and psychological discomfort. Journal of Environmental Psychology 2010;30:533–541.
[10] Chang C-Y, Chen P-K. Human response to window views and indoor plants in the workplace. HortScience 2005;40:1354–9.
[11] Farley KM, Veitch JA. A room with a view: A review of the effects of windows on work and well-being 2001.
[12] Gladwell VF, Brown DK, Barton JL, Tarvainen MP, Kuoppa P, Pretty J, et al. The effects of views of nature on autonomic control. European Journal of Applied Physiology 2012;112:3379–86.
[13] Kim J, Wineman J. Are windows and views really better. A Quantitative Analysis of the Economic and Psychological Value of Views Lighting Research Center, Rensselaer Polytechnic Institute, New York 2005.
[14] Li D, Sullivan WC. Impact of views to school landscapes on recovery from stress and mental fatigue. Landscape and Urban Planning 2016;148:149–58.
[15] Heschong L, Wright RL, Okura S. Daylighting impacts on human performance in school. Journal of the Illuminating Engineering Society 2002;31:101–14.
[16] Christoffersen J, Johnsen K, Petersen E, Hygge S. Post-occupancy evaluation of Danish office buildings. Publications de La Commission Internationale de l'éclairage (CIE) 1999;133:333–7.
[17] Lam, William. Perception and lighting as formgivers for architecture. McGraw-Hill; 1977.
[18] CIE. Discomfort Glare in the Interior Working Environment. Vienna: Commission Internationale de l'Eclairage; 1983.
[19] Day JK, Futrell B, Cox R, Ruiz SN, Amirazar A, Zarrabi AH, et al. Blinded by the light: Occupant perceptions and visual comfort assessments of three dynamic daylight control systems and shading strategies. Building and Environment 2019;154:107–21. https://doi.org/10.1016/j.buildenv.2019.02.037.
[20] Wienold J, Christoffersen J. Evaluation methods and development of a new glare prediction model for daylight environments with the use of CCD cameras. Energy and Buildings 2006;38:743–57.
[21] CEN. 17037, Daylight of Buildings. Brussels, Belgium: European Committee for Standardization; 2018.
[22] Jakubiec JA, Reinhart CF. The `adaptive zone'–A concept for assessing discomfort glare throughout daylit spaces. Lighting Research and Technology 2012;44:149–170.
[23] Wienold J, Iwata T, Sarey Khanie M, Erell E, Kaftan E, Rodriguez RG, et al. Cross-validation and robustness of daylight glare metrics. Lighting Research & Technology 2019;51:983–1013.
[24] Empa - NEST - SolAce n.d. https://www.empa.ch/web/nest/solace (accessed January 28, 2022).
[25] Saint-Gobain. MicroShade® Redefining Solar Shading 2017.
[26] Glaströsch. Swisslamex Coolshade, Protection solaire grâces aux lamelles insérées dans l'assemblage du verre feuilleté de sécurité. Swisslamex Coolshade 2012. https://www.glastroesch.ch/fileadmin/content/images/service/broschueren/franzoesisch/bro_swisslamex_coolshade_fr.pdf (accessed January 28, 2022).
[27] Kostro AG, Geiger M, Scartezzini J-L, Schueler A. Towards microstructured glazing for daylighting and thermal control. Proceedings of CISBAT 2011 - CleanTech for Sustainable Buildings, Lausanne, Switzerland: EPFL; 2011.
[28] Gong J, Kostro AG, Schüler A. Towards novel glazing with seasonal dynamics based on micro compound parabolic concentrators. Proceedings of the ISES Solar World Congress 2017, 2018.
[29] Florio P, Tendon X, Fleury J, Costantini C, Schueler A, Scartezzini J-L. Performance Assessment of a nZEB Carbon Neutral Living/Office Space and Its Integration into a District Energy-Hub. Energies 2022;15. https://doi.org/10.3390/en15030793.
[30] Foldbjerg Rasmussen Helle. MicroShade™ provides daylight and view out in the new Confederation of Danish Industry' building in Copenhagen. REHVA Journal 2013. https://www.rehva.eu/rehva-journal/chapter/microshadetm-provides-daylight-and-view-out-in-the-new-confederation-of-danish-industry-building-in-copenhagen (accessed April 25, 2022).
[31] Gong J, Kostro A, Motamed A, Schueler A. Potential advantages of a multifunctional complex fenestration system with embedded micro-mirrors in daylighting. Solar Energy 2016;139:412–25.
[32] Pierson C, Wienold J, Jacobs A, Bodart M. Luminance maps from High Dynamic Range imaging: calibrations and adjustments for visual comfort assessment. 2017.
[33] Wienold J. Evalglare–A new RADIANCE-based tool to evaluate daylight glare in office spaces, 2004.
[34] Wienold J, Andersen M. Evalglare 2.0–new features faster and more robust hdr-image evaluation, 2016.

IOP Conf. Series: Earth and Environmental Science 1099 (2022) 012024 doi:10.1088/1755-1315/1099/1/012024

Physiological effects of white and coloured LED lighting for elderly people

Y Oe[1], Y Miura[2] and Y Akuzawa[2]

[1] Department of Architecture, Faculty of Science and Technology, Tokyo University of Science, 2641 Yamazaki, Noda-shi, Chiba, Japan
[2] Department of Architecture, Graduate School of Science and Technology, Tokyo University of Science, 2641 Yamazaki, Noda-shi, Chiba, Japan

yukioe@rs.tus.ac.jp

Abstract. The aim of this study is to investigate the physiological stress effects of light colour on white and coloured LED lighting in the elderly people. The laboratory experiment was conducted on 19 elderly subjects and analysed using heart rate variability. The results showed that the 2700 K and 6500 K conditions, which significantly decreased HR and significantly increased HF, were considered to be relaxing tendencies. Based on the experimental results of previous studies, comparisons with young age groups showed that there were no differences in tendency according to light colour between age groups, but that the tendency to relax may be weakened in elderly people.

1. Introduction

Coloured light (light in the region outside the vicinity of the Planckian locus) is expected to be used in a variety of situations with LED lighting, which can adjust its colour easily. It is increasingly being used in spaces where relaxation is required, such as restaurants and delivery rooms. However, it is necessary to consider stress to coloured light as it differs from the light environment to which we are exposed in daily life. With the discovery of iPGC cells (intrinsically photosensitive retinal ganglion cell) in the 21st century, it has become important to clarify the effects of light exposure on humans not only from a psychological perspective but also from a physiological one. In recent years, in addition to the findings on the illuminance and correlated colour temperature of white lights (light on the Planckian locus) in young subjects [1][2][3], the other studies have also reported about primary light colours (R, G and B) with LEDs [4][5][6][7][8]. In our previous study, we also examined the physiological effects of the colour and illuminance of white lights, primary colour lights, and mixed primary colour lights, as well as psychological stress [9].

In addition, as the population ages (especially in Japan), there is an increasing need to design lighting environments that are suitable for the elderly, whose visual characteristics are declining. Several studies have already examined lighting for the elderly from a physiological perspective. Izsó et al. and Wang et al. both reported on the effects of white lights with various illuminance and correlated colour temperature on elderly subjects using heart rate [10][11]; Laufer et al. exposed the elderly to red and blue light and found a slight increase in parasympathetic activity with blue over red [12]. However, these studies did not compare results with younger subjects under the same experimental conditions. The ultimate aim of this study is to clarify the effects of different light colours (including mixed colour lights) and intensities on physiological and psychological stress in order to utilize coloured LED lighting for a wide range of age groups. As an initial study, we investigate the physiological stress of different primary colour lights on elderly people at the same illuminance in this paper. Comparisons are also made with the results of previous experiment carried out in our laboratory target at young people.

2. Experimental method

2.1. Experimental room

Figure 1 shows the plan and cross-sectional view of the experimental room. The dimensions of the room were 3370mm in width, 2400mm in depth and 2700mm in height. The experimental space was assumed to be a living room, so two sofas and one table were placed. To eliminate the influence of external light, no windows were installed. The LED luminaire was placed on the wall so that enough intensity of light reached the subject's eyes. Two floor lamps were placed in preparation before the start of the experiment and were switched off at all times during the experiment.

2.2. Subjects and lighting conditions

The subjects were 19 people (17 males and 2 females) with normal colour vision, aged between 65 and 82 years (average: 73.5 years old). The colour vision of the subjects was confirmed by the Ishihara Colour Test Chart. The essential conditions for subjects to participate were that they must be non-smokers, consent to heart rate and blood pressure measurements, no heart disease and no regular medication affecting the autonomic nervous system. The diseases and medications were judged by the medical doctor. In addition, the subjects were asked to do the following four things before the start of the experiment: 1) not to drink alcoholic beverages within 24 hours, 2) not to consume caffeine-containing beverages within 6 hours, 3) not to eat within 2 hours, and 4) not to perform strenuous exercise within 6 hours. The experiment was conducted from October 2021 to March 2022.

Six colour lighting conditions were set: three conditions of 2700K, 4000K and 6500K were used for white light, and three conditions of monochromatic light (R, G and B) were used for coloured light. The illuminance in front of the eyes (height from floor level: 1100mm) in each lighting condition was 53 lux (CL-500A, KONICA MINOLTA). Figures 2 and 3 show the chromaticity and spectral characteristics of each lighting condition, respectively. The u'v' chromaticity diagram was used because it can show the most even colour differences corresponding to the light source colours.

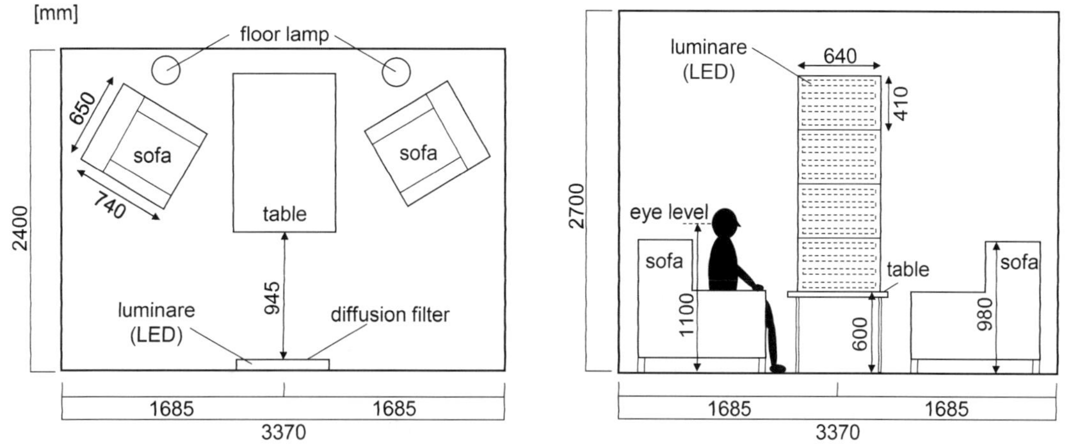

Figure 1. The plan and cross-sectional view of the experimental room.

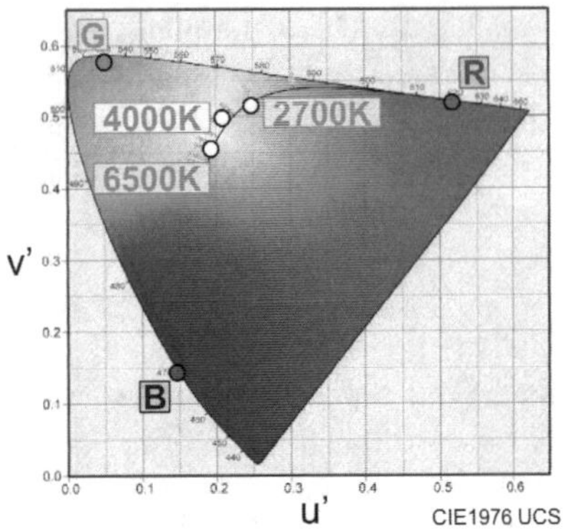

Figure 2. The chromaticity of lighting conditions.

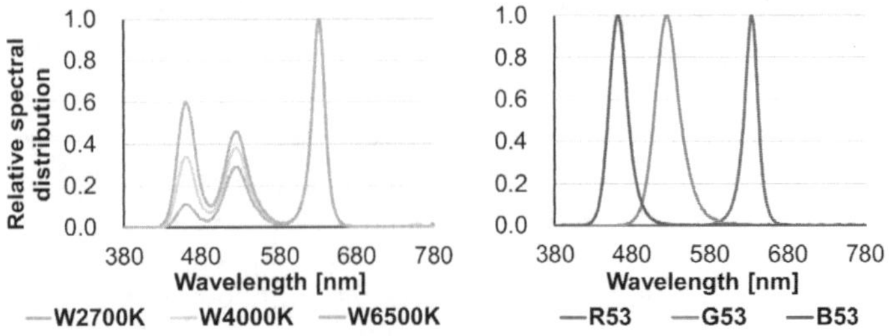

Figure 3. The spectral characteristics of lighting conditions.

2.3. *Experimental procedure*

The experimental procedure is shown in figure 4. The subjects wore a heart rate monitor (RF-ECG2, GM3) on their chest during the experiment and were exposed to two lighting conditions per experiment. The Subjects were instructed not to look directly at the light source. The experiment consisted of the following five phases: 1) 6 minutes of rest in the dark (dark adaptation, 1st), 2) 1 minute of adaptation under first lighting conditions, followed by 8 minutes each of KAPLA (French building blocks), Sudoku (Number puzzles), and sketching (dark adaptation after each task; 2nd, 3rd, 4th), 3) dark adaptation (5th), 4) the same as phase 2 under the second lighting condition, 5) dark adaptation. The experiment was conducted for a total of about two hours each time, with the experiment starting at either 8.00 a.m., 11.00 a.m. or 2.00 p.m. The day of week and the period of time on each subject's experiment were always fixed. The content of this experiment was based on medical advice and was approved by the Ethics Committee of Tokyo University of Science.

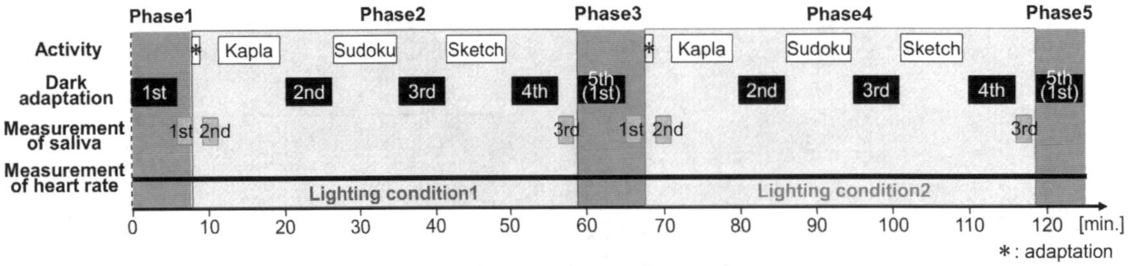

Figure 4. The experimental procedure.

3. Results

Following previous research [13][14], the function of the autonomic nervous system was confirmed from heart rate variability. The heart rate (HR), which reflects the workings of the sympathetic and parasympathetic nerves, and the high-frequency component (HF) were used. If HR did not increase after exposure to the lighting condition, the subject was considered to be stress-free, and if HR decreases and HF increases, the subject was considered to be relaxed. In this paper, the HR or HF at the 5th dark adaptation divided by the HR or HF at the 1st dark adaptation was used in the analysis.

3.1. *Physiological responses of elderly people*

Figure 5-1 shows the comparison results of HR and HF before and after exposure to the lighting conditions. HR decreased and HF increased in all lighting conditions, indicating a tendency to reduce stress after exposure to the lighting condition. In particular, the 2700 K and 6500 K conditions, where

HR significantly decreased and HF significantly increased, can be considered as relaxing tendency. There were no significant differences between lighting conditions and no effect of light colour was observed.

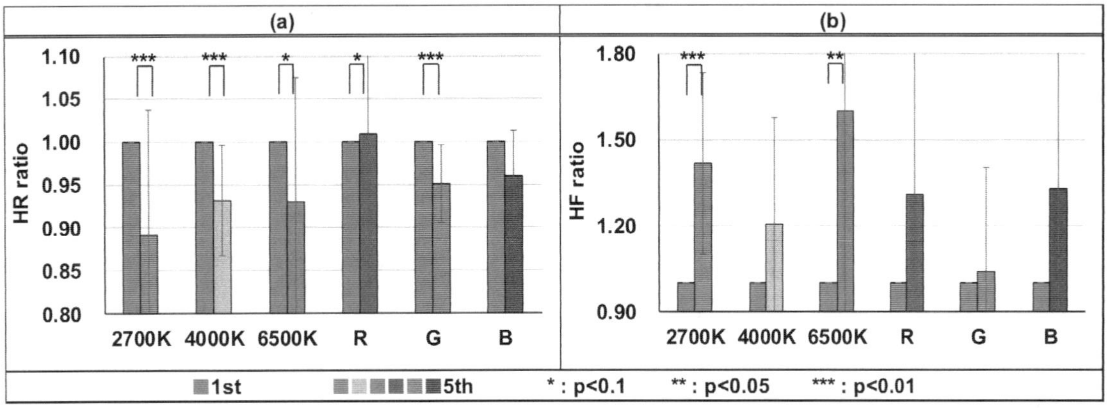

Figure 5-1. The comparison results of HR and HF before and after exposure to the lighting conditions on elderly people (a: HR, b: HF).

3.2. Comparison of physiological responses between age groups
This section describes the results of a comparison with previous experiments conducted on young people in previous studies. The experimental procedure was the same as described in chapter 2, and the subjects were 55 people (23 males and 32 females) with normal colour vision. The results of HR and HF on the young subjects are shown in figure 5-2.

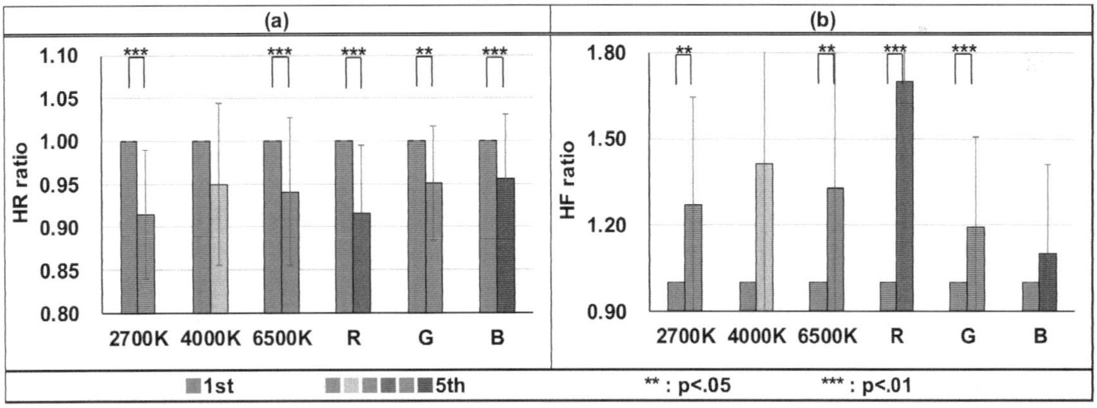

Figure 5-2. The comparison results of HR and HF before and after exposure to the lighting conditions on young people (a: HR, b: HF).

The young people tended to relax under the three white light conditions (2700K and 6500K), red and green lights, indicating that the elderly people have fewer relaxing lighting conditions than the young people. This may be due to a decrease in retinal illuminance and cone sensitivity, which may have reduced the physiological effects of lighting by making it more difficult for older people to perceive it. Furthermore, significance difference tests between age groups showed that HR and HF were not significantly different between age groups in all lighting conditions, and no differences were found between light colours. It was hypothesized that there would be differences in results between age groups in blue light due to reduced transmission of short-wavelength light in the lens with ageing, but the hypothesis was rejected. As described in the technical report published by CIE, there is a process by which humans compensate for changes in retinal stimulation caused by lens aging in order to maintain a stable white point throughout life [15]. This mechanism may have influenced the results of this experiment.

4. Conclusion

In this paper, differences in physiological responses by light colour caused by age-related changes in visual characteristics were investigated. There were no differences in the tendency according to light colour between age groups, but the tendency to relax may be weakened in the elderly people. It was suggested that this was influenced by age-related changes. It is not possible to conclude that the effects are due to age-related changes, so experiments will continue in the future with the aim of securing the number of subjects and examining other illuminance conditions. In addition, the correspondence with psychological assessment will be discussed.

References

[1] Mingli L, Songtao H, Zhu M, Ring L, Song X and Hongyu G 2020 Research on work efficiency and light comfort based on EEG evaluation method *Build. Environ.* **183** 107122

[2] Chao L, Limei S, Xiaotong J, Yalin Z, Xi M, Chao J and Weijun G 2022 How correlated color temperature (CCT) affects undergraduates: A psychological and physiological evaluation *J. Build. Eng.* **45** 103573

[3] Ruijun C, Meng-Chun T and Yaw-Shyan T 2022 Effect of Color Temperature and Illuminance on Psychology, Physiology, and Productivity: An Experimental Study *Energies* **15** 4477

[4] Christian C, Mirjam M, Szymon K, Kurt K, Roland S, Peter O, Selim O and Anna W 2005 High Sensitivity of Human Melatonin, Alertness, Thermoregulation, and Heart Rate to Short-Wavelength Light *J. Clin. Endocrinol. Metab.* **90(3)** pp 1311–6

[5] Chang-Jin C, Kyung-Sao K, Chul-Min K, Se-Hong K and Whan-Seok C 2011 Reactivity of heart rate variability after exposure to colored lights in healthy adults with symptoms of anxiety and depression *Int. J. Psychophysiol.* **79** pp 83-8

[6] Daniela L, Lu W, Ingrid G and Gerhard L 2013 The Influence of New Colored Light StimulationMethods on Heart Rate Variability, Temperature, and Well-Being: Results of a Pilot Study in Humans *Evid. Based Complementary Altern. Med.* **2013** 674183

[7] Jesus M, Miguel A L, Diego A R, Maria J S and Francisco P 2017 Blue lighting accelerates post-stress relaxation: Results of a preliminary study *PLoS ONE* **12(10)** e0186399

[8] Felix S, Timo H, Andreas J M, Martin W and Ursula W 2017 Effect of short-term colored-light exposure on cerebral hemodynamics and oxygenation, and systemic physiological activity *Neurophotonics* **4(4)** 045005

[9] Miura Y, Oe Y and Yoshizawa N 2021 Examination of Psychological Effects and the Additivity Assumption in the Stress Evaluation of Chromatic LED Lightings *Proc. CIE 2021* pp 73-9

[10] Izsó L, Láng E, Laufer L Suplicz S and Horváth Á 2009 Psychophysiological, performance and subjective correlates of different lighting conditions *Light. Res. Techonol.* **41** pp 349-60

[11] Yajing W, Haijing H and Gang C 2020 Effects of lighting on ECG, visual performance and psychology of the elderly *Optik* **207** pp 1-9

[12] Laufer L, Lang E, Izso L and Nemeth E 2009 Psychophysiological effects of coloured lighting on older adults *Light. Res. Technol.* **41** pp 371-8

[13] Billman G 2013 The LF/HF Ratio Does Not Accurately Measure Cardiac Sympatho-vagal Balance *Front. Physiol.* **4(26)** pp 1-5

[14] Berntson G G, Cacioppo J T and Quigley K S 1991 Autonomic determinism: the modes of autonomic control, the doctrine of autonomic space, and the laws of autonomic constraint *Psychological Rev.* **98(4)** pp 459-87

[15] CIE 2017 *Lighting for Older People and People with Visual Impairment in Buildings* **227** p 9

Acknowledgments
We are grateful to D. Dumortier for collaboration on the early stages of this work. The authors would like to thank T. Suzuki and M. Oyoshi for an assistance with the experiments.

The circadian impact of computer monitors with different color configurations

A Huguet-Ferran[1,2], D Kántor[1], S Hernández[2] and B Garrido[2]

[1] Enlighting Technologies, Kumux, carrer Baldiri Reixac 4, Parc Científic, Torre R, planta 2, B1, Barcelona, Spain
[2] Departament d'Enginyeria Electrònica i Biomèdica, carrer Martí i Franquès 1, planta 2, Facultat de Física, Universitat de Barcelona, Barcelona, Spain

adria.huguet@kumux.io, daniel.kantor@kumux.io, shernandez@ub.edu, blas.garrido@ub.edu

Abstract. The objective of this study is to find out if it is possible to achieve the circadian levels recommended by different published lighting guidelines using different color configurations in a text editor. By measuring 25 monitors of different sizes, the circadian impact is evaluated at a standard distance using two different parameters: Circadian Stimulus and Melanopic Equivalent Daylight Illuminance. After analyzing the results, we see that during the night Circadian Stimulus is, in most cases, within the recommended range, but the same is not true for Melanopic Equivalent Daylight Illuminance. In addition, the circadian impact of the f.lux software has been analyzed and compared with two different cases of color configurations in a text editor.

1. Introduction

The human circadian system is related to numerous physiological, biochemical, and psychological variables, directly related to people's health. Without the presence of external elements, such as light, experiments show that the circadian rhythm has a duration of about 25 hours, as opposed to 24 hours a day. The impact of light helps to regulate and synchronize circadian rhythms with our environment, readjusting them to the daily 24-hour cycle [1].

At the same time, modern human beings live a large part of our time enclosed in spaces without access to natural light. Artificial light, in general, is static with no regulation according to external daylight conditions. This can lead to an imbalance in circadian rhythms, as we lose reference to the outside world.

Several studies have been published linking light exposure to the regulation of melatonin, a hormone related to circadian rhythms. Essentially, the sunlight that reaches our retina helps to synchronize our internal circadian rhythms with the outside world, telling our body what to do at any given moment. An imbalance in circadian rhythms can lead to health problems, increasing the predisposition to mood disorders, including impulsivity, mania, and depression [2]. If exposure to light is not correct, in the long term, a disruption of the light-dark cycle can lead to other problems, such as weight gain, obesity or diabetes [3], as well as certain forms of cancer [4], among others.

Institutions such as Lighting Research Center (LRC) have carried out studies in different environments and have published guidelines with a series of recommendations on how to correctly use light, from a circadian point of view [5]. In this case, a parameter developed by LRC itself is used: the circadian stimulus (CS). CS is a parameter related to the suppression of melatonin due to a light stimulus incident on the human eye [6]. LRC itself has published calendars that relate a time of day to a CS value. A case applied to office lighting is shown below:

Table 1. Recommended CS value based on the time of day in an office. EOB refers to End of Business Day.

Local time	CS
07:00 – 16:00	0.3
16:00 – 17:00	0.3 → 0.2
17:00 – 19:00	0.2
19:00 – 20:00	0.2 → 0.1
20:00 - EOB	0.1

On the other hand, other authors suggest the adoption of other parameters such as Melanopic Equivalent Daylight Illuminance (mEDI), which describes the light-sensitivity of human physiological response [7]. In a similar way, the circadian impact of light reaching the human eye is evaluated and a calendar that relates mEDI values to the time of the day has been defined:

Table 2. Recommended mEDI value depending on the time of day.

Local time	mEDI (lx)
Daytime	> 250
Evening	< 10
Nighttime (vision)	< 10
Nighttime	< 1

When assessing the circadian impact of light, it is important to take into account the spectral shape of the light source and the luminous flux incident on the eye, i.e., the illuminance. Both CS and mEDI take these two factors into account, although the equations defining them are different. Generalizing, at the spectral level, both cases are mainly related to the amount of blue contained in the incident light, although the effective circadian response is different in each case. Other longer wavelengths (green, yellow, red...) also affect the circadian system, although their impact is much more limited. To see the lack of direct relationship between CS and mEDI, we have analyzed and compared a set of 718 commercial white LEDs, as can be seen in figure 1:

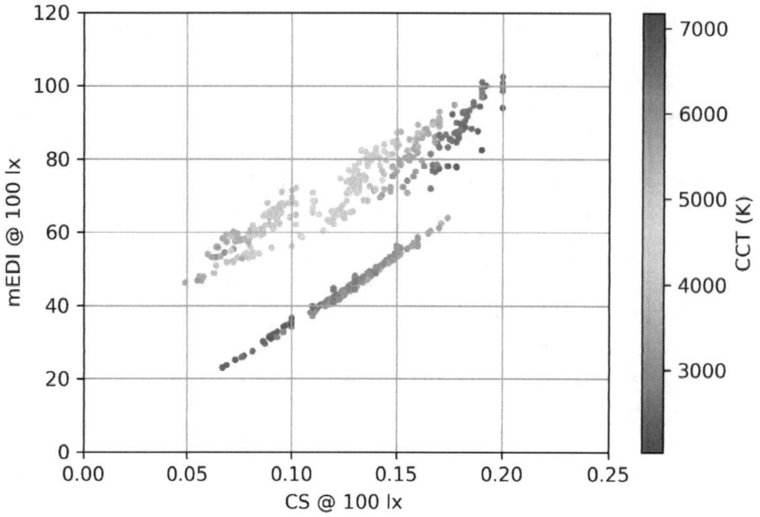

Figure 1. Relation between mEDI and CS at 100 lx for 718 commercial white LEDs.

The data for the 718 LEDs of the figure 1 have been collected through the information provided by different LED manufacturers and is available on the spectral design platform developed by Kumux. All the information has been standardized and numerous colorimetric and circadian parameters have been calculated, including mEDI and CS.

Often only the illumination coming from outdoors (natural light) or indoors (from luminaires) is taken into consideration. However, studies have shown that prolonged exposure to light from a computer monitor has an impact on melatonin secretion in adolescents and can suppress melatonin by a 30%, approximately [8].

To minimize this effect, some initiatives have arisen, such as f.lux, a computer program which varies the warmth of the colors of a computer screen depending on the time of day [9]. Thus, as night approaches, the computer screen on which this software is installed begins to redden, while during the day it retains its original colors.

This article presents a new approach to the problem. In this case, to find out the potential impact, a text editor called Visual Studio Code (VS Code) is used, which allows full customization of the colors of the graphical interface. VS Code has been chosen because it is one of the most widely used programs for software developers, although there are other very popular editors (VIM, Emacs...) that also allow the customization of colors, being a commonly used tool among programmers. In a similar way to f.lux, two different color configurations have been created and spectral and illuminance measurements have been taken for both, comparing the acquired values with the recommended CS and mEDI values shown in table 1 and table 2.

2. Method

A study on the impact of computer screens at the circadian level is conducted. For this purpose, two different color schemes of the VS Code text editor have been created, providing two different scenarios with potentially low (with a dominance of dark colors) and high (with a dominance of white colors) melanopic impact.

Twenty-five displays of different sizes, technologies and time of use (aging) have been analyzed to find general patterns that can be used for future studies. This analysis includes circadian calculations based on CS and mEDI parameters, as well as a comparison with f.lux software.

2.1. Location

The study was carried out at the Facultat de Física of the Universitat de Barcelona and at the Barcelona Science Park. In total, two offices with access to natural light and one laboratory without access to the outside were analyzed. However, the calculations did not take into account the contribution of illumination beyond the light from the monitors themselves.

2.2. Experimental equipment

In this case, we have used a UPRtek MK250S Premium spectrometer. It is a portable spectrometer with an integrated camera with a wavelength range of 380 to 780 nm, 1 nm resolution and an illuminance measurement range from 1 to 150,000 lx. In this case, the spectrometer provides the spectral power distribution (SPD) and the illuminance (in lx) of the measured screen, the two parameters needed to calculate CS and mEDI.

To perform the measurements, the spectrometer was mounted on a tripod with an adapter created with LEGO Technic, as shown in figure 2:

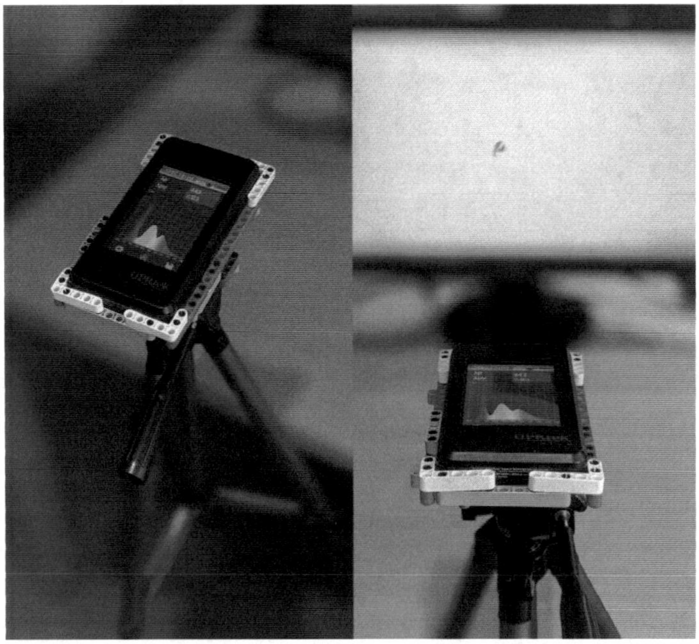

Figure 2. Spectrometer mounted on the tripod making measurements on a monitor.

2.3. Data acquisition

The acquired data comes from 25 different monitors. These monitors differ in manufacturer, size, technology (TFT, LCD and LED) and aging. To perform the measurements, two different images (scenarios) are displayed on the different screens: one with white colors dominance and one with dark colors dominance. In other words, the same image with a white background and the same image with a black background is displayed on all screens, and spectral measurements of both are then taken. In all cases, the monitors have been configured to provide the maximum possible brightness (100%). Other options such as blue dominance have been discarded, because although the circadian response of the eye depends on the amount of blue light impinging the eye, in a computer screen, the white color already includes the blue color at maximum power. For clarification, the white generated by the computer is composed of a vector of red, green and blue components, all at maximum. This fact implies that the maximum circadian effect that can be achieved coincides with the white color at maximum power.

Moreover, as mentioned above, the measurements were taken with a spectrometer mounted on a tripod. Since the spectrometer has a built-in camera, the measurements have been taken pointing to the center of the screen. The distance of the measurements to the screen has been set at 68 cm from the center, as this is the working distance recommended by some studies [10]. The height has been set at 1.20 m, which corresponds to the recommended height in the mEDI schedule of table 2.

For each of the 25 monitors, three different measurements were carried out: a dark measurement with the screen turned off but with the corresponding ambient light, a measurement with black color dominance, and a measurement with white color dominance on the screen. Thus, by subtracting the dark measurement from the two measurements with the screen on, it is possible to obtain the clean spectral and illuminance values of the monitors in each case, suppressing the contribution from ambient light. Figure 3 shows the two images displayed by the screens with black and white color dominance:

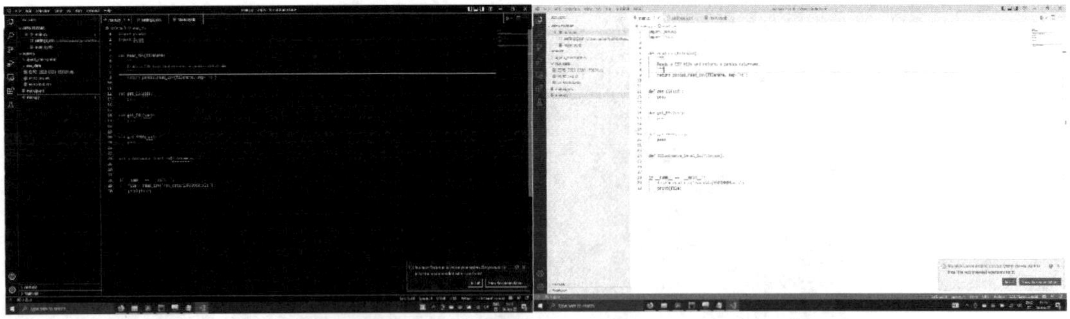

Figure 3. Screenshot of the VS Code Graphical User Interface with black color dominance (left) and with white color dominance (right).

2.4. Data analysis

For each measurement, the spectrometer generates a spreadsheet with spectral information that includes both SPD and measurements of basic colorimetry, horticulture and some other non-visual parameters, in addition to the illuminance value. CS and mEDI parameters have been evaluated from the acquired spectra using a simple script to perform the required calculation.

The script is developed in Python 3, and its main function is to read the set of spreadsheets derived from the spectrometer measurement and perform the calculations of the CS and mEDI parameters. To perform the calculations of these parameters, the Luxpy library has been used [11], which allows a fast implementation of the necessary equations. In these calculations, the dark values are subtracted from the values of the two-color configurations described. In this way, the values of the light coming from the monitors analyzed are obtained.

An analysis was performed in 3 parts. In addition to calculate CS and mEDI parameters, illuminance was also measured. Illuminance can be employed as a reference for checking the dependence on the diagonal of the monitor and its relation with the monitor area (i.e., quadratic dependence with the diagonal). Actually, as the diagonal of the monitor increases or decreases, the number of emitting points and the effective illumination area scales quadratically with the diagonal. The illuminance measurements of the 25 monitors with the white colors dominance configuration can be seen in figure 4:

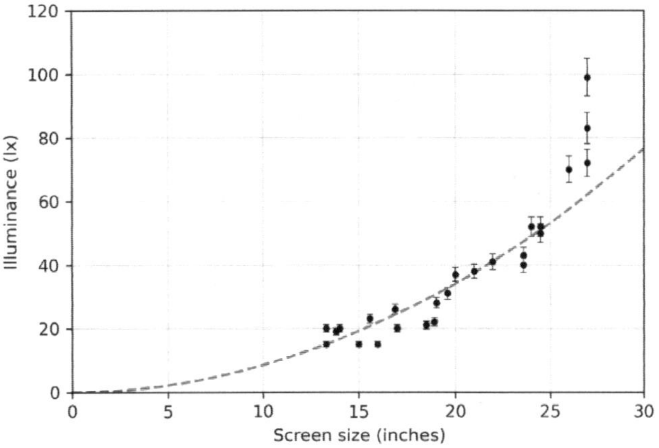

Figure 4. Illuminance (lx) vs screen size (inches) of the monitor in the white color dominance configuration. A quadratic function (dotted line) intersecting with the coordinate origin has been drawn as a reference.

In figure 4 an error of 2 cm has been estimated in the experimental measurement at 68 cm. This is approximately a 6% error in the illuminance value. This means that, to the human eye, such errors are only noticeable when the illuminance is high, as can be seen in the error bars of figure 4. On the other hand, the auxiliary quadratic function, of the type $f(x) = a{\cdot}x^2$, often does not fall within the error margins of the measurements, which means that there are other effects or other sources of error that were not considered. The discrepancies can be originated by the differential factors between monitors mentioned above, such as light emission technology, screen ratio, or aging, which can cause a decrease in the luminous flux of the screen.

In the case of the dark color-dominant configuration, no appreciable illuminance value could be measured. In other words: there is no perceptible light emission that circadianally impacts the human system. For the larger displays, the difference in illuminance measurements between the dark color dominance configuration and the screen off is much less than 1 lx, as can be seen in figure 5. Since CS and mEDI parameters depend on the illuminance value, if the illuminance has a zero value, CS and mEDI will also be zero. On a practical level, the dark color dominance setting is equivalent of having the screen turned off.

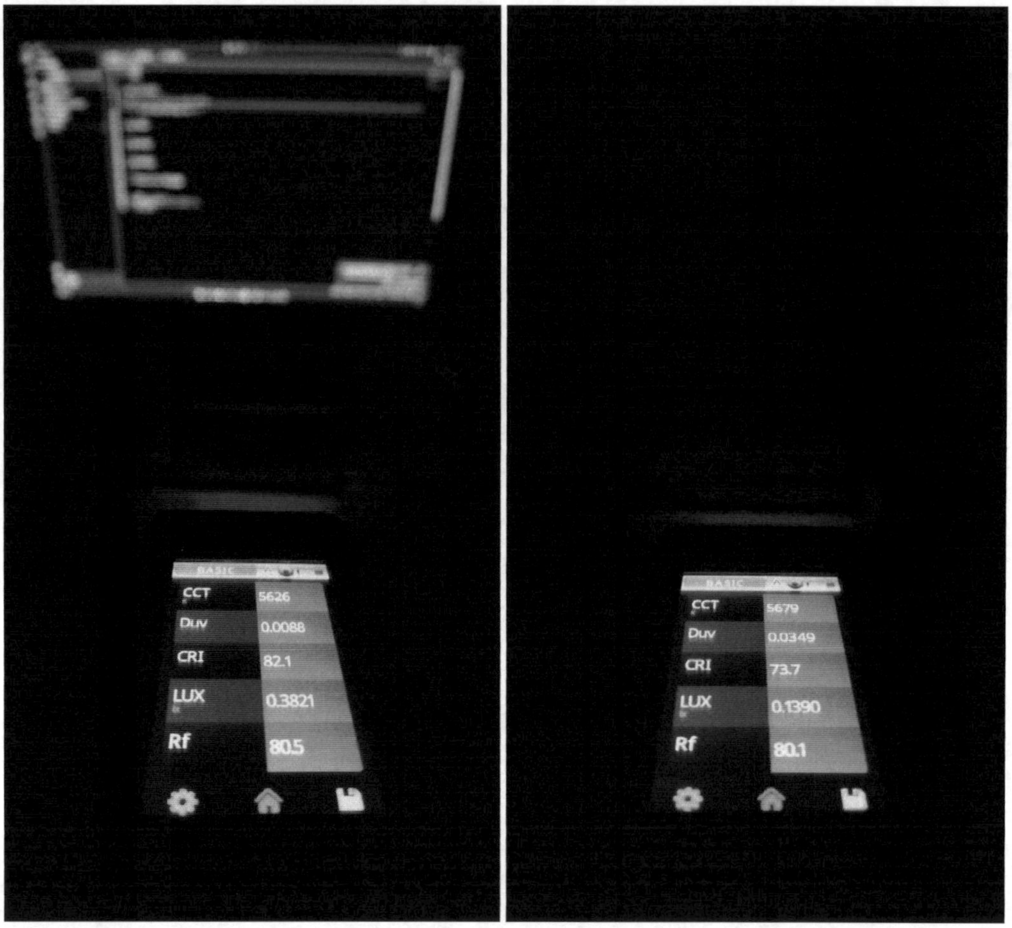

Figure 5. Photograph of the values measured by the spectrometer for a 27-inch screen in dark color dominance configuration (left) and with the screen turned off (right). It can be seen that the difference in illuminance is 0.24 lx.

3. Results

Similar to the case of illuminance, the dependence of CS and mEDI parameters on the screen size is visualized in figure 6. Due to the imperceptibility of the illuminance in the dark color configuration, only the net white color configuration, resulting from subtracting the ambient light (screen turned off measurement) has been taken into account.

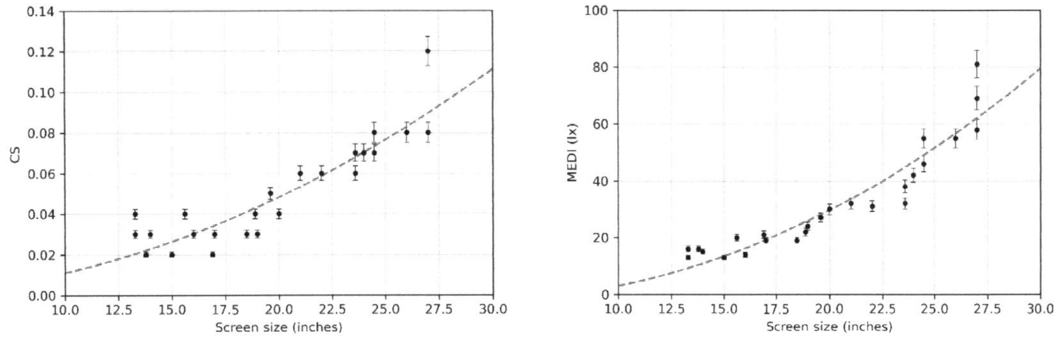

Figure 6. CS (left) and mEDI (lx) (right) *vs* screen size (inches) of the monitor in the net white color dominance configuration. A quadratic function (dotted line) intersecting with the coordinate origin has been drawn as a reference in both cases.

Similarly to the illuminance behavior with the screen size (figure 4), in the case of CS and mEDI we also find points that fall outside the expected values. The reasons for this are identical to those previously mentioned. Nevertheless, by comparing the values obtained with the values in table 1 and table 2, we can draw the following conclusions:

- To achieve the recommended circadian impact during daylight hours (CS > 0.3 or mEDI > 250 lx) a large screen is required. Following the reference trend lines in figure 6, the intersection between the recommended values and the screen size occurs at 50-inches for both CS and mEDI.

- To achieve the recommended circadian impact during nighttime hours (CS < 0.1 or mEDI < 1 lx) the conclusion is not clear for the net white color dominance configuration. On the one hand, if we stick to the LRC recommendations, we see that all but one of the screens analyzed meet the condition of CS < 0.1. Only large screens can have a circadian effect greater than necessary. However, if we look at the mEDI recommendations, none of the results fall below 1 lx. In fact, in the case of mEDI, none of the results meet the more restrictive conditions, related to evening values and nighttime activities requiring vision (mEDI < 10 lx).

We are going to analyze whether, in the case of mEDI, using software such as f.lux it is possible to fall below the recommended limits in the white color dominance scenario.

3.1. Comparison with other software

The f.lux software has been installed and spectral measurements of the mEDI parameter (the most restrictive during nighttime hours) have been performed at the same distance as in the previous cases. The results of the relative reduction of mEDI can be seen in figure 7:

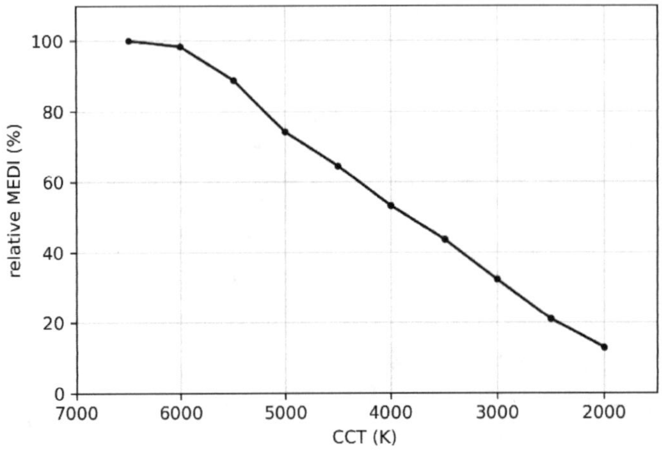

Figure 7. Relative mEDI values as a function of the CCT of the screen controlled by the f.lux software. Measurements were taken on a 27-inch screen.

The f.lux software allows automatic control of the display's CCT, which varies according to geolocation and time of day. In this way, the screen maintains its original colors during daylight hours and turns orange (with the absence of blues) as night approaches. With this control, as can be seen in figure 7, a progressive reduction of mEDI is achieved, reaching 92% at its most extreme point, when the CCT of the display is minimal.

Applying the 92% reduction of mEDI provided by the f.lux software to the set of displays analyzed in figure 6, we see that all of them would meet the evening condition (mEDI < 10 lx). However, although the smaller ones come close, there is no display that achieves the nighttime value of mEDI < 1 lx.

4. Discussion

Analyzing the results in depth, we have seen that a number of conclusions can be drawn about the circadian impact of the computer monitors considering that there is no ambient light:

- Having a black background in a text editor, at the circadian level, is practically equivalent to having the screen turned off. It was not possible to measure a relevant illuminance of screens with the dark color dominance setting, as it is too small, implying that the circadian impact is also null.

- To achieve a high circadian impact (CS > 0.3 or mEDI > 250 lx), a large screen (> 50-inches) or a combination of smaller screens is required. In addition, in a text editor such as VS Code, it is necessary to combine it with a white color-dominant background. However, a possible daytime scenario where there is only illumination due to screens is extremely rare.

- At night, according to the two guidelines analyzed, there is no consensus on the potential impact of the screens. On the one hand, if we follow the recommendations of the CS parameter, we see that in the vast majority of cases (small and medium-sized screens) there is no problem, even if a white-dominant background is used. However, if we follow the recommendations of the mEDI parameter, we see that no display meets the requirements in the white color dominance setting. For this reason, it is necessary to change the luminance of the display with some software resources that allow setting darker colors or with less blue content.

- As it can be seen in figure 1, there is no direct relationship between CS and mEDI. This can lead to problems such as the one described in this article: while according to the CS guideline, during the night, the requirements set out in table 1 are met, the same is not true for mEDI (table 2). This can lead to confusion, and it is necessary that scientifically there is a clear methodology to follow.

- A study of the f.lux software has been conducted, and while it is absolutely true that it can greatly reduce circadian impact, it is not as effective as other strategies, like changing the background color. Changing the background color to black results in zero circadian impact, while using the f.lux software results in a 92% reduction.

For this reason, as a general recommendation, it is proposed to use white-dominant backgrounds during the day and black-dominant backgrounds at night. The potential circadian impact of the screen, especially during the day, is not very high, although depending on conditions it may help to reach the limit of CS > 0.3 or mEDI > 250 lx. During the night, on the other hand, we should limit exposure to white colors as much as possible in computer screens. Specifically, in the case of CS, in most cases the CS > 0.1 value is not reached, but possible contributions from external illumination to the monitor must be taken into account. Thus, in the case of displays it is very important to control both the potential output spectrum and the luminance, since a low luminance value will hardly imply a high circadian impact.

In a similar way, the conclusions can be extrapolated to today's intelligent lighting systems. There are numerous systems that allow for CCT tuning or brightness variation, although in order to have a high circadian impact, both parameters must be controlled simultaneously. Moreover, these variations must take into account factors such as geolocation, type of activity or time of day and date of year, among others. All these parameters have a huge influence on the amount and type of illumination entering the retina of the eye, so automated solutions are needed to allow automatic control of the circadian impact due to artificial lighting systems.

Future studies are needed with a larger sample of monitors to differentiate technologies or aging, among other parameters. In this way, it will be possible to draw more appropriate conclusions about the correct use of monitors during the day and at night.

Acknowledgements
Thanks to the people of Facultat de Física of the Universitat de Barcelona.

References

1. Kerkhof GA. Inter-individual differences in the human circadian system: A review. Biological Psychology. 1985 Mar 1;20(2):83–112.
2. Salgado-Delgado R, Tapia Osorio A, Saderi N, Escobar C. Disruption of circadian rhythms: a crucial factor in the etiology of depression. Depress Res Treat [Internet]. 2011 [cited 2022 Apr 25];2011. Available from: https://pubmed.ncbi.nlm.nih.gov/21845223/
3. Depner CM, Stothard ER, Wright KP. Metabolic consequences of sleep and circadian disorders. Curr Diab Rep [Internet]. 2014 [cited 2022 Apr 25];14(7). Available from: https://pubmed.ncbi.nlm.nih.gov/24816752/

4. Samuelsson LB, Bovbjerg DH, Roecklein KA, Hall MH. Sleep and circadian disruption and incident breast cancer risk: An evidence-based and theoretical review. Neurosci Biobehav Rev [Internet]. 2018 Jan 1 [cited 2022 Apr 25];84:35–48. Available from: https://pubmed.ncbi.nlm.nih.gov/29032088/
5. Healthy Living [Internet]. [cited 2022 Jan 25]. Available from: https://www.lrc.rpi.edu/healthyliving/
6. Rea MS, Figueiro MG, Bierman A, Bullough JD. Circadian light. Journal of Circadian Rhythms [Internet]. 2010 Feb 13 [cited 2022 Apr 25];8(1):1–10. Available from: https://jcircadianrhythms.biomedcentral.com/articles/10.1186/1740-3391-8-2
7. Brown TM, Brainard GC, Cajochen C, Czeisler CA, Hanifin JP, Lockley SW, et al. Recommendations for daytime, evening, and nighttime indoor light exposure to best support physiology, sleep, and wakefulness in healthy adults. PLOS Biology [Internet]. 2022 Mar 1 [cited 2022 Apr 25];20(3):e3001571. Available from: https://journals.plos.org/plosbiology/article?id=10.1371/journal.pbio.3001571
8. Figueiro M, Overington D. Self-luminous devices and melatonin suppression in adolescents: http://dx.doi.org/101177/1477153515584979 [Internet]. 2015 May 6 [cited 2022 Apr 25];48(8):966–75. Available from: https://journals.sagepub.com/doi/abs/10.1177/1477153515584979
9. f.lux: software to make your life better [Internet]. [cited 2022 Apr 25]. Available from: https://justgetflux.com/
10. Charness N, Dijkstra K, Jastrzembski T, Weaver S, Champion M. Monitor Viewing Distance for Younger and Older Workers: http://dx.doi.org/101177/154193120805201965 [Internet]. 2008 Sep 1 [cited 2022 Apr 27];3:1614–7. Available from: https://journals.sagepub.com/doi/abs/10.1177/154193120805201965
11. Smet KAG. Tutorial: The LuxPy Python Toolbox for Lighting and Color Science. LEUKOS - Journal of Illuminating Engineering Society of North America. 2020 Jul 2;16(3):179–201.

IOP Conf. Series: Earth and Environmental Science 1099 (2022) 012026 doi:10.1088/1755-1315/1099/1/012026

Dynamic Lighting and Natural Ventilation in Patient Rooms at New Psychiatry Bispebjerg

C Volf [1, 3], K Martiny[1], PM Petersen[2], D D Corell[2], C Dam-Hansen[2]

[1] Region H, New Interventions in Depression (NID-Group), Edel Sauntes Allé 10, 2100 Copenhagen E, Denmark,
[2] Department of Electrical and Photonics Engineering, Technical University of Denmark, Frederiksborgvej 399, 4000 Roskilde, Denmark

[3] Corresponding author e-mail: carlo.volf@regionh.dk.

Abstract
This paper presents preliminary results of analyzes of the indoor environmental quality (IEQ) in a full scale rotatable 1:1 mockup of a patient room. As initial part of the planning of the indoor environment at a new mental hospital, the project seeks to uncover the natural differences and limitations of the building, caused by the geographical orientation. Examining the outdoor diurnal variations and their effect on the indoor environment, in order to create a balance between on the one hand, daylight and artificial lighting and on the other hand the demand for passive cooling, using natural ventilation. The analysis discuss potential health aspects of daylight and fresh air and proposes a new strategy for dynamic artificial lighting, including natural ventilation to supplement mechanical ventilation. In this article we uncover daylight conditions based on two main parameters; 1) the geographic orientation (N, S, E and W) and 2) the season (summer and winter). Based on the preliminary data and the regional daylight conditions, the article proposes a new way to plan the indoor environmental quality called Lattitude and Orientation Compensated Architectural Lighting (LOCAL).

1. Introduction
Planning of hospitals dates centuries back in time and is deeply rooted in old traditions of architectural planning and healing architecture (1, 2), traditions which originally revolved around creating a healthy environment around the patients. Considerations on health of the patients have been a pivotal point in the planning of hospitals, in a balance of course, with other considerations, such as economy, operation costs, etc. Or rather, in a symbiosis with these considerations, since a balanced, healing environment directly influenced a wide range of other factors, such as recovery time, treatment of patients (3, 4, 5) together with the general performance and overall efficiency of a hospital over time (6).
Over the years the planning of hospitals has changed dramatically. Originally, hospitals were planned in rural and natural settings with plenty of natural light and fresh air. Nowadays however, hospitals often are built in densely populated urban settings. As a consequence of this, new hospitals are more compact, maximizing the occupied building area and the footprint of the building, through simple high rise constructions. In this process, high ceilinged rooms and slender building typologies originally built to optimize daylight, one-sided natural ventilation and cross ventilation, has been replaced with a practice based on mechanical ventilation and artificial lighting, both crucial elements for a balanced indoor environment quality in compact, deep buildings with lower ceiling heights, no longer built to optimize daylight and natural ventilation. In the postwar period, the so-called *technological hospitals* emerged.

These hospitals were based on all-time low energy prizes during the 1950´s and 1960´s using electricity to a wide range of purposes, such as heating, cooling, lighting, ventilation, etc. Together with medical development and improvements in treatment, this typology created new standards for the architectural planning of hospitals (7).
In recent years, focus on optimizing the energy performance of hospitals, furthermore has led to increased requirements for insulation and air-tightness of the building as well as requirements for energy consumption for heating and lighting. In Denmark, the current building requirements BR18 (8) were made even stricter for hospitals, since they should meet "future requirements", actually never applied to the building code and today only functioning as a voluntary low-energy class. These strict requirements have had a major impact on the planning of the new hospitals, the indoor environmental quality being planned in a more technological way compared to earlier practice. Focus on avoiding over-temperatures in relation to the requirements for thermal indoor climate, often compromises sufficient daylight, sufficient window area, glass quality, etc.

The conflict between requirements, the thermal indoor climate and the desire for optimal lighting conditions seems poorly resolved when not taking into account the dynamic, diurnal and annual variations in the outdoor climate and changing needs over time. In this respect, this study is a step towards reestablishing these dynamic variations over time. In doing so, the article puts the energy planning phase in a longer perspective, not only thinking energy efficient hospitals, but all-efficient hospitals in a total operation phase. Including the benefits of a healing environment – which potentially may exceed the savings in the planning of the energy performance. The latter only amounting to 4% of the total energy and climate costs, with salaries (57%), medicine (15%) and maintenance (14%) covering almost 90% of total costs (9).

2. Case study, New Psychiatry Bispebjerg

The project studies a specific new, large hospital building New Psychiatry Bispebjerg. The hospital consists of three 3-storey buildings. Building 51, 52 and 53. This article focus on Building 51, having a total of 72 single patient rooms, all equipped with separate toilet/bathroom, see Figure 1.

2.1. Lighting system

In Building 51, the planned DALI-controlled dynamic LED lighting system consists of two ceiling lamps as general lighting (tunable white), supplemented by bathroom lighting, bed lighting and TV lighting (constant light intensity and CCT), see Figure 1. The lighting system is planned to be managed at luminaire level, each luminaire having its own IP address.

2.2. Ventilation system

Building 51 is mechanically ventilated via six VAV pressure-controlled section regulators for the overall air distribution as well as motorized zone dampers controlled via PIR, CO2 and temperature sensors in the individual patient rooms. Each unit is equipped with a heat exchanger having a projected temperature heat recovery of 0.8. The units are also equipped with cooling coils that can lower the temperature by a few degrees when maximum airflow cannot keep the desired temperature set point. The ventilation ducts are projected for a maximum air velocity of 4 m/s in order to keep low noise levels and at the same time have energy efficient air transport. The VAV indirectly determines the air temperature since the supply air has a temperature of. 19 - 21 °C. All rooms have VAV, and there are dampers at both inlet (Ø125 mm) and extraction unit (Ø160 mm) per patient room.

IOP Conf. Series: Earth and Environmental Science 1099 (2022) 012026 doi:10.1088/1755-1315/1099/1/012026

Figure 1. Visualization of New Psychiatry Bispebjerg 2020. Building 51, 52 and 53 each consists of 72 patient rooms identical to the 1:1 mockup room. Building 51 (right) is planned as Phase 1 in the construction of New Psychiatry Bispebjerg.

3. Method

Based on the New Psychiatry Bispebjerg, this paper raises two research questions in order to come up with advices on how to plan operation and control of lighting and ventilation in the best way, taking into account season and the orientation of the patient rooms.

3.1 Research questions

- Can dynamic LED-lighting help distribute light through the day and through the year?
- Can natural ventilation help improve the overall indoor environmental quality?

The advices include supplementing and adjusting daylight with tunable LED lighting, time- and daylight control of LED lighting, energy saving, using natural ventilation as supplement to mechanical ventilation to help cope with overheating and balance indoor temperatures and external heat loads dependent on orientation,

3.2 A rotatable 1:1 mockup

In order to study aspects of the performance of the patient rooms, a full-scale rotatable mockup was build on the building site. Data was collected for four geographical orientations E, S W and N, both during the summer and winter period. Computer simulations were used to analyze how the patient rooms would perform in the real hospital. The rotatable 1:1 mockup was placed on the upper (7th) floor deck of an existing parking building. All monitoring consisting of consecutive recordings of temperatures, daylight illuminances and spectral light distribution in the mockup was carried out. All measurements were conducted over two periods, during the summer period on 02.07.2018 - 02.09.2018 and during the winter period 10.12.2018 - 03.02.2019, respectively.

The study collected data over one coherent year, collecting and analyzing data from representative weeks during the summer and winter period, respectively. The time plan over the experimental setup is shown in Table 1. Measurements were supplemented with photographic recordings of the luminances and the light distribution in the patient room. All tests and measurements in this study were supplemented with simulations in BSim, a simulation program used as part of the Danish Building Requirements.

Table 1. Time schedule for measurements in the rotatable 1:1 NPB mockup. The table shows dates for the four geographical orientations of the Mockup, for the summer and winter period, respectively.

Period of year / Orientation	Summer	Winter
East	16.07.2018 – 06.08.2018	24.12.2018 – 07.01. 2019
South	09.07.2018 – 16.07. 2018	28.01.2019 – 06.02. 201
West	02.07.2018 – 09.07. 2018	08.02.2019 – 19.02. 2019
North	21.08.2018 – 02.09. 2018	28.11.2018 – 17.12. 2018

3.3 Sensors for data collection

A number of different sensors were installed for measurement and monitoring of the conditions inside and outside the 1:1 mockup throughout the measurement period of one year. Light sensors monitored the incident light both outside and inside the mockup. A combination of spectral sensors, (SPARK, Ocean Optics) and illuminance sensors (Wireless tags pro ALS, Cao Gadgets LLC) were used for the light measurements. The light sensors were calibrated at DTU Fotonics to spectral irradiance in the range 400-700 nm (W/m^2 nm) and illuminance (lux), respectively. A combination of a spectral sensor and an illuminance sensor was used to measure the incident light through the window and on the wall opposite the bed. An illuminance sensor was placed on the wall above the bed. Indoor climate sensors (IC-meter, IC-Meter) were used to continuously monitor the indoor climate parameters. These included temperature (°C), relative air humidity (%), CO_2 (ppm) and noise level (dB(A)). As a reference sensor for the available daylight, an illuminance light sensor was installed on the roof, measuring horizontal illuminance. For outside temperature measurement, an IC-meter was placed under the mockup. The setup of all sensors is illustrated on the patient room plan as shown below in Figure 2.

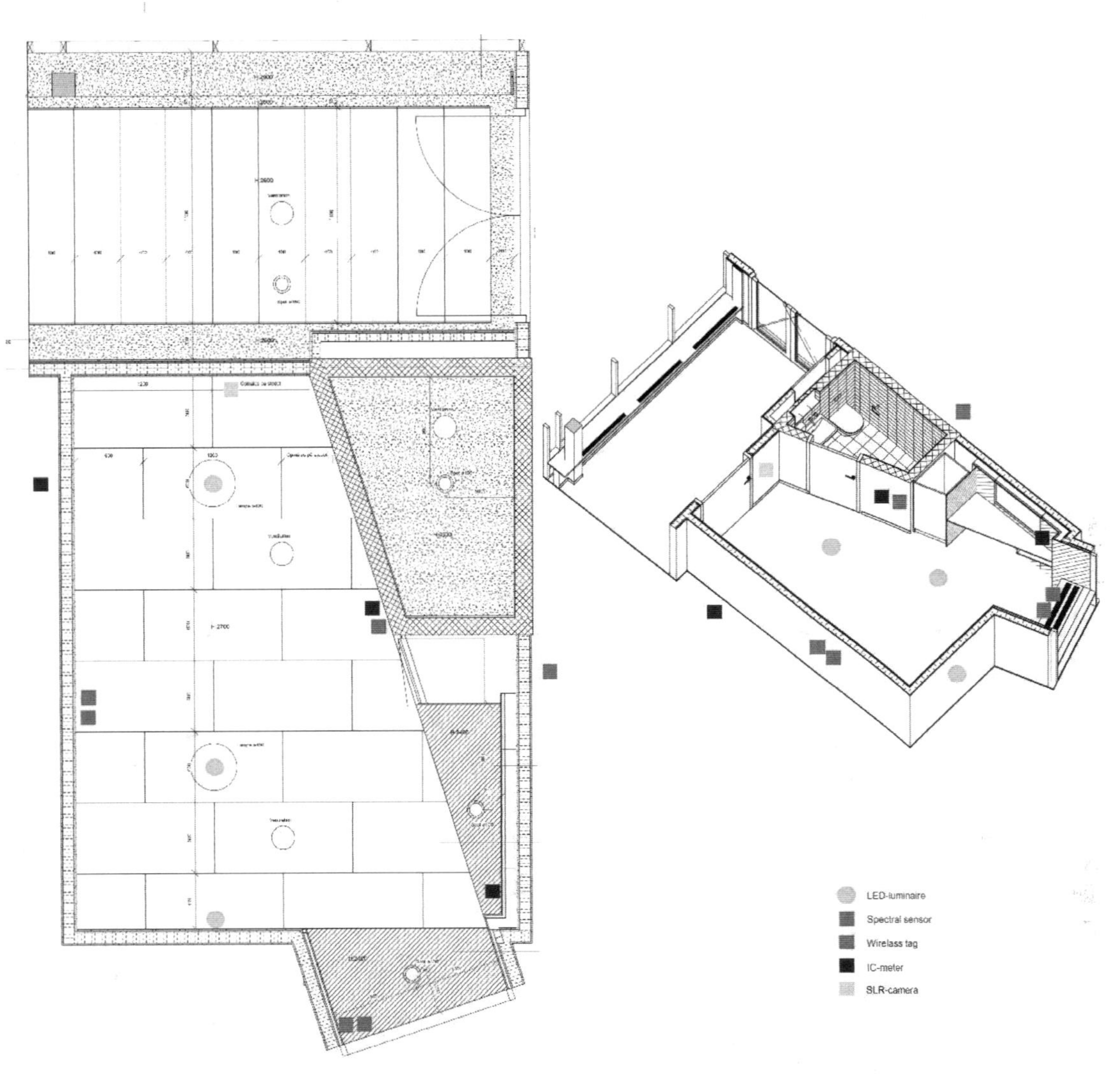

Figure 2. Plan of the patient room illustrating the location of the installed LED luminaires and all sensor. Note that the window direction deviates from the façade orientation by 18 degrees.

3.4 Photographical representation of light

Photographic recordings were performed in the 1:1 mockup NPB using two radio-controlled SLR-cameras, Canon EOS 550D camera. The recordings were made at summer solstice (21.06.2018) and winter solstice (21.12.2018) for each orientation of the façade W, S, E and N, respectively. Recordings were made every second hour as time-lapse photography through one consecutive day for each orientation. Recordings were made in two ways, using a) a fixed shutter-speed and b) a fixed aperture. In this way the estimated light levels and the perceived light distribution corresponding to the luminance was recorded both summer and winter. For all photographic recordings, both daylight and artificial lighting were recorded.

4. Results

In this section, registrations and measurements from the rotatable 1:1 mockup are presented. Starting with a photographical representation of the lighting conditions in the patient room for each different geographical orientation, summer and winter, respectively. These recordings are followed by luminance and spectral measurements of lighting conditions (10, 11) in the patient room for each geographical orientation, summer and winter. Finally, measurements of temperatures for each different geographical orientation are presented. The latter only for geographical orientations having solar exposure and only for the summer period, since focus was on the demand for natural ventilation as active cooling during the summer period.

4.1 Photo series of mockup

The photographical representation of the lighting conditions showed a marked difference between, on the one hand, what we experience when we occupy a patient room, Fig 4, and on the other hand, the amounts of light we actually receive when occupying the patient room, Fig 3. The results showed large differences between E, S W and N during the day. N and W receiving little daylight during the morning, and E and S receiving little daylight during the evening. Large differences were observed between the summer and the winter period. In the winter period, E received morning daylight and S received daylight during most part of the day, while N and W received very little daylight. Only 8 hours of daylight during the winter was observed during the winter period, while 16 hours of daylight was observed during the summer period at 56th N lattitude.

Figure 3. Photographical representation of the light levels in the 1:1 mockup NPB every second hour during the day from 05 am to 11 pm for each of the four world corners, E left), S, W and N (bottom) in summer 21.06.2018 (right) and winter 21.12.2018 (bottom).

IOP Conf. Series: Earth and Environmental Science 1099 (2022) 012026 doi:10.1088/1755-1315/1099/1/012026

Figure 4. Photographical representation of the perceived daylight and artificial light distribution every second hour during the day in the 1:1 mockup NPB from 05:00 to 23:00 for each of the four corners of the world, E, S, W and N in summer 21.06.2018 (top) and winter 21.12.2018 (bottom).

4.2 Light measurements

Illuminance was measured over a four week period during the summer and the winter period, over which the orientation of the 1:1 mockup was rotated as described in Table 1. Results of the illuminance measurements included the diurnal variation of light in the mockup for each of the four orientations.

For both summer and winter measurements, data showed that most daylight was provided in the W-facing orientation of the room in the afternoon, whereas it occurs in the middle of the day for the S-facing orientation and in the morning for the E-facing orientation. For the N-facing orientation, a low illuminance is seen throughout the day. These results are in line with the earlier findings shown in the photo series of mockup.

To get a better picture of the amount of morning and evening light available in the mockup and the relationship between morning and evening light, data was analyzed over the four week for each orientation, during both summer and winter period. The morning light was defined by the light integral from 06:00 to 14:00 and the evening light was defined by the light integral from 14:00 to 22:00. These were given as

$$I_{morning} = \int_{06:00}^{14:00} E(t)dt, \qquad I_{evening} = \int_{14:00}^{22:00} E(t)dt,$$

where *E(t)* was the illuminance at time *t*. The light integrals were given in [lux hours]. The ratio of the morning to evening light integral was given by:

$$Ratio = \frac{I_{morning}}{I_{evening}}$$

The measured illuminance as a function of time during the measuring period for summer and winter, respectively, was applied to this analysis of morning and evening light integrals. The ratio between morning light and evening light was calculated for all the days studied.
A ratio greater than 1 thus corresponds to most light in the morning hours and a ratio of less than 1 corresponds to most light in the evening hours. It is observed that there are large variations in the total light integral from day to day for all orientations and that the total light integral is lowest for the north orientation. It is observed that the ratio for south, east and north are all larger than 1, while it is less than 1 for the orientation towards west. It provides information of the variations in these parameters due to weather changes over the day and from day to day.

These calculations of the measured ratios were performed for all four orientations of the 1:1 mockup, in order to get a better overview of the variations in the measured morning to evening light ratios. Mean values are shown in the bar diagram in Figure 5 for each orientation for each of the three positions inside the room; at window, at bed and opposite bed, respectively. The marked error bars show the standard deviation of the measured ratio for each week/orientation.

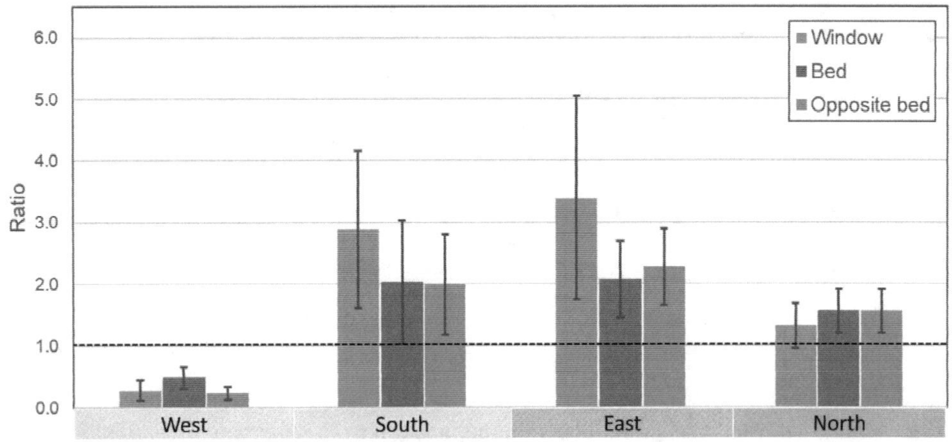

Figure 5. Mean values of the measured ratio of morning and evening light integral for each of the four orientations of the mockup in the summer period. The ratio is shown for the three positions in the mockup. The error bars show the standard deviation of the measured ratio for a week. The dashed line indicates the ratio of 1 between morning light and evening light.

It is seen from Figure 5 that the ratio was lowest for the W-facing room, having values around 0.2-0.3, which means that there were appr. 3 – 5 times more evening light compared to morning light. For the other orientations, this ratio was over 1, meaning that the morning light integral was higher when compared to the evening light integral. For South and East, this ratio was observed to be around 2 and a slightly higher at the window with a value of 3, however with a higher standard deviation. For the North orientation this ratio was around 1.5 for all positions inside the mockup.

The corresponding analysis for the measured ratio between morning and evening light was performed for the winter period. In Figure 6 the mean values and standard deviations are shown for the three positions inside the room.

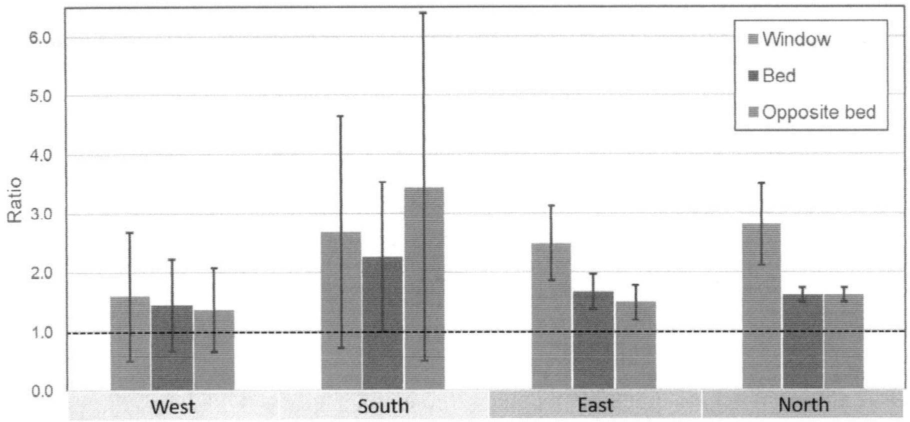

Figure 6. Mean values of the measured ratio of morning and evening light integral for each of the four orientations of the mockup in the winter period. The ratio is shown for the three positions in the mockup. The error bars show the standard deviation of the measured ratio for a week. The dashed line indicates the ratio of 1.

The measured ratio shows large variations in the winter period for the west and south orientation of the room. Although the mean values are larger than 1 for all orientations, the variation covers values higher and lower than 1 in the west orientation. This means that there are days with more evening light than morning light in west-facing patient rooms. For the east and north orientation there are little variation, and all measured ratios are higher than 1, with a value of approx. 1.6 for near the bed. South facing rooms have most morning light and the highest ratio of 2.2 for near the bed.

4.3 Temperature measurements

In order to compare differences in diurnal over-temperatures during the summer period, data from the summer period data was collected for each sunny orientation. The mockup was oriented with the façade facing three corners of the world one week at a time, first facing W, then S, then E. The course of indoor temperature, outdoor temperature and difference between indoor and outdoor temperature was recorded. Data showed a very warm period the first 3 weeks facing W, S and E, respectively.

The purpose of the temperature measurements was to analyze the influence of the orientation of the room on the variation of the indoor temperature, especially during hot summer days. These variation of and the differences between indoor and outdoor temperatures namely influence when and how much natural ventilation can reduce the need for mechanical ventilation throughout the day and the night. Since solar radiation was the only heat load in the mockup, the temperature variations only reflected the variations of the solar load over the day. From the measurements of light transmitted through the window (described in section 3) the solar load could be approximately determined from the light efficacy, which was set to 109 lumen/W (mixed efficacy of direct radiation and diffuse sky radiation). The calculated graphs of solar loads in W/m^2 on the selected days are shown in Figure 7. The graphs also were corrected for the ratio of heat transmittance (g-value) over light transmittance, LT (0.67/0.83 = 0.81). It should be noted that when the facade was facing South (180°), the window actually deviated 18 degrees towards East, due to the architectural design, see Figure 3, hence the actual window orientation was 162°.

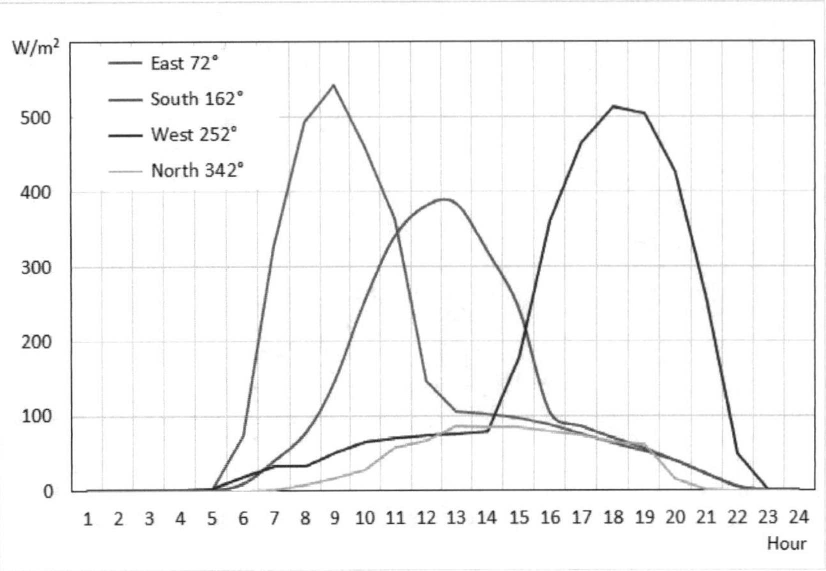

Figure 7. Solar radiation load W/m^2 in the mockup during the summer period. The solar radiation (W/m^2) on the selected days is determined from the light measurements (lux) the estimated light efficacy, (109 lumen/W), and corrected for ratio of heat transmittance over light transmittance.

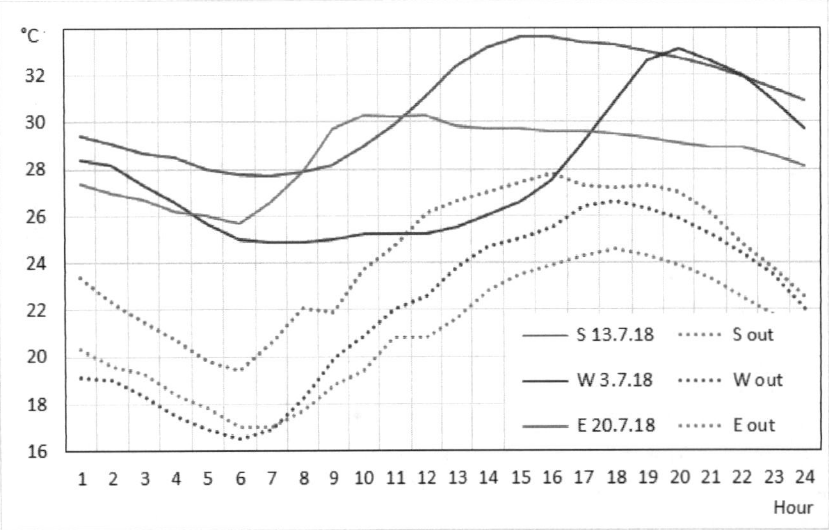

Figure 8. Measured indoor temperatures during the summer period for each orientation (shown by dates) and corresponding outdoor temperatures on the same dates.

Figure 8 shows the measured indoor temperatures on selected days for the sunny façade orientations West (3.7.2018), South (13.7.2018), and East (20.7.2018) together with the corresponding outdoor temperatures for the 3 days (out).

As shown, the outdoor temperature differed on the three selected days, which of course also effected the indoor temperatures. Comparing the graphs in Figure 7 clearly shows how the indoor temperature rises when the solar radiation hits the window. It can be seen, that although the temperature in the W-facing orientation starts to rise later, at 13-14 o'clock (daylight Saving Time), the maximum value becomes somewhat higher when compared to the other orientations. This indicates that the need for ventilation is increased because of the solar radiation. At the same time, the data show that outdoor air temperature using natural ventilation, effectively can supplement mechanical ventilation, to maintain comfort temperatures throughout the day and especially throughout the night during the summer period, see Figure 8.

5. Conclusion

In this article we have presented data and observations based on investigations carried out in a rotatable 1:1 model of a patient room, as part of the planning of the New Psychiatric Hospital Bispebjerg. Through on-site measurements and analysis of light and indoor environmental quality, the study has documented conditions in and around a patient room. Data included illuminance, spectral irradiance, indoor and outdoor temperatures, collected for all four orientations E, S, W and N from summer 2018 to winter 2018/2019. Measurements and data were supplemented with photographical representations of light conditions for each orientation, during the day and during the summer and winter period, respectively. Based on these data and results, the study is able to make the following main-conclusions:
All in all, fundamental differences in lighting and thermal conditions in the patient room were found, depending on the geographical orientation and the season of the year. For the summer period, morning to evening light ratios were larger than 1 for all orientations except for the W-facing rooms. A ratio of 0.2-0.3 in W-facing rooms means that there are appr. 3 - 5 times more evening light than morning light. In S-facing rooms this ratio was 2, meaning that there was twice as much morning light as evening light. In N-facing rooms this ratio is around 1.5.
For the winter period, the morning to evening light integrals generally were much lower than during the summer period. The meant value of the ratio remained larger than 1 for all orientations, with S-facing rooms having most morning light and the highest ratio of 2.2 and E- and N-facing rooms having the lowest ratio of approx. 1.6. However, large variations were observed for the W orientation, for which ratios were both larger and lower than 1. This meant that there were days with more evening light than morning light in W-facing patient rooms.
For the N-facing patient room a lower illuminance was observed throughout the day and throughout the year, below 10.000 lux near the window during winter and below 20.000 lux around the clock during the summer. Generally, the spectral light measurements indicated that E- and S-facing patient rooms perform better than W and N-facing patient rooms all year around. The results showed that E- and S-facing patient rooms provide the highest melanopic EDI response during the morning and the lowest melanopic EDI response during the evening, potentially facilitating sleep initiation and consolidation better. However, these latter potential health benefits are to be investigated further in another project.

Based on the preliminary results presented in this article, we intend to develop and implement a new planning tool for lighting, called Lattiude and Orientation Compensated Arhitectural Lighting (LOCAL) at New Psychiatry Bispebjerg. LOCAL as a dynamical light planning concept for patient rooms will help compensating for the discoveries from our investigation of a patient room in this study.
LOCAL as a new planning tool is dividing the day into a morning period (06 – 14) and an evening period (14 – 22), setting two scenarios, one morning scenario with recommended higher light levels, and one evening scenario with lower recommended dim-light levels. To assess the differences in light doses during the day, a ratio of light in the morning period (06 – 14) over light in the evening period (14 – 22) will be introduced. A ratio <1 means that the cumulated light dose (light integral) of evening light is higher than the cumulated light dose of morning light. A ratio of >1 means that the cumulated dose of morning light is higher than that of the evening light.

The novelty of LOCAL as a planning tool is, that it is dividing the day into two scenarios; a morning scenario (06-14) and an evening scenario (14-22), respectively. The morning scenario differing significantly from the evening scenario. In this way the planning of lighting takes advantage of the daylight differences during the day in order to optimize the electrical lighting and contribute to a good indoor climate that better supports variations around the clock and across seasons.

Based on the preliminary results presented in this article, we intend to implement natural ventilation as a supplement to mechanical ventilation in all 72 patient rooms. The preliminary results indicate that supplementing with natural ventilation can help cooling the hospital building passively in the summer period. Measurements of the temperatures in the mockup, in general showed that indoor temperatures follow the solar radiation transmitted through the window, creating excessive temperatures late in the afternoon, highest in the W-facing patient room. During the summer period, the need for ventilation is closely connected to the solar radiation load and therefore dependent on the geographical orientation of the patient room. Natural ventilation is an effective and free source to reduce overheating in the patient room because the outdoor temperature – during daytime and especially during night time – is lower than the indoor temperature, when compared to the typical inlet temperature from mechanical ventilation. Natural ventilation not only can reduce the need for mechanical ventilation and save energy, it also may provide a cleaner air, giving a feeling of fresh air, closer contact to nature and better control of the indoor environment. All in all, these preliminary results indicate that it makes sense to take account of the differences caused by the geographical orientation and seasonal changes when it comes to planning dynamic LED lighting and supplementing with natural ventilation. We expect that dynamic LED-lighting, together with natural ventilation, will have an overall positive therapeutic effect on patients suffering from depression, providing a better sleep, less anxiety in patients, etc.

In a coming project, we will study the effect of these interventions. The project is expected to include a total of 1.000 patients and results are expected to be published in 2025.

6. Acknowledgements

This study was supported by Elforsk, PSO 350-050 and Aase og Ejnar Danielsens Fond. Furthermore, experiments were implemented with help and funding from Region H Psychiatry.

References

[1] Volf C, Natural Light, Natural Air and Natural Surroundings – in Different Hospital typologies 1910 – 2010. 2016

[2] Volf C. Light, Architecture and Health – a Method. Aarhus School of Architecture. 2013

[3] Beachemin, KM & Hays, P. Sunny hospital rooms expedite recovery from severe and refractory depressions. Journal of Affective Disorders. 1996

[4] Ulrich, R. "View Through a Window may Influence Recovery From Surgery." Science 224: 420-421. 1984

[5] Lewy, AJ, Bauer VK et al. Morning vs evening light treatment of patients with winter depression. Arch Gen Psychiatry. 1998

[6] Krzytow et al. Depressed Patients Hospitalized in Southeast-Facing Rooms Are Discharged Earlier than Patients in Northwest-Facing Rooms.2016

[7] A Adams. Medicine by design - The Architect and the Modern Hospital, 1893–1943.

[8] Trafik- Bygge- og Boligstyrelsen 2018. Bygningsreglement.dk. https://bygningsreglementet.dk/

[9] A publication, Glostrup Neurohabiliteringscenter 2012

[10] CIE S 026/E:2018. CIE System for Metrology of Optical Radiation for ipRGC-Influenced Responses to Light, https://www.techstreet.com/standards/cie-s-026-e-2018?product_id=2030705

[11] Http://cie.co.at/publications/position-statement-non-visual-effects-light-recommending-proper-light-proper-time-2nd, last visited 15-11-2020.

IOP Conf. Series: Earth and Environmental Science 1099 (2022) 012027 doi:10.1088/1755-1315/1099/1/012027

Identifying Nurses' Perception of a Lighting Installation in a Newly Built Hospital

Schledermann, Kathrine (1)(3), Bjørner, Thomas (2), Mullins, Michael (2), and Hansen, Torben (3).

(1) Aalborg University Copenhagen, Denmark, kas@chromaviso.com; (2) Aalborg University Copenhagen, Denmark; (3) Chromaviso, Denmark.

Kas@chromaviso.com

Abstract. Prior to the installation of new integrative lighting systems, perceived experiences by nurses of the existing lighting were collected as a baseline study through interviews at two newly built hospital units in Denmark. This paper addresses the advantages and barriers that the current lighting application presents for the nurses' work-practice. The study found that ten out of twelve of the interviewed staff at the Neurointensive Care Unit (neuro-ICU) and Postanesthesia Care Unit (PACU) prefer to work in relatively dark conditions. The interviews revealed how light and darkness are perceived as important to the staff for entraining the circadian rhythm of the patients; this was manifested in their working routines. An identification and description of nurses' experience will guide and inform future implementation of integrative lighting systems at neuro-ICU and PACU to ensure patients' sleep quality and recovery, improve visual comfort for the staff, enhance the degree of perceived usefulness of the lighting systems and facilitate technology acceptance. The study concludes that there is a need for implementing differentiated lighting design for the different units and contexts in the hospital, in contrast to a 'one-size fits-all' lighting implementation.

1. Introduction

Health-care work environments have received growing attention in research during the past decades [1]. This can be explained by the significant effects of the physical indoor environment on patients and staff, when designed correctly e.g., the right light at the right time. It is already well documented how light not only enables us to adequately see in specific work environments, but also its importance for physical and psychological functioning [2–5]. In previous studies, work environments for hospital staff have been discussed in relation to carrying out nightshift work, as light has been shown to have an impact on health and well-being (e.g., quality of sleep) of staff, reduction in occurrence of errors and productivity increase [6–11]; additionally, shift work and work at night have gained increasing focus in lighting research due to the interest in preventing work-related, long-term diseases such as breast cancer and fatigue, which have been linked to nightshift schedules in recent papers [7,12–16]. There is thus a documented need to improve safety, performance and health in the workplace by reducing circadian misalignment [9,17–19], particularly for shift workers. As a response to the issue at hand, researchers have stressed the need for attention to the design of lighting for individuals that do shift work and sleep at atypical hours, towards achieving a positive effect on performance and circadian regulation. Past research has also highlighted individual differences in light sensitivity and preferences, suggesting implementing complementary individual lighting solutions at work [9,17]. The focus of past research has to a large extent been on non-image forming (NIF) effects of light; consequently, there are few studies applying a qualitative framework of methods to staff's perception of lighting. As previously argued [20], this might be explained by the complexity of applying such methods to real-life cases, and the difficulty of studying lighting conditions in real-life situations, where field research at hospitals can present the researcher with difficult access and ethical issues. The number of studies focusing on nurses' perception of lighting in health care environments is however, growing steadily [21,22]. For instance, McCunn et al. [24] conducted a content analysis of open-ended responses, concluding that controllability of brightness levels was among the most valued attributes of lighting, together with better task lighting. In addition, Davis et al. [23] found that nurses working in a 'contemporary environment-of-care lighting system' (zone divided luminaires with separate control, night-time lighting, dimming) compared to a 'traditional environment-of-care lighting system' (simple switching controls and without dimming) perceived the lighting to be of higher quality, reported fewer complaints from patients and needed less additional light sources to perform tasks. Based on their analysis, suggestions were made for future implications for practice. Although some, albeit few, previous papers deal with these implications, there is a need to obtain a deeper understanding of the practical challenges that lighting systems have for work situations in hospital units. On this basis, the research question underlying this study was how do nurses working at a Neurointensive Care Unit and Postanesthesia Care Unit use a lighting system and adjust it to their and the patients' needs? The novelty of this paper lies in its identification of the advantages and barriers that the lighting system poses for the work of the nurses during different working hours. Moreover, the paper presents a qualitative field-study conducted at a newly build hospital. An identification and description of nurses' experience can guide and inform future implementation of integrative lighting systems at the neuro-ICU and PACU to ensure patients' sleep quality and recovery, improve visual comfort for the staff, enhance the degree of perceived usefulness of the lighting systems and facilitate technology acceptance. Perceived usefulness and technology acceptance is defined in the study according to the technology acceptance model [23]. Perceived usefulness relates to whether the lighting is beneficial for tasks and workflow, that it has a positive impact on job efficiency and that it supports the care of patient and their needs. Visual comfort is related to the judgement of brightness and colour temperature of a room, and are concepts related to appraisal and visual capabilities [24,25]. Thus, visual comfort often affects perceived usefulness, as it can affect the staff's capability to perform tasks with varying degrees of accuracy. Visual comfort is also connected to perception of adequate light or having enough options to adjust the lighting. In this paper, it refers to whether the nurses perceive the light to be adequate for their ability to perform a task; visual comfort relates primarily to illuminance but can also refer to light settings.

2. Setting and Lighting Installation

The study was conducted at a Neurointensive Care Unit (neuro-ICU) and a Postanesthesia Care Unit (PACU) located in a new facility under the Neuroscience Centre that is part of Copenhagen University Hospital, Rigshospitalet in the Capital Region of Denmark. The two units opened in September 2020 and are located on the second floor which are adjacent to each other with the capacity of 12 ICU-beds, 8 daytime and 4 night-time PACU-beds. The neuro-ICU treats patients with acute brain and nerve injuries caused by, for example, strokes, brain tumours and traumatic brain injuries. In the neuro-ICU, the patients have single-bed rooms and are monitored 24-hours a day, whereas at PACU, the rooms are designed to accommodate multiple patients with up to five beds, as the patient turnover is more frequent with a maximum of one night observation. After half a year in service, a renovation of the lighting was commissioned requesting an integrative lighting solution that supports the working environment and entrains the biological 24-hr sleep-wake cycle of patients and staff. This paper presents findings of the nurses' experience of the lighting prior to this renovation. The interior of the department comprises a dark, blue-grey linoleum floor and white-painted walls. The rooms have large window sections that allow daylight to illuminate the rooms during the day. When compared with a standard patient-room, the neuro-ICU is a highly technical environment containing three computer screens and four machines attached to the patient that constantly radiate light and sound. The nurses have a working station with a computer and screen at one end of the room, where they can document their work and the treatment of patients.

Illustration 1. Daytime [DT], Night-time [NT], Postanesthesia Care Unit [PACU], Neurointensive Care Unit [ICU

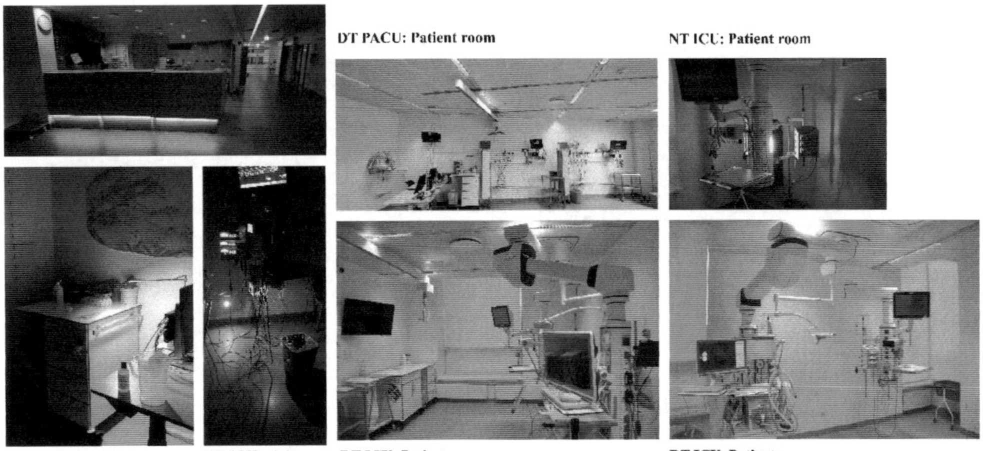

The lighting consists of six linear profiled luminaires by Zumtobel (Tecton C LED4000-840 L2000 WB, Dali controlled) fitted in the ceiling (K= 4000, CRI= min. 80) with five options for control, an adjustable table-lamp (model: Kelvin Edge by Flos, K=2700/3200, CRI=90) with three possible settings, and an examination lamp (Maquet Lucea, K=4500, CRI=95). Around the bed, there are in addition three light sources (uplight, horizontal downlight 1m from the floor and vertical illumination at the height of the bed's headboard) integrated in a critical-care pendant (Maquet Moduevo) developed by Maquet, Getinge. Additionally, the nurses have a pencil light (flashlight) of the brand Abena (K=3000, Ra=99.5). The hallway lighting consists of a continuous LED by Zumtobel (Tecton C LED2000-840 L1000 WB, Dali controlled) located on the left side of the ceiling. There are three open workspace areas in the department, which have spotlights (model: Pleiad G3 CMP) by Fagerhult fitted in the ceiling and, at the desk, Flos table lamps (Kelvin Edge). Zumtobel LED was also used for the medicine dispensary, cleaning and storage spaces. Below each patient bed, a small square-shaped luminaire was fitted in the wall, intended for orientation light at night (Berker B.7) that was intended for night-time use. The ceiling luminaires had four options for adjusting illuminance: 25% intensity, 50% intensity, a manual dimming option where the lighting increased in steps (see table 1 for ratio of min. and max intensity), a night function (luminaire behind bed) and an option to switch it off. At night and in the evening, the nurses often used 25%, the desk lamp, the flashlight, the luminaire behind the bed and/or the uplight on the critical-care pendant. The nurses also manually adjusted the illuminance themselves.

Figure 1 shows the distribution of illuminance (lx) of the ceiling mounted linear profile luminaires set to 25% intensity in a standard neuro-ICU patient room (5.8*5.12*2.80m), measured with a Konica Minolta CL-500A spectrometer. 37 measured points were taken at floor level with 9 cm from floor to the light sensor. The measurements were taken horizontally with approximately 0.6 m between measured points. No daylight was present, as the measurements were taken after dark. The glass windows between the room and hallway were covered by white, shade blinds. The window facing the streets were also covered by blinds. Lights from three screens, though on standby mode, and a narrow glass window in the door radiated additional light in the room. The average illuminance of the room were 109 lx and correlated colour temperature (CCT) of 3986 K with a uniformity of 0.4. Besides these, measurements were taken at three identified work-zones: 1) the medical cart 2) the desk and 3) by the bedside (table 1). Measurements at the bed were taken on the right side of the headboard, where the medical equipment and monitors are located, and the nurses often worked from this side of the bed. The respirator was located on the left side of the bed, as well as an extra working station with a computer screen.

Table 1. Photometric measurements of the ceiling luminaires and desk lamp.

Location	1. **Medical cart:** Horizontal measurements taken at 95cm from floor and 45cm to wall		2. **Desk:** Horizontal measurements taken at 76 cm from floor to desk		3. **Bed (right side of headboard):** Vertical Measurements taken at 173cm from floor. Looking toward patient with head tilted in a 25-degree angle.	
Setting	**Lx**	**CCT**	**Lx**	**CCT**	**Lx**	**CCT**
25% intensity	77	3992	118	3984	81	4018
50% intensity	163	4001	238	3995	136	4023
Min. intensity	3	4081	4	4047	2	4647
Max. intensity	324	4007	481	3997	370	4019
Only light from desk lamp	3*	2831	3*	2868	-	-
	*Desk lamp pointed towards medical cart giving ambient illuminance		*Desk lamp positioned behind computer screen. A setup that the nurses chose to create indirect light)			

Fig 1. Illuminance measurements of ceiling luminaires in a standard neuro-ICU patient room (2021/09/01 at 21:45-00:30). The grey boxes illustrate the position of six profile luminaires. The grey circle on the desk illustrates the position of the desk lamp. Numbers 1-3. refer to the measurements displayed in table 1.

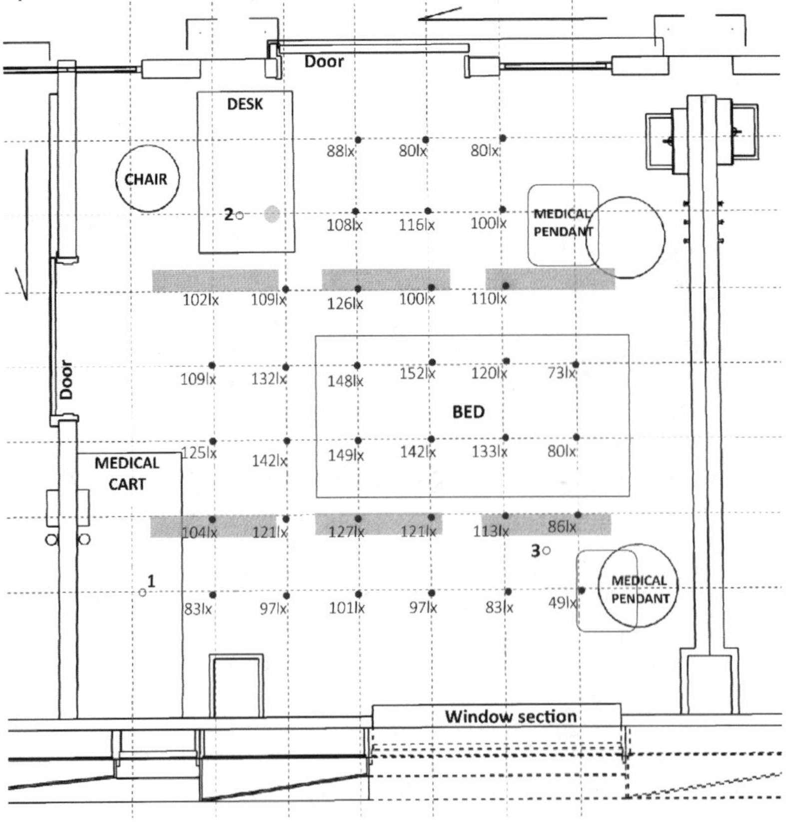

3. Method

3.1. Participants

The participants in the study were healthcare staff working in the neuro-ICU or PACU at Copenhagen University Hospital. 12 nurses (1 male, 11 females) out of 20, were recruited by a researcher employed at the neuro-ICU. All 12 nurses did not have fixed working hours, but had rotating day, evening and nightshifts, except one nurse who worked only day and evening shifts. The participants were selected based on inclusion/exclusion criteria. The recruitment was based on qualitative quota sampling [26], where the recruiter divided the participants into two pre-defined groups: 25-39 years (n=10) and above 40 years old (n=10). For eligibility in the study, the participants had to be able to communicate about their work-routines and work environment in Danish or English, had worked at the institution for a minimum of two months at the point of inclusion, worked a minimum of 25 hours per week at the department, regularly worked either day, evening and nightshift, or a combination of these, and were willing to participate in more than one interview. The participants knew about the upcoming re-installation of lighting.

3.2 Ethics

All participants in this study are anonymized by ID numbers, meaning that all data are labelled with appurtenant ID numbers, time of data collection and date. We followed the SRQR standards for reporting qualitative research [27]. Prior to voluntary participation, all participants completed an informed consent form that included the right to withdraw at any time, the right to refuse to answer the questions and a guarantee of the participants' anonymity. We applied special ethical considerations for access, permission and data handling within the healthcare sector. As the data was collected during the covid-19 pandemic, safety measures were applied, and guidelines were upheld. All sensitive information about patients from the interviews and observations was anonymised; likewise, names and information of colleagues or external people e.g., family and friends. The work carried out by the staff was respected, with granted permissions to observe and allowance for interruptions for attention to patients or colleagues. In these instances, the interviews and audio recorder were paused. The researcher making observations attempted to be as unobtrusive as possible, by respecting when there were situations where observations were not appropriate. This study received ethical approval (ID2021- 020-00540) by an ethics committee at Aalborg University and the research protocol was described and approved by the head of the Neuroscience Centre at Copenhagen University Hospital.

3.3 Procedure

The study was conducted between 29th May and 1st July 2021. Semi-structured interviews focused on the staff's experience with the lighting and were conducted in a patient room, either empty or with the patient present, depending on the nurses' preferences. The method employed was similar to [20], a study in a nursing-home context. The questions in the interview started with an introduction where the interviewee was asked to introduce themselves, i.e., education, what shifts they typically work, their work experience, etc. Then the interviewee was asked in brief to explain a typical day, including tasks and routines. This question was raised for the interviewer to get acquainted with the context and flow of working at a neuro-intensive care unit. The next set of questions focused on the lighting, i.e.: "During a shift day/evening/night, what lighting sources and settings do you typically use?", followed by: "Why?", "What is your experience of the lighting?" and "Do you notice the lighting?". The interviewee was also encouraged to give examples of when they used the different fixtures or settings, e.g., what lighting sources they used for different tasks, how they affected the task at hand, the difference in their use of lighting at different times of day, whether they adjusted the illuminance level, etc. Questions about the experience of controlling the lighting were also raised, together with a discussion on the different light settings and the shading system. The interview often ended with questions about improvements, e.g., whether there was something they were missing, or whether they felt they had the lighting they needed to complete their tasks and the patient's needs.

3.4 Data analysis

The transcription process consisted of two phases; firstly, the audio-file from interviews were auto-transcribed using the software Sonix; and secondly, the text was read while listening to the audio simultaneously. The interview material was coded in positive and negative comments about the lighting system [28]. These comments were divided in categories, summarised and are presented in table 2.

4. Findings

The study found that an overarching theme was that the patients' needs act as a driver or motivation for the nurses' use and adjustment of the lighting. In particular, five themes were found identifying advantages and barriers to the existing lighting system (table 2). The identified barriers relate primarily to mornings, evenings and especially night-time, while the nurses could rely on natural daylight during the day.

The nurses' choice of light setting was related to the patient's well-being, to care and create a comfortable environment for them and their recovery. In parallel, the staff struggled with their own visual needs and execution of tasks, due to inadequate light settings or not having enough lighting. It was challenging for the staff to find a light setting that fitted both the patients' and their own needs; consequently, the staff prioritized the patients' well-being. This conundrum was present primarily in the early morning hours and night-time. The interviews showed that the nurses were deeply focused on creating the optimal environment for the patient's recovery, especially their sleep, regardless of whether the patient was unconscious (theme 4 and 5). For this reason, few light sources were employed despite the large variety in tasks that are performed during evenings and nightshifts; this was criticised by the staff (theme 2). A night-shift nurse explained it as:

"Even if we do have the curtains down, then it is rather bright in these patient rooms, with this light (referring to the light sources in the room). [...] Or else, it is with the small flashlights we use, because it is important that it is dark... Also, I have a bit of light from the computer and such, but it is primarily the pupil light we use or else we have a small night lamp behind the bed that we can switch on and that gives enough light" [ID2, nightshift].

The nurses expressed that the use of the sparse light sources and dim light settings was primarily to shield the patients from bright light, in an attempt to prevent disturbances to the patients' sleep-wake cycles. The latter is especially important as it is common for patients admitted to an ICU to experience delirium [29]. Two of the nurses also pointed out that the reason they only used few of the luminaires was in lack of introduction to the lighting and the various options: *"One of the reasons why I don't use much of the lighting is due to the lack of introduction to the lighting, such as how do we switch it on, when is it appropriate to have it switched on, what is the purpose of the lighting, and how is it intended to be used?" (ID11, night shift).* Even though the rooms had multiple light sources and settings to choose from, the staff found it challenging to set the lighting to the desired level. The nurses expressed the need for something in-between their current settings (theme 4). This related to the option for setting the desired illuminance and CCT. A nurse explains the light settings and how she perceives them: *"you have options, but then after all not really, because the lighting outcome is not great. There isn't really anything in-between, either we have the lighting switched off entirely or we have it on. The difference between 25% and 50% is not that evident" (ID3, evening shift).*

In the Postanesthesia Care Unit, where multiple patients were present, this experience also related to having either the entire room lit up or having no light at all. The nurses expressed lacking an option to control the luminaires individually (dividing the room into zones) or having a lamp near the headboard to avoid having to disturb all the patients when checking up on one individual (theme 4). As a result, the medical flashlight was used if the task was not too complicated. The nurses critiqued this part of the lighting system, as it put them in a situation where they had to decide whether to turn on all the luminaires for the task and thus potentially disturb multiple patients. The medical flashlight was commonly used as a light source: Eleven out of twelve nurses interviewed mentioned that they primarily used a medical flashlight (also referred to as the 'pupil light' that is meant for checking pupil response), the light on the medical vertical pendants or the desk-lamp, when performing tasks after dusk, e.g., for checking the amount of urine in the drainage bag, checking the patient's skin or checking the medical dispersion system that is connected to the patient by a tube. The desk lamp was appraised by the nurses due to its warmer colour temperatures, that they could rotate the luminaire and tilt the light source to a desired angle (theme 2).

Additionally, the nurses switched off the computer screens after use, and found rather creative methods to shield the patients from light, putting a pillow cover over the machines, adding coloured post-it notes on the night lamp behind the bed to dim it or angling the examination lamp over the patient's bed towards the floor away from the patient for some ambient light.

Besides adjusting the illuminance in the room with the electrical lighting, they used internal curtains. However, the internal curtains had not been a part of the building interior from the very start (automatic external solar blinds were the only shading option). The complaints about the curtains were not merely about creating a dark environment or blocking out heat but were also about creating privacy for the patients, as the rooms in this section of the hospital had partly glass walls and were located opposite to residential buildings or towards the hallway that had glass walls.

4.1. Light cycles

Another critique of the ceiling light was of it being static or, in other words, it's lack of being dynamic. ID12 referred to their previous department where the lighting in the morning automatically changed in illuminance and CCT – "like a sun rising" *(ID12, dayshift)* (theme 4). Likewise in the afternoon (4pm), the nurses stated they needed more soft lighting. Although they appreciated having a dimmer for controlling the illuminance, the dimming function provided were experienced as limited. ID3 explained that they preferred the previous lighting system (that they had in their previous department) more than the new one, because of having more options to grade the lighting and make it dimmer than with their current lighting system.

4.2. Visual performance

Another explanation for the sparse use of light was connected to the amount of experience with certain tasks: some tasks the nurses know 'like the backs of their own hands', and as such are not tasks that require much light. As a nurse working at the neuro-ICU expressed: *"You quickly get used to perform your routines in darkness or with dim light" [ID7, nightshift].* Besides creating an optimal environment for the patients, the use of dim or no light at night was also connected to the staff's own visual perception of the lighting; they found it comfortable to work in darkness as the bright and cold light gave them tired eyes and headaches (theme 2).

The nurses wanted to have an adequate amount of light at the area of the patient. For example, the examination lamp (providing high levels of illuminance) above the bed was described as being useful for technical tasks, such as attending wounds, applying stitches, checking the drain attached to the patient's skull or inserting a catheter, etc. ID1 explained that they have an adequate amount of lighting for these kinds of tasks.

However, five nurses were unsatisfied with lighting for tasks at the medical chart (theme 1). These tasks include finding medical kits, preparing liquid bags, drawing medicine into syringes, food formulars, etc. The PACU nurses used the desk lamps instead of the ceiling light, when for instance managing medical kits at the stationary cart located in one the corner of the room; however, it is not at every medical chart that they had a desk lamp at close range.

As a result, they worked in darkness or used the flashlight. ID3 explains her routine: *"Actually, in the rooms we use them [desk-lamps] to create a little more like ... to have more light I mean. When we have to go to the room to treat the patient, then we turn on the lights [the desk lamps] near the medical cart opposed to turning on the big light [ceiling luminaires]. That gives a tiny bit of light in there, and we turn it off again once we leave the room"[ID3, evening].* ID8 also confirms this and explain that often the only option is to turn on the ceiling luminaires because they must check that they have the correct medicine, which results in stressing the patients. The practice of using the desk lamp to illuminate the room was also addressed by neuro-ICU-nurses who sometimes raised the desk-lamp to shine light towards the medical cart. The interviews also revealed that the nurses found the lighting intuitively, however, in some interviews the comments were both positive and negative. For example, ID3 explained that it intuitive to control the lighting, but at the same time found the light settings did not match her expected outcome.

4.3. Hallways and work environment

Compared to the neuro-ICU, where staff spend a major part of their shifts in one patient's room, PACU-nurses spend a large part of their shift in the open office environment, situated between the hallway and the patient rooms. The PACU nurses criticized the hallway lighting (theme 4), as they were unable to control the lighting in their part of the department. The hallway lighting was controlled for the entire department at the farther end of the neuro-ICU. The open office space lighting and hallway lighting was described as very bright and was pointed out as an issue because of the glass walls between the hallway and patient rooms, where the lighting due to its high illuminance created discomfort for the patients. Moreover, staff working late wished to turn off the lighting in the hallway and have the dimmest lighting in the open office space. As ID3, 8 and 10 argued, if they needed more light, they would employ the desk lamps.

Table 2. Perceived advantageous and barriers of the installed lighting system through accumulated codes from interview statements.

Themes		**Perceived Advantageous** (neuro-ICU and PACU)	**Perceived Barriers** (neuro-ICU and PACU)
1.	**Visual performance**	Having bright light to read is good (ID3).	Not having light for tasks at the medical cart (issue at night) (ID1, 5, 7, 8, 10)
		In the patient room it is bright enough for most tasks. The flashlight is suitable for tasks such as checking the pupil response or drain can be done with the flashlight (ID2, 9, 11)	
2.	**Light at night**	The desk lamp is useful as it is flexible and can be angled to how we find it suitable (ID2, 3, 4, 11, 12). "My good friend at night" and it is sufficient for most tasks at night as you can adjust the angle (ID11).	Difficult to make the room dark despite having curtains down. (ID2, 3, 5, 6, 7)
			At night, the lighting is too bright and too cold (ID3, 5, 7, 8, 11, 12). It makes you feel hungover (ID11) and causes headaches (ID8).
3.	**Perceived ease of use**	It is easy to use and intuitive (ID1, 2, 3).	Having lighting controls that do not match the description (ID3) and the controls are not logical (ID8). The controls are sensitive: you need to be careful when pressing a setting if you do not want a room that is very bright (ID11).
4.	**Light adjustment**	Being able to adjust the intensity of the light is appreciated (ID1, 2, 3, 11) including the hallway that has been dimmed (ID3). For example, when a new patient is admitted, they turn on the brightest light to make sure	Not having the option to set individual lighting for the patients in the Postanesthesia Care Unit (light zones) (ID3, 8) [exclusive to PACU]
			Lack of an option to dim the light or turn it off in the hallway and work environment around dusk and at night.

		nothing is critical with patient they must attend to, then they dim it.	Staff working nightshifts find it uncomfortable to work in and set it to the lowest value (ID3, 6, 7, 8, 10) [exclusive to PACU]
			Missing functions that are in-between. The lighting controls either provide too much light or too little. In the evening, staff lack the possibility to illuminate the room without having to turn on the ceiling luminaries (ID3, 5, 8, 12). The light sources on the medical pendants are considered too bright and shine directly on the patients (ID2, 12). The electrical light is the same all day around, it does not change during the day. You do not experience the sun slowly rising it is from pitch dark to fully lit (ID12).
5.	**Atmosphere**		The lighting does not support the creation of warm context for e.g., a family member visiting a patient, family members saying farewell to a loved ones (LED candles are used instead) (ID1) or a hospitalized kid who are scared (ID8).

5. Conclusion

The analysis of twelve interviews identified possible factors that may influence the nurses' experience of their work environment at a modern hospital department from 2020. The nurses' experience of lighting for a good work environment, for example a dark room, can be related to their ability to provide care for the patients, who are the centre of most of their tasks; and secondly, the possibility to carry out the tasks as intended (visibility). The interviews revealed that the nurses working on nightshifts are frequently placed in a quandary, where they must choose between adjusting the light for their own visibility to conduct the task or for the patient's well-being. Similar findings [30] revealed difficult visual task performance for the nurses at night, balancing both the task performance and the patient's well-being. Some nurses also revealed how they became used to working in darkness or dim lighting. The study found that the nurses regularly were obliged to either turn on bright ceiling luminaires or work without any additional light in the room. As a result, the flashlight is often employed for performing tasks. This inclination to use the flashlight or desk lamp as a substitute for the ceiling light has also been found in other studies [18], but it is also interesting that the nurses are missing light settings between either on or off. The nurses use dim illumination at night and try to create darkness in the rooms, however, they also point out the challenge of creating dark environments in the hospital due to devices, windows, shading and light pollution from streetlights. The interviews showed that the staff were discontented and frustrated with the current lighting installation, however, they also revealed that the nurses appraised the option to dim the lighting and having a luminaire that can be tilted and adjusted to a specific angle. The findings show the importance of including health-care staff in the early phases of planning and the strength of evaluating the installation to create sustainable work environments that support health and wellbeing of patients and healthcare workers where the lighting is adapted to their needs. Further studies are required to investigate if the findings in this study also apply to other types of hospital departments or to other types of lighting systems. Based on the findings, we have listed some recommendations for practice implementation:

- Introduce the staff to the lighting and teach them how to use the lighting, when it is appropriate to use, what the different light settings are intended for and how light can affect the 24hr sleep-wake cycle.
- Design lighting that is tailored to the different department's needs
- Zone divided lighting, e.g., for patient rooms with multiple beds, individualized lighting is important to minimize the disturbance of other patients and for staff who might have different preferences
- Having lighting for task-areas used for handling medical kits and medicine (e.g., medical cart)
- Having light settings with different CCTs to accommodate different tasks and needs (e.g., the need for creating calm atmospheres)
- Consider indirect light to minimize glare and for tasks at night that require minimal visibility
- Having curtains for windows and glass walls to block out streetlighting at night and to create a sense of privacy
- Having task lighting that is focused by the bed but also with some ambient and soft illuminance
- Position and distribution of light fixtures should be carefully considered. Carefully consider various rooms that is frequently occupied including medicine rooms and workstations
- Designing a dynamic lighting profile that automatically entrains the circadian rhythm of patients and staff. Including light setting(s) (long-wavelength light) designed specifically for nightshift workers, with minimum disturbance of circadian rhythm and a high CRI

Acknowledgement
The researchers would like to thank the Neuroscience Centre for providing access to the staff and collaboration on the research. We would like to thank all participants and Laura Krone Larsen for her assistance with recruitment of participants and initiating access to conduct interviews and Ane Søby Eskildsen and Claus Puggaard at Chromaviso for assisting with communication to the department.

References
[1] Al horr Y, Arif M, Katafygiotou M, Mazroei A, Kaushik A and Elsarrag E 2016 Impact of indoor environmental quality on occupant well-being and comfort: A review of the literature *Int. J. Sustain. Built Environ.* **5** 1–11
[2] Boyce P R, Veitch J A, Newsham G R, Jones C C, Heerwagen J, Myer M and Hunter C M 2006 Lighting quality and office work: Two field simulation experiments *Light. Res. Technol.* **38** 191–223
[3] Smolders K C H J, De Kort Y A W and Van den Berg S M 2013 Daytime light exposure and feelings of vitality: Results of a field study during regular weekdays *J. Environ. Psychol.* **36** 270–9
[4] Veitch J A, Stokkermans M G M and Newsham G R 2013 Linking Lighting Appraisals to Work Behaviors *Environ. Behav.* **45** 198–214
[5] Ru T, Kort D, Chen Y A W H J and Zhou & 2018 NIF effects of illuminance and correlated color temperature of office light on alertness, mood, and performance across cognitive domains Citation for published version (APA) **149** 253–63
[6] Jensen H I, Markvart J, Holst R, Thomsen T D, Larsen J W, Eg D M and Nielsen L S 2016 Shift work and quality of sleep: effect of working in designed dynamic light *Int. Arch. Occup. Environ. Health* **89** 49–61
[7] Griepentrog J E, Labiner H E, Gunn S R and Rosengart M R 2018 Bright environmental light improves the sleepiness of nightshift ICU nurses *Crit. Care* **22** 295
[8] Figueiro M G, Hunter C M, Higgins P A, Hornick T R, Jones G E, Plitnick B, Brons J and Rea M S 2015 Tailored lighting intervention for persons with dementia and caregivers living at home *Sleep Heal.* **1** 322–30
[9] Lowden A and Kecklund G 2021 Considerations on how to light the night-shift *Light. Res. Technol.* **53** 437–52

[10] Chen Y, Broman A T, Priest G, Landrigan C P, Rahman S A and Lockley S W 2021 The Effect of Blue-Enriched Lighting on Medical Error Rate in a University Hospital ICU *Jt. Comm. J. Qual. Patient Saf.* **47** 165–75
[11] Schulmeister L 2005 Ten simple strategies to prevent chemotherapy errors. *Clin. J. Oncol. Nurs.* **9** 201–5
[12] Dall'Ora C, Ball J, Recio-Saucedo A and Griffiths P 2016 Characteristics of shift work and their impact on employee performance and wellbeing: A literature review *Int. J. Nurs. Stud.* **57** 12–27
[13] Xie Y, Tang Q, Chen G, Xie M, Yu S, Zhao J and Chen L 2019 New insights into the circadian rhythm and its related diseases *Front. Physiol.* **10**
[14] Costa G 2010 Shift work and health: Current problems and preventive actions *Saf. Health Work* **1** 112–23
[15] Applebaum D, Fowler S, Fiedler N, Osinubi O and Robson M 2010 The Impact of Environmental Factors on Nursing Stress, Job Satisfaction, and Turnover Intention *J. Nurs. Adm.* **40** 323–8
[16] Vetter C 2018 Circadian disruption: What do we actually mean? *Eur. J. Neurosci.*
[17] Hadi K, DuBose J R and Ryherd E 2016 Lighting and Nurses at Medical–Surgical Units: Impact of Lighting Conditions on Nurses' Performance and Satisfaction *Heal. Environ. Res. Des. J.* **9** 17–30
[18] Davis R G, McCunn L J, Wilkerson A and Safranek S 2020 Nurses' Satisfaction With Patient Room Lighting Conditions: A Study of Nurses in Four Hospitals With Differences in the Environment of Care *Heal. Environ. Res. Des. J.* **13** 110–24
[19] Foster R G 2021 Fundamentals of circadian entrainment by light *Light. Res. Technol.* **53** 377–93
[20] Schledermann K M, Bjørner T and Hansen T S 2021 Danish Nursing Home Staff's Perceived Visual Comfort and Perceived Usefulness of a Circadian Lighting System *Proceedings of ACM International Conference on Information Technology for Social Good (GoodIT '21), September 9–11, 2021, Roma, Italy. ACM, New York, NY, USA,* (New York: ACM) p 6
[21] Davis R G, McCunn L J, Wilkerson A and Safranek S 2020 Nurses' Satisfaction With Patient Room Lighting Conditions: A Study of Nurses in Four Hospitals With Differences in the Environment of Care *HERD* **13** 110–24
[22] McCunn L J, Safranek S, Wilkerson A and Davis R G 2020 Lighting Control in Patient Rooms: Understanding Nurses' Perceptions of Hospital Lighting Using Qualitative Methods *Heal. Environ. Res. Des. J.* **14** 204–18
[23] Holden R J and Karsh B-T 2010 The Technology Acceptance Model: Its past and its future in health care *J. Biomed. Inform.* **43** 159–72
[24] Veitch J A, Newsham G R, Boyce P R and Jones C C 2008 Lighting appraisal, well-being and performance in open-plan offices: A linked mechanisms approach *Light. Res- Technol* **40**
[25] Boyce P R 2014 *Human Factors in Lighting* (CRC Press. Taylor & Francis Group)
[26] Bjørner T 2015 *Qualitative methods for Consumer Research* ed H Persson (Hans Reizel)
[27] O'Brien B C, Harris I B, Beckman T J, Reed D A and Cook D A 2014 Standards for Reporting Qualitative Research: A Synthesis of Recommendations *Acad. Med.* **89** 1245–51
[28] Báldy I D, Hansen N and Bjørner T 2020 How to design and evaluate a serious game aiming at awareness of therapy skills associated with social anxiety disorder *ACM Int. Conf. Proceeding Ser.* 156–62
[29] Barbateskovic M, Krauss S R, Collet M O, Larsen L K, Jakobsen J C, Perner A and Wetterslev J 2019 Pharmacological interventions for prevention and management of delirium in intensive care patients: a systematic overview of reviews and meta-analyses *BMJ Open* **9** e024562
[30] Graves E, Davis R G, DuBose J, Campiglia G C, Wilkerson A and Zimring C 2020 Lighting the Patient Room of the Future: Evaluating Different Lighting Conditions for Performing Typical Nursing Tasks *HERD* 1937586720972078

IOP Conf. Series: Earth and Environmental Science 1099 (2022) 012028 doi:10.1088/1755-1315/1099/1/012028

Evaluating an integrative lighting design for elderly homes – a mixed methods approach

Anna-Kristin Bochnia[1,2], **Senja Maarit Ruohonen**[1,3], **Mihkel Pajuste**[1], **and Ellen Kathrine Hansen**[1]

[1]Department of Architecture, Design and Media Technology, Aalborg University Copenhagen, Denmark

[2]anna@bochnia.de
[3]senja.ruohonen@gmail.com

Abstract. The experience of light in a space and the physiological effect of light combined is complex to study. Methods which combine parameters from the fields of the biological, visual and atmospheric effects of light through a combination of both qualitative and quantitative data collection on site, seem more important than ever to validate the potentials of integrative lighting. A case study, conducted in two elderly homes in Copenhagen and Aarhus, in Denmark, is used to evaluate a mixed methods approach, assessing an integrative lighting design solution. Luminaires with circadian settings are implemented as a supplementary lighting with the aim of supporting various objectives, including biological stimulation and a homely atmosphere in the elderly homes. To evaluate the implementation of the lighting, a convergent mixed methods approach is used, collecting quantitative and qualitative data. In parallel to collecting quantitative measurements, such as melanopic equivalent daylight illuminance (mEDI) and photopic lux, semi-structured interviews based on questionnaires and card-sorting, on the same subjects, are carried out. To evaluate the approach for evaluating the implementation of this integrative lighting design, three criteria are introduced: a) integration, b) flexibility and c) validity. The integration of results enabled to compare the different methods and findings to identify and study correlations, differences or contradictions. That was conducted across the three topics: biological, visual, and atmospheric effects of the integrative lighting design solution. A framework for a holistic approach for evaluating lighting in elderly homes was defined, with an indication of the need for further development in an iterative process.

1. Introduction

A holistic lighting design evaluation requires the inclusion of multiple effects of light. The complexity how light affects people both physically and physiologically makes it particularly difficult to conduct studies. The multidimensionality of lighting design has been investigated with different methodologies to evaluate and identify the most important factors. According to the current state of science, the biological effect of light on humans, such as melatonin suppression, should be considered in particular [1], as well as factors such as visual comfort [2] and perceived atmosphere [3]. However, studies on the effect of light are often investigated one factor at a time and primarily in laboratories [4]. This study is therefore intended to help assess the multidimensionality of a lighting design solution based on the factors related to the biological, visual and atmospheric effects of light. In this context, conflicting factors pose a particular challenge for evaluation. How can this be dealt with if two of the factors produce conflicting results? Is a scoring of the factors necessary and if so, how can this be achieved?

Lighting design that encompasses these considerations is described as integrative lighting, defined by CIE [5]. Integrative lighting is an essential topic for the AAU Lighting Design Research Group. It also touches upon the core idea of the master's program – designing with lighting transdisciplinary, across multiple fields of knowledge and expertise to understand light as a multidimensional design element [6]. It is hypothesized, that an integrative approach, which includes the areas of biological, visual and atmospheric effects of light can form a holistic evaluation, applying methods from quantitative and qualitative research. This leads to the first criterion against which the approach will be tested: *(1) Integration. The methods must integrate the various effects of light. The methods should incorporate biological, visual, and atmospheric effects of light, as well as allow for integrative data analysis.*

Due to the similar interest in light and health and its effect on humans in a holistic sense, the AAU Lighting Design Research Group was invited to be part of the project and evaluate a new integrative lighting solution for eldercare homes. While the lighting solution in question had already been tested in lab settings by other parties, AAU participated in the project as a conductor of a case study. Studies have demonstrated, that case studies can be important for evaluating a lighting design solution, as knowledge can be drawn from setting up tests in *"specific spaces under specific lighting conditions"* [4].

Designing lighting for eldercare homes bears a significant responsibility, as light is known to influence health and well-being, for instance the effect on the circadian rhythm [5]. Residents of eldercare facilities are especially susceptible for disturbances in circadian rhythm, since many of them are physically restricted, and therefore have limited access to daylight, which is known to be a zeitgeber of circadian regulation [7]. Similar to the research mentioned, the municipalities in Copenhagen and Aarhus had also found in their preliminary studies, that there is a need for supporting the well-being of residents in eldercare facilities by supporting the circadian rhythm with biological stimulation, and by creating visually pleasant atmospheres in their homes. Together, the two municipalities created a tender in which the needs of the elderly residents and the staff working in these facilities were described [8]. Thereafter the tender was released, and the bids of different lighting design companies were assessed. Lyhne Design with their product *SolMate* [9] was chosen and a detailed evaluation of their suggested integrated lighting design solution, further referred to as ILD solution, was planned.

Referering to Schledermann et al. [10] an evaluation in nursing homes is not without challenges. The complexity and heterogeneity of the facilities make it hard to conduct research in the nursing home itself, due to the great complexity of dealing with residents in sensitive situations. Implementing the new lighting can also present challenges related to acceptance, adaptation and usability of the technology, as Schledermann et al. discovered. Lastly, the number of samples depends on the condition of the participants, especially the residents. It is often necessary to rely on the residents' contact person(s) from the given facility to reflect on certain behaviours. To ensure that the evaluation takes into account these needs for flexibility, the second criterion was established. *(2) Flexibility. The methods should offer a high degree of flexibility in their application and design. The methods must be able to be carried out in an eldercare facility and allow valid data collection in different locations, circumstances and/or lighting conditions.*

A combination of both, qualitative and quantitative methods, seems more important than ever to validate the potentials of integrative lighting. For instance, the study conducted by Schledermann et al. demonstrates the value of applying mixed methods to a case study in the field of healthcare, a nursing home [11]. According to Creswell and Plano Clark [12], mixed methods can be defined as different designs, which can be included in a specific framework as a core design. It is of particular value in the field of lighting design to combine different methods and to draw conclusions from them, as *"[a] mix of qualitative and quantitative studies, each chipping away at a single point, can start to provide answers"*, as Albers [13] states. An approach including multiple methods can help to reinforce, understand, or question the data collected. Defined by Creswell and Clark, *"[i]ntegration is the point in the research procedures where qualitative research interfaces with quantitative research"* [12]. Though the integration differs depending on the type of mixed methods design, four key considerations can generally be made to enable the integration of data analysis and interpretation of the chosen topics, as proposed by Creswell and Clark: integration intent, integration data analysis procedures, representation of integration results, and interpretation of integration results [12]. Another advantage of mixed methods is the possibility to include not only quantitative and qualitative methods, but also multiple different factors. As the evaluation of experiments in the Double Dynamic Lighting innovation project [4] shows, the diversity of mixed methods designs [12] allows for the inclusion of a wide range of factors. Depending on the design, the factors can stand next to each other and be evaluated simultaneously or sequentially build on each other.

Lastly mixed methods can provide validity, even where topics occur where *"it has been a challenge [...] to quantify the thing which cannot be measured through conventional measurement techniques"* [14]. Often described as an issue [15], the quality in mixed methods approach can still be discussed, e.g., based on the criteria from O'Cathain [16]. Building on the Good Reporting of a Mixed Methods Study (GRAMMS) [17] O'Cathain provided a set of evaluation criteria, from which, four key criteria can be derived to ensure the validity and quality of the study: rigorous data collection and analysis, intentional integration, organisation into specific research design and framing within theory [12]. This leads to the final criterion for the evaluation of an integrative light design - the validity of the conducted study. *(3) Validity. The methods must demonstrate their validity for scientific research. The methods must ensure the quality and thus the validity of the study conducted.*

Consequently, the following question is asked: *How can a mixed methods approach be applied for a holistic evaluation of an integrative lighting design in elderly homes?* To evaluate the mixed methods approach, three criteria have been formulated, about the topics of *(1) integration*, *(2) flexibility* and *(3) validity* of the methods.

2. Method

To evaluate the integrative lighting design solution, a case study, using a convergent mixed methods approach, is carried out. The case study is part of the phase 2 of the evaluation design process to evaluate and develop the ILD solution for elderly homes as illustrated in **Figure 1**. As the different phases of this study unfold, active integrated conclusions, about the ILD solution and the use of the mixed methods approach, can be drawn based on the results previously obtained.

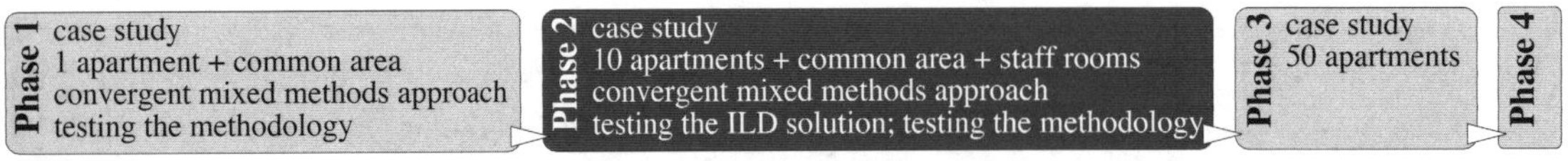

Figure 1. Evaluation design process – Integrated Lighting Design

As *"[...] in areas such as education and nursing, the phenomena studied are complex and tensions arise over the efficacy of both qualitative and qualitative research"*, stated by Gilbert [18], the simultaneous collection of quantitative measurements and qualitative data on the same subject, allows to compare these data and gives an overview of whether correlations, differences or contradictions exist. While the scope of the case study is to study integrative lighting [5], metrics for both visual and non-visual effect are applied. Although non-visual effects could refer to a large variety of biological human responses [19], in this paper the measurements of non-visual effects of light are focused on the qualities of light that would have an effect on circadian rhythm and sleep [7]. The chosen metric for studying the non-visual effects of light is melanopic daylight illuminance (mEDI), as it is based on an expert scientific consensus [7] and recommended by CIE [20]. The visual effect of light could refer to a vast array of perceived qualities of light, such as colour, light distribution, spatial pattern and contrast ratio [4]. In this study, as the lighting design solution to be evaluated takes place in an indoor workplace, the chosen approach for quantifying the visual effect is to study the illuminance on horizontal planes of the relevant task areas, with a reference to the standard, DS/EN 12464-1-1:2021. For ensuring a valid quantitative data collection professional equipment is used and calibrated. Further, measurement points are decided to be similar that the resulting data will be comparable between each other by limiting the amount of measurement heights and directionalities.

As the perceived atmosphere of light is not straightforward measurable [14], qualitative methods are introduced. Semi-structured interviews are a key element towards grasping the participants' experience. This type of interviews offer the possibility to ask questions of different types, depending on the nature of the participants and the context [21]. The interviews are based on questionnaires using the tool SurveyXact [22] in order to enable precise and standardized data collection. This tool also provides the flexibility to send the questionnaires online, if interviews with the healthcare workers on-site due to the unpredictable conditions in the nursing homes is not possible [18]. To verify and expand the collection of the data, contact persons interviews are introduced additional to the interviews with the residents. As a projective technique for the interviews, card sorting is used to help the participants to elaborate the queried topics [21].

In addition, observations are planned during data collection to better compare data sets later and to account for possible irregularities or open issues that may arise due to the dynamics of daylight and possibly differentiated lighting situations [4]. The data sets are first analyzed separately and then compiled in a joint display [12]. The combination of the data based on the three selected subjects allows the reinforcement, understanding and questioning of the results. The previously described methodology is illustrated in **Figure 2.**

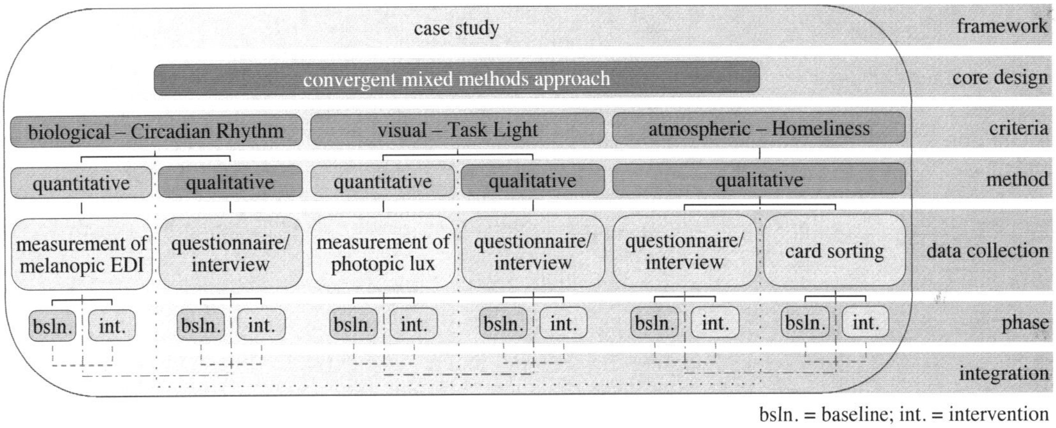

Figure 2. Methodology chart based on Creswell & Plano 2018

3. The Case Study

The evaluation process of the new ILD solution was designed in co-operation between AAU, Lyhne Design, and the municipalities of Copenhagen and Aarhus. Regarding the tender materials from the municipalities, the main approach in developing the new ILD solution has been to design the lighting for eight defined target groups or areas [8], in the eldercare facilities.

Based on the defined target groups and their needs, and the introduced criteria of biological, visual and atmospheric effects of light, three topics and their research questions have been introduced to evaluate the ILD solution. The first topic, corresponding with the theme of biological effects, is the effect of the ILD solution on the circadian rhythm: Does the ILD solution support the circadian rhythm of the participants? Second topic, considering the theme of visual effects, is task lighting: Does the ILD solution provide sufficient light for (healthcare) tasks? And thirdly, the feeling of homeliness, for evaluating the perceived atmosphere of the new lighting design, is discussed: Does the ILD solution support the participants' feeling of homeliness?

The resulting data of the quantitative and qualitative methods will be evaluated against the targets listed in the municipalities tender material [23]. The case study was conducted in Danish and translated to English for this research paper.

3.1. The Integrative Lighting Design Solution

The new ILD solution consisted of three different types of light fixtures, presented in **Figure 3**, which were designed to be used as complementary lighting, supporting the lighting that existed prior to the implementation of the new fixtures. These existing luminaires consisted of varying types of light fixtures, both residents' personal fixtures, as well as several types of luminaires from the facilities.

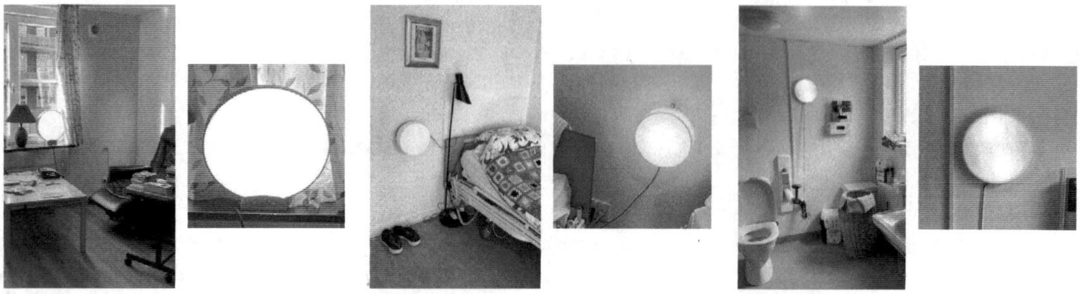

SolMate NightMate BathMate

Figure 3. The new light fixtures implemented in the ILD solution

One of the new fixture types, SolMate, had a predefined circadian lighting setting built in. SolMate was programmed for a dark setting (<3mEDI) from 9p.m. to 6:30a.m. From 8:45a.m. until 3:30pm the setting was at its brightest. In between these times, the light was dimming gradually from bright setting to dark setting during the evening and vice versa in the morning. SolMates were designed not only as tabletop version, but also as wall mounted, ceiling recessed and hanging pendants. In the apartments only the tabletop version was used, while in the common and staff areas all four types were installed. The other two fixtures, NightMate and BathMate, were intended for use only during evening and night-time, and had no circadian profile settings.

3.2. Experimental Set-up

The total experiment period took place from the 5th until the 29th of April 2022. Baseline study, prior to the implementation to the new ILD solution was carried through from the 5th to the 8th of April 2022 both in Aarhus and Copenhagen. After the baseline study, the new ILD solution was implemented, and the intervention study took place from the 26th to the 29th of April 2022. The new lighting was implemented 18 days before the intervention study in Aarhus and for nine days before the intervention study in Copenhagen. Five residents in Copenhagen and five in Aarhus participated in the study, as well as four healthcare workers from Copenhagen and six from Aarhus.

For the intervention period, SolMates were implemented in two resident apartments in Aarhus and Copenhagen, resulting in four apartments in total. Two staff rooms in Aarhus and one staff room in Copenhagen were equipped with SolMates, as well as one common area per each city. The number and placement of the SolMates varied in each space. NightMate was implemented in three apartments in Aarhus and two apartments in Copenhagen, the placement of the fixture being in close proximity to the bed. BathMates were installed in four toilets in total, with two in Aarhus and two in Copenhagen. Overall, the ILD solution was implemented minimum of one and maximum of four rooms per floor. In detail set-up of the study has been documented, however excluded due to the length of the paper.

3.2.1. Quantitative Methods. The mEDI was always measured from the field of view of the observer, with a 180° hemisphere sensor, the main direction of gaze being the angle 0 [20]. The measurement device was a spectrometer by GL Spectis (1.0 touch + flicker). In case of seated position, the measurements were taken from the vertical plane in the height of 1.2m, and in the case of a standing, from the vertical plane in the height of 1.6m [7]. The measurements for the scenario of a resident laying in a bed were taken from the height of 1.2m in an angle of 45° during daytime, referring to a typical view angle from a raised hospital bed [24]. In the night-time, the measurements from the beds were taken from the horizontal plane facing upwards in the height of 1.0m. In the apartments where SolMate fixtures were installed, the measurements were taken five times a day (at 9am, 12am, 3pm, 6pm and 9pm). In the baseline period the measurements were taken with daylight only and with daylight and the house lighting combined. In the intervention period, the lighting was measured both with daylight only and with daylight and the new lighting combined. In the apartments, where the ILD solution consisted only of fixtures intended to be used at night, the mEDI value was recorded only at 9pm, once with the luminaire turned off and once with the luminaire turned on. Task lighting was measured as illuminance on relevant task areas, using Voltcraft lux meter MS-200. All the task lighting measurements were taken at 9pm, as it was in the interest of the case to provide task lighting at evening and night. For the measurement, all the other light sources were turned off, except for the fixtures implemented in the ILD solution. With the light fixtures that were dimmable, the value was recorded both with dimmed down and dimmed up settings.

3.2.2. Qualitative Methods. For the qualitative research, semi-structured interviews were conducted, using card sorting as a projective technique. The interviews were based on questionnaires specific to the topics and target groups addressed. SurveyXact [22] as a data collection platform was utilized to record the interviews. For the card sorting, ten cards were employed in which the ratio between positively and negatively perceived words was balanced. The selected cards were based on previous studies [4,11] on the perception of light in relation to the feeling of homeliness. The chosen cards read as follows: glary, soft, cosy, bright, dark, sharp, sterile, sufficient and insufficient. Each interview lasted 10-20 minutes. All the ten resident participants were planned to be interviewed, but in the test period only six residents were found to be cognitively fit to participate in the interviews. Card sorting could not be applied to any of the resident interviews. In the test period, nine daytime and one night-time healthcare workers were interviewed. In cases of resident interviews in which the resident was estimated to be potentially agitated by the interview situation, a healthcare worker was present to monitor the interview situation. In the cases where the healthcare workers could not be interviewed in person, a questionnaire was sent for them, including the same questions that were included in the interviews. In total four healthcare workers filled out the questionnaire online. In addition to measurements and interviews conducted, observations of the placement of the participants (where and how they were positioned in the room), and light conditions in the space were noted down each time the researchers entered a space to conduct a study. The distances from the observer to the luminaires were measured with a measurement tape, and the direction of gaze and placements of luminaires were noted in a floor plan sketch. Observations of the daylight inflow were noted in text, describing the weather condition, and whether direct sunlight was present in the space. Additionally, for the intervention period, if other lights than the ILD solution were used simultaneously, they were registered in the observations. Additional observations of the interview situations were made regarding the quality of the interview, noting down whether the used techniques seemed to confound the participants or if they were understood in a way appropriate for the research.

3.3. Integration of the Results

Detailed results of the evaluation process have been collected and analysed in the case study, and the summary of the results within each topic is presented in **Table 1**. The summaries are provided both as numerical values regarding how many of the measurement positions reached the target values, and as number and tendency of responses per each interview question. The target values referred to are based on the tender material provided by the Municipalities [23].

The inspection of the summarized results reveals correlations, differences, and contradictions in the data, both within the topics and across. A contradiction within a topic can be found in the results regarding circadian rhythm of the residents, where the measurements point mostly towards positive values, while the interview data shows no positive change. Therefore, the result on the topic of circadian rhythm is inconclusive regarding the residents. The effect of light on the circadian rhythm of staff is as well inconclusive since no qualitative data was collected.

IOP Conf. Series: Earth and Environmental Science 1099 (2022) 012028 doi:10.1088/1755-1315/1099/1/012028

<table>
<tr><th></th><th></th><th>Circadian Rhythm
Is circadian rhythm supported by the ILD solution?</th><th>Task Lighting
Does the ILD solution provide sufficient light for tasks at night-time?</th><th colspan="2">Homeliness
Does the ILD solution support the feeling of homeliness?</th></tr>
<tr><td rowspan="2">Residents</td><td>Quantitative Results</td><td>Daytime avg. >250 mEDI:
Apartments: 3/4 positions
Common areas: 0/2 positions

Night-time avg. <3mEDI:
Apartments: 6/6 positions
Bathrooms: 1/2 positions</td><td>Numerical target values for the task lighting for residents were not defined in the project.</td><td rowspan="2">Qualitative Results</td><td rowspan="2">Fixture design:
2/6 positive feedback
2/6 unclear
2/6 negative feedback

Experience of the new light:
0/4 positive experience
4/4 negative experience

Willingness to keep the new luminaires:
0/5 willing to keep
3/5 don't know
2/5 not willing to keep</td></tr>
<tr><td>Qualitative Results</td><td>Experience of good night sleep:
0/4 improvement
3/4 no change
1/4 deterioration

Waking up at night:
0/7 improvement
5/7 no change
2/7 deterioration

Wandering at night:
1/1 no change</td><td>Residents for whom bathroom lighting was installed did not use bathroom at nights and therefore could not response the questions.</td></tr>
<tr><td rowspan="2">Healthcare workers</td><td>Quantitative Results</td><td>Daytime avg. >250 mEDI:
Staff rooms: 2/5 positions

Night-time avg. <3mEDI:
Staff toilets: 1/2 positions
Staff rooms: 5/5 positions</td><td>>5lx for minor tasks:
Resident beds: 3/4 positions
Staff rooms: 5/5 positions

>100lx for major tasks:
Resident beds: 0/4 positions
Staff rooms: 2/5 positions

>10lx for tasks in bathroom:
Resident toilets: 0/2 positions</td><td rowspan="2">Qualitative Results</td><td rowspan="2">Fixture design:
4/9 positive feedback
4/9 unclear
1/9 negative feedback

Experience of the new light:
1/7 positive experience
4/9 unclear
2/7 negative experience</td></tr>
<tr><td>Qualitative Results</td><td>Interviews regarding the effect of the light on the circadian rhythm of the staff were not conducted.</td><td>Perceived as sufficient. However, the respondent referred to all the apartments in general, and not specifically to the ones with the new ILD solution. It could not be confirmed, how the new lighting solution was perceived.</td></tr>
</table>

Table 1. Joint display of summarized results

The photopic lux measurements on the relevant task areas indicate, that lighting was partially sufficient during night-time for the healthcare workers to carry out healthcare tasks. The qualitative data is inconclusive, and therefore cannot support nor contradict the quantitative results. The integrated results on the topic of task lighting for the residents are inconclusive, since the participating residents did not have need for task lighting at night, and target values for the task lighting for the elderly had not been set.

The topic of homeliness was only ever planned to be studied with qualitative methods, and therefore the conclusion is drawn based on results from the qualitative methods alone. There is no indication in the results that the ILD solution would have supported a feeling of homeliness for the residents. In contrast, the healthcare workers expressed mostly positive feelings towards the new lighting and light fixtures. The interviews indicate, that the ILD solution was supporting the feeling of homeliness for the healthcare workers.

4. Discussion

In this chapter it will be discussed whether the study succeeded in meeting the set criteria of *(1) integration, (2) flexibility* and *(3) validity*.

The intentional **integration** of the results in different test phases, methods and topics was partly achieved. Regarding the integration across topics, as an example, the results on the topic of homeliness can be used to reflect upon the results on circadian rhythm and task lighting. For instance, the measured illuminance levels for sufficient task lighting were not met at night for care tasks for the staff, according to the target values set by the municipalities [23]. Similarly, a citizen expressed discomfort of the new lighting, experiencing too low light levels in the evenings. A contradiction and therefore conflict in the results can be found between the measurements of circadian rhythm (mEDI values) for healthcare workers at night-time, where the ILD solution supports the circadian rhythm of the workers, and the task light at night-time, not being sufficient for carrying out the tasks, when the ILD solution runs on the predefined setting. Looking into the integration within one topic a challenge was met regarding the comparability between the quantitative and qualitative data and results. A notable example of this is the data on night-time task lighting for the elderly, where questions were asked about the usage of the bathroom, but no numerical target values for this task or other night-time tasks for the elderly were defined and thus measured. This illustrates the importance of ensuring that comparability between the different methods is maintained in later phases of the study. Successful comparable data from the quantitative and qualitative methods were, however, obtained with regard to the circadian rhythms of the elderly. The interview questions referred to all times of the day and the measurements were taken throughout the day. A deficiency can be found in the comparability regarding the results for the circadian rhythm of elderly in relation to the baseline and intervention periods. The analysed mEDI measurements, due to time restrictions in the analysis process, do not reveal how the light has been before the implementation of the new lighting, while the interview results are based on the change between the baseline and intervention periods.

Having found contradictions in the results, leads to a search for potential explanations. The reason for contradictions might lie in the set target values, methods used for evaluation, or in the lighting design solution itself, such as the programmed light curve over the day. Some of the limitations and challenges mentioned might have been avoided through a simplified evaluation design, for example by focusing the study on a single topic and fewer research questions. With a single focus topic, resources could have been directed more efficiently towards refining the implementation of the chosen method within the short study period, rather than spreading the researchers' attention across multiple topics. If the research had been based solely on one method or topic, however, many opportunities for improvement would have remained undiscovered. Developing methods for transdisciplinary research is commendable, because finding inconsistencies between data sets shows that the design may be functional in one aspect but undesirable in another.

Another option to improve the evaluation design can be an extended implementation period. An indication of an unfavourable design emerged, for example, in the resident interviews on the topic of homeliness, where several residents expressed their discontent with the new design. A plausible explanation for this could be the very short implementation period, as the healthcare staff pointed out that new things are often difficult for the elderly to accept. Therefore, the result led to looking for ways to improve both the design and the evaluation methods.

In terms of **flexibility** by following a mixed-methods approach, the simultaneous collection of quantitative and qualitative data disrupts the everyday life of the participants for a shorter period than conducting several independent studies. It also allows the number of samples to be reduced as they can be validated with each other in the process of integration. In this study, it was vital to evaluate the lighting situation in the real settings, where the citizens were exposed to the lighting, e.g., measurements were taken from the positions the residents seated themselves. This led to a more open approach comparing different settings on the same topic with each other and allowed a greater sample size.

Another hurdle is the number and scope of interviews that could be collected with the residents. Supported by Schledermann [10], this depends on the condition of the residents and can lead to incomplete data sets under certain circumstances. The chosen method allowed to combine data sets of different residents with the same profile. In addition, inspired by the same study, data on the residents could be collected from their contact persons instead, if the resident was not capable on answering the questions themselves, due to cognitive or other limitations. The interviews with the contact persons of the residents were also essential in cases where the resident was able to be interviewed, as this allowed for further validation of the collected data. For example, in an interview a resident expressed having woken up often at night in the study period, and this information was confirmed by the same statement made by their contact person. The collection of valid data is further supported using a mixed-methods approach, in that a baseline interview on the effect of the old lighting on circadian rhythms was compared with the effect of the new ILD solution. The methods chosen also allowed to collect data regardless of the heterogeneous schedules of the health care workers. As the interviews are based on questionnaires, they could be sent online when necessary, as it occurred during the baseline study in Aarhus. The flexibility of the application led to more data and therefore more validity in the overall study. In some cases, losing participants compromised the validity of the results. This happened in the evaluation of the night-time task lighting for the healthcare workers, when only one night-time worker was reached to answer the questionnaire.

The **validity** of the study was ensured by organising the study into the specific research design of a case study as a framework and the convergent mixed methods approach for the core design. The evaluation design is based on state-of-the-art practice and the chosen methods were embedded in theory, the origins of which have been explained in the introduction. Points for further research were noted for some methods regarding their adaptability from theory to real-life situation, although particular attention was paid to the workflow for interviewing techniques, including card sorting, and the rigorous measurement of the mEDI values. During data collection, some situations arose that led to compromises in the implementation of the methods, as illustrated in the following example regarding the measurements of mEDI values. While the theory states that the mEDI is measured from the observer's field of view, it was found that in practice it can be difficult to determine the participant's field of view. Some participants were constantly on the move from one space to another, while others had severe physical restrictions, yet the measurement technique was the same for all. One to three most relevant field-of-view positions were established for each participant, on which the measurements were based. It was found that some measurement positions reflected the participant's actual light exposure well, while in some cases the need for a more accurate method of estimating a particular participant's light exposure in future work became apparent. Other methods that can help with more accurate measurement of mEDI over the course of a period, such as the use of a wearable sensor, can also be considered when evaluating the ILD solution.

As discussed earlier, the evaluation of integration of results shows that it can add to knowledge and increase the validity of the study. However, in this study, discrepancies occasionally occurred that reduced the validity of the results by excluding the possibility of integration. The validity of a study, though, can be judged not only by the results, but also by the possibility of replication. Through careful planning and documentation, similar studies can be repeated. However, due to the special characteristics of a case study in elderly homes with numerous changing factors, such as uncontrollable use of the spaces and artificial lighting by the participants and thus light exposure, it can never be set up exactly the same. This should not be seen as a disadvantage, as case studies, mentioned in the introduction, can provide rich data and new knowledge from real-life environments.

5. Conclusion

In conclusion, through the rigorous collection and evaluation of quantitative, as well as qualitative data in response to the research questions posed on circadian rhythm, task lighting and homeliness, this approach allowed the three topics of biological, visual, and atmospheric effects of the new ILD solution to be combined and evaluated in integration. The joint display demonstrated correlations, differences, and contradictions in the results of the quantitative and qualitative data collected.

The way this study was developed and executed, incorporating the convergent mixed methods design, addressed the needs and challenges involved in conducting the study in the specific context of nursing homes. This approach was framed in relevant theory and therefore took multiple criteria into consideration to demonstrate the validity and quality of the study. The mixed methods approach allowed to evaluate an integrative lighting design, to reinforce, understand and question the collected data and to further develop and adapt the methods in the process to the challenges.

It was found that a mixed methods approach can lead to answering the posed question and help finding correlations, differences, or contradictions. The results led to new knowledge that can be used for the iterative process of developing the ILD solution and evaluation in the next phases that might later lead to answer the overall research question in this specific case and in other cases related to integrative lighting design.

The formulation of the criteria for *(1) integration*, *(2) flexibility*, and *(3) validity* was valuable in the approach to the study and highlighted the most important issues to be considered in the evaluation design. This paper can potentially contribute to other research by presenting an approach that could be applied to other lighting design cases where more than one factor needs to be addressed and investigated in field studies.

6. Acknowledgments

Thank you to our colleagues from the Lighting Design Research Group at AAU, Mimi Ravn, Georgios Triantafyllidis and Kathrine Schledermann for the support in the conduction of the study and sharing knowledge on the methods of evaluating an integrative lighting design solution. Many thanks also to Anita and Morten Lyhne from Lyhne Design, Anders Lundin from Belid, as well as Lene Vad Jensen and Matilde Rodríguez Gammelgaard from the Municipalities of Copenhagen and Aarhus for the opportunity and support with the project.

References

[1] Lucas RJ, Peirson SN, Berson DM, Brown TM, Cooper HM, Czeisler CA, et al. *Measuring and using light in the melanopsin* age. Trends Neurosci 2014;37:1–9. https://doi.org/10.1016/j.tins.2013.10.004.

[2] European Standards. EN 12464-1 *Light and lighting - Lighting for work places - Part 1: Indoor work places* 2021.

[3] Böhme G. *Architektur und Atmosphäre*. 1. Aufl. München: Wilhelm Fink; 2006.

[4] Hansen EK, Bjørner T, Xylakis E, Pajuste M. *An experiment of double dynamic lighting in an office responding to sky and daylight: Perceived effects on comfort, atmosphere and work engagement*. Indoor Built Environ 2021:1420326X2199119. https://doi.org/10.1177/1420326X21991198.

[5] CIE. *Position Statement on Non-Visual Effects of Light - Recommending Proper Light at the Proper Time*, 2nd edition 2019.

[6] Ellen K. H, Mullins M. *Lighting Design: Toward a synthesis of science, media technology and architecture*. In: Thompson EM, Education in Computer Aided Architectural Design in Europe, Northumbria University, editors. eCAADe, Northumbira University, Newcastle: 2014, p. 613–20.

[7] Brown TM, Brainard GC, Cajochen C, Czeisler CA, Hanifin JP, Lockley SW, et al. *Recommendations for daytime, evening, and nighttime indoor light exposure to best support*

physiology, sleep, and wakefulness in healthy adults. PLOS Biol 2022;20:e3001571. https://doi.org/10.1371/journal.pbio.3001571.
[8] Aarhus Kommune og Københavns Kommune. *Kontraktbilag 1: Behovs- og Procesbeskrivelse, Innovationspartnerskab n.d.*
[9] Lyhne Design. *SolMate*. 2020.
[10] Schledermann KM, Mullins MF, Sjörgen M, Bech-Larsen P, Flyvholm A. *Implementing Circadian Rhythm Lighting – Two Health-Care Case Studies*. vol. 8, Kunnskapssenteret St. Olavs Hospital, Trondheim, Norway: SINTEF Academic Press; 2021, p. 194–202.
[11] Schledermann KM, Bjørner T, Hansen TS. *Danish Nursing Home Staff's Perceived Visual Comfort and Perceived Usefulness of a Circadian Lighting System*. Proc. Conf. Inf. Technol. Soc. Good, Roma Italy: ACM; 2021, p. 91–6. https://doi.org/10.1145/3462203.3475881.
[12] Creswell JW, Plano Clark VL. *Designing and conducting mixed methods research*. Third Edition. Los Angeles: SAGE; 2018.
[13] Albers MJ. *Quantitative Data Analysis—In the Graduate Curriculum*. J Tech Writ Commun 2017;47:215–33. https://doi.org/10.1177/0047281617692067.
[14] Joshi A, Kale S, Chandel S, Pal D. *Likert Scale: Explored and Explained*. Br J Appl Sci Technol 2015;7:396–403. https://doi.org/10.9734/BJAST/2015/14975.
[15] Fàbregues S, Molina-Azorín JF. *Addressing quality in mixed methods research: a review and recommendations for a future agenda*. Qual Quant 2017;51:2847–63. https://doi.org/10.1007/s11135-016-0449-4.
[16] O'Cathain A. *Assessing the Quality of Mixed Methods Research: Toward a Comprehensive Framework*. SAGE Handb. Mix. Methods Soc. Behav. Res., 2455 Teller Road, Thousand Oaks California 91320 United States: SAGE Publications, Inc.; 2010, p. 531–56. https://doi.org/10.4135/9781506335193.n21.
[17] O'cathain A, Murphy E, Nicholl J. *The Quality of Mixed Methods Studies in Health Services Research*. J Health Serv Res Policy 2008;13:92–8. https://doi.org/10.1258/jhsrp.2007.007074.
[18] Gilbert T. *Mixed methods and mixed methodologies: The practical, the technical and the political*. J Res Nurs 2006;11:205–17. https://doi.org/10.1177/1744987106064634.
[19] Houser K, Boyce P, Zeitzer J, Herf M. *Human-centric lighting: Myth, magic or metaphor?* Light Res Technol 2021;53:97–118. https://doi.org/10.1177/1477153520958448.
[20] CIE. CIE S 026/E:2018 CIE *System for Metrology of Optical Radiation for ipRGC-Influenced Responses to Light*. International Commission on Illumination (CIE); 2018. https://doi.org/10.25039/S026.2018.
[21] Bjørner T, editor. *Qualitative methods for consumer research: the value of the qualitative approach in theory and practice*. Kopenhagen: Hans Reitzels Forlag; 2015.
[22] Ramboll. SurveyXact. Olof Palmes Alle 20, DK-8200 Aarhus N, Denmark: Ramboll; n.d.
[23] Aarhus Kommune og Københavns Kommune. *Kontraktbilag 1a: Teknisk grundlag for biologisk lys n.d.*
[24] Babilon S, Beck S, Khanh T. *A field test of a simplified method of estimating circadian stimulus*. Light Res Technol 2021:147715352110446. https://doi.org/10.1177/14771535211044664.

Designing Natural Atmosphere in Office Environment Through Daylight Responsive Dynamic Lighting

Verda Sigura[1,2], Ellen Kathrine Hansen[1], Henrik Clausen[3]

[1] Aalborg University, A. C. Meyers Vænge 15, 2450 Copenhagen, Denmark.
[2] University of Copenhagen, Karen Blixens Plads 16, 2300 Copenhagen, Denmark.
[3] Fagerhults Belysning AB, Åvägen 1, SE-566 80 Habo, Sweden.

E-mail: verda.sigura@jur.ku.dk

Abstract. Often the criteria for designing office lighting is to increase efficiency through a bright and evenly distributed lighting. Yet, research has found that this static lighting leads to an unstimulating, boring, and dull luminous environment that supports neither task focus, nor a pleasant atmosphere. The goal of this paper is to create a closer connection to a natural light environment by designing an office lighting that inspired by light as natural phenomenon creates a lively and more nature-based atmosphere. We developed a lighting design with focus on analysing the perceived atmosphere as a metric for a dynamic lighting design for a meeting room of a real estate company in Denmark. Through a mixed method approach, we explore the design factors that determine the desired natural atmosphere. Our design aims at enhancing the function of the meeting room as a formal space where negotiations are held by 1) presenting the identity of the company to the business partners; and 2) supporting employees' sense of belonging to the company culture. As a result, our lighting design creates a lighting hierarchy in line with the architectural elements of the space, complementing daylight through electric lighting in a dynamic way, accounting for both sky-type and daylight inflow.

1. Introduction

Atmosphere can be defined in many ways, as expressed by architect Peter Zumthor: "I enter a building, see a room, and in the fraction of a second- have this feeling about it." [1]. Pallasmaa [2] argues that atmosphere precedes other perception and exists beyond the Aristotelian five senses. He compares the concept of Genuis loci, spirit of place to atmosphere, as they both encompass the unique identity and characteristics of a space. As human perception has evolved from being outside in ever changing patterns of daylight cycles it is essential to understand these dynamics when we design the indoor atmospheres that seem pleasant. However, it is challenging to meet these atmospheres in an office space, as offices are typically considered as formal and neutral spaces, with the function of providing an effective working environment. Throughout mid-20th century, the aim of office managers was to ensure high visibility to facilitate workers' task performance through bright lighting, with even distribution on a horizontal surface, in the context of a transparent and efficient atmosphere in the workplace [3]. This trend continued until nowadays and a lot of workplace lighting research focuses directly on factors like attention span, focus and other various criteria of productivity. However, due to the COVID-19 pandemic, as the line between homes and offices blurred, some companies that have a human centered work culture begun to seek ways to transform their office spaces via new learnings of this era. We would like to illustrate this trend with a few examples. Research and design lab Space10 renewed their office space to adapt to the "new normal" by creating an office space that is tailored around its users' needs, providing more flexibility, fluidity, and variety [4]. The company managed these changes via reshaping materiality, colours, spatial re-organization and of course light: all the necessary tools to create a desired atmosphere in the workplace. Similarly, design consultancy firm IS IT A BIRD created a homely atmosphere in their Copenhagen office [5].

Although the culture of each company is unique and hence calls for a unique identity catered to its needs, these examples demonstrate that the ideology of designing workplaces based on productivity alone is no longer valid. Both of the aforementioned companies discovered that adopting an oblique approach to productivity by creating a positive atmosphere perception actually allows employees to do their best while boosting their well-being. For corporate offices, as in the case study of this paper, considering the degree of formality that the company culture requires, it might not be plausible to create a totally home-like atmosphere. Yet, it would still be highly valuable to create a more natural luminous environment to inspire and comfort the users of the space with reference to the natural variations we have a preference for. More specifically, dynamic lighting could be the key answer to human need for natural variation, to ensure pleasant atmospheres and well-being. We argue that, despite cultural and individual differences, design solutions inspired by our common evolutionary traits that define our connection to nature, lead to a more universal perception of atmosphere. According to Pallasmaa [2], atmospheric qualities stem from a degree of irregularities and vagueness and that a static design, perfect form, and formal cohesion work against the creation of atmosphere. In his writings, this notion also connects to the idea that dynamic spaces and atmosphere create each other, which, intuitively supports the use of dynamic lighting to create atmospheres. Building on the above, for the context of this study we explore the following questions: what does the atmosphere of an office space communicate? Which acts does it invite its occupants to? How do we create an atmosphere where people enjoy spending time and working? What are the lighting qualities of an inviting and pleasant workplace?

In this paper, we aim to design an indoor office lighting that creates a lively and natural atmosphere-similar to the feeling of lighting associated with sunlight filtering through the trees under a blue sky. A sustainable and innovative lighting solution for meeting rooms is proposed, with a lighting design that creates a lighting hierarchy in line with the architectural elements of the space and complements daylight through electric lighting in a dynamic way while taking sky-type and daylight inflow into consideration. The proposed solution is validated through solid methodology, simulations and initial feedback received upon presentation to the case company.

2. Methods

A mixed-method approach [6]that combines qualitative and quantitative methods has been used for this design project. First, a user survey provided input at the research stage. The "Perceived Atmosphere Questionnaire" developed by Vogels[7] has been used as a basis to understand the affective evaluation of the space by the users. To achieve a more comprehensive picture, the complete survey has been designed to start with an open-ended question providing qualitative data on participants' general evaluation of the lighting in the space, followed by modified "Perceived Atmosphere Questionnaire" and "Visual Appearance Questionnaire" (with identifying terms that fit the evaluation of an office space). From the 38 atmosphere terms [7] mentioned, 12 were used for this study: cosy, motivating, personal, intimate, formal, stimulating, relaxed, lively, detached, boring, clinical, tense. The visual appearance terms have been selected among the ones previously used in Double Dynamic Lighting Research [8]. The following 9 terms have been used to understand visual appearance of lighting: comfortable, sufficient, task-focused, natural, contrasting, dim, bright, glary, disturbing. the participants rated each adjective's applicability on a 7-point Likert scale: 1- Not applicable at all; 7- Very applicable. Due to Covid-19 restrictions, only 6 out of 25 employees completed the survey. The qualitative side of the user research has been supported by two semi-structured interviews, one with a user of the space (in person interview), the other with the Business Area Manager of the company that is responsible of the administration of the building where the office in case is located (phone interview). Second, researchers' own observations of atmosphere, visual appearance of light and architectural elements in the space have been documented. Third, the results of Double Dynamic Lighting Research have been investigated [8–10] and those that fit the current project (described under section 4.1. Design Criteria), have been transferred. Fourth, light measurements, sun path analysis and DIALux Simulations have been conducted both to understand the daylight availability and the movement of daylight throughout various seasons and to test researchers' own design ideas in connection to lighting standards. Finally, the three elements of lighting [11] – Focal glow, Ambient luminescence, and Play of brilliants – have been used as a design tool.

3. Analysis of the space

The object of our study is the biggest meeting room of the office of Freja Ejendomme – a real estate company owned by the Danish State. The office is located on the 15th floor of a 19-storey building called Codanhus - the tallest commercial office building in Frederiksberg, Copenhagen in Denmark. The Meeting Room 1 has three west oriented windows.

Figure 1. Current lighting Freja Ejendomme- Meeting Room 1 -04.05.2021.

3.1. Light Measurements

In order to analyse the lighting situation of the space, two consecutive measurements have been conducted: first, only daylight; second, only current electric lighting. For all the lux measurements, Testo 540 luxmeter (uncalibrated) has been used. The measurements revealed that according to EU Standards [12], daylight alone does not provide enough light in the room, especially on the task area (horizontally) and for lighting of faces (vertically at 1.2m- Eye level for seated person) of users sitting on the side of the table, with their back and side facing windows. The light levels provided with electric lighting alone is mostly inadequate both for task lighting (horizontally) and lighting of faces (vertically at 1.2m-- Eye level for seated person). The light distribution is uneven as well. In Copenhagen, these are especially problematic during winter, as sunset can start as early as 15:30pm.

3.2. Materials

For diffuse materials, reflectance (ρ) was estimated using a Testo 540 Luxmeter. Illuminance has been measured 15 cm distance from the material's surface, first toward the surface (reflected light) and after away from it (surface incoming light) as done in previous DDL studies [10] For non-diffuse materials, such as glossy and specular, this method cannot be used and due to the lack of access to measuring device such as a gloss meter, a visual method has been used to classify the reflectance type of these materials. From different angles, photos of the materials have been made and the presence of a directional distribution of light has been assessed. The reflectance of ceiling paint (0.8), most of the wall areas (0.8 and 0.7 for the beige and white walls respectively) are high values that increase the light levels in the space by reflecting most of the incoming light. On the other hand, the reflectance of dark grey painted wall (0.3) and mixed grey carpet (0.1) are low values that decrease the light levels by absorbing most of the incoming light. The glossy beige laminate table, glossy white windowsill paint and specular light grey ceiling panels are also high in reflectance whereas, the glossy dark grey fabric of the curtains has a low reflectance. The dark red/brown colour of chairs and medium brown colour of wood plant pots contribute to creating warmer reflections of light in an area mostly designed with muted/cold colours.

3.3. Windows

In order to understand the properties of daylight input into the room, transmission of the window glass has been investigated. To understand the transmission and coating of the glass, illuminance, the spectral power distribution, and CRI of the daylight has been measured first pointing spectrometer directly to daylight with windows open, second touching the spectrometer to the glass with windows closed (Table 1). The measurements have been conducted with the spectrometer: AsenseTek Lighting Passport (ALP-01) (Uncalibrated). The interview with the building administration informed us that the windows have been covered with a film in order to reduce the heat inside the building. This kind of filter is known to filter the long wavelengths (red and infrared parts of the spectrum). The windows are double-glazed, and a neutral colour of glass has been used. The glass, accordingly, absorbs some light: the ratio between the direct daylight illuminance and the daylight illuminance from glass is 2000/4300 = 0.5. Furthermore, the comparison of Spectral Power Distributions (Figure 2) shows that the glass absorbs light at some of the shorter (violet-blue) and some of the longer (yellow-red) wavelengths and lets the mid length (blue-green) wavelengths pass.

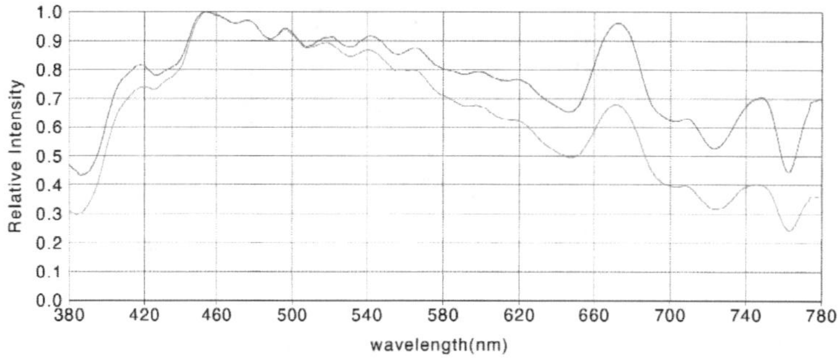

Figure 2. Comparison of SPD for Direct Daylight (blue) and Daylight through Glass (orange).

Table. 1. Comparison of Daylight Qualities- Direct and through Glass.

	CRI (Ra)	CCT	Illuminance
Direct Daylight	100	6300K	4300lux
Daylight through Glass	95	7200K	2000lux

3.4. Sun Path

The 3D sun path diagrams, using the tool by Marsh [13], have been made for the following three dates: December 21, 2021- Winter Solstice, the shortest day with sunrise at 08:37, sunset at 15:38 (Total daylight hours: 07:01); March 20, 2021- Vernal Equinox, day and night are of equal length with sunrise at 06:12, sunset at 18:23 (Total daylight hours: 12:11); May 4, 2021- The day that researcher has physically been to the space to do the light measurements and observations with sunrise at 04:20, sunset at 19:53 (Total daylight hours: 15:33) The diagrams (not reported here) demonstrated that the Meeting Room 1 with a west façade is illuminated by the sun only during the afternoon. It also points out to the stark differences in sun elevation and total daylight hours throughout the seasons as a characteristic of Scandinavia.

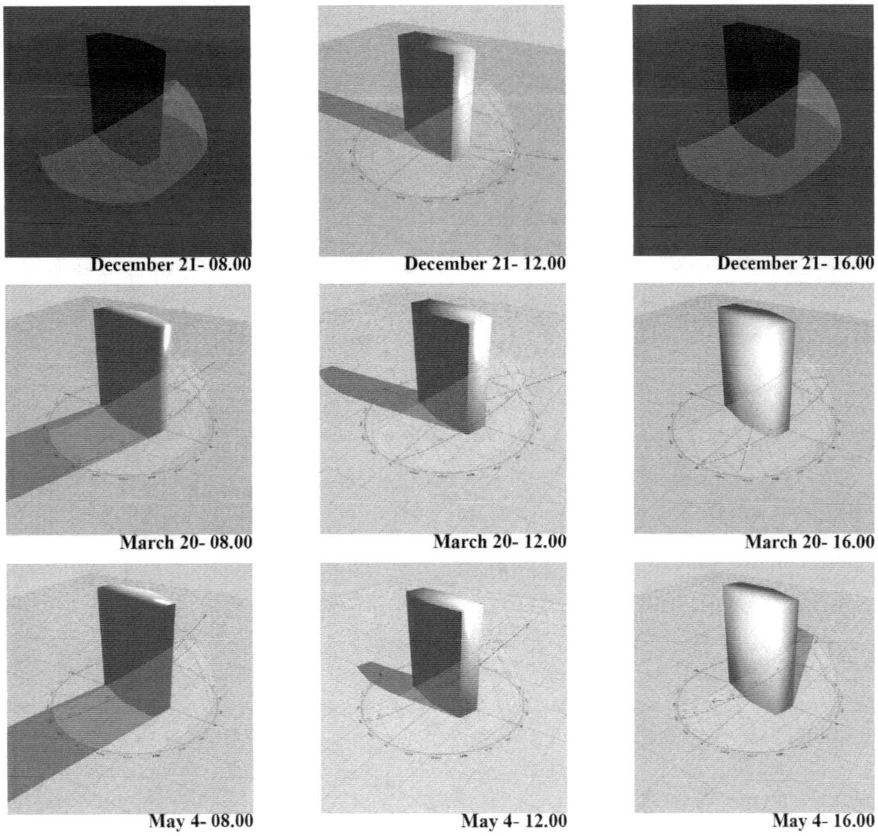

Figure 3. Sunpath Diagrams for Codanhus building west façade.

3.5. Daylight Calculations

Each window has a glass area of 143 cm by 136 cm and total glass area of all three windows equals to 18% of the floor area. However, as stated in BR18, there is a need to adjust the glass area for any reduced light transmittance [14]. As mentioned earlier the visible light transmittance has been calculated as 0.5, which leads to a readjusted glass area/floor area ratio of 8.3%, which is less than minimum 10% required to document sufficient access to daylight by BR18 [14]. The daylight factor has also been calculated through DIALux Software and has been found to be 0.8 %. Tregenza & Wilson [15] argue that when average daylight factor from side windows in an office setting is around 1%, the room has a gloomy appearance and there is a harsh contrast with the view outside. In turn, complementary electric lighting during daylight hours can help to balance daylight variations. Furthermore, as the space is facing west, even when it gets sunlight in the afternoon, the light modelling values are not sufficient. In order to have a good understanding of when the light requirements are not met with daylight alone, daylight simulations have been run in DIALux Software and it has been found that there is a need for electric light in all cases. All the simulations have been run for 2020 (December 21, 2021- Winter Solstice) and 2021 (March 20, 2021- Vernal Equinox) & May 4, 2021- The day that researcher has physically been to the space to do the light measurements and observations). For May 4, both overcast and clear sky situations have been run in simulations but for March 20 and December 21, only the overcast sky scenarios have been run as it corresponds better with the probability for overcast sky (70% and 80% respectively for those dates in Copenhagen) [16][17]

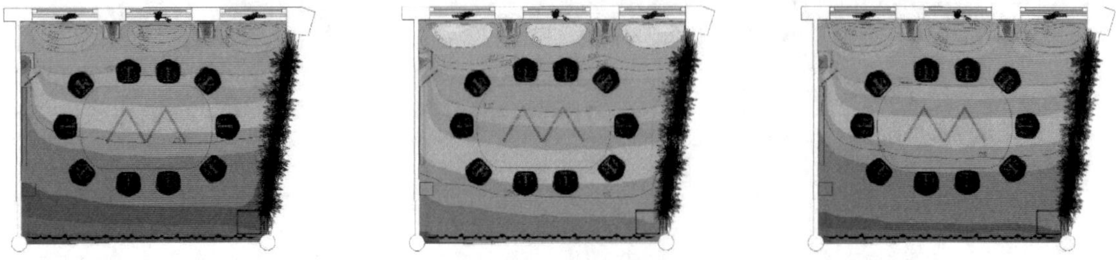

March 20- Morning (Overcast Sky)
Unmet light requirements:
No requirement is met

March 20- Noon (Overcast Sky)
Unmet light requirements:
Grey wall, area of activity: standing presenter, windows wall, visual task area and immediate surroundings

March 20- Afternoon (Overcast Sky)
Unmet light requirements:
All room main surfaces except green wall, area of activity: standing presenter, visual task area and immediate

Figure 4. Illumination simulations (Lux levels) for daylight availability for the Meeting Room 1 at 0.8m height (Simulation for March 20- Overcast Sky is demonstrated as a sample from all simulations conducted)

3.6. User Survey Results & Interviews
The survey results for "Visual Appearance" and "Perceived Atmosphere" are illustrated in Table 2. The number corresponding to each term is the average score it received on a 7-point Likert scale: 1- Not applicable at all; 7- Very applicable.

Table. 2. Average scores for Visual Appearance and Perceived Atmosphere metrics.

Visual Appearance		**Perceived Atmosphere**	
Natural	4,0	Boring	4,7
Comfortable	3,5	Relaxed	4,7
Contrasting	3,2	Cosy	4,3
Sufficient	3,0	Formal	4,2
Task-focused	2,7	Clinical	3,8
Disturbing	2,7	Personal	3,3
Glary	2,6	Stimulating	3,2
Bright	2,5	Motivating	3,0
Dim	2,3	Detached	3,0
		Intimate	2,8
		Lively	2,7
		Tense	2,7

Survey results and interviews show that the current lighting solution is insufficient and not enough task-focused. This is marked by the high scores on "Relaxed" and "Boring", and the low scores on "Stimulating", "Lively", and "Motivating". During the open-ended question, the participants restated this assessment. Participant 6 described the lighting as "maybe too relaxed", Participant 5 expressed the overall lighting experience as "Not good for working, Often very dark. A bit cold". Finally, Participant 3 described it as "A bit dull/sleepy, especially during winter". The user of the space that has been working at Freja Ejendomme for 8 years – in the semi-structured interview – described his perception of the current lighting as "lazy, dull, and sleepy"; adding that they would like "more light". He mentioned that the company tried to replace the light bulbs in ceiling panels in three meeting rooms,

IOP Conf. Series: Earth and Environmental Science 1099 (2022) 012029 doi:10.1088/1755-1315/1099/1/012029

keeping the lumen output constant but changing the colour temperature, to see if this would give the impression of "more light" but the general perception is that it hasn't been a success. The discomfort is exacerbated by the lack of manual controls that would allow the users to regulate the lighting.

4. Design

The previous analysis has shown that the current lighting in the space fails at being stimulating, task-focused, motivating and lively. As such, our goal is to design a task-focused, motivating and lively lighting that complements architectural elements in the space. At the same time, we strive to enhance the function of the meeting room in two ways: 1) by representing the company identity to the business partners while negotiations are held during meetings; 2) by supporting employees' sense of belonging to the company culture.

4.1. Design Criteria

Delving into outcomes of most recent double dynamic lighting research [8], the following criteria have been adapted for this project: 1) Neutral ambient lighting of less than 5000K for overcast sky and cooler ambient lighting of more than 5000K for clear sky; 2) Combination of direct warm light with diffuse cool or neutral light to achieve a natural feeling [8–10]. In fact, having some degree of direct lighting, as opposed to only diffuse lighting, increases socialization in the space [10]. However, a balance is needed: too little direct light causes a loss of details whereas too much causes harsh shadows [9].

To adhere with standards accepted in Denmark, the lighting in meeting rooms should provide a (average) maintained illuminance value of 200lux for visual task area [12] , Unified Glare Rating should not exceed 19, uniformity should not fall below 0.6 and the CRI (Ra) should be minimum 80 and the lighting should be adjustable. In addition, (average) maintained illuminance value should be higher than 50lux on the walls and 30lux on the ceiling, both having a uniformity of at least 0.1 [18] Although, the Danish Annex to European Standard EN 12464–1 deems 200lux acceptable for visual task area in meeting rooms, to provide a better visual environment, the values in EU Standard are taken as design criteria: minimum 500 lx level on the task area and 300lux in the immediate surroundings [18]. Standards don't specify rules on illuminance for peripheral surroundings yet, according to our investigations, it is suggested to have at least a third of the illuminance of immediate surroundings, for peripheral surrounding, which in this case is 100lux [19]. For the purpose of this project, the definitions of all three areas are as follows: task area- 70cm inward from the edges of the table, all around; the immediate surroundings- 50cm outward from the edges of the table; peripheral surroundings- starts at a distance of 0.5m from the room's walls. In addition, as the space is a meeting room, there is a need to calculate mean cylindrical illuminance at 1.2m above floor level- for people seated around the meeting table to enable good visibility of faces. To ensure good light modelling, the ratio of the cylindrical illuminance to horizontal illuminance at any point (at 1.2m is a good height) is advised to be between 0.3 and 0.6 [10] [18].

4.2. Design Concept

4.2.1. Overview. Our design is guided by Kelly's [11] three elements of lighting- Focal glow, Ambient luminescence, and Play of brilliants. The dynamic aspect of the design is achieved through matching the colour temperature of diffuse component of electric lighting to the sky condition. As the sensor technology for programming dynamic lighting reacting to sky condition automatically is not yet commercially available, the proposal for implementation of the lighting solution has been formulated as follows: one person in the company is in charge of controlling the lighting through an application on a tablet located in the meeting room. Programming of the lighting is done via DALI lighting control system. Diffuse lighting is programmed to be 5500K for clear sky and 4000K for overcast sky condition. Direct lighting is programmed to be 3000K for both scenarios. For no daylight situation, the same scenario as overcast sky automatically applies, without need for choice by users.
The room is divided into 3 light zones: Light zone A- Meeting table, Light zone B- Green wall, Light

zone C- Gray wall. Aforementioned CCT changes on light zones have been done manually in the point-in-time simulations and are presented in Figures 5 and 6. These light zones can be controlled independently from each other, in order to offer maximal flexibility. For instance, during a presentation using the screen, users can turn off Light zones B and C to decrease general light levels and provide focus on the screen while keeping Light zone A brighter to meet the requirements of reading and writing. On daily use, a light level sensor automatically dims the light according to daylight availability and ensures light levels described in the criteria are met. The sensor also functions as a presence sensor and turn the lights automatically on upon detection of users.

Table. 3. Summary of overall design concept.

Design Challenge	Design Intention	Design Solution
The lighting in the space is described as not task focused by users.	How to support primary function (task) of the space: holding meetings?	Create Focal glow via a direct and warm light forming a pool of light around the meeting table.
As the meeting room is located on 15th floor of a high-rise building, it doesn't get any light reflected from the ground to the ceiling inside the space. The light comes from above but floor at the space doesn't reflect it due to dark grey/black carpets that absorb the light.	How to make a feeling of sky light in the space? How to create a lighting that will accentuate/complement the contrast between the grey carpet and white ceiling?	Create Ambient luminescence by lighting up the ceiling complementing the sky type in terms of CCT.
The wall that the screen is mounted on is painted grey which absorbs the light and as the space is facing west, this wall gets the least daylight.	How to light the grey wall so that the tone of grey is perceived beautifully, instead of trying to go against the grey (dark colour) by trying to make it bright?	Create Ambient luminescence by amplifying the (blue) undertones of grey.
The lighting in the space is described as unstimulating, boring, demotivating and dull by users.	How to make a lighting design that excites the optic nerves and creates interest but doesn't harm the plant growth?	Create visual interest and increase alertness of users by Play of brilliants, lighting up the plant wall in a natural way so that it comes to life as a decoration element/piece of art, amplifying the green colour with warm light reminiscent of sun rays.

4.2.2. Development. Our design is inspired by light as natural phenomenon. For Light Zone A- Meeting Table, to create a focal glow around the meeting table, we took inspiration from the following manifestations of focal glow: sunset, campfire, moon. This effect has been achieved through use of pendants. For Light Zone B- Green Wall – the ambient lighting – to light the ceiling in a way that emphasizes and complements the architectural features of the space, we took inspiration from natural/artificial skylights. The triangular zones created through the presence/division/absence of ambient lighting on the ceiling are in line with the geometry of the space which has a corner on the side of the green wall. This has been achieved using illuminated ceiling made of textile. For the accent lighting, to create a play of brilliants that has a natural look and feeling, we took inspiration from the concept of "Komorebi" which describes "Sunlight filtering through trees" [20].This effect was achieved with the use of spotlights. For the interplay of direct and diffuse lighting, to create soft shadows of plants, we took inspiration from the shadows of plants, created by natural daylight and sunlight of late afternoon hours. For Light Zone C- Gray Wall, the skylight concept of Light Zone A's diffuse component has been implemented as well. Sketches and inspirational images for each design element per light zones are shown in figure 5. Visualizations of the final lighting plan are shown in figure 6.

IOP Conf. Series: Earth and Environmental Science 1099 (2022) 012029 doi:10.1088/1755-1315/1099/1/012029

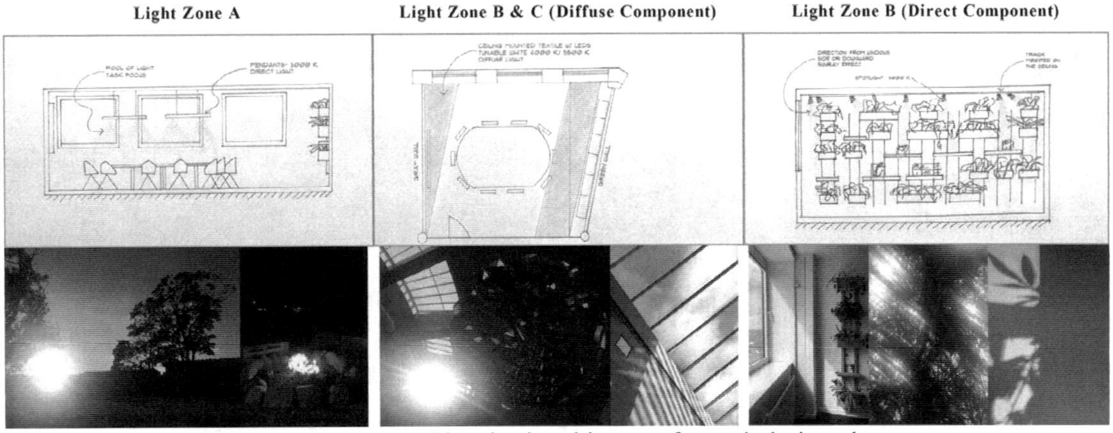

Figure 5. Sketches and inspirational images for each design element.

Figure 6. Visual representation of the Final Lighting Solution for Meeting Room 1- Various sky conditions.

4.2.3. Testing. In order to understand how the lighting design performs according to the aforementioned lighting standards under criteria section, we used DIALux to test varying light scenarios in relation to room main surfaces, visual task areas, areas of activity and glare (UGR). Three main lighting scenarios have been chosen for further investigation: December 21- 08:00 (No daylight), March 20- 12:00 (Overcast sky), May 4- 12:00 (Clear sky). When we assess the values that don't match the target, light modelling at the areas of activity (meeting table) stand outside the target range of [0.3 – 0.6]. Yet, this range is explained to be as a "rough guide" in the standard [18]. For no daylight and overcast sky scenarios, the values are quite close to the target values of 0.29 and 0.66 respectively; while for clear sky condition the value is 0.88, which is due to more diffuse light levels through increased daylight intake. For further work, light modelling qualities can be fine-tuned through site testing. All the UGR levels are below the max value of 19, except one value of 19.1. Figure 7 illustrates light distribution on the workplane height for each light scenario. Overall, we argue that our lighting design meets most criteria successfully.

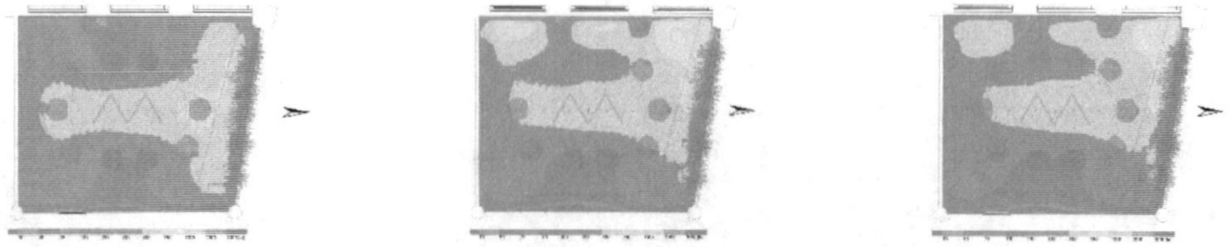

Figure 7. Illuminance simulation of the Final Lighting Solution at 0.75m height; Images from left to right: December 21- 08:00 (No daylight); March 20- 12:00 (Overcast sky); May 4- 12:00 (Clear sky)

5. Discussion

This study followed a design process guided by a combination of qualitative and quantitative methods to address a complex and intangible issue as atmosphere and at the same time to support the function and unique character of the space. The main methods that have been used included: User survey on Perceived Atmosphere and Visual Appearance; semi-structured interviews with a user and building administration; researchers' own personal observations of atmosphere, visual appearance of light and architectural elements in the space; input from the results of Double Dynamic Lighting Research (becoming part of design criteria); testing through simulations for daylight and electric light (to adhere and go beyond EU Office Lighting Standards) and finally Kelly's three elements of lighting- Focal glow, Ambient luminescence, and Play of brilliants, as a design tool. When the users said they wish for "more light" in the space, the usual approach would be just to make existing lighting brighter (i.e., double the lux levels) but previous Double Dynamic Lighting Research already proves that this approach would not actually solve the problem but would just add more flat light. In connection following criteria from Double Dynamic Lighting Research has been put in place: neutral ambient lighting of less than 5000K for overcast sky and cooler ambient lighting of more than 5000K for clear sky and a combination of direct warm light with diffuse cool or neutral light to achieve a natural feeling. From the perspective of a lighting designer, the aim has been to transform the space described as unstimulating, boring, demotivating, and dull by users to a stimulating, lively, motivating and task focused natural luminous environment by using the methods summarized above.

The study has certain limitations. First, due to Covid-19 restrictions, the access to the space has been limited to a duration of one day visit which made it impossible to do mock-ups and site testing. Therefore, there is a definite need for site testing to fine-tune the design, in case of realization. Second, due to difficulties reaching all employees, it was only possible to receive feedback for the user survey for 6 out of 25 employees. Nevertheless, considering each survey result pointed out to some "need to be improved" points that have been supported with interviews and complemented researchers' own observations of the space, a clear picture of the user perspective is believed to be obtained. Third, to make sure the intended atmospheric qualities and functions are accomplished, future work should include receiving feedback from a large sample of users through photorealistic renderings, beyond the initial presentation of the design solution made to the company and in the case of realization of the design concept, conducting a post occupancy user survey. Fourth, future research would benefit from employing climate-based daylight modeling to enrich the input on dynamic nature of daylight. Last but not least, a specific light scenario for daylight transition hours (in the morning and afternoon) hasn't been included in the scope of this study as this is a quite large area of research/design on its own and the season that the study has been conducted, the sunrise and sunset hours have been beyond work schedule hours.

6. Conclusion

This project started with an investigation of complex interplay among factors affecting the perception of atmosphere and pointed out to the effect of cultural and individual differences in perception. We argue that, despite cultural and individual differences, design solutions inspired by our common evolutionary traits that define our connection to nature, lead to a more universal perception of atmosphere. This approach has guided us through the entire process. This study dwelled on perceived atmosphere as a metric for a dynamic lighting design of an indoor environment, with electric lighting complementing daylight, to achieve a natural, lively, and pleasant atmosphere in the space. While making the desired atmosphere, the lighting design solution for this case study should also succeed in creating a lighting hierarchy in line with the architectural elements of the space and complementing daylight through electric lighting in a dynamic way, taking sky-type and seasonal changes in daylight inflow into consideration. It is reasonable to conclude that the solution will work as intended to create a better work environment; first, the design solution has been achieved through a mixed-method approach providing support for the design with anthropological research methods (surveys and interviews), well-established architectural lighting principles and adaptation of criteria resulting from extensive testing through long periods of time and in various office settings in the context of previous Double Dynamic Research Projects and other relevant literature; second, the design solution has been simulated in 3D with real climate and location data across different seasons and times of the day and been tested both visually through renderings and for standards through light calculations.

References

[1] Zumthor Peter 2006 *Atmospheres : architectural environments, surrounding objects* (Basel: Birkhäuser Verlag AG)
[2] Pallasmaa J 2014 Space, place and atmosphere. Emotion and peripherical perception in architectural experience *Lebenswelt. Aesthetics and philosophy of experience.* **0** 230–45
[3] Petty Maile M, Rappaport N, Kleinman K, Kahn A, Fizer N and Kahn A 2008 Scapes 7 School of Constructed Environments Parsons the New School for Design
[4] Anon 2021 Reimagining the Office for 'The New Normal' | SPACE10
[5] Thomsen R Is your office livable?
[6] Creswell J W and PlanoClark V L 2017 *Designing and conducting mixed methods research /* ed J W Creswell and V L Plano Clark (Los Angeles: SAGE)
[7] Vogels I 2008 Atmosphere Metrics: a tool to quantify perceived atmosphere *International Symposium Creating an Atmosphere*
[8] Hansen E K, Bjørner T, Xylakis E and Pajuste M 2021 A field experiment of double dynamic lighting in an office responding to sky and daylight: Perceived effects on comfort, atmosphere, and work engagement. *Indoor and Built Environment* **0** 1–20
[9] Hansen E K and Mathiasen N 2020 Dynamic lighting balancing diffuse and direct light *Proceeding of ARCH19: Building for Better Health - Research and innovation in architecture and urban design for care and health* p 10
[10] Hansen E K, Pajuste M and Xylakis E 2020 Flow of Light: Balancing Directionality and CCT in the Office Environment *LEUKOS - Journal of Illuminating Engineering Society of North America* **00** 1–22
[11] Kelly R 1952 Lighting as an Integral Part of Architecture *College art journal* **12** 24–30
[12] Anon 2015 *Lys og belysning - Belysning ved arbejdspladser - Del 1: Indendørs arbejdspladser = Light and lighting - Lighting of work places -Part 1: Indoor work places.* (Charlottenlund: Dansk Standard)
[13] Anon PD: 3D Sun-Path
[14] Agency H 2018 BR18 Executive order on building regulations **2017**
[15] Tregenza P and Wilson M 2011 *Daylighting, architecture and lighting design.* (New York (NY): Routledge)
[16] Anon Average Weather on March 20 in Copenhagen, Denmark - Weather Spark

[17] Anon Average Weather on December 21 in Copenhagen, Denmark - Weather Spark
[18] Anon 2011 *Lys og belysning - Belysning ved arbejdspladser - Del 1: Indendørs arbejdspladser = Light and lighting - Lighting of work places - Part 1: Indoor work places.* (Charlottenlund: Dansk Standard)
[19] Anon Principles for the working area - Fagerhult (International)
[20] Anon Komorebi - Wiktionary

IOP Conf. Series: Earth and Environmental Science 1099 (2022) 012030 doi:10.1088/1755-1315/1099/1/012030

Colour of reflected light: a determinant of sustainable spatial function

M Bhatia

Msc in Architectural Lighting Design from KTH Royal Institute of Technology (2019); Bachelors in Architecture from Chandigarh College of Architecture (2014), India

Email: mbhatia.ld@gmail.com

Abstract. According to the works of W.R. Ott, we are an 'indoor species', spending on an average, 87% of our time in closed environments [1]. Our long interaction with indoor surroundings, creates the need for such spaces to be sustainable, bringing factors affecting user behaviour and consequent usage patterns under the microscope. This is an empirical study investigating if reflected light carries a component of colour taken from pigmented surfaces and if this colour of reflected light has any influence on usability of a space and its user pattern. The study is conducted in domestic interior environments of a student housing in Stockholm under artificial light conditions that were static and constant. The period of study was limited to weekends to observe the user behaviour in conditions of free will. It is a comparative analysis between two dining spaces having identical architectural and lighting attributes, with colour of walls as the only change in the two study sets. There were two separate user groups of similar age groups, similar needs and similar time schedule. The study includes four stages of observation and data gathering by adopting qualitative and quantitative methodologies. The effects of colour on the perception of space and light and hence usability of the two spaces cannot be negated completely, even if more study is needed to prove the same. The results can be used to question the photometric instruments in their inability to detect such aspects of reflected light. This also helps to underline the need for involvement of a qualified lighting designer in the making of spaces from the beginning till the end for interpreting the effects of the interactions between light and pigmented surfaces. The findings may be considered a stimulus for developing a set of standards for colour and coloured reflected light in a given environment as a measurable parameter of light for sustained usage of indoor environments.

1. Introduction

Since times immemorial, humans have endeavored to create indoor environments in which they can feel comfortable and safe. According to the 'The National Human Activity Pattern Survey (NHAPS) conducted by NEIL E KLEPEIS and W.R. Ott, we are an 'indoor species', spending on an average, 87% of our time in closed environments [1]. It is in the indoor environment, that we live, work, play, and rest in. Compromised indoor environment of a building has a direct effect on the comfort, health and productivity of the occupants [2]. Suffice to say that the quality of the indoor environments created by us, leading to its efficient and sustained use, becomes a matter of concern. From this point of view, the indoor environment quality is directly linked to efficiency and sustainability of the indoor spaces. And hence there is need to study and do continuous research on the factors that affect the quality of these spaces.

The term 'indoor environmental quality' (IEQ) is a domain that encompasses diverse factors like indoor air quality (IAQ), lighting comfort, lighting quality, thermal comfort, acoustics, services, ergonomics and other related ones. The World GBC's Health & Wellbeing Framework lays down its second Principle for a Healthy, Sustainable Built Environment as 'Prioritize comfort for building user'[3]. Its sub principle 2.2 says we should 'Maintain exemplary lighting for occupant wellbeing' through not just daylight but artificial, too - one that is appropriate for space use [3]. Additionally, its sub-principle 2.4 states that 'The built environment can be directly responsible for a range of human health and comfort influencers that are outside traditional building design conscience. Wider comfort indicators include olfactory, ergonomic, and visual comfort.' [3] It also lists colour as one of the many 'Design and operational factors that can produce the additional 'flourish' factor', the others maybe interior design, aesthetics, character, layout and functionality, space, access to views, nature, and greenery.' [3]

It is with reference to above that the author makes a point that artificial light and colour are intrinsic to human perception, visual comfort and hence critical to the sustained use of a space. While the above-mentioned framework lays down lighting as an established determinant of IED and sustainable spatial function through the importance of illumination levels, daylight and glare control, its direct relation with colour is ignored. Most of the green building rating systems around the world and sustainability practice manuals studied by the author have tended to do the same.

This paper aims to present a study of

- The color of light reflected from differently colored surface in two nearly identical indoor environments (similar in character, layout, functionality and space) in a student accommodation.
- Compare the space usage pattern in the two environments.

With the assumption that visual comfort and human perception of a space creates sustained and longer use of the space by a greater number of people, in conditions of free-will, the author will then go on to suggest a co- relation between the above and present a conclusion.

2. Methodology

The study follows both qualitative and quantitative methods of observation and data collection.

2.1. Area and parameters of analysis

The study is carried out analysing shared dining areas of two separate corridors in the student accommodation of Lappis. Two almost identical common dining areas of two separate corridors in the student accommodation of Lappis, Stockholm, Sweden were chosen as the area of study. All the users were students doing post-graduation degrees. To create best opportunity for exercising free will in choosing the duration of time spent in the area, two weekends were chosen to carry out the study. Since the study is related to the effect of artificial light only, the observations were made after 1800hrs on Fridays, (23rd November, 30th November 2018), after 1600hrs on Saturdays and Sundays (24th -25th November, 1st -2nd December 2018).

2.2. Quantitative analysis

As the lighting condition under study is artificial light and hence constant, the measurement of both the spaces were taken on Friday 23rd November, 2018, and not on other days.

2.2.1. Observation. The author and a helper observed and recorded the frequency and duration of usage of people in the two dining areas over the stated study time period.

2.2.2. Measurements. Two types of measurements were carried out. First, the chosen dining area were measured physically and a layout drawing prepared of that. Second, the artificial photometric data measurements that were taken with spectral photometer. Three horizontal measurements and three vertical measurements taken were in each of the dining areas. The results were then compared to find any patterns or consistencies in the measured photometric data. Colour temperature of light, wavelength of light and illuminance were taken as parameters to see if there is a measurable colour difference in the light conditions of the two spaces.

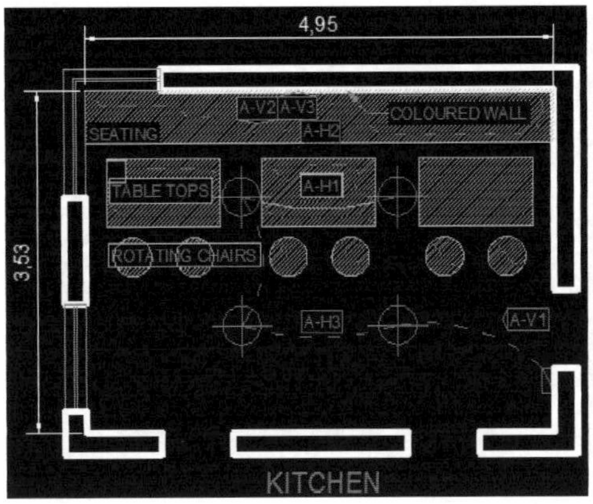

Figure 1. Plan of Dining area 1 with green wall

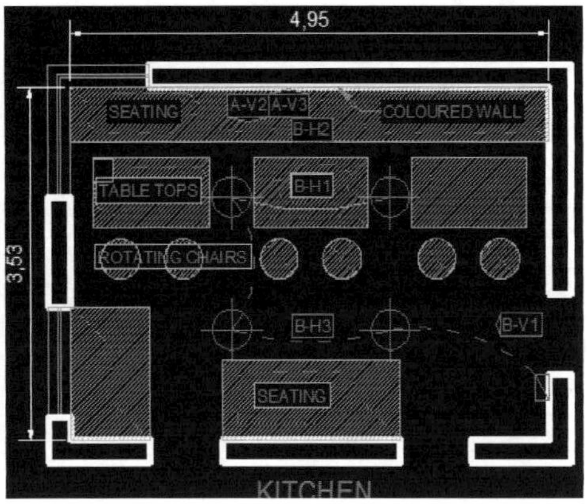

Figure 2. Plan of Dining area 2 with red wall

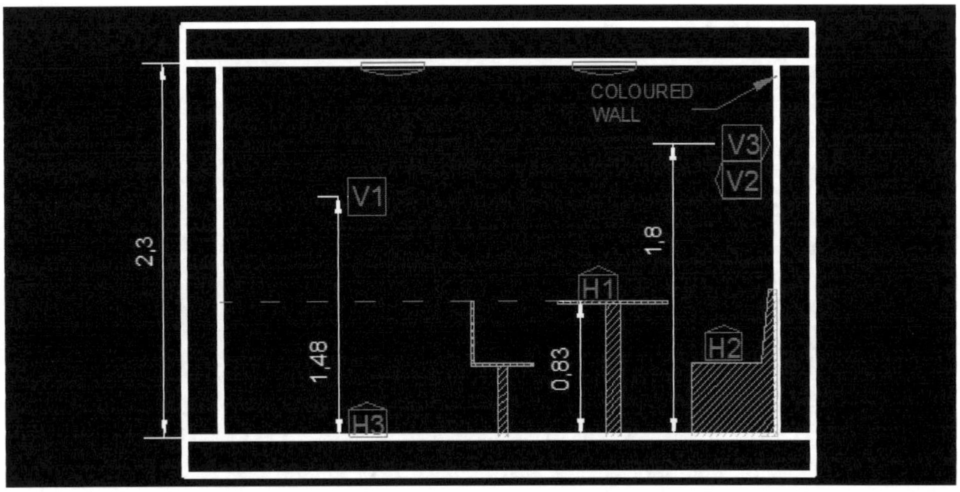

Figure 3. Typical section of the dining areas

2.3. Qualitative analysis

2.3.1. Visual analysis. Both spaces are analysed visually for significant differences in layouts and lighting.

Figure 4. Dining area 1 with green wall

Figure 5. Dining area 2 with red wall

2.3.2. Questionnaire. A Google form was created to seek opinion on the following:

- Rate the likeability of the space in general on a scale from 1-9 (1 as not like it at all and 9 as love it)
- Rate the likelihood of having their meals there on weekends on a scale from 1-9 (1 as least likely and 9 as most likely)
- Pick the perceived colour of the white walls in the two spaces on a pictorial scale of colour temperatures from 1-9 (1 as 1000K and 9 as 9000K)

Figure 6. Pictorial scale for choosing perceived colour of white

24 persons with various nationality backgrounds in age group of 24-29 (14 females, 10 males) agreed to participate in the survey questionnaire. 12 used one corridor dining space and 12 used the second corridor dining space.

3. Results

3.1. Observation

The red walled dining area had 18.2% more no. of persons/hour visiting the space than the green walled dining area. The red walled dining area had 22.7% more human activity hours than the green walled dining area.

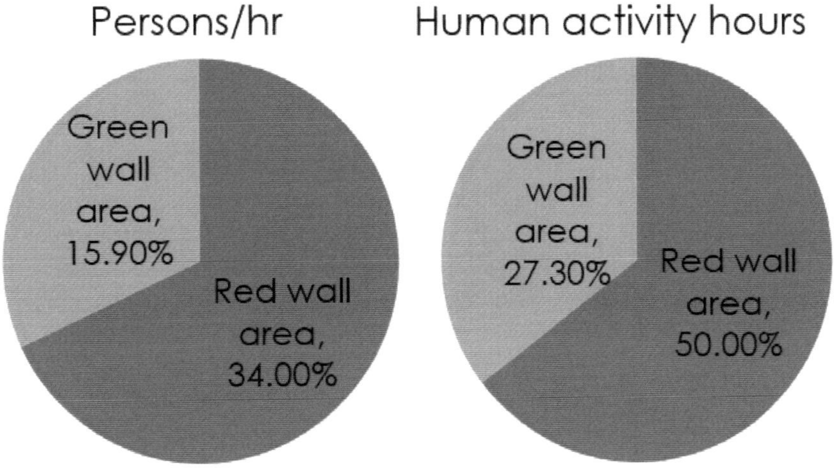

Figure 7. Comparison of the two spaces on the basis of 1. Persons /hour visiting the places, and, 2. Human activity hours that the two spaces support in the given period of study

3.2. Measurement

The spectral distribution curves in both the areas show almost identical distributions typical to an LED light source as seen in figure 8 and figure 9. The Illuminance, wavelength, CRI and colour temperature (CCT) measured at 3 vertical and 3 horizontal positions, is almost identical in both the areas.

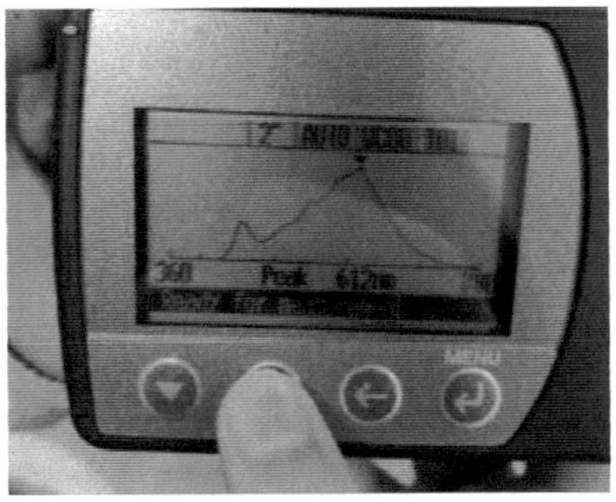

Figure 8. Dining area 1 with green wall

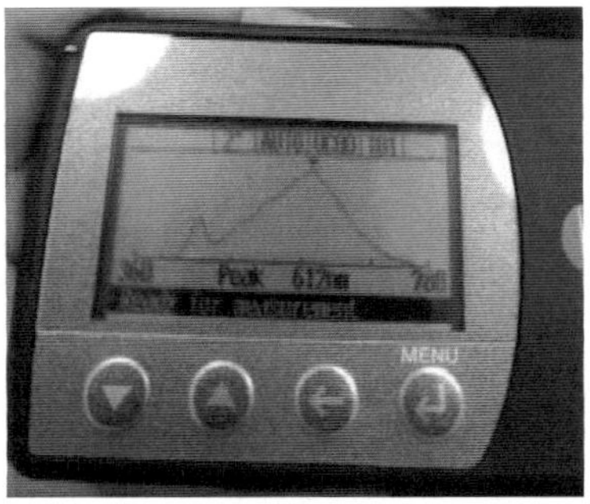

Figure 9. Dining area 2 with red wall

3.3. Visual analysis

Visually both the spaces have identical sizes, layouts, lighting systems, relative position of the wall with respect to the space and ease of accessibility with respect to placement in architectural layout.

3.4. *Questionnaire*

15% less users want to eat in the green coloured room, than in the red coloured room as shown in figure 10. General likeability of the green area is almost 17% more than the red area as shown on figure 11.

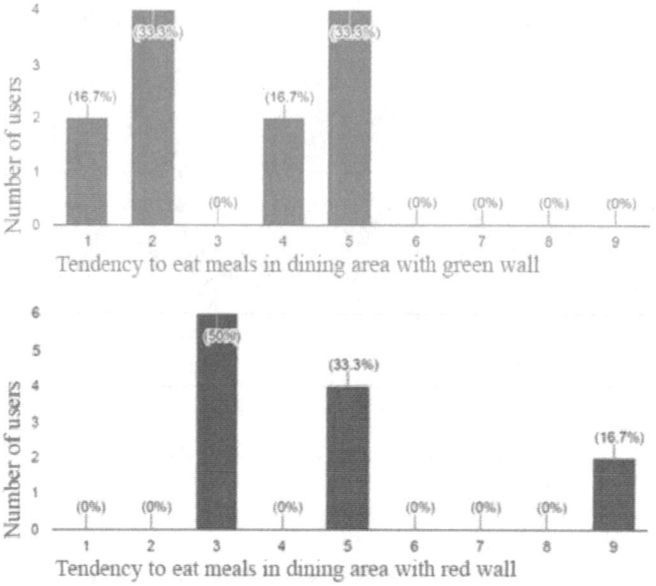

Figure 10. Tendency to eat meals in the dining area on weekends, as recoded from responses to the questionnaire

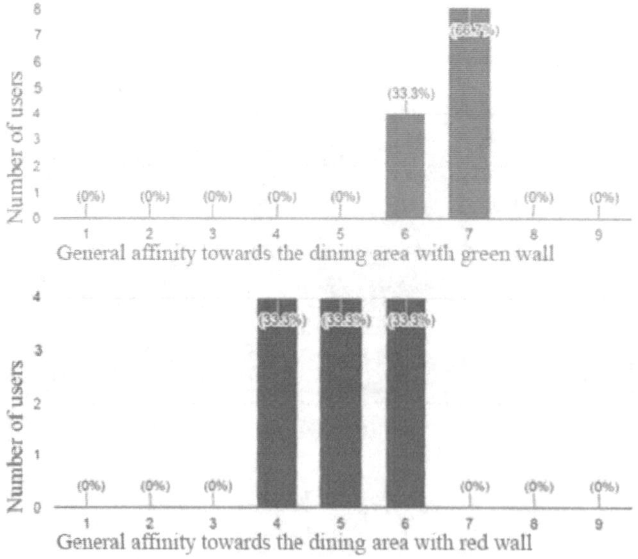

Figure 11. General likeability of the area, as recoded from responses to the questionnaire

In the green walled room, 50% people perceive the white walls to be coherent with 6000K colour temperature of white light as shown in figures 12, 14 and 50% people perceive the white walls to be coherent with 4000K colour temperature of white light in the red walled room as shown in figure 13, 14.

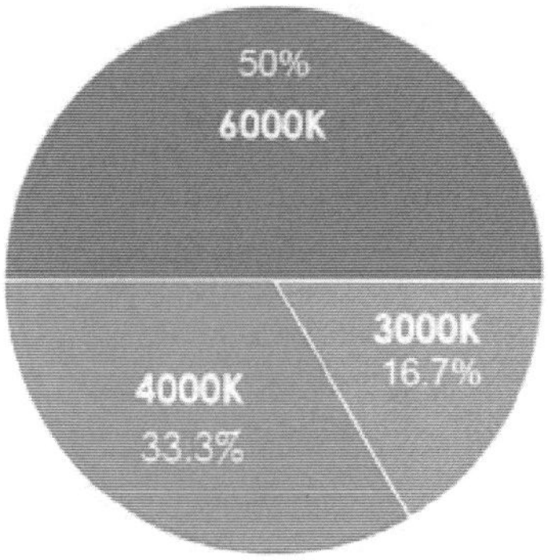

Figure 12. Perceived white colour opinions in green walled area

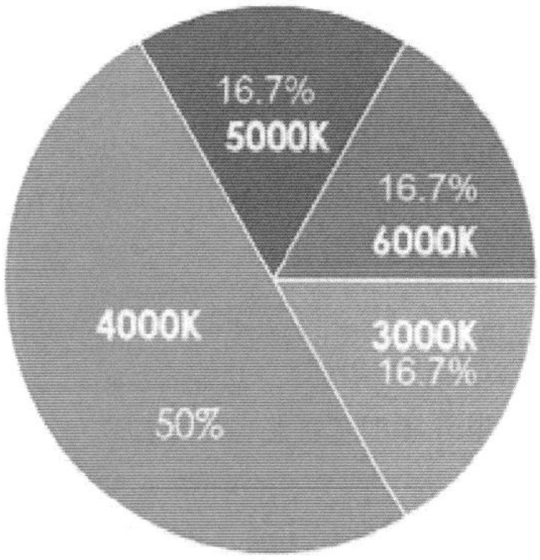

Figure 13. Perceived white colour opinions in red walled area

Figure 14. Pictorial selection of most common choice in both the areas

4. Discussion

Keeping all the parameters same, only the colour of a single wall is changing and we have been able to observe that this influences the decision of an individual to spend time in that space. Even though green is a bright colour, it does not invite the user. The white walls too in this room appear to be perceived on the colder side of white light colours. At the same time, the red wall created the perception, of warmer toned white walls. This could be because of the light bouncing off the coloured wall as shown conceptually in figure 15.

Figure 15. Colour changing the perceived colour

Generally, too, reds are known to stimulate adrenaline and increase appetite while greens are known to have a calming effect. Even though the observations confirm this, the reference to context is very important for the colour used as "the observer perceive a light environment as a whole, including both light character and colour setting" [4]. It is intertwined with our environment and with light. Hence it must be considered in totality, when designing with colour and with light.

5. Conclusion

The effect of colour in a space has a significant effect on our emotional mindscape. With the light measuring instruments, the effect is still immeasurable. Therefore, according to the instruments, the two spaces are identical and should not have any change in the effect it has on the users and the pattern of usage. But the study shows that colour does have profound effect on perception, leading to an emotional response. This in turn has an effect on users' decision of using the space and the usage pattern of the space. It goes on to prove the well discussed fact that colour shapes our responses. The study brings forth the immeasurable ambient colour of reflected light that affects the overall atmospheric quality.

6. Suggestions for further research

It would be interesting to study the activity levels in these rooms if daylight was considered criteria and how the perception of these colours changed with the shift from daylight to artificial light at night. The instruments designed to measure various aspects of light are designed to treat light as being devoid of colour. Colour that has proven effect on human psychology and behaviour is not taken as a measurable entity of light. Just as Johanna Enger is developing methods to design standards for the perceptual qualities of light [5], maybe colour and its physical and perceptual effects too need a set of standards, and this paper may provide the necessary fillip to spur research in that direction.

References

[1] Klepeis N and Ott W R (2001) The National Human Activity Pattern Survey (NHAPS): a resource for assessing exposure to environmental pollutants. *J Expo Sci Environ Epidemiol 11*, 231–252. https://doi.org/10.1038/sj.jea.7500165

[2] Alhorra Y, Arif M, Katafygiotou M, Mazroei A, Kaushik A and Elsarraga E (2016) Impact of indoor environmental quality on occupant well-being and comfort: A review of the literature. *International Journal of Sustainable Built Environment, Volume 5*, Issue 1, June 2016, Pages 1-11, https://doi.org/10.1016/j.ijsbe.2016.03.006

[3] https://worldgbc.org/principle-2-prioritise-comfort-building-users

[4] Enger J (2018) Visual Concepts and Sensory Experience of Light Environments. https://www.researchgate.net/publication/325106260_Visual_Concepts_and_Sensory_Experience_of_Light_Environments

[5] Boork M, Enger J, Hiller C and Wendin K (2018) *User-centric measures of perceived light quality: development of tools to promote energy efficient lighting.* As presented at BEHAVE 2018, 5th European Conference on Behaviour and Energy Efficiency, Zurich, 6-7 September 2018

The school lighting innovation dilemma

Tove Karlsson (1), Jonas Kjellander (2), Åsa Machado (3), Olle Strandberg (3), Peter Götlind (4) and **Reine Karlsson (5)**

Lighting designer, InsideLight AB (1), Senior architect and lighting designer, Sweco (2), Development Manager for school architecture, Sweco (3), ICT specialist, Malmö city (4), Facilities Manager, The Centre for Economic Sciences, Uppsala University (5) Prof. EcoDesign, InsideLight AB (6)

Corresponding author: reine.karlsson@insidelight.se

Abstract. This paper presents a qualitative descriptive case study that aims to build awareness about the need for improvement of the indoor environment in classrooms and to exemplify how renewal-oriented processes can be initiated. The study focuses on innovation processes that have resulted in both completely new designs and further spread of innovative solutions. The reason for the need to improve is that too many pupils and students have problems with concentration, relaxation and recuperation. The number of diagnoses such as Autism and ADHD is growing. One factor which significantly influences the pupil's mental wellbeing is the daily physical environment. The single aspect which may be improved in reasonable time, at a reasonable cost, is the lighting design of classrooms. An underlying reason for the habit to continue installing insipid uniform static "light-carpets" is the lack of awareness about how lighting affects well-being and health. As with innovation in other industries, it is an innovation dilemma that the vast majority of the people who sell and procure classroom lighting tend to prefer the established solutions that they are used to. The article describes how Malmö's early investment in Human Centric Lighting and the unique indoor environment of the Study hall in the Centre for Economic Sciences at Uppsala University are used as role models for lighting design in classrooms. The case study combines the authors' own observations and earlier interviews with users. Malmö's temporal light variation and parts of the Study hall's lighting design are used in more than 30 classrooms, in Malmö, Stockholm Business School and Iggesund. The article concludes with suggestions on how today's lighting design knowledge and technology can be combined to provide more advanced adaptations to the varying lighting needs of pupils, students and teachers.

Key words: Classroom lighting, Health, Spatial light variation, Human Centric Lighting, Environmental psychology

1. Introduction

A high number of children suffer from psychological distress. Every four years the Swedish Public Health Authority studies health indicators for different ages. From 1986 to 2018 the percentages of 15-year-old girls declaring that they have experienced mental disorders has increased: from 13 to 41% for feeling low, from 12 to 36% for nervousness and from 18 to 34% for sleep disorders [1]. The number of persons diagnosed with neurodevelopmental disorders like Autism and ADHD is also increasing.

IOP Conf. Series: Earth and Environmental Science 1099 (2022) 012031 doi:10.1088/1755-1315/1099/1/012031

While the physical environment is only one aspect that carries influence on their well-being it is relevant to note that 80 percent of the information we receive about our environment is taken in through our eyes and the visual system [2]. Light affects how we perceive and understand our surroundings, improved lighting in classrooms is a cost-effective measure to enhance student well-being by design. The combination of todays' SSL, the new knowledge on light and health and the ongoing large-scale replacement of fluorescent tubes ought to be an excellent opportunity for improvement of classroom environments. However, Swedish school actors routinely continue using uniform light distribution in almost all classrooms. The ambition with this paper is to investigate the lack of action and to encourage and provide guidance for initiation of innovative transformation of lighting design in Swedish classrooms.

This paper presents a qualitative descriptive case study that aims to support buildup of awareness about today's school lighting situation and to exemplify how renewal-oriented improvement processes can be initiated. An innovation process here denotes a goal-oriented series of efforts and activities. To enhance the ability to overcome the resistance to change, it is crucial to combine innovative ways of working and utilization of today's lighting design knowledge.

The paper also discusses whether and how the lighting design in classrooms can be improved to better support well-being and health. The lighting design chapter can be read as a conceptual light-related "map" of environmental psychology knowledge. The theoretical overview presents some underlying sensory and psychological mechanisms behind people's interpretations of environments and situations.

We describe the processes which resulted in noted radical lighting design improvements for two educational premises. Kjellander has designed the Study hall at the Centre for Economic Sciences at Uppsala University, which received the Swedish Lighting Award 2019. The city of Malmö introduced user-adapted Human Centric Lighting already in 2015. The Study hall in Uppsala and the classroom in Malmö has served as role models and inspiration for start-up of innovative improvement processes. The process that started with the Study hall has inspired lighting design improvements at Stockholm Business School and in Iggesund. The city of Malmö has installed temporal light variation in 30 classrooms and Human Centric Lighting in three nursing homes. Götlind and Strandberg have on site insights about the installations at Uppsala University and in Malmö city respectively. The paper also builds on professional inside experiences from a city's procurements and Sweco's market development for student centered lighting design. The students' age is 13-15 years for the classroom in Malmö, 6-9 years in Iggesund and most students at Ekonomikum and Stockholm Business School are 16-20 years or older.

2. The Innovation Dilemma for Transformational Change

Innovation aims for value-enhancing deployment of new ideas. This means that the basic "new idea", e.g. the lighting design knowledge already exists. Invention and innovation are two different concepts. To invent something means to discover a new thing. To innovate means to make value enhancing use of already existing ideas and methods.

However, most of the companies' and society's interest in innovative development is focused on how to increase the business volume. From user value and sustainability point of view it is important to strengthen the ability "to do the right things", i.e. to improve the effectiveness.

Established businesses tend to focus on *efficiency* and quantitative growth. Business development for new solutions is more related to *effectiveness* and qualitative improvement. The difference between effectiveness and efficiency is fundamental to value creation and sustainable development. Efficiency (to do the things right) is about producing whatever somebody has chosen to produce as quickly and cheaply as possible. When you limit the perspective to efficiency, it does not matter if you produce for a throwaway society. Effectiveness (to do the right things) emphasizes the fundamental importance of making the product / solution / architecture truly useful to and appreciated by users.

For a century the main goal for the lighting sector has been to sell more light (efficiency). All incandescent lamps gave the same kind of light. The combination of today's LED and control technology enable radical improvements of the ability to get the right light, in the right place, at the right time (i.e.

to improve effectiveness) [3]. One innovation hurdle is that it is crucial and difficult to make the customers interested in new (unfamiliar) solutions.

> "Businessmen will have to learn to practice "marketing" as an innovative force in itself. They have to learn that the truly new does not, as a rule satisfy demands that already exist. It creates new expectations, sets new standards, makes possible new satisfactions. 'Innovative marketing' therefore creates markets. New technology always needs new markets which were not even conceivable until the new technology created new demands (Drucker 1969)" [4]

Christenson's bestseller *The Innovator's Dilemma* [5] has brought to the fore that the resistance to change also is caused by people within companies and organizations in the established business system. It is a natural human preference to prefer to continue to use the solutions, the technology, the working methods and established documents that one already works with or have an established role in. People tend to continue to relate to the mindsets they are used to.

3. Lighting design and environmental psychology

Environmental psychology is multidisciplinary and focuses on how people affect and are affected by the social and physical environment [6]. One basis for environmental psychology theory is how our evolution in nature has affected man as a species [7 & 8]. One main aspect is the circadian variation between light and darkness and the variation of spatial light distribution. The EU project SSL-erate has shown that "Light affects our well-being and health much more than most people realize" [9].

The characteristics of light in nature is our best guidance in lighting design for wellbeing and health. Attention should not be stimulated by simply raising the level of diffuse dim lighting [10]. Human Centric Lighting aims to combine visual, emotional and biological needs [11]. We do not see without light, but neither do we see and feel without contrasts, shadows, direction and darkness. Sufficient darkness is vital for accent and functional lighting to be well experienced [12].

In school, pupils are expected to focus on defined tasks and avoid expression of inappropriate emotions. According to environmental psychology theory, directed attention needs regular and frequent temporary relief by contemplative fascination, such as a fire or moving water [13 & 14]. Fascination is important to avoid mental fatigue and to stimulate curiosity [15]. Lighting of vegetation promotes mental recuperation by observing beautiful scenery (fascination), deepened visual depth (change of environment) and emphasis on nature (shifted context) [16]. Accent lighting can highlight elements in the classroom and thereby stimulate fascination.

The Prospect and Refuge theory is based on human development by natural selection. The quest for food while simultaneously avoiding risks - to see (prospect) without being seen (refuge) - is a challenge for all animals. If we lack views and/or refuge, we tend to feel stressed or uncomfortable [15]. When reading people tend to prefer cozy seating [17]. To be overly exposed by lighting in longer sessions tends to be stressful.

Environmental psychology emphasizes that it is vital to feel as a part of one's environmental context, to feel included. One central brain structure to this end is the Hippocampus which organizes the flow of sensory input to build memories and the utilization of earlier memories for interpretation of the situations we meet [18]. It is vital to interpret one's environment, including facial expressions of other individuals, which is facilitated by combined directional overhead and lateral lighting. Concentrated light from above as the single light source will generate harsh shadows and make faces appear threatening or unnatural [10]. The spectrum affects whether and how the cognitive system can interpret moods, as the oxygen level affects skin hues [19]. Doubts about one's own ability to interpret what's happening creates worries and uncertainty about one´s capabilities. Such a mind-loop may be self-reinforcing [20]. Visual problems to interpret the surroundings increases the risk for stress, burnout problems, anxiety and psychological distress.

4. Human Centric Lighting for classrooms in Malmö

The starting point for Malmö's decision to improve the school lighting was that trusted influential persons at the technical development department convinced the city to join the application for the EU project SSLerate. The ambition included a sustainable social development ambition with a similar aim as the WHO definition of health; "Health is a state of complete physical, mental, and social well-being and not merely the absence of disease or infirmity." [21] The project received funding and Malmö started to look for opportunities to make a full-scale installation of Human Centric School Lighting.

A classroom was selected and organizational settings established. A user adapted light-variation scheme was developed in a master project [22]. The project combined interviews about pupil's and teacher's preferences and scientific knowledge on circadian effects. The automatic control of the light aimed for slow hardly noticeable (i.e., undisturbing) variations. The design process was iterative.

The initial goal was to have a period of wake-up light in the morning. The teachers wanted to have periods with harmonious lighting in the morning and before and after lunch, because it tended to be messy in the classroom at those times. Furthermore, the color temperature is low in the afternoon to increase alertness without suppressing melatonin [23], see Figure 1. Finally, the automatic variation scheme is combined with a manual control panel for selection among predefined scenarios.

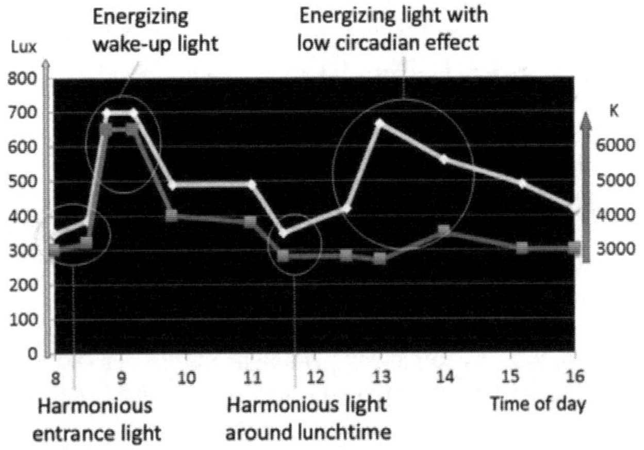

Figure 1 Light variation scheme for Malmö's first classroom installation of "Human Centric Lighting" by means of temporal light variation. The diagram shows the automatic variation of brightness and color temperature, during a school day. The intense wake-up light is on for 30 minutes in the morning. As an average for the day the scheme is aiming for a relatively low level of lighting. All light changes are slow and each one take at least 20 minutes.

It may appear sufficient to manually adapt the light to different educational situations. However, the studies [22] indicated that teachers did not change settings and that lengthy periods of activating light caused headaches for some of the pupils. Consequently, the control system has a function for automatic return to the lighting scheme.

Figure 2 shows a classroom with traditional fluorescent tubes and the classroom with Human Centric Lighting using 60x60 panels. In both cases the light distribution is uniform.

Figure 2 Traditional classroom lighting (left) and Human Centric Lighting (right) with 60x60 panels, temporal light variation and 5000K color temperature when photographed. Photo Olle Strandberg

When the project started Malmö city had an ambition to find turnkey suppliers and Strandberg invited both lighting and control system companies to present their experience with the suggested type of light variation. No company was found with experience of combining the desired lighting and control. Consequently, the city's technical development department organized the installation of tunable white luminaires and the control system. Strandbergs's key objective in the dialogues to build interest within the city was to try to awaken feelings that touch the heart.

Interviews with the teachers and students indicate that the installation and light variation are appreciated. [22] Sleep studies indicate that the installations have a positive circadian entrainment effect [24]. Strandberg emphasizes that Malmö is using tunable white 60x60-panels panels with high quality diffusers and that they do not have problems with glare. LED light has a reputation of being too intense and glary. Lighting designers tend to dislike LED and 60x60 panels. The reason seems to be the disapproval of only roof lighting and uniform static light distribution. This is highly relevant in relation to the emotional significance of the visibility of accent lighting effects.

The EU-project SSLerate attracted international media interest and then local media became very interested in the installation in Malmö. The Nobel Price 2017 related to the importance of circadian mechanisms [25] also increased the interest in temporal light variation. The city introduced user-adapted circadian light variation in 2015 and has up until now installed temporal light variation in 30 classrooms and Human Centric Lighting with more homelike lighting in three nursing homes.

5. The Study hall at the Centre for Economic Sciences, Ekonomikum, Uppsala University

The university print shop was closed down and Ekonomikum was left with two empty basement premises. Meanwhile there was an urgent need for more study areas. The aim was to create an attractive environment despite being in a basement without daylight, actually in an active bomb shelter.

The vision was to create a room with interior and lighting design to support various study situations with diverse seating positions and study tools, digital as well as analogue. Lighting should be possible to adjust to suit personal preferences and contribute to a calm, contemplative and pleasant atmosphere. The furnishing together with the lighting signalize a quiet space for mainly individual reading and working situations. Successively lifted floor levels from center to sides offers a good overview.

Figure 3 The Study hall at The Centre for Economic Sciences, Uppsala University Photo: Mikael Silkeberg

Surrounding a central fireplace there are 14 specially designed armchairs. The fireplace is a point of focus visible from nearly all seats in the room with light from the gas-flames creating a cozy atmosphere and a relaxing distant eye-rest between long straining sessions of looking at books and screens.

The circular seating is also intended for conversation occasions, with optimal physical conditions and a size not larger than enabling to hear and to read facial expression and body language across the circle. Gathering around the campfire is obviously the archetype of meeting and quality conversation in all cultures through history.

The design of the Study hall reuses some of the design elements from Matildelund preschool, awarded with the Swedish Lighting Award 2012 [26], in particular the Reading-refuges. One of Kjellander´s ambitions is to understand architecture from the child's perspective, i.e., related to the child's own perceptions and understandings. A refuge which attracts children is pleasant also for adults.

The design process was an iterative and interactive process that included dialogue with customers and user representatives and about the different partial solutions that suppliers suggested. Götlind and Kjellander made an initial draft that was presented to the board of Ekonomikum. The proposal was appreciated, but the investment cost was too high. An application for central funding from the university's strategic funds was made. In order for the application to be granted, the project needed to be anchored at all levels within the organization – a time-consuming but necessary process to reach the goal. The time from idea to decision was about three years. After another three years, the premises were inaugurated. During the dialogues, Kjellander presented his vision to the users and suppliers so they could better articulate their needs and deliveries respectively. The installation received the Swedish Lighting Award in 2019. The jury's motivation:

> The Study halls lighting design demonstrates advanced insights about lighting interaction with color, shape and material. The project has been carried out with great passion and expertise on study environments, and belief in the young individual and his/her rightful claim to dignity. Function, sensory experience and atmosphere are integrated by the ergonomic and innovative treatment of light. Darkness plays an obvious role in the design and creates focus. With great courage and creativity, the lighting designer stages original means of expression for experiencing safety, familiarity and wonder to enhance the feeling of well-being in this unique learning environment. [27] *Translation Tove Karlsson*

The Study Hall is very popular and also attracts students from other parts of the university. It is often rather filled and many students arrive early to catch their favorite seats. The cleaners want to delay the opening to 07:00 am, to be able to clean.

Götlind has observed that the mood is more harmonious also in other parts of Ekonomikum. The users of the study hall seem to spread the experienced harmony. Pictures from and descriptions of the Studyhall are used as vital parts in the marketing of the Centre for Economic Sciences. The interest in the Study hall influences the impression of Uppsala University. Through national and international study visits, the premises have provided inspiration for planning of other educational facilities.

Kjellander's design of the Study hall aims for a holistic combination of architecture, interiors, lighting design and equipment that meets the users different and varying functional and emotional needs. The CIE concept "Integrative Lighting" also suggest early integration of lighting and architecture [28]. Some of Kjellander's design concepts for the Studyhall are reused for a Conversation-room, Figure 4 and in classrooms, Figure 5, at Stockholm School of Economics and for current installations in a classroom at Iggesundsskolan, Figure 6. These installations also have a user-centered design ambition. Some design concepts that have been reused are the low level of general lighting, the spatial variation of the light, the round shape, the fire to attract spontaneous attention, the reading refuges and the variety of seating opportunities.

The Conversation-room and the classrooms in figures 5 and 6 all have a low level of general lighting. The Conversation-room has 12 seats and is primarily intended for dialogue.

Figure 4 The round Conversation-room at Stockholm School of Economics. Photo Jonas Kjellander

Some of the classrooms at Stockholm School of Economics, figure 5, have curved desks which provide a more social and interactive atmosphere compared to normal rows of desks where students are facing the backs of others. The lighting will be improved, aiming for spatial variation and highlighting the art cabinets. The study compartments are designed as in the Study hall at Uppsala University.

Figure 5 Classroom at the Stockholm School of Economics. The curved interior design (left) as well as the wall with individual reading refuges (right) are inspired by the Study hall in Uppsala. Photo Åsa

The lighting design for the classroom at Iggesundsskolan includes a panel for manual selection between lighting for different forms of activities, e.g. reflective dialogue and focused individual work, Figure 6

Figure 6 Iggesundsskolan lighting scenarios that aims to provide suitable conditions for the pupils Reflections on their studies (left) and for Focused individual work (right). Photo: Jonas Kjellander

IOP Conf. Series: Earth and Environmental Science 1099 (2022) 012031 doi:10.1088/1755-1315/1099/1/012031

6. Discussion

The initial installation in Malmö has a scientific basis in the knowledge about the awakening effect of intense blue light and was developed in dialogue with researchers in an EU-project. The lighting design for the Study hall at the Centre for Economic Sciences has received the Swedish Lighting Award. Strandberg and Götlind have on site insights from contacts with the usage of the facilities. During the Ekonomikum design process Kjellander was working four years with gradual buildup of mutual understanding, about the holistic interior design. Malmö's temporal light variation was designed in a master project that included dialogues and interviews with the teachers and pupils [22]. This paper does not compare the lighting design for these installations. The overarching ambition is to describe the broadening innovative deployment processes that have been inspired by the interest in the initial installations.

The intended transformation of school lighting is dependent on a spectrum of different kinds of innovation processes. At one side there is a need to get started with radical improvements, as exemplified by the Study hall and the first installation in Malmö. At the other side of the spectrum there is a need to deploy further copies of earlier innovative solutions, like Malmö's additional installations of classroom lighting with temporal light variation. To build interest in a new kind of solution it is important to have access to full-scale installations that can be demonstrated. It is also very valuable to have access to persons that are using and can tell about their own experience of the new kind of solution. Most school lighting installations does not involve a lighting designer nor dialogue with the users of the installation. Teachers etc. are not used to be involved in preparations of specifications for procurement of lighting.

> As a team leader for the city's implementation of new ICT, I am convinced that there is great technical potential to flexibly vary the light in user-adapted ways. The innovation dilemma is that it is difficult to involve teachers etc. in dialogue about the varying goals for the lighting in the classrooms. *Olle Strandberg*

To build mutual understanding between designers and user representatives there is need for open dialogue. To enable further development, it is interesting to combine parts from an already introduced solution with renewal oriented use of further improvement ideas and new technology.

The intended transformation of school lighting is dependent on a spectrum of different kinds of innovation processes. At one side there is a need to get started with radical improvements, as exemplified by the Study hall and the first installation in Malmö. At the other side of the spectrum there is a need to deploy further copies of earlier innovative solutions, like Malmö's additional installations of classroom lighting with temporal light variation. To build interest in a new kind of solution it is important to have full-scale installations that can be demonstrated. It is also very valuable to have accesses to persons that are using and can tell about their own experience of the new kind of solution. Most school lighting installations does not involve a lighting designer nor dialogue with the users of the installation. Teachers etc. are not used to be involved in preparations of specifications for procurement of lighting.

One key factor that initiated and enabled the innovative processes for the Malmö schools and related to the Study hall was that experienced trusted professionals took own initiatives. To motivate people to engage as change agents it is not sufficient to describe the advantages of the new solution. It is also necessary to clarify the shortcomings of the established standard. However, it is difficult to awaken interest in light variation among persons that are used to think that uniform stable light is preferable.

A basic characteristic of life is the constant oscillations between wakedness and sleep, work and rest, as well as taking in and digesting information. The 24-hour circadian cycle is one aspect. Environmental psychology stresses that we have a need for more frequent variation between activating and restorative experiences. In comparison to the living natures multifaceted variations, one might say that the uniform static lighting that is used in almost all classrooms gives a feeling that the pupil's environment is dead.

The ambition to provide the right light, in the right place, at the right time, is difficult to combine with today's standards and habits. It is easier to specify and procure a traditional lighting solution. One critical aspect of standardization is the influence on innovation. Standardization like the CIE concept

Integrative lighting may be important to open up for deployment of new solutions. On the other hand, established standards tend to impede deployment of renewal oriented solutions.

Renewal can be driven by industrial companies investing in development of more products to increase their market. Tunable white has given the opportunity to change CT and intensity, which is used for the classroom installations in Malmö. However, this is hardly affecting spatial light variation, since light distribution is a design variable [29]. Producers only invest in improving their own product portfolios. Lighting design companies are normally much smaller and sell their competence directly to the end customers. In most classroom projects no lighting designer is involved. The user involvement in the first Malmö classroom installation and for the Study hall in Uppsala has built awareness. There is a need to develop renewal-oriented lighting design concepts and descriptions which touch the heart.

It is a social sustainability issue that most schools continue to use class-room lighting that is neither stimulating, nor recuperative. The investment cost for user adapted light variation is practically negligible in comparison to the long-term costs when pupils get stuck in developmental problems. During the SSLerate project Strandberg often said that "if Human Centric Lighting can save only one pupil from lifelong exclusion from society, it is definitely worthwhile to invest in significantly better lighting".

6.1. Possible further design developments

The authors' observations and the jury's Lighting Award motivation for the Study hall show that significant emotive advantages have been achieved by combination of individually controllable luminaires, low general lighting, spatial light variation, the starry sky roof lights and the fire that attracts spontaneous attention. The lighting design and an advanced interior design with round shape, reading refuges and a variety of seating opportunities has been developed in concert. The lighting equipment is a minor part of the cost and parts of the Study hall lighting design can be adapted for classrooms. The Study hall demonstrates a rich holistic mixture of inspirational ideas. The emotional pleasure of being in the Study hall shows that it is possible to create attractive room, also in windowless basements – in fact it seems as if this falls more naturally when darkness can be used as a starting point.

Among lighting designers, it is well known that general lighting must be limited in order for accent lighting to be well experienced. There is a positive correlation between the psychological advantages of reduced general lighting and energy saving. Furthermore, eye problems among children and adolescents, partly caused by spending long periods watching a screen at close range, are increasing. This indicates that there is too little sensory stimulus to refocus and observe more distant objects. One way of dealing with this is to create sightlines towards accent-lighted sensory-stimulating details. The fireplace and the fiber optic star ceiling, in the Study hall, captures spontaneous attention and let the eyes rest.

One possible further development is to make combined use of various aspects from different installations, e.g., accent lighting with the kind of lighting zones applied in Malmö and at Iggesund. Analogous to the lux and CT variation in figure 1 it is technically feasible to achieve a periodically stronger visualization of certain objects by coordinated temporal variation of accent lighting, functional lighting and general lighting.

7. Conclusion

The main innovation dilemma for classroom lighting is that the lighting design knowledge on how to vary the light to improve emotional wellbeing is not used. When nothing specific is stated about light variation the general practice is to install uniform light-carpets with relatively high illumination levels.

The presented installations and processes show that it is achievable to make innovative full-scale school installations with radical lighting design improvements. Malmö's first full-scale installation of Human Centric Lighting awoke an interest to make additional installations. The Study hall is receiving successively more visitors. The increasing number of related installations shows that full-scale role models can inspire start-up of innovative deployment processes. Renewal oriented installations are important as showcases to build awareness about the relation between light and health.

References

[1] Folkhälsomyndigheten 2018. *Skolbarns hälsovanor i Sverige 2017/18*. Grundrapport

[2] Starby L. (2006) *En bok om belysning, för planering av belysningsanläggningar*. Ljuskultur

[3] Fagerhult (2022) Our view of Human Centric Lighting https://www.fagerhult.com/knowledge-hub/human-centric-lighting/our-view-of-human-centric-lightning/ (downloaded 20220420)

[4] Maciariello J.(2009) Marketing and innovation in the Drucker Management System J. of the Acad. Mark. Sci.

[5] Christensen C. (1997) *The Innovator's Dilemma: When New Technologies Cause Great firms to fail*. Harvard Business Review Press

[6] Roethlisberger F.J., & Dickson W. (1939) *Management and the Worker*. Cambridge, Mass.: Harvard University Press.

[7] Hildebrand, G. (1999) *Origins of architectural pleasure*. University of California Press

[8] Kaplan S. & Kaplan R. (1989) Kaplan, Stephen och Kaplan Rachel. 1989. *The experience of nature: A psychological perspective*. Cambridge: Cambridge University Press

[9] Vick, A. et.al. (2016) *Lighting for Health and Wellbeing in education, work places, nursing homes, domestic applications and smart cities*. EU: SSLerate FP7-ICT-2013-11-619249.

[10] Taylor l. & Socov E. (1974) The movement of people toward lights. *Journal of the Illuminating Engineering Society*, 3(3), s.237-241

[11] Lighting for People (2016) What is Human Centric Lighting? http://lightingforpeople.eu/human-centric-lighting/ (downloaded 220420)

[12] Wänström Lindh, U. (2018) *Ljusdesign och rumsgestaltning*. Lund: Studentlitteratur

[13] Bell, Paul A.; Greene, Thomas C; Fisher, Jeffery D.; Baum, Andrew. 2001. *Environmental psychology. Fifth edition.* Taylor & Francis group: New York

[14] Ulrich, R. (1984) View through a window may influence recovery from surgery. *Science, 224, 420-421*

[15] Appleton, J. (1996) *The experience of landscape*. Revised edition. Chichester: 1996

[16] Nikunen H. & Korpela K.M. (2009) Restorative lighting environments *J Light and Visual Environment*, The Illuminating Engineering Institute of Japan, 33(1), s.37-45

[17] Karlsson T. (2021) Reading-Refuge – Harmonious indoor environments for reading in libraries. *Master in Library and information science, Uppsala, Uppsala University*

[18] Østby H. & Østby Y. (2018) *Att dyka efter sjöhästar - hur vårt minne fungerar, hjälper och sviker oss*; Nordstedts

[19] Changizi, M. (2009) The Vision Revolution: How the Latest Research Overturns Everything We Thought We Knew About Human Vision. Dallas: Benbella Books

[20] Paulsen, R (2020) *Tänk om - en studie i oro*. Albert Bonniers Förlag

[21] Karlsson R. et.al. (2016) *Sustainability issuses for* EU: SSLerate FP7-ICT-2013-11-619249.

[22] Karlsson T. (2015) Human Centric School Lighting - Evidence based design of light characters and light variation. *Master in Architectural Lighting Design, Stockholm, KTH*

[23] Figueiro M. (2015) Chronobiology & non-visual effects of light, *Light Symposium, Stockholm*

[24] Eriksson S. & Wingren K. (2016) . The effect of dynamic LED light intervation on sleep quality of adolescent students. *Master in Biomedical Engineering, Lund, Lund University*

[25] Nobel Assembly (2017) Nobel prize in Physiology or Medicine to Hall, Rosbash and Young, for molecular mechanisms controlling the circadian rhythm, Karolinska Institutet, Stockholm

[26] Ljuskultur (2012) Svensk förskola vinner Nordiskt ljuspris (downloaded 2022-05-04) https://news.cision.com/se/ljuskultur/r/svensk-forskola-vinner-nordiskt-ljuspris,c9337652

[27] Rokka M. (2019) Vinnare av svenska Ljuspriset – Ekonomikum läsesal. Ljuskultur. Maj. https://ljuskultur.se/artiklar/ekonomikum-vinner-svenska-ljuspriset/ 2022-05-04

[28] Stanley Lemon, C. (2021) Human-centric lighting moves closer to integrative future. *LEDs Mag*.

[29] Boyce P. (2014) Edi.: Light distribution – a missing variable. *Lighting Res. Technol*. V. 46:2014

A review of lighting research in educational spaces

S Angelaki[1], U Besenecker[1] and C B Danielsson[2]

[1] Division of Lighting Design, Department of Architecture, KTH Royal Institute of Technology, 100 44 Stockholm, Sweden

[2] Department of Architecture, KTH Royal Institute of Technology, 100 44 Stockholm, Sweden

angelaki@kth.se, ute.besenecker@arch.kth.se, christina.bodin.danielsson@arch.kth.se

Abstract. The present paper summarises tendencies related to research in lighting for school environments from 1980 to 2020. Methods and tools used to evaluate and analyse both daylight and electric lighting are presented. The studies in this literature review were grouped in four decades and are presented chronologically. The review suggests a shift from on-site evaluations in the early decades, both when using qualitative and quantitative research methods, to software-based research and analysis in more recent studies. during the past decade, there appears to be a general increase in both software and user-based techniques for evaluation and design of educational spaces. However, it is interesting that the lighting layout in classrooms does not show significant changes during the four decades since it follows the same grid pattern regardless of the changes in architectural, design and teaching tendencies.

1. Introduction and background

Lighting in schools is not a new research topic, and lighting research in classrooms is not only present on recent studies. Several topics have been investigated related to lighting in schools. For example, there have been prior investigations related to lighting in schools with the focus being either on energy efficiency (1-3), performance, mood and cognition (4, 5), or perception and visual comfort (6-8). The topics presented in research are often related to research interests and tendencies, based on the scientific developments, such us the existence of the ipRGCs, or technological advancements, such as the use of digital tools, of each decade. Consequently, tools and methods are either being further developed or new tools and methods are presented. In this review we are interested in how the reported studies measured outcomes in their areas of investigation and correlated them with specific lighting characteristics such as, for example, the light levels, the light distribution and the correlated colour temperature of the electric lighting. The methodology and tools or equipment presented in each study are summarized and presented throughout this paper.

1.1. Objective of paper

This paper's main goal is to present the methodology and the tools used in lighting-related research in educational spaces throughout the past four decades, to discuss parameters related to lighting design and to identify knowledge gaps and potential areas for future research. The paper discusses the various parameters affecting and defining, in many cases, the direction and primary focus of selected lighting-related research papers within the selected time frame. The various methods and tools used for gathering data in school environments are the focus of this review. Papers corresponding to research performed during each decade from 1980 to 2020 were collected and grouped chronologically. For collecting the papers, the following databases were used: Scopus, Web of Science, Google Scholar and JSTOR.

1.2. Methods

For this literature review, papers written from 1980 to 2020 were collected using the following databases: Scopus, Web of Science, Google Scholar and JSTOR. The search string in the different databases was defined as follows: *("light" OR "lighting" OR "artificial lighting" OR "electric lighting" OR "daylight" AND "school" OR "primary education" OR "classrooms" AND "performance" OR "visual comfort" OR "energy consumption" OR "energy" OR "behaviour" OR "perception").* Due to the limited number of papers in the period being investigated, the papers chosen were not limited to research performed in the Nordics, even though this paper is part of a research project situated in Sweden. The review does not focus on research performed in specific locations but on each paper's chosen methods and tools. After each search, the resulting papers were further processed to eliminate papers that did not investigate school environments. The additional keywords that were used to filter out papers not related to schools were: [*(school(s), classroom(s), educational environment(s), school building(s)*]. Papers related to research performed for educational spaces at a university level were excluded because the authors' current research project focuses on younger pupils. Both daylight and electric light studies were included in the review process, and in some cases, both aspects were investigated in a single project. Furthermore, only peer-reviewed articles or conference papers and reviews have been included in the process.

One limitation needs to be pointed out related to the changing in terminology. There are shifts in keywords and research interests that are not the same during each decade. An example of that is the use of the term *"classroom"* during the first three decades, which is currently being replaced by the terms *"educational environment"*, *"learning environment"*, or *"innovative learning environment"*.

2. History of lighting in school buildings

Lighting in educational spaces has been studied in different frequencies along the years. In the following categorisation, the studies are presented in four decades from 1980 to 2020. The methods and tools used to measure and evaluate light are presented. Furthermore, the main areas of investigation are pointed out to find research tendencies along the years.

2.1. The decade 1980-1989

The review process for the first decade, from 1980 until 1989, includes six articles after the described elimination process. During this decade, not many research projects focused primarily on school lighting. A combination of projects investigating school lighting and office lighting suggests that similar practices were being applied in these spaces, and only a limited number of projects work primarily with schools. In terms of methods, light meters for registration of illumination levels and time-lapse photography (9-11) are the dominant tools for registering and documenting the light. The collected data in the case of time-lapse photography are analysed frame by frame to find differences between the investigated scenarios. For outcome measures studied, one paper examines behavioural changes and environmental characteristics through questionnaires (12). Another article explores the potential connection between sickness leaves and fluorescent luminaires with spectral variations and illumination levels in the classroom (13), which is the only article from the chosen ones in which lighting is linked to health. In two of the papers, the area of investigation focuses on the pupils' preference for windows and daylight (12, 14). And then, there are two studies comparing energy consumption of classroom lighting with different fluorescent tubes (11, 13). One of the main arguments for investigating lighting in school environments was that energy consumption generated by lighting was up to one-third of the yearly energy costs. Therefore, energy savings related to lighting were considered a priority for the schools, and a supported research topic.

It is essential to mention that the published studies that did not administer perception or performance tests to collect data, such as the studies focussing on measuring energy consumption, did not include the pupils' or teachers' active participation in the investigation of the lighting. Two out of six studies used methods allowing the teachers' and students' participation through interviews and questionnaires (11, 12). Furthermore, in all cases, the lighting changes refer to ceiling mounted lighting fixtures, and the lighting layout typically follows a grid configuration to achieve the required lighting levels, existing published guidelines and standards.

2.2. The decade 1990-1999

The following decade covers articles from the years 1990 to 1999. Seven studies met the selection process described above. All of the ones considering educational facilities at university level have not been included. Similar to the prior decade, there is performance-oriented research linking performance rates of learning activities to lighting. Of these studies, one focused on electric lighting (15), or as described in this case, artificial lighting, two more investigated the effect of daylight on performance (16, 17), and one study combined both electric light and daylight (18). Regarding the variables to which lighting aspects were linked, two research papers focussed on health and development-related aspects by collecting data related to sick leaves throughout a predetermined time and registering body growth and stress hormone levels (15, 18).

During this decade, furniture layout variations emerged in classrooms, and, therefore, research related to how lighting should be designed in response. Specifically, using direct observations as a method, Schreiber describes and explains how lighting can highlight certain corners in the classroom (19). The creation of "pools of light", or concentrated light, mentioned by referencing Moore's report *"Recommendations for child care centres"(20)*, can support different activities taking place in the space.

Comparisons of conventional fluorescent lamps, with their distinct limitation in light spectral power distribution, with so-called "full-spectrum" fluorescent luminaires with ultraviolet light supplements and high-pressure sodium vapour lamps are presented (15, 21). The comparisons primarily aimed at exploring the relationship between the different luminaires in terms of spectrum and illumination levels and performance of learning activities, body growth or cognitive and physical development, but the studies also mention the aspect of energy efficiency in comparison to the conventional fluorescent luminaires. One of the presented articles focuses on the energy consumption aspect without investigating any other factor, resulting in the outcome that properly designed educational spaces in which natural light is used throughout the majority of the year could save up to 80% of energy costs compared to using incandescent lighting fixtures (22). Two of the presented studies focused both on electric lighting, daylight availability and school performance (16, 18). In both cases, daylight is defined as an environmental factor that can potentially influence pupils' performance. The authors, Kuller and Lindsten, were comparing daylit classrooms with windowless classrooms where only electric light is present, while the authors Nicklas and Bailey only focus on daylight availability during teaching hours and its connection to pupils' performance without considering electric light (16).

In this decade, the equipment used to collect data slightly varied compared to the previous decade. Photo registration and aerial photography were used to collect data and analyze the space occupancy. Equipment related to light measurements was used to register illuminance levels and changes in illumination when comparing different luminaires. When exploring light in connection to performance related to learning activities, tests were handed out to pupils to evaluate potential changes in their performance, based on math, reading tests or visual tasks performed under certain illuminationlevels or attention and memory in certain subjects. It is essential to mention that school architecture underwent alterations related to architecture and interior spaces simultaneously (open-plan schools, wing schools, modular schools). Even though the architectural and interior elements were changing the lighting schemes remained similar to the prior applications. Therefore, an additional variable was present regarding space characteristics compared to the last decade. Software used for calculating energy consumption was mentioned for the first time (22).

2.3. The decade 2000-2009

The following decade spans from 2000 to 2009, and the elimination process resulted in the presentation and analysis of ten papers and reports. Five out of the ten studies investigated classroom daylight-related aspects (23-27). Of the remaining papers, four focussed on the electric light (28-31), and the last paper focused primarily on design elements and how they influence the students' outcomes (32). In this study, lighting was a parameter investigated as one of several design elements. In five studies, the central aspect linked to lighting was energy efficiency and reduced yearly energy costs (23, 24, 27-29). During these years, six studies were performed without the involvement of any user participation, neither on behalf of the teachers nor the pupils. In these studies, the primary tool for data collection and space evaluation was software-based.

In one literature review summary paper, the aspect of daylight has been highlighted in individual studies (27). In this paper the descriptions and analyses of pupils' and teachers' health, performance, teachers' and pupils' attendance and energy efficiency were provided. Two of the presented studies focused on the effect of daylight on pupils' performance; they showed considerable development in perfromarce and increased student test scores in classrooms with high daylight intake (26, 27). These studies support that efficient design can result in more optimal daylight conditions, enhance students'/pupils' performance, and result in efficient systems for energy consumption and associated costs. The remaining three studies did not include users' participation but used software and measurements of various factors , for example to evaluate the space and the current indoor conditions one of which was lighting (23-25).

Kruger and Zannin explored how the environmental factors of noise, light and temperature can potentially influence classroom teaching activities (25). In this case, the comparison of traditional classrooms with ones where light shelves (high reflective horizontal surfaces placed on windows' upper part for directing daylight into a room) were placed, taking into account the difference in orientation and daylight intake, resulted in improved lighting/visual comfort in the classrooms with the light shelves. Traditional classrooms were described as rectangular spaces with a desk layout oriented towards the whiteboard. It was noted that those traditional classrooms still had thermal comfort issues even after installing the light shelves. Data collection related to light, noise and temperature showed increased performance rates in pupils and teachers after installing light shelves, resulting in increasing the daylight intake.

Two of the studies focusing on electric lighting investigated only aspects of energy consumption throughout the year relating it to energy costs (28, 29), while of the remaining two, one focused on visual discomfort (31) and the other focused on learning-based activities in the classroom and the use of digital screens (30). This last study explored parameters of lighting linked to digital screens while pointing out that designing lighting in classrooms should also consider the additional light sources that are part of the teaching/learning activities (laptop screens, monitors, projectors).

The methodologies used for data collection in the listed studies show a greater variation compared to the previous decades and a slight tendency towards evaluation processes based on simulation without users' participation. Four studies were based on software and evaluation of simulations, and only one of them included on-site measurements. Software simulations allowed the exploration of a greater variety of cases and alternatives, but they did not include feedback or data collected from the teachers or students. Two studies used on-site measurements of light levels, investigating different combinations of electric lighting and blind position in the classroom (25, 31). Three more studies collected data based on tests linked to evaluation of learning processes and activities, thus including the pupils in the data collection process (26, 30, 32). Only one of the studies included both the teachers and students to get feedback on the lighting conditions in the classroom by using questionnaires as a form of data collection (30). Tanner and Uline collected data from students using a Likert-scale questionnaire to investigate the impact of various design elements in school environments and their relation to pupils' performance (32).

The overall impression from the studies presented above is that the simulation-related projects focus on standards and regulations to evaluate lighting conditions and lighting levels. At the same time, a common denominator in these studies is the energy efficiency aspect of lighting. The use of software simulations eliminated feedback from the pupils or teachers who are the main users of the space. Photo registration seemed to be replaced by simulations gradually. Tests addressed to students, focusing on attention, memory and performance of specific activities and courses depending on each research project, are still the primary source of data collection in performance-based studies, similar to the previous decades.

2.4. The decade 2010-2020

The last decade of the review covers the years from 2010 until 2020. During this decade, a considerable increase in the number of published studies and reviews related to school lighting is apparent. The number of papers in this decade is more than twice the number of papers from 2000 to 2009. It is not clear, however, whether that means that research in the subject increased overall, or just the possibility and motivation to publish. Out of the total number of papers and by following the same elimination process used for the prior decades, the twenty-four papers are included in this article. The papers will be described first based on whether they explored daylight or electric lighting, and which other variables were looked at. The next part will focus on methods and tools, and the last part will present the role of user participation. At first, it needs to be mentioned that four of the papers are review articles summarising and investigating different aspects of school environments related to lighting. Some of the selected papers presented here are also part of the analysis in these review articles.

During this decade, daylighting and electric lighting were in focus, with daylight-related research being included in fourteen out of twenty-four studies, meaning that more than half of the papers are investigating daylight related aspects in school classrooms (3, 33-44). The number of studies that focussed on electric lighting in school classrooms is twelve (1, 36, 42, 45-53). Only two studies investigated both daylight and electric lighting in school classrooms (36, 42). In all other studies, both natural light (or daylight) and electric lighting were mentioned, however, only focussing one of the two aspects. Most daylight studies were linked to energy efficiency, stating that the electricity a school building consumes throughout the year can be considerably reduced if better design practices, are in place. This argument has been used in previous decades as well. More specifically, half of the studies investigating daylight focussed on the energy efficiency aspect, while the corresponding number for the electric lighting studies is three out of fourteen studies. Other areas of investigation are visual comfort, students' performance and behaviour, environmental and thermal comfort, and the perception of space. More specifically, four studies investigated the relationship between students' performance and lighting. Three of these studies explored the correlation between daylight scenarios and their effect on students' activities (36, 39, 44). Two studies presented the interrelation between electric lighting and performance (36, 46). Two studies, both investigating aspects of electric lighting, focussed on students' attitude and behaviour, (51, 52). Visual comfort, specifically eye fatigue linked to glare, was addressed in three studies; all of these investigated daylight, evaluating various possible solutions such as the installation of light shelves, different types of glazing, window size and orientation. (35, 43, 54). One of the studies focused on occupants' spatial perception in response to the electric lighting in space, and more specifically the illumination levels on the task areas (46). Environmental comfort, related to visual and thermal aspects, and electric lighting, in terms of perceived light levels in space, was the focus in another one of the described studies(47).

The methods found in these papers are different from the ones used during the previous decade. There is an increasing combination of qualitative and quantitative methodologies, regarding the use of both measurement-based practices, such as illuminance levels, spectral distribution and daylight analysis, as well as methods dealing with the users' feedback and evaluation of the space, such as questionnaires and tests for students. Even though some studies rely entirely on simulation software for spatial assessment and light levels calculations, a limited but considerable number of examples re-introduce user-based techniques for spatial evaluation. On-site measurements are also present, and the schemes according to which these measurements are performed are more complex and detailed than in the previous two decades. This could be explained by the technology advancing, making available a wider variety of affordable field measurement equipment to measure and calculate various light parameters.

Most of the selected studies used more than one methodology for assessing the space, resulting either in lighting related recommendations or design proposals. Seven of the here presented studies used software simulations for assessing the space, collecting data related to light attributes, such as illuminance levels and daylight availability and evaluating lighting design proposals (3, 34, 40, 43, 47, 53, 54), and eleven out of twenty-four articles had calculation schemes for on-site calculations and evaluation of the classroom (1, 3, 35, 39-41, 46, 52-55). Qualitative methodologies used and described in the published studies were questionnaires linked to perceived light levels, various tests for students' assessment, direct observations of the activities and teaching processes and the photo registration of the space. Photo registration, however, was presented as a methodology only in one study (50), and direct observations were used to collect data in three studies. Out of these three papers, one investigated daylight and electric lighting using direct observations (42). The second paper focussed on daylight, and the last one explored only electric lighting (47). Tests were used to evaluate performance and attention in five studies and were linked both to changes in daylight and electric lighting (36, 44, 45, 48, 56). In five articles, questionnaires were used to collect data. Three of them investigated the attributes of environmental comfort and electric lighting (36, 46, 47), and the remaining two were investigating the attributes of students' performance related to daylight (44, 54). One of the main findings of reviewing the papers during this decade is the relatively low level of participation of the primary users of the space, meaning the teachers and students. Considering that four out of twenty-four articles are reviews, the resulted number of design-based papers is twenty. Nine out of these studies, which accounts for almost half of the papers, used non-participatory-based evaluation of classrooms and educational spaces. This means that the users' feedback was not part of the evaluation or design process; the proposals were based entirely on the calculations performed by the software in response to the available guidelines at the time of the study.

3. Discussion and summary

This paper explores and describes the research tendencies found in lighting-related research in school environments during the past four decades. The review involves studies performed in schools from kindergarten to high school level. Both daylight and electric lighting are investigated during each decade as illustrated in figure 1. The first outcome of this review is that the number of available published studies considerably increased after 2010—the studies matching the inclusion criteria during the period from 2010 to 2020 were more than double compared to the previous decades. This might be either due to an increased interest in the subject as a topic or research, or due to a general increase in publishing practices.

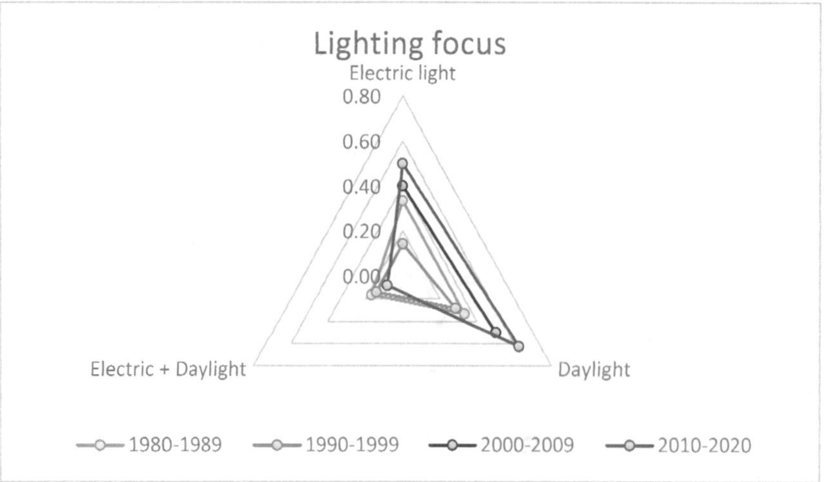

Figure 1. Graph illustrating the focus in relation to electric lighting or daylight during the four decades, from 1980 to 2020.

Throughout all four decades, daylight is primarily linked to energy efficiency. The main argument related to studying daylight is that it can considerably reduce the electricity consumed during the academic year. Therefore, it is more financially viable and cost-beneficial. During the first two decades, the evaluation processes related to daylight utilized on-site measurements and time-lapse photography. After 2000 the use of simulation software for calculation and evaluation of educational spaces gradually replaced prior practices. Simulation-based evaluations allow the exploration of various scenarios and design solutions under different sky conditions. From 2010 to 2020, simulations and on-site measurements were used in a considerable number of studies. Additionally, the evolution of specific software and the integration of weather data and lighting standards within the software offers the opportunity of evaluating more spaces through a faster process. However, we observed that such proposed solutions were based upon the lighting standards at the time of the study without considering the users' feedback.

Another method used more or less often throughout every decade is the use of light level and light spectrum measurement equipment, referring to luminance and illuminance meters, spectrometers and other similar devices as mentioned in the studies presented above. Light measurement equipment has advanced through the years becoming smaller and more available and affordable for field work. Subsequently, the methods to which spectral measurements are defined and performed are more refined closer to 2020. From 1980 to 1989, lighting measurements concerned mainly light levels and contrast ratios and not necessarily spectrum. Lighting equipment is widely used during the first two decades and then again after 2010 as seen in figure 2. The broader use of software-based evaluations characterizes the third decade. From 2010 to 2020, lighting equipment has been used, along with equipment registering noise and temperature levels. It is gradually, since 2000, that more environmental factors have been studied along with lighting.

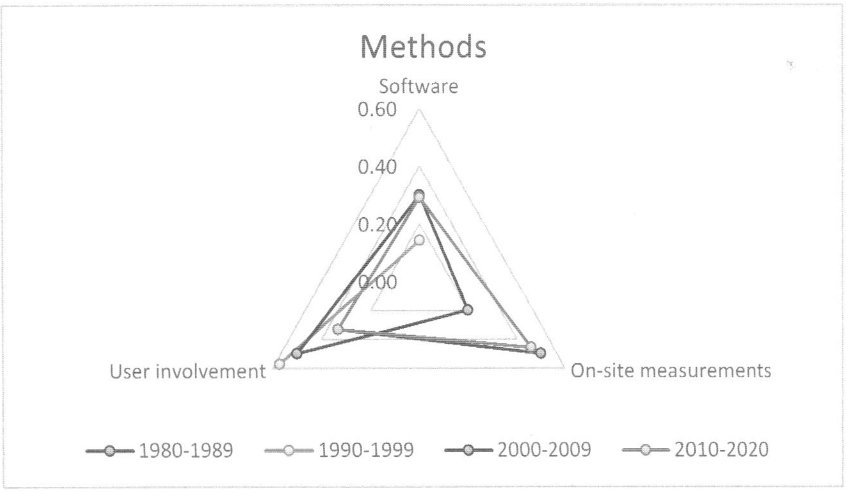

Figure 2. Graph illustrating the tendency among three different methods for each separate decade.

Regarding the use of electric lighting, a shift from fluorescent fixtures to LED light sources took place from 2000 to 2010, and more studies were present after 2010. During the first two decades (1980 to 2000), fluorescent fixtures were compared to full spectrum fluorescent lights, and the main attribute studies were investigating was the light source spectrum. During the following years the argument was, in many cases, related to the decreased energy requirements LEDs had, but studies related to visual comfort and students' performance enhanced the shift to LED sources. Starting 2010 LED lights gradually started to replace fluorescent fixtures in school classrooms.

Attention and performance have been studied throughout the four decades, and the tools used for their evaluation were performance and attention tests, such as the d2 attention test. In addition, behaviour and attitude were studied through observations and interviews with teachers and students using primarily questionnaires. It is mainly during the first two decades (1980-2000) that participatory methods were used for spatial evaluation and evaluation of lighting qualities. Photo registration and time-lapse photography were used in three of the studies presented in this review until 1989, but only one study mentioned photo registration as part of its methodology in the last decade.

Furthermore, user-based methods and tools that have been used during the first two decades (1980-2000) and were gradually replaced by simulation-based calculations. During the last decade (2010-2020), considering that the number of studies was greater, less than half of the studies included methodologies that engaged the users. In most of these studies, the area of investigation focused on students' performance.

One of the common denominators regarding electric lighting for the reviewed studies throughout all decades is that they all investigated different light sources and fixtures that were ceiling mounted following the same regular grid pattern and ambient distribution. No significant changes were found in terms of light topology in the classrooms, the two main typologies are either rectangular or square panels, or linear pendants. Even in cases where the design or furniture layout deviated, the lighting layout appears to follow the same grid pattern without exploring alternative design solutions or placements of the lighting in the space. Additionally, it was mainly brightness levels and spectrum that have been tested in relation to lighting. Other attributes, such as density of light fixtures or variations in light distribution have not been tested as variables.

Reviewing the published research, suggests the that testing different lighting layouts would fill a current gap in research while supporting the layout of existing elements, such as furniture layouts, in the classrooms that do not necessarily follow a grid pattern. Furthermore, since in today's classrooms not all pupils have the same needs and do not necessarily use the same tools simultaneously (books, laptops, tablets), there is also use of various working spots, and the need/opportunity for lighting to support working in different locations throughout the space (57-59). Although architecture and design have changed based on teaching and learning tendencies, classroom lighting design has remained almost the same for the past forty years, and it is time to change that.

Acknowledgements:
This work is part of a project financially supported by the Bertil & Britt Svensson's Foundation for Lighting Technology and the Swedish Energy Agency.

References

1. Li DHW, Cheung KL, Wong SL, Lam TNT. An analysis of energy-efficient light fittings and lighting controls. Applied Energy. 2010;87(2):558-67.
2. Rebec KM, Klanjsek-Gunde M, Bizjak G, Kobav MB. White LED compared with other light sources: age-dependent photobiological effects and parameters for evaluation. Int J Occup Saf Ergon. 2015;21(3):391-8.
3. Delvaeye R, Ryckaert W, Stroobant L, Hanselaer P, Klein R, Breesch H. Analysis of energy savings of three daylight control systems in a school building by means of monitoring. Energy and Buildings. 2016;127:969-79.
4. Knez I. Effects of Indoor Lighting on mood and cognition. Journal of Environmental Psychology. 1995(15):39 - 51.
5. Hygge S, Knez I. Effects of Noise, Heat and Indoor Lighting on Cognitive Performance and Self-Reported Affect. Journal of Environmental Psychology. 2001;21(3):291-9.
6. Liu Q, Huang Z, Li Z, Pointer MR, Zhang G, Liu Z, et al. A field study of the impact of indoor lighting on visual perception and cognitive performance in classroom. Applied sciences. 2020;10:1-17.

7. Noda L, Lima AVP, Souza JF, Leder S, Quirino LM. Thermal and visual comfort of schoolchildren in air-conditioned classrooms in hot and humid climates. Building and Environment. 2020;182.
8. Fakhari M, Vahabi V, Fayaz R. A study on the factors simultaneously affecting visual comfort in classrooms: A structural equation modeling approach. Energy and Buildings. 2021;249.
9. Hunt DRG. The Use of Artificial Lighting in Relation to Daylight levels and Occupancy. Building and Environment. 1979;14:21-33.
10. Levy AW. Lighting Controls, Patterns of Lighting Consumption, and Energy Conservation. IEEE TRANSACTIONS ON INDUSTRY APPLICATIONS. 1980;IA-16.
11. Fletcher D. Effects of Classroom Lighting on the Behavior of Exceptional Children. EEQ. 1983;4:75-89.
12. Stewart DM. ATTITUDES OF SCHOOL-CHILDREN TO DAYLIGHT AND FENESTRATION. Building and Environment. 1981;16(4):267-77.
13. London WP. FULL-SPECTRUM CLASSROOM LIGHT AND SICKNESS IN PUPILS. Lancet. 1987;2(8569):1205-6.
14. Dunn R, Krimsky JS, Murray JB, Quinn PJ. Light up Their Lives_A Review of Research on the Effects of Lighting on Children. The Reading Teacher. 1985;38:863-9.
15. Hathaway WE. Effects of School Lighting on Physical Development and School Performance. The Journal of Educational Research. 1995;88:228-42.
16. Nicklas MH, Bailey GB. Analysis of the Performance of Students in Daylit Schools. North Carolina; 1996.
17. Heschong L. Daylighting in Schools: An Investigation into the Relationship between Daylighting and Human Performance. Detailed Report, Academic Achievement. 1999;140.
18. Kuller R, Lindsten C. Health and Behaviour of children in classrooms with and without windows. Journal of Environmental Psychology. 1992;12(4):305-17.
19. Schreiber ME. Lighting Alternatives: Considerations for Child Care Centers. Young Children. 1996;51(4):11-3.
20. Gary T. Moore, Carol Gee Lane, Ann B. Hill UC, Tim McGinty, , Frederick A. Jules, Marie H, Hollenbeck, et al. Recommendations for Child Care Centers. Wisconsin-Milwaukee: Center for Architecture and Urban Planning Research Books; 1996.
21. Edward E. Green, Cook PF, Bolt L. Fitting New Technologies into Traditional Classrooms: Two Case Studies in the Design of Improved Learning Facilities. Educational Technology. 1996;36:27-38.
22. M. Santamouris, Balara CA, E. Dascalaki, A. Argiriou, Gaglia A. Energy Consumption and the potential for energy conservation in school buildings in Hellas. Energy 1994;19:653-60.
23. Ho M-C, Chiang C-M, Chou P-C, Chang K-F, Lee C-Y. Optimal sun-shading design for enhanced daylight illumination of subtropical classrooms. Energy and Buildings. 2008;40(10):1844-55.
24. Mistrick R, Sarkar A. A study of daylight-responsive photosensor control in five daylighted classrooms. Leukos. 2005;1(3):51-74.
25. Krüger EL, Zannin PHT. Acoustic, thermal and luminous comfort in classrooms. Building and Environment. 2004;39(9):1055-63.
26. Heschong L, Wright RL, Okura S. Daylighting impacts on human performance in school. Journal of the Illuminating Engineering Society. 2002;31(2):101-+.
27. L. Edwards, Torcellini P. A Literature Review of the Effects of Natural Light on Building Occupants. 1617 Cole Boulevard
Golden, Colorado 80401-3393: National Renewable Energy Laboratory; 2002.
28. Daniel J. Fonseca, Kirtikumar B. Bisen, Midkiff KC, Moynihan GP. An expert system for lighting energy
management in public school facilities. Expert Systems. 2006;Vol. 23:194-211.

29. Perez YV, Capeluto IG. Climatic considerations in school building design in the hot-humid climate for reducing energy consumption. Applied Energy. 2009;86(3):340-8.
30. Ramasoot T, Fotios S. Lighting for the Classrooms of the Future.
Electronic classrooms: A new challenge for school lighting guidance. Light & Engineering. 2009;17:62-70.
31. Winterbottom M, Wilkins A. Lighting and discomfort in the classroom. Journal of Environmental Psychology. 2009;29(1):63-75.
32. Tanner CK, Uline CL. Effects of school design on student outcomes. Journal of Educational Administration. 2009;47(3):381-99.
33. Ramli NH, Masri MH, Zafrullah M, Taib HM, Hamid NA. A Comparative Study of Green School Guidelines. Procedia - Social and Behavioral Sciences. 2012;50:462-71.
34. Pellegrino A, Cammarano S, Savio V. Daylighting for Green Schools: A Resource for Indoor Quality and Energy Efficiency in Educational Environments. Energy Procedia. 2015;78:3162-7.
35. Secchi S, Sciurpi F, Pierangioli L, Randazzo M. Retrofit Strategies for the Improvement of Visual Comfort and Energy Performance of Classrooms with Large Windows Exposed to East. Energy Procedia. 2015;78:3144-9.
36. Abdolreza Gilavand Mohammadreza Gilavand SG. Investigating the Impact of Lighting Educational Spaces on Learning and Academic Achievement of Elementary Students. International Journal of Pediatrics. 2016;4.
37. Bellia L, Fragliasso F, Stefanizzi E. Why are daylight-linked controls (DLCs) not so spread? A literature review. Building and Environment. 2016;106:301-12.
38. Costanzo V, Evola G, Marletta L. A Review of Daylighting Strategies in Schools: State of the Art and Expected Future Trends. Energy and Buildings. 2017;7(4).
39. Costanzo V, Evola G, Marletta L, Pistone Nascone F. Application of Climate Based Daylight Modelling to the Refurbishment of a School Building in Sicily. Sustainability. 2018;10(8).
40. Nocera F, Lo Faro A, Costanzo V, Raciti C. Daylight Performance of Classrooms in a Mediterranean School Heritage Building. Sustainability. 2018;10(10).
41. Calama-Gonzalez CM, Suarez R, Leon-Rodriguez AL. Thermal and Lighting Consumption Savings in Classrooms Retrofitted with Shading Devices in a Hot Climate. Energies. 2018;11(10).
42. Vásquez NG, Felippe ML, Pereira FOR, Kuhnen A. Luminous and visual preferences of young children in their classrooms: Curtain use, artificial lighting and window views. Building and Environment. 2019;152:59-73.
43. Bakmohammadi P, Noorzai E. Optimization of the design of the primary school classrooms in terms of energy and daylight performance considering occupants' thermal and visual comfort. Energy Reports. 2020;6:1590-607.
44. Baloch RMM, Maesano CN, Christoffersen J, Mandin C, Csobod E, Fernandes EO, et al. Daylight and School Performance in European Schoolchildren. Int J Environ Res Public Health. 2020;18(1).
45. Barkmann C, Wessolowski N, Schulte-Markwort M. Applicability and efficacy of variable light in schools. Physiol Behav. 2012;105(3):621-7.
46. De Giuli V, Da Pos O, De Carli M. Indoor environmental quality and pupil perception in Italian primary schools. Building and Environment. 2012;56:335-45.
47. De Giuli V, Zecchin R, Corain L, Salmaso L. Measured and perceived environmental comfort: field monitoring in an Italian school. Applied Ergonomics. 2014;45(4):1035-47.
48. PJC Sleegers, Moolenaar N, Galetzka M, A Pruyn, Sarroukh B, Zande Bvd. Lighting affects students' concentration. Lighting Research & Technology. 2013(45):17.
49. Peng S, Chen Y, Tang X, Heynderickx I. Lighting For Projector Use In School Classrooms To Improve Visibility. IEEE. 2014.
50. Georgieva D, Schledermann KM, Nielsen SML, Hansen EK. Designing User Centred Intelligent Classroom Lighting. In: Brooks AL, Brooks E, Vidakis N, editors. Interactivity, Game

Creation, Design, Learning, and Innovation. Lecture Notes of the Institute for Computer Sciences Social Informatics and Telecommunications Engineering. 229. New York: Springer; 2018. p. 314-23.
51. L. Morrow B, M. Kanakri S. The impact of fluorescent and led lighting on students attitudes and behavior in the classroom. Advances in Pediatric Research. 2018.
52. Pulay A, Read M, Tural E, Lee S. Examining student behavior under two correlated color temperature levels of lighting in an Elementary School classroom. Educational Planning. 2018;23:53-69.
53. Doulos LT, Kontadakis A, Madias EN, Sinou M, Tsangrassoulis A. Minimizing energy consumption for artificial lighting in a typical classroom of a Hellenic public school aiming for near Zero Energy Building using LED DC luminaires and daylight harvesting systems. Energy and Buildings. 2019;194:201-17.
54. Michael A, Heracleous C. Assessment of natural lighting performance and visual comfort of educational architecture in Southern Europe: The case of typical educational school premises in Cyprus. Energy and Buildings. 2017;140:443-57.
55. Hartstein LE, Tuzikas A, Karlicek RF. The Impact of Dynamic Changes in Light Spectral Power Distribution on Cognitive Performance and Wellbeing. Leukos. 2019;16(4):289-301.
56. Keis O, Helbig H, Streb J, Hille K. Influence of blue-enriched classroom lighting on students' cognitive performance. Trends in Neuroscience and Education. 2014;3(3-4):86-92.
57. Rands MLG-T, Ann M. The Room Itself Is Active: How Classroom Design Impacts Student. 2017:9.
58. Marchand GC, Nardi NM, Reynolds D, Pamoukov S. The impact of the classroom built environment on student perceptions and learning. Journal of Environmental Psychology. 2014;40:187-97.
59. Kerstin S. Corridors, Classrooms, Classification – The impact of school layout on pedagogy and social behaviours. Designing for the future of schooling: Routledge; 2018.

Design Principles for Natural Lighting

P Thursfield [1], R vd Ven[2]

[1] Light Experience Design Lead, Signify, Eindhoven, The Netherlands
[2] Lighting Designer, Signify, Eindhoven, The Netherlands
E-mail: paul.thursfield@signify.com

Abstract. Natural day light has important qualities that we need for our physical comfort, biological health, and psychological wellbeing. Modern lighting addresses our physical need for light and human centric lighting our biological needs, but both lack qualities of natural daylight we need. The Design Principles for Natural Lighting references Biophilic design theory and focuses on our experience of light and its psychological effect through our senses. A model is proposed for composing natural light experiences practically and various examples assessed. NatureConnect is presented as a purposely designed lighting system based on the design principles and its approach discussed.

1. Introduction

People are naturally attracted to nature and sunlight. They seek out the sun when on holiday and enjoy sporting pursuits such as sailing, cycling, and running in nature. The homes people value most have views on natural settings, particularly overlooking water, and the sea. To unwind or revive themselves, people go outside for a walk and a breath of fresh air. This fact that we find the outdoors and natural daylight invigorating is not surprising, after all we have evolved as a species living in natural environments for millennia. It is only in recent history with the dawning of the industrial age that people migrated from the countryside and an outdoor life to live and work in cities and a predominantly indoor lifestyle. According to the United Nations [1] it is now estimated that 82% of Americans and 74% of Europeans live in built-up urban environments and a study sponsored by U.S. Environmental Protection Agency [2] found that Americans spend 87% of their day living indoors! Indoor services such as sanitation, ventilation, heating, and lighting have grown as this migration progressed and have improved our basic health and life expectancy. However, it has also led to a new wave of lifestyle diseases and mental health disorders. There is a growing consensus that indoor conditions are inherently unnatural for us and are poor for our long-term health and wellbeing [3]. Reconnecting with nature, fresh air and natural light are important to addressing our modern ills with the introduction of new standards such as the Well building standard [4] and theories such as Biophilia [5] and Biophilic Design [6]. Human centric lighting is a solution that attempts to solve this, but only addresses one dimension, our biological needs. To create a complete solution, we must address all three dimensions, physical comfort, biological heath and last but not least our psychological wellbeing. To do this we must bring all the qualities of nature into our living and working spaces. This paper focuses on the missing qualities of nature experienced through light, the psychological effects, and the way we can design them.

2. The purpose and function of lighting

The early years of electric illumination was predominantly driven by the needs of industry to lengthen the workday into the night and to provide sufficient illumination levels needed for safe and efficient working conditions inside buildings. The decorative possibilities of lighting also emerged during this time using light to illuminate architecture, interiors, and merchandise for the purpose of beautification and attracting people's attention. The lighting design profession arose to satisfy the needs for good quality lighting that combined functional performance and attractiveness. Another more specialized lighting design profession also emerged in the theatre, music and movie industries using lighting as a scenic, narrative, and emotional device. Today these functions of light are well known, and the principles of application well understood. However, we are entering a new era where the health and wellbeing benefits of artificial lighting and its application is emerging with solutions such as human centric lighting.

2.1. Light for health and wellbeing

Even though the health and wellbeing benefits of natural light have been known and used since antiquity when the Greeks used solariums to enjoy sunlight and its health-giving properties, it was not until 2007 that science finally proved the existence of light sensitive receptors in the human eye that influence our biorhythm [7]. The use of 'human centric lighting' has developed over the last decade with measurable biological outcomes but has still to be widely adopted, which the authors believe is in part because there is not one but three aspects that are important for health and wellbeing:

2.2. Physical Comfort

Our eyesight performs optimally under the broad spectrum of natural daylight, and the contrast between direct lit areas and shadows are important for our perception of depth and shape. We have all experienced how quickly we can become disorientated in fog or a white-out while skiing on snow and how colours disappear in the twilight after the sun has set. We also do not like too much light direct light in our eyes which we experience as glare, such as very bright sunlight reflecting off water. The right qualities of light can affect how physically comfortable we feel [8] and helps us to perform activities optimally.

2.3. Biological Health

The non-visual 'third receptor' in the human eye is sensitive to the electromagnetic spectrum with peak response in the blue zone at 480 nm, stimulating the production of hormones that govern our biological clock and determining our state of alertness or preparedness for sleep [9]. These wavelengths are found in abundance in natural daylight and absent during the night aiding our readiness for rest and sleep. Artificial light used in offices is a poor source of these wavelengths during the day leading to the erosion of healthy circadian rhythms that has multiple effects on health and wellbeing [3].

2.4. Psychological Wellbeing

The psychological effects of light are well known in decorative and theatre lighting where our emotional response to certain kinds of light effects can be exploited to create the desired mood and atmosphere. Many of these effects are derived from experiences found in nature such as the golden hour at sunset that has an instant relaxation effect and is strongly associated with social interaction at the end of the day. In the same way a bright sunny morning immediately lifts our spirits with the prospect of a beautiful day and gives us an energy boost when we see the bright blue sky and warm sunlight. The constantly changing dynamics of nature such as the relaxing dapples of sunlight in the shade of trees and the sparkles on water are also very influential on our state of mind and energy levels as set out in the attention restoration theory [10].

It is this last category the authors have researched in detail, as the way we perceive, and experience natural light and nature has great potential for benefiting people's health and wellbeing.

3. Our sense of light

Like our sense of hearing, smell, taste and touch, our sense of light, our sight is an important part of how we experience our surroundings. Our senses are so innate and effortless to us that we are not necessarily aware of how influential they are. We do not just see, but also feel and know about the world around us because of the visual input of light. This is strongest when it comes to natural environments, for which we have special affinity, first proposed by Edward O Wilson, who's theory of Biophilia [5] meaning 'the love of living things', describes how we have a psychological and even genetically determined need for nature. These ideas were further elaborated into a general theory of Biophilic Design [6] for architecture and recently Terrapin Bright Green, a consultant on workspaces published the *14 Patterns of Biophilic Design* [11] identifying the ways we can exploit our affinity for nature to design better environments for our health and wellbeing. A companion document also sets out the economic value that Biophilic Design approach can generate [12].

These ideas inspired the authors to explore how we might practically use artificial light to obtain these positive benefits for people living and working indoors far from natural daylight. Using the principles of Biophilic design we identified three fundamental sensations that are essential for light to reconnect us with the processes and features of nature and therefore its experience:

3.1. The Sensation of space – distribution of light in the environment

Light needs surfaces and objects to become manifested and the way light interacts with our surroundings helps us determine the depth, shape, and texture through the contrast between light and dark areas in our vision. As mentioned previously we become aware of this when fog and mist reduce our ability to orientate ourselves due to the lack of contrast and features. Light changes the apparent size and shape of spaces leading us to see our surroundings differently depending on the time of day and is often used in architectural lighting to shape our experience of indoor spaces.

3.2. The sensation of time – the changing qualities of light over the day

The elevation and direction of the sun and the colour of the sky tell us about the time of day and even the season due to the movement of the earth and the sun. These changes are slow and gradual and come with clear signatures such as the shape and length of shadows and the changing gradient of colours in the sky at sunrise and sunset. These aspects also differ depending on where we live on earth, such as the very long twilight found in Scandinavia, or the very rapid arrival of nigh in equatorial regions.

3.3. The sensation of energy – the constantly changing dynamics found in nature

The seemingly chaotic and random movements found in nature derived from the interaction between sunlight, the weather, and the environment, give us the sensation of mood and atmosphere from energetic liveliness to peaceful tranquillity. Signatures such as sparkles of light on water or the filtered light through leaves and the resultant dapples of light under trees make us aware of the conditions around us. Natural environments are constantly changing and never repeating but very familiar to us because of the recognizable patterns and frequencies that are explicitly different from the noise and pollution in manmade environments [10].

Though each of these elements can be identified, isolated, and analysed it is the goal of this paper to show how these elements are utilized to create complete compositions or experiences that are richer than the sum of their parts. It is when the elements come together that the magic of nature truly comes to life.

4. Designing Natural Lighting experiences

The first coherent principle for designing with artificial lighting was proposed by Richard Kelly in a lecture called 'Lighting as an Integral Part of Architecture' in 1952 [13]. He describes three elements; *ambient luminescence*, *focal glow* and *play of brilliants*, each explained with observations stemming from natural experiences 'figure 1'. He also describes their application using the analogy of how an artist uses watercolours to paint a landscape, balancing the use of broad washes of colour (ambient luminescence), contrast (focal glow) and detail (play of brilliants) to create a composition. There can be different combinations of the three elements resulting in varying but equally pleasing outcomes.

"Ambient luminescence is the uninterrupted light of a snowy morning in the open country. It is foglight at sea in a small boat, it is twilight haze on a wide river where shore and water and sky are indistinguishable."

"Focal glow is the shaft of sunshine that warms the end of the valley. Focal glow draws attention, pulls together diverse parts, sells merchandise, separates the important from the unimportant, helps people see."

"Play of brilliants is sunlight on a fountain or a rippling brook. Play of brilliants excites the optic nerves, and in turn stimulates the body and spirit, quickens the appetite, awakens curiosity, sharpens the wit...."

Figure 1. Three elements of lighting, Richard Kelly, 1952

Though Kelly never explicitly mentions the idea of natural lighting or health and wellbeing as an outcome of his approach he does address the idea of people's experience, and his three elements have a rough correlation with the sensorial aspects of space, time, and energy. However, it should also be recognized that his approach was created at a time when lighting design was about lighting architecture, as the title of his lecture suggests. The simplicity and flexibility of Kelly's model in application is difficult to supersede as an approach to lighting design, so perhaps this explains why no fresh set of principles have emerged since the 1950's.

4.1. The language of light – composing with light

Besides the specific elements identified by Kelly it is his analogy to watercolour painting that is most intriguing. He could have easily made the same analogy to other creative professions such as musicians, cinematographers, chefs, or writers. What all creative professions have in common is a set of elements such as 'words' that are brought together using design principles such as 'grammar' to achieve narrative composition that is meaningful and engaging to a broad audience. When we hear the voice of the 'narrator' in our minds, we are not only able to understand the story but also appreciate the manner with which the story is told. All creative acts involve a language and grammar of a kind, and light is no exception conveying a strong meaning and feeling, i.e., we see, feel, and know things from light. Painters know the language of light very well as they use light and shadow in paint to create an illusion of place, time of day, the atmosphere, and emotion of subjects. Painting with light has many parallels with this so it's not surprising that Kelly used this as his analogy, and as in painting, if we observe the natural world, we can use our chosen medium of light to create an impression of nature.

4.2. Principles for natural lighting

To create experiences of natural lighting we need a set of design principles, our 'grammar', to guide the creation process both intellectually and practically. During the process of prototyping natural light experiences, the authors developed a layer model 'figure 2' to guide the composition of natural light experiences. Comprising four levels, each layer represents a different scale of space, time, and energy, each layer ever closer to the observer. The approach is not entirely new as a similar approach can be found in constructing VR environments where reality is re-created based on physical models of a real environment. What makes this approach unique is its application to artificial lighting in a real-world setting 'figure 3', where the objective is to create the impression of sunlight entering an indoor space that might lack any real connection with outdoors.

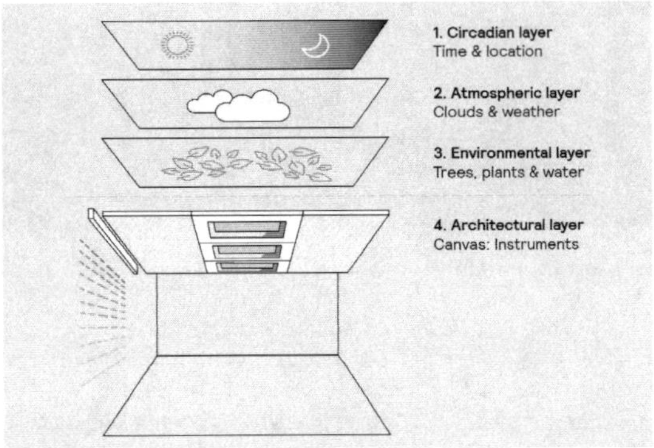

Figure 2. Layer model natural light, Authors, 2019

Figure 3. NatureConnect concept, Authors, 2019

4.2.1. Circadian layer. The long cycle of the day and the seasons, connecting us with the slow cycle of change of sunrise, daylight, sunset, twilight, moonlight etc. and the change of season from spring to summer, autumn and finally winter, orientating us to the time of day and season.

4.2.2. Atmospheric layer. The changing conditions of the weather from a heavily overcast, dull, and chilly winters day through to a warm sunlit summer's day, making us aware of the changeability of the weather and its many moods from relaxed and lazy to exciting and even fury.

4.2.3. Environmental layer. The way natural surroundings filter daylight, from the playful dapples and shadows of trees to the sparkles of sunlight dancing on water, and the reflections that come from the refracted light that make us aware of the living qualities of nature.

4.2.4. Architectural layer. The way light enters a building, via windows, skylights and other architectural features that connect us with the outdoors, offering us a prime vantage point or (prospect) while offering protection against dangers (refuge).

4.3. Putting the natural lighting principles into practice

The earliest examples of this layer model approach can be found in *trompe l'oeil*, first seen during the ancient Greek period, to create illusory spaces in architecture. This technique of 'fooling the eye' reached its peak in the frescos of Renaissance architecture as seen in 'figure 4'. All the methods of light and shadow, forced perspective and foreshortening are being utilized to create a believable architectural oculus with an illusory space beyond containing cherubs, animals and people gazing in. Despite the convincing rendering of the illusion, it is only partially convincing due to the inherent limitations of paint that cannot reproduce the contrast and brightness of real light. Also, its lack of depth means that *motion parallax*, the perception that objects that are closer appear to move faster than objects that are further away, immediately renders the image as flat and constrained to the surface on which the image is rendered. Despite these limitations it is still captivating.

Figure 4. Oculus on the ceiling of the Spouses Chamber, castle of San Giorgio in Mantua, Italy

Figure 5. The Weather Project, Tate Modern, Olafur Eliasson, 2007

More recently there are examples of artists and designers exploring natural experiences in their work. To name two notable examples we should start with Olafur Eliasson's monumental installation at the Tate Modern, 'figure 5' of a setting sun that made him famous and highlighted his ideas about the human experience in art and his frequent use of natural phenomena often using light. This work triggered a most unusual response from visitors, such as lying on the floor to sunbathe and dancing in what could be seen as a form of sun worship! This experiential and emotional connection with an artwork is also found in the work of James Turrell, 'figure 6 & 7' who uses daylight as his medium to create framed experiences. Turrell describes his objective as encouraging people to slow down and take time to observe the quality of light in nature and appreciate its ever-changing beauty. What connects these two artist's works is use of natural phenomena that stimulates the feelings and experience of the audience rather than an expression of the artist and his technique. Eliasson for instance sees himself as a co-producer of reality where his work is only complete with the involvement of the audience [14].

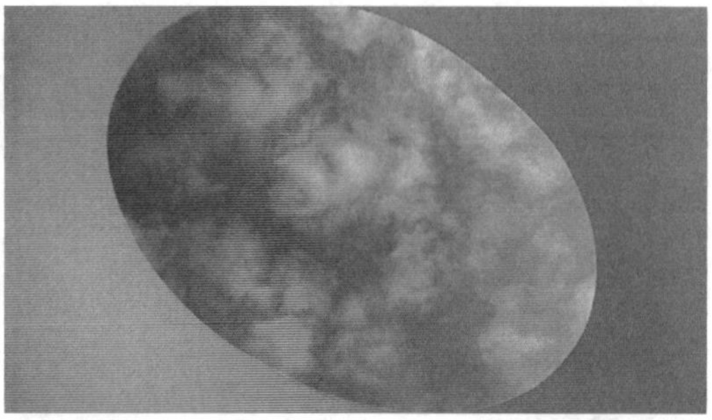

Figure 6. Skyspace, 2011, James Turrell, Tremenheere Sculpture Gardens, Cornwall, UK

Figure 7. Stonesky, 2005, James Turrell, Nappa Valley, California

The earliest example of creating a natural lighting effect for a work setting is probably Naoto Fukasawa's installation shown in 'figure 8'. Fukasawa creates an abstract blue sky using back projection on a horizontal screen suspended over a worktable with a separate spotlight providing the sensation of warm sunlight spilling over the chair. In many ways the illusion is not very realistic but creates the impression of sunniness through the effective use of the colour contrast between the blue light of the diffused sky and the warm white directional spotlight to emulate sunlight. Another example of creating the illusion of natural light is Daniel Rybakken's installation *The subconscious effect of daylight* in 'figure 9' a series of experiments that creates the illusion of sunlight that falls into a room seemingly from a hidden window, striking the wall opposite. The illusion is very convincing, though in reality it is made using an LED backlight embedded in the wall and does not change shape or white point with the time of day.

Figure 8. Personal Skies, Naoto Fukasawa, 'Workspheres' exhibition 2001 MoMA, NY

Figure 9. Subconscious effect of daylight, Daniel Rybakken, 2007

Working from these references and observations of real skylights the authors and colleagues identified the various important signature elements needed to create a convincing skylight illusion. The sequence of images in 'figure 10' are as series of renderings in which the first is a simple blue sky with recessed window frame. The second introduces the reflection of the glass plane that deepens the apparent depth of the sky. The third image in the sequence introduces shadow line between the sunlit and shadow side of the window frame that gives the impression of the sun's direction and elevation.

Figure 10. Skylight renderings series 1, Authors 2022

The second series of images in 'figure 11' show the changing aspect of the skylight illusion as time passes. The first image shows a cloudy skylight where the sunlight shadow has disappeared, the second a sunset effect with golden light and coloured sky with long sun shadow and the last image of a night sky with moonlight shadow effect.

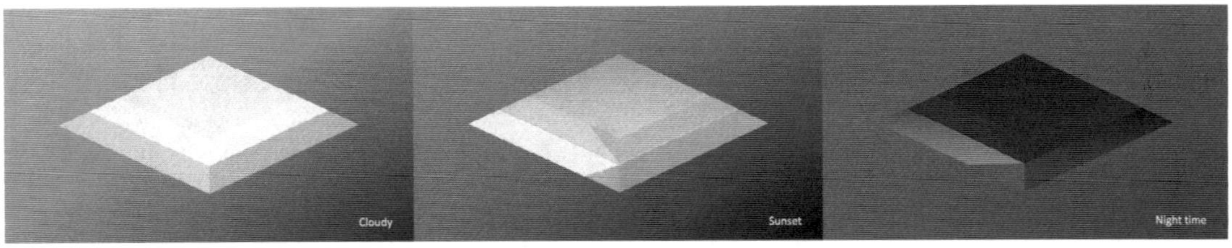

Figure 11. Skylight renderings series 2, Authors 2022

These studies were refined through various prototypes into commercial solution called NatureConnect 'figure 12' where a series of skylights are brought together with other lighting elements to create a meeting room with a natural feeling of sunlight and connection with outdoors. Surrounding the skylights are daylight panels that bring white tuneable illumination and the LightScape wall washer that makes a sky gradient with changing colours and vertical illumination that enlarges the apparent size of the room. The lighting system changes dynamically throughout the day emulating a day rhythm 'figure 13' from bright and energizing midday sun to the golden tones of sunset, and the cool darks of twilight.

What is not immediately apparent from images of the system is the immersive quality of light in the room that literally makes the room feel sunny, even in the absence of any windows. Though it is impractical to create the brightness of sunlight, which can be up to 100,000 lux on a bright day, it is in fact unnecessary and undesirable for meeting the office requirements of glare with UGR<19 and people's thermal comfort. What is more important is creating the feeling of sunlight, for which 1000 lux seems to be more than adequate and ensuring colour rendering of CRI>90 with no discernible flicker, as found in natural daylight. However, besides the technical characteristics, the most important trigger for the sunniness experience seems to be the simultaneous presence of cool sky tones contrasted against warm splashes of sunlight that varies correctly for the time of day. Learning to 'fine tune' and 'play' the orchestra of light 'instruments' to balance both technical criteria and a convincing experience of changing daylight over the day was the biggest challenge as there were no precedents or examples to follow and is still the subject of further exploration.

IOP Conf. Series: Earth and Environmental Science 1099 (2022) 012033 doi:10.1088/1755-1315/1099/1/012033

Figure 12. NatureConnect lighting system, Authors 2022

Figure 13. NatureConnect lighting system at different times of the day, Authors 2022

5. Conclusions

We believe that the NatureConnect product has successfully balanced the competing demands of good visual comfort, with sufficient melanopic lux levels for biological health and the unmistakably feel of natural daylight and sunniness that is positive for psychological wellbeing. The solution meets and in fact exceeds many of the criteria the very best office lighting standards such as the Well Building. Feedback from those who regularly work in environments with NatureConnect lighting report positive benefits of living and working under what feels like real sunlight, and though the illusion of natural daylight in NatureConnect is far from perfect, the authors have learnt that perfect fidelity is not entirely necessary. All that is required for the illusion to work is something that is reasonably analogous to the real thing. In fact, attempts achieving hyper real effects often end with an uncanny feeling that is quite undesirable. If we return to the analogy with painting, the objective instead is to create an impressionistic composition where the mind recognizes what is intended but is free to fill in the details, as this seems more pleasing.

It is also unnecessary to re-create the incredibly high light levels found outdoors on a sunny day which would make the lighting system unaffordable to run and is in fact uncomfortable for workers in an office. Being asynchronous with outdoors is also another advantage, where users are free to decide what they need on a given day. If I am living in Norway in the depths of winter, being able to dial up some extra sun time when it is twilight all day is highly desirable. I may also wish to have a less sunny effect at times to relax or make a presentation.

What are clearly lacking in the NatureConnect system are the faster dynamics identified by the design principles for natural lighting. The natural variations found in nature, such as seasonal variations, sun variations due to cloud, and nature textures such as dynamic shadows and dapples of leaves are among the missing elements that could add further cues that give a connection with nature. Early experiments indicate that subtle ongoing variations have an impact on the impression of naturalness and are strongly associated with natural environments. These variations seem not to be distracting but rather relaxing and energizing simultaneously, in what is referred to as *attention restoration* in Biophilic theory [10]. The authors continue to develop further ideas based on these observations and the further application of the design principles for natural lighting.

6. References

[1] United Nations. 2018. World Urbanization Prospects, Revision 2018: The keyfacts. *United Nations*, https://population.un.org/wup/Publications/

[2] Klepeis, N E et al. 2001. The National Human Activity Pattern Survey (HHAPS): A Resource for Assessing Exposure to Environmental Pollutants. *Lawrence Berkeley National Laboratory*.

[3] Boubekri M, Cheung I N, Reid K J, Wang C H, Zee P C. 2014. Impact of windows and daylight exposure on overall health and sleep quality of office workers: a case-control pilot study. *J Clin Sleep Med*. **10**(6) pp 603-611

[4] IWBI 2022. WELLv2, Q2, 2022. https://v2.wellcertified.com/en/wellv2/overview.

[5] Wilson, Edward O. 1984. *Biophilia*. (Cambridge, MA: Harvard University Press) ISBN 0-674-07442-4

[6] Kellert S F, Heerwagen J H and M L Mador Eds. 2008. *Biophilic Design: The Theory, Science & Practice of Bringing Buildings to Life*. (Hoboken, NJ: John Wiley & Sons)

[7] Zaidi F H, et al. 2007. Short-wavelength light sensitivity of circadian pupillary, and visusal awareness in humans lacking an outer retina. *Current Biology*. **17** (24), pp2122–8.

[8] Smolders K C, de Kort Y A, Cluitmans PJ. 2012. A higher illuminance induces alertness even during office hours: findings on subjective measures, task performance and heart rate measures. *Physiol Behav*. **107**(1) pp 7-16

[9] Foster RG, Hankins MW. 2007. Circadian vision. *Curr Biol*. **17**(17) R746-R751

[10] Kaplan R, & Kaplan S. 1989. *The experience of nature: A psychological perspective*. (Cambridge: Cambridge University Press)

[11] Terrapin Bright Green. 2014. *The 14 Patterns of Biophilic Design: Improving Health and Wellbeing in the Built Environment*. (Terrapin Bright Green, LLC)

[12] Terrapin Bright Green. 2014. *The Economics of Biophilia*. (Terrapin Bright Green, LLC)

[13] Kelly R. 1952. Lighting as an Integral Part of Architecture. *College Art Journal*, **12** (1), pp 24-30. doi:10.2307/773361

[14] Eliasson O. 2010. The reality we live in is our own construction. *Spiegel Online*. https://www.spiegel.de/international/zeitgeist/danish-artist-olafur-eliasson-the-reality-we-live-in-is-our-own-construction-a-691464.html

Analyzing Natural Lighting Conditions from the Perspective of Biophilic Design in Indoor Office Environments

Gulsah Dogan Karaman (1), Ayse Nihan Avci (2)
(1) Çankaya University Faculty of Architecture Department of Interior Architecture, Turkey, gdogan@cankaya.edu.tr; (2) Çankaya University Faculty of Architecture Department of Interior Architecture, Turkey.

Abstract

The term "biophilic design" is recognized by some designers who coined a word for designing instinctively using fresh understandings, reasons, and views, which helps reconnect us with the natural world in the indoor environment. Within the various patterns that connect humans with nature, natural lighting provides an orientation to the day and night according to the sun's location and cycles, which is essential for human health and wellness. The supply and quality of natural lighting, including its influence on mood, have been extensively investigated and written about for many years in various locations and the overlapping science and design disciplines. This paper analyzes the biophilic design approach to mood for the natural lighting conditions that change according to the position of the office environments in winter. The selected two indoor office environments were located in the Faculty of Architecture at Çankaya University on Main Campus, on the second floor of the building. Office-1 overlooked the design studio, preventing it from taking natural lighting directly, and Office-2 directly penetrated the natural lighting. Natural lighting-based photos were taken from the selected offices during three periods of the day. The photos and a survey were directed at the two groups: Group 1 consisted of the students who took the Natural and Artificial Lighting Course, and Group 2 students who did not take this course. The experimental instruments analyzed their moods to evaluate natural lighting conditions.

Keywords *Natural light, Biophilic Design, Mood, Office Environment*

1. INTRODUCTION

1.1. Natural Lighting

Natural light is light received on Earth from the sun and a mixture of direct, atmospheric, and indirect light. It is distinguished by its variety [1]. Sunlight has a broad spectrum, with a peak in the blue-green and a cut-off in the ultra-violet at 290 nanometers (nm), due to the Earth's atmosphere acting as a filter. The atmosphere intercepts shorter wavelengths, with the ozone layer playing an important role. Natural light falling on a desk in space consists of three elements: direct sunlight from the sky, light from external surfaces, and light from internal surfaces. It also varies from time to time and from site to site in amount, spectrum and distribution.

The experience of natural lighting is vital to human health, well-being, productivity, motivation, and performance, allowing orientation to the day, night, and seasons in response to the position and cycles of the sun [2]. Natural lighting awareness may also aid movement and wayfinding and provide comfort and satisfaction. Natural lighting may take on aesthetically pleasing shapes and patterns beyond mere exposure through the creative interplay of light and shadow, diffuse and changeable light, and the integration of light with spatial qualities. Natural lighting may be introduced deep into interior rooms using glass walls and clerestories, reflective colors and materials, and other interior design strategies.

1.2. Natural Lighting and Biophilic Design in an Indoor Environment

Kellert [3] argues that nurturing man's intrinsic passion for and connection to the environment, called biophilia, is critical to humans' well-being in the modern world. Humans are biologically evolved to respond positively to interaction with nature, and this relationship can help them achieve intellectual, emotional, and spiritual fulfillment [4]. There is a design strategy called "biophilic design" to create a connection with nature in an environment.

People spend most of their time indoors [5]. Feeling nature in the interior directly, indirectly, and integrated with the space can also be considered a crucial feature of the biophilic design strategy that sustains the human focus [2]. According to Kellert et al. [6], biophilic design identifies two basic dimensions in an indoor environment. The first fundamental dimension of biophilic design is an organic or naturalistic dimension, described as shapes and forms in an indoor environment that directly, indirectly, or symbolically represent the intrinsic human attraction to nature.

The second fundamental dimension of biophilic design is a place-based or vernacular dimension, described as structures and landscapes that link to the culture and ecology of a location or geographic area. These two dimensions are followed by six biophilic design elements in an indoor environment: Environmental features, natural shapes and forms, natural patterns and processes, light and space, place-based relationships, and evolved human-nature relationships. Figure 1 represents the biophilic design's two basic dimensions and six design elements.

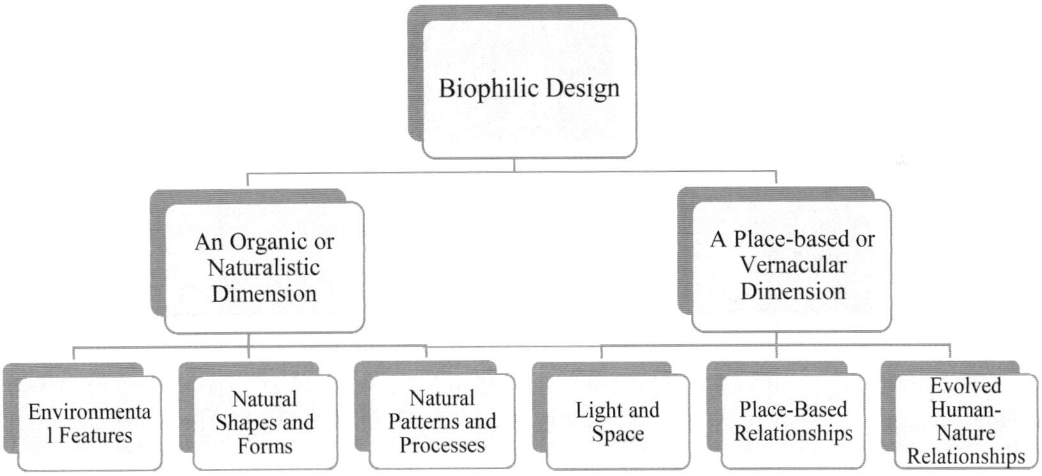

Figure 1. The Two Basic Dimensions and Six Design Elements of the Biophilic Design [6]

The biophilic design closely relates to physical, sociological, and psychological well-being. Humans' exposure to nature is thought to benefit their psychological health. Therefore, it is possible to create an atmosphere that encourages well-being by adopting restorative methods that boost mood and increase mental health, lowering stress. Natural substances such as natural lighting can help people cope with stress and improve their mood by reducing anxiety [3].

Natural lighting contains all hues and positively impacts human productivity and psyche. Light and shadow create an effect that emphasizes or helps differentiate one object from another. Reflected light reduces eye strain by balancing glare and producing a soft impression. Warm light makes people feel protected, and peaceful spatial diversity boosts mental and emotional activity [7]. To sense the natural balance and harmony, the inside-outside areas must connect the inside with natural lighting [3].

According to the World Health Organization, one in every ten people in the world has a mental illness, and two in every five people in Turkey are in a negative mood [8]. According to the studies, simply looking at nature photos can boost mood, working memory, and cognitive abilities [9]. Despite mounting evidence of links between nature interaction and stress reduction in urban settings [10], little is known regarding the impact of natural integration in indoor environments [11]. In the workplace, incorporating natural components into the setting promotes mental health advantages such as better focus, less stress, and improved mood [12, 13, 14]. From that perspective, humans evolved under natural, diurnal light conditions, so natural lighting and its processes should be preferred and beneficial. The hue and angle of natural lighting in the interior influence the room's mood at different times of the day [15].

Moods are seen as subjective emotional states [16] and mood augmentation is regarded as a primary method of stress reduction [13]. Biophilic design is mostly connected with stress reduction results, and there are several studies that show the biophilic design and mood connection [17, 18]. The initial five minutes of nature exposure result in the biggest boosts to mood and self-esteem [19]. Mood as a sub-title associated with biophilic design is assessed in the study with Positive and Negative Affect Scale (PANAS).

2. METHOD

2.1. Participants

The volunteered participants for the study consisted of 72 students from the Faculty of Architecture, Çankaya University, Ankara, Turkey: thirty-six took the Natural and Artificial Lighting Course and were in the 4th class, and another half of them were in the 1st class who did not take this course. This course was included in the 2nd class curriculum, so they were chosen from the 1st and 4th classes. Their ages ranged from 18 to 31.

2.2. Experimental Setting and Lighting Conditions

This paper aims to analyze the natural lighting conditions on mood from a biophilic design approach perspective according to the two different positioned office environments. The selected two indoor office environments were located in the Faculty of Architecture at Çankaya University / Main Campus. They were on the second floor of the building. Çankaya University is on 39°49' North Parallel and 32°33' East Meridian. The weather was mostly cloudy and the air temperature varied between -3°C and 8°C. Figure 2 shows the sun paths of the Çankaya University, on the 1st, 14th, and 28th of February in 2022.

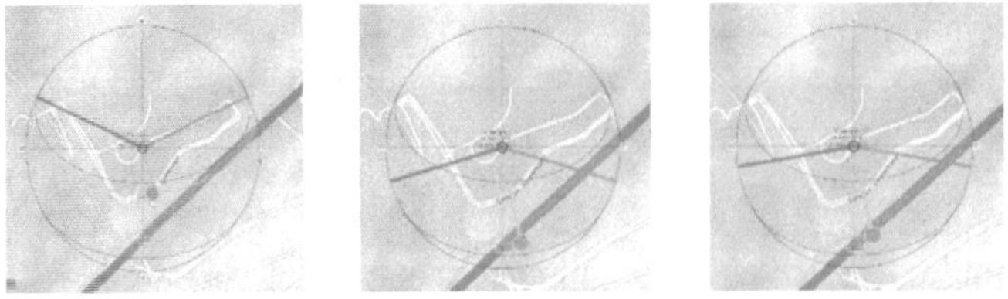

Figure 2. Sun Paths of the Çankaya University, Main Campus

An online survey is prepared to implement the intended comparison. Participants were required to reply the online survey in the design studio that was on the ground floor. It was 225.92 m2 (h: 3.30), with eight vertically positioned windows on one of the northwest-facing walls, white-painted walls, a white ceiling, and concrete on the floor. The working table near the first window was chosen for the experiment to provide the required illuminance level. It is white, and its dimensions are 120 x 80 cm. Figure 3 presents the part of the design studio. Accordingly, it was determined that the desired illuminance levels could be achieved for a reading task (500 lx) between 12:00 and 1:00 p.m. in winter.

Figure 3. The Desk that the survey conducted

2.3. Experimental Instruments and Procedure

The Positive and Negative Affect Schedule (PANAS) [20] was adopted to assess the natural lighting conditions in terms of biophilic design in indoor office environments. This scale is one of the scales used to measure the subjective mood. PANAS consists of 20 items. These 20 items cover the evaluation of ten positive and ten negative moods out of 5. Positive mood scales are interested, excited, strong, enthusiastic, proud, alert, inspired, determined, attentive, and active; negative mood scales are distressed, upset, guilty, scared, hostile, irritable, ashamed, nervous, jittery, and afraid. The PANAS test was also directed to the participants for 3-hour periods in which the illuminance level was measured in order to measure how 2 different offices that benefit from natural lighting or not, which is the main purpose of the study, affect the mood with a biophilic design approach. The selected two indoor office environments were located in the Faculty of Architecture at Çankaya University on Main Campus, on the second floor of the building. They were approximately 11 m2. Office 1 overlooked the design studio, preventing it from taking natural lighting directly, and Office 2 directly penetrated the natural lighting. Figure 4 shows the location of the offices in the building. Figure 5 and 6 represent the natural lighting-based photographs were taken from the selected offices during three-day periods (10:00 a.m., 1:00 p.m., and 4:00 p.m). A Luminance Meter, 'Konica Minolta CL-70F', was used to measure the three types of natural lighting characteristics, which were illuminance level (IL), color rendering index (CRI), and correlated color temperature (CCT) for each room. Table 1 presents the values of these lighting characteristics.

a) b) c)

Figure 4. The Location of the Offices in the Building a) Partial plan of the second floor (not to scale), b) View of the Office 1 (taken from the studio opposite), and c) Façade view of the Office 2 (taken from the outside)

a) b) c)

Figure 5. Natural lighting-based photos taken from the Office 1 in three periods of the day a) 10:00 a.m, b) 1:00, and c) 4:00 p.m

a) b) c)

Figure 6. Natural lighting-based photos taken from the Office 2 in three periods of the day a) 10:00 a.m, b) 1:00, and c) 4:00 p.m

IOP Conf. Series: Earth and Environmental Science 1099 (2022) 012034 doi:10.1088/1755-1315/1099/1/012034

Table 1. The Values of the Natural Lighting Characteristics in Selected Offices

	Lighting Characteristics		
Office 1	IL (lx)	CRI (Ra)	CCT (K)
10:00	59.7	90	4548
1:00	78.2	89.5	4479
4:00	63.4	89.3	4395
	Lighting Characteristics		
Office 2	IL (lx)	CRI (Ra)	CCT (K)
10:00	222	96.1	5999
1:00	637	95.6	5279
4:00	1280	95.3	5014

IL: Illuminance, **CRI:** Color rendering index, **CCT:** Correlated color temperature

Seventy-two students volunteered in the Faculty of Architecture. Data were collected between the 1st and 28th of February in 2022. According to the pre-scheduled timetable, participants were invited to the test environment (see Figure 2). The desired illuminance levels could be achieved for a reading task (500 lx) between 12:00 and 1:00 p.m. in winter. So, the experiment was carried out only at the specified time interval. They were then seated on a stool and allowed to place their laptop on the working table. First, they were informed about the aim of the research and the ethical assurance. Their access to the Google Survey prepared for the study was provided. It consisted of 11 questions. The first five questions were related to the class, age, gender, attending the natural and artificial lighting course, and knowledge of biophilic design. First, four questions had two options; the question related to the knowledge of biophilic design had the 5-point Likert scale. The other six questions include the photos of the offices with PANAS scales. To analyze their subjective mood according to the natural lighting conditions of the office environments, students were required to evaluate the photos with a 5-point Likert - PANAS scale. In addition, they were asked to interpret the experiment to see its positive and negative sides.

3. RESULTS AND DISCUSSION

IBM SPSS Statistics 23.0 software was used to analyze the collected data statistically. The Shapiro-Wilk test was used to evaluate the data for normality. Mean and standard deviation statistics were performed to make some observations about the distribution of the items. It was considered sufficient for normal distribution that the skewness and kurtosis values were between +3 and -3 [21, 22]. According to the results, it was accepted that the scale scores showed a normal distribution. Parametric methods were used in the analyzes. Independent samples t-test was utilized to investigate the relationship between sample groups (class and gender) with the PANAS Scale according to the natural lighting conditions of the offices in terms of biophilic design.

Table 2. Descriptive Statistics of the Data

Average of the items of the PANAS scale	N	Mean	SD
Office 1_10:00_Ave.	72	2.22	0.53
Office 1_1:00_Ave.	72	2.21	0.66
Office 1_4:00_Ave.	72	2.28	0.71
Office 2_10:00_Ave.	72	2.25	0.69
Office 2_1:00_Ave.	72	2.37	0.65
Office 2_4:00_Ave.	72	2.55	0.67
Office 1_10:00_Ave._P	72	1.65	0.59
Office 1_10:00_Ave._N	72	2.78	0.78
Office 1_1:00_Ave._P	72	2.36	0.80
Office 1_1:00_Ave._N	72	2.07	0.83
Office 1_4:00_Ave._P	72	2.48	0.84
Office 1_4:00_Ave._N	72	2.09	0.91
Office 2_10:00_Ave._P	72	2.33	0.89
Office_2_10:00_Ave._N	72	2.16	0.94
Office 2_1:00_Ave._P	72	2.77	0.88
Office 2_1:00_Ave._N	72	1.97	0.82
Office 2_4:00_Ave._P	72	3.28	0.91
Office 2_4:00_Ave._N	72	1.83	0.91

Ave.: Average, **P:** Positive items, **N:** Negative items

While analyzing the data, the averages of the adjectives in the PANAS test were taken in this study. Accordingly, the three periods of the day and the adjectives measuring positive and negative subjective moods were grouped according to the three periods of the day. Table 2 shows that the participants generally used the positive side of the rating scale to evaluate the natural lighting conditions with the PANAS Scale. Independent samples t-test was utilized to investigate the relationship between sample groups. For Office 1 and Office 2, the results indicated no statistically significant differences between the 1st and 4th classes of the students for all scales in all periods (t (70) = -1.644, p = 0.105; t (70) = -0.053, p = 0.958; t (70) = -0.634, p = 0.528; t (70) = 0.478, p = 0.634; t (70) = 0.289, p = -0.774; t (70) = -0.478, p = 0.634). No significant difference was also found when mood scales were evaluated separately as positive and negative items (t (70) = -0.777, p = 0.440; (t (70) = -1.628, p = 0.108; (t (70) = 0.604, p = 0.548; t (70) = -0.662, p = 0.510; t (70) = 0.264, p = 0.792; t (70) = -1.244, p = 0.218; t (70) = 0.742, p = 0.461; t (70) = 0.000, p = 1000; t (70) = 1.253, p = 0.214; t (70) = -0.876, p = 0.384; t (70) = 0.051, p = 0.959; t (70) = -0.762, p = 0.449). Table 3 presents these results.

Table 3. Results of the Independent Samples t-Test

Office	Side of the Scale	Time	*p*
Office 1	P+N	10:00	0.105
Office 1	P+N	01:00	0.958
Office 1	P+N	04:00	0.528
Office 2	P+N	10:00	0.634
Office 2	P+N	01:00	-0.774
Office 2	P+N	04:00	0.634

Office 1	Side of the Scale	Time	*p*	Office 2	Side of the Scale	Time	*p*
	P	10:00	0.440		P	10:00	0.461
	N	10:00	0.108		N	10:00	1000
	P	01:00	0.548		P	01:00	0.214
	N	01:00	0.510		N	01:00	0.384
	P	04:00	0.792		P	04:00	0.959
	N	04:00	0.218		N	04:00	0.449

P: Positive items, **N:** Negative items

According to the gender, there was a statistically significant difference for the 10.00 a.m. natural lighting condition t (70) = -3.042, p = 0.003) in Office 1. The evaluations of male groups (M= 2.53) were slightly higher than those of female groups (M= 2.06), and they felt slightly more intense emotions than female groups. However, male students generally evaluated the illuminance levels of the offices from photos in the range of 1-3, and their moods were not intensely affected. There were also statistically significant differences for the 1.00 p.m. and 4.00 p.m. natural lighting conditions (t (70) = -2.414, p = 0.018; t (70) = -2.351, p = 0.022) in Office 2. The evaluations of male groups (M= 2.62; M= 2.81) were slightly higher than those of female groups (M= 2.24; M= 2.42). Table 4 presents these results.

Table 4. Results According to the Gender (1)

Office / PANAS Item	Time 10:00	Time 01:00	Time 04:00
Office 1 / All Items	p = 0.003		
Office 2 / All Items		p = 0.018	p = 0.022

* It was evaluated by taking the average of all positive and negative items.

However, there were also significant differences found when mood scales were evaluated separately as positive and negative items. In other words, for Office 1, there were statistically significant differences on positive items at 10:00 a.m. (t (70) = -3.640, p = 0.001; M= 1.48, M= 1.98) negative items at 1:00 p.m. and 4 p.m. (t (70) = -3.498, p = 0.001; M= 1.84, M= 2.52), (t (70) = -2.193, p = 0.032; M= 1.93, M= 2.42). For Office 2, there were statistically significant differences on negative items at 1:00 p.m. and 4 p.m. (t (70) = -2.509, p = 0.014; M= 1.80, M= 2.30), (t (70) = -3.476, p = 0.001; M= 1.58, M= 2.32). The evaluations of male groups were slightly higher than those of female groups. The results indicated that the male participants felt slightly more intense emotions than the female participants. Table 5 presents these results.

Table 5. Results According to the Gender (2)

Office / PANAS Item		**Time**	
	10:00	**01:00**	**04:00**
Office 1 / Positive Items	p = 0.001		
Office 1 / Negative Items		p = 0.001	p = 0.032
Office 2 / Positive Items			
Office 2 / Negative Items		p = 0.014	p = 0.001

* It was evaluated separately as positive and negative items.

4. CONCLUSION

This study aimed to investigate the natural lighting conditions in terms of biophilic design in an indoor office environment. In the study, photos were taken at three different times of the day (10.00, 01.00, and 04.00) from two different office room types, one of which receives direct natural light and the other facing the studio, and questions were asked to the 1st and 4th-year students of the interior architecture department by using these photographs in the survey. As a result of the study, the effect of knowing the biophilic design and taking the lighting course was not determined. The main surprising result of the study was in terms of gender differences in the results. There was a significant similarity between the answers given by the male students, a remarkable similarity in the answers of the female participants, and a significant difference between the answers of the two genders.

This study might be limited in terms of participants, yet this study is a preliminary study that provides the basis for enlarged research. In future studies, the effects of gender on mood in terms of biophilic design and lighting can be examined in more detail by working with a more extensive study group.

REFERENCES

[1] Boyce, P R 2003 *Human factors in lighting* Crc Press.
[2] Kellert, S and Calabrese, E 2015 *The practice of biophilic design* Terrapin Bright LLC.
[3] Kellert, S R 2005 Building for life: Designing and understanding the Human-Nature connection (2nd Edition) Island Press.
[4] Soderlund, J and Newman, 2015 Biophilic architecture: A review of the rationale and outcomes. *AIMS Environmental Science* 2 (4) pp 950–969.

[5] Cramer, J S and Browning, W D 2008 Transforming building practices through biophilic design *Biophilic design* Hoboken: Wiley pp 335-346.
[6] Kellert, S R, Heerwagen, J and Mador, M 2011 *Biophilic design: the theory, science and practice of bringing buildings to life* John Wiley & Sons.
[7] Dogan Karaman, G and Selçuk, S A 2021 Restorative design approach for elderly people: Revisiting biophilia hypothesis and universal design principles in assisted living facilities *Gazi University Journal of Science Part B: Art Humanities Design and Planning* 9 (1) pp 19-28.
[8] Turkstat 2019 *İstatistiklerle yaşlılar* Available at http://www.tuik.gov.tr/PreHaberBultenleri.do?id=33712.
[9] Berman, M G, Jonides, J and Kaplan, S 2008 The cognitive benefits of interacting with nature *Psychological science* 19 (12) pp 1207-1212.
[10] Van den Berg A E, Hartig T and Staats H 2007 Preference for nature in urbanized societies: stress, restoration, and the pursuit of sustainability. *Journal of Social Issues* 63 pp 79-96.
[11] Yin J, Yuan J, Arfaei N, Catalano P J, Allen J G and Spengler J D 2020 Effects of biophilic indoor environment on stress and anxiety recovery: a between-subjects experiment in virtual reality *Environment International* 136 p 105427.
[12] Hartig, T, Mang, M and Evans, G W 1991 Restorative effects of natural environment experiences *Environment and behavior* 23 (1) pp 3-26.
[13] Kaplan, S 1995 The restorative benefits of nature: Toward an integrative framework. *Journal of Environmental Psychology* 3 (15) pp 169-182.
[14] Leather, P, Pyrgas, M, Beale, D and Lawrence, C 1998 Windows in the workplace: Sunlight, view, and occupational stress. *Environment and behavior* 30 (6) pp 739-762.
[15] Dalay, L 2020 The Impact of Biophilic Design Elements on the Atmospheric Perception of the Interior Space *Uluslararası Peyzaj Mimarlığı Araştırmaları Dergisi (IJLAR) E-ISSN: 2602-4322*, 4 (2) pp 4-20.
[16] Gardner, M P 1985 "Mood states and consumer behavior: a critical review", Journal of Consumer Research, Vol. 12 No. 3, pp. 281-300.
[17] Shibata, S and Suzuki, N 2004 Effects of an indoor plant on creative task performance and mood *Scandinavian journal of psychology* 45 (5) pp 373-381.
[18] Brown D K, Barton J L and Gladwell V F 2013 Viewing nature scenes positively affects recovery of autonomic function following acute-mental stress *Environmental Science & Technology* 47 (11) pp 5562-5569.
[19] Barton J and Pretty J 2010 What is the best dose of nature and green exercise for improving mental health? A multi-study analysis. *Environmental science & technology*, 44 (10) pp 3947-3955.
[20] Watson, D, Clark, L A, and Tellegen, A 1988 Development and validation of brief measures of positive and negative affect: the PANAS scales. *Journal of personality and social psychology* 54 (6) p 1063.
[21] Groeneveld, R A and Meeden, G 1984 Measuring skewness and kurtosis. *Journal of the Royal Statistical Society: Series D (The Statistician)* 33 (4) pp 391-399.
[22] Hopkins, K D and Weeks, D L 1990 Tests for normality and measures of skewness and kurtosis: Their place in research reporting *Educational and psychological measurement* 50 (4) pp 717-729.

IOP Conf. Series: Earth and Environmental Science 1099 (2022) 012035 doi:10.1088/1755-1315/1099/1/012035

Shining Light on Colors: Using a Puzzle Game to Teach Additive Color Theory

Dovilė Perednytė

E-mail: dpered18@student.aau.dk

Katarzyna Ewa Flig

E-mail: kflig17@student.aau.dk

Kristinn Bragi Garðarsson

E-mail: kgarda18@student.aau.dk

Malte Elkær Rasmussen

E-mail: marasm18@student.aau.dk

Sara Munk

E-mail: smunk18@student.aau.dk

Simonas Čeponis

E-mail: scepon18@student.aau.dk

Henrik Schønau Fog

E-mail: hsf@create.aau.dk

Abstract. Game-based learning is rapidly transforming the educational landscape by providing engaging supplementary tools to students around the world. A problem area where game-based learning has not been empirically studied is the subject of additive color theory. Can a puzzle video game increase players' knowledge of additive color theory? This paper describes how a puzzle game can be designed in order to teach this topic. The puzzle game prototype *Temple of Starlight* was developed, and an evaluation of its effectiveness at teaching was performed. The evaluation was done online following a within-subjects study design, where the participants answered questions about additive color theory before, immediately after, and a week after playing through the game prototype. Two analyses were performed: first one comparing the amount of correct answers from before playing the game and right after playing; and the second one comparing answers from before playing and a week after playing. Results showed a significant increase in knowledge of additive color theory, both right after ($N = 37$, $p = <.001$) and a week after ($N = 30$, $p = <.001$) playing through the game prototype. This suggests that the game prototype could be used as valuable supplementary material for teaching additive color theory in learning environments.

1. Introduction

In recent years, digital games have increased in popularity, especially for teens and young adults [1]. Consequentially, there has been fast growth in the educational games industry as well [2; 3]. The idea of using games in an educational setting is not a novel concept, and it has long been acknowledged that games are excellent at increasing and retaining motivation – something traditional teaching approaches often lack [4]. Additionally, games add to the effort of transferring learning to a more sustainable digital structure, where updates or changes to the learning content can be implemented and distributed smoothly.

Among existing educational games, science is one of the more popular topics [5]. This may be explained by a growing need for more active instructional methods to be used particularly for science teaching [6]. Furthermore, game-based learning appears especially beneficial in this area, as games can create interactive virtual environments where students can freely get hands-on experience with the learning topics.

One science topic that could benefit from a more interactive and contextualized teaching method is the additive color property of light. This topic is also becoming increasingly important in today's highly digitalized world, where most things are presented and designed for light based displays, in the form of screens. A study by Martini et al. [7] shows that students often struggle with gaining intuitive understanding of this theory, and indicates that this struggle may stem from misconceptions or lack of knowledge about the differences between additive color theory and the subtractive color theory, as the latter is commonly taught early on in kindergarten art classes relating to the mixing of various pigments. Such a misconception may create a barrier for many and ultimately demotivating them when learning about lighting design or light related art. Expanding the understanding of additive color theory will therefore be the basis of this study, with an investigation of whether game-based learning can prove useful for mediating knowledge on this topic. This raises our research question:

To what extent can learners' knowledge of additive color theory be increased by a video game?

2. Background

2.1. Color Mixing

What humans recognize as various colors are different wavelengths of light from a part of the electromagnetic spectrum called the visible spectrum, which covers the interval of wavelengths that are perceivable to the human eye [8].

The visible spectrum stems from the types of photoreceptors in the eyes [9]. Our main color perception is based on one type of photoreceptors called cones, which come in three variations, each sensitive to a specific wavelength range. These three wavelength ranges define what we consider to be the primary colors of light: red, green, and blue.

When humans perceive non-primary colors, it is the brain interpreting the combinations of different cones' signals. This serves as the basis of additive color mixing [8], seen on the left side of Fig. 1. Additive color theory is mainly applied when working with digital screens, since these display color by emitting red, green and blue light in different combinations. It is otherwise useful in the art, design and digital industries as it teaches about color perception and formation.

The other type of color mixing concerns subtractive colors. Subtractive color theory is often taught in relation to mixing paints, and refers to how physical objects absorb and reflect different wavelengths of light [8].

From the model of subtractive color mixing (Fig. 1), the primary and secondary colors have been switched compared to the additive colors. This is because the subtractive color model focuses on which of the three primary light colors are being absorbed, and therefore not perceived. For example, if we perceive a box to be red, this means that its surface material is reflecting the red light that is hitting it back at our eyes, while any potential green and blue light is being absorbed. As such, if said red box were to be lit only by a cyan light, then we would perceive it

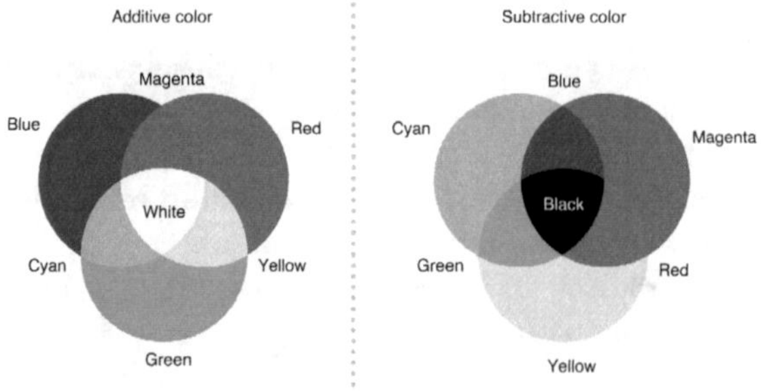

Figure 1. Model showcasing additive and subtractive color mixing [10].

as black, as no light would be reflected off of it [8].

2.2. Transformational Games

Transformational games are experiences that change the player by modifying behavior or improving a personal skill [11]. For a game to be effective at transforming players, the desired change has to both transfer and persist. For example, if the player becomes skilled at a game's mechanic that requires color mixing, it cannot be inferred that a successful transformation has taken place. Transfer requires the player to be able to apply color mixing outside the game's context. Moreover, persistence requires the player to be able to blend colors even some time after the game experience has ended.

2.3. Game-based Learning

The transformational framework suggests that educational games or learning games can be considered a part of the umbrella term "transformational" [11]. In related research, various terms have been used to describe the application of games or game elements in an educational context. While some terms may be used to describe a varying focus on the educational part compared to the game part, all terms are usually used to refer to learning in a game-like manner [12]. In this paper we will be using the terms "game-based learning" and "educational games" interchangeably to refer to games that are designed as learning tools.

It has long been accepted that the activity of play is important for a child's development [12], and the psychologist Vygotsky expressed the importance of play as a leading activity determining children's volitional, emotional and cognitive development [13]. Currently, the use of games in an educational context is a well-known and established concept [14]. The significance of play in early education has also been addressed [15], as well as the usefulness of games in primary education [16]. Nevertheless, educational games are seldom used as a learning tool in higher education, although there seems to be no evidence of any natural decrease in the desire for playing or the benefits hereof as we get older [12].

Game-based learning is rarely thought to be a substitute for traditional formal education but rather a supplement [12]. Educational games are often developed to teach only a limited set of learning goals. Nevertheless, there are multiple arguments for including game-based learning in an educational context, and it could prove useful in conjunction with other teaching methods or as a starting point for discussion [17]. There have already been a number of successful implementations of educational games and studies have indicated a promising use of games for learning purposes [18].

One important argument for game-based learning is its ability to motivate students due to their inherent entertainment factor [19; 12]. For example, games that are used for entertainment purposes are able to motivate players through rewards (points, trophies etc.), or interesting and enjoyable game mechanics and activities that would contribute towards high situational interest [20]. Similarly, enjoyment has been shown to be a large influential factor for learning effectiveness [21; 22; 23], as well as for motivation [24; 25]. This aligns with more general learning theory, stating that if a learning activity is enjoyable it will have a positive affect on performance and competences [26]. The motivational factor is also closely related to players' engagement, which can be facilitated in educational games, for instance, by interactions, visuals, narratives, mechanics or by having the game adapt to player's individual capabilities [12]. The capability of games to engage players is very useful, as engagement in an educational game can have a positive effect on learning [27].

Another benefit of game-based learning is that with digital games it is possible to have a safe environment, where players are allowed to make mistakes without having large negative consequences [12]. This also makes it possible to situate a student in a simulated position that would not be possible for them in reality. Investigations by Barab et al. showed that students who interacted with a game were much more motivated to participate in the classroom activities than those who learned in a traditional classroom [28].

2.4. State of the Art
Following one of the pillars from the transformational framework (section 2.2), this section will investigate prior works related to additive color theory and games.

Looking at games aimed specifically at teaching either or both additive and subtractive color theory, there are *True Colors* [29] and *Color Mixing* [30]. These games are structured very similarly; they both present a palette of available colors and specify a color that must be mixed. This same structure of teaching is offered by the *RGBox* [31], a physical device illustrating the mixing of additive colors using a line of LEDs.

All three examples serve their purpose of providing a space for potentially learning and/ or practicing how to mix different colors using either additive or subtractive color theory. However, it could be argued that the entertainment value they provide is low, as they do not offer much variation in gameplay or difficulty. This could be because development teams for educational games often lack resources to make the game engaging. As such, these games likely struggle to keep the players' attention for long enough to properly teach the theory they present [32].

3. Prototype Design
The transformational framework was used to establish different aspects of the transformational factor of the game in pre-production (see section 2.2).

We developed a prototype in the game engine Unity [33], which is a first-person puzzle game called *Temple of Starlight* that uses additive color mixing as the core of its light-focused mechanics.

The game has the player go through 19 levels, in which each level has a puzzle for them to solve in order to proceed to the next. These puzzles are constructed from two core mechanics based on colored light:

- Colored objects become visible and gain physical properties only when they are lit up by a color of light that would be reflected by the object. E.g. a magenta box would be visible under blue and/or red light, but not under green or no light (see Fig. 2).
- Power nodes can be activated by shining a combination of lights on them to match the node's color. E.g. a cyan power node will be activated when hit by both green and blue light simultaneously (see Fig. 3).

Figure 2. Screenshots of a magenta door under blue light (left), under a combination of red and blue light (center), and while unlit (right).

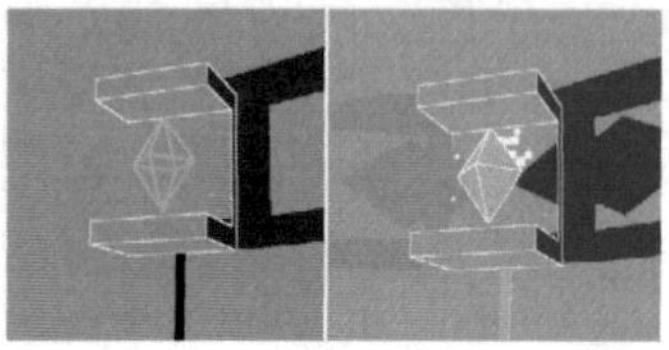

Figure 3. Screenshots of a cyan power node when it is inactive under green light (left) and when active under a combination of blue and green light (right).

3.1. Game Progression

All levels are designed so that the exit starts out blocked by a solid colored door (as seen in Fig. 2). The player then moves around in each level, activating and deactivating various lights, and changing their color in order to solve the puzzles, resulting in the door being unblocked.

While progressing through the levels, the player is gradually introduced to the different mechanics, providing a smooth learning curve. It is important to create an optimal experience, where the difficulty of the task (in this case, a level of the game) relates to the player's skill [34]. Optimal experiences increase the enjoyment when playing educational games, which in turn can positively affect learning [27].

3.2. Game Visuals

The game's visual style was designed around the core focus that the light and its colors should be clearly visible and stand out from the environment. As a result of this, the environment itself is kept in a single darker color. Since one of the game's core mechanics is based on making colored objects appear and disappear using the lights, the various ornamental objects in the levels were kept the same dark color as the environment itself to prevent the players from being confused about which objects were important for the puzzles. To ensure that the environment and objects were still identifiable, white wireframe-like outlines were added along all edges. All of this can be seen in Fig. 4.

Deliberate changes were also made to the visual appearance and physical behavior of the lights. The light modifications were done to ensure that the lit areas were clearly defined, and that the color mixing was kept to clean secondary colors and white only (as seen in Fig. 4). This means that if two red lights and one green light are all shining at the same area, the color will still be yellow instead of orange. This was mainly done to limit our domain concepts (section 2.2) to reach a more realistic scope.

4. Methods

After iterating on the game prototype for this study, it could then be used in an attempt to answer the research question:

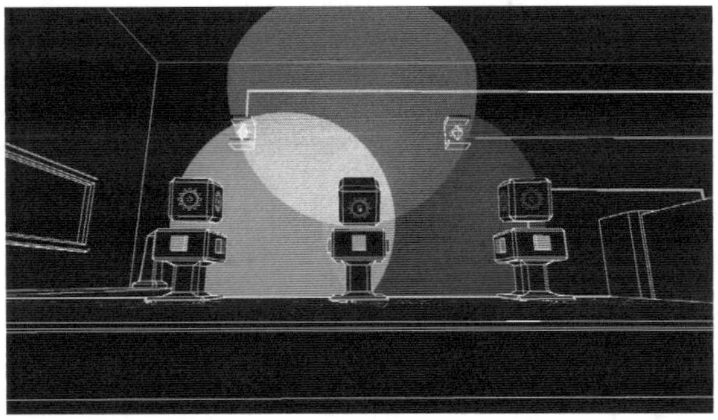

Figure 4. Screenshot showing off what color mixing looks like in the game.

To what extent can learners' knowledge of additive color theory be increased by a video game?

Following the transformational framework (section 2.2) we established the player transformation *"the player intuitively knows what color an object reflects based on what color it is and what light is shining,"* and we could conduct scientific experiments as part of our assessment plan. This also suggested the hypotheses:

H0: The puzzle video game prototype does not increase the players' knowledge of additive color theory.

H1: The puzzle video game prototype increases the players' knowledge of additive color theory.

To evaluate the player transformation, we did a repeated measures experimental design with the following experimental variables:

- **Independent variable** – time since completing the game.
- **Dependent variable** – level of knowledge of additive color theory.

4.1. Design of Experiment

Based on the variables, we defined three different conditions:

(i) Answered questions about additive color theory without (or before) playing through the game (later referred to as pre-gameplay).

(ii) Answered questions about additive color theory shortly after playing through the game (later referred to as post-gameplay).

(iii) Answered questions about additive color theory a week after playing through the game (later referred to as follow-up).

To better distinguish the change in additive color theory knowledge, the same participants should both play through the game and answer all of the questions. This suggests that ideally, a within-subjects study design should be followed, where each participant produces one result for each condition [35]. Within-subjects study design also allows to minimize confounding variables, as the baseline of knowledge of additive theory is captured when the participant fills out the questionnaire prior to playing the game. It can then be compared to the results of the questionnaire after the play-through.

In this study's case, the 'carry-over' effect [35] is minimized by ensuring that the participants do not get feedback on their questionnaire answers. We also avoid giving questions that could suggest answers or provide hints to upcoming questions.

4.2. Participants
Audience & context is one of the pillars in designing a transformational game (section 2.2), and it is essential in any experimental design. The most important aspect of the target group for the project was to distinguish colorblind participants (the prototype was not adapted for colorblind accessibility). These characteristics could impact the outcome of the study and serve as additional confounding variables; colorblindness could act as a barrier (section 2.2). This information, however, was gathered via some additional questions during the test procedure.

Participants for the test procedure were recruited online in several ways:

- Contacted personally, following convenience and snowball sampling methods.
- Invited to participate in the study via various websites and social media sites.
- Distributed to students through the help of teachers.

As part of the participant gathering effort, a chance to win different video games was advertised.

4.3. Measures
A questionnaire was prepared for the purpose of getting an understanding of the participant's level of knowledge about additive color theory. The questionnaire begins with demographic questions, asking about age, gender, nationality, and if they were colorblind. A follow-up questionnaire was required to evaluate if the game is able to fulfill the transfer and persistence aspects of transformational games (as discussed in section 2.2).

After the demographic questions, the core of the questionnaire consisted of nine questions of varying difficulty regarding additive color theory.

- All of the theory questions had the same eight answer options (Red, Green, Blue, Yellow, Magenta, Cyan, Black, White).
- Images were used to present the questions and make them more understandable without giving away the answers.
- The participants did not receive feedback as to whether their answers are correct or incorrect.

Three sets of the additive color theory questions were used that had the same amount of questions and the same structure and phrasing, but described different colors instead. For counterbalancing, each time a participant was doing a questionnaire, one of the three sets was randomly assigned.

4.4. Procedure
The whole procedure was performed online. The participants were directed to a website. There, all of the information and materials (questionnaire, consent form and a link to the video game prototype) were presented. If needed, the researchers could be contacted via provided e-mail address.

The starting page of the website provides information about the study, data handling, instructions regarding withdrawal from the study, and requirements to participate. The procedure then went as follows:

(i) The questionnaire was presented (described in section 4.3), where the participants first filled out demographic information and e-mail address.
(ii) Participants proceeded to the first additive color theory questionnaire set.
(iii) Game prototype link was provided. Participants were instructed to play through it and continue with the questionnaire once the game was finished.
(iv) The second additive color theory questionnaire was presented.

(v) Participants were presented with open-ended questions related to additional usability feedback (visuals, audio, and gameplay), their continuation desire, and issue reports.

If the participants agreed to answer the follow-up questionnaire, they were then contacted via their provided e-mail address after a week. The follow-up questionnaire only consisted of additive color theory questions. Those participants also received an opportunity to enter a video game key giveaway.

5. Results

Two separate data analyses were performed, since fewer participants responded to the follow-up questionnaire seven to nine days after their initial participation in the study.

5.1. *Experiment 1*

For Experiment 1, the first two conditions were compared (see section 4.1). The two conditions have the same number of participants, since they were completed one after the other. The responses to the questionnaires were treated as ordinal data. Each participant had a score from 0 to 9, based on whether they correctly answered each of the nine questions; a correct answer increased their score by 1, and an incorrect or partial answer did not increase the score at all. The study had a total of 43 participants. Some of the data had to be discarded, since several participants did not fully play through the game. Data from one colorblind participant was separated, and was not used in the results analysis. Among the 37 valid participants, there were 27 males and 10 females, aged 16-30 (M = 22.78, SD = 3.19). Participant nationalities were Lithuanian (N = 11), Danish (N = 10), Polish (N = 9), and 6 other. The average of the scores was found for each condition: The average pre-gameplay questionnaire score was 4.32 (SD = 1.84). The average post-gameplay questionnaire score was 5.73 (SD = 1.56). The one-tailed Wilcoxon signed-rank test indicated that the increase in questionnaire scores was statistically significant, $T = 452$, $p = <.001$, with a large effect size (CLES = .73).

5.2. *Experiment 2*

Experiment 2 was performed on a portion of the participants (out of the initial 37) which had successfully gone through all three of the conditions. Experiment 2 analysis was performed to compare the first and third condition. Valid data used for the results analysis was gathered from 30 participants (out of the 37 participants from Experiment 1). Among 30 valid participants, there were 21 males and 9 females, aged 16-30 (M = 23.07, SD = 3.16). Participant nationalities were Lithuanian (N = 9), Danish (N = 9), Polish (N = 6) and 6 other. The average of the scores was found for each condition: The average pre-gameplay questionnaire score was 4.57 (SD = 1.77). The average follow-up questionnaire score was 5.77 (SD = 1.65). One-tailed Wilcoxon signed-rank test for the longer-term effect indicated that there was a statistically significant increase of additive color theory questionnaire scores, $T = 33$, $p = <.001$, with a large effect size (CLES = .70).

6. Discussion

6.1. *Findings*

The results showed a statistically significant increase in the participants' measured knowledge. With this, the null hypothesis can be rejected. This greatly increases our confidence that the game could serve as supplementary educational material to teach individual about additive color theory. Having only a week between the gameplay and the follow-up questionnaire is a relatively shorter time in relation to educational topics. However, the results suggest that the game is capable of fulfilling the two main purposes outlined by the transformational framework (section 2.2); change to both transfer and persist.

6.2. Subtractive Color Theory

As shortly described in the Color Mixing section 2.1, people tend to confuse additive and subtractive color theory since they are usually taught to mix pigment at a younger age. Upon deeper inspection of the data, we noticed a tendency in pre-gameplay responses, as 12 participants answered that either **red, blue and yellow** or **cyan, magenta and yellow** are the primary colors, and 12 participants answered that mixing all the colors together would result in black. These two sets of 12 participants partially overlap.

One possible concern from teaching additive color theory through our game, is whether it is communicated clearly enough that this color mixing specifically applies to light. This concern stems from the notion that if it is not clear that light mixes differently than pigments, the game might cause players to merge their color mixing knowledge together; potentially causing more confusion later on.

6.3. Real-world Application

In order to be fully beneficial in an educational context, the game should be a supplement of lectures on this topic. By itself, the game lacks theoretical explanation of why the colors of light mix in such manner, and does not provide a theoretical explanation of the physical properties of light. As such, the game mainly serves to create an intuitive understanding of how additive colors mix, but needs to be played alongside additional teaching of the theory, in order to convey the full scope of knowledge needed for use in real-world applications. Although our solution is very limited in what it teaches, it can be used to break a barrier many may experience when learning about colors of light, which is a fundamental concept in lighting design and visual art.

Conclusion

Game-based learning has shown to be a promising supplement to traditional learning methods, and especially within the field of science, educational games may be beneficial for providing students with hands-on experimentation with the learning topics. This paper set out to investigate if game-based learning can be used as an effective learning approach for additive color theory – a topic of which students often appear to lack intuitive understanding. Transformational games and color mixing were researched in order to find a way of designing such an educational game. Based on the research, we developed a prototype of a first-person puzzle game.

An evaluation was conducted online on 37 participants. 30 of the participants also took part in a follow-up study a week after, where their knowledge of additive color theory was assessed once more. The results showed that a significant increase in the knowledge of additive color theory was found both immediately after playing the game and after a week had passed since playing it. The game was well-received by the majority of the participants, showing that such a game could potentially be used as supplementary material when teaching the topic of additive color theory.

Finally, we hope to inspire other game developers to make games that have some kind of transformational value. It is a general issue within the field that there is a lack of established methods to evaluate transformational games [36], and only by coming together and making a stronger community around it can we hope to achieve advancements and further knowledge in the field.

References

[1] A. Lenhart, J. Kahne, E. Middaugh, A. R. Macgill, C. Evans, and J. Vitak, "Teens, video games, and civics: Teens' gaming experiences are diverse and include significant social interaction and civic engagement.," *Pew internet & American life project*, 2008.

[2] Verified Market Research, "Educational games market size and forecast," 2020. Online; accessed 14/12/2021; `https://www.verifiedmarketresearch.com/product/educational-games-market/`.

[3] IMARC, "Game-based learning market: Global industry trends, share, size, growth, opportunity and forecast 2021-2026," 2021. Online; accessed 14/12/2021; `https://www.imarcgroup.com/game-based-learning-market`.

[4] M. Prensky, "Digital game-based learning," *Computers in Entertainment (CIE)*, vol. 1, no. 1, pp. 21–21, 2003.

[5] E. A. Boyle, T. Hainey, T. M. Connolly, G. Gray, J. Earp, M. Ott, T. Lim, M. Ninaus, C. Ribeiro, and J. Pereira, "An update to the systematic literature review of empirical evidence of the impacts and outcomes of computer games and serious games," *Computers & Education*, vol. 94, pp. 178–192, 2016.

[6] M. Riopel, L. Nenciovici, P. Potvin, P. Chastenay, P. Charland, J. B. Sarrasin, and S. Masson, "Impact of serious games on science learning achievement compared with more conventional instruction: an overview and a meta-analysis," *Studies in Science Education*, vol. 55, no. 2, pp. 169–214, 2019.

[7] B. Martini, M. Tombolato, and R. D'Ugo, "Primary colors as a source of possible misconceptions: an insight into teaching and learning about color," *Cultura e Scienza del Colore-Color Culture and Science*, vol. 11, no. 02, pp. 25–33, 2019.

[8] T. B. Moeslund, *Introduction to video and image processing: Building real systems and applications.* Springer Science & Business Media, 2012.

[9] S. Yantis and R. A. Abrams, *Sensation and Perception*, vol. 2nd edition. Worth publishers, 2017.

[10] Apple Developer Archive, "Color spaces." [Archive page], 2005. Accessed Nov. 17, 2021.

[11] S. Culyba, *The Transformational Framework: A process tool for the development of Transformational games.* Carnegie Mellon University, 2018.

[12] S. Greipl, K. Moeller, and M. Ninaus, "Potential and limits of game-based learning," *International Journal of Technology Enhanced Learning*, vol. 12, no. 4, pp. 363–389, 2020.

[13] L. S. Vygotsky and M. Cole, *Mind in society: Development of higher psychological processes.* Harvard university press, 1978.

[14] J. Piaget, *Play, dreams and imitation in childhood.* Routledge, 2013.

[15] G. S. Ashiabi, "Play in the preschool classroom: Its socioemotional significance and the teacher's role in play," *Early Childhood Education Journal*, vol. 35, no. 2, pp. 199–207, 2007.

[16] T. Hainey, T. M. Connolly, E. A. Boyle, A. Wilson, and A. Razak, "A systematic literature review of games-based learning empirical evidence in primary education," *Computers & Education*, vol. 102, pp. 202–223, 2016.

[17] M. Pohl, M. Rester, and P. Judmaier, "Interactive game based learning: advantages and disadvantages," in *International Conference on Universal Access in Human-Computer Interaction*, pp. 92–101, Springer, 2009.

[18] B. Snow, "The potential for game-based learning to improve outcomes for nontraditional students," *Muzzy Lane Software Report, funding from Bill and Melinda Gates Foundation*, 2016.

[19] J. L. Plass, B. D. Homer, and C. K. Kinzer, "Foundations of game-based learning," *Educational Psychologist*, vol. 50, no. 4, pp. 258–283, 2015.

[20] J. I. Rotgans and H. G. Schmidt, "Situational interest and academic achievement in the active-learning classroom," *Learning and Instruction*, vol. 21, no. 1, pp. 58–67, 2011.

[21] T. M. Connolly, E. A. Boyle, E. MacArthur, T. Hainey, and J. M. Boyle, "A systematic literature review of empirical evidence on computer games and serious games," *Computers & education*, vol. 59, no. 2, pp. 661–686, 2012.

[22] G. Hookham and K. Nesbitt, "A systematic review of the definition and measurement of engagement in serious games," in *Proceedings of the australasian computer science week multiconference*, pp. 1–10, 2019.

[23] C. Steiner, P. Hollins, E. Kluijfhout, M. Dascalu, D. Albert, W. Westera, and A. Nussbaumer, "Evaluation of serious games: A holistic approach," in *ICERI2015: 8th annual International Conference of Education, Research and Innovation*, pp. 4334–4342, IATED Academy, 2015.

[24] N. Iten and D. Petko, "Learning with serious games: Is fun playing the game a predictor of learning success?," *British Journal of Educational Technology*, vol. 47, no. 1, pp. 151–163, 2016.

[25] E. Fokides, P. Atsikpasi, P. Kaimara, and I. Deliyannis, "Factors influencing the subjective learning effectiveness of serious games.," *Journal of Information Technology Education*, vol. 18, 2019.

[26] M. Csikszentmihalyi, K. Rathunde, and S. Whalen, *Talented teenagers: The roots of success and failure.* Cambridge University Press, 1997.

IOP Conf. Series: Earth and Environmental Science 1099 (2022) 012035 doi:10.1088/1755-1315/1099/1/012035

[27] J. Hamari, D. J. Shernoff, E. Rowe, B. Coller, J. Asbell-Clarke, and T. Edwards, "Challenging games help students learn: An empirical study on engagement, flow and immersion in game-based learning," *Computers in human behavior*, vol. 54, pp. 170–179, 2016.

[28] S. A. Barab, M. Gresalfi, and A. Ingram-Goble, "Transformational play: Using games to position person, content, and context," *Educational researcher*, vol. 39, no. 7, pp. 525–536, 2010.

[29] ABCYa, "True colors - color theory." [Game], 2021. Accessed Nov. 17, 2021.

[30] Novel Games, "Color mixing." [Game], 2019. Accessed Nov. 17, 2021.

[31] G. Bílá, "Rgbox." [Electronic device], 2019. Accessed Nov. 17, 2021.

[32] T. Fullerton, *Game design workshop: a playcentric approach to creating innovative games.* AK Peters/CRC Press, 2019.

[33] Unity Technologies, "Unity: 2021.1.20f1." [Game Engine], 2021. Accessed Dec. 16, 2021.

[34] M. Csikszentmihalyi and I. S. Csikszentmihalyi, *Optimal experience: Psychological studies of flow in consciousness.* Cambridge university press, 1992.

[35] A. Field and G. Hole, *How to design and report experiments.* Sage, 2002.

[36] G. Petri and C. G. von Wangenheim, "How to evaluate educational games: a systematic literature review," *Journal of Universal Computer Science*, vol. 22, no. 7, pp. 992–1021, 2016.

IOP Conf. Series: Earth and Environmental Science 1099 (2022) 012036 doi:10.1088/1755-1315/1099/1/012036

Sociological approach to the territorialization of night-time atmospheres

Houel Nicolas (1), Potard Saïg (2)

(1) Associate Researcher at the AAU Laboratory - ENSA Nantes, France, founder of the company L'Observatoire de la nuit

(2) Head of Public Lighting, Light Signals and Telecommunications Networks Division of Brest Métropole, France

Nicolas.houel@gmail.com, saig.potard@brest-metropole.fr

Abstract. This article presents a case study conducted in the metropolis of Brest, France in 2022, designed to support public policy decision-making by integrating the knowledge and needs of citizens during the different stages of a night-time urban planning project, around the notion of synchronization of the city's rhythms. It presents three innovations: (1) Actors: coordination of public and private actors, common vocabulary, cartographic representations of assets, introduction of participatory approaches, activation of citizen knowledge in political expertise. (2) Sociology: participative modalities, objectification of citizens' knowledge, definition of their decision-making perimeter, (3) Public lighting policies: territorial and social strategy of light sobriety, regularization and empowerment of the network of actors. The conclusion of this article will focus on the limits, added values and development axes of nocturnal urban sociology as a support for decision-making in public lighting management and renovation strategies, and will open up the mobilization of the rhythms and temporalities to be mobilized in order to restore balanced nocturnal ambiences between lighting, which is essential, and darkness, which is vital.

1. Introduction

Since 2018, public lighting in France has been subject to a regulatory framework of mandatory application, underlining the urgency of a shift towards its sobriety[1]. Its determinants are the reduction of energy consumption and light pollution[2] and the preservation of natural night environments[3]. Guides, labels and actions now qualify darkness as a resource[4]. In parallel, sociological initiatives are opening up[5], presenting isolated participative approaches, often mobilized in a political framework seeking to verify a proposal for an atmosphere that is already underway.

In the follow-up of energy and environmental awareness, we propose to deepen the territorial role of the societal awareness of the relation to night-time ambiences. Today, the user migrates from a passive posture, developed in front of a safety artificial lighting[6], then advertising[7], towards a role of actor of the reduction of light pollution[8].

However, the notions of citizen knowledge are rarely mobilized in the framework of operational lighting projects [9]. Only some scientific studies explore the fields of individual perception [10]. Finally, the notion of rhythms [11] plays a decisive role: rhythms of technological innovations, political mandates, regulatory changes, renovation of heritage and the daily life of regular or occasional night-time users. This is where our problem lies: Can the identification of ordinary and extraordinary rhythms and uses become a support for decision making in political and technical strategies for artificial lighting sobriety?

The main motivation of the project lies in the political decision support of the territorialization strategies of public lighting (regimes, forms, photometries, atmospheres). It is part of the operational approach of the Schéma Directeur d'Ambiances Nocturnes (SCAN), or Nocturnal Ambience Coherence Scheme (NACS) initiated by Brest Métropole. Its methodology works in three phases:

- Diagnosis of the lighting assets, help in determining the political, technical and social issues pursued by the community,
- Territorialization of the issues by opening them up to the expertise of experts[12] and users,
- Definition of a strategy for the creation and renovation of lighting assets determined in time and space and represented by an evolutive digital cartographic support.

We will present this case study by taking as a starting point the ambition of luminous sobriety which has become essential in view of the energy, environmental and societal breaking points encountered today in artificial lighting practices.

Through the study of these elements, we will observe the transformation of urban and rural nocturnal ambiences towards the concept of the black grid. The latter will be observed as a form of imposed restoration of darkness in inhabited areas, for the benefit of energy sobriety and respect of the environment. We will question the decisional, operational and innovation temporalities, which are nowadays decisive in the maintenance of the lighting practice framework. We will propose an adaptation towards a concept that could be unifying, that of the rhythm of the city and its uses.

2. Sobriety issues in Brest Metropole

The challenges of sobriety expected from street lighting are becoming prominent in France. Since 2018, they have been the subject of specific national regulations to reduce light pollution emissions[1]. In 2021, an annual public report[2] was published highlighting the lack of an ambitious and long-term vision for the management of this heritage, whose obsolescence represents a significant source of energy consumption, which can definitely be improved. Faced with these observations, a growing number of authorities have embarked on the study of renovation strategies that are often unilateral (energy/environment) and prioritize the extinction of lights in more or less extensive areas of their territories and according to time slots that are most often determined without analysing local night-time activities.

These new practices, which are virtuous in terms of the objective of sobriety, are however out of step with the priority service of street lighting: to accompany individuals moving about in the public space at night. Determining these new street lighting practices involves considerable work, both in terms of cost and time. To undertake them without a detailed analysis of local night-time rhythms and uses, both daily and occasional, presents, from our point of view, the non-negligible risk of an evolution

[1] We refer to the ministerial order of December 27, 2018 on the prevention, reduction and limitation of light pollution. [URL] https://www.legifrance.gouv.fr/loda/id/JORFTEXT000037864346/ - last consultation on 06/08/2022.

[2] We refer to the annual public report of the Cour des Comptes, published on 18 March 2021. [URL] https://www.ccomptes.fr/fr/documents/55052 - last consultation on 06/08/2022.

of a public service - street lighting - which has been renovated but which is nevertheless out of sync with its initial objective: to accompany uses.

Here, sobriety as a territorial strategy is understood with the help of the developers and users of the places concerned. It is part of a system where considerations related to the preservation of energy and the environment are balanced according to the expertise of both administrators and users, according to a bottom-up logic where the expertise of uses comes to complete and enhance the technical and political expertise. Our case study, located on the territory of Brest Metropole, France, presents the first steps of this methodology.

2.1. *Brest métropole*

In 2022, the Brest metropolitan area will have more than 36,000 light points. These are spread over 8 municipalities and provide a public service to nearly 208,500 inhabitants, i.e. nearly one light point for every 6 inhabitants. Given this vast territory, the metropolis is pursuing a global approach to the operation of lighting and a homogeneous strategy for taking needs into account. Historically committed to an approach based on sobriety, Brest Metropole presents a technical management based on night-time cut-off and dimming of public lighting. This configuration, initially defined empirically by knowledge of the territory and its uses, is now evolving according to the observation of real traffic conditions and feedback from citizens, in order to get closer and closer to the reality of uses.

2.2. *Actors and coordination*

The metropolis is responsible for the creation, development, management and maintenance of public space in the community and municipal domains and its facilities. Political governance is carried out by the elected vice-presidents responsible for services to the population and territorial proximity.

The “street lighting” unit is part of the "Public Lighting, Illuminated Signals and Telecommunications Networks" Division, which is part of the "Roads, Networks and Infrastructures" Department of the "Public Space and Environment" division of Brest Metropole.

The “street lighting” department is made up of an engineer and two technicians, who develop the community's strategy by proposing asset renovation programs and monitoring the operation of the network in the best possible conditions. In order to maintain control over operations, the community has chosen to delegate management by relying on two public contracts with purchase orders, one for new work and another for maintenance.

2.3. *The project for Nocturnal Ambience Coherence Scheme (NACS)*

Brest Metropole has begun studies to develop a Nocturnal Ambience Coherence Scheme. This initiative is fully in line with action no. 56 of the Brest Metropole's territorial climate and energy plan: the pursuit and enhancement of the energy optimization of the public lighting network. The NACS continues to define local policies in terms of public lighting, which it develops in light of contemporary energy and environmental issues. It is interested in political, technical and technological innovations, which it territorializes by integrating a sociological approach conducted by action research protocols and the development of participatory survey methods dedicated to nighttime environments, in order to meet the objective of “lighting right”.

NACS proposes a 3-step methodology. The first step proposes a transversal approach to the various departments involved in the development of public space in order to understand the needs and levers of action of each. At the same time, it allows to measure the different levels of political and technical complexity of public space planning, and to set up a shared vocabulary, around the issues of night time ambience, between political decision makers and public engineers and planners. These exchanges take the form of collaborative workshops, from which a general street lighting strategy oriented towards sobriety emerges. We believe that this first strategy is out of sync with the habits of the territory's night-time users, who are not consulted during its development. Therefore, a second step called territorialization is proposed. This step aims at identifying and valorizing the daily and occasional night activities of the users in professional and recreational situations. Like step 1, it takes place in the form of workshops, this time with citizens, and allows the establishment of a shared vocabulary on a larger scale. The result is a so-called territorial strategy, where the night-time atmosphere is sober and adapted to the activities of the users.

However, this work is carried out without any experiential form of the imagined atmospheres. The NACS proposes a final stage of experimentation of the future ambiances. This last one can be done in real conditions or in a simulator, and aims at presenting the future spatialized environments to the decision makers, technicians and citizens. The experimentation finally allows to represent and to value the political and technical commitments towards a reasoned lighting (fig. 1).

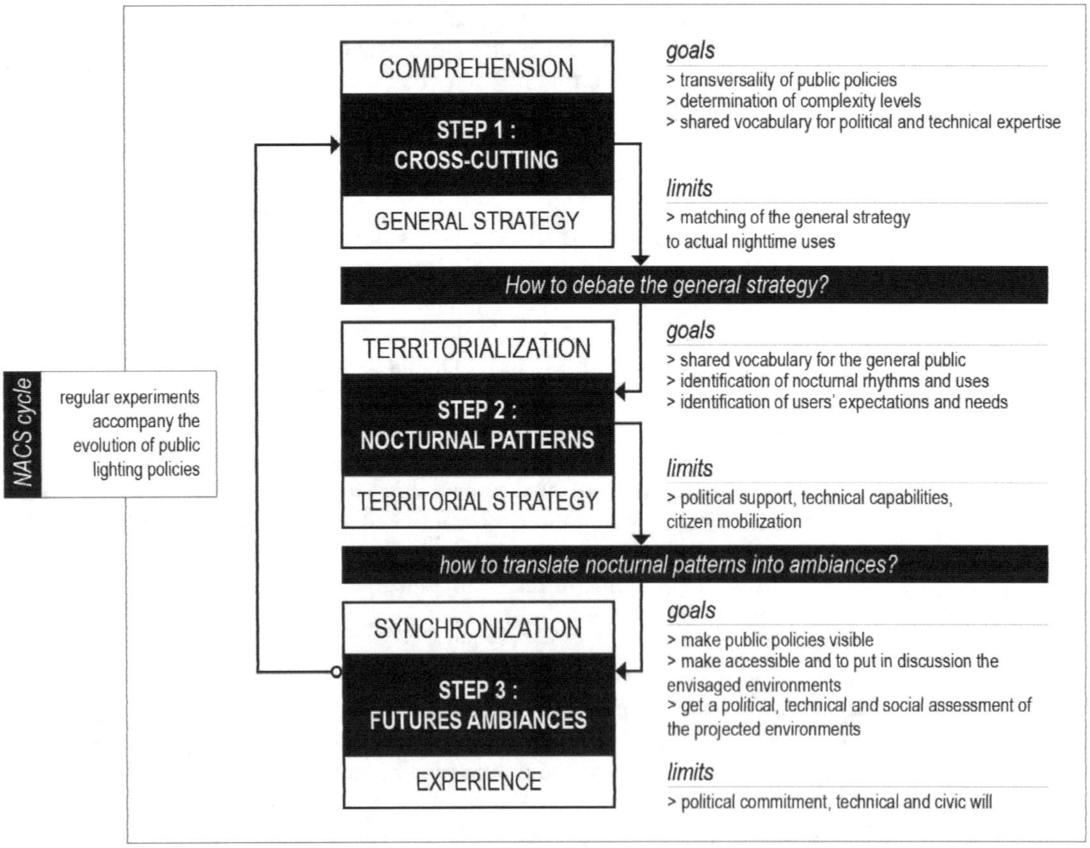

Figure 1 - Diagram of the NACS methodology: synchronize the political and technical expertise of artificial lighting with the uses in order to produce adapted and evolving night atmospheres

The objective of providing the right lighting for the city continues to be developed in the precise definition of the lighting service, taking into account all of the challenges facing public spaces, whether they are societal, urban, energetic, environmental, economic, tourist or even innovative. For Brest Metropole, designing the right public lighting involves a global approach to these issues, with a system of varying and evolving priorities depending on the spatial and temporal contexts, understood under the notion of rhythms.

3. From black frames to nocturnal rhythms

On a national scale, these dynamics of sobriety are today motivated by a dominant concept, although still awaiting a precise and shared definition, of the black frame. The environmental backlash to the luminous generosity of the second half of the 20th century, the nocturnal landscapes of the 2020s are seeing their ambiences determined by a quantity of artificial lighting measured as close as possible to the respect of the environment, participating in the emergence of these black frames, geographical interpretations locally superimposed on the green and blue frames, sometimes enhanced by atlases of biodiversity, exceptionally augmented by rhythmic and temporal emergences. These black frames, widely debated in academic and operational circles, are currently making their way towards the notions of dark frames, where lighting is moderated, and light frames, where lighting is preserved with regard to identified uses on the ground.

The grid approach is a first form of local response to the challenges of preserving biodiversity. Nevertheless, it still seems to be insufficiently supported by the necessary identification of the fauna and flora species of the places concerned, nor of their rhythms (permanent or seasonal presence). Thus, the grid approach presents geographical strategies that must now be associated with rhythmic strategies for flora, fauna and humans.

Through this territorial approach of mutualizing space and time indicators, one of the major evolutions of the future of urban nights is presented, that of a correspondence to the ordinary and extraordinary uses of nocturnal space-time. The city of Brest, with its years of participative experience, is committed to experimenting with the development of a strategy for harmonized nighttime atmospheres for all users in its territory.

4. Territorialization: enhancing the transversality of nighttime urbanism

The harmonization of a public service to its users is understood under the notion of territorialization, in this case of night-time environments, affordable at least under two aspects:

1. In political science, territorialization consists in identifying and valorizing territorial specificities, through an approach to public policies that encourages interdepartmental cooperation.
2. In the academic community, the geographer Claude Raffestin explores "nocturnal territorialities", understood as our daily, ordinary practices and uses of nocturnal spaces.

In France, where the development of public space is organized in silos (green spaces, roads, traffic lights, public tranquility, urban quality, new mobility, public lighting, etc.), the question of night-time environments is regularly treated in a secondary, primarily technical manner, except in specific cases of large-scale urban development. Skills, knowledge, budgets and meeting time are reduced to a strict minimum, to the benefit of a reflection on daytime developments and associated atmospheres.

Our study explores a methodology for cross-cutting issues related to street lighting. We propose to approach aspects 1 (political science) and 2 (academic) from a similar angle: meeting and sharing knowledge with planners and decision-makers (aspect 1) and then with panels of professional or private night-time users (aspect 2). This approach will allow us to present to decision-makers and users the different levels of complexity now encountered in the management of street lighting, and in parallel to grasp, from their expertise, the levels of complexity they encounter in the administration of their own sectors of:

- Expertise (decision-makers/planners),
- Uses (users).

This step, entitled common vocabulary, will allow us to set all the participants at an equal level of knowledge (vocabulary, issues, needs, expectations). It is carried out in the form of interviews and is considered to be the preliminary stage in the development of the NACS of the metropolis.

5. Brest Metropole: the levels of collaboration in the service of the territory's nocturnal resilience

In Brest, the Nocturnal Ambience Coherence Scheme is an innovative, qualitative and sensitive approach to nocturnal spaces and temporalities. Mobilized in the design of lighting environments, it reveals their complexity and accompanies the management of change through a collaborative modality deployed in 3 stages: A preliminary stage of transversal diagnosis; a first phase of territorialization, called general strategy; a second phase of territorialization, called territorial strategy.

5.1. Transversal diagnosis

The transversal diagnosis maps the lighting data of Brest Metropole as well as all the physical indicators of the public space: road network, centrality poles, natural spaces, school sectors, or tourist points of interest. It allows the first indications of complexity to emerge between the historical roles attributed to artificial lighting and contemporary aspirations for environmental respect, energy and social resilience.

5.2. General strategy

A first phase of territorialization, known as the general strategy, initiated by interdepartmental cooperation. It aims at a broad definition of the needs and expectations of each department of the public space department in terms of night-time ambience. This first step encourages the de-siloing and allows to reveal the richness of the current stakes fed by reasoned practices of artificial lighting. It opens the way to the definition of a macro-scale strategy, oriented towards lighting sobriety and transcribed via an interactive digital interface within which each collaborator can contribute his or her specific expertise to the NACS.

5.3. Territorial strategy

A second phase of territorialization, called territorial strategy, conducted through collaboration with the users of the public night space. The goal is to transmit, share and refine the general strategy through the real expertise of the territory's users. This third step enhances energy and environmental considerations through a social approach, and guides the evolution of cultural representations towards reasoned nocturnal atmospheres. The result is a precise cartography of expected atmospheres, allowing the implementation of operational documents prioritized in time and space. At the same time, it allows us to apprehend, directly from the lived space, the expectations and needs of individuals in terms of lighting ambiences. All these elements are grouped together in a specific digital cartographic support designed for the occasion, the NACS interface.

6. NACS interface: decision support, collaboration and digital transition of night strategies.

This collaborative approach is associated with an exclusive digital mapping tool, dedicated to the nocturnal strategy and to the wider dissemination of its issues. It is presented as a unique support for awareness, projection and collaboration on the topics of light transition, projected atmospheres and technical and political decisions to be taken. The device encourages synergy between all decision-makers, actors and users of the public night space, and facilitates decision support through a dynamic interactive system accessible to everyone.

Developed to provide a scalable and collaborative decision support system, NACS has the following values:

- The concept of a Scheme indicates a framing document for the lighting strategy on a metropolitan scale, which can accompany its implementation over the long term,

- The coherence approach is understood as spatial, temporal and of uses, on the scale of all living beings evolving outside at night. The nocturnal environment is adapted in order to respect a balance between human, fauna and flora activities according to ordinary and extraordinary cyclical criteria,
- The projection of the Ambiences places the planning of the development in a qualitative logic, where the photometric indicators adapt to the needs and activities, and not the opposite,
- The mobilization of the term nocturnal encourages the reintegration of a temporality, the night, and its natural resource, darkness. It encourages policy documents to move away from considerations exclusively motivated by a predefined amount of artificial lighting.

The case study analyzed in Brest is part of an exploratory approach to the nighttime environments experienced by citizens. In this respect, the collaborative approach is conducted directly from the digital interface, offering a transition from an existing panel of highly technical tools to a new generation didactic interface, dedicated to political, strategic and social considerations. Regarding the latter, the collaborative approach encourages the essential evolution of the user's posture, from spectator-consumer to actor of the nightlife of his territory, by providing a fine and daily expertise of use otherwise considered inaccessible.

7. Transversal diagnosis and general strategy of Brest Metropole

The transversal diagnosis of the nighttime atmosphere is carried out by mobilizing the following territorial data: public lighting heritage; road network; architectural and urban enhancement areas; historical monuments; natural areas; green and blue networks; schools. The data is obtained via the internal services of the metropolis, or online via an Opendata platform[3].

7.1. Transversal diagnosis of Brest Metropole

Individually, these indicators represent uses and rhythms on the ground and require a certain amount of lighting, but cannot, on their own, become the exclusive determinants of a night-time strategy. When they are crossed, we obtain clues, particularly in terms of conflict between two rhythms and uses. For example, according to the normative and regulatory texts, a main traffic route requires a specific minimum amount of lighting. On the contrary, some natural areas expect little or no lighting. What strategy should be adopted when a major roadway crosses or borders a natural area: that of preserving a potentially unused lighting service or that of preserving the environment at the risk of exposing human beings to less comfortable atmospheres? At this point, the collaborative methodology, as much in the establishment of a shared vocabulary as in the identification and enhancement of nighttime rhythms, will be able to objectively and operationally accompany the choice of decision-makers and planners regarding the parameters of the expected nighttime ambiences. Among these parameters, we find, in order of importance:

1. Should lighting be used? If the answer is no, it is recommended to remove the equipment, or at least to remove the structural components (lamp, driver, electrical box),
2. If it is necessary to light, it is then necessary to know the current state of the equipment and its environment (road network, built environment, natural areas, etc.).

The acquisition of these elements is done by consulting, preparing and representing all the physical databases available on the territory. This step allows the transversal diagnosis to be carried out, which will determine the next step, called the general strategy.

[3] For this study, we mobilized the Opendata Brest Métropole platform. [URL] : https://opendata.brest-metropole.fr/pages/accueil/ - last consultation on 06/08/2022.

IOP Conf. Series: Earth and Environmental Science 1099 (2022) 012036 doi:10.1088/1755-1315/1099/1/012036

7.2. General strategy of Brest Metropole

The general strategy is developed around the crossing of territorial indicators. It mobilizes the photometric criteria determined in the ministerial order of 27 December 2018[4] and offers a wider metropolitan distribution. Thus, the color temperatures of the sources adapt to the biodiversity environments, and lighting and extinction times to road categories (fig.2).

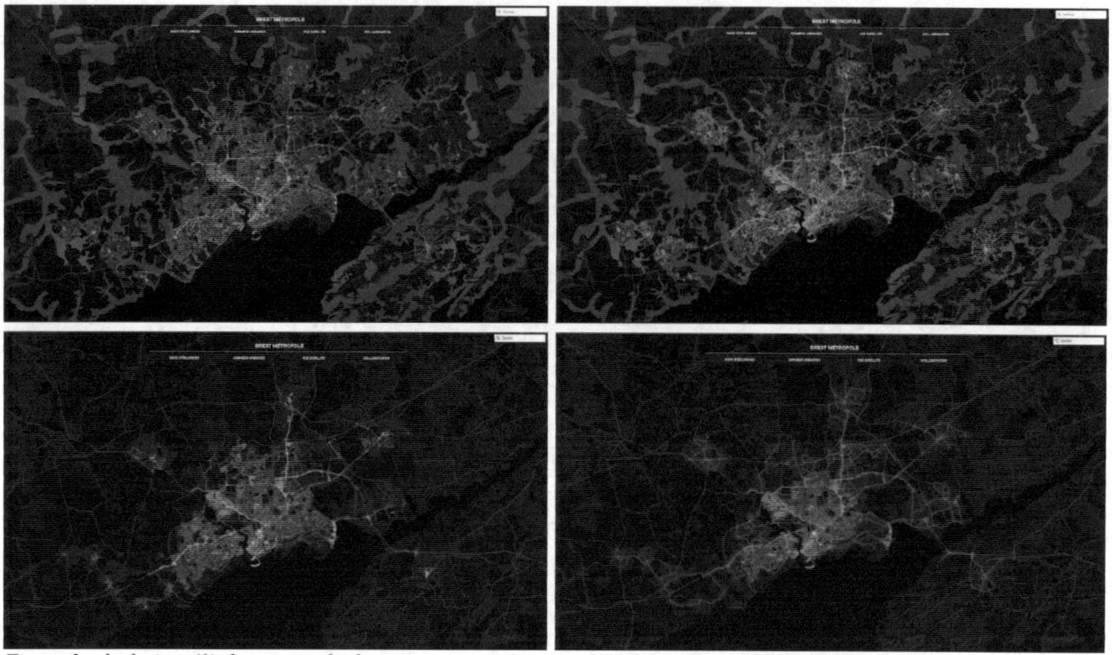

Figure 2: clockwise: (1) diagnosis of color temperatures (orange: 2200°K; yellow: 3000°K; cyan: 4500°K+), green and blue flat area: green and blue screen. (2) color temperature distribution strategy (orange: 2200°K; yellow: 3000°K), the lowest color temperatures are located near the natural areas. (3) midnight power level diagnosis (blue: 0%; cyan: 10%; green: 30%; orange: 60%; yellow: 100%). (4) power variation strategy at 00h (blue: 0%; cyan: 10%; green: 30%)

The diagnostic process allows us to measure the important disparities between urban, natural and ambient structures. The color temperatures are distributed without visible pattern, and the dimming regimes do not seem to follow a precise territorial logic. Different factors appear to be at work:

- The evolution of lighting technologies: mercury vapor, now banned, high pressure sodium, with orange lighting, and metal halide, with yellow lighting.
- The creation and renovation of the lighting heritage, concentrated in certain areas and at certain periods.

At this point, the notion of rhythm raises the question of the phenomenon of desynchronization. The latter seems to occur between the major rhythms of the administration of public lighting, namely:

- Political, normative and regulatory decisions and their timetables.
- Technical and technological innovations encountered and implemented over the decades,
- The operations of creation and renovation of the heritage,

These three desynchronized rhythms produce ambiances scattered over the metropolis. The color temperatures, regularly linked to lighting technologies, do not respond to precise territorial structures

[4] We refer to the ministerial order of December 27, 2018 on the prevention, reduction and limitation of light pollution. [URL] https://www.legifrance.gouv.fr/loda/id/JORFTEXT000037864346/ - last consultation on 06/08/2022.

IOP Conf. Series: Earth and Environmental Science 1099 (2022) 012036 doi:10.1088/1755-1315/1099/1/012036

but rather to the political need to produce a lighting service that meets normative requirements, regardless of the resulting ambiance in terms of visual comfort. The light levels are determined according to a classification of motorized roads whose daily and seasonal cycles of use are mostly unknown. The recommendations made by NACS meet the current regulatory requirements, namely a reduction in light levels and color temperatures in the area of biodiversity. However, the knowledge in terms of fauna and flora species is weak, making the recommendations a partial response to environmental preservation issues.

7.3. Limitations of the general strategy
The desynchronization of the rhythms of decision, innovation and operation produces a nocturnal territory with resolutely heterogeneous atmospheres. Technical, political and regulatory priorities dominate and erase the mobilization of the criteria of the rhythms of users, fauna, flora and human beings. The general strategy proposed in Brest corresponds to a harmonization of the luminous ambiances with the contemporary environmental and energetic determinants but, like historically encountered practices, always remains at a distance from the needs and uses on the ground. However, this initial strategy phase allows for the building of a common language with the city's elected officials, who acquire a more detailed knowledge of the trajectory and challenges of artificial lighting. This initial pooling of expertise (political and technical) encourages rigorous decision-making that is adapted to the expectations of the energy and environmental transition of artificial lighting practices. Finally, this collaborative work allows to engage the territorial strategy with planners and citizens, through the NACS interface.

8. Brest Metropole's territorial strategy
The territorial strategy is considered as the element of synchronization of all the rhythms of decision, innovation and operation related to artificial lighting, around priority rhythms: nocturnal fauna, flora and human activities daily or cyclical.

8.1. Synchronization around the ordinary and extraordinary rhythms of public space
The identification of these uses is, in the case of Brest Metropole, determined by the collaboration of the different departments of the public space (interdepartmental collaboration) and by the mobilization of citizen panels (nocturnal territorialities). In order to facilitate the stages of contribution, representation and restitution of the participants, the digital interface of the NACS proposes a section for contribution and consultation of the transmitted data (fig. 3).

Figure 3 - On the left, the NACS interface enables a contribution to be made simultaneously with the general strategy observation. On the right, the contributions can be consulted in the form of pop-up displaying the content of the contribution.

8.2. First contributions to the territorial strategy of Brest Metropole
In May 2022, the head of the public lighting department, Saïg Potard, accompanied a participatory initiative on public space, entitled exploratory walk of women in Kerourien. He collects the opinions of the mobilized women users and transcribes them via the NACS collaboration interface. At the same

time, he encourages the collaborators of the other departments of the metropolis to use the digital interface to analyze the general strategy and to comment on it when points of evolution are noticed. These initial comments are summarized in Table 1.

Observation	Process	Typology
Issues with respect to biodiversity.	inter-service collaboration	Environment
Seen from the metro, the entrance to Kerourien is very dark: scary, not feeling like going to the area.	citizen participation	Representation
Porch on Amiral de Gueydon Street is well lit and safe. Women use this route because it is safer than the porch on Père Ricard Street.	citizen participation	Representation
Reassuring lighting, especially with the supermarket open late.	citizen participation	Representation
Playground "le rond" not lit + dark atmosphere of the Père Ricard street: rethink the lighting of this place.	citizen participation	Representation
Père Ricard street : lack of light / Review the light on the main street.	citizen participation	Use
Basketball area not lit / Possibility of lighting until 8pm?	citizen participation	Use
Unlit playground: request for playground lighting.	citizen participation	Use

Tableau 1 - Contributions from the interdepartmental collaboration and citizen participation processes. Contribution from the Parks and Gardens department indicates a problem of respect for biodiversity. 7 citizen contributions provide observations of more or less comfortable atmospheres from the street.

8.3. Contribution analysis

We propose a classification of contributions by process (inter-service collaboration; citizen participation) and by typology (environment; representation; use). The typology environment focuses on the unidentified conflict areas between lighting practices and biodiversity preservation issues, in order to concretize the environmental transition of lighting practices. The representation typology gathers the citizens' contributions related to the feeling of security and well-being towards the lighting presence, considered as local cultural representation criteria. The use typology underlines the users' expectations in terms of lighting of some recreational equipment (playground), unusable in the evening due to lack of lighting.

8.4. Continuation of the territorial strategy

Undertaken in 2021 at the initiative of the head of the public lighting department in Brest, Saïg Potard, the NACS is simultaneously proceeding with the stages of general and territorial strategies. Framed by the rhythms of decision-making, the next participatory stages remain to be determined in terms of time, space and methodologies to be used. The SCAN approach is intended to be part of a long time perspective, dedicated to understanding the daily and occasional rhythms of all the territory's users, in order to adapt the political, technical and technological choices that are essential to a responsible practice of public lighting, thought out by and for the users.

8.5. The NACS methodology in the evolution of Brest Metropole's lighting policies

The general strategy of the SCAN is approved by a coordination committee made up of elected vice-presidents in various fields: territorial climate air and energy plan, services to the population, territorial proximity, urban planning, development, sensitive natural spaces, active mobility, mobility (public transport), urban tranquility, citizen participation.

The sharing of the challenges of each public policy, the pooling of their knowledge, thanks to the digital cartographic tool, allows the analysis of their respective impacts on street lighting and their collective weighting debated. This defines the general strategy of the SCAN. The territorial strategy is based on the field expertise of the users of the public space and also of the local elected officials (mayors of the communes and districts of the city of Brest) which contributes to the definition of the general strategy. Each street lighting renovation or development project complies with the SCAN guidelines and includes the politically validated recommendations in terms of atmosphere: illumination level, color temperature, nighttime dimming profile of the illumination level.

Regarding the nighttime atmosphere resulting from the NACS recommendations, a post-renovation field survey is currently being considered in order to collect the perceptions of users before and after the evolution of the nighttime atmosphere.

9. Discussion

Light sobriety, through its social approach, is a political project. Driven by the notion of rhythms, this in-depth strategic work is part of a temporal process motivated by criteria of decision-making, financial, technical and innovation capacities, all grouped under the dominant notion of uses. The clues collected through the NACS open up priority operational actions (regulations, energy, environment) and socio-cultural approaches embedded in the constancy of a decisive philosophical approach: nocturnal atmospheres are able to meet complex and qualitative sensitive information, to open up to a quantitative and qualitative balancing and to reveal historical and new nocturnal atmospheres, where the rhythms of lighting, users and darkness cohabit in space and time.

The methodology undertaken in Brest Metropole hopes to ensure the transition from a top-down method, where the user is still absent from the collaborative loop, to a bottom-up method, where he is a contributor to the strategies. Its insertion is envisaged once the general strategy has been politically validated, with the aim of testing it in the field and determining the territorial scale, adapted to the needs and activities of each place. In this sense, and to use Arnstein's scale, we envisage a transition from tokenism to partnership. However, one of the most important unknowns lies in the individual desire to be mobilized around a question of ambiences with resolutely acquired fundamentals and potentially uncomfortable changes.

This methodology is being deployed in parallel in other French communities and underlines the desire of decision-makers and planners to reduce the amount of artificial light in order to achieve substantial energy savings, greater respect for biodiversity and to place their renovation programs in an accessible and controlled time frame. The collaborative approach and the digital tools make it possible to consider the simplified evolution of strategies for the nighttime ambience of specific territories, in line with political and technical developments, and with current and future cycles of daily, seasonal and generational use, with the aim of consolidating the dynamics of sober lighting, adapted to the uses and temporalities of the territory.

After decades of progress and innovation in favor of widely illuminated territories, and without any other consideration than the rhythms of economic growth and the extension of the duration of professional and recreational activities, the criteria of energy savings and environmental preservation open up unexpected social considerations to lighting practices. These considerations could potentially lead to the development of responsible practices of artificial lighting, motivated by a sensitive and qualitative expectation of night landscapes, shared and consolidated by the expertise of the users.

In terms of rhythms, which temporalities will be needed to rediscover balanced nocturnal atmospheres, whose dark resource will eventually present physiological, psychological and philosophical virtues?

References

[1] Houel, N. (2020) La pédagogie de la sobriété lumineuse, thesis
[2] Falchi, F.; Cinzano, P.; Duriscoe, D.; Kyba, C.; Elvidge, C.; Baugh, K. ; Portnov, B. ; Rybnikova, N.; Furgoni, R. (2016). The new world atlas of artificial night sky brightness. Science Advances.
[3] Sanders, D., Frago, E., Kehoe, R. et al. (2021) A meta-analysis of biological impacts of artificial light at night. Nat Ecol Evol 5, 74–81.
[4] Challéat S., (2019), Sauver la nuit. Comment l'obscurité disparaît, ce que sa disparition fait au vivant, et comment la reconquérir, Paris, Premier Parallèle, 296 p.
[5] Franchomme, M., Hinnewinkel, C. (2019). Acceptabilité sociale de la trame noire : exemple de la Métropole Européenne de Lille.
[6] Paquot T. (2000). Le sentiment de la nuit urbaine aux XIXe et XXe siècles. In: Les Annales de la recherche urbaine, N°87. Nuits et lumières. pp. 6-14.
[7] Deleuil J.M.; Toussaint J.Y. (2000). De la sécurité à la publicité, l'art d'éclairer la ville. Les Annales de la Recherche Urbaine, PUCA, pp.52-58.
[8] Zielinska-Dabkowska, K.; Xavia, K.; Bobkowska, K. (2020). Assessment of Citizens' Actions against Light Pollution with Guidelines for Future Initiatives. Sustainability. 12. 4997.
[9] Chelkoff, G.; Bardyn J.L. (1990) Une approche qualitative de l'éclairage public. [Rapport de recherche] 17, CRESSON; GEG, pp.167.
[10] Deleuil J.M. (2009) Eclairer la ville autrement, expérimentations et innovations en éclairage public. Suisse. PPUR.
[11] Lefebvre H. (2004) Rhythmanalysis, space, time and everyday life. Bloomsbury Academic.
[12] Zielinska-Dabkowska K. ; Bobkowska K. (2022) Rethinking Sustainable Cities at Night: Paradigm Shifts in Urban Design and City Lighting. Sustainability. 14(10):6062.

An examination of reflected glare prediction method based on luminance distribution

M Sagawa[1], A Maeda[1], Y Mizokami[2], M Takashima[3], Y Oe[4] and N Yoshizawa[4]

[1] Department of Architecture, Graduate school of Science and Technology, Tokyo University of Science, 2641, Yamazaki, Noda-shi, Chiba-ken, 278-8510 Japan
[2] Department of Imaging Sciences, Graduate School of Engineering, Chiba University, 1-33, Yayoi-cho, Inage-ku, Chiba-shi, Chiba-ken, 263-8522 Japan
[3] The National Museum of Western Art, Tokyo, 7-7, Ueno-Koen, Taito-ku, Tokyo, 110-0007 Japan
[4]Department of Architecture, Faculty of Science and Technology, Tokyo University of Science, 2641, Yamazaki, Noda-shi, Chiba-ken, 278-8510 Japan

yoshizawa@rs.tus.ac.jp

Abstract. Explicit evaluation methods and suppression techniques for reflected glare have not been clarified in museum lighting. The purpose of this study is to elucidate the occurrence mechanism of reflected glare and to make use of its knowledge for future exhibitions in museums. In this paper, for the first step, we focused on the several factors, which were assumed to have an influence on the intensity of the reflected glare, such as the balance between ambient light levels for an exhibition room and target light levels for paintings, the Munsell value of paintings, and the amount of varnish coating. We conducted a subjective experiment in which the reflected glare on the replica paintings was evaluated by subjects aged in their 20s in an experimental room that was assumed to be an actual exhibit space. Based on the results of this experiment, we confirmed that the degree of reflected glare in museums could be roughly estimated from luminance distribution, reflectance, and glossiness of the paintings.

1. Introduction

Reflected glare on the surface of exhibits in museums often interferes with an appreciation of paintings and must be suppressed to create a good environment for viewing art. Curators who exhibit artwork also cite reflected glare as a serious problem, and clear indicators are needed. Currently several standards or guidance on museum lighting suggest the methods to suppress reflected glare based on geometric aspects. Reflected glare can be avoided by devising the position and direction of spotlights, or tilting the paintings toward the viewer[1][2]. Figure 1 shows the recommended position of light sources to avoid occurring the reflected image of luminaire on the displayed object in Japanese Technical Guide for Museum Lighting[3]. On the other hands, suppression methods that take into account other factors are not fully elucidated at the moment. Possible influencing factors on reflected glare are as follows - Munsell values, surface roughness, and materials of paintings, the optical characteristics of display cases, and the balance between ambient light levels for an exhibition room and target light levels for paintings. There have been several metrics to evaluate discomfort glare of indoor electric lighting and daylighting[4][5][6][7][8][9][10], however these metrics does not deal with reflected glare directly. The final purpose of this study is to elucidate the overall mechanism of reflected glare in museum lighting. In this paper, as a first step, we focused on following three factors: the balance between ambient light levels for an exhibition room and target light levels for paintings, the Munsell value of paintings, and the amount of varnish coating, and examined whether degree of reflected glare could be predicted based on these factors.

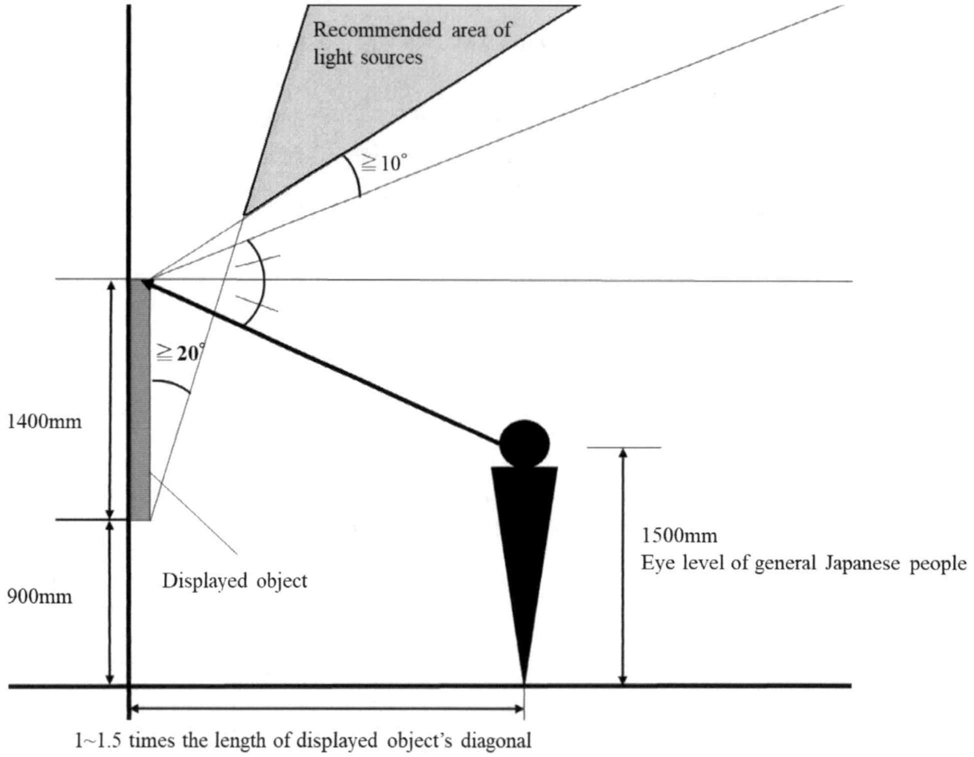

Figure 1. Recommended area of light sources.

2. Experimental method

We conducted the subjective experiment to understand how the effect of reflected glare changes by three experimental variables described in the previous section. There were 18 subjects (8 males and 10 females) with normal colour vision, and all subjects were students in their 20s. Colour vision was confirmed by Ishihara chart.

2.1. Experimental room

The experimental room as shown in figure 2 was designed assuming an actual art museum. The dimensions of the room were 2900 mm in width, 3000 mm in depth and 2400 mm in height. All interior surfaces of this room were painted white (N9.5) and its mean reflectance is 85.0%. In order to produce the reflected glare on the paintings, the luminaire configuration was intentionally designed. However, since the position of the lighting equipment and the direction of light incident on the paintings were fixed due to the experimental room constraints, these factors were not included in the experimental variables. The eye level was 900 mm above the floor and the installation height of the spotlight was 1200 mm. The luminaire with wide beam light distribution was used as an ambient light (blue-phosphor white LEDs, Ra:97, CCT:3000K, duv: -0.004) and the spotlight was used as a target light on the painting (purple-phosphor white LEDs, Ra:93, CCT:3000K, duv: -0.005). The spotlight was equipped with the trimming cutter to illuminate only the painting surface in a rectangular shape.

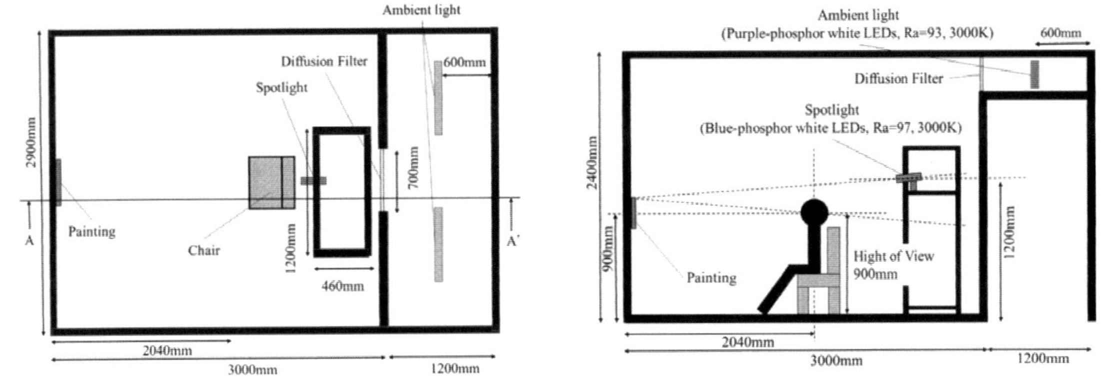

Figure 2. Experimental room (a: Plane section b: A-A' Cross section).

2.2. *Experimental conditions*

Lighting conditions (table 1) were created by varying the ratio of the painting illuminance by the spotlight to the one by the ambient light. There were 12 lighting conditions that could be adjusted with the luminaire output in a combination of three levels of illuminance on the painting (100lux, 150lux, 200lux) and five levels of illuminance ratio (0.03, 0.11, 0.25, 0.43, 1.00), which was defined as the illuminance on the painting by the spotlight divided by the illuminance by ambient light. All illuminance measurements were taken at the centre of the painting. In order to confirm the effect of Munsell value and gloss on reflected glare, nine replica paints were prepared by combining three levels of Munsell values (black, grey, white) and three levels of varnish application (no varnish, one layer and two layers of varnish) on the painting surface. The surface of paintings was painted with one colour. The replica paints were created under the supervision of the conservator at the National Museum of Western Art, Tokyo, and we took a method of applying commercially available opaque oil tube paints and varnishes to a ground-painted canvas. The colour of oil tube paint was chosen to be achromatic. The information of the replica paint colours are shown in table 2. The mean reflectance of these paintings was measured at 9 points with a spectrophotometer (CM-700d, KONICA MINOLTA). In the same way, the mean glossiness was measured with a glossmeter (GM-60, KONICA MINOLTA). The dimensions of the paintings were 410mm in width and 318mm in height. There were a total of 108 experimental conditions.

2.3. *Experimental procedure*

The experimental procedure was as follows: 1) The subject adapted to the light environment (50lux at the centre of painting) for 10 minutes after entering the experimental room, 2) The experimenter set a experimental condition, 3) The subject adapted to the condition for 30 seconds and then evaluated the painting, 4) The process 2) and 3) were repeated. The experimental conditions were presented randomly for each subject, and all 108 conditions were divided into three separate experiments. Each experiment takes approximately 2 hours. The subjects evaluate the whole painting and designated areas within the painting. The designated areas in the painting are shown in figure 3, at the top and bottom. This paper deals only with questions related to reflected glare on the upper designation area where the most reflected glare occurred. Table 3 shows the evaluation items used in this paper. The rating scales were a monopolar scale of 0 to +6.

Table 1. Lighting conditions.

Illuminance on the painting by the spotlight [lux]	Illuminance on the painting by the ambient light [lux]	Illuminance on the painting [lux]	[a]Illuminance ratio
5	195	200	0.03
5	145	150	0.03
20	180	200	0.11
10	90	100	0.11
15	135	150	0.11
40	160	200	0.25
20	80	100	0.25
30	120	150	0.25
30	70	100	0.45
60	140	200	0.45
45	105	150	0.45
50	50	100	1.00

[a]Illuminance ratio= Illuminance on the painting by the spotlight/ Illuminance on the painting by the ambient light.

Table 2. Replica paint conditions.

Painting colour	Munsell Value	Chroma	Reflectance	Glossiness (No varnish/One layer of varnish /Two layers of varnish)
Black	1.72	0.25	2.47	0.133/1.21/2.22
Grey	5.92	0.20	28.4	0.489/1.31/2.22
White	9.51	0.38	88.0	0.444/1.49/3.04

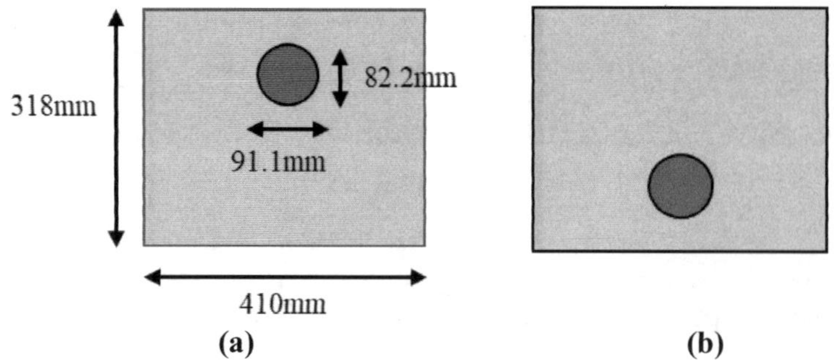

Figure 3. Designation area in the painting (a: upper designation area b: lower designation area).

Table 3. Evaluation items for reflected glare.

Occurrence of reflected glare
Dazzling caused by reflected glare
Uncomfortable view caused by reflected glare
Difficult to distinguish colours because of reflected glare

3. Results

3.1. The effects of experimental variables on reflected glare

In this paper, the luminance ratio was calculated as the mean luminance of the area where the reflected glare occurred (the upper designation area) divided by the mean luminance of the other area on the painting, and was used in the analysis described below. The luminance ratio ranges from 0.44 to 2.46. We will only report the result on "Occurrence of reflected glare", since the other three evaluation items had almost similar relationship with the physical quantities, such as the luminance ratio. Figure 4-1 shows the relationship between the evaluation and the luminance ratio by the Munsell value on the painting. It was confirmed that the effect of reflected glare increased as the luminance ratio increased at all Munsell value levels. In the case of paintings with the same Munsell value, there were no clear differences due to the amount of varnish. Figure 4-2 shows the relationship between the evaluation and the luminance ratio by the mount of varnish. Reflected glare scarcely occurred in any levels of Munsell value on the painting without varnish. In the one-layer and two-layer varnish, the degree of reflected glare was higher at lower Munsell value under low luminance ratio conditions below 1.2, however, when the luminance ratio exceeds a certain value, the relationship between perceived reflected glare and the luminance ratio appeared almost the same among different Munsell values. There would also be a certain luminance ratio at which the degree of reflected glare reaches the peak.

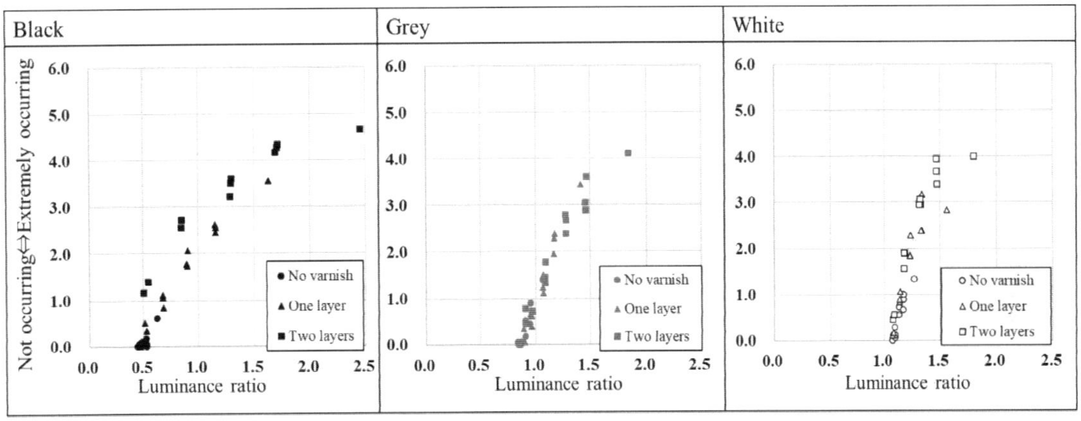

Figure 4-1. Relationship between reflected glare and luminance ratio by Munsell value.

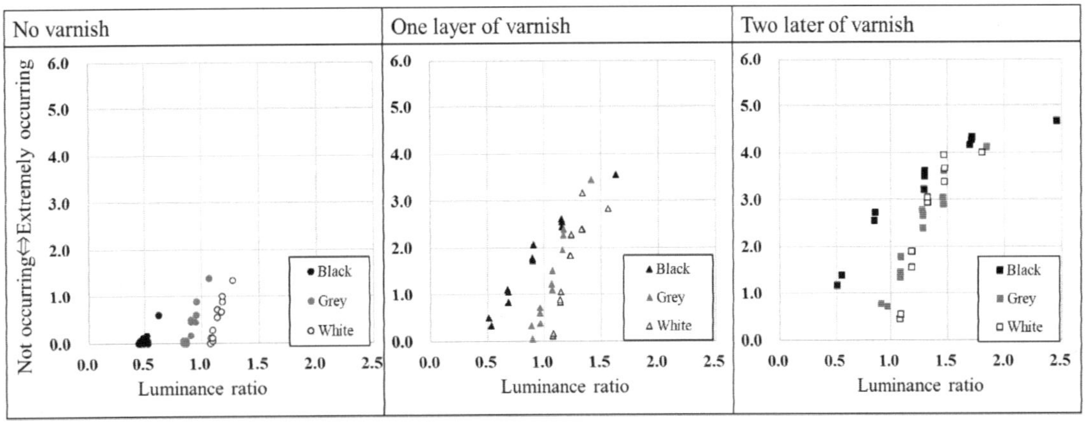

Figure 4-2. Relationship between reflected glare and luminance ratio by varnish.

3.2. Prediction of reflected glare by physical quantity

A single linear regression analysis was performed using the mean evaluation value of the reflected glare occurrence as objective variable and the luminance ratio as the explanatory variable. The luminance ratio could be assumed as a physical quantity that covers the effects of three experimental variables mentioned in the previous section. This analysis could not construct a regression model with a high coefficient of determination (R^2=0.61, Figure 5-(a)). This result might be due to the fact that the luminance ratio did not encompass all the effects of glossiness and reflectance. Therefore, we performed a multiple regression analysis with three explanatory variables, adding the reflectance and glossiness of each painting. As a result, we obtained a regression equation with higher coefficient of determination (R^2=0.82, Figure 5-(b)). These results suggest that the degree of reflected glare was difficult to predict from the luminance distribution of the painting surface only, but might be predicted to some extent from three physical quantities including reflectance and glossiness on the painting. Furthermore, since the plot distribution of figure 5-(b) reach a peak, there is a possibility that the effect of reflected glare saturates when the luminance ratio exceeds a certain value.

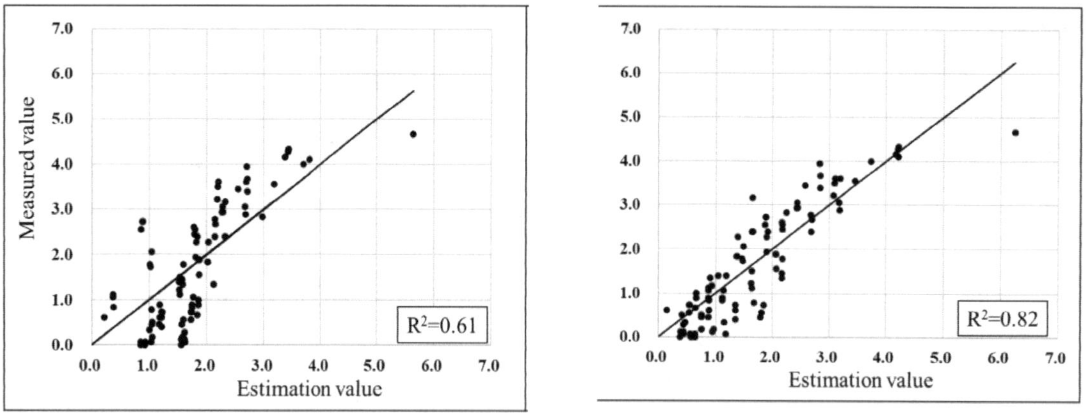

(a) **(b)**

Figure 5. Prediction of reflected glare (a: single regression analysis b: multiple regression analysis).

4. Conclusion

In this study, as a first step in developing an evaluation method for reflected glare in museum, we confirmed the possibility of predicting its degree from a regression model that uses the luminance distribution, the reflectance, and the glossiness of the painting surface as variables. The next step is to consider a non-linear regression model for reflected glare to account for the possibility that the effects of reflected glare saturate. In addition to the physical quantities focused on in this paper, it is necessary to continue further experiments by adding other factors that would affect reflected glare, such as the characteristics of the display case and the surface properties of the paintings. Furthermore, it is necessary to examine the luminance distribution used in the analysis not only within the painting surface, but also in a more extended range including the exhibition wall. In future work, we will also examine whether the current results obtained from the controlled experiments can be applied to actual paintings and exhibition spaces in museums.

References

[1] CIBSE 2015 *Lighting for the built environment LG8: Lighting for museums and art galleries* (Norwich: Page Bros Group) chapter 4 pp 20–4, chapter 8 pp 56–9
[2] Christopher C 2007 *Light for Art's Sake* (London: Routledge) pp 164–5
[3] Sano C, Yoshizawa N, Kinoshita S, Kubo k, Kotani A, Suzuki J, Takashima M, Tsukada M, Nagato S, Nakamura Y, Fujiwara T, Hotta R, Matsuoka H, Mizokami Y, Yamauchi T, Yoshida

N, and Yoshitsuka N 2021 *Technical Guide for Museum Lighting* (Tokyo: The Illuminating Engineering Institute of Japan) pp 36–7
[4] Holladay L 1926 The fundamentals of glare and visibility *J Opt. Soc. Am.* **12(4)** pp 271–319
[5] Luckish M and Guth S K 1949 Brightness in visual fid at borderline between comfort and discomfort *Illuminating Engineering* **44** pp 650–70
[6] Commission Internationale de l'Eclairage 1995 Discomfort Glare in Interior Lighting (Vienna: CIE) *CIE Publication* **117**
[7] Kasahara T, Aizawa D, Irikura T, Moriyama T, Toda M and Iwamoto M 2006 Discomfort glare caused by white LED light sources *J. Light Vis. Environ.* **30** pp 95–103
[8] Hopkinson R 1972 Glare from daylighting in buildings *Appl Ergon* **3** pp 206–15
[9] Tokura M, Iwata T, and Shukuya M 1996 Experimental study on discomfort glare caused by windows part 3 Development of a method for evaluating glare from a large light source *J. Archit. Plann. Environ. Eng., AIJ* **489** pp 17–25
[10] Nazzal A 2001 A new daylight glare evaluation method Introduction of the monitoring protocol and calculation method *Energy Build.* **33** pp 257–65

IOP Conf. Series: Earth and Environmental Science 1099 (2022) 012038 doi:10.1088/1755-1315/1099/1/012038

Relationship between photon flow and the perception of light field under simple lighting environments

Otori Kaira[1], Ken Komazawa [1], Roland Schregle [2] and Nozomu Yoshizawa[1,*]

[1] Department of Architecture, Faculty of Science and Technology, Tokyo University of Science, 2641, Yamazaki, Nodashi, Chiba, 278-8510, Japan
[2] School of Engineering and Architecture, CC Building Envelopes and Civil, Engineering, Lucerne University of Applied Sciences and Arts, Werftestrasse 4, 6002 Luzern, Switzerland

[*] Correspondence: yoshizawa@rs.tus.ac.jp

Abstract. In our previous study we proposed a new method, "photon flow", to visualise the light propagation in space. Theoretically the photon flow depicts the physically correct light distribution, however, it was yet unclear whether people could correctly read the physical light field in actual spaces by observing the photon flow. Therefore, in this study we verified the relationship between the light field perceived in the real space (visual light field) and the one estimated by observing the photon flow. The experiment was conducted using the method of a white sphere, on which shade represents the light field at that point. We clarified the characteristics of the perceived light field in a simple rectangular space with one spotlight or two. The perceived light field when observing the photon flow was basically similar to the physical light field when there is one spotlight in the space. However, under the more complex scene, the subjects sometimes could not predict the physical light field well when seeing the photon flow. Basically, the perceived light field when observing the photon flow was close to the visual light field than expected. In future work we would like to clarify under which conditions people can estimate the correct physical light field by observing the photon flow and compare the difference between experienced and naïve observers.

1. Background and purpose

Two-dimensional indexes such as illuminance/luminance distribution are commonly used in lighting environment evaluation, however, these indexes can merely present the result of light distribution and cannot directly express the three-dimensional process of light propagation in space. Photon flow, which was introduced in the previous study[1)], shows lighting environment in three dimensions at arbitrary positions in space, and it can illustrate the physical light field itself. Whereas Koenderink[2)] clarified that there are physical and visual light fields, it is yet unclear which light field will be visually perceived by observers when seeing the photon flow. If people can easily read the physical light field by observing the photon flow, it could be directly used as one of the helpful tools in the lighting design. If this is not the case, we need to clarify which information of light fields the photon flow will convey and how to effectively present the photon flow for designers. The purpose of this study is, as a first step, to clarify what kind of light field is estimated when naïve observers see the photon flow.

2. Methods

Subjects first observed a real space illuminated by a spotlight or two and estimated its light field. Then, they observed the photon flow of the same space, which was calculated using Radiance, and estimated the light field again. This experiment employed the white sphere method proposed by Kartashova)[3], in which subjects answered the shade on a virtual white sphere, which was not installed in the actual space, to capture the perceived light field. This perceived light field is referred to as "visual light field (VLF)".

2.1. Experiments to estimate the visual light field using a real space

2.1.1 Space Overview. The actual experimental space used in this research is shown in figure 1. The ceiling and walls were uniformly white at N9.5, and the floor surface was red (Y, x, y) = (10.9, 0.56, 0.29) or gray with the same reflectance. The light source used in this room is shown in Table1. In the simulation of the light environment by Radiance, the officially published IES data was used.

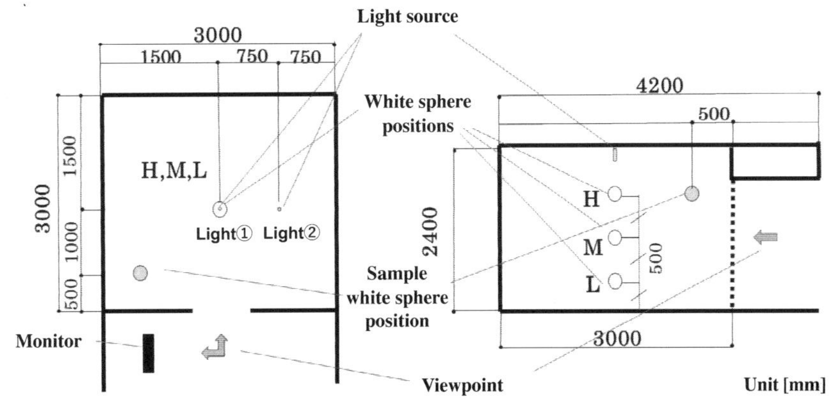

Figure 1. Floor plan (left) and cross section (right) of the experimental space.

Table 1. Light source specification.

Model number	Colour temperature [K]	Dimming Angle [°]	Dimming rate [%]	Model number	Colour temperature [K]	Dimming Angle [°]	Dimming rate [%]
① NTS01179WRZ1	2700	19	6	② NTS01189WRZ1	2700	31	10

Figure 2. Real space (Gray floor and Red floor) observed by subjects.

2.1.2 Observation conditions and Experimental procedure. Experimental factors and levels are shown in Table 2, with a total of eight conditions combining the following factors and levels. The subject observed the experimental space illuminated by the spotlight and judged the light field in the space. The experimenter displayed the rendered image of a white sphere, which was prepared in advance for every grid point in the room, on the monitor . Then the subject was asked to compare it with their own estimation of the shade on the virtual white sphere which was supposed to be at the position instructed by the experimenter. Subjects judged using a three-point scale, whether the shade presented on the monitor was correct/undecided/incorrect for the virtual white sphere (hereinafter referred to as consistency judgement). Total 20 subjects in their 20s participated in the experiment.

Table 2. Observation conditions in real space.

Floor colour	Light condition	White Sphere height		
	SpotLight	500(L)	1000(M)	1500(H)
Gray	1 Light ①	L Condition_Gray	M Condition_Gray	H Condition_Gray
	2 Light ①②		M2 Condition_Gray	
Red	1 Light ①	L Condition_Red	M Condition_Red	H Condition_Red
	2 Light ①②		M2 Condition_Red	

2.1.3. Making of the white sphere images for consistency judgment. Table 3 shows the monitor used in the experiment. Adobe Photoshop was used to present the white spheres on it. To correctly produce the colour and luminance of the white sphere image presented on the monitor, all materials inside the experimental space were measured with a spectrophotometer (Table 4). The CIE standard light source A was used as the observation light source, since the correlated colour temperature of the spotlight used in this experiment was 2700[K].

Table 3. Monitor specification.

Name	Apple Display XDR	Reference Model	Pro Display XDR(P3-1600nit)
Finish	Nano-texture Grass	Color Temperture	D65
Resolution	6016×3384	Whie point	P3
Pixel Pitch	218ppi	Gamma setting	2.2
Output Briightness	0.005~500(cd/m2)	Dimensions	71.8×41.2cm

Table 4. Material information.

	X	Y	Z
Gray floor	17.0	15.7	6.0
Red floor	39.0	19.1	1.1
Wall & Celling	97.2	88.5	30.6

The von Kries model[4] was used for the transformation of colour adaptation and the Sharp RGB Colour Space[5] for rendering. The XYZ values of the CIE standard light source A were converted to the Sharp RGB colour space, then converted to the Radiance RGB values. The experimental space had high reflectance and high interdiffusion components, therefore the photon mapping algorithm was used for the rendering. The parameter settings were shown in Table 5.

Table 5. List of parameters used in the calculations.

Parameter Setting									
-ap	500	-ar	4096	ds	0.02	-dj	0	-pj	0.5
-ad	1	-ad	4096	-dc	0.17	-ps	1		
-aa	0.1	-as	2048	-dt	0.05	-dp	2048		

Subjects evaluated the rendered image displayed in P3 colour space on the monitor. Therefore, the colours of the rendered images in the Sharp RGB space were converted to those in the P3 space. The

RGB values converted to P3-space colours were multiplied by 1/2.2 (γ value) to obtain the final input values. The viewing angle of the consistency judgment images when presented on the monitor was set be the same as that when the white sphere was actually placed in the real space.

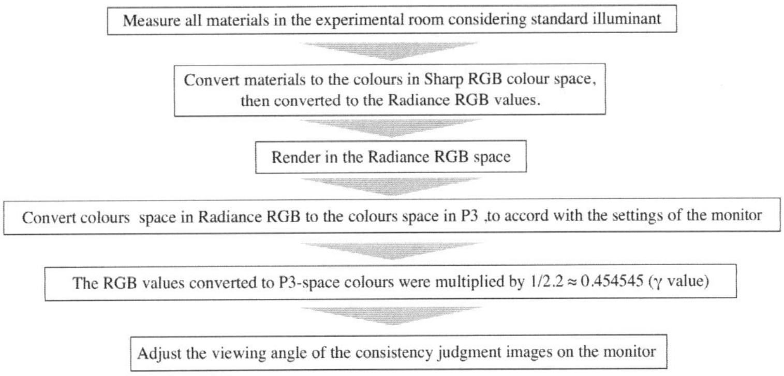

Figure 5. Flow chart for making consistency judgment images (white sphere images)

2.2. *Experiments to estimate the light field using photon flow*
Instead of the real space, subjects observed on the monitor the photon flow that faithfully reproduces the same space and estimated the shade on the white sphere as in the experiment conducted in the real space described in section 2.1.

2.2.1. Photon flow expression. Since the radiant flux of each photon is the same value, the luminance of all photons must be uniform on the monitor. Some images of photon flow are shown in figure 6. Note that the colours have been changed to make them easier to see in this paper.

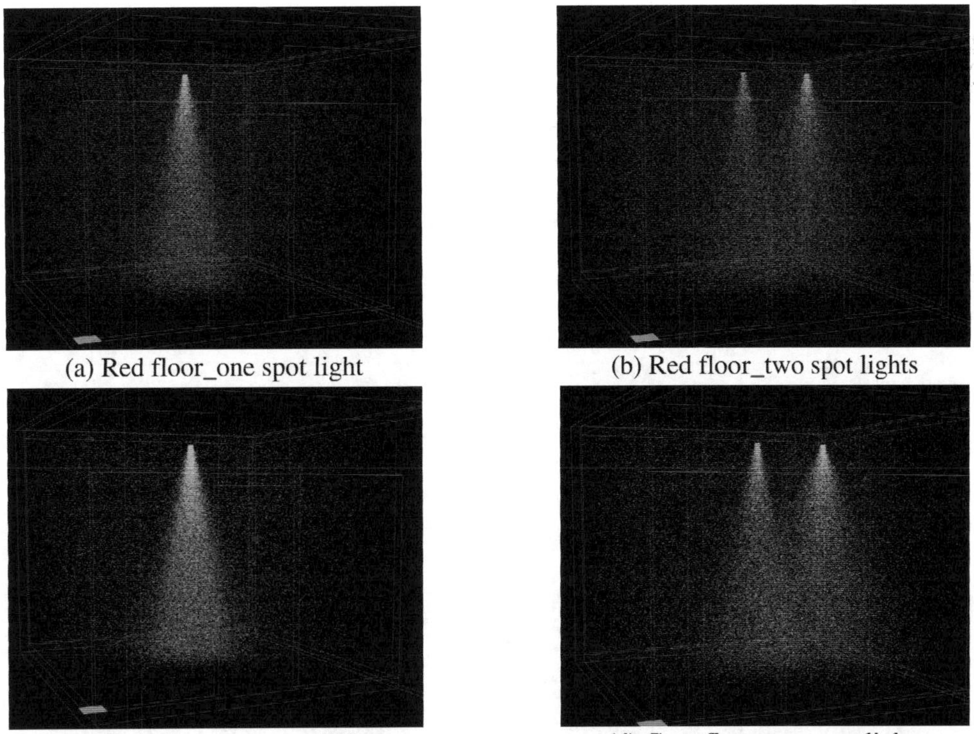

(a) Red floor_one spot light
(b) Red floor_two spot lights
(c) Gray floor_one spot light
(d) Gray floor_two spot lights

Figure 6. Photon flow images.

2.2.2. Experimental conditions, experimental procedures, and subjects. The experimental conditions, experimental procedures, and subjects were the same as in the real space experiment. Since it was difficult to read the absolute brightness of the white sphere, which corresponded to the photon density, we simultaneously presented the rendered image of a sample white sphere which was located at the corner of the experimental space as a reference for estimating the shade on the virtual white sphere.

3. Results

To clarify the relationship between the perceived light field when observing the photon flow and the physical light field (PLF), we used four indices (directionality/intensity/diffuseness/Colour difference) to confirm how well subjects could evaluate the PLF. In this paper, the colour difference was the RMSE of ΔE_{00} (CIE DE 2000). If the subject answered "Correct" to the consistency judgment image at the correct position (= the white sphere position), the evaluation was 0 for all indices. If the subject answered "Correct" for the consistency judgment image at the positions other than the correct position, we evaluated each indicator based on how much the selected image differed from the image at the correct position (For example, for directionality, we calculated the angle between "the illumination vector's direction of the consistency judgment image at the correct position" and "the direction of the image on which the subject answered "Correct").

Figure 7 was created by the following steps. (I) Relative values were calculated for all consistency judgment images to the maximum value of the index. (II) The relative values at the position where the subject answered "correct" were totalled for each condition. The total value was then divided by the number of selected images and the average value was calculated. (If a subject answered "correct" for consistency judgment images No.1 [relative value: 0.3], No.5 [relative value: 0.8], and No.10 [relative value: 0.1], the value in this step is (0.3+0.8+0.1)/ 3 = 0.4). (III) For all subject, STEPs (I) and (II) were implemented, and the average of the results of all subjects was calculated for each condition. (IV) Finally, figure 7 summarizes the results of step (III) for each condition. The black dotted line in the graph is the baseline for relative evaluation of the results for all conditions. This time, half of the maximum value for each indicator was used as the baseline. In other words, if the condition of which value for each indicator was less than the baseline, the perceived light field was regarded as being close to the physical light field.

$$\text{Directionality}=\cos^{-1}\frac{\overrightarrow{E_v}\cdot\overrightarrow{E_p}}{|\overrightarrow{E_v}||\overrightarrow{E_p}|} \quad (1)$$

$$\text{Intensity}=\frac{|\overrightarrow{E_v}|}{4}+\sim E \quad (2)$$

$$\text{Diffuseness}=1-\frac{|\overrightarrow{E_v}|}{4E_{scalar}} \quad (3)$$

$$\text{Colour difference}=\sqrt{\frac{1}{n}\sum_1^n(\boldsymbol{\Delta E_{00}})^2} \quad (4)$$

where

$\overrightarrow{E_v}$: Vector representing the direction of light at the white sphere position

$\overrightarrow{E_p}$: Vector representing the direction of light at positions other than the white sphere position

$|\overrightarrow{E_v}|$: Magnitude of the vector representing the direction of light at the white sphere position

$|\overrightarrow{E_v}|$: Magnitude of the vector representing the direction of light at the position other than the white sphere position

$\sim E$: Symmetric illumiance

n : Number of pixels of the consistency judgment image

ΔE_{00} : Color difference of each pixel of the consistency judgment image between the white sphere position and other positions.CIE DE 2000

3.1 Conditions: PLF could be read

Total 8 conditions were classified into four groups by cluster analysis based on the indicators' value.

The groups were Group 1: H Condition_Red/Gray, Group 2: M/L Condition_Gray, Group 3: M2 Condition_Gray, and Group 4: M/L/M2 Condition_Red.

Figure 8 summarizes the results of the graphs in figure 7, especially for Group 2. All indicators showed that the PLF could be relatively well perceived in Group 2.

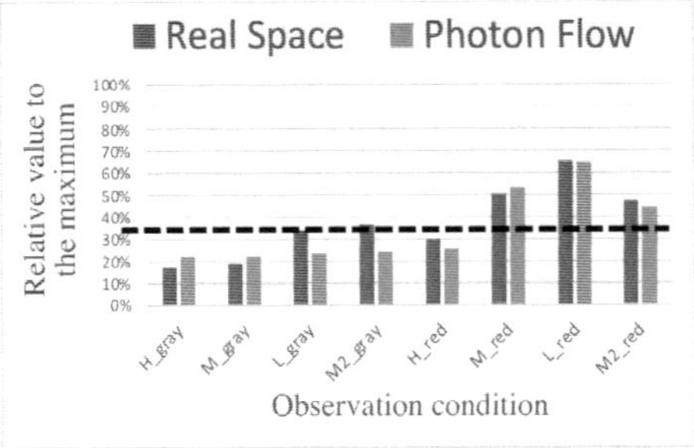

1.Directionality

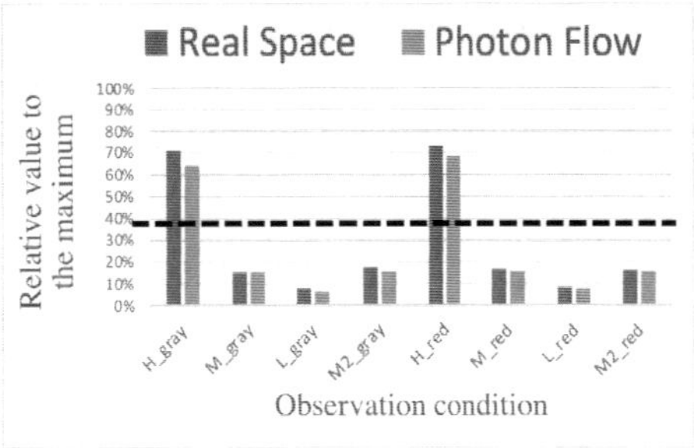

2.Intensity

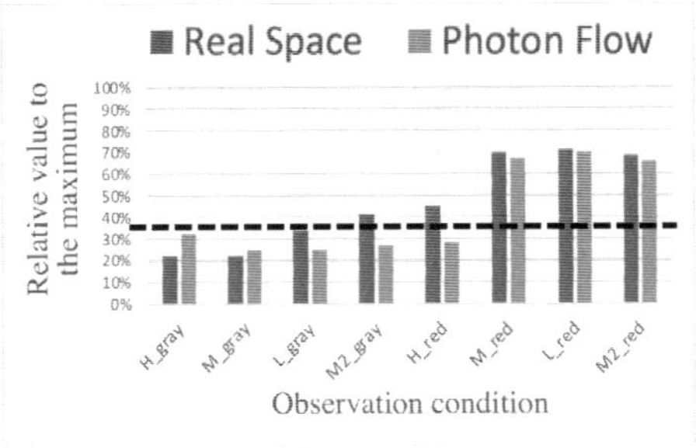

3. Diffuseness

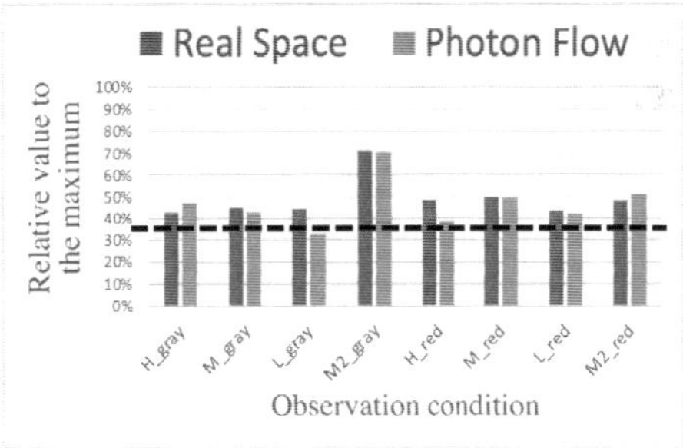

4. Colour difference

Figure 7. Relative value of the indicator calculated for each condition to the maximum value.

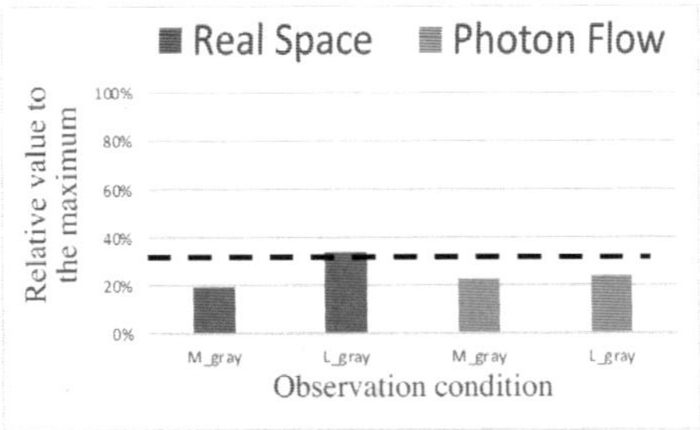

1.Directionality

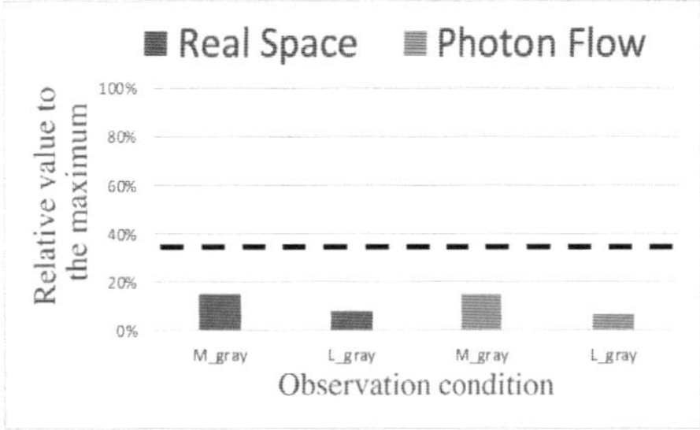

2.Intensity

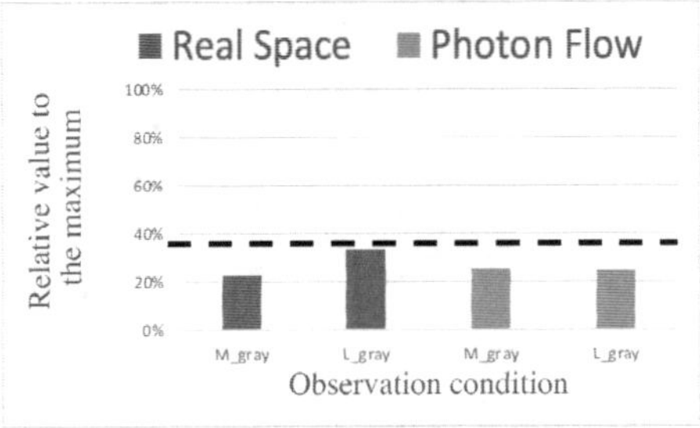

3. Diffuseness

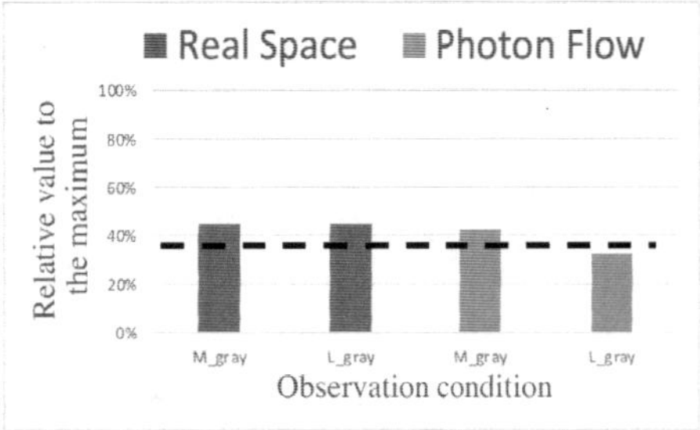

4. Colour difference

Figure 8. Relative value of the indicator calculated for each condition to the maximum value in Group .2

IOP Conf. Series: Earth and Environmental Science 1099 (2022) 012038 doi:10.1088/1755-1315/1099/1/012038

3.2 Conditions: PLF could not be read

Figure 9 summarizes the results of the graphs in figure 7, especially for Group 4. Most of the indicators indicated that the PLF could not be well perceived in Group 4. Figure 10 shows white sphere images on the gray floor and red floor conditions. Figure 11 is the logarithmic luminance maps of the white sphere image in figure 10. There is little difference between the luminance of the simulated white spheres on the red and gray floor. There is a possibility that, in the red floor conditions, subjects judged the contingency of the images just by the fact that the bottom areas of the white spheres on the red floor were tinted red, regardless of the directionality on them. Further investigation is required on this issue in future.

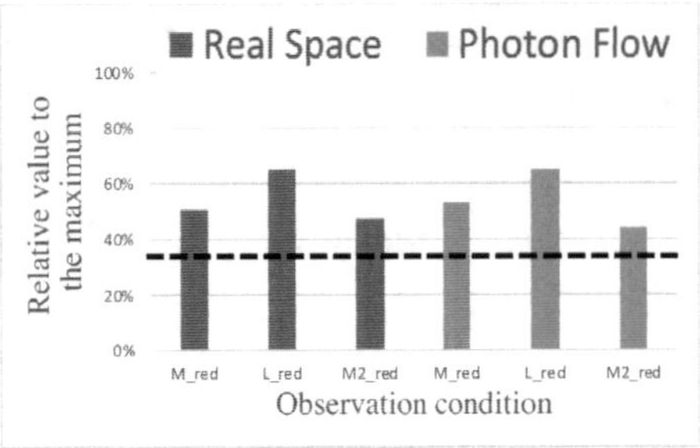

1.Directionality

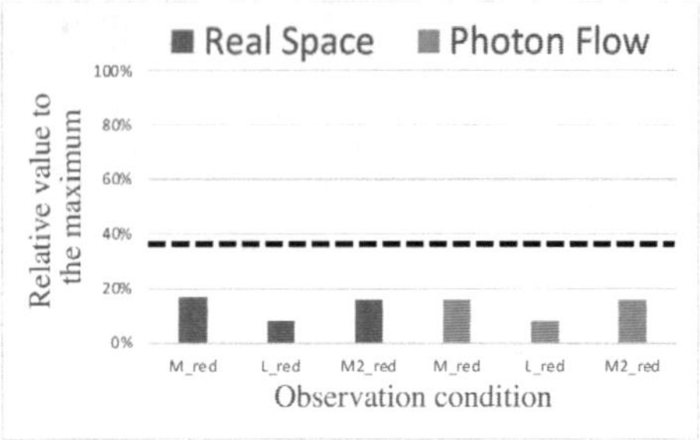

2.Intensity

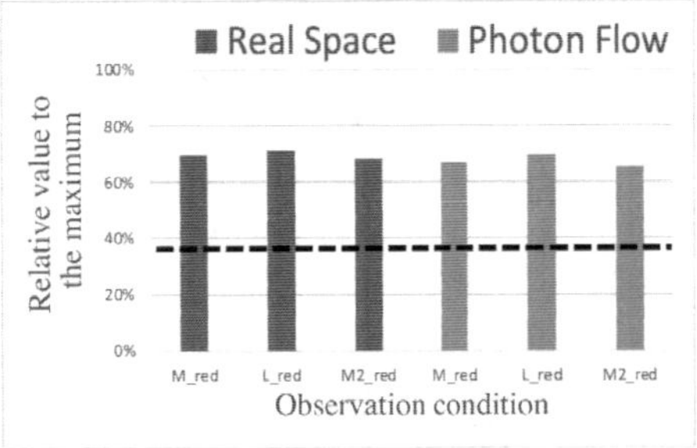

3. Diffuseness

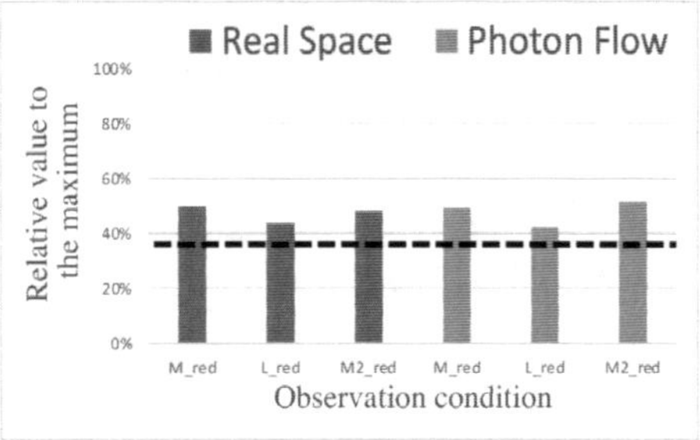

4. Colour difference

Figure 9. Relative value of the indicator calculated for each condition to the maximum value in Group 4

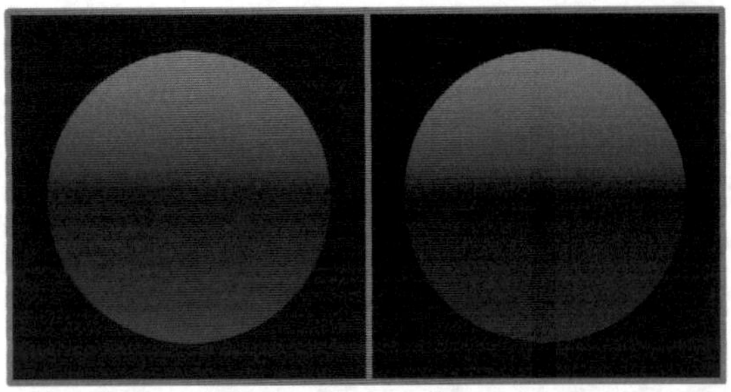

Correct position

(X, Y, Z) = L(1500, 1500, 500)

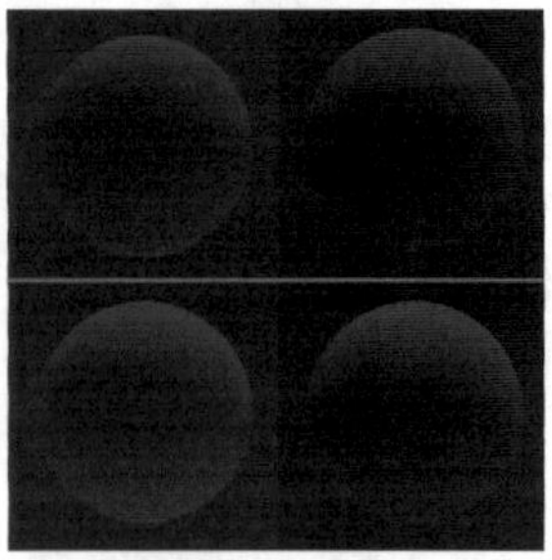

Incorrect position

(X, Y, Z) = (1500, 1000, 500)

(X, Y, Z) = (1500, 1000, 250)

Figure 10. The consistency judgment images of the white sphere position, Red floor (Left) and Gray floor (Right) under one spotlight.

Correct position

(X, Y, Z) = L(1500, 1500, 500)

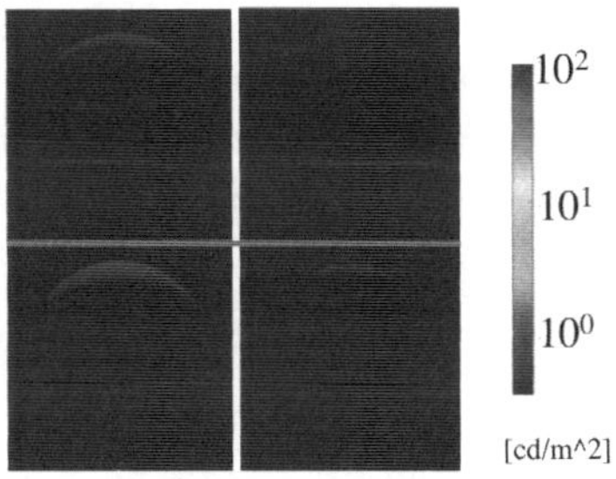

Incorrect position
(X, Y, Z) = (1500, 1000, 500)
(X, Y, Z) = (1500, 1000, 250)

Figure 11 Luminance maps of the consistency judgment images of the white sphere position, Red floor (Left) and Gray floor (Right) under one spotlight.

Each sphere in figure 12 shows the percentage of subjects who answered "Correct" on the consistency judgment image at the position in M2 Condition_Red. The larger the sphere size is, the more subjects responded with "Correct". In this condition, the white sphere was virtually placed at the intersection of the horizontal line M and the vertical Z axis. From these graphs we can see that the perceived light fields by observing the real space and its photon flow were quite similar in this condition, and most subjects answered "Correct" at positions slightly shifted to the positive direction of X-axis

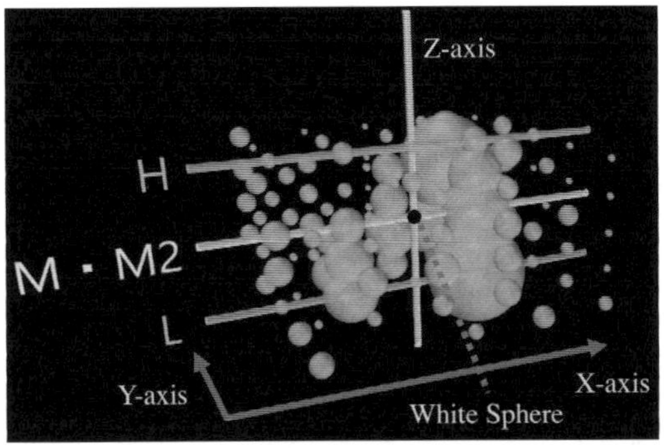

Real space

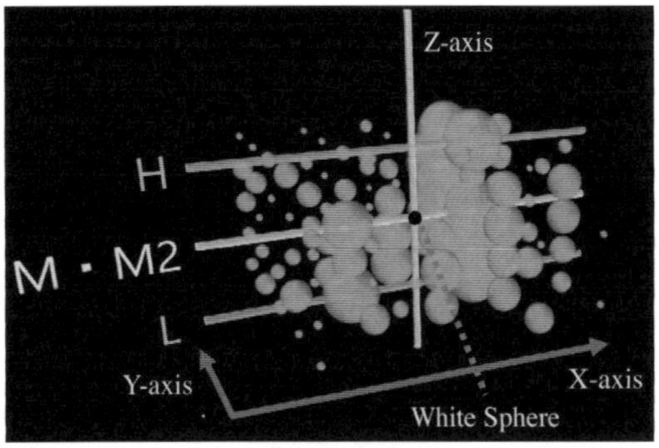

Photon flow

Figure 12. Response distribution in M2 Condition_Red: Real space (Left) and Photon Flow (Right)

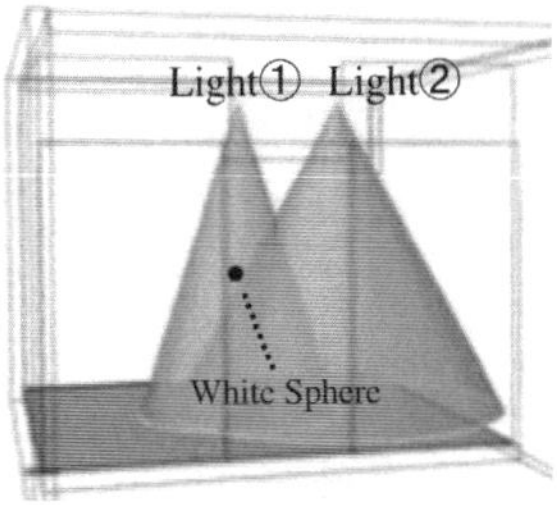

Figure 13.
The light distribution angles of the light sources

At the correct position (X, Y, Z) = (1500, 1500, 1500)

At a position shifted to the right of the correct position (X, Y, Z) = (1500, 2000, 1500)

Figure 14. Consistency judgement images: M2 Condition_Red.

from the correct position. As shown in figure 13, the white sphere position M is just on the boundary of the light distribution angle of the light source ②. There is a possibility that subjects overestimated the effect of the light source ② when seeing the real space and the photon flow, and the white sphere images which were illuminated from the upper right (Right figure in figure 14) seemed for them to be at the correct position.

3.3 Correspondence with visual light field

Figure 15 shows the relationship between the percentage of respondents who answered "Correct" when observing the real space and the one when observing its photon flow in all conditions. The darker the green plot is, the larger the number of consistency judgment images included in this plot is. The graph shows that the perceived light field by observing photon flow was generally close to the VLF. This tendency is common for each group, which indicates that the photon flow could convey the information on the physical light field for naïve observers only when they can read the physical light field by observing the real space.

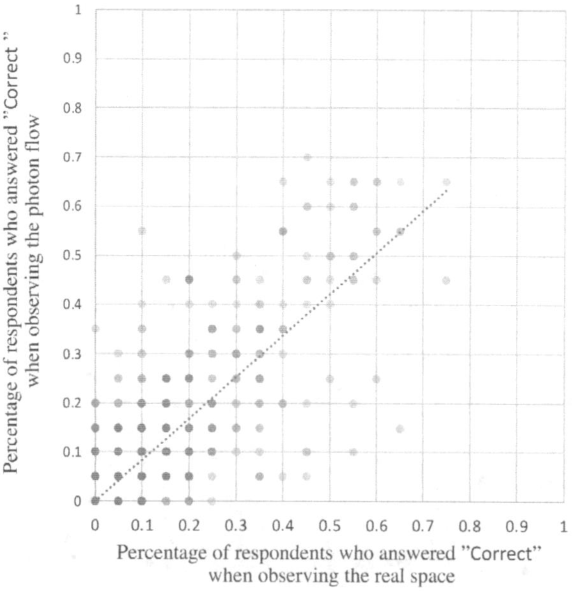

$y = 0.843x$
$R^2 = 0.755$

Figure15. Percentage of respondents who answered "Correct" when observing the real space/photon flow in all conditions.

4. Conclusion

In this study, the light field perceived by observing the photon flow was similar to the physical light field when there was a simple spotlight in the space. However, under more complex scenes, the subjects could not read the physical light field readily by looking at the photon flow. Besides, the light field perceived by naïve observers seeing the photon flow is generally close to the visual light field (VLF).
In the future works, we would like to clarify under which conditions people can estimate the correct physical light field by observing the photon flow and compare the differences between experienced/professional people and naïve observers. The final goal is to develop a useful tool for architectural and lighting designers to understand the light field more easily in their design stages.

References

[1] Okamoto T 2019 Case Studies of a Three-Dimensional Expression of Coloured Light Flow Using Volume Photon Mapping *29th CIE SESSION* DOI 10.25039/x46.2019.PO146.

[2] Koenderink JJ, Pont SC, van Doorn AJ, Kappers AML, Todd JT 2007 The visual light field **36** pp1595–1610.

[3] Kartashova T 2016 The global structure of the visual light field and its relation to the physical light field, *J. Vis.***16** pp14-5 .

[4] von Kries J 1902 Festschrift der Albrecht-Ludwig-Universität [Translation: D. L. MacAdam 1970 Chromatic adaptation (MIT Press, Cambridge)].

[5] Ward G, Eydelberg-Vileshin E 2002 Picture Perfect RGB Rendering Using Spectral Prefiltering and Sharp Color Primaries *Thirteenth Eurographics Workshop on Rendering(2002).*

IOP Conf. Series: Earth and Environmental Science 1099 (2022) 012039 doi:10.1088/1755-1315/1099/1/012039

Artificial light(ing) or electric light(ing)?

Federico Favero (1), Ute Besenecker (1)

(1) KTH Royal Institute of Technology, School of Architecture and the Built Environment, Division of Lighting Design, Stockholm, Sweden

E-mail: ffavero@kth.se; uteb@kth.se

Abstract. Researchers and designers use the words "artificial" or "electric" to describe lighting products, design, or research related practices, and there appear to be differing opinions about which is the more appropriate term. Generally, there are challenges with a common use of language and vocabulary in interdisciplinary research and this might be also valid for design and research in lighting design across different disciplines. The authors were educated in opposing practices of using "electric" lighting vs "artificial" lighting; this started a discussion and the conceptualization of this article. The paper explores, summarizes and discusses through literature review and a survey the concepts described and conveyed by both terms in different disciplines. Interestingly we could find differences among and between disciplines and professional backgrounds. This might indicate that the education and nomenclature in the field influences the use of terms. We found a tendency to refer to light sources either in terms of the energy used to generate the light, e.g. electric light or gaslight, but also in terms of the effect that it evokes, e.g. candle light is defined natural. Generally, a common lighting glossary could be developed through continuous discussion and studies. As today's complex questions are discussed in interdisciplinary teams, a common language might promote effective communication and stimulate sustainable solutions.

1. Introduction

Researchers and designers use the words "artificial" or "electric" to describe lighting products, design, or research related practices, and there appear to be differing opinions about which is the more appropriate term. Generally, there are challenges with a common use of language and vocabulary in interdisciplinary research [1] and this might be also valid for practice and research in lighting design across different disciplines.

The authors were educated in opposing practices of using "electric" lighting vs "artificial" lighting; this started a discussion and the conceptualization of this article. The objective for the paper is to explore, summarize and discuss through literature review and survey: What concepts are described and conveyed by both terms? How are both terms defined and contextualized in different disciplines?

Language contributes to constructing concepts of reality; how we describe people and the environment around us can influence perception and appraisal [2] [3], not just the other way around. For example, in current usage, one might appraise the terms "natural lighting" as positive, and "artificial lighting" as negative. As a matter of fact CIE deprecates the use of the term "artificial lighting" in favour of "electric lighting" [4]. However, one of the etymologies of artifice and artificial is the Latin artificium ('art' + 'make'), which translates into "made or achieved with human skill" [1]. While artificial is not a synonym for artistic, what constitutes "artificial" lighting should be discussed. For example, it could be used for "human-made" lighting in all its aspects [5]. But where to draw the line when describing the lighting design in a spatial environment? Would daylighting design, or the implementation of a fireplace, be considered "artificial" lighting or "natural" lighting? Moreover, is "electric" lighting always "artificial" lighting? The term "electric" lighting, meaning "powered by electricity", acknowledges how the light is generated, and for "electric light", similar to "gaslight", "firelight", or "sun/daylight", the definition is clearly tied to the type of energy used to generate the visible spectrum that we perceive as light. To develop the discussion, the distinction of the terms lighting and light are added, with, for example, "lighting" indicating the act of illumination of a spatial environment, and "light" describing the general physical phenomenon and the visual sensation that is evoked. Light as defined [4][6] is a human sensation (tied to vision) triggered by the combination of electromagnetic energy in the visual spectrum (380 nm - 780 nm) and our sensory organs, eye and brain. Is there "artificial" (human-made) lighting, but no "artificial" light? And is "artificial" lighting the strategy of combining different light sources (powered by electricity, gas, fire, the sun, or other) into one term for human-made lighting practices? With this starting point, the paper looks into literature from different disciplines working with light and lighting and explores the use of the terms "artificial" and "electric". The disciplines included were chosen based on the authors' experience with the multidisciplinary field of architectural lighting as well as the literature search database records. From the broad knowledge areas of the Social Sciences, Humanities and Engineering, we included contributions from the fields of architecture, lighting design, and lighting research (engineering and design). From the knowledge areas of Medicine, Biology and Agricultural Sciences, lighting related contributions from the fields of biology and ecology, health and neuroscience were added.

2. Methods

2.1. Literature search

We performed a literature search to gain insights in how the terms "electric lighting" and "artificial lighting" were used and defined. Scopus (https:
www.scopus.com) was chosen for its multidisciplinary database and for the possibility to select subject areas. We searched the following query: ("artificial lighting" OR "electric lighting" AND LIMIT-TO (SUBJAREA "ENGI" OR "ENER" OR "AGRI" OR "ENVI" OR "SOCI" OR "MEDI" OR "ARTS" OR "MULT" OR "NEUR" OR "IMMU" OR "PSYC"). The countries of the search were limited to the countries of the participants of the survey, see below section, to relate to similar cultural contexts. The query produced 1969 papers published within the time period from 1962 until August 2022, subdivided in 11 subjects areas in the Scopus catalogue. We excluded papers that did not involve studies of lighting in the environment, e.g. silicon chips development or electrochromic glazing. The 5 more frequent keywords including the terms light and/or lighting found among these documents are listed in Table 1. We further selected the four most cited papers in each of the following four subject areas, which represent a broad multidisciplinary view on the areas of interest for light and lighting: Engineering, Medicine, Agricultural sciences and Biology, and Social Sciences. The final list encompasses 16 publications, 4 publications in each of the 4 queried fields of research. See Tables 1 and 2 for findings. We preferred human studies to animal studies although these dominate the field of Agricoltural sciences and Biology.

[1] "artificial, adj. and n." OED Online, Oxford University Press, December 2021, www.oed.com/view/Entry/11211. Accessed 12 February 2022.

2.2. Survey

We asked 57 professionals/ experts working with light(ing) from selected countries (UK, USA, Italy, Germany, Switzerland, Sweden, Denmark, Norway, Netherlands, Japan, Brasil) to answer 10 questions in an online survey on the Phonic platform (http: www.phonic.ai). We hand-picked participants among speakers of the past Light Symposium and in our network, aiming at a balance of gender and age. We contacted them via e-mail and asked whether they could distribute it to colleagues. These professional backgrounds were represented: lighting design, lighting research, architecture, biology, medical/health related research, and neuroscience. A total of 32 participants completed the survey, 18 had design background, of which seven work also in research or education, and the rest in design practice. Fourteen participants had background in medicine or biology. The gender was balanced (Female=16), and the median age was 44 years. The participants worked in the field of light and lighting for a median of 15 years, twelve had more than 20 years of experience. We clustered participants' answers into design (i.e. professionals with a background in lighting design and architecture) and non design background.

2.2.1. Questions The complete 10 questions are presented below (in parenthesis the available multiple choices):

- Which of the following describes your profession best? ('Daylighting expert, 'Lighting designer, 'Integrative designer, 'Lighting researcher, 'Lighting scientist, 'Architect', 'Biologist', 'Neuroscientist', 'Anthropologist', 'Other: please specify in the next page')
- Please describe your profession
- How many years have you worked in the field of light and lighting?
- In your work, do you make a distinction between the terms light and lighting? If yes, how and when do you use each one?
- How do you refer to the light source that you work with the most? ('Daylight ', 'Electric light', 'Natural light', 'Artificial light', 'Digital light', 'Analogue light', 'Other: please specify in the next page')
- Could you please give an example and describe some qualities of the light source you were referring to in the prior question?
- How would you refer to lighting a space with luminaires? ('Daylighting ', 'Electric lighting', 'Natural lighting', 'Artificial lighting', 'Digital lighting', 'Integrative lighting', 'Analogue lighting')
- How would you refer to lighting a space with both openings to the outside [e.g. windows or skylights] and luminaires? ('Daylighting ', 'Electric lighting', 'Natural lighting', 'Artificial lighting', 'Digital lighting', 'Integrative lighting', 'Analogue lighting')
- How would you characterize these items: 'Surface-mounted fluorescent fixture', 'Torch', 'Window', 'Ceiling-mounted LED spotlight', 'Fireplace', 'Sunlight projected from a window frame', 'Gas lamp', 'Skylight', 'Mobile phone', 'Fire', 'Candle', 'Halogen Task lamp', 'Moonlight', 'Sunlight', 'Projected image'. (categories: 'Artificial', 'Electric', 'Natural')
- Please provide any additional comments you may have

3. Findings

The findings to date were divided by methods, literature search first and survey later. We summarized data in tables and figures and described the results in the following sections.

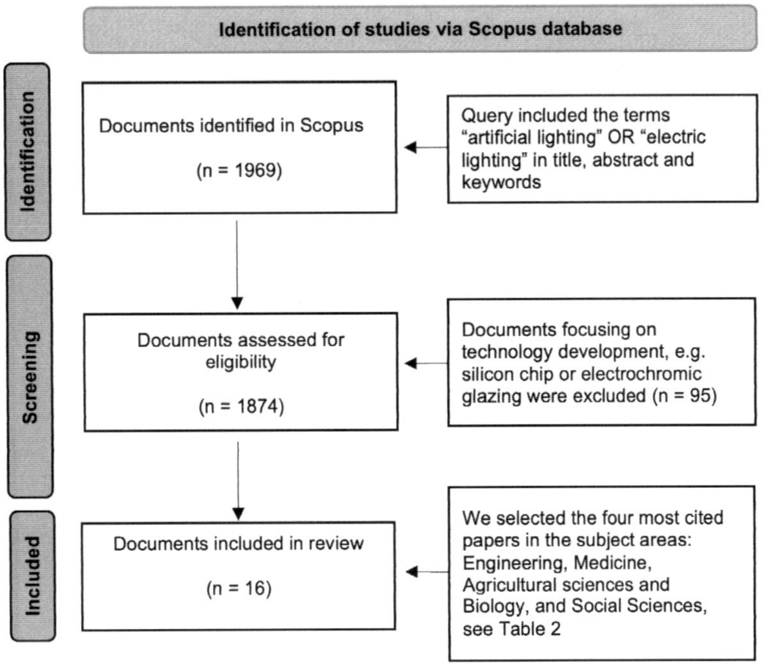

Figure 1. Flow diagram adapted from PRISMA 2020 (https://prisma-statement.org)

3.1. Literature search

The search revealed that the term electric lighting is present in more than double the amount of documents than the term artificial lighting, see Table 1. Engineering is the dominant subject area with more than 1200 documents and in this subject area authors refer predominantly to electric lighting in the last twenty years, see Table 2. Daylighting is present with more than 200 documents while in literature the term natural (light or lighting) is included in less than 20 documents. We could notice a predominance in use of artificial lighting in Biological and Social sciences, see Table 2. In Medicine authors use the terms interchangeably.

We selected references from 5 papers where the terms electric and artificial lighting are contextualized and they provide examples of the concepts to which the authors refer to, see Table 3. For instance, in Engineering, electric lighting might refer to fixtures, for instance "occupants did not switch off the electric lighting" [7] as well as the electric lighting system. In Biology artificial is often intended as a stimulus that "alters natural light regimes" and "influences biological systems"[8]. This same concept can be expressed as electric lighting by authors when referring to lighting in Architecture, e.g. "Illumination from electric lighting in the built environment is quite different from solar radiation in intensity, spectral content, and timing during the 24-hour daily period" [9]. The authors of one paper use both terms "Today's modern society is exposed to artificial electric lighting in addition to the natural light-dark cycle" indicating that all electric lighting is artificial [10].

3.2. Survey

The findings from the survey questions in tables and figures are reported in the sections below. We clustered the results depending on the participants' background: one cluster includes architects and designers (n=18), the other researchers, medical doctors or biologists (n=14). Table 6 collects transcriptions from audio recordings and written answers to open questions of

IOP Conf. Series: Earth and Environmental Science 1099 (2022) 012039 doi:10.1088/1755-1315/1099/1/012039

Table 1. Terms including light and/or lighting found in literature search from 1962 until today with the search "artificial lighting" OR "electric lighting"

Total documents	Electric lighting	Lighting	Artificial lighting	Daylighting	Light
1874	723	360	319	240	150

Table 2. Terms in literature search

Title	Author(s)	Year	Term	Subject area
Use of model predictive control and weather forecasts for energy efficient building climate control	Oldewurtel, F. et al. [11]	2012	Electric	Engineering
Adding advanced behavioural models in whole building energy simulation: A study on the total energy impact of manual and automated lighting control	Bourgeois, D. et al. [12]	2006	Electric	
Occupant preferences and satisfaction with the luminous environment and control systems in daylit offices: a literature review	Galasiu, A.D. and Veitch, J.A. [7]	2006	Electric	
The use of artificial lighting in relation to daylight levels and occupancy	Hunt D.R.G [13]	1979	Artificial	
Sources of stress in captivity	Morgan, K.N. and Tromborg, C.T. [14]	2007	Artificial	Biological sciences
The ecological impacts of nighttime light pollution: A mechanistic appraisal	Gaston, K.J. et al.[8]	2013	Artificial	
Plant productivity in response to LED lighting	Massa.G.D. et al.[15]	2008	Electric	
Street lighting changes the composition of invertebrate communities	Davies, T.W. et al. [16]	2012	Artificial	
Melatonin-induced temperature suppression and its acute phase-shifting effects correlate in a dose-dependent manner in humans	Deacon, S. and Arendt, J.[17]	1995	Artificial	Medicine
Light in the built environment: Potential role of circadian disruption in endocrine disruption and breast cancer	Stevens, R.G. and Rea, M.S. [9]	2001	Electric	
Ocular photoxicity	Roberts, J.E. [18]	2001	Artificial	
Dim light at night disrupts molecular circadian rhythms and increases body weight	Fonken, L.K. et al. [19]	2013	Electric	
When innovations meet institutions: Edison and the design of the electric light	Hargadon, A.B. and Douglas, Y.[20]	2001	Electric	Social Sciences
Lighting and discomfort in the classroom	Winterbottom, M. and Wilkins, A. [21]	2009	Artificial	
A simplified method to estimate energy savings of artificial lighting use from daylighting	Krarti, M. et al. [22]	2005	Artificial	
The impact of light and colour on psychological mood: A cross-cultural study of indoor work environments	Küller, R. et al. [23]	2006	Artificial	

the survey that also inspired threads in the discussion section.

3.2.1. Light and Lighting We summarize findings from the question: 'In your work, do you make a distinction between the terms light and lighting? If yes, how and when do you use each one?' in Table 4. The participants seem to agree on the fact that light refers to the phenomenon and the quality of a source (or radiation/stimulus/inspiration) while lighting refers to the application of light in space, which includes process, discipline, design.

3.2.2. Lighting a space The participants refer to work with these sources of light, in ascending order: daylight, electric and artificial, see Figure 2. The majority of designers refer to artificial light as the source they work with but a handful switches to electric lighting when asked their term for "lighting a space with luminaires", see Figure 3. Participants without design background refer more consistently to electric lighting for the action of lighting a space with luminaires. Interestingly this category of respondents is not at all consistent answering this question: "how would you refer to lighting a space with both openings and luminaires?". Here research background respondents divide themselves in between daylighting, electric, artificial and integrative lighting. Designers chose predominantly the term integrative lighting, see Figure 3.

Table 3. Quotations from literature as examples of terms in context

Extract	Authors	Subject area
"The review is structured around two themes: studies that examined the preferred physical and luminous conditions in daylit office environments; studies that investigated the occupant satisfaction and acceptance in relation to the control of *electric lighting* and window shading in daylit offices." "The authors also noted that sometimes occupants did not switch off the *electric lighting* even when the indoor daylight illuminance was rather high because "they failed to notice that it was on""	Galasiu, A.D. and Veitch, J.A. [7]	Engineering
"Changes in light regime can be characterized as changes in the spatial distribution, the timing and the spectral composition of *artificial light* sources. As human communities and lighting technologies develop, *artificial light* increasingly encroaches on dark refuges in space, in time, and across wavelengths"; "we propose a framework that focuses on [..] the ways in which *artificial lighting* alters natural light regimes (spatially, temporally, and spectrally), and the ways in which light influences biological systems, particularly the distinction between light as a resource and light as an information source"	Gaston, K.J. et al. [8]	Biological sciences
"Illumination from *electric lighting* in the built environment is quite different from solar radiation in intensity, spectral content, and timing during the 24-hour daily period. Humans evolved over millions of years with the day-night pattern of solar radiation as the primary circadian cue. This pattern maintained a 24-hour rhythm of melatonin release, as well as a host of other physiological rhythms including the sleep-wake cycle. *Electric lighting* in the built environment is generally more than sufficient for visual performance, but may be inappropriate for the maintenance of normal neuroendocrine rhythms in humans; e.g., insufficient during the day and too much at night. Lighting standards and engineering stress visual performance, whereas circadian function is not currently emphasized."	Stevens, R.G. and Rea, M.S. [9]	Medicine
"[In]1878 September Edison begins working on incandescent illumination, recording in one of his notebooks: "Object, Edison to effect exact imitation of all done by gas, so as to replace lighting by gas, by *lighting by electricity*" (Basalla, 1988: 48)""Edison's success lay in developing a system of *electric lighting* that adapted and integrated each of these components and, as important, that gained rapid acceptance and provided the model for subsequent development of the technology."	Hargadon, A.B. and Douglas, Y. [20]	Social Sciences
"Some authors place most emphasis on the importance of *daylighting*, but the need for integrated systems of day- and *artificial lighting* is broadly accepted"	Winterbottom, M. and Wilkins, A. [21]	Social Sciences

Table 4. Quotations from survey participants on difference between "light" and "lighting"

Profession	Quotation/Answer
Lighting designer and educator	Light (luz) as the entity, the visual element. Lighting (iluminação) as the discipline and the process.
Lighting designer and lighting researcher	I use light when I refer to both daylight and electric light. And lighting or electric lighting when referring to artificial light.
Lighting researcher	Light is a general concept describing the light as such, but lighting is only about man created light, mainly by electricity, but also by other means such as fire.
Lighting designer, architect and educator	[I refer to] Light as a physical phenomenon (natural or artificial) and lighting as a design solution
Lighting designer, architect and educator	light = natural light lighting = light designed in a functional and aesthetic way
Lighting designer, architect and educator	Light for the phenomena, Light is the bridge between human and space, Lighting is what you need to get light
Lighting researcher and psychologist	Light is the physical stimulus, the optical radiation to which organisms respond. (Organisms are mostly humans, but can include other animal species or plants.) Lighting is the means by which humans deliver light.
Lighting designer, architect	I use light as a term for the light itself, regardless the source or the situation. I use lighting as a term for how the light is "set" in a space

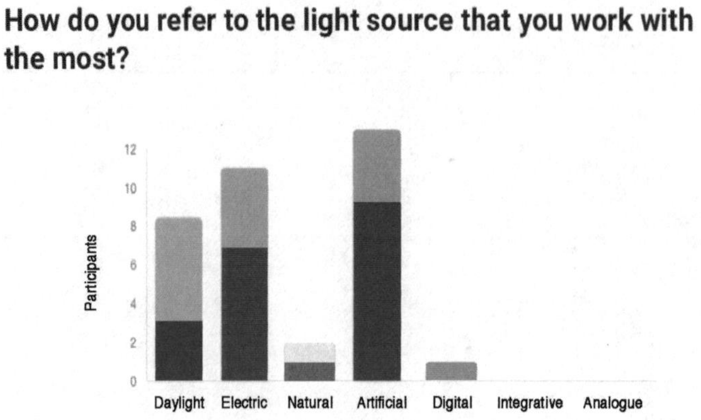

Figure 2. Participants' responses to the question: how would you refer to the light source that you work with the most? This question allowed multiple answers. The darker shade represents the participants with design background, e.g. 9 participants with design background refer to work mostly with artificial light, 7 with electric light.

3.2.3. Terms used to identify objects and phenomena We divided the 15 objects or phenomena that we asked about, into four categories, see Table 5: luminaires, devices (phones and projected images), combustion sources or fixtures, daylight phenomena. This last category is the only one where there is absolute consensus among respondents. Answers to the other categories are rather mixed independently from the respondents' background. Responses in the luminaires and devices categories are split between artificial and electric, combustion sources between natural and artificial. Participants without design background predominantly characterize the first three sources in table 5 as electric, designers' responses are split between artificial and electric.

4. Discussion

Both review and survey, provided an overview that we could summarise as follows.

Different disciplines do use different terms for lighting, see Table 2, Figure 3 and Table 5. What was a discussion topic between colleagues turned into a more general finding and possibly a conversation maker among professionals related to light and lighting.

Regarding these two terms, the survey helped to tease out the difference between them. Most participants use light and lighting differently and we could find consensus on a sort of definition of the two terms. Summarising the comments from the survey, 'Light' is the visual phenomenon, the source of radiation, but also of inspiration, which for many relates to natural light or daylight; 'Lighting' refers to the application, the design of light, including the process, frames and conditions of this action or activity.

We do not have consensus about the terms artificial and electric. Some participants respond to use the two terms interchangeably and the literature search highlighted that the two terms coexist in the same subject areas. Biologists indicate artificial lighting as the concept of human made light affecting biological/ecological systems. Engineers use electric lighting for a system that integrates luminaires and components in the built environment, in contrast to other, previous, lighting, such as gas lighting: "Edison to effect exact imitation of all done by gas, so as to replace lighting by gas, by lighting by electricity", see Table 3. Nowadays, electric lighting is probably the most prevalent in the general concept of artificial lighting. However, as one researcher mentioned, see Table 4, artificial lighting can also relate to fire, for example. And there might be future sources that use other sources of energy and develop other systems, which

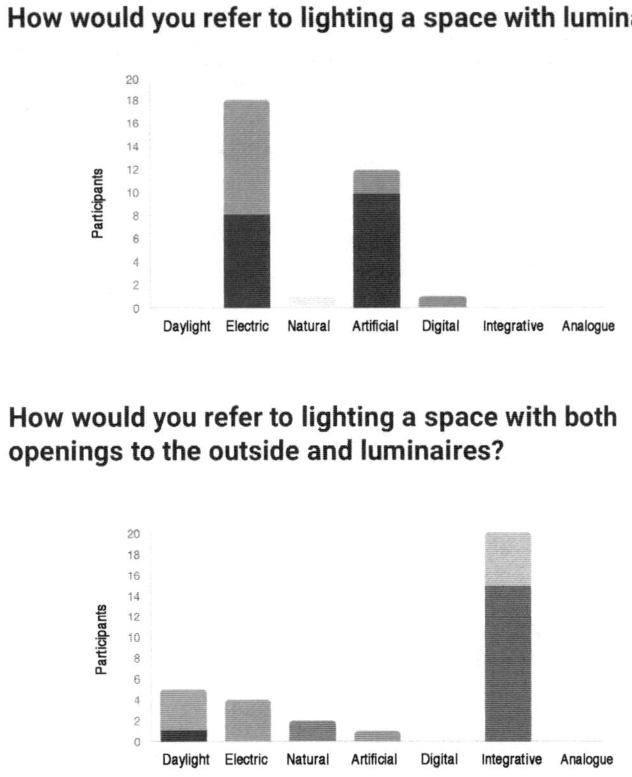

Figure 3. Participants' responses to the question: 'how would you refer to lighting a space with luminaires?' above, and 'how would you refer to lighting a space with luminaires and openings to the outside?', below. These questions did not allow multiple answers. The darker shade represents the participants with design background, e.g. 10 participants with design background refer to artificial lighting in the figure above.

might also be part of the bigger category of artificial lighting. In this sense the two definitions, artificial and electric, coexist and complement each other.

Results from both survey and literature search suggest that professionals identify light sources by their character and/or by the source of energy or method that produces light. This duality of interpretations between energy and character of the effect manifests mainly with combustion sources, for example candles. Candle is referred as either natural or artificial independently by the professional background, see Table 5. Our interpretation is that some define candle a human-made product, fuelled by an artificial energy source; others might characterize candle by the similarity of the light they produce to the one produced by fire and thus identify them as natural, or simply refer to the fact that candles *feel* natural. We spotted this trend with another artefact (fire-place), but also interestingly with fire. Not all combustion sources follow the same trend, in fact participants characterize gas-lamp as artificial. This might have to do again with the character of the light, which is rather cold and ghostly depending on the gas-mantle that glows in a gas-lamp. One participant comments about devices that use daylight as light source, solar tubes, that "They rather feel like some artificial lighting because it's not electric", see Table 4. Therefore even a natural source like daylight can produce an artificially appearing effect. A designer and educator states that a future definition of natural light should be "light that is

Table 5. Terms used for lighting devices, fixtures, phenomena

Term	Artificial	Electric	Natural
`Surface-mounted fluorescent fixture`	8 (7)	22 (7)	
`Ceiling-mounted LED spotlight`	9 (8)	22 (10)	
`Halogen task lamp`	9 (8)	21 (11)	
`Mobile phone`	13 (10)	17 (7)	
`Projected image`	24 (15)	8 (4)	
`Fire`	9 (3)		21 (16)
`Fireplace`	13 (6)		17 (13)
`Candle`	14 (7)		16 (12)
`Gas lamp`	26 (15)	1 (1)	3 (3)
`Sunlight projected from a window frame`	1		31
`Window`	1 (1)		31
`Sunlight`	1		31
`Moonlight`			32
`Skylight`			32

In parenthesis the amount of answers from participants with design background, e.g. 8 participants, of which 4 with design background, characterized 'Projected image' as 'Electric'.

Table 6. Quotations from survey participants

Profession	Quotation/Answer
Lighting designer and educator	"I would really love to live in a world where Natural Light would mean something more profound, as light that is in harmony with nature, independently from its sources... but that's an Utopia, I guess..."
Lighting researcher and psychologist	"Your choice of survey questions does reveal inconsistencies in lighting terminology. (I use lighting here to refer to the field of study.) However, I think your forced-choice of three categories on the previous page was rather limiting and does not fully capture the issues. "Gas lamp" in particular fits poorly in any of these, because the word "artificial" can mean "fake" or "counterfeit", which these objects are not - but neither are they electric or natural."
Daylighting expert and Lighting researcher	" There are natural light sources that are not considered daylight but they would look very artificial even if they come from daylight. I'm not speaking about fire [but] for example to solar tubes, this type of systems which they don't really feel as daylight. They rather feel like some artificial lighting because it's not electric."
Lighting designer and researcher	"The CIE considers "integrative lighting" to be that which combines light impacts on health together with light for vision and comfort. It likely also includes daylighting."
Daylighting expert, Architect and Lighting researcher	"In the question about light from both windows and luminaires [..] I would call this condition 'mixed lighting' or similar (e.g., using my own vocabulary, both daylight and artificial lighting). I would like to also note that I use the terms 'daylight' and 'natural light' interchangeably, and the same for 'artificial' and 'electric' lighting. For me, 'artificial' encompasses both electric and simulated light (e.g., from renderings)."

in harmony with nature, independently from its sources". These observations find support in established lighting analysis approaches that integrate visual and physical characteristics, e.g. [24] and contemporary investigations in semiotics [25] or atmosphere [26].

Language is a very powerful tool which influences appraisal. We used the term artificial in the survey as made or achieved with human skill. We discovered by both survey and literature search that the use of the word artificial is strongly avoided among Northern American researchers. Perhaps, as a participant noted in her final comments, "because the word 'artificial' can mean 'fake' or 'counterfeit', which these objects are not", see Table 6. It could be interesting in a further research to identify whether artificial relates more to designed, human-made or to unnatural and counterfeit.We excluded 'torch' from the results summarised in table 5 because multiple participants found the definition ambiguous, for instance "I took the word torch to mean flashlight, but that's probably just the difference between British and American English". Most participants with design background refer to "integrative lighting" to define "lighting a space with luminaires and openings" while researchers provided multiple answers. One researcher participant provides the current definition of the International Commission on Illumination: "The CIE considers "integrative lighting" to be that which combines light impacts on health together with light for vision and comfort. It likely also includes daylighting.", see Table 6. We don't know how many respondents were familiar with this definition, in any case we could observe that designers prefer this term to the others presented.

Our ambition was to discuss whether there could be one set of definitions that would enable both researchers and designers from different disciplines and backgrounds to find common ground, even while using different terms in their texts. In the current status of the paper we could find differences and started identifying the needs for definitions to be included in a common lighting glossary. This common lighting glossary can be an effective tool to improve communication in interdisciplinary research and a source of inspiration for the lighting design field.

4.1. Limitations and future thinking

Our objective was to start a discussion and raise awareness about the terms used in the field of lighting and the concepts underlying these terms. The intent was neither to offer a definitive answer to the use of terms in the lighting design profession, nor to suggest the use of one or the other. During this study we had to make decisions that had an effect on the findings presented. We are aware that the criteria of paper selection by citation rate might over-represent the dominant subject area (Engineering). At the same time this did not over-represent older papers, in fact most papers presented in Table 2 belong to the last twenty years. We think the survey, which was more focused on lighting researchers and designers, well complemented the information found through the literature search. A more specific future search in the fields of Architecture and Design could offer additional, specific insights into aspects related to these professions and related current/contemporary research. In this regard we see LS22 as an opportunity to involve the audience into the discussion and a second step of the investigation to expand on the current results. In addition, while the authors consider it a necessary, practical starting point to focus this international topic on the English language, it could be interesting and revealing to expand the study into other languages where definitions and sensibilities might differ.

5. References

[1] Bracken LJ and Oughton EA. 'What do you mean?' The importance of language in developing interdisciplinary research. *Transactions of the Institute of British Geographers* 2006; 31(3): 371–382. DOI:10.1111/j.1475-5661.2006.00218.x.

[2] Barrett LF. *How Emotions Are Made.* New York: Houghton Mifflin Harcourt, 2017. ISBN 9780544133310.

[3] Landau AN, Aziz-Zadeh L and Ivry RB. The influence of language on perception: listening to sentences about faces affects the perception of faces. *Journal of Neuroscience* 2010; 30(45): 15254–15261. DOI: 10.1523/JNEUROSCI.2046-10.2010.

[4] CIE. CIE S 017/E:2020. Technical report, International Commission on Illumination, Vienna, 2020. DOI:10.25039/S017.2020. URL `https://cie.co.at/publications/ilv-international-lighting-vocabulary-2nd-edition-0`.

[5] Weisbuch C. Historical perspective on the physics of artificial lighting. *Comptes Rendus Physique* 2018; 19(3): 89–112. DOI:10.1016/j.crhy.2018.03.001. URL `https://doi.org/10.1016/j.crhy.2018.03.001`.

[6] ANSI/IES. LS-1-21 Lighting Science: Nomenclature and Definitions for Illuminating Engineering, 2021. URL `https://www.ies.org/standards/definitions/`.

[7] Galasiu AD and Veitch JA. Occupant preferences and satisfaction with the luminous environment and control systems in daylit offices: a literature review. *Energy and Buildings* 2006; 38(7): 728–742. DOI: 10.1016/j.enbuild.2006.03.001.
[8] Gaston KJ, Bennie J, Davies TW et al. The ecological impacts of nighttime light pollution: A mechanistic appraisal. *Biological Reviews* 2013; 88(4): 912–927. DOI:10.1111/brv.12036.
[9] Stevens RG and Rea MS. Light in the built environment : Potential role of circadian ... *Cancer Causes and Control* 2001; 12: 279–287. URL `https://link.springer.com/content/pdf/10.1023%2FA%3A1011237000609.pdf`.
[10] Moreno CR, Vasconcelos S, Marqueze EC et al. Sleep patterns in Amazon rubber tappers with and without electric light at home. *Nature Scientific Reports* 2015; 5(December 2014): 1–11. DOI:10.1038/srep14074.
[11] Oldewurtel F, Parisio A, Jones CN et al. Use of model predictive control and weather forecasts for energy efficient building climate control. *Energy and Buildings* 2012; 45: 15–27. DOI:10.1016/j.enbuild.2011.09.022. URL `http://dx.doi.org/10.1016/j.enbuild.2011.09.022`.
[12] Bourgeois D, Reinhart C and Macdonald I. Adding advanced behavioural models in whole building energy simulation: A study on the total energy impact of manual and automated lighting control. *Energy and Buildings* 2006; 38(7): 814–823. DOI:10.1016/j.enbuild.2006.03.002.
[13] Hunt DR. The use of artificial lighting in relation to daylight levels and occupancy. *Building and Environment* 1979; 14(1): 21–33. DOI:10.1016/0360-1323(79)90025-8.
[14] Morgan KN and Tromborg CT. Sources of stress in captivity. *Applied Animal Behaviour Science* 2007; 102(3-4): 262–302. DOI:10.1016/j.applanim.2006.05.032.
[15] Massa GD, Kim HH, Wheeler RM et al. Plant Productivity in Response to LED Lighting. *HortScience* 2008; 43(7): 1951–1956. DOI:10.21273/HORTSCI.43.7.1951. URL `https://journals.ashs.org/view/journals/hortsci/43/7/article-p1951.xml`.
[16] Davies T, Bennie J and Gaston K. Street lighting changes the composition of invertebrate communities. *Biology Letters* 2012; 8(5): 764–767. DOI:10.1098/rsbl.2012.0216.
[17] Deacon S and Arendt J. Melatonin-induced temperature suppression and its acute phase-shifting effects correlate in a dose-dependent manner in humans. *Brain Research* 1995; 688(1-2): 77–85. DOI: 10.1016/0006-8993(95)96872-I.
[18] Roberts J. Ocular photoxicity. *Journal of Photochemistry and Photobiology B: Biology* 2001; 64(2-3): 136–143. DOI:10.1016/S1011-1344(01)00196-8.
[19] Fonken L, Aubrecht T, Meléndez-Fernández O et al. Dim light at night disrupts molecular circadian rhythms and increases body weight. *Journal of Biological Rhythms* 2013; 28(4): 262–271. DOI: 10.1177/0748730413493862.
[20] Hargadon AB and Douglas Y. When Innovations Meet Institutions : Edison and the Design of the Electric Light Author. *Administrative Science Quarterly* 2001; 46(3): 476–501.
[21] Winterbottom M and Wilkins A. Lighting and discomfort in the classroom. *Journal of Environmental Psychology* 2009; 29(1): 63–75. DOI:10.1016/j.jenvp.2008.11.007. URL `http://dx.doi.org/10.1016/j.jenvp.2008.11.007`.
[22] Krarti M, Erickson P and Hillman T. A simplified method to estimate energy savings of artificial lighting use from daylighting. *Building and Environment* 2005; 40(6): 747–754. DOI:10.1016/j.buildenv.2004.08.007.
[23] Küller R, Ballal S, Laike T et al. The impact of light and colour on psychological mood: A cross-cultural study of indoor work environments. *Ergonomics* 2006; 49(14): 1496–1507. DOI:10.1080/00140130600858142.
[24] Liljefors A. *Lighting -Visually and physically.* Stockholm: KTH Royal Institute of Technology – School of Architecture, 1999.
[25] Schielke T. The Language of Lighting : Applying Semiotics in the Evaluation of Lighting Design. *LEUKOS* 2019; 15(2-3): 227–248. DOI:10.1080/15502724.2018.1518715. URL `https://doi.org/10.1080/15502724.2018.1518715`.
[26] Stokkermans M, Vogels I, de Kort Y et al. A Comparison of Methodologies to Investigate the Influence of Light on the Atmosphere of a Space. *LEUKOS - Journal of Illuminating Engineering Society of North America* 2018; 14(3): 167–191. DOI:10.1080/15502724.2017.1385399. URL `https://doi.org/10.1080/15502724.2017.1385399`.

Challenges in lighting education: a recommended practice

Paulo Sergio Scarazzato (1), Jéssica Cristine da Silva Fonseca Matos (2), Ana Judite Galbiatti Limongi França (1), Taísa Dóccosse Pavani (1)

(1) University of São Paulo. School of Architecture and Urbanism, Brazil; (2) University of Campinas. School of Civil Engineering, Architecture and Urban Design.

pasezato@usp.br

Abstract. Although lighting is a theme usually loved by architects, there are few studies about appropriate pedagogical strategies focusing on its teaching in architectural undergraduate courses. This subject rarely is present in conferences, symposia, and similar events regarding architectural education. Due to historical mistakes, lighting education is often conducted in such courses with no link to the architectural design process. Regarding daylighting, it not always goes beyond solar geometry and insolation control, and electrical lighting is addressed to the discipline of electrical installations, strictly based on a calculation approach. It urges to present lighting to students as an integral and indissociable part of the architectural space itself. Also, it is essential to support a process of learning based on observation and laboratory experimentation. This paper discusses these issues and presents a pedagogical proposal favouring drawing with light, both daylighting and electrical lighting, since the first steps of the architectural design process. An electronic tool yet under development and destined to help in decision-making related to daylighting in the initial phases of design development is also presented. For a sustainable and safe future in a world in rapid transformations due to technological advances and pandemic events, lighting education deserves to be exhaustively studied by educators and architecture schools worldwide. Sharing knowledge in a network seems to be the best way for such an objective.

Keywords: *Architectural Design Process; Lighting; Teaching-learning; Pedagogical Practice*

1. Introduction

For centuries, the teaching of architecture was carried out at the construction site, where the elders transmitted the skills of the craft to the younger ones, and there are no known records about how the classes were organized. There is no information too regarding how issues inherent to the appropriation of daylight were discussed. What is known, from material testimonies that have survived over time, is that the ancient Egyptians, Greeks, and Romans used daylighting with ingenuity, especially in religious buildings, and that both Greeks and Romans carried out the design successfully. of their cities to ensure healthy conditions, depending on the solar trajectories.

Vitruvius' Treatise, "De Architectura", is the only written source from antiquity that has come to our days, which brings some explicit clues about architects' education. According to Vitruvius, the architect must be disciplined and cult, have a solid education, be versed in various subjects without, however, being a specialist in any of them, and receive both theoretical and practical knowledge. Regarding questions about light and lighting, he warns that architects must know the *"rules of astrology and the sky"*, that is, the solar trajectories and the luminous climate.

In the Middle Ages, the architect's education was still offered essentially on the construction site, and controlled by corporations, which even offered apprentices access to ancient texts, translated into the vernacular. However, writings of the time on light were restricted to philosophical reflections. The treatise "De Luce" (On Light) by Robert Grosseteste (1168 – 1253), for example, helps to understand the culture of the time, in which the search for perfection – also applicable to architecture – had a strong symbolic connotation, influenced that it was for Christianity.

After Vitruvius, it was only in the 15th century that a second treatise on architecture was published, *"De re aedificatoria"*, by Leon Battista Alberti, which, according to some scholars, would have arisen from the need to question the then-current practice, and for the establishment of guidelines for the future professional exercise of the architect. However, that work does not address lighting issues, because Alberti is an *"architect of surfaces or textures, he gives up the effects of light and presents three-dimensional characteristics to take advantage of the flat composition, which he almost always exercised superbly"* (STRÖHER, 2006).

The teaching-learning methods adopted in today's undergraduate architecture courses derive from pedagogical proposals that emerged in the 19th century in Europe and the United States. However, none of them tackled lighting issues.

In the architecture course of the School of Fine Arts in Paris, created in 1816, the central axis of that course was the design of buildings, presented with such rigor that it thwarted any initiative of creativity, which must be the essence of the practice of architecture. Shapes and proportions should follow strict rules that guarantee harmony from the point of view of established principles. In such a context, windows, skylights, or any other openings in walls or roofs were designed to meet, above all, formal requirements, which recurrently resulted in pieces with dimensions more than necessary to ensure adequate lighting conditions. Usually, internal blinds were responsible for controlling the solar access and glare. There was no concern regarding the orientation of facades as a function of the apparent trajectories of the sun.

In the United Kingdom, unlike in France, the architecture course at the Architectural Association (AA) offers the students the freedom to exercise their creativity. Still, daylighting and electrical electric lighting issues were never – and are not – explicitly mentioned in his undergraduate course. The same occurred in Germany and in the United States, where the teaching of architecture did not include lighting aspects.

Although the systematization of formal architectural education dates from the early 19th century, the incorporation of contents related to daylighting and electrical lighting teaching into its curricula only happened at the end of that century due to the development of sanitation concepts. This scenario led to the introduction of issues related to the healthiness of buildings and electric lighting. However, the teaching approach was still poor and incomplete once there was no integration of the daylighting and electric lighting solutions during the design process. From the 20th century until the first decades of the 21st century, it is possible to note some understanding regarding the importance of an integrated approach (BOYCE, 2006). But still as a matter of theoretical or complimentary content.

In the middle of the 20th century, the availability of electric lighting sources associated with low energy costs and no attention to environmental issues resulted in a progressive abandonment of good daylighting practices. Simultaneously, because most architects had no understanding of the potential of electric lighting in valuing the built environment, they often delegated its design solutions to electrical engineers. This scenario greatly influenced the state of lethargy in which the teaching of daylighting and electric lighting lasted in the architect's education. Not even the resumption of awareness of the importance of daylighting from the 1980s onwards, resulting from the two oil crises of the previous decade, and the consolidation of the concept and the adoption of pro-sustainability actions provoked curricular changes to make the teaching of lighting integrated with the teaching of design. This finding leads to the need to promote a reflection on how to review the teaching of design in undergraduate architecture courses that include lighting issues. This paper focuses on such a matter and presents a viable didactic proposal that has a good potential for a new direction in design teaching in such courses.

2. Lighting on architect's education

The claim that teaching lighting in architect education is often disconnected from teaching design could be confirmed by a Systematic Literature Review (SLR) that carried out screening for related papers on the Capes Periodical Portal (<www.periodicos.capes.gov.br>, accessed 2022.04.01) with the following keywords: architecture AND lighting AND "undergraduate course" AND (pedagogy OR "teaching-learning" OR "pedagogical practice"). The filter was established between the years 2002 to 2022, with an interval of the last 20 years.

The screening returned 23 papers. The first phase of analysis of titles and abstracts excluded 12 articles, as they were outside the area of knowledge of the review. The second phase of complete reading of the articles excluded 5 more articles because they do not deal with teaching architecture, much less lighting in architecture courses (Figure 1).

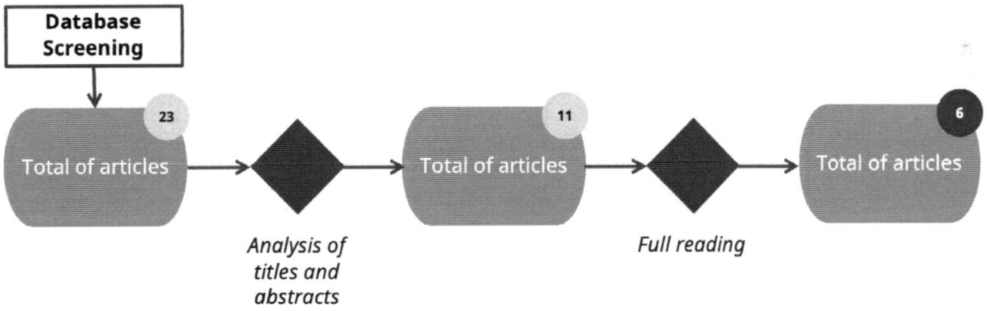

Figure 1. RLS Protocol.
Source: the authors.

The selected papers were then organized into three groups: knowledge areas related to architecture; teaching in architecture; and teaching lighting in architecture/for architecture students. The group with the most significant number of articles that make up this review contains studies that discuss general aspects of architectural education. Table 1 highlights the most relevant contributions and t issues to the discussion that regards the teaching of lighting.

Table 1. List of the main contributions to this study found after assessing the papers.

Authorship	Contribution to this study
Feijoó *et al.* (2021)	The authors evaluated the results of implementing web-based teaching tools during the pandemic. The main findings were (1) some interactions between student and teacher are desirable; (2) the gamification tools and the use of tablets increase students' interaction; (3) it is desirable to make the essential (theoretical) content available in videos before the classes, support debates and promotes dynamic interactions.

Continue

IOP Conf. Series: Earth and Environmental Science 1099 (2022) 012040 doi:10.1088/1755-1315/1099/1/012040

Continuation

Maghool *et al.* (2018)	The authors discussed learning theories in architecture and the inclusion of Virtual Reality in teaching and presented some quantitative data: - 33% of the architecture students are sensory learners. On the other hand, most teachers used intuitive teaching methods. - 82% of the students are visual learners, and the courses are based on auditory material. Virtual Reality (VR) can be an alternative to deal with these divergences between types of learners and teaching methods. The authors evaluate a VR tool and bring some advantages of this model: - 3D visualization allows the students to perceive the actual scale of architectural components in a three-dimensional environment. - high level of interaction. - the realistic environment with its real-time lighting and reflection calculations offers a high degree of immersion. - the possibility of time control or rewind allows the students to go back or forward the construction process. - option to include extra audio, video, and images while inspecting the 3d model. - lower implementation cost compared to an actual visit to the construction site.
Smith (2011)	The researcher presented a pedagogical proposal for subjects with theoretical content based on the understanding that texts are the essential means of teaching theory. Despite that, the gathered content must be evidenced by assessing the performance of existing built environments and benchmarking design solutions. Also, there is a proposal of a course based on three main components: reading, writing and classroom discussions.
Makakli (2019)	The author discusses the STEAM concept - an adaptation of STEM (science, technology, education, and mathematics) with the inclusion of the letter A for arts which here represents creative thinking and design - in architectural education since architecture is the intersection between technology, art, and science. In this context, the design studio is the foremost teaching-learning environment for teaching architecture - it has been the application of STEAM in the project. The author concludes that the contributions of Virtual Reality to the teaching of design are beneficial and should be included not as a tool but as a platform to open new perspectives of space.

Source: the authors.

The only article that deals with the teaching of lighting in undergraduate architecture (FARIA, 2014) is also the most significant for the discussion proposed: The researcher deals with digital inclusion in the learning of daylighting, using free software such as DIALux, Sketchup, and Radiance. According to him, when teaching daylighting, the professor must stimulate the student to perceive the effects of daylight on the project. The researcher also notes that when teaching daylighting, professors should stimulate students to feel the variation of effects resulting from it. Furthermore, he argues that teaching should take place in an environment familiar to future professionals, through drawings and not in numerical form or algorithms. The results should illustrate the design expectations regarding lighting in space and not graphs and tables.

Faria (2014) performed a series of experiments using the various software and plugins aimed at simulating daylighting and concluded that these came to complement what the tools for estimating the amount of daylight did not provide: the visualization of the model illuminated by sunlight. Therefore, these tools can give to the students a better understanding of the characteristics of daylighting interaction with the built environment and related visual effects in the space, thus improving the design solutions concerning openings placement and other envelope components.

A web search concerning disciplines that address lighting provided information about the curricula of several architectural undergraduate courses. The survey results were tabulated and analysed using

IOP Conf. Series: Earth and Environmental Science 1099 (2022) 012040 doi:10.1088/1755-1315/1099/1/012040

descriptive statistics. Themes related to lighting design in the architectural context were found in topics that mostly talk about visual, thermal, and acoustic comfort. Table 2 shows the lighting terms used worldwide for naming the disciplines. The search compiled information from thirty-seven colleges in all five continents: ten are from South America, six from North America, nine from Europe, six from Asia, one from Africa, and five from Oceania. Of those, thirty colleges have information on undergraduate courses, and the other seven colleges are either graduate courses or had no curriculum information available on the institution webpage.

Table 2. Keywords used in the lighting discipline

Issue	Language	Keyword
Climate Project	Dutch	Klimaatontwerp (Climate Design).
Environmental Design	English	Environmental, Environmental Design, Environmental Technologies, Context.
Environmental Comfort,	Portuguese	Conforto Ambiental, Iluminação (Environmental Comfort, lighting).
Environmental Comfort	Spanish	Acondicionamiento Ambiental, Instalaciones, Luminotecnia; (Environmental conditioning, Installations, Lighting technology).

Source: the authors.

The results suggest a common understanding that lighting issues are essential for architects' education. Nonetheless, lighting teaching often is still more focused on quantitative than sensitive aspects. Experimenting with the possibilities and effects of light has been little explored.

A comparative analysis of the lighting discipline syllabus from thirty architectural courses around the world revealed that approximately 53% of the best ones have content about lighting, covered by disciplines related to aspects of thermal and acoustics comfort, ventilation, and others (Table 3). Only two among them have specific lighting disciplines on the undergraduate level. Among seventeen countries selected, only Brazil has universities that offer specific lighting disciplines in their architectural undergraduate courses. The survey showed that in approximately 93% of selected universities, lighting studies are more consistently offered only in the specialization courses.

Table 3. Disciplines dedicated to discussing lighting in undergraduate architecture courses

	Bachelor's degree in architecture	Time to degree		Name of the discipline(s)
Argentina	Universidad de Buenos Aires (UBA)	**6 years**	Yes	Instalaciones III: Luminotecnia. Acustica, Instalaciones eléctricas, Instalaciones termomecânicas.
Australia	Curtin University	**3 years**	Yes	ARCH3006 Environmental and Building Systems in Architecture and ARCH1025Architectural Science in Context.
	RMIT University	**3 years**	No	-
	The University of Melbourne	**3 years**	No	-
	The University of Sydney	**3 years**	Yes	BADP3001: Designing for Environmental Quality.
Brazil	Universidade de São Paulo (USP)	**5 years**	Yes	AUT0282 -– Conforto Ambiental 1 – Fundamentos. **AUT0284 -– Conforto Ambiental 2 – Iluminação.**
	Universidade Estadual de Campinas (Unicamp)	**6 years**	Yes	**AU121 Teoria e Projeto VI -– Iluminação Natural e Artificial na Arquitetura**; **AU146 -– Teoria Projeto VI : Ilum. Nat. E Art.l na Arquitetura** and **AU021 Metodologia de Projeto VI: Ilum. Nat. E Art. Na Arquitetura**.
	Universidade Federal do Rio de Janeiro (UFRJ)	**5 years**	Yes	FAT231 -– Conforto Ambiental I.
Canada	University of British Columbia	**4 years**	Yes	ARCH 513 (3) Environmental Systems and Controls I.
Chile	Pontificia Universidad Católica de Chile (UC)	**6 years**	Yes	Instalaciones, Sist. Y Proyecto Arquitectónico AQC0300.
	Universidad de Chile	**6 years**	No	-
China	Tsinghua University		-	*No information about the curriculum or description of the discipline was found.*
	Tongji University		-	*No information about the curriculum or description of the discipline was found.*
Colombia	Universidad de los Andes	**4, 5 year**	No	-

Continue

Continuation

	Universidad Nacional de Colombia	**5 years**	No	-
Denmark	Aalborg University			***Avaliable only in the master program.***
Finland	Aalto University	**3 years**	No	-
Hong Kong	The University of Hong Kong		-	*No information about the curriculum or description of the discipline was found.*
Italy	Politecnico di Milano	**3 years**	No	-
Japan	The University of Tokyo		-	*No information about the curriculum or description of the discipline was found.*
Netherlands	Delft University of Technology	**3 years**	Yes	BK1TE1 klimaatontwerp and BK2TE3 Klimaatontwerp.
New Zealand	The University of Auckland	**3 years**	Yes	ARCHTECH 210: Environmental Design 1.
Peru	Universidad de Piura (UDEP)	**5 years**	No	-
	Universidad Nacional de Piura	**5 years**	Yes	AR2330 Acondicionamiento Ambiental I e III.
Singapore	National University of Singapore (NUS)	**5 years**	No	
South Korea	Seoul National University	**4,5 years**	Yes	Architectural Environmental Planning 2.
Spain	Universidad de Jaén (UJA)			***Avaliable only in the master program.***
	Universidad Politécnica de Madrid	**3 years**	Yes	Acondicionamiento Ambiental Y Habitabilidad.
South Africa	University of Cape Town	**3 years**	-	*No information about the curriculum or description of the discipline was found.*
Switzerland	ETH Zurich – Swiss Federal Institute of Technology	**3years**	No	-
United Kingdom	University College London (UCL)	**3 years**	Yes	Environmental Design (BARC0086).
	University of Cambridge – Cambridge.	**3 years**	Yes	Paper 5: Fundamental principles of environmental design.
United States	Columbia University	**4 years**	No	-
	Harvard University	**4 years**	No	-
	Massachusetts Institute of Technology (MIT)	**5 years**	Yes	4.401 / 4.464 Environmental Technologies in Buildings.
	University of California, Berkeley (UCB)	**4 years**	Yes	Either ARCH 140 Energy and Environment.
	University of California, Los Angeles (UCLA)	**2years**	No	-

Source: the authors.

3. Discussion

Several authors defend the teaching of professional practice in undergraduate courses (Brandston, 2010; Freire, 2004; Gropius, 1977; Nicol, Pilling 2000; Lara, Marques, 2015; Salama, 2016; Scarazzato, 2018; Schön, 2000; Polião, 1999). In architectural undergraduate courses, studio design is the locus par excellence for such a practice. There, students can apply the acquired knowledge to the theory needed for the building design process (Habib *et al.*, 2021; Maghool *et al.*, 2018; Makakli, 2019, Matos, Scarazzato, 2017).

It is necessary to involve students as active agents of their own apprenticeship, and the use of interactive tools represents a good alternative for such a target (Habib *et al.*, 2021, Feijoó *et al.*, 2021). The distribution of didactic support material before classes allows for the introduction of the theme and encourages discussions, with more interactive sessions of class debates, equalizing knowledge and better understanding by all students (Feijoó *et al.*, 2021).

Faria (2014) notes that a way to promote the desirable involvement of the students is by using computer simulation software. However, as referred here, specific methodologies for teaching lighting with the support of digital tools still need to be further developed. Also, it is necessary to establish practices to insert and encourage the use of light as an essential aspect for modelling the built environment and not only as a checklist item to meet design standards and recommendations.

4. Proposed and Applied Didactic

Aiming to develop the perception regarding the relations between the design solutions and the visual quality in the built environment (MERCIER, MERCIER, 2014), the students are introduced to light from images of daylighting and electrical lighting solutions. Then they are asked to search for references for what they consider good and bad applications of light, with justifications for their understanding. The results are exposed and discussed in class.

The second step consists in sensitizing the eye by observing scenes under daylighting or electrical lighting conditions, at the choice of each student. This activity comprehends two draws: graphite pencil on a white sheet of paper and white colour pencil on a matte black car. The last one represents the total absence of light (Figures 2 and 3).

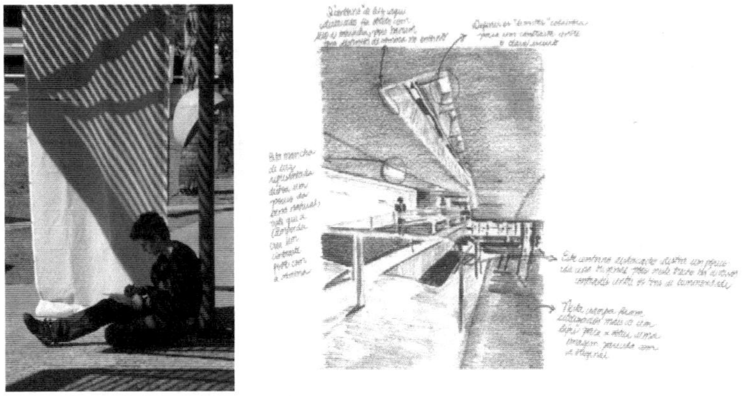

Figure 2. Left. Student drawing by observation. **Right.** Draw by observation. Pencil on white paper
Source: the authors.

Figure 3. Draw by observation. White colour Pencil on matte black card.
Source: the authors.

In sequence, students learn to use High Dynamic Range images (HDR)[1] to evaluate quantitatively and qualitatively the distribution of daylighting and electrical lighting inside or outside buildings[2] (Figure 4). At the end of these exercises, students start to pay more attention to issues of lighting in their design process, albeit in an embryonic form. There is no distinction or preference between daylighting and artificial lighting. Light, in its essence, is what matters.

Figure 4. Photo of an actual architectural room taken by an undergraduate student
and its HDR image, in false colours.
Source: the authors.

[1] software used to generate HDR is denominated **Light Photo Analyzer**, currently in Beta version, and available on the web <https:// www.liv.ic.unicamp.br/HDR/>. The Light Photo Analyzer software is derived from the WebHDR software, by Axel Jacobs (for more information, access: jaloxa.eu/webhdr).

[2] HDR images are excellent for quantitative analysis of daylighting, but not for quantitative analysis of artificial lighting.

The fourth stage of apprenticeship consists of making physical scale models to observe daylighting. Students are generally very motivated at this stage, even because physical scale models are familiar to them (Figures 5 and 6).

Figure 5. Physical scale model of Barragán House-studio crafted by architecture's undergraduate students. Source: the authors.

Figure 6. Architecture's undergraduate students preparing and observing physical scale-models Source: the authors.

As an unfolding of the use of the physical scale models as a teaching-learning tool, it is confident to use the HDR technique to evaluate the indoor daylighting distribution. Figure 7 shows an example of the use of HDR imaging applied to a scale model as an evaluation method for fenestration solutions during the design process.

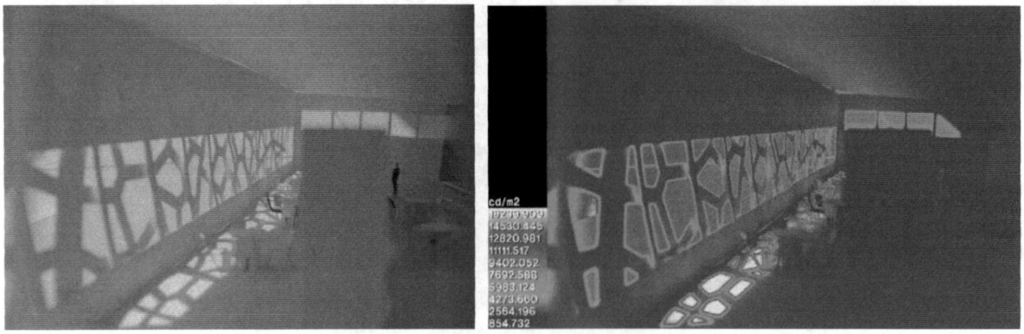

Figure 7. Photo of physical scale model crafted by architectural undergraduate students and its HDR image in false colours
Source: the authors.

Related to electrical lighting students first develop conceptual sketches. Based on concepts learned, standards requirements, and the best practices related to lighting computational modelling, the students develop skills in computational simulation (Figure 8).

Figure 8. Conceptual sketch (left) and electrical lighting simulation of a room modelled by a student with the use of DIALux EVO software (right).
Source: the authors.

Once introduced to lighting, it is time for students to have contact with guidelines for the inclusion of lighting in the design process. They receive information about the possibilities of daylighting and its control, including criteria for dimensioning and spacing of illuminating openings. Regarding electrical lighting, they learn about different ways to illuminate, light sources, luminaires, and control devices. Photometric magnitudes and units, as well as calculation methods and simulation tools, come last. This step corresponds to a maximum of 20% of the total course load.

In the last exercise, students develop a medium-complexity design, in teams of five students, with an emphasis on daylighting and electric lighting. For this exercise, students use a tool called Daylight Design[3] which uses the concepts of design patterns from the urban scale to building details and furniture. It aims to aid students or any design professional in making decisions related to daylight. Its actual configuration brings the sun-path diagram and design guidelines for 330 Brazilian cities, according to the Brazilian Bioclimatic Zoning Norm (NBR 15 220 - 3). It contains sixty-five daylight patterns related to the inclusion of daylighting in the design process. Two are possibilities for applying the tool in teaching daylighting: (i) use it in design studio exercises and (ii) as a theoretical basis for the analysis of referential projects.

The Daylight Design tool is under development, and it is being tested during lighting classes of architectural undergraduate courses (Figure 9). To better understand its possibilities and limitations, the students voluntarily answer a survey. The tool is freeware and it's available at <www.daylightdesign.com.br> (accessed 2022.08.09).

[3] The tool is still under development by one of the authors as part of a future Ph.D. The thesis research has been supported by The São Paulo Research Foundation (FAPESP) – Process Number 2017/24479-4

Figure 9. Screens of the Daylight Design: (1) Home page; (2) sun-path diagram.
Source: the authors.

Figure 10 shows the proposed teaching-learning flowchart. It highlights every learning phase of the process and its related practical activities.

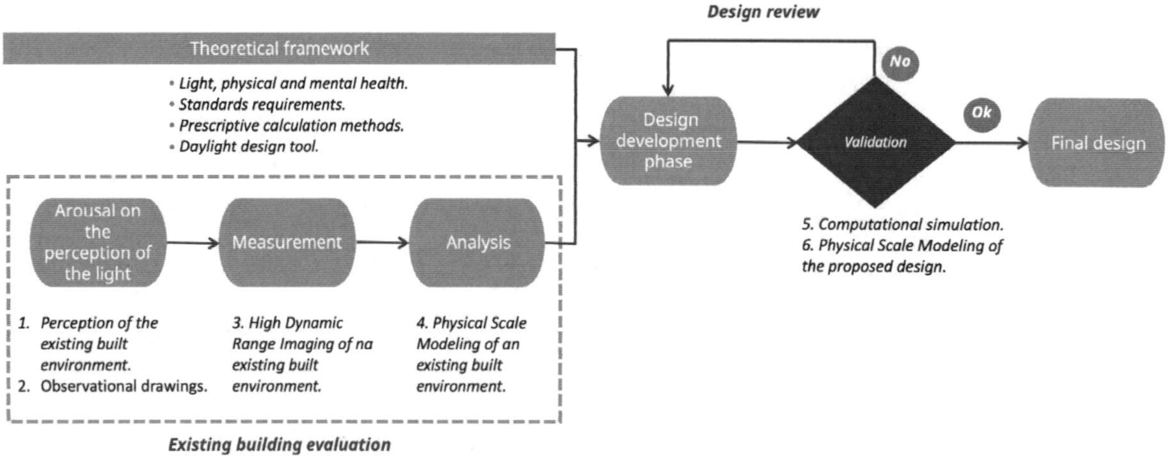

Figure 10. Teaching-learning process flowchart.
Source: the authors.

The presented didactic proposal has been applied in two Brazilian universities, in disciplines, respectively with fifteen weeks and six class hours per week, and fifteen weeks and four class hours per week.

5. Final considerations

Considering the understanding that lighting is an essential issue in creating the built environment (BAEZA, 2010), the teaching-learning process should occur both in the studio and beyond the classroom. This is a critical aspect once learning how to think about the built environment's complexity is the central element of the architect's education (SALAMA, 2009). Therefore, the architect in training must have opportunities for experimentation and awareness. The sensorial experience, given by the set of stimuli offered to the user, is fundamental for the perception of the built environment, as well as the promotion of their health and well-being (BÖHME, THIBAUD, 2016; ZUMTHOR, 2006).

Lighting applied to architecture is a complex subject. So, it shall consider the evaluation of tasks performed, their features, duration, users' gender, age, habits, and expectations. Thus, lighting design in the architectural context is not only theoretical nor just qualitative but shall present the integrated comprehension of the design problem. An integrative lighting design requires a vision encompassing daylighting and electrical lighting, as well as its technical, biological, and perceptive aspects, in addition to an understanding of the specific aspects of each group of users.

One can conclude that the teaching-learning process of architecture, including lighting, should not be limited to applying a set of procedures. On the contrary, it is essential to be aware and provides a means for students to track and continually use technologies and processes and respect design requirements (LAWSON, 2001).

Finally, in a post-pandemic scenario, it will be necessary to rescue the commitment to appropriate human shelter conditions that can contribute to improving and maintaining the physical and mental health of buildings users and visitors. Habits acquired during the pandemic and now consolidated, such as remote work and studies and prolonged periods at home, should be incorporated.

Certainly, lighting education has new challenges. Thus, the non-visual aspects of light and safety in lighting are two new topics that should be part of the syllabus of lighting disciplines of undergraduate - and graduate - architecture courses going forward.

Acknowledgments
The Daylight Design toll presented in this paper has the support of The São Paulo Research Foundation (FAPESP) - Process Number 2017/24479-4.

References

[1] BAEZA, Alberto Campo. (2010) Architectura sine luce nulla architectura est: sobre la luz. In: Cosa Mentale. Paris: [s.n.].
[2] Böhme, Gernot; Thibaud, Jean Paul. (2016) The Aesthetics of Atmospheres. Oxon: Routledge.
[3] Boyce, Peter R. (2006) Education: the key to the future of lighting practice. In: Lighting Research & Technology, v. 38. n. 4. p. 283–291.
[4]. Brandston, Howard M. (2010). Aprender a Ver: a essência do design da iluminação. Trd. Paulo Sergio Scarazzato. 1ª ed. São Paulo: De Maio.
[5] Faria, João Roberto Gomes. (2014) Inclusão digital no ensino de iluminação natural. In: Paranoá: cadernos de arquitetura e urbanismo, n. 11, p. 123–133.
[6] Feijóo, Juan Carlos Mosquera; Suárez, Fernando; Chiyón, Isabel; Alberti, Marcos García. (2021) Some Web-Based Experiences from Flipped Classroom Techniques in AEC Modules during the COVID-19 Lockdown. In: Education sciences, Basel, v. 11, n. 5, p. 211.
[7] Freire, Paulo (2004) Pedagogia da Autonomi: saberes necessários à prática educative. São Paulo: Paz e Terra.
[8] Gropius, Walter. Bauhaus: nova arquitetura (1977). São Paulo: Perspectiva.
[9] Habib, Maki K; Nagata, Fusaomi; Watanabe, Keigo. (2021) Mechatronics: Experiential Learning and the Stimulation of Thinking Skills. In: Education sciences, Basel, v. 11, n. 2, p. 46.
[10] Lara, Fernando; Marques, Sonia. (2015) Quidi Novi? Architectural education dilemmas in the 21st century, Austin: Nhamerica.
[11] Lawson, Bryan. (2001) How Designers Think: the design process demystified, Oxford: Architectural Press.
[12] Maghool, Sayyed Amir Hossain; Moeini, Iradj, Seyed Hossein; Arefazar, Yasaman. (2018) An Educational Application Based on Virtual Reality Technology for Learning Architectural details: Challenges and Benefits. In: ArchNet-IJAR, Cambridge, v. 12, n. 3, p. 246.
[13] Makakli, Elif Süyük. (2019) STEAM approach in architectural education. In: SHS web of conferences, Les Ulis, v. 66, p. 1012.
[14] Matos, Jéssica; Scarazzato, Paulo. (2017) A iluminação natural no projeto de arquitetura: revisão sistemática da literatura. In: PARC Pesquisa em Arquitetura e Construção, Campinas, SP, v. 8, n. 4, p. 249–256.
[15] Mercier, Kim; Mercier, Paul. (2014) Architecture for light. New York: IES.
[16] Nicol, David; Pilling, Simon (2000). Changing Architectural Education.New York: Spon.
[17] Polião, Marco Vitruvio (1999). Da Arquitetura. Trad. e notas Marco Aurélio Lagonegro. São Paulo:

Hucitec; FUPAM.
[18] Salama, Ashraf M. (2009) Transformative Pedagogy in Architecture & Urbanism, Oxon: Routledge/Taylor & Francis Group.
[19] Salama, Ashraf M. (2016). Spatial Design Education. New York: Routledge.
[20] Scarazzato, Paulo Sergio (2018). Questões de Luz na Formação do Arquiteto. Tese (Livre-Docência). São Paulo: Faculdade de Arquitetura da Universidade de São Paulo. Available at: <https://www.teses.usp.br/teses/disponiveis/livredocencia/16/tde-04112019-163717/pt-br.php>, accessed 2022.08.09.
[21] Schön, Donald A. (2000). Educando o Profissional Reflexivo: um novo design para o ensino e a aprendizagem. Porto Alegre: Artmed.
[22] Smith, Korydon H. (2011) Architectural theory in the undergraduate curriculum: a pedagogical alternative. In: International journal of technology and design education, Dordrecht, v. 23, n. 1, p. 117–128.
[23] Ströher, Ronaldo De Azambuja. Lições Albertianas para a Teoria e a Prática da Arquitetura Contemporânea. PhD. Thesis. Porto Alegre: UFRGS. School of Architecture, 2006. Available at: <https://lume.ufrgs.br/handle/10183/7681>, accessed 2022.08.09.
[24] Top universities. QS World University Rankings by Subject 2021: Architecture & Built Environment, Topuniversities, 202. available at: <https://www.topuniversities.com/university-rankings/university-subject-rankings/2021/architecture-built-environment>, accessed 2022.04.14.
[25] Zumthor, Peter. (2006). Atmospheres: architectural environments; surrounding objects. Basel.

IOP Conf. Series: Earth and Environmental Science 1099 (2022) 012041 doi:10.1088/1755-1315/1099/1/012041

Light4Health eLearning Course: health research for interior lighting design. Re-thinking design approaches based on science

KM Zielinska-Dabkowska[1], L Godley[2], F Kyriakidou[3], U C Besenecker[3], G Triantafyllidis[4]

[1] GUT LightLab, Faculty of Architecture, Gdansk University of Technology, Narutowicza 11-12, 80-233, Gdansk Poland;
[2] Thomas Jefferson University, 4201 Henry Avenue, Philadelphia, PA 19144 USA;
[3] Division of Lighting Design, Department of Architecture, KTH Royal Institute of Technology, Osquars backe 5, 114 28 Stockholm, Sweden;
[4] Lighting Design Lab, Aalborg University Copenhagen, AC Meyers 15, Copenhagen 2650, Denmark
Corresponding author's e-mail: k.zielinska-dabkowska@pg.edu.pl

Abstract. This paper presents the results of 'Light4Health' (L4H), a three-year EU Erasmus+ Strategic Partnership grant project (2019-2021), which investigated, systematized and taught health-related research on the impact of natural and artificial light on human health and well-being relevant to indoor lighting design. The objective was to re-think evidence-based lighting design approaches for residential, working/educational, and healthcare spaces, in order to develop a novel cross-disciplinary eLearning platform, that intersects lighting design and current peer-reviewed health research through a select combination of the most relevant research, methods, and tools. The content was developed through teaching workshops with international researchers, teachers, and students. Participating students were introduced to the application of relevant research and new metrics, in order to produce creative lighting design proposals. Students were able to inform lighting design approaches that support health and well-being without compromising creativity via the tools and methods developed through the health sciences. With this project, participants and consortium members have narrowed the gap that exists between research and the practice/application of lighting, by translating research from complex scientific jargon into various tools for designers to use. The knowledge gained, was consolidated into an open-access online curriculum for international lighting design students, educators and professionals via the free eLearning Moodle platform (https://course.light4health.net/).

1. Introduction

In 1935, the Swiss architect Le Corbusier was already highlighting "the struggle for light" as one of the emerging planning problems of the modern world [1]. Today, densely filled, tall buildings in cities mean access to daylight is even more restricted.

The invention of electric lighting allowed inhabitants to spend more time in enclosed spaces, in order to perform necessary visual tasks long after the sun set. Nevertheless, the latest research indicates that a lack of exposure to proper, quality light has immense effects on people's health, well-being, and productivity [2-4]. Therefore, an unprecedented paradigm shift has been taking place in the lighting community, which involves moving beyond energy efficiency towards quality light for well-being, and from a vision/perception-oriented lighting approach to one that is also oriented towards human biology. The key application of health research knowledge in the lighting design profession, is for the development of lighting to be an element of preventive healthcare. It balances the visual, emotional and biological benefits of lighting on humans; promotes good vision, and simultaneously satisfies human emotional and biological needs; also taking into account that light stimulates biological effects on psychology and physiology. Given that every human is constantly exposed to light, the introduction of lighting that is based on research could have a significant social and economic impact for the EU and beyond.

Additionally, lighting practitioners, responsible for designing artificial lighting plans for indoor spaces, have been overwhelmed with recently developed health-related lighting metrics by scientists and researchers. Lighting practitioners require clear guidance on which metrics to apply and the reasons for applying them, in order to focus on making lighting safer and healthier [5]. The Health Research-Based Innovative Open Educational Resources and Tools for Lighting Design Students and Professionals LIGHT4HEALTH project was developed to respond to these challenges. Sponsored by Erasmus+, the intra-university project involved experts from Lighting Design, Light and Health Research, Neurology, Architecture, and the Built Environment from six internationally renowned institutions including the UK (University of Wolverhampton - UOW), Denmark (Aalborg University - AAU), Germany (Hochschule Wismar - HSW), Sweden (Royal Institute of Technology- KTH), Russia (ITMO University), and the USA (Thomas Jefferson University - TJU).

The first objective of this project was to re-think existing lighting design approaches for residential, working/educational, and healthcare spaces, by developing a cross-disciplinary eLearning platform, that incorporates information from current peer-reviewed health research into lighting design education via the selection of a combination of relevant and appropriate methods and tools. These were informed by evidence-based research in Neurology, Photobiology, Neuroendocrinology, Neurobehavioral Studies, Psychophysiology of Perception, as well as aspects of Environmental Psychology, which all aimed at fostering accessibility and the implementation of the curriculum across Europe and overseas. Researchers from Thomas Jefferson University's Light Research Program, which included Dr George Brainard and his team, joined this project as experts to advise on the physiological effects of light [5-9]. They also provided guidance on the use of appropriate lighting metrics, which includes tutorials on the use of CIE α-opic Toolbox [10]. Thanks to this knowledge, lighting professionals can be confident with science informed circadian protocols. Teachers from the partner universities were responsible for efforts that included the development of the curriculum, preparation of pre-recorded lectures and exercises, reviews of student's work, and the creation of a teachers' guide.

The second objective was to meet the growing needs of the lighting industry to fill the gap between research, education and application, and to better connect them. As there are significant growth and employment opportunities in the lighting sector that have not yet been undertaken, the lighting industry, together with EU legislators, aims for the growth of the European lighting market through more intelligent and human oriented lighting that supports health and well-being. At the same time, educational programs about this topic in Europe and overseas are falling behind, failing to produce the required number of graduates. To fulfill these needs, we developed the course content for students in graduate Lighting Design programs as well as professional lighting designers. The project envisaged the involvement of various groups, such as: students enrolled in summer school workshops, adult learners/practitioners, teachers in fields related to lighting design, developers of study programs, employers at the regional, national and European level, the lighting industry, education certification, healthcare authorities, and local communities/general public.

2. Research questions

Three key research questions were explored and investigated during the course development to achieve expected project objectives and to help position and operationalize the findings in relation to the following identified issues:

1. How to apply the outcomes of peer-reviewed research about the effects of light on human health and well-being to current design education via an opensource teaching platform?
2. What are the most appropriate metrics currently being used to measure the impact of light on health and well-being, specifically for indoor environments?
3. How to predict and evaluate whether an interior lighting design supports the well-being of users?

This paper focuses solely on the project and the implementation of knowledge, however it does not answer the specific design-related questions that have been investigated.

3. The general structure of the L4H eLearning course

This online course aims to bring the latest applicable health research findings to lighting design education. The three course goals are:

- Highlight the significance of physiological and psychological health research,
- Enhance the quality and relevance of participants' applicable knowledge and skills,
- Equip graduates with current and relevant research, design, and skills for the lighting industry.

The development of the L4H eLearning course consisted of five steps (Figure 1).

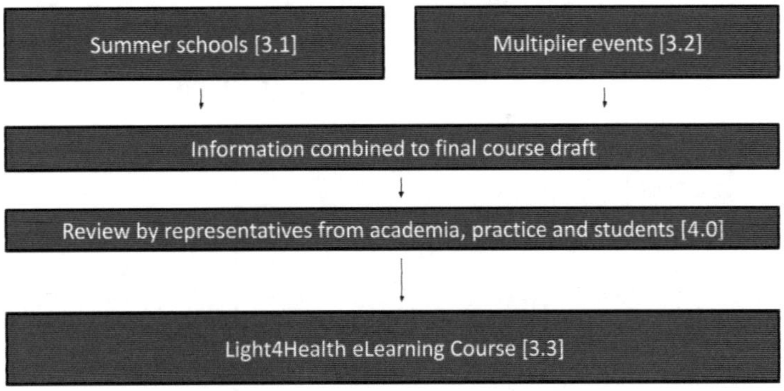

Figure 1. Overview of the L4H project

3.1. Summer Schools (Workshops)

Three innovative teaching modules were validated through test-runs in the format of transnational Summer Schools at partner organizations, resulting in first-hand experience and extensive feedback data to improve structure and navigability of the resulting eLearning platform. The workshops took place with a total of 91 students and 14+ teachers in various locations such as: Wolverhampton 2019 (in person); Hochschule Wismar 2020 (online and in person) and Aalborg University 2021 (online). All had different topics, timeframes, and due to the COVID-19 pandemic, it was necessary to shift from in-person summer schools to remote classes with different teaching methodologies (see Figure 2 for more information).

The workshops helped to develop curricular and applicable knowledge in both the conceptual and practical aspects of health research, as well as the lighting design for a variety of environments. Project

work was the central pillar of each of the schools. It served to structure the understanding of health research methods so that students could develop a critical approach to what is known about the impact of different lighting solutions on humans in specific environments. Students were introduced to new metrics and software for measuring and simulating possible impact on health, in order to then apply these methods in their design process and concepts. This required students to experiment with research methods and metrics in various settings, which allowed them the opportunity to conceptualize their ideas and consider appropriate design solutions. Additionally, interactive quizzes and sample task assignments were provided for each summer school module to assess the students' understanding of the material.
Feedback from the students at the end of each workshop helped to refine consecutive workshops and the eLearning platform.
Example comments:
"It was a really great opportunity for all of us lighting designers to attend such a valuable workshop like this. It offered not only the knowledge and materials, but also the practical ways to practice and deliver what we learned through the platform."
"Thank you for this great course, it was a pleasure to be part of it. I wish that I could have been part of the first course as well. The data provided is more than useful and interesting, as was the feedbacks from a wide range of professionals. I will definitely join the next course and strongly recommend it to others."
"I think it is a very good opportunity to me to be introduced to different methods to design lighting schemes with a mindset that focused on the impacts for health, and I believe this experience and knowledge will be useful for my career."

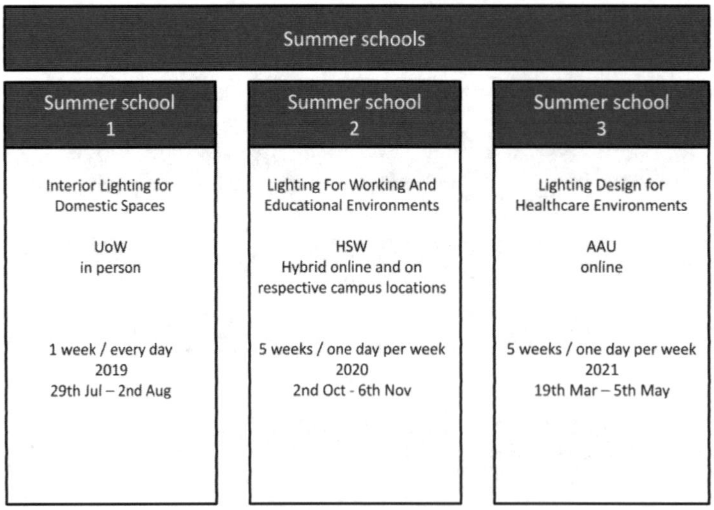

Figure 2. Summary of course content of the three Summer Schools

3.2. Multiplier events
Multiplier events (MEs) are required with all Erasmus+ projects in order to share the intellectual results of the project with a larger audience throughout the timeframe of the grant. Four MEs were conducted with a total of 300 external participants. The selection and involvement of participants depended on the host partner and the phase of the project. All MEs were advertised in social media and through partners' organizations, as well as by individual invitation. Arc lighting magazine (the foremost publication for lighting in architecture) was very supportive in disseminating the project through its newsletter as well.

3.2.1. Multiplayer Event #1 – L4H Public Workshop "Healthy Lighting for Working and Educational Environments" hosted by HSW

This event took place in March 2021, during the Light Symposium Wismar 2020/21 (LSW 2020/21) - a three-day online forum. This symposium delivers a state-of-the-art outline of how daylight and electric light affects the physical and mental health, efficiency, and performance of human beings. The L4H consortium partners used the opportunity to introduce the L4H project to a wider audience. The major target audience for this workshop was a mixture of: academia (educators and researchers), the lighting practice (designers), and the lighting industry (manufacturers). The workshop program included a Light4Health project presentation that discussed Summer School 1 and 2, and an introduction to Consortium partners. This was then followed by a panel discussion with 14 invited members of the Light4Health Team. The event was streamed online and it had an international reach, hosting 354 participants.

3.2.2. Multiplayer Event #2 – Light4Health Online Conference – Hosted by UOW

This event took place in May 2021, as an online conference aimed at students and academics, as well as employers. All L4H partners participated as speakers. The morning session gave an overview of the project so far (each of the three modules and summer schools), an introduction to factoring lighting in the design of buildings, with two case studies. In the afternoon session, speakers discussed lighting standards in domestic housing in the UK, the psychological influence of lighting, and there was also a talk about improving and installing lighting for astronauts for NASA. Each of these sessions provided an opportunity for feedback, discussion and Q&A. Over 150 people accepted invitations to attend and register; 22 external participants attended. The virtual format enabled participation from an international audience.

3.2.3. Multiplayer Event #3 – Launch of Light4Health eLearning Course "Health Research for Lighting Design" – Hosted by KTH

August 2021 – This course launch included a Light4Health project presentation with an introduction to consortium partners, a short summary of all the summer schools, and a summary of the previous multiplier events. This was followed by a panel discussion with six invited panelists who had reviewed the material for the online course. Feedback from reviewers was discussed. This was followed by questions from the audience. The event concluded with the live launch of the online course. Participants were also shown how they could register for the course, and were given a tour through the eLearning course on the Moodle platform. The event had an international reach due to being streamed online. In total 123 people attended the event (of which 108 were external participants). The major target audience for this workshop was a mixture of: academia (educators and researchers), the lighting practice (designers), and the lighting industry (luminaire manufacturers).

3.2.4. Multiplayer Event #4 – Online Conference "Transfer of Knowledge: Closing the Gap between Lighting Research and Lighting Design" – Hosted by ITMO

August 2021 – An online event to present the recently launched Light4Health educational eLearning course and eLearning platform. There was also a demonstration of relevant case studies and a discussion with the invited speakers about topics related to light and health. This event was streamed online and it had an international reach with 148 people registering for the online conference and 77 people attending, 70 of whom were external participants.

3.3. L4H eLearning course

The L4H eLearning course available on the Moodle platform, includes seven folders. The first is a course description, syllabus, and consortium information, the second folder contains material for teachers in the form of a Teachers' Guide and sample task assignments, and folders 3-7 contain five educational Modules with 28 video lectures and pdf slides, as well as reading reference lists and interactive self-assessment quizzes (Figure 3).

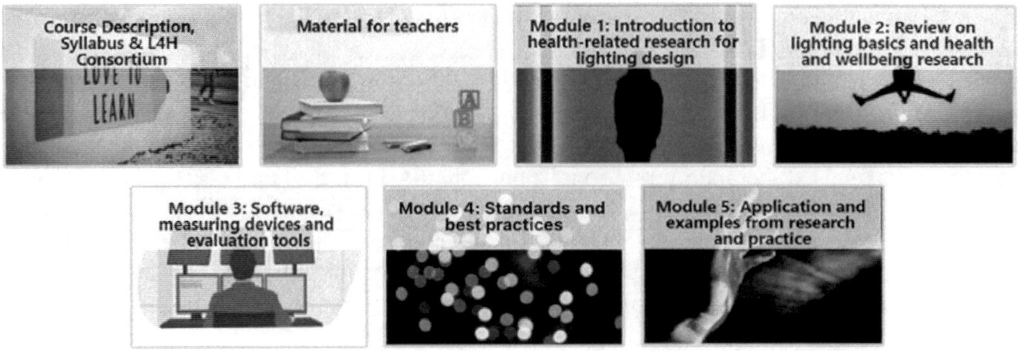

Figure 3. The L4H online course content as viewed on the Moodle eLearning platform

The development and organization of the content for the final L4H eLearning platform were based on content that was developed for the previous workshops, feedback and assessment from faculty and students, and the restructuring of the previous content. The course was designed to either function as both independent lectures and lessons, or as a complete course covering 1) a historical background about health- related research for lighting design, 2) an overview of lighting basics, and physiology related to light, 3) a range of tools for measuring, applying, and evaluating lighting metrics, 4) standards and best practices for lighting for health, and 5) their application in the form of case studies.

A more detailed description of the content of the five educational modules is provided in Table 1.

Table 1. Five Course Modules with detailed content

Module 1:	Introduction to health-related research for lighting design: Provides participants with a short history about light and health, and why we need to consider health as part of lighting design. The effects of light on the human body are also covered with an introduction to aspects of the physics of light and the physiology of vision.
Module 2:	Review on lighting basics and health and well-being research topics: Provides a review of lighting basics and lighting-related aspects of health. Included is daylight in architecture, the impact of daylight on health, and, in general, the neuroendocrine, neurobehavioral, and circadian effects of light on the human body. The psychology of light is also introduced; more specifically, how light is used as a visual trigger for psychological and behavioral impact. In addition, the topics of glare and flicker are discussed.
Module 3:	Software, measuring devices and evaluation tools: This module investigates software, measuring devices, and evaluation tools. Metrics are presented which can be used to assess the potential for the physiological impact of lighting, and different software and measurement tools are discussed. In addition, some assessment techniques are presented for the subjective impressions of a space. Finally, a tour is given of the Jefferson Lighting Research Lab.
Module 4:	Standards and best practices: Provides an overview of standards and good practices. Metrics for daylight evaluations are presented, and the "Manchester Recommendations" for healthy daytime, evening, and nighttime indoor light exposure are introduced. Moreover, design integrations are discussed. This involves employing measurements and design criteria for physiological impact and visual perception.

Module 5:	Application and examples from research and practice: Provides application examples and case studies from research and practice. In this context, several examples of light and health research are presented. This includes project examples from workspaces, and educational and healthcare environments. Specialty applications (e.g., space travel, users with autism), are also included, as are concepts of spectral modeling for light and health considerations, double dynamic lighting, biophilia, and information on therapeutic lighting applications.

4. Evaluating the L4H project content

An external quality assurance of the intellectual outputs of the project was important to its credibility. This involved three levels: 1) Qualitative evaluation by the L4H summer schools students; 2) Qualitative evaluation by the external reviewers/experts, and 3) Quantitative evaluation of the effectiveness of the online learning platform as an educational product. Formal validation and provision of the European Credit Transfer System (ECTS) was not possible due to the framework of a public open access platform, however each of the course modules was developed to be equivalent to 1 ECTS.

4.1. Pre-Launch evaluation by the L4H summer schools' students participants

Three innovative teaching workshops resulted in first-hand experiences and extensive feedback data to further improve the structure and navigability of the eLearning platform. The students participating in the workshops piloting the tools provided valuable feedback, including: that they learned many things that were unavailable elsewhere, they gained inspiration for their master thesis, they were introduced to new tools and software to evaluate lighting with an innovative approach that would be 'really meaningful in the professional field', and that working in a team with 'different cultural, educational background, and experience, had so much impact on the information I had and my way of thinking'.

Students had the opportunity to broaden their networks and get to know companies, academics and software as well as peers at partner universities. Exposure to scientific research to inform design was evaluated and shown to expand the skills and employability of students. Over 95% of survey respondents (21) reported that the experience of attending a L4H summer school session was a valuable experience overall, and 100% (22) said that the knowledge gained can be integrated into their studies and/or design practice. Other concrete impacts on L4H 'graduates' were increased confidence from completing the course and working in international teams.

4.2. Pre-Launch evaluation of the eLearning platform by the reviewers/experts

A draft of the final course content was reviewed and validated through 38 international expert reviewers including academics in the fields of lighting and health research, and professional experts in architecture and lighting design, who had access to the draft course materials for one month. The reviewers provided comments via a detailed online questionnaire. These comments were then used to make necessary improvements to the final course. Supportive feedback and valuable constructive criticism were provided on the pedagogical aspects and content of the draft; generally, there was a willingness to replicate the course. The collected feedback has been documented and it was used to make improvements to the course prior to the course launch, and logged to further inform possible updates of future content. 73% of the reviews by academics, indicated they were likely/highly likely to incorporate the materials in their teaching. Reviewers from professional lighting practices, architects and lighting manufacturers, expressed a strong interest in using the materials to inform their activities.

Academic reviewers included representatives from Parsons University (USA), Jönköping University (Sweden), Kongsberg University (Norway), Mount Sinai Medical Center (USA), Aarhus (Denmark), Build Research Institute (Denmark), Konstfack University (Sweden). Reviewers from professional lighting practices included: Global Designers Arup, The Lighting Practice (USA), Licht Kunst Licht (Germany), Liska (Iceland), Lightsphere (Switzerland) and Hoare Lea (UK).

Example comments:
"Students will clearly benefit from adding the topic of light and health to their lighting education."
"This course material could serve as a clear incentive to extend offerings and use this course as a resource, rather than having to develop material at each university."
"It provides engaging in-depth lectures on highly specific topics with field implications from the researchers engaged with the subject matter. This gives students unique insights into the topic in a medium that is very accessible."
"This gives a good foundation on light and health directly from the researchers involved. This could be a highly useful teaching tool, either as scaffold for a course or as supplement material to an array of lighting courses."

4.3. Post-Launch evaluation of the eLearning platform as an educational product
More than 221 users have registered to the L4H eLearning platform (https://course.light4health.net/) since the Final Course was launched at the end of August 2021. Users are mainly students, but there are also teachers and professionals. This number is in addition to the 117 users of the prototype during the summer school pilots.
The views of the L4H eLearning platform are more than 3500, including the views for the Modules, Teacher's Guide and project description. More specifically, Figure 4 shows the number of views per module for lectures, reference list and quizzes per module, while Figure 5 shows the percentage of views per continent. It is important to note that the high number of views for the self-assessment quiz, as well as the use of the online course all around the world.

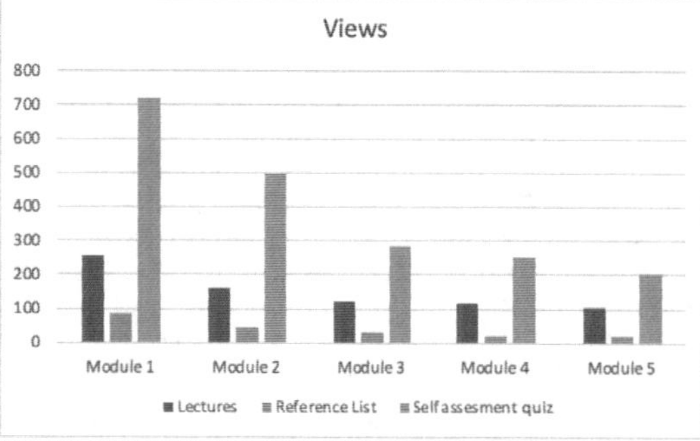

Figure 4. Number of views per module content

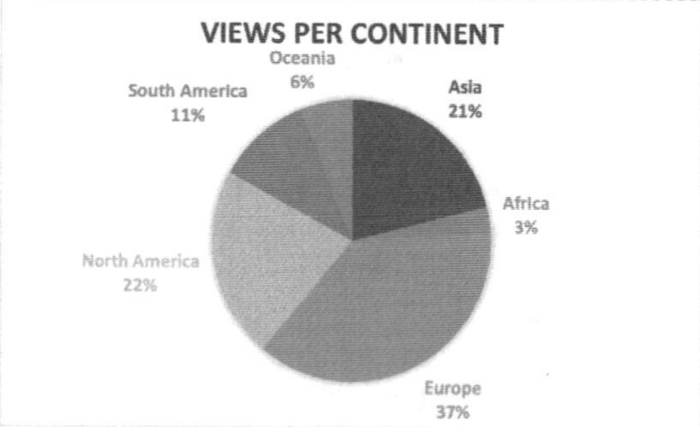

Figure 5. Percentage of views per continent

5. Findings and Potential Impact
Key innovative findings about this lighting design education project are grouped into the following categories below. They can inspire other researchers and teachers in lighting design.

5.1. Approach
- Research Driven Learning - students were shown how to access and interpret relevant and current peer reviewed papers from scientific research as a foundation for their designs.
- Interdisciplinary Learning - sharing education and research expertise from six organizations across multiple countries, universities, and departments that incorporates knowledge from Photobiology, Neuroscience research, Architecture, Psychology as well as Lighting Design, opened discussion to enable innovative solutions that reaches across diverse fields rather than silo-based learning.
- Team-Based Learning - teamwork plays an important role in the designing and construction of the built environment. The feedback from architects indicated having a base understanding of

Light4Health principles and tools, can facilitate team work and help ensure that lighting knowledge is not lost or compromised when engaging with the architect or those who commission buildings. Student feedback indicated the experience of working in teams to realize their final design project, demonstrated to them that design is rarely achieved, and that they need to develop a network in order to aid their future practice.

5.2. Application

- Applied Research - the application of scientific research directly to the design practice is not at all common in the area of architecture and lighting design. Students were shown how to go beyond studying the research, and to apply it to their own design processes.
- Adaptable to various educational platforms – the course is very flexible and versatile in that it can be taught in-person or remotely (or a combination of the two), or self-taught with assessment tools that can be used in both contexts by students or for ongoing training by professionals in the lighting industry.
- Accessible across professions - an innovative theme that emerged during the piloting and external review, was the real potential of the project to support and enhance teamwork between professionals beyond just lighting design (architects, engineers, surveyors, etc.) and those who commission construction or remodel works, highlighting where it is necessary to incorporate healthy lighting design. The online and modular nature of the course, clarity with which the topics are presented and the varying backgrounds and different perspectives of the partners that contributed to the course, makes it accessible as a resource to educate the wide range of practitioners and stakeholders that will contribute to the future development of healthy lighting.

5.3. Assessment

- Project Review – The curriculum provides a clear approach for both students and professionals, in how to apply research, as well as the methodology and tools that are needed to evaluate the impact of the final design. In addition to this, interest from the target groups (including academics, lighting practitioners/industry and students) was strong, as demonstrated by the 300 external participants across the five multiplier events (the target was 155), with a high proportion of international attendees. Finally, the goal for the official project website (https://www.light4health.net/) was 300 visits. At the time of writing this report, there have been more than 1000 views.
- Participant Review – Through extensive student and consortium partner feedback, not only was there continual refinement of the L4H project, but also room for further development and collaboration.
- External Review – The broad range of external reviewers confirmed the need and the value of the overall project and L4H platform for use across academic and professional fields. Manufacturers that attended the launch event considered it of utmost importance that they are invited into academia and that there is more discussion between lighting manufacturers, researchers, architects and designers so knowledge can drive innovation, and that this project has improved this dialogue.
- Platform Review – Since the launch in August 2021, data continues to be collected as per the overall use of the platform.

5. Future studies and course developments

We are aware that research on the effects of lighting on human health is a new and rapidly evolving field, and that subsequent investigation and its complexity demands further discussion to better understand it. This requires financial support to conduct independent research, the application of appropriate research tools and the development of new methodologies for measuring and predicting impact. In addition, how such tools are best implemented in the design processes is still a new field of investigation. We acknowledge that the L4H course content, while wide-ranging, does not encompass the full extent of knowledge as health-related lighting design research continues to grow. Thus, the current course content is subject to refinements, updates, and revisions in future iterations. Further studies are needed so more knowledge can be integrated. Also, in addition to indoor lighting, outdoor lighting with increased global light pollution has been recently considered a potential origin of certain chronic diseases in human beings. We hope this project inspires further investigation by the health research community and that this also involves a close collaboration with the design community to inform health-conscious design processes and its implementation in the built environment.

6. Conclusion – Enacting change

This work highlights the timely recognition of the significance to health and well-being of indoor environments where people spend most of their time. The emphasis on the importance of quality lighting in the built environment indicates a shift toward a more holistic approach.

The Light4Health project's impact includes many aspects – both within the EU and internationally, while also providing benefits to society and the economy at a local, regional and national level. Through intensive cross-border and cross-disciplinary cooperation, the project has triggered collaborations and produced results that are already in use at the university level across many countries, and are likely to develop further. The project will help to prepare graduates for the lighting industry and related employers in Europe and around the world, thus contributing to more research-driven, knowledge-based, competitive economies.

We have estimated that approximately 10,000 people were reached directly or indirectly by the L4H project activities. These include: (1) those involved in developing the course content - either through the contribution of their expertise or review; this group encompasses students and graduates of partner universities, lighting professionals, academics, architects, and lighting manufacturers; (2) External experts and stakeholders; (3) Summer School participants and those who accessed the material contained in the first three modules tracked as through the eLearning platform and the users of the final L4H course measured since its launch at the end of August 2021 (4) External participants in the Multiplier Events; (5) those involved through social media, the project website, recruitment for summer schools, publications such as press releases and articles, television and radio broadcasts, and other forms of promotion and dissemination of the project.

With this project participants and consortium members have narrowed the gap between research and the practice/application of lighting, translating research from complex scientific jargon into tools for designers to use. Through the use of the L4H education eLearning platform, we hope the lighting community will acknowledge the importance of health and wellbeing as a major principle in engineering and the design of light and lighting systems.

References

[1] Zielinska-Dabkowska, KM and Xavia K 2019 Protect our right to light. *Nature*, **568,** 451–453.
[2] Münch,M. et al. The Role of Daylight for Humans: Gaps in Current Knowledge. 2020 *Clocks & Sleep.* **2**, 61-85.
[3] Changing perspectives on daylight: Science, technology, and culture. Science/The American Association for the Advancement of Science, Custom Publishing Office Washington, DC, 2017.
[4] Zielinska-Dabkowska, K.M. Make lighting healthier. 2018 *Nature*, **553**, 274–276.
[5] Brown T et al. Recommendations for Healthy Daytime, Evening, and Night-Time Indoor Light Exposure. Preprints 2020, 2020120037.
[6] Wright KP et al. Entrainment of the human circadian clock to the natural light-dark cycle. 2013 Curr Biol., 23, 16, 1554-1558.
[7] Vetter CP et al. A Review of Human Physiological Responses to Light: Implications for the Development of Integrative Lighting Solutions, 2022 *LEUKOS* **18**(3), 387-414
[8] Commission Internationale de l'Eclairage (CIE). CIE Position Statement on Non-Visual Effects of Light. Recommending proper light at the proper time, 2nd ed,. CIE Publication: Vienna,

Austria, 2019; Available online: https://bit.ly/2NysTq0 (accessed on 14 February 2022).
[9] Lucas JR et.al. Measuring and using light in the melanopsin age. 2013 *Trends in neurosciences* **37**(1), 1-9.
[10] Schlangen LJM (2020, Apr 2). Toolbox, user guide and video to support the use of the international standard CIE S 026:2018. CIE. http://cie.co.at/news/launch-cie-s-026-toolbox-and-user-guide

Acknowledgements:
This work was financially supported by EU Erasmus+ Strategic Partnership grant agreement number: 2018-1-UK01-KA203-048246.We thank all our colleagues Ellen K. Hansen, Paul Hampton, Bipin Rao, Ezekiel Chinyio, Dmitrii Ingi, Valeriia Lukinskaya, Natalia Bystriantseva, Iris Molendijk, Michael F. Rohde, George Palamas, Emmanouil Xylakis, Federico Favero, and Natasha George, as well as the reviewers from academia and the professional lighting practice for the form of the final course, who provided insight and expertise which greatly assisted the research.
We would like to acknowledge the contribution from the Jefferson Light Research Program, specifically Dr. George Brainard, Dr. John Hanifin, and Benjamin Warfield for their input with research and Melanopic EDI lighting metrics. Lastly, we thank our international Summer Schools' students as well as external speakers and partners for their participation and creativity which allowed us to develop and test this course.

Envisioning Community Engagement in Smart Lighting Design

Leni Schwendinger[1] **and Philip R Ross**[2]

[1] International Nighttime Design Initiative, United States, leni@nightseeing.net

[2] Studio Philip Ross, the Netherlands, mail@studiophiliross.nl

Abstract. Smart Lighting is widely viewed as a means to reduce energy use, streamline city services and lower maintenance costs of public lighting. These are valid perspectives and useful pursuits, but they leave a blind spot that needs urgent attention: how the application of Smart Lighting technologies could affect and benefit people in public space. Illumination as it is currently deployed is mostly utilitarian. Creative and innovative opportunities are often overlooked. It is critical for designers to reclaim their role in this technological sea-change to combine the operational benefits of Smart Lighting with benefits for people on the ground. Involving the community should in our view be part of this process. Our goal is to unearth techniques and design approaches so that Smart Lighting becomes a flexible tool for community-informed night experiences.

1. Background and questions

What if citizens' nighttime public-space use—whether night-shift workers and/or a mother traversing the high street after dark—was the highest consideration when designing a Smart Lighting system? What would the system look like if illuminated scenography echoed and enhanced night-use zones such as transit hubs, markets, parks, institutional, cultural, and hospitality districts? What techniques could help designers include a citizens' perspective in a Smart Lighting design process? Here, these questions are examined. And based on our experience in practice and examples from literature, we highlight techniques, concepts, and capabilities that are instrumental in bringing the citizens' perspective into the Smart Lighting design process.

1.1. Who is the "community"

This paper focuses on community engagement to ascertain who has a stake in the nighttime environment and who makes final decisions. Community involvement increases local support, reduces complaints, and provides a sense of accomplishment and ownership of the lighting project. In parallel, a sense of neighborhood belonging is developed.

A series of questions is proposed to define the project community's population.

Is the community the same as stakeholders? Are they composed of residents? Business owners? Elected officials? Local NGO organizations and associations? Who makes decisions, influences decisions and/or participates in decision-making? When it comes down to funding and implementing design outcomes, which are installed? *Who Represents "The Community"* elaborates further [1].
Throwing a wider net, *Urban Learning* [2], a European Union's Horizon project is centered on energy programs. They indicate that stakeholders are "administrative departments, utilities, enterprises, scientific institutions, and civil society representatives".

This paper makes a distinction between **stakeholders** – authorities; decision-makers with power, such as elected officials and business owners – and **community** – on the ground residents, shop employees, and local associations.

Smart Lighting design teams should implement participatory local research to understand the composition of both stakeholders and community in the project district – to make equitable choices to involve citizens and officials.

2. Techniques and approaches

The authors highlight several techniques and approaches that contribute to a community-informed Smart Lighting design process. These approaches are derived primarily from our practices and research, supplemented with the *Guidance Ethics Approach* [3]. New and speculative elements are posited to fit the demands and opportunities of Smart Lighting.

2.1. Uncover existing conditions: Shades of Night

An "existing conditions" survey by engineers, architects, and landscape architects, among others, is a standard phase that launches a public space project. Similarly, a Shades of Night (SoN) study [4] underpins lighting strategy. Through mixed methods, night fears and after-dark assets, desires for the future, and city data (transportation, transit, crime, open/shut commercial hours, etc.) are addressed and analyzed. See Figure 1.

The SoN methodology is conducted throughout the night, from dusk to dawn. Small teams document observations and conduct interviews that are time stamped.

Quantitative: SoN loosely follows the Gehl protocol [5] to measure the circulation of people, bikes, and motor vehicles as well as their behavior, such as lingering, walking with shopping or strollers, and directionality. Other examples are time-based measures like business's open and closing hours. Lighting measures include light levels from both public and privately provided light and identifying illuminated shop windows.

Qualitative: Intercept interviews are conducted throughout the night to ascertain people's destinations, for example, heading toward a bar, transit, or nightshift work, and their perceptions of the illuminated area.

The Product: A basis for lighting improvements including Smart Lighting is produced, which is a living schedule of lighting control in the area. "Living" emphasizes that the control schedule is instituted with evaluative review periods and changed if/when needed.

Figure 1. Shades of Night Analysis Diagram. For existing conditions studies, this diagram illustrates a broad framework with which to measure lighting and human activities from dusk to dawn

2.2 Co-design and local participation

There are many levels of community engagement and consultation--from presentations to side-by-side collaborative design.

A co-design imaginary for a social housing complex is proposed. This section describes a methodology for full community engagement using a site modeled from a courtyard. The purpose: to guide area design scenarios based on local needs and desires for the future. The engagement program would include capacity-building by teaching skills and hands-on activities. A Citizen Advisory Lighting Corps (CALC) is envisaged. The CALC working party would be composed of a variety of ages, interests, and other diverse demographics that mirror the subject area. This citizens' group would work alongside the Design Team. The local city council, municipal representatives, and a university research/planning division would be instrumental in this effort.

Here an added skill set is employed: community-based planning and design, which necessitates collaboration between design professionals, locals, and stakeholders. Specific literature is available for lighting designers to explore co-design and connectivity to layers of society [6]. Benefits include:

- Spatial character is identified: existing and aspirational conditions
- Capacity building: professionals and community members plan "scenarios" together
- Engagement of community in creative decision making
- Community co-evaluation of lighting success over time to underpin future programming adjustments

Figure 2 illustrates a before and after scenario sited on a public housing campus. Qualitative social research has already taken place through interviews, workshops, and discussions. It is revealed that after dark, women are afraid to use the diagonal path although it is a practical shortcut to the corner grocery store. The basketball court, although it has functional lighting, it is never switched on because of the nuisance: noise and congregation. Additionally, the entire plaza is empty of amenities such as quality lighting, greenery, comfortable seating, path paving, or cultural amenities. A Smart Lighting pilot is proposed, which, is co-designed by community and professionals to include new landscape and lighting.

For the duration of the pilot, a full-on evaluation of the site design, testing of timed basketball illumination, and other programmed lighting are agreed upon by CALC. Prospective areas for evaluation after the new design is implemented are listed below.

Path: Evaluation is undertaken of path usage during times of day and night after new paving and lighting are installed.

Basketball Court: Testing would include trial periods for illumination. For example, for one or two weekend nights lights would be turned on for a certain duration with assessments taking place at regular meetings. The lighting is programmed and can be changed.

Poet's Corner and Sitting area: Other timed experiments and evaluations are conducted for the Poet's Corner and Sitting Area, with varied times of brightness, color and on-off as part of the test process.

Community informed public design generally requires multiple disciplines. In the example of Figure 2, we argue for the incorporation of street and landscape design.

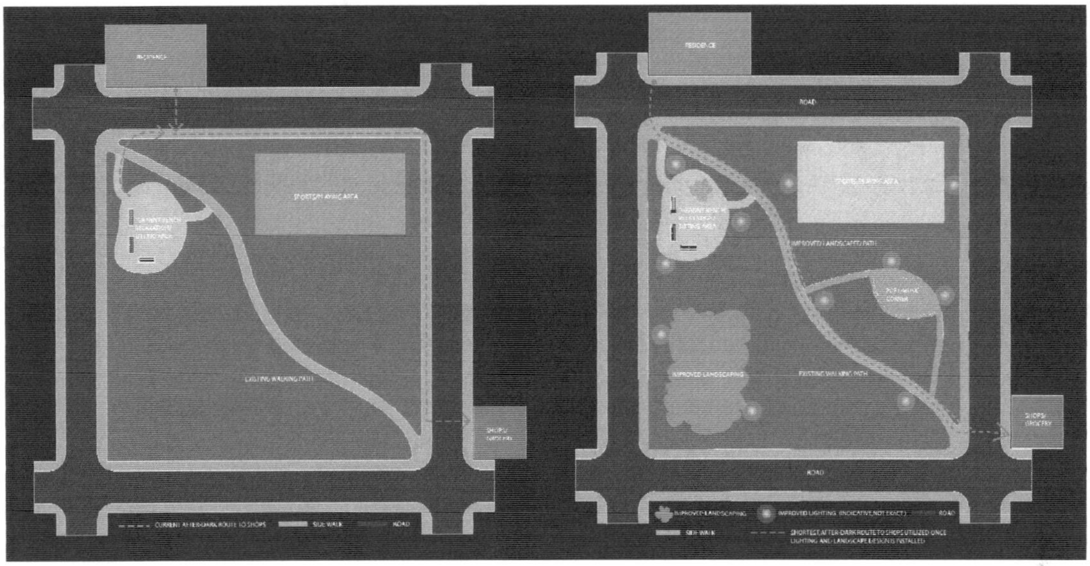

Figure 2. Pilot Site Diagram: Left / Existing, Right / Improved pilot. This example site is a typical US social housing courtyard. Through interdisciplinary design, e.g., lighting, landscape, grocery trips, cultural events and seating after dark are made possible for tenants. Diagrams by Fatima Terin

2.3 Continuous evaluation.

Involving the community does not end when a system is designed and realized. Once a system is installed, interactions take place on a day-to-day basis. People may notice the dimming effects of Smart Lighting, they may learn about the kinds of data that are gathered, and may adapt their behaviours to the enhanced environment. The designer must remain involved in this 'adoption' process alert to unanticipated responses, negative opinions, and aware of new opportunities that may surface. How to keep the community engaged in the process of designing, realizing and maintaining a Smart Lighting system?

A helpful technique for a continuous conversation with the community in relation to technology development is the Guidance Ethics approach [3]. Figure 3 shows the three stages of this iterative approach.

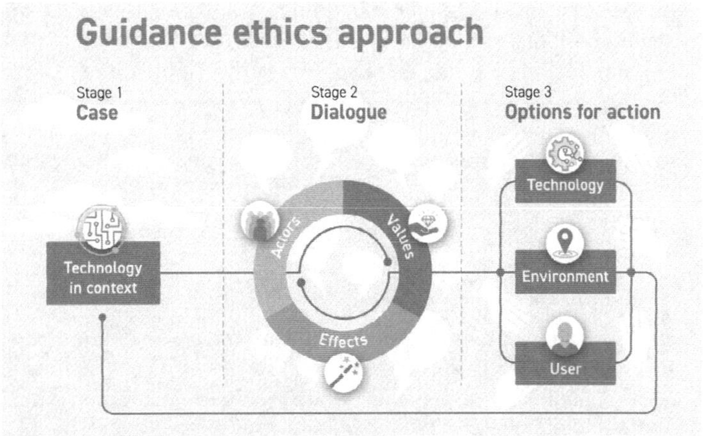

Figure 3. The three-step iterative process of the Guidance Ethics approach [3], reprinted with permission.

Stage 1 is about gaining or improving an understanding of the technology in its context, and what its use could mean for people. For example, in the case of a Smart adaptive dimming, this stage would encompass a clear description of what the system senses, how it responds, what the light does, etc.

Stage 2 is about dialogue with the relevant parties and investigating what the possible effects of the technology in its context are. For this, the relevant actors – community and stakeholders -- need to be involved, for example, the residents in a neighborhood, the people biking or walking or driving there, maintenance engineers, law enforcement, etc. Guided conversations uncover important values to underlie the design. This purpose of this discussion is to learn about positive, negative, direct and indirect effects for individual, social and society.

Stage 3 is about action, based on the results of Stages 1 and 2. In some cases the initial design will be adapted to meet principles agreed upon about technology and values. For example, if cameras are deemed too invasive, they could be replaced by infrared sensors. These are initial patches. After the in-depth explorations in Stages 1 and 2, the community realizes changing local traffic rules – like lowering the speed limit, or street design, a "road diet" or asphalt-art painting has higher value than a lighting system adaptation.

The Guidance Ethics approach is iterative. After installing the system, the cycle of Stages 1, 2 and 3 can be revisited over time. The implications and results of the system are clearer at that point, and the community conversations can be more concrete. Additionally, context or community changes may have occurred, requiring specific adjustments. This approach is not about end results, but rather improving a system step by step. Note that such an iterative Smart Lighting design process also requires that maintenance and design grow together.

2.4 Light Sketching

Another illustration of community participation is the Light Sketching for Placemaking method [7]. This project investigates how Smart Lighting could contribute to the quality of public space as experienced by its residents, drawing inspiration from the Placemaking movement [8]. Essentially, community dialogue about light is enabled *through light*. Smart Lighting offers a wide range of inventive lighting possibilities, such as color accents, dynamics and different light distributions. This allows people without training to imagine the potential of these possibilities for their after-dark environment and express what they would prefer.

To enable a creative and nuanced exchange about the fresh Smart Lighting possibilities with residents, the second author created the Light Sketching system. This system's custom luminaires include numerous individually-controlled RGBWW LED modules that enable a wide variety of light and color patterns. A tablet interface enables people to draw light and color patterns around the luminaires as easily as finger-painting. Each drawing action on the tablet interface translates immediately into cast light. In Light Sketching sessions, the existing public lighting is turned off, and the Light Sketching luminaires take over. Participants are asked to sketch light in their own environment to best support the activities they desire to undertake there.

Figure 4 shows results from a Light Sketching in a small neighborhood square in Eindhoven, Netherlands.

Figure 4. Four light sketches made by residents at their own neighbourhood square. The top left shows the normal street lighting situation. Photos by Bart van Overbeeke.

The Light Sketching for Placemaking project demonstrates a creative, active, visible tool set. Active learning of this sort leads to nuanced discussion. For example, qualities like glare, dynamics, adaptively, color temperature, light distribution were discussed in this example, whether or not the technical terms were known. Surprisingly dynamics and interactive scene selection were brought forward by residents – the basics of Smart Lighting. These results point toward possible benefits of Smart Lighting for local communities. Light Sketching is a way to engage people creatively in a design process.

2.5 Create Urban night scenarios – Learning from theatre

This discourse aims to enthuse the reader about advanced atmospheric, localized, nighttime illumination through LED and intelligent, or Smart, controls in urban settings.

A process is envisioned. The Design and community CALC Team would evolve a timeline for lights on, lights dimmed, lights off. Timing, color usage, and lighting levels would be installed with review periods to evaluate and change as needed.

In staged productions, the illumination designer works with the production team to intensify the scenic design with atmospheric effects. Lighting directs the viewers' gaze for the duration of changing scenes, actor's roles and story.

Additionally, lighting provides time references. This is technically achieved by dimming or in theatrical lingo, "fades" and "cross-fades". Atmospheric transformations such as sunrise, evening's end, inclement or fair-weather fade in and out defining the character of space, time, and narrative.

The play begins when lights fade to black. In comparison, the real-life city scenario night is the dark canvas into which we generate volumes of light to intensify the character of place.

Lived settings, unlike theatre, embrace real people traversing public space. The relatively recent ability to dim outdoor lights is key to translating scenography into city life. Hybridizing time-based technology with urban planning opens up a new approach. Thus, smart and dimmable lighting tools can visually externalize use areas such as town centers, institutional, residential, and entertainment districts.

Subscribing to the notion that urban lighting would be compelling and joyful if it takes a page from theatrical design has several parts:

1. Identifying and characterizing a use area with the full range of brightness, dark and color
2. Making use of temporality–dimming in real time– attracting visitors to linger in public space or signaling the end of active night, for example, through a reduction of lighting
3. Mirroring the collaborative process of theatre by co-design: by incorporating the input of stakeholders and citizen users

Fully utilizing the dimming capacity of Smart/intelligent lighting reinforces placemaking after dark.

Theatre design techniques have been applied to Figure 5, *Dreaming in Color*. This public art installation by the first author is located on the Seattle Center campus. It frames a passageway from the street leading to campus destinations such as a large fountain, sports arena, museums, and parks. The walkway threads between an opera house and a rehearsal hall. The art concept was developed to convey a sense of performance from the indoor stage into the public realm. Illumination is focused on nine vertical screens which intentionally spills onto the ground plane to cast light upon visitors. The result is an immersive, rhythmically changing, 3-dimensional color field. A dimming system facilitates the rhythmic fades and transitions.

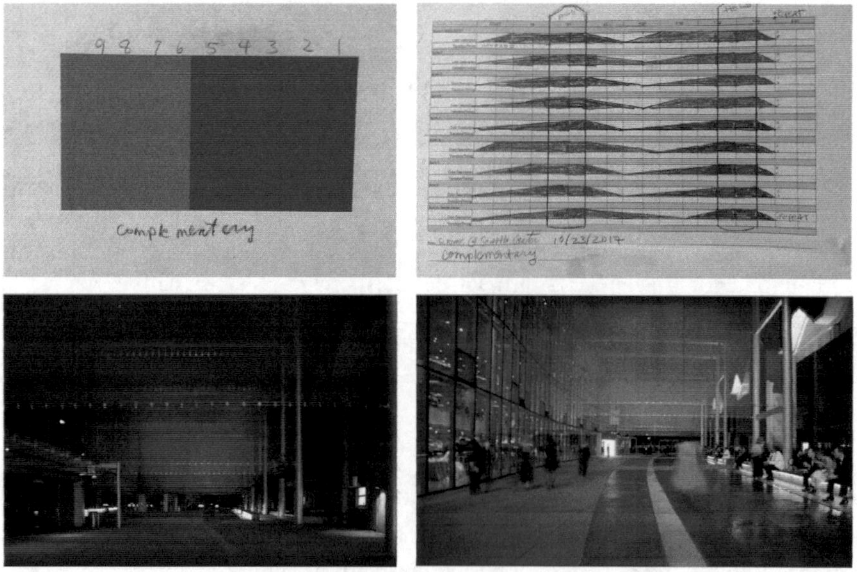

Figure 5. *Dreaming in Color*, a public art installation, Seattle, Washington, United States. **Top right and left:** The artist's working drawings to plan out colour palettes over time. On the left, the colour palette, on the right, borrowing from a musical score structure, the colours simultaneously play out over tempo with fades added. **Bottom right and left:** Moment photos of the light cast onto 9-metal mesh scrims and the ground plane. Photo left by AJ Epstein, Photo right by Gustafson Guthrie Nichol

3. Discussion

The authors posit that today's primary rationale for LED conversion and Smart Lighting — longevity, energy saving, lowered maintenance costs — are not enough. Creative atmospheres are possible in our city streets for citizens' safety and joy. Which atmospheres will they be and by what process will they be designed?

The underlying ambition is to integrate design into Smart. The authors suggest that the objective of Smart Lighting is not merely functional – to consider all-inclusive design. This design vision would meet a range of human needs – spaces for social relationships, creativity, discovery, and a sense of belonging. The richest design is a two-way learning process involving designer and neighborhood.

Several engagement processes from one-time workshops to ongoing participation by locals, designers, and municipalities have been described to underpin a new holistic Smart Lighting design.

The approaches and techniques described in this paper, are primarily drawn from the authors' practice and experience. However, Smart Lighting as a generally accepted *design form* is uncharted territory. Exciting, hard work is needed to develop the approaches further through practical application. The authors have provided points of departure and an invitation for lighting designers to use, evaluate and expand on them in their Smart Lighting practice.

To utilize Smart Lighting's full potential, the lighting designer's awareness and skill set must broaden – ranging from technical to organizational. Illumination designers should acquire basic competencies in IT and grasp the potential of temporal, dynamic layers that Smart Lighting allows. Additionally, for success in the engagement practices suggested here, community involvement and outreach training or consultation with experts may be necessary. These competencies will integrate and enhance specialist collaboration.

Finally, this paper outlines twin approaches: local participation, and time-based creative design. These are often challenging individually, and even more so when combined. Collaborators may include overlapping municipal agencies, which are often siloed, cooperation from electrical engineers who may feel uncomfortable with creative solutions, and other naysayers unable to see beyond traditional street lighting. However, light is malleable, and people are resilient and curious. By harnessing these forces, let's embrace safe and inspiring nights.

Keywords
Co-design, Cities, Citizen Participation, Community Engagement, Smart Lighting, Lighting Design

References

[1] Talking Headways: A Streetsblog podcast 2022 Who Represents "The Community" Part 1 and 2. https://streetsblog.libsyn.com/episode-374-who-represents-the-community-part-1 and https://streetsblog.libsyn.com/episode-375-who-represents-the-community-part-2, retrieved April 20, 2022
[2] Urban Learning: Stakeholders 2022 http://www.urbanlearning.eu/toolbox/stakeholders/, retrieved April 20, 2022
[3] Verbeek P P and Tijink D 2020 *Guidance Ethics Approach,* The Hague: ECP
[4] Schwendinger L, Besenecker U, Appleman K, Terin F and Planning4Places 2020 *Municipal Smart City Street Light Conversion & Evolving Technology Guidebook*
[5] Gehlpeople Public Life Data Protocol 2022 https://gehlpeople.com/tools/public-life-data-protocol-beta/, retrieved April 20, 2022
[6] Entwistle J and Slater D 2019 Making space for 'the social': connecting sociology and professional practices in urban lighting design *Br J Sociol* **70** 2020-41
[7] Ross P R and Kalinauskaite I 2020 Sketching Smart Light for Placemaking, *ILI Magazine* **13** 16-9
[8] Project for Public Spaces 2022 www.pps.org, retrieved April 15, 2022

IOP Conf. Series: Earth and Environmental Science 1099 (2022) 012043 doi:10.1088/1755-1315/1099/1/012043

Urban-Centric Lighting Task Group: Tactical Lighting as an innovation strategy

A Barberá[1], B Teixidó[2]

[1]Senior Lighting Designer, Anoche Lighting Design; Founder, Lupercales – Lighting Tactics, Barcelona, Spain
[2]Project Manager, Artec Studio; Founder, Berta Teixidó Studio, Barcelona, Spain

E-mail: abarbera@anoche.eu
E-mail: berta@bertateixido.com

Abstract. The present article describes a perspective on tactical urban lighting and its power as a source for disruption and innovation in the field of urban lighting. It is the result of an applied research that is currently being carried out by a task group formed by technicians, planners, managers and designers from different private companies and public institutions with responsibility and expertise in public lighting and urban planning. The group objective is to explore innovative methods and state of the art technology implementations as a practical application of the evidence-based knowledge in the field of artificial lighting for the sake of health and well-being. The definition of health is extended, as the concept of well-being includes the urban space as an ecosystem. There is also a fundamental source of awareness that comes from social sciences about how the built environment design promotes well-being, or else constitutes a hazard for health. After a definition phase for the methodologies and the technology that would need to be developed, the task group is focused on developing new products and planning a prototype installation on a real playground. A tactical, lightweight intervention in an urban green area in the Metropolitan Region of Barcelona is designed to implement, test, and communicate these innovative methods. Once the experiment is validated, a collaboration with public authorities will allow to introduce innovative perspectives in street lighting. Having received support from local's regional government innovation programmes, through a competitiveness strengthening initiative, the project is partly funded by the European Commission.

1. Introduction

Urban-centric lighting is the research topic of an Urban Lighting task group. The group was formed to carry out research about the influence of urban exterior lighting in people's health. The Task Group is promoted and managed by CICAT, a Spanish Cluster for innovation in the sector of lighting. The cluster members come mainly from the lighting manufacturing industry, and also include universities, and other related companies, including independent lighting designers.

This applied research is being carried out by technicians, planners, managers and designers from different private companies and public institutions, with responsibility and objective markets centered in public lighting and urban planning.

The main purpose of this project is to evaluate and discriminate objective parameters that will define the lighting design and its features, choosing among the available technological innovation those that are properly grounded on scientific evidence.

At the same time, there is research and innovation that come from disciplines such as ecology, environmental psychology, critical urban studies, and sociology, that are strictly necessary to nurture with sense and meaning design decisions in the urban realm. They have given the knowledge for establishing critical approaches such as gender perspective and environmental justice awareness, among others. Such approaches should be included in the toolset for the building of an innovative methodological framework. This why we need to identify and evaluate inclusive lighting strategies on urban space and avoid those perspectives that can drive to an unequal or unjust result that will exclude because of gender, race, age, or socio-economic status.

The task group is collaborating with Barcelona City Council's office for mobility, to implement experimental lighting projects in some of the temporary arrangements of public space that they are managing. This opportunity spaces are usually former car transit areas now reclaimed for pedestrian and civic use. They remain in a transient state while they receive the official and definitive transformation project. Joining forces with municipality will allow the group to experiment the framework with temporary lighting interventions that apply the results of the research on a busy, full-scale, urban scenario.

2. Notes for a holistic approach to a well-being definition in urban environment

To start with, it is detected that an extended definition of health needs to be agreed, in order to drive the evolution of the research and set proper foundations for the methodology. This definition of health would comprise, for human beings, such "a state of physical, mental and social well-being, and not only the absence of disease and infirmity", as defined by World Health Organization, no matter how vague or generic this definition will appear for our purposes [1]. The scope of our definition of such well-being includes other living species, and of course, the urban space as a whole ecosystem, both natural and social.

From the biological perspective, the framework must apply the knowledge and scientific evidence about the non-visual effects of lighting on mental and physical health and has the parameters that involve this interaction identified and quantized. This effects on health and behavior have been studied and demonstrated on humans, animals, insects, and plants, so there is a founded consciousness about the harmful effects of light at night on living organisms. The framework will have to be aimed to minimize such disruptions.

The perspective of social well-being is based on subjective perceptions and also on the human-space interaction dynamics. Affection, subjective perception of trust, sense of belonging, among others, are successful indicators for such values. We include under this title the perspective offered by the environmental justice movement [2], as an analytical toolset to address situations involving inequality in the public realm.

There is also an important ecological framework to consider. The need for more and better access to nature and green areas, also in cities, is influencing urban planning evolution, so this ecological perspective will become more necessary as the naturalization of urban spaces advances. This is critical in the case of children, where research tells us that regular time outdoors is essential for their healthy development [3]. A proper lighting approach should take special care not to disturb the interaction between humans and nature. The experience of darkness at night, and the vision of the night sky, is considered as a vital and beneficial access to nature. Also, allowing for an extended time outdoors in winter season using the appropriate outdoor lighting can substantially benefit the equitable access to nature for children in urban parks and green areas.

3. Research questions

The initial questions that motivated the research, have been categorized to better understand the complexity of the dynamics and tensions that affect public space, in order to organize the context.

a) Methodological approach: How innovation in urban lighting can benefit from temporary, experimental projects in urban space?
b) Human health and equity: Is it possible to identify a set of tools appliable to urban lighting design that is objective and represents an evolution in terms of health and environmental justice for our cities' habitants?
c) Ecosystemic approach: How can a set of guidelines guarantee the minimum biological interference with the non-human living organisms in an urban environment?
d) Connectivity: How much smart does a city lighting system need to be for such environmental and health purposes?
e) Darkness and access to nature: What will be the evolution in the intersectional spaces between urban and natural spaces?

4. Theoretical framework for strategical thinking: A living-centered map for urban lighting planning

The group has summarized the available, state-of-the-art, sources of relevant knowledge that will influence the design decisions regarding the urban intervention, and that relate to the research questions:

4.1. Methodological approach: Tactical Urbanism and urban lighting

Short-term street interventions are frequently interpreted as a performative action by users, and work as starting point for interaction between people and places [4][5]. They are recommended as a practice for making better long-term planning decisions in urban space, and immediately influences a more positive perception of it [6]. This is also true when public space interventions affect only lighting. They are proven to be relevant actions in terms of social engagement and communication strategy [7].

During the COVID-19 pandemic, a considerable amount of innovative temporary urban infrastructures helped to partially deal with the challenges that the situation brought up. Contemporary urban planning approaches that are focused on environmental health and sustainability, agree in the need of enhancing the quality of life in the city through eliminating car transit and gaining space for pedestrian use. There is a new demand for soft mobility and community-based urban services that addresses the convenience of distributing mixed-uses and proximity services, avoiding the concentration and monoculture of uses, with an arrangement that will result in the availability of all the services that contemporary city life needs within walkable or biking distance [8].

Figure 1. Barcelona's Superblock: Tactical Urbanism applied to a project of urban retrofit.

The superblock model that cities like Barcelona are applying impacts mobility, reducing car transit, and establishing reduced speed transit areas [9]. As a way of introducing these changes, the government is using the codes of temporary interventions (also known as Tactical Urbanism), to understand the viability of the new design and communicate the project to its neighbours and stakeholders [10]. The pilot program has come to an end, and now a major renovation is in progress, an evolution of the initial concept that will establish green mobility infrastructures (the "Green Axis") that go beyond the superblock scale.

The fact is that the innovations applied to urban space configuration has not found a response with a similar evolution in terms of urban lighting. Street lighting that focuses on the demands of traditional mobility has been dominant over the years and has created urban nightscapes that are both unhealthy and unappealing. This is the present situation in most of our cities: uniform, high light levels aimed to deliver visual needs of high-speed transit of vehicles. When mobility is challenged in such a way that speed is drastically reduced and traffic density is expelled from the city centres and residential neighbourhoods, such visual features are no longer needed. The reduction in speed comes together with a reduction in the requirements of the visual task. The time that we spend inside the space is longer. Lighting strategies deserve to be challenged, at least with the same strength. When more time is dedicated to experience places, visual needs are different: there is enough room for adaptation to lower light levels, to enjoy the balance between light and shade, even darkness, without risk of having an accident.

4.2. Human health and equity: Non-visual effects of artificial lighting
Human-Centric Lighting has framed design strategies and technology developments to address the non-visual effects of lighting in humans inside the built environment. The discoveries around the potential of lighting as a circadian stimulus [11], combined with a growing awareness about how the way a space is designed has effects on human health and wellbeing, have caused an evolution that has been focused mainly on the field of interior lighting, as the built interiors are the places where most of the artificial light is applied to humans [12].

4.3. Human health and equity: Environmental Justice framework
It is empirically demonstrated that significant levels of light pollution are associated to lower income neighbourhoods [13]. There is social inequity in exposure to hazardous light levels at night, and dark skies are considered as an amenity that is not within reach for everybody, while crime prevention or reduction is the main excuse for over lighting, without an empirical demonstration of its benefits for it, except for a topical (and arguable) sense of perceived safety when we are in a more lit environment. Here the confusion lays between our ability to read and understand the surrounding space (quality) and the amount of light (quantity). The reduction of such threat for health is also an urgent claim for social justice.

Tackling gender inequities in urban lighting design will include giving visibility to women in public space, and to traditionally gender-associated activities, such as caring and nurturing children and elder people [14]. On the other hand, granting physical and mental security can be sustained through an overall perception and understanding of the environment, that results in a global perception of safety.

4.4. Ecosystemic approach: Light pollution and circadian interference reduction
The adoption of highly efficient LED technology in public lighting is just a technology replacement, and not a solution to the extreme amount of light emission and light pollution of the night sky. In fact, due to the better energy efficiency and the better access to the technology, the adoption of LED has impacted negatively in terms of light pollution, as new installations appear, or greater luminous power is applied without significant increases in power demand. Without a change in methodology, a shift in the practice that is grounded on theory and experimentation, non-visual effects on light, circadian disruption and other environmental threats will not be deactivated. This will mean, at least, a better optic control, light levels limitation, light levels reduction in curfew times, and the limitation of the spectral content of emitted light.

4.5. Connectivity: Lighting control
Some of the innovation oriented to the connectivity of public lighting is devoted purposely on control and surveillance technology. This results on hybrid devices where the light side is only one feature

among others. The technology and connectivity that is supposedly planned to enhance management and efficiency, is also able to constitute a platform for sensitive data gathering. The urban green space (or common space) should be free of such technological threats. As Linnea Tillet explains [15], lighting systems do not need such functionalities to properly work, as 'street/pedestrian lighting in no way requires any of these services nor are they a necessary component of more efficient LED light sources.' The lighting professional 'needs to be clear with his/her client and, by extension, their communities' when considering this technology. There are strategies to smartly save energy and limit light pollution through simple, unmonitored control networks. So, the new green deal for fighting the climate change does not justify at all such highly technological deployments and the potential amount of data exchange.

4.6. Darkness and access to nature: The growing need for a darker green

Target number 7 of the Goal 11 of the UN Sustainable Development Goals 2030 states: 'provide universal access to safe, inclusive and accessible, green and public spaces, in particular for women and children, older persons and persons with disabilities'. The policy of developing and providing proximity green spaces will therefore mean the *naturalisation* of the urban space, or the *urbanisation* of the natural environment. In both cases, the green areas will behave as darkness repositories at night, for the sake of the vegetation and fauna, but also to enhance human well-being and reduce light pollution. This is a concept that has been developed by Roger Narboni at Agence Concepto, under the global name of 'dark infrastructure' [16].

The above considerations constitute the framework that is guiding the design decisions and the development of the technology that will be applied to the experimental intervention.

5. Lighting tactics: Benchmarking the technical requirements

Once the framework is set, a beta-test installation is agreed in a controlled environment. This beta-project will be installed in Parc del Molinet, a green public space just nearby the Besòs Riverside Park, that creates a natural frontier between the cities of Barcelona and Santa Coloma de Gramanet. This is an area of medium-low-income population. The restoration of the Besòs riverbanks has reclaimed a natural space that is now a successful asset for the neighbours.

Figure 2. Parc del Molinet is an urban infrastructure inaugurated in 1989. The lighting was renewed once since that date, but LED technology is still not applied.

The park was built and is currently owned and maintained by AMB (Area Metropolitana de Barcelona), and it was designed by Arriola and Fiol Arquitectes. The AMB is the public administration of the metropolitan area of Barcelona, a large urban conurbation made up with 36 municipalities. AMB is a supra-local institution that, among other attributions, manages, plans, and makes urban and mobility projects at a metropolitan level. As they have not made the transition to LED lighting on their infrastructure, they are very interested also in exploring the framework to apply it in the future, mostly for parks and green areas, a scope that is covered by our research. For the task group this applied exercise will serve as an opportunity for testing both the technology and the methodology, in a relatively controlled environment, at urban level but with a lower density and more flexible requirements.

5.1. Lighting objectives: visibility and space representation

Following the guidelines described previously, the installation consists of luminaires with tunnable spectrum, controlled by sensors in a closed and autonomous network that works unmonitored. The areas intervened are:

1. The background pergola: this longitudinal space limits the park and works as a background, with the vegetation growing as a vertical – and horizontal – frame. From a gender perspective, it is important to be able to see this space from far when approaching, and to be seen once you have reached the place.
2. The children playground: this space is surrounded by tables when accompanying people and such can rest and stay. Both the playground and the tables area were unlit. It was important to give the possibility of using these spaces in the winter afternoons, by installing the adequate lighting.

5.2. Spectral tuning and blending

The software used for Spectral tuning summarizes the objective parameters of a light source in relation with measurable qualities, such as color rendering, CCT, and circadian properties. It can work with combinations of different LED sources, that are modelled by their spectral power distribution. Equivalent Daylight Illuminance (MEDI) is the parameter that we have chosen to guide the spectral contents of the resultant mixed-spectrum. A prototype set of lighting fixtures developed on purpose for the project, have been set to use two pairs of combinations: a mix of PC-Amber and 3000K LED arrays, and 2200K and 3000K LED arrays are mounted in standard urban fixtures, controlled by wireless network devices.

Figure 3. The prototypes developed blend two different light sources: for both protecting living species and providing good colour rendition, depending on the use of the space.

5.3. Light levels

The main strategy follows the concept of darkness as an asset, as part of the infrastructure. For the sake of a healthier environment, lighting is applied only where, when, and why it's needed.

Instead of completely switching off lighting when no presence is detected, a background level of spectrally tuned lighting that minimizes Circadian effects on humans and fauna, is set. This basic situation is delivering and average illuminance of 10 photopic lux in PC-Amber light, which corresponds to less than 1 lux MEDI.

Figure 4. Commissioning of the Luminaires of the pergola in Parc del Molinet.

This strategy means that visitors would be able to perceive the structure of the space when they are approaching it, and will be accompanied by whiter, better rendering lighting when they decide to stay at the place. This way, biological hazard is minimized, and only present (reduced to its minimum expression) when is absolutely needed. This means, in terms of MEDI, that a maximum of 10 lux MEDI is recommended at dusk. Using a combination of PC-Amber and 3000K white light blended source, this will allow us to have up to 15 photopic lux with only 7 lux MEDI (Figure 5).

COLORIMETRIC

The Color Rendering Index is a quantitative measure of the ability of a light source to reveal the colors of various objects.

CCT (K)	2786
Duv	-0.0047
Rf	83
Rg	96
CRI	80
R9	3
COI	6.74
x	0.4454
y	0.3948

NON-VISUAL

The Color Rendering Index is a quantitative measure of the ability of a light source to reveal the colors of various objects.

M/P ratio	0.489
MDER	0.443
CS @15 lux	0.019
Cla @15 lux	14
EML @15 lux	7
MEDI @15 lux	7
440-490nm %	0.1

POWER

The Color Rendering Index is a quantitative measure of the ability of a light source to reveal the colors of various objects.

Efficiency (lm/W)	83
Electrical Power (W)	55.28
Luminous Flux (lm)	4576

Figure 5. Summary output for the Kumux Spectral design software, representing a spectral blend of the prototype's two light sources that minimizes circadian effects while keeping color rendering and a reasonably good efficiency.

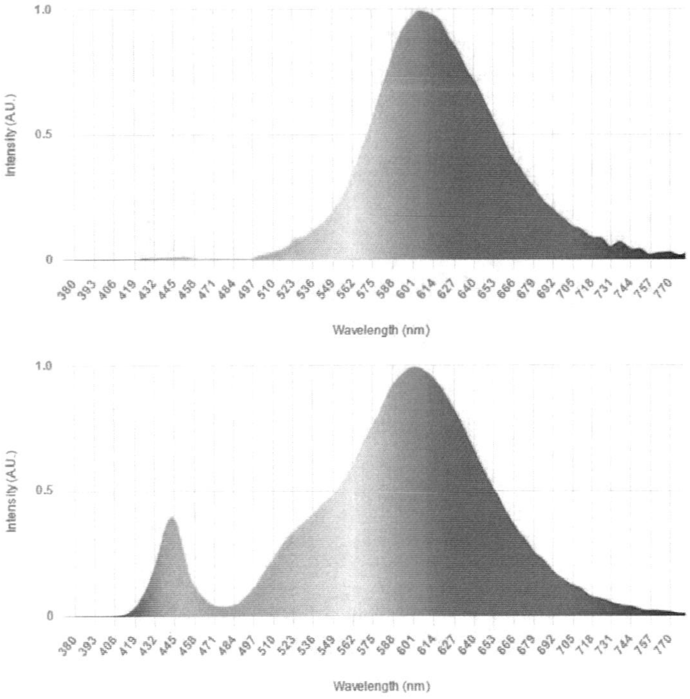

Figure 6. Spectral Power Distribution measured on-site for the PC-Amber and the White Blend at 3000K

5.4. Control strategy

The control network is fed with the signal that comes from presence detectors that will activate a change in the basic lighting when people come to the vicinity of the lighted area. When nobody stays around or passes by the installation, lighting is restricted in level and spectral composition, so it will minimize hazard for humans and other living species.

Covering open areas with IR sensors is not always a feasible objective. The solution tested in the children playground area has been using a single, short-angle sensor (30 degree), that the user has to activate, placing him or herself under a beam of light that represents the sensor beam. Instructions to activate the white light will be displayed on metal plates installed on the lighting poles. This way, white light will be triggered only on demand, allowing for a personal interaction with the space.

This strategy means that visitors would be able to perceive the structure of the space when they are approaching it, and will be accompanied by whiter, better rendering lighting when they decide to stay at the place. This way, biological hazard is minimized, and only present (reduced to its minimum expression) when is absolutely needed.

The main feature here is that the control network can work unmonitored and disconnected from any hub or repeater, so the data generated never is ‘spilled’ out of the place. These autonomous wireless networks will allow for complex programming, and they will provide the exact amount of automation for energy efficiency, circadian disruption avoidance and light pollution purposes.

The use of smooth transitions between lighting states will provide people with a gentle mood of the place, as light is controlled for creating a better experience of the space, and not expresses merely control or surveillance intentions. The ‘white light’ state is set to stay for 30 minutes before going back to the more environmentally friendly PC-Amber spectrum.

6. Provisory conclusions and further development

The research project is ongoing, and it is planned to be fully operative and running in this first placement from September 2022. The basis for the framework is established, and the results will be gathered and evaluated in the following months.

Figure 7. Lighting improves colour rendering and illuminance level on the horizontal plane when motion sensors are activated in the pergola area.

The group has consolidated its internal dynamics and has succeeded in the challenge of conceptualising, designing, planning and executing a full exterior lighting project workflow. The next phases will include analysing the outcomes and communicating the results. After this first prototype is fully operative and analysed, pros and cons established, the group is going to be ready for sharing the strategies with AMB technical department, and Barcelona City Council.

Some conceptual challenges will be addressed when interacting with technicians and planners, that are mainly focused on the use of a certain level of commissioned darkness (justified with the reduction of traffic speed, and the activation of higher levels with sensor strategies). The deployment of wireless local networks for lighting control, integrating the use of presence sensors, is a discussion topic that will have to be worked to persuade public authorities of its feasibility and reliability. Some local authorities are reluctant to the use of sensors and devices alike, although dimming is already established and accepted, it is not in a responsive manner or on-demand.

The positive results in the beta test will serve as a foundation for the conversation, as it will be the political intention to address the sustainability challenges that the present is bringing to the attention of technicians, designers, and planners.

References

[1] Boyce, P R 2010 The impact of light in buildings on human health *Indoor and built environment* **19** https://doi.org/10.1177/1420326X09358028

[2] Colsa Perez, A et al 2015 Evolution of the environmental justice movement: activism, formalization and differentiation *Environmental research letters* **10** https://doi.org/10.1088/1748-9326/10/10/105002

[3] Louv R 2010 Last child in the Woods *Atlantic Books*

[4] Killing Architects Urban Tactics – Temporary interventions + long term planning https://issuu.com/alisonkilling/docs/urbantactics_tempinterventions_longtermplanning/8

[5] Street Plans Collaborative Tactical Urbanism Vol. 1 and Vol. 2 https://issuu.com/streetplanscollaborative/docs/tactical_urbanism_vol.1 https://issuu.com/streetplanscollaborative/docs/tactical_urbanism_vol_2_final

[6] Roe J, Barnes L, Napoli N J, Thibodeaux J 2019 The Restorative Health Benefits of a Tactical Urban Intervention: An Urban Waterfront Study *Frontiers in Built Environment* **14** https://doi.org/10.3389/fbuil.2019.00071

[7] Barberá A, Bas I, Portaluri P 2014 *Lupercales Project: Lighting Up Urban Tactics* IALD Enlighten Europe Conference Berlin

[8] Moreno, C, Allam, Z, Chabaud, D, Gall, C, Pratlong, F 2021 Introducing the "15-Minute City": Sustainability, Resilience and Place Identity in Future Post-Pandemic Cities *Smart Cities* **4** 93-111 https://doi.org/10.3390/smartcities4010006

[9] Mueller N et al, 2019 Changing the urban design of cities for health: The superblock model *Environmental International* **134** https://doi.org/10.1016/j.envint.2019.105132

[10] Barcelona Superblocks Program https://ajuntament.barcelona.cat/ecologiaurbana/en/bodies-involved/citizen-participation/superblocks

[11] Brainard G C, Hanifin J P, Greeson, J M, Byrne B, Glickman G, Gerner E, Rollag M D 2001 Action spectrum for melatonin regulation in humans: evidence for a novel circadian photoreceptor Journal of Neuroscience **21** 6405–6412 https://doi.org/10.1523/JNEUROSCI.21-16-06405.2001

[12] Brown T M et al 2022 Recommendations for daytime, evening, and nighttime indoor light exposure to best support physiology, sleep, and wakefulness in healthy adults *PLoS Biology* **20** 3 https://doi.org/10.1371/journal.pbio.3001571

[13] Nadybal S M, Collins T W, Grineski S E 2020 Light pollution inequities in the continental United States: A distributive environmental justice analysis *Environmental research* **189** https://doi.org/10.1016/j.envres.2020.109959

[14] Col·lectiu Punt6 2019 Urbanismo feminista. Por una transformación de los espacios de vida *Virus Editorial*

[15] Tillett L 2022 Thinking Ecologically *Tillett Lighting Design Associates webpage* https://www.tillettlighting.com/practice

[16] Narboni R, 2015 *Recent evolutions in lighting master planning* Proceedings of 2015 PLDC 5th Global Lighting Design Convention

Dark Adaptation in Urban Environments: An Innovative Design Framework for Pedestrian Lighting

Chiara Ester Amoruso[1], **Mette Hvam Larsen**[2], **Mette Hvass** [3], **Georgios Triantafyllidis**[3], **Ellen Kathrine Hansen**[3]

[1]Licht Kunst Licht AG, 10997 Berlin, Germany,
[2]LEDVANCE, 2630 Høje-Tåstrup, Danmark,
[3]Lighting Design Lab, Aalborg University, 2450 Copenhagen, Denmark.

Corresponding author: chiara.e.amoruso@gmail.com

Abstract. Outdoor lighting has mainly been approached as an engineering subject throughout its history. The current urban lighting requirements are established upon factors which mainly refer to motorised traffic dynamics. As a consequence, excessive lighting levels are found in the city environments with a negative impact on the human as well as the environmental sphere. By synthesising scientific knowledge on human visual system and outdoor lighting theories, this research seeks to formulate strong arguments to point at the lack of human scale in the contemporary practice and to question the adequacy of the standard requirements for human vision in nighttime settings. The knowledge gained from scientific studies and ongoing research shows how the nighttime visual experience of pedestrians is still an untapped dimension, unveiling the lack of defined guidelines and methods to approach the design of urban lighting through the lens of this target. With a focus on soft mobility, the aim of the study is to address dark adaptation as a human factor in the design of urban lighting and to provide evidence on the human eye's ability to perform in dim conditions. Important findings from the literature review were used as requirements to formulate a methodology to approach the design of pedestrian lighting in urban environments. Strong conclusions could be drawn on the figure of dark adaptation and important lighting requirements were identified to facilitate this process in the condition of soft mobility. Even though further research and experimentation are required to implement dark adaptation in the design of urban lighting, this paper's goal is to open up new perspectives in the field, shaping the vision for the future of cities with reduced urban illumination.

1. Introduction

Outdoor lighting in urban environments has mainly been approached as an engineering subject throughout history, with its main purpose of regulating conflict between motorised and non-motorised traffic, to additionally providing society with a sense of safety and security and boosting the economy after dusk [1]. The set of requirements for urban lighting are regulated by the so defined "class" categories, which are established upon a set of parameters mainly referring to motorised traffic dynamics, regardless of the presence or absence of motorised vehicles in the city space of interest. In fact, the methods employed to design lighting for different targets than motorists, such as pedestrians (class-P) are generally a smaller-scale imitation of traffic route methods [2,3]. Many experts are questioning the correctness of defining lighting classes by means of parameters that are highly related to motorised traffic, which do not find any correlation within urban areas that are limited to low speed traffic such as bicycles and walkers [4].

Furthermore, a number of contemporary issues related to the human as well as the environmental sphere gravitate around the misuse of lighting in urban environments [5,6]. While standards and regulations have become embedded in the urban infrastructure allowing governance of spaces and urban life after dark, today they pose big challenges to the lighting design practice because the way they classify and rule different urban areas no longer meet the demands of contemporary planning [7]. In the context of this discussion, many authors and researchers have been investigating the impact of outdoor lighting on the nocturnal environments as well as the way the human visual system responds in nocturnal scenarios, in order to use this knowledge to improve urban lighting applications. Vision science explains how the human eye's sensitivity response to light changes according to brightness levels [8]. This defines three modes of vision respectively depending on high light levels conditions (daytime/photopic vision), dark conditions (nighttime/scotopic vision) and dark/intermediate light levels conditions (mesopic vision). Moreover, many studies have proved that the way humans perceive brightness at daytime completely differs from nighttime [8]. In addition, modern measuring system devices and the established units are based on the eye response curve of photopic vision (high light levels), meaning that the measured brightness does not correspond to the perceived brightness at nighttime [9]. Following this line, new research is urgently needed to understand that we are using excessive lighting that may not be needed, especially when it concerns pedestrians and soft mobility, where the visual needs are not as demanding as the vision of drivers. Besides high energy consumption, overly bright urban areas have caused issues of lighting pollution in many contemporary cities, which impact negatively on the biological nocturnal environment [5,10]. Therefore, alternative approaches are needed to frame and solve this environmental problem.

By collecting, reviewing and synthesising scientific knowledge on the topic of dark adaptation, this paper seeks to promote evidence on the human eye's ability to perform in dim conditions and to formulate a proposal for a novel approach to pedestrian lighting providing designers with a set of requirements to take into consideration to facilitate the process of dark adaptation.

1.1. Background work

Zooming in urban spaces, it is possible to experience in different ways the consequences deriving from the misuse of lighting such as excessive brightness levels, poor targeted lighting, glare and trespass which have all shown to influence people' stress level and to disrupt nocturnal ecosystem patterns [5,12]. Additionally, the aforementioned lighting class categorization provides requirements which generally fulfil specific areas' needs, regardless of the surroundings, and therefore this non-comprehensive approach results in great disparity of lighting levels throughout the urban environment. Pedestrian crossings or passages in junctions between trafficked roads and residential areas or urban parks are just a few examples of the spatial context this paper refers to.

Figure 1. Examples of dramatic variations of lighting levels in urban scenarios.

It can be truly challenging to classify all the possible cases from the urban environment in which incongruencies of the different lighting class requirements generate unwanted results, specifically as excessive brightness against naturally dark areas. In the attempt to identify possible spatial characteristics of such cases, it becomes important to additionally take into account the different features shaping the context of each case, such as the social and cultural traits amongst others. Brightness as a tool is differently understood and managed across socio-cultural contexts. The individual experience of brightness and darkness does not only rely on the physiological visual response, but expectations are influenced by the cultural background and prior experience [11].

However, as case studies have been suggesting, a domestic scale of urban lighting favours the presence of people in an urban space, which naturally promotes a collective sense of pride and safety [13]; this contradicts the governance formal control, which justify higher brightness because it allows a higher degree of surveillance that in turn increases the perceived safety amongst people [14]. In recent years lighting engineers have been able to measure the impact of brighter illumination on road safety and light pollution, yet there is still little knowledge on how to measure the social embodied and affective impacts of higher brightness [15]. In spite of how much the context may differ across the various spatial, social and cultural backgrounds, the issue of excessive brightness is present at different scales in the urban environment. Therefore the intent of this paper is to tackle the problem by proposing a comprehensive approach which addresses the eyes' potential of dark adaptation as a design objective.

As research has proved, the exposure to dramatic variations of brightness levels as found in urban areas can influence the inherent ability of the eye to adapt to dark environments [6]. This consideration emerges from collecting knowledge on the way lighting levels influence the state of adaptation of the human eye, which determines the vision mode [8]. Exposure to bright light levels causes blindness or temporary visual discomfort (technically known as discomfort or disability glare) when suddenly entering a dark area, compromising the process of adaptation. Additionally, with the technological advancement of light sources, urban lighting has tended to switch to smaller but brighter sources, implemented at different scales. This has increased the brightness exposure of pedestrians while walking through urban areas [16]. This important matter reconnects to the acknowledged lack of human centric approaches in the design of urban lighting.

1.2. Research question

The knowledge gained from scientific studies and ongoing research shows how the nighttime visual experience of pedestrian mobility is still an untapped dimension, unveiling the lack of defined guidelines and methods to approach the design of urban lighting through the lens of this target. With the goal of creating a new approach to this topic, a research question was formed which this study aims to answer: *How can we define a lighting design methodology to facilitate dark adaptation for pedestrians transiting from bright to dark areas?*

The following sections will unfold the intentions of this paper by firstly describing the methods and search strategy deployed for the literature review, which is the backbone of this study. Secondly, it will provide the reader with fundamental knowledge synthesised from the literature topics and it will identify essential parameters to finally implement in the design process. The resulting design framework will propose guidelines to develop lighting solutions in support of pedestrian dark adaptation. The paper will also discuss practical considerations in the implementation of the design framework to then finally open up suggestions for further development of this research.

2. **Methodology**

The methodology employed in this work should aim at identifying, selecting, processing, and analysing information about the topic of dark adaptation, with the interest of developing a design framework guidelines of lighting for nocturnal urban wanders. To this result, an exploratory research model [17] is followed, trying to explore and investigate dark adaptation. An explanatory study aims to explain how certain phenomena function. In other words, explanatory research is an attempt to connect different ideas to understand the nature of cause-and-effect relationships. It analyses problems and explains how variables interact with each other. In this context, a literature review [18] for the

phenomenon of dark adaptation is conducted. A literature review provides a critical analysis of the relevant contributions that have been made in the field of dark adaptation. During the literature review, the theoretical foundation for the research is identified, the novelty and relevance of the research are identified, and the focus, research questions, and design framework guidelines are clarified and refined. In order to answer the research question and frame the process of adapting to dark environments and the shift of brightness, three separate inquiries were conducted to maximise the chances of finding relevant studies: Dark Adaptation, urban artificial lighting qualities for pedestrians, and one for Scotopic and Photopic terminology. Literature was searched by publications indexed in Researchgate, Scopus and ISI Web of Science and more, using keywords such as dark adaptation, artificial lighting at night and additionally using the key words photopic and scotopic. The terms were searched in titles, abstracts, and keywords of the papers. The desired papers were refined to include peer-reviewed search articles and articles in press, short surveys, books chapters, and books. The searches were performed in two different periods of time: spring of 2020 and winter 2022. The goal was to identify studies that investigate urban lighting practice with particular attention to human factors. All studies were examined by title and abstract in order to determine if the studies satisfied the eligibility criteria. If they had a valid number of academic references with professionals or somewhat recognized authors and included the research keywords, the full texts were examined. The content of the studies was analysed to map out the function of the eye and the process of adapting to dark environments when travelling from a lit environment. The data was used to form a lighting design framework for artificial lighting at night in urban environments.
The aim with this literature review is to better understand the function of the human vision and its modes. Furthermore, to raise the discussion of facilitating more human-centred sustainable nocturnal lighting design in the future. Important findings from the review of existing knowledge were gathered to formulate a proposal for a novel approach to address dark adaptation in the design of urban lighting for pedestrians transiting from bright to dark areas.

3. Review on adaptation mechanism

In order to understand the factors involved in the process of adaptation, it is of course important to explore the human eye as a complex mechanism; eyesight is only one of the five senses we perceive within all environments. The human visual system depends upon a few essential components of the eye. Figure 2 illustrates the eye's structure with its photoreceptive components [19]. The photoreceptor cells are responsible for how humans perceive light and furthermore adapts to darkness. The photoreceptors placement within the retina structure at the back of the eye distinguishes the different visual field regions foveal and peripheral of which cones and rods are respectively responsible [8].

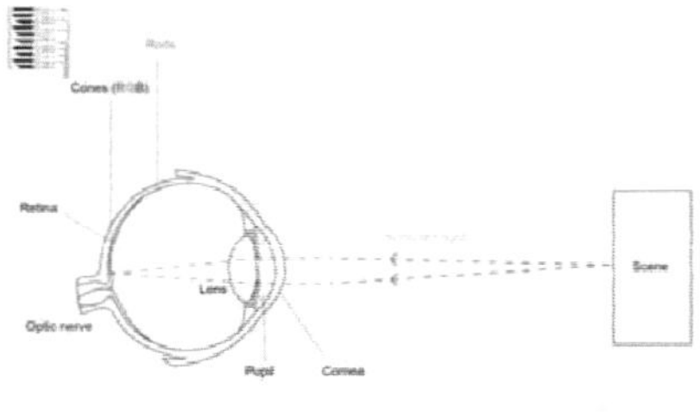

Figure 2. Human eye structure.

3.1. The vision modes

The state of the different conditions of the photoreceptors are referred to as photopic vision mode, scotopic vision mode and mesopic vision mode [20].

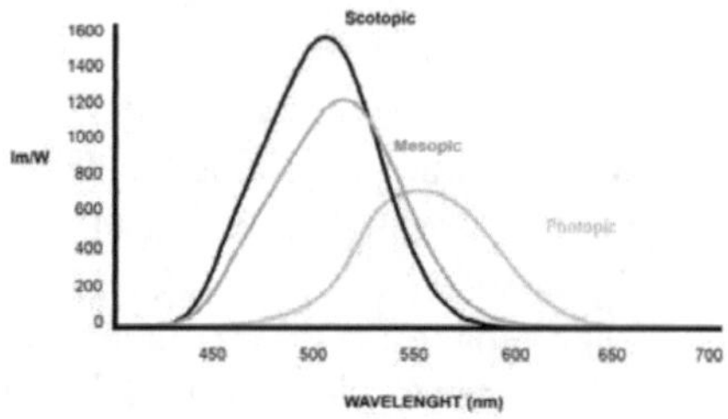

Figure 3. Vision modes sensitivity curves.

When the luminance levels range from 0.001 to 3 cd/m2, the transitional state from photopic to scotopic modes is referred to as mesopic vision. In these illuminance ranges, which are those usually found in urban nocturnal settings, both photoreceptors are engaged in the retinal response to luminance, enabling humans with peripheral and foveal perception [8]. It is important to stress at that point that due to a different presence in the number of the photoreceptors in the retina (92 million rods and 4.6 million cones), rods are still the most influential for the eye's sensitivity response in the mesopic mode [20]. This means that considering the rods' higher sensitivity to brightness, studies state that, provided of not being impaired, any person regardless of the age can detect trip hazards on the ground with only less than 1 lx horizontal photopic illuminance [3].

3.2. The dark adaptation process

Dark adaptation occurs when the retina photoreceptors sensitivity adapts to lower illuminances, and this process consists of four factors. Firstly, the intensity and duration of the pre-adaptation light. Secondly, size and orientation of the retina contribute to the process of dark adaptation. Thirdly, the spectral distribution of the pre-adaptation luminances and, lastly, the regeneration of rhodopsin [21]

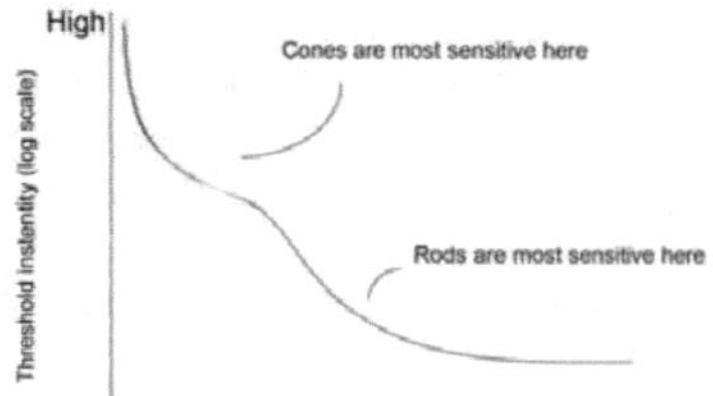

Figure 4. Graphical representation of dark adaptation process.

The degree of influence of the pre-adaptation exposure largely depends on the brightness level as well as directionality. Under most conditions light adaptation generally is an extremely rapid phenomenon, and humans adjust to the levels shift within seconds [21]. The photoreceptors engage in the process fairly quickly compared to rods, as rods need the photopigment rhodopsin to be produced and get activated. When transitioning to darkness after exposure of high or prolonged luminnances, a proportion of the visual pigment in the photoreceptors gets "bleached", therefore it can take several minutes to return to normal visual sensitivity. This very slow recovery in which rods are engaged, is called 'dark adaptation' [21]. Following this line, research shows that the level of pre-adaptation exposure determines the state of the adaptation. In urban environments, the exposure to artificial lighting can reach light levels up to 150 lux at nighttime, showing an increasing negative impact on the biological life of humans and animals [6]. Even though important information on the mechanism of dark adaptation is found in scientific studies, this knowledge fails to be addressed in the contemporary urban lighting practice, determining a new problem area which calls for further investigation and new approaches.

3.3. Lighting qualities for dark adaptation

Understanding the behaviour of the human eye in outdoor dim conditions is essential to determine the visual needs of active people at nighttime and how lighting can be used to aid the accomplishments of tasks within a specific context. Nowadays, lighting is used for a range of purposes in the outdoor environment. Lighting in cities is aimed at people and property 'safety and security', as well as emphasising the architectural and historical identity, to attract people and to create meaning throughout their spaces. Though, a big portion of this realm is concerned with road installations for motorised vehicles and other alternative means of navigating through the urban environment. This consideration leads to identifying different targets which the lighting in the urban spaces is designed for. According to the condition of experience of the urban space, whether it is static or dynamic, and whether it has different speeds in the condition of mobility, the design of urban lighting can be aimed at motorists or pedestrians and soft traffic users, as well as it could be targeted at different categories together. As specified in the paper introduction, the target focus for this research are pedestrians. Based on the understanding of this target visual needs in the condition of navigation, this section provides a set of parameters which call for attention when designing lighting for this target.

3.3.1. Topology. Before identifying the quantitative parameters of lighting to consider as tools for shaping the nighttime visual environment within any given urban space, it is important to consider the physical arrangement of the lighting sources. The physical placement of the lighting is defined by the distance between sources and mounting heights and it is highly determined by the lighting requirements of the area, usually defined by the class and the spatial geometry of the environment.

In outdoor settings, the selection of luminaires will be determined by the specific context of application, as an urban area may have different requirements in relation to whether it is a highly motorised trafficked area or a fully pedestrian zone and so on. As explored in the analysis on the human visual system with its different vision modes, an important piece of information is that the main photoreceptor engaged in the scotopic and mesopic vision is the rod [8]. Photoreceptors responsible for visual-forming effects *(rods and cones)* have different placement within the eye's retina [20]. Rods photoreceptors are placed in the peripheral part of the retina, improving peripheral vision in scotopic illuminations. Since all photoreceptors are equally engaged in the process of mesopic vision, the peripheral region becomes more sensitive due to its rods' high sensitivity to brightness [19]. Research also shows that photoreceptors cones are placed in the center of the retina, enabling in photopic settings sharp vision with high details discrimination in the foveal region [8]. Through the visual experience under mesopic vision, the foveal and peripheral views are shifting in which is being most dominant according to the illumination levels. Following these claims, it can be stated that because of the high sensitivity of rods cells, there should be no glare exposure, therefore creating visual comfort in dark environments. Extreme contrast between the visual field regions would bleach the photopigment rhodopsin and impairing the photoreceptors in adapting to lower levels.

3.3.2. Colour appearance and S/P ratio. Colour appearance, as a perceptual concept, is an essential quality of light which contributes to the way space and objects are perceived. According to the spectral composition of a lighting source, the human eye's photoreceptors respond differently to specific light wavelengths, influencing the visual perception process. The parameter which quantifies the spectrum of a lighting source, which will also influence its colour appearance, is the **Spectral Power Distribution (SPD)** [19].

The literature review on the human visual system shows how the photoreceptors rods and cones determine the state of the retina and the vision's mode, and they have different sensitivity responses to the wavelengths composing the light spectrum [8]. In mesopic ranges (0.001 cd/m2 — 3 cd/m2), where both cones and rods and engaged in the process of visual perception, the great contribution of the rods alters the sensitivity response to certain wavelengths, influencing the perception of the luminous environment [8]. Following the scientific knowledge on the rods' high sensitivity to short wavelengths, the lighting within a dark area can be reduced to very low brightness levels when working with the blue enriched spectrum, as the rods are most sensitive and most perceptive in these spectral conditions. An important parameter which indicates the effectiveness of light under mesopic condition is called S/P ratio, as in the ratio between the scotopic and the photopic lumens of a source. This value gives indication of the source efficiency in terms of lumen output according to both photopic and scotopic spectral and spatial response [9]. If the ratio value is 1, the light source performs just as good under scotopic conditions as photopic. The higher this number the better the eye performs under the light source.

3.3.3. Levels. In relation to lighting levels, the first observations to be made regard the dual dimension of the context - bright environment against dark environment. The project scope gravitates towards the idea that in dark environments visual comfort can be fully achieved providing previous adequate exposure to low light levels. Therefore, it is essential to understand the relations between the two areas and the degree by which the bright area impacts the experience of the dark environment. As proved by research, the exposure to different high light levels shows to influence the time required by rods to get engaged in the process of visual perception when entering dim environments [6].

Aiming at supporting dark adaptation and at smoothing the transition from the bright to the dark environment, a first step to approach the problem is to analyse the lighting configuration in the bright area and the way this affects the process of dark adaptation for soft traffic users. In relation to this, *illuminance levels* play a fundamental role in understanding this issue, but it is also important to specify that whenever referring to lighting levels, topology and spectrum become interconnected variables. The research work developed by Steve Fotios on pedestrian street lighting states that reduction on illuminance levels recommended by the Standards can be achieved by providing the use of lighting source with high S/P ratio, i.e. source with high blue content in the spectral power distribution. Additionally, lighting topology has a role in determining the distribution of luminance across the space. A regular layout with specific distancing and heights will provide a more uniform distribution with continuous visibility. Fotios experiments show that low illuminances ranging between 0.10 to 0.62 lx are adequate for hazard detection for pedestrians regardless of the age, whereas illuminance levels between 1 and 2 lx are sufficient for cyclists urban navigation, providing the paths are not extremely hazardous [3]. These values become important data to be considered in the design framework, which addresses the need for creating a new approach for slow traffic lighting in urban environments [2,4,16,22].

4. Design framework proposal

The review on the human visual system proved that specific human factors can be integrated in the definition of a framework useful for designing urban lighting to aid the process of dark adaptation for pedestrians transiting urban areas with variations from lit to dark environments. In this context, findings from the analysis provide three essential requirements for the lighting design framework, translating existing data from scientific knowledge into design tools. Considering the high sensitivity of the peripheral vision in dim settings, it is important to provide a physical arrangement of the lighting which minimises glare exposure within the extension of the field of view. Furthermore, due to the rod cells' high sensitivity to brightness and contrast, illuminance levels should be low at approximately 1 lux and uniformly distributed at ground level. Additionally, as research proves, rod cells are most sensitive to short wavelengths of the spectrum, meaning that the average illuminance can be minimised if using blue enriched SPD lighting with a high S/P ratio [4], which indicates that the lighting source is efficient both in photopic and scotopic response. Ultimately, three criteria of lighting topology, spectrum and level become the specific lighting requirements for the new design framework:

- **Lighting Topology:** The lighting is placed to provide a consistent visibility throughout the space and to avoid extreme contrast between zones;
- **Lighting Spectrum:** The sources S/P ratio is high, with short-wavelength enriched spectrum and high CRi;
- **Lighting Levels:** The illuminance level is very low and lighting is uniformly distributed to reduce contrast.

This section covers and explains the steps of proposal to a new framework to approach the design of lighting for pedestrians in the context of urban areas with variations from bright to dark environments. The proposed lighting design framework becomes a new toolkit for designers sharing the goal of improving the nocturnal experience of the contemporary urban environment for walkers, tackling the inadequacy of excessive brightness compromising the visual experience of the dark environment.

4.1. *Measure*

The primary task of the designer is to understand the degree by which exposure to high light levels in bright urban areas affects dark adaptation time when entering dark areas. The more the adaptation process is delayed the longer the users will need sufficient but lower levels to reduce the discomfort effect experienced in dark areas.

The first step of the framework suggests to map the critical points of lighting exposures and create a transitional sequence to collect quantitative data about the bright intervals the walkers are exposed to. The defined transitional sequence gathers important data within the observer's luminous experience along the path, providing information on where the light/darkness shift occurs. These bright intervals from the transitional sequence are defined as pre-adaptation exposures, meaning the bright exposure before undertaking the dark path, determining and potentially influencing the process of dark adaptation. The designer is suggested to gather the following quantitative measurements from the critical points: Lux values at ground level, luminance exposures and source spectral composition.

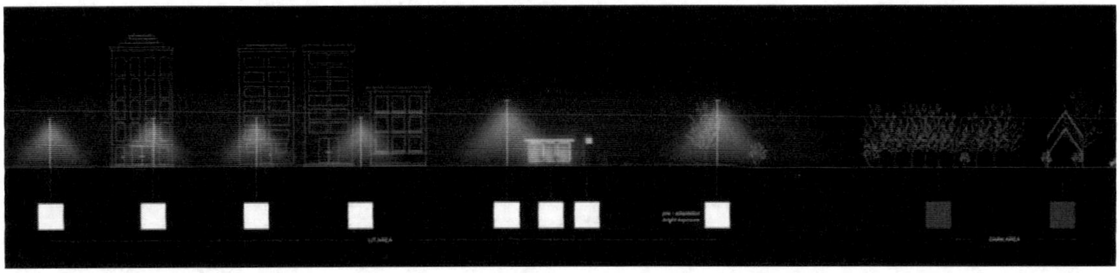

Figure 5. Illustration showing the process of transitional sequence definition.

4.2. *Test*

The second step suggests reconstructing the scenario of the pre-adaptation exposures in a simulated environment. The aim of the testing phase is to measure the way the exposure to brightness influences the time required to adapt to the dark environment. Subjects should be exposed to a light scene, simulating the real context where the sequence of pre-adaptation exposures is measured. The specific testing setting and procedure require focus on different aspects, such as the time of exposure and the position of the subject in relation to the lighting source. After the bright exposure, the second action would be to monitor the time required by the subject to perform a task-related response like detecting a hazard on the ground. Therefore, each subject will be standing in a simulated dim environment and exposed to bright pre-adaptation light as measured in the real scenario. The luminance will be distributed uniformly onto the retina by means of a source placed at the top of the subject. After the exposure to the sequence of luminance, the response time will be measured. Research argues that different levels of exposures affect the adaptation curve, resulting in different values of dark adaptation time.

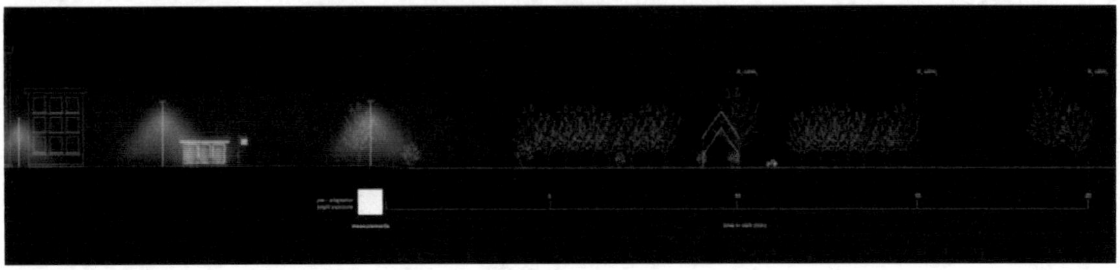

Figure 6. Illustrations showing dark adaptation delays in relation to different bright exposures.

4.3. *Translate*
The final step suggests a design approach which translates the measured adaptation time into spatial distance. According to the area usages and users, an estimation of average walking speeds can be converted into distance, defining the area of interest in need of a lighting scheme to aid pedestrians in the process of dark adaptation. The distance is a value measured in metres at which the designer will implement the newly developed lighting scheme to guide and support the users in the process of adaptation to the dark environment.

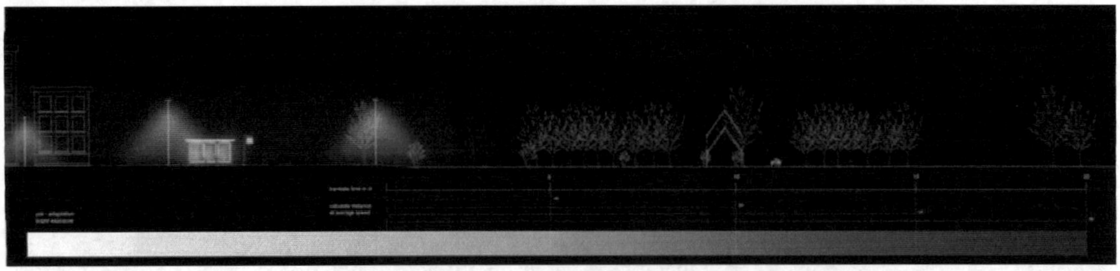

Figure 7. Illustration showing the process of time/distance translation

At this final stage of the process, the designer will reconnect to the knowledge gained in the review of the subject of dark adaptation, and the three success criteria mentioned in Section 4 will be applied as a design tool to improve the urban lighting transition.

5. Limitations and discussion
In spite of this research goal of contributing with practical design guidelines, the applicability of the design framework has its limitations, which the authors aimed at identifying in the following section. Firstly, it is pertinent to raise questions on whether dark areas are a rare scenario or not in urban environments, as excessive lighting is a general issue arising from the urban lighting practice. If this

scenario would be true, then the interest in implementing the design framework would be low.
While city centres shine from clusters of intense lighting, brightness spreads out across the urban environment, covering wide areas with layers of thousands of streetlights, advertisement's screens, architectural lighting and so on. In the contemporary image of urban nights, the highest contrast that can be identified in this scenario seems to be solely found at the edges between the urban environment and natural areas. At this point, it is logical to question whether it is possible to identify a clear taxonomy of areas where darkness is preserved by zooming in the urban environment. The process of mapping out urban cases of lighting transition is an essential operation in order to understand the possibility of application of the developed design framework and how this could be a valuable stepping stone for further research applied in the world of outdoor lighting
Furthermore, it is essential to consider that different cultural and social backgrounds may result in a different understanding of darkness and consequent use of brightness at night. Further investigation and the application of the design framework are required in real scenarios to establish what specific context could benefit or not from the proposed guidelines.

Regarding the framework steps, several limitations of the testing phase must be discussed. The chosen test scenario of the proposed framework refers to a task-related response experiment, in which the subjects need to give note when a hazard is detected on the ground in the dark after being exposed to the simulated pre-adaptation exposures. It can be questioned whether such a hybrid approach is a correct setting for a reliable study. A collection of test hypotheses should be considered for a more comprehensive approach. For instance, the task-related response could be distinguished between object or movement detection, depending on the hypothesis to be confirmed or rejected. Continuing on this line, a focused experimentation should be formulated around the subjects of face recognition and expression discrimination, which becomes an important task for pedestrians to discern movement and intentions of other space occupants [4]. Adding on the test discussion, the difference in the sensitivity of the cones and rods and their placement in the retina lead to question whether the topology of the pre-adaptation exposures have different impacts of the dark adaptation time. According to the testing set up suggested in the design framework, pre-adaptation luminances are exposed to the retina uniformly, allowing the authors to understand how the visual system responds to the exposure of specific luminance values in a simulated environment. In a real urban scenario, luminance values are usually not distributed uniformly across the visual field, and therefore it must be questioned whether different distributions onto the retina can somehow affect the stimuli responses. When testing the bright exposures in a controlled environment, it could potentially be considered if the pre-adaptation luminances should be defined into two types, peripheral and foveal pre-adaptation exposures, and monitor whether they would give different results or not. The luminance maps generated from the bright scenario are a useful tool for understanding luminance distribution in terms of foveal and peripheral exposures. Moreover, it would be pertinent to examine the medical eye´s condition of the target group, as different age, gender and medical condition targets might have different visual responses.
The flexibility of the design framework in applying testing results can also be discussed in relation to the length of the path of interest. In an ideal urban scenario with extended dark areas, a sufficiently long path would provide enough distance for the dark adaptation to be aided by the framework lighting requirements. In a more realistic urban scenario with small dark areas, it can be stated that the scope of the framework should be to minimise the soft traffic visual discomfort caused by the brighter environment when approaching the dark path where levels dramatically drop.

In essence, from the big picture of the city environment, the urban areas presenting high contrast from brightness to darkness where the framework could be successfully applied seem to be found more often at the edges with the natural areas, as mentioned before. Additional considerations should be stressed on whether the lighting requirements suggested to support nighttime vision and aid the process of dark adaptation can potentially impact wildlife patterns in urban environments. Spectral content of the applied lighting system as well as the integrated topology in the spatial context should be carefully evaluated in order to minimise negative impact on animals and plants from the natural dark area. Finally, reconnecting to the unwanted consequences arising from incomprehensive approaches, the design phase's main focus is the prevention or reduction of the negative impacts of excessive brightness, which is an ever growing contemporary issue due to excessive illumination in big cities. Many outdoor lighting researchers and practitioners have shown resistance for how the standardised approaches and their unclear design principles somehow promote a misuse of brightness in urban environments [23]. By drawing attention to existing scientific knowledge on the eye´s dark adaptation and its design potential in the context of pedestrian lighting, the framework proposes a way to resist the pre-definition of design needs for slow traffic areas exclusively by means of standards and regulations. Since a collective resistance to the challenges posed by these regulations has been widely shown, it is clear that a reformulation and reassessment of lighting planning principles is yet needed. For instance, the redefinition of lighting classes for slow speed areas according to different weighting factors than traffic volumes and motorised vehicle presence, such as specific physical characteristics, brightness exposure of neighbouring areas and adaptation process.

6. Conclusion and future work

Despite existing scientific studies on the human eye abilities, this knowledge still fails to be addressed in most contemporary urban lighting applications.

As a result of synthesising knowledge on human vision, three main findings are converted into three success criteria, which constitute the foundation of new lighting requirements for dark urban areas in need of illumination, facilitating the process of adaptation to dim settings for nocturnal urban wanders. Based on the information gained on the figure of dark adaptation, it could be concluded that the process demands time to be completed, consequently requiring sufficient distance to be available in most dark urban paths. This has led to questioning the applicability of the proposed design framework in urban context. However, these reflections frame the key point of this study, shaping a vision for the future encouraging coherent lighting master plan for modern cities, with no excessive and unnecessary use of lighting and with the consideration of its negative effect on the human and environmental spheres. The overarching vision of this research is to deploy the lighting design framework on a bigger urban scale, potentially triggering a domino effect of reduced lighting levels throughout the city environment. By addressing dark adaptation as a human factor for non motorised navigation, this research aims at providing evidence on the human eye's ability to perform under dim conditions, giving a new perspective on the outdoor lighting design practice and contributing to a human-scale-oriented urban development. By tackling important human factors in various design disciplines within the urban environment, many contemporary issues affecting the human as well the environmental sphere can be framed and potentially solved, strengthening and sustaining livability in future cities. In relation to this, opportunities for future experimentation arise and interesting directions could be undertaken to strengthen dark adaptation as an important factor in the design of urban lighting. A future development of this research would initially see the conduction of a protocoled testing phase in a real urban scenario to collect scientific evidence on the figure of dark adaptation and the parameters which affect it, opening channels for new studies.

Future directions to be considered for this new research area are the following:

- Combining design-oriented approach and medical examination of the eye in the study of dark adaptation. Due to the physical/chemical nature of dark adaptation within the eye anatomy, the use of medical equipment would aid the production of scientific data, supporting the task-based approach to study the subjects' response in a spatial dimension.
- Consider the possibilities offered by VR as a digital tool for advanced design-oriented testing, studying subjects' responses in a simulated condition of navigation through an enhanced reproduction of urban scenarios.

7. References

[1] Brandi, U and Geissmar- Brandi, C 2006, *Light for cities - Lighting Design for Urban Spaces, A Handbook*, Birkhauser - Publishers for Architecture, Basel, Boston, Berlin

[2] Fotios, S., Gibbons, R. 2018 *Road lighting research for drivers and pedestrians: The basis of luminance and illuminance recommendations*, Lighting Research & Technology, 50(1), 154-186.

[3] Fotios S., Boyce P., 2015 *Understanding lighting for pedestrians*, Discussion Document, The Institute of Lighting Professionals, Regent, House, Rugby

[4] Fotios, S. 2020 A review of design recommendations for P-class road lighting in European and CIE documents - part 1 : parameters for choosing a lighting class. Lighting Research & Technology, 52 (5). pp. 607-625. ISSN 1477-1535

[5] Longcore, T.; Rich, C. 2004 E*cological Light Pollution*, Front. Ecol. Environ, *2*, 191–198

[6] Grabusic et al, 2019 Review Light Pollution, Circadian Photoreception, and Melatonin in Vertebrates, Sustainability 2019, 11(22)

[7] Edensor T 2017 *From Light to Dark: Daylight, Illumination and Gloom*. Minneapolis, MN and London: University of Minnesota Press

[8] Boyce, P. 2003 *Human Factors in Lighting*, 2nd Edition. 2nd ed. London: CRC Press

[9] Lighting Industry association 2013 Lighting Industry Association Ltd, 2013. *SP Ratios And Mesopic Vision*. Lighting Industry Association Ltd

[10] International Dark-Sky Association, *2020 International Dark-Sky Association*. 2020. International Dark Sky Association | International Dark-Sky Association

[11] Lam, W. M. C. 1977 Perception and lighting-as formgivers for architecture McGraw-Hill Inc. NY

[12] Falchi F, Cinzano P, Elvidge CD, et al. 2011 *Limiting the impact of light pollution on human health, environment and stellar visibility*. DOI: 10.1016/j.jenvman.2011.06.029 Journal of Environmental Management 92(10): 2714–2722

[13] Newman O, 1996 *Creating Defensible Space*. Washington, DC: U.S. Department of Hous- ing and Urban Development Office of Policy Development and Research

[14] Jacobs J, 1993 *The Death and Life of Great American Cities*. New York: The Modern Library.

[15] Shaw R, 2014 *Street-lighting in England and Wales: New technologies and uncertainty in the assemblage of street-lighting infrastructure*. Environment and Planning A 46: 1–16

[16] Fotios S, Johansson M. *Appraising the intention of other people: Ecological validity and procedures for investigating effects of lighting for pedestrians*. Lighting Research and Technology 2019; 51(1): 111-130.

[17] Stebbins, R. A. 2001 *Exploratory research in the social sciences*, Vol. 48 Sage

[18] Snyder, H. 2019 Literature review as a research methodology: An overview and guidelines. Journal of business research, 104, 333-339

[19] Tregenza, P. and Loe, D. 2014 *The Design Of Lighting*. 2nd ed. New York: Routledge Taylor and Francis Group London and New York

[20] Zele, Cao, 2015 Zele, A. and Cao, D. 2015 *Vision under mesopic and scotopic illumination* Frontiers in Psychology

[21] Lamb, T. and Pugh, E., 2006. *Phototransduction, Dark Adaptation, and Rhodopsin Regeneration* The Proctor Lecture. Investigative Ophthalmology & Visual Science, 47(12), p. 5138

[22] Fotios, S., Uttely J. 2019 *The influence of road lighting on cyclists numbers and safety*, DOI: 10.25039/x46.2019.OP12, CIE x046:2019, CIE, Vienna

[23] Ebbensgaard, C. L. 2019 *Standardised difference: Challenging uniform lighting through standards and regulation,* https://doi.org/10.1177/0042098019866568, Queen Mary University of London, UK

IOP Conf. Series: Earth and Environmental Science 1099 (2022) 012045 doi:10.1088/1755-1315/1099/1/012045

Amber is not yet the white knight for sustainable outdoor light

L C Radetsky

DesignLights Consortium, 10 High Street Suite 10, Medford, MA 02155, USA

lradetsky@designlights.org

Abstract. Non-white light (NWL), and specifically yellow-orange or amber light sources, are often posed as a solution for environmentally sensitive areas and for astronomical sky glow. Lighting requirements published by various institutions ranging from public advocacy groups to governmental regulators demonstrate little agreement on NWL spectral thresholds and metrics. A recent market and performance analysis of NWL policies and products found that current international guidelines are not aligned, the terminology currently used by the lighting industry is inconsistent, and NWL LED products would likely not meet performance requirements for luminaire efficacy or colour rendition set by existing certification bodies.

1. Background

Human-produced light at night from electric light sources is often noted by the initialism ALAN, meaning anthropogenic light at night or artificial light at night. ALAN is increasing year over year [1,2], penetrates marine ecosystems [3], and is exacerbated by increasing market penetration of "blue-rich" LED sources producing consistent short wavelength radiation in the 400 – 500 nm (violet-blue) range (see for example, [4]). As of 2018, LED outdoor lighting had an estimated penetration rate of 46 - 50% [5], and it is estimated that the demand for per-capita light consumption has not been saturated [5,6]. LED lighting offers many benefits over incumbent technologies, including higher efficacies, and better optical and temporal control. However, the increased amount of short-wavelength radiation in phosphor-converted- (pc)-white LED spectral power distributions (SPDs) can be deleterious at night, especially if high CCT LED products are used and/or the application area is over lit.

Municipalities, energy efficiency programs and other outdoor lighting decision-makers can better support their energy reduction goals and abide by dark sky policies and ordinances, by aligning with third-party performance requirements for specific categories of outdoor LED fixtures, such as those set by the DesignLights Consortium (DLC) LUNA Version 1.0 Technical Requirements [7]. Specifiers seeking to fulfill the light pollution and trespass requirements of LEED and WELL building programs, or to follow application guidance in the joint International Dark Sky Association (IDA)-Illuminating Engineering Society (IES) Model Lighting Ordinance, may also be interested in these types of products.

However, many performance specifications, including LUNA, apply only to white-light LED outdoor products with CCTs at 3000K or below and do not include non-white light (NWL) LED luminaires deemed appropriate for environmentally sensitive wildlife areas, such as phosphor-converted (pc-) amber and direct emission (de-) amber products.

This paper provides an overview of a recent whitepaper from the DLC [8] that reports the state of the science and current recommendations for NWL light sources in outdoor lighting applications, and performance considerations relative to existing performance specifications. Next steps are suggested to address gaps in existing research, standards, and guidelines that would enable holistic performance specifications for NWL products in the future.

2. Environmental Impacts of White Light vs. NWL products

Outdoor LED lighting offers a range of benefits over incumbent technologies such as high-pressure sodium (HPS) and low-pressure sodium (LPS) fixtures, including higher efficacies, more choices of SPDs, better control of the optical distribution, more granular dimming control, and instantaneous on/off responses. The increased amount of short wavelength radiation in these LED SPDs, however, has been linked to several deleterious impacts, particularly if applications use high CCT products and/or areas are excessively lit. The known negative impacts from light at night, at meaningful doses and times, can include:

- Disruption in the circadian systems of animals and plants [9–12];
- Disorientation of wildlife such as sea turtles and migrating birds (the latter suffering significant mortality from collisions with illuminated buildings) [13–16];
- Harm to both diurnal and nocturnal insects [17,18]; and
- Increases in pathogenic risks [19,20].

Against this backdrop, NWL sources that eliminate short wavelength radiation have surfaced as a potential strategy to limit light pollution and other negative effects of anthropogenic/artificial light at night (ALAN). While HPS and LPS lamps have been common for outdoor lighting historically (Figure 1), two types of "amber" outdoor LED fixtures - phosphor-converted- amber (pc-Amber) and direct emission- amber (de-Amber) - are becoming readily available (Figure 2), and a third market category of "amber" LED chips near the Planckian locus with better colour rendition (pc-LEDs) is emerging (Figure 3). Pc-Amber and Pc-LED luminaires are marketed as replacements for HPS luminaires. De-Amber luminaires are marketed for use in coastal communities with sea turtle nesting beaches.

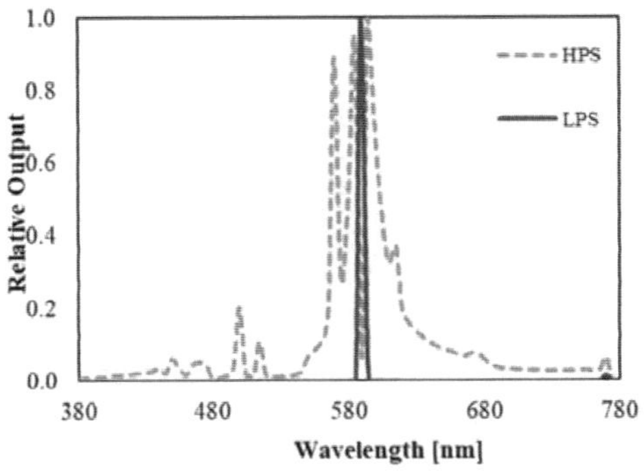

Figure 1. SPDs for representative HPS and LPS lamps

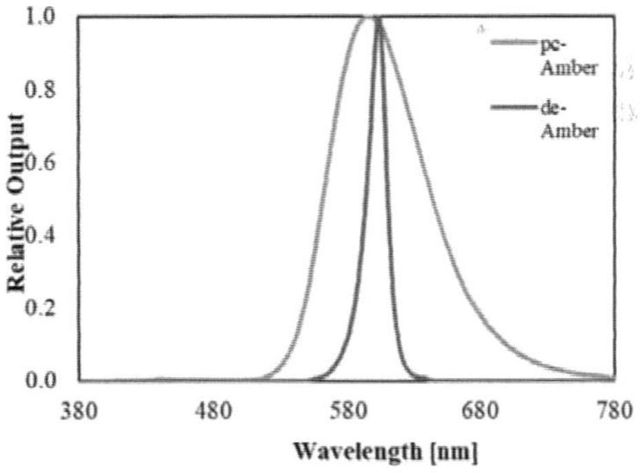

Figure 2. SPDs for representative pc-Amber and de-Amber LEDs

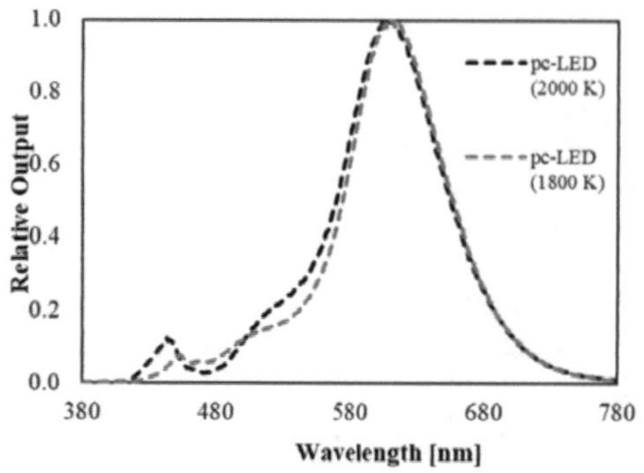

Figure 3. SPDs for two pc-LEDs near the Planckian locus at 1800K and 2000K

3. Lack of Standardization

As researchers and stakeholders seek to describe the capabilities and features of NWL products, however, the lack of standardization in metrics and measures is clearly apparent. A market review evaluated the lighting requirements published by various institutions ranging from public advocacy groups to governmental regulators found little agreement on NWL spectral thresholds and metrics (Table 1). Furthermore, there is wide variation and lack of standardization in LED luminaire and chip terminology. The term "amber" is a catch-all used for many LED luminaires that encompasses pc-Amber products, de-Amber products, and in some cases, pc-white products marketed as amber. Some luminaire manufacturers use CCT values (e.g., 1800K) to describe their products' chromaticity. Some others use the underlying chip technology as designations or application terms such as "Turtle Safe Lighting", "Turtle-Friendly Lighting", or "Coastal Wildlife Amber".

4. Method

An understanding of the performance of NWL products is needed to support efforts to reduce the astronomical and ecological impacts of ALAN while also considering energy efficiency and lighting quality characteristics. NWL light sources were evaluated by their luminaire efficacy, colour rendition, and sky glow potential. Lumen and chromaticity maintenance were also evaluated [8], but not reported here for the sake of brevity.

To understand commercially available outdoor luminaire efficacies, photometric data (e.g., .ies files) were downloaded in Q3 2021 from websites of commercial North American lighting manufacturers. Luminaires were also categorized by either 1) the CCT keyword in the .ies files, or 2) the manufacturers' catalog number nomenclature. Where the manufacturer reported CCTs of 1400K – 2000K, luminaires were categorized as pc-Amber.

Colour rendition measures were calculated for commercially available de-Amber, pc-Amber, and pc-LED chips, and an HPS and LPS lamp for reference. Calculated measures included measures from ANSI/IES TM-30-20 (R_f, R_g, $R_{f,h1}$, and $R_{cs,h1}$)[21], CIE 13.3-1995 (R_a and R_9)[22], and The Total Light Source Error Score (R_d)[23].

Because there is not yet a standard prescribing sky glow metrics, the DLC used the Relative Sky Glow (RSG) metric, which was computed using an unlocked version of the DOE Sky Glow Comparison Tool [24] to supplement [25]. RSG was calculated for the LED and HID products mentioned previously. RSG was also evaluated for 12,245 additional simulated SPDs that were created using random linear combinations of commercially available pc- and de- LED SPDs and binned according to their chromaticities.

IOP Conf. Series: Earth and Environmental Science 1099 (2022) 012045 doi:10.1088/1755-1315/1099/1/012045

Table 1. Spectral astronomical or ecological light pollution recommendations or requirements from various bodies

Organization or Regulating Body	Spectral Recommendation or Requirement
IDA International Dark Sky (IDA) Fixture Seal of Approval (FSA) – For Commercial Luminaires	Light sources shall have a maximum CCT of 3000K
IDA – For Residential Luminaires	Light sources shall have a maximum CCT of 3000K
IDA – "Innovation"	No more than 7% of visible emissions in 380-520nm
IDA – "Wildlife Tuned"	Sea turtle specific spectrum: 0% less than 565nm
FWC - Florida Fish and Wildlife Commission (FWC) Wildlife Lighting Certification Program	0% radiation below 560 nm Short wavelength light sources, PC Ambers, RGBs, dual lighting boards, and colour change options are not acceptable.
HCC - Hawaii County Code	Lamps with less than 2% radiation between 400 nm and 500 nm have no operation restrictions, other new lighting technologies are prohibited or must be switched off from 11 PM to sunrise (Class 1 lighting). 1931 CIE xy chromaticity coordinates outside of any of the traffic signal colour boxes as defined by ITE ST-052 500/AGS-PM/1105.
Smart Outdoor Lighting Alliance (SOLA) Community Friendly Lighting (CFL) Program	Less than 25% radiation between 430 nm and 530nm. This is typically achieved by light sources with CCTs ≤ 3000K.
Soft Lights Standard	Maximum CCT of 2200K for business and residential areas. CCT between 1000 K and 1800 K for sensitive and rural areas. Maximum CCT of 2700K for headlights.
Low Impact Lighting Standard	CCT ≤ 2200K AND less than 6% radiation below 500 nm. If average illuminance is below 5 lux: CCT ≤ 2700K AND less than 10% radiation below 500 nm.
Australian Guidelines for Outdoor Lighting	Percent radiation below 500 nm "as low as possible" Radiation above 680 nm "should also be avoided" LEDs with a nominal CCT ≥ 3000K should not be used in ecologically sensitive areas. Warm white LEDs with a nominal CCT ≤ 3000K or lower usually radiate a very low blue component and can be recommended from a health and ecological point of view.
City of Flagstaff, AZ	Preferred Source – LPS lamps and Narrow-Spectrum Amber LEDs
The UK Parliament's All-Party Parliamentary Group (AAPG) for Dark Skies	CCT ≤ 3000K. CCT ≤ 2400K in certain protected areas such as nature reserves and national parks.
French decree of 27 December 2018 relating to the prevention, reduction and limitation of light pollution	CCT ≤ 3000K. Additional requirements for protected areas such as nature reserves and parks: CCT ≤ 2700K for the "built environment" of towns and villages, ≤ 2400K otherwise.

5. Results

Results are shown in Table 2, and in Figures 4-6. While many of the organizations shown in Table 1 use a maximum CCT to limit the spectral element of astronomical or ecological light pollution, there are disadvantages to this. First, chromaticity boundaries for exterior LED products with CCTs under 2200K are not defined by any standards organization, so the use of low CCTs alone, does not result in consistent technology selection. Second, promulgating NWL CCTs without additional spectral requirements will not result in consistent specifications. For example, Table 2 shows that pc-Amber sources with similar CCTs have colour rendition and relative sky glow performance that do not vary monotonically with increasing CCT, while de-Amber sources with overlapping CCTs have much worse colour rendition but improved relative sky glow. Third, the analyses indicate that using CCT to limit the relative sky glow due to spectrum is not as effective as using an alternative metric, the S/P ratio, as CCT could result in large residual RSG values. Depending on the SPDs selected, the RSG for pc-white products could vary by up to three times for the same nominal CCT (Figure 4), but much less so for NWL products. The residual values for RSG as a function of S/P ratio are small (Figure 5).

5.1. Luminaire Efficacy performance

The DLC requirements were used as a threshold for luminaire efficacy, applying 105 lm/W for outdoor SSL products. An efficacy allowance is available for products with CCTs below 3000 K, resulting in a threshold efficacy of 100 lm/W for lower CCT products. Using commercially available photometric data for outdoor "amber" roadway/area LED luminaires, it was found that:

- Very few pc-Amber products could meet the 100 lm/W threshold efficacy requirements.
- No de-Amber products could meet the 100 lm/W threshold efficacy requirements.

5.2. Spectral impacts on colour rendition and relative sky glow reduction

With regards to colour rendition and reduction of relative sky glow, it was found that:

- LPS and de-Amber sources, as well as evaluated pc-Amber sources, all produced lower RSG than HPS.
- LPS and de-Amber sources had the worst colour rendition.
- Pc-LED products had better colour rendition than other NWL sources but increased RSG.
- There are spectral trade-offs for each type of NWL product evaluated, and specifiers will have to find a balance between reducing RSG and having good colour rendition. No NWL light source outperformed all the others in every aspect.

Table 2. CCT, colour rendition, RSG and S/P Ratio values of some representative NWL LED chips and HID lamps.

Metric	pc-Amber				de-Amber			pc-LED		HID	
	Mfr. A	Mfr. B	Mfr. C	Mfr. D	Mfr. A	Mfr. B	Mfr. D	Mfr. C	Mfr. C	HPS	LPS
CCT	1775	1873	1735	1724	1314	1619	1728	1998	1828	2041	1721
R_f	42	48	64	50	2	1	1	76	78	42	0
R_g	40	48	66	54	0	0	0	95	91	62	0
$R_{cs,h1}$	-31%	-29%	-22%	-32%	-76%	-82%	-82%	-16%	-17%	-47%	-99%
$R_{f,h1}$	35	40	54	36	0	0	0	69	67	13	0
R_a	37	41	58	36	-13	-28	-29	73	72	19	-46
R_9	-114	-108	-58	-115	-340	-417	-428	-27	-26	-221	-496
R_d	84	76	60	84	636	592	568	16	16	40	312
RSG	0.51	0.58	0.69	0.50	0.18	0.25	0.27	1.25	1.21	1.00	0.23
SP Ratio	0.40	0.45	0.50	0.39	0.17	0.23	0.25	0.79	0.76	0.63	0.22

For the 12,245 composite LED SPDs, it was found that:

- CCT is a poor predictor of RSG (Figure 4), whereas the Scotopic/Photopic (S/P) Ratio was the strongest predictor of RSG (Figure 5).
- RSG is not predicted by colour fidelity metrics such as CIE R_a (Figure 6) or TM-30 R_f.

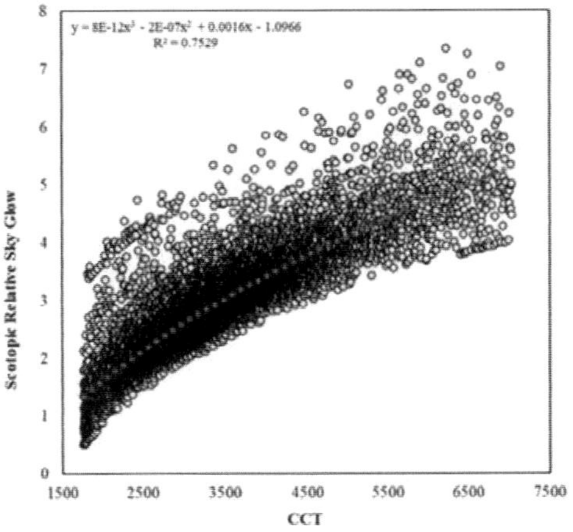

Figure 4. RSG as a function of CCT for 12,245 composite SPDs. The red dotted line is a polynomial regression line. CCT is not a good predictor of sky glow, as at each CCT a wide range of RSG values are possible, up to a factor of 3x different.

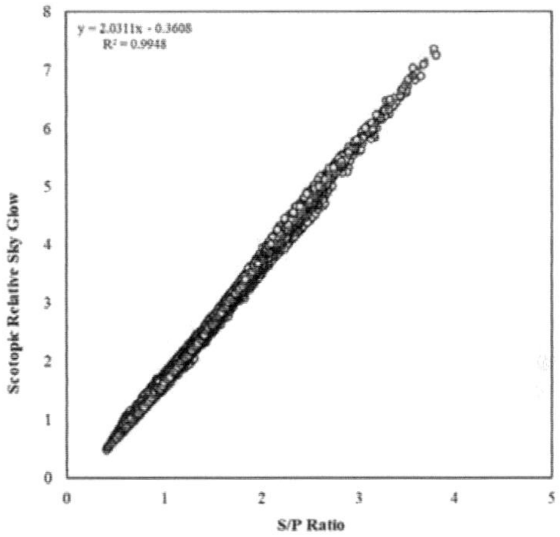

Figure 5. RSG as a function of S/P ratio for 12,245 composite SPDs. The red dotted line is a polynomial regression line. S/P ratio was a very strong predictor of RSG.

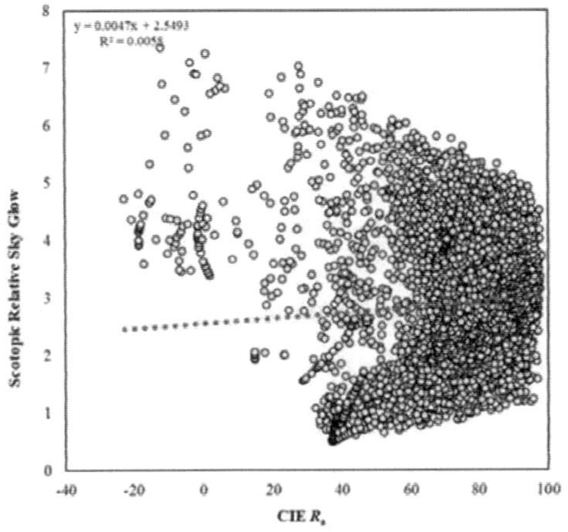

Figure 6. RSG as a function of the colour fidelity metric CIE R_a for 12,245 composite SPDs. The red dotted line is a polynomial trendline. RSG is not predicted by CIE R_a. At each R_a value a wide range of RSGs is possible, up to a factor of 6x different.

6. Call to action

The lighting industry needs an integrated approach to solving the unintended negative consequences of light pollution on the natural environment. Nominal designations must be developed and standardized by consensus, because chip and luminaire manufacturers are currently doing this independently, leading to inconsistent colour naming and nominal designations and unintentional confusion across the industry.

Luminaire manufacturers should also show the LED SPD graph on their specifications sheet to help both specifiers and regulators understand the product spectrum.

Lighting SDOs are encouraged to address this problem to help manufacturers communicate and differentiate their products, to help specifiers and other stakeholders understand a product's colour characteristics, to help regulators communicate what products are allowed or disallowed in their regulations, and to help researchers publish species-specific data in ways that can be applied practically, to deliver better lighting.

Stakeholders utilize certified product lists and/or standards to ensure that LED luminaires can be reliably and consistently measured and evaluated worldwide. Existing lighting standards must be updated to include NWL sources so that stakeholders can evaluate these products using a consistent framework. To appropriately address these types of luminaires in the future, necessary developments include:

- Standardized chromaticity boundaries for NWL products, including "amber," "red-orange," "red,";
- Standardized terminology and naming conventions;
- Standardized nomenclature that encompasses the totality of optical radiation to which non-human taxa are sensitive;
- Guidance on colour rendition thresholds for mesopic applications including NWL sources;
- Standardized reporting requirements for light source spectral power distribution; and
- Standardized calculation procedures for computing astronomical sky glow and other negative impacts of ALAN.

As efforts get underway to develop uniform standards for NWL LED products aimed at mitigating the negative impacts of ALAN, stakeholders and SDOs across the industry should get involved to ensure that high quality outdoor lighting products achieve energy optimization, and minimize light pollution, provide appropriate visibility for people and limit negative impacts to the environment.

Acknowledgments
The DLC's efficiency program members of the DLC provided funding for this project. Tony Esposito with Lighting Research Solutions worked as a consultant to the DLC on this project, and produced the figures and tables shown. Adrian Martin contributed to the luminaire efficacy analyses.

References

[1] Kyba C C M, Kuester T, Sánchez de Miguel A, Baugh K, Jechow A, Hölker F, Bennie J, Elvidge C D, Gaston K J and Guanter L 2017 Artificially lit surface of Earth at night increasing in radiance and extent *Sci. Adv.* **3** e1701528

[2] Sánchez de Miguel A, Bennie J, Rosenfeld E, Dzurjak S and Gaston K J 2021 First Estimation of Global Trends in Nocturnal Power Emissions Reveals Acceleration of Light Pollution *Remote Sens.* **13** 3311

[3] Smyth T J, Wright A E, McKee D, Tidau S, Tamir R, Dubinsky Z, Iluz D and Davies T W 2021 A global atlas of artificial light at night under the sea *Elem. Sci. Anthr.* **9** 00049

[4] Hung L-W, Anderson S J, Pipkin A and Fristrup K 2021 Changes in night sky brightness after a countywide LED retrofit *J. Environ. Manage.* **292** 112776

[5] Elliott C and Lee K 2020 *Adoption of Light-Emitting Diodes in Common Lighting Applications* (US Department of Energy Office of Energy Efficiency and Renewable Energy)

[6] Tsao J Y and Waide P 2013 The World's Appetite for Light: Empirical Data and Trends Spanning Three Centuries and Six Continents *LEUKOS* **6** 259–81

[7] DesignLights Consortium 2021 *LUNA Technical Requirements Version 1.0* (Medford, MA)

[8] DesignLights Consortium 2022 *Non-White Light Sources for Nighttime Environments* (Medford, MA)

[9] Grubisic M, Haim A, Bhusal P, Dominoni D M, Gabriel K M A, Jechow A, Kupprat F, Lerner A, Marchant P, Riley W, Stebelova K, van Grunsven R H A, Zeman M, Zubidat A E and Hölker F 2019 Light Pollution, Circadian Photoreception, and Melatonin in Vertebrates *Sustainability* **11** 6400

[10] Aulsebrook A E, Connelly F, Johnsson R D, Jones T M, Mulder R A, Hall M L, Vyssotski A L and Lesku J A 2020 White and Amber Light at Night Disrupt Sleep Physiology in Birds *Curr. Biol.* **30** 3657-3663.e5

[11] Falcón J, Torriglia A, Attia D, Viénot F, Gronfier C, Behar-Cohen F, Martinsons C and Hicks D 2020 Exposure to Artificial Light at Night and the Consequences for Flora, Fauna, and Ecosystems *Front. Neurosci.* **14**

[12] Zheng Q, Teo H C and Koh L P 2021 Artificial Light at Night Advances Spring Phenology in the United States *Remote Sens.* **13** 399

[13] Witherington, B.E. 1997 The problem of photopollution for sea turtles and other nocturnal animals *Behavioral Approaches to Conservation in the Wild.* (Cambridge University Press.)

[14] Salmon M 2003 Artificial night lighting and sea turtles *Biologist* **50** 163–8

[15] Van Doren B M, Horton K G, Dokter A M, Klinck H, Elbin S B and Farnsworth A 2017 High-intensity urban light installation dramatically alters nocturnal bird migration *Proc. Natl. Acad. Sci.* **114** 11175–80

[16] Van Doren B M, Willard D E, Hennen M, Horton K G, Stuber E F, Sheldon D, Sivakumar A H, Wang J, Farnsworth A and Winger B M 2021 Drivers of fatal bird collisions in an urban center *Proc. Natl. Acad. Sci.* **118**

[17] Giavi S, Fontaine C and Knop E 2021 Impact of artificial light at night on diurnal plant-pollinator interactions *Nat. Commun.* **12** 1690

[18] Owens A C S and Lewis S M 2018 The impact of artificial light at night on nocturnal insects: A review and synthesis *Ecol. Evol.* **8** 11337–58

[19] Kernbach M E, Cassone V M, Unnasch T R and Martin L B 2020 Broad-spectrum light pollution suppresses melatonin and increases West Nile virus–induced mortality in House Sparrows (Passer domesticus) *The Condor* **122** duaa018

[20] Kernbach M E, Martin L B, Unnasch T R, Hall R J, Jiang R H Y and Francis C D 2021 Light pollution affects West Nile virus exposure risk across Florida *Proc. R. Soc. B Biol. Sci.* **288** 20210253

[21] IES 2019 *ANSI/IES TM-30-20. TECHNICAL MEMORANDUM: IES METHOD FOR EVALUATING LIGHT SOURCE COLOR RENDITION* (New York, NY: Illuminating Engineering Society)

[22] CIE 1995 *CIE 13.3-1995. Method of measuring and specifying colour rendering properties of light sources* (Austria: Commission internationale de l'éclairage, CIE Central Bureau)

[23] Esposito T and Houser K 2019 A new measure of colour discrimination for LEDs and other light sources *Light. Res. Technol.* **51** 5–23

[24] Pacific Northwest National Laboratory (PNNL) Sky Glow Comparison Tool Version 1.0. PNNL-SA-138348

[25] Kinzey B R, Perrin T E, Miller N J, Kocifaj M, Aube M and Lamphar H A 2017 *An Investigation of LED Street Lighting's Impact on Sky Glow* (Pacific Northwest National Lab. (PNNL), Richland, WA (United States))

IOP Conf. Series: Earth and Environmental Science 1099 (2022) 012046 doi:10.1088/1755-1315/1099/1/012046

The Moon as a Light Source

New sustainable ways of lighting up cities at night

Camilla Monberg Rathsach[1] and Mette Hvass[1]

[1]Aalborg University, Copenhagen, Denmark. Department of Architecture, Design and Media Technology

Corresponding author: crathsach@gmail.com

Abstract. In our well-lit cities today, light pollution has become an increasing issue and serious threat of natural darkness and the view to a night sky full of stars. Darkness is not only a natural part of the day- and night-time cycle, but it is a basic need for all living creatures. And as light takes over in our night-time environment, this causes issues for both wildlife, nature, and human health. This project thematizes these issues by introducing new perspective on how to create a balanced relationship between human needs for lighting and natural needs for darkness. Through the case of Anholt island, the darkest place in Denmark, this is done by investigating human perception in low-level lighting and the Moon as a natural light source, that can be incorporated in a lighting design. The result is a dynamic lighting scheme, which incorporates the Moon's cycle by turning the electrical light 'on' and 'off' depending on the significance of moonlight. This both in a functional and aesthetical way, which not only protects the darkness but enhances the atmosphere of it, while at the same time supporting human needs for lighting in an urban context.

1. Introduction

According to the International Dark Sky Association, light pollution is characterized as an inappropriate or excessive use of electrical light hindering us to see and experience the night as dark [1]. Nearly all living organisms, including humans, have evolved under a natural rhythm of day and night, and as lighting becomes brighter and more extensive, this distinction becomes blurred [2]. This results in disruption of the human circadian rhythm, causing several health effects, such as cancer, depression, diabetes, and sleep disorders [3, 4]. It also affects the behavioral and population ecology of organisms in their natural habitats including foraging, reproduction, migration, and communication [5]. With the increasing issues of light pollution, it is important that we start searching for new ways of lighting up outdoor spaces at night. We need to create a more balanced relationship between the human needs for lighting and the natural needs for darkness. This is not only to minimize the impact on human health and the natural environment but to simply bring back the experience of the night. Before electrical outdoor lighting, complete darkness characterized the night-time environment together with an undisturbed view of the sky full of shining stars. While the sight of the Milky Way and the experience of moonlight on a full Moon night was a natural part of the night-time environment, these natural wonders are today obscured by light pollution. This means, that one must travel far out of urban areas and into the more untouched places on Earth to experience this fully.

The following article is an attempt to challenge the way we light up cities at night and to provide new perspectives on how we can balance the delicate relationship between light and darkness. Through a case on Anholt Island, the darkest place in Denmark, this is investigated with a focus on the Moon, nature's own light source, and how this can be used as a design tool when designing sustainable outdoor lighting. By investigating the darkest place in Denmark, the intention is to produce knowledge, which can be used when designing lighting at Anholt but also to create an awareness, which can be used in more densely populated urban areas. In larger cities the Moon is visible on the sky, but the human eye is not able to experience moonlight on urban surfaces because the brightness level from urban lighting is too high. The demands for visibility and safety are higher because of a higher density of people and traffic. But studies show that if brightness levels are balanced between a lit zone and the surrounding lit zone, brightness levels can be lowered. With a holistic approach to our urban lit zones, visibility can be increased by avoiding contrast. Studies show that less contrast leads to a more relaxed atmosphere and an increased feeling of safety because visibility is increased [6, 7]. Lower, more balanced lighting levels can possible also lead to natural dark night experience such as moonlight phase shifts and starry nights, in bigger urban environments than the city of Anholt.

2. Aim

The aim of the project is to investigate how moonlight can be incorporated in an outdoor architectural lighting design to protect the natural darkness, while at the same time supporting human needs for lighting in the town scape of Anholt. The aim is, that this can provide new ways of thinking and designing outdoor lighting, which can lead to sustainable lighting solutions in the future – not only in Anholt, but in cities around the world.

3. Background

The study of this article springs from a Lighting Design Master Thesis project at Aalborg University in Copenhagen. It began in summer 2020 with a field trip to Anholt Island where observations in the town area and interviews with the inhabitants were conducted. Anholt is a small Danish Island with around 130 inhabitants. It is located in the Kattegat Sea, right in between Denmark and Sweden. It is around 50 km. away from the mainland on each side, and it is therefore completely undisturbed by light pollution. The article reveals findings from a case study of Anholt Church, which is located in Anholt town in the centre of the island.

Walking through Anholt town at night, one is overwhelmed by how dark the night really can be. There is only a handful of streetlamps along the narrow streets and a few wall lamps on public buildings to signal the entrance. A bit of lighting is sensed from inside of people's private homes, but besides this, it is completely dark. Looking up, the stars are visible. Not just the ones forming the big Chariot, as we are lucky enough to experience in the rest of Denmark, but also the Milky Way. And the rest of the universe – that is how it seems at least. For a person growing up in urban areas, this simply feels unreal. It is not surprising, that this natural experience of darkness is one of most important elements for the inhabitants living here. They value the darkness deeply; it is part of their identity, and they fight to protect it. Going for a walk to look at the night sky was mentioned by the inhabitants as one of the most appreciated activities. Here they find 'calmness' and 'contemplation' as the informants describe it. Interviewing the inhabitants about their relationship with darkness, it was furthermore interesting to discover, that moonlight plays a significant role in this experience. One informant describes it: "(…) there are bright nights, semi-bright nights, and dark nights. If the Moon is there, either half or full, then you do not need any electrical lighting. Most of the nights are bright. They are illuminated by the Moon and contours are visible. It is completely dark when it is new Moon, and when it is cloudy. This natural change of light and darkness is something I enjoy". Even though the inhabitants enjoy the darkness, they expressed a certain unsatisfaction with the town area when it is dark. This seemed to be connected to the streetlighting, which is described as "obscuring the view to the night sky" and "destroying the impressions of natural lighting and darkness". Besides the experience of lighting being too excessive, the informants also expressed a lack of it. One informant explained that: "It is simply too dark to navigate just a little bit". Another informant also described that the lack of lighting in the town creates a feeling of the place being 'empty' and 'boring' in the darkness.

As a result of the fieldtrip to the island, Anholt church was chosen as a case study for the project based on its social- and visual function in the town. The church is one out of only a few public buildings and it is mentioned as an important local meeting point. Here, the community meets for holidays, anniversaries, concerts, and other communal events. Besides being a central meeting point for the community, the building is literally in the centre of the narrow town. It is small and simple in its

architectural style, but with its unique placement, it is a beautiful landmark visible from the near distance. Through an architectural lighting design, the church has the potential to not only provide visual identity during the day but also during the dark hours. Right now, the church has no electrical light at night, and is, like the rest of the town, 'hidden' in the darkness. By creating a lighting design for this specific place, it would be possible to support visual orientation and atmosphere in the near surroundings of the town, which could hereby lead to a more enjoyable night-time environment for the people living here.
Moonlight became the core of the project based on the local knowledge and experience; that it has exceptional lighting qualities in the darkness providing enough visibility for people in Anholt to see efficiently without any electrical lighting. This specific insight led to the study of how the lighting design at Anholt church can incorporate and interact with moonlight. This is to protect the experience of the night, while at the same time supporting the inhabitants needs for lighting in Anholt town.

4. Design process

The design process involved three steps in the order of 1) a literature review of guidelines on how to minimize light pollution and an extensive literature review of the Moon's phases, illumination levels and human visual perception of it, 2) visualization of lighting scenarios, and 3) an online survey. Each step will be elaborated in the following sections.

4.1. Literature review

4.1.1 Light Pollution Guidelines. As astronomer David L. Crawford explains, avoiding light pollution and the negative effects caused by it, doesn't mean that we need to turn off all electrical outdoor lights at night [8]. Instead, research shows that there are four overall principles to follow; *need, spectrum, intensity,* and *direction,* which together can minimize the negative impact of outdoor lighting [8, 9, 10, 3]. 1. Need: Humans are the only creatures that want light at night [10], and it is therefore recommended to only leave it on and install it, when/where there is a human need for it [9]. 2. Spectrum: Controlling the color spectrum is another essential recommendation. The general rule is to avoid any light that has emissions in the ultraviolet spectrum and adjacent short wavelengths [9]. This has to do with the fact, that short wavelengths of the spectrum disrupt the circadian rhythms in wildlife (and humans) and compared to longer wavelengths this attract many species, which then gets "trapped" in the light [9, 10, 11, 12]. The shorter, blue, wavelengths are also more likely to be scattered by the atmosphere and returned to Earth as sky glow [11]. For these reasons, it is recommended to avoid using light with wavelengths shorter 480 nm ("cold white" or "blue" light) and use amber and "warm white" light with a color temperature of 3000 K. or lower [12]. 3. Intensity: Regarding the intensity of light, it is essential that this is kept as low as possible to avoid disturbing species in the area of the lighting as well as reducing the reflection of light in the atmosphere [9]. The lighting has to correspond to the minimum amount of light required for the given task [8, 9], but in case of designing lighting for natural area, it is recommended to not solely rely on standards created by professional societies since these are generally developed for urban/suburban areas with little to no regards for wildlife. Instead, the efforts should be made to reduce the intensity of lights and still achieve the desired function [9]. 4. Direction: Managing the direction of light emissions is important to reduce impacts to natural lands and species, and to prevent light from shining directly towards the sky [9]. To do this, the light must be directed carefully, so that it will only shine on the target to be illuminated. Besides this, the light source must not be visible at large distances in any direction beyond the target area, and any illumination (including floodlight) must be directed downwards [12]. To reduce glare, it is recommended to install a 'full/sharp cut-off' luminaire, which do not shine any direct light above a horizontal plane that passes through the bottom of the fixture [10].

4.1.2 The Moon. The Moon is a complex phenomenon constantly changing its appearance throughout the month. To incorporate moonlight in a lighting design an extensive literature review of it was therefore conducted. It is important to note, that when we talk about moonlight, it is not the Moon itself that produces it. The light we experience is instead light from the Sun, which is reflected towards the Earth from the surface of the Moon. The Moon is not always visible on the night sky. It appears at different times, sometimes not at all, and it looks different in both size and shape. This has to do with the natural phases of the Moon. Half of the Moon's surface is always illuminated by the sun. However, the side of the Moon that faces the Earth, the 'near side', only becomes visible to us on Earth through eight different stages [13]. These stages change from one day to the next because of the Moon's orbital motion around the Earth, which takes 27,32 days. When the Moon has completed the eight-stage cycle, it starts all over again [14]. During the first phase, defined as the 'New Moon', the Moon occurs when the Sun and Moon are on the same side of the Earth, which makes it more or less invisible to us in the night sky [15]. As the Moon continues into to the next phases of 'waxing', the illumination from the Sun increases on the surface of the Moon and it slowly becomes brighter until it reaches its peak at 'full Moon'. From here, the illumination decreases as the Moon continues into the phase of 'waning' and finishes its cycle at 'new Moon' again [14].

It is estimated that the light of the full Moon is approximately half a million times less bright than of the sun [14, 16]. However, moonlight provided by full Moon is enough illumination for the dark-adapted human eye to see efficiently in darkness and hereby navigate properly in the night-time environment [17, 18, 8, 19]. The exact amount of light provided, is still uncertain and it is something that scientists continue discussing [20, 9]. In research by Christopher Kyba et al., it is announced, that many existing publications report incorrect values for full Moon illuminance varying between levels of 0.5 lux and all the way up to 2.2 lux [20]. According to Kyba et al., these levels of illumination are far too high and unrealistic, and based on their study of the Moon, they conclude that the full Moon, even at its highest elevation, will only produce between 0.05-0.2 lux on the ground [20]. At rare moments, when the Moon is directly at zenith, the full Moon illumination can be up to 0.32 lux. But as Kyba et al. note, this is only at optimal atmospheric conditions, for a few hours each year, in the tropics [20]. Research by Travis Longcore and Catherine Rich provide the same results of full Moon illumination at a maximum level of 0.3 lux, in rare scenarios, and they furthermore write that in most cases it is closer to 0.1 lux on the ground [21], this is also concluded in research by R. Dick [10]. Regarding the rest of the Moon's cycle, research show that the illumination is significantly less than at its peak at full Moon. Two days right before the full Moon, and two days right after, the illumination is only ½ of the maximum it reaches at its full stage [18]. In the later phase, at 'quarter' Moons, in both periods of 'waxing' and waning, is around 0.03 lux on the ground, and finally it is down to around 0.0013 – 0.008 lux at first and last stage of 'crescent' [21].

Not only does moonlight provide visual orientation at night, but it also creates a unique visual experience of the night-time environment. Despite the fact, that moonlight is a warmer orange-red correlated colour temperature (CCT) than sunlight, around 4100K CCT [22, 23], it is, by the dark-adapted eye, perceived as a cold silvery light with a blue tinge [20]. As Marco Ciocca and Jing Wang describes it: "(…) everything bathed in it [moonlight] looks the same yet different, cold and peaceful." [22]. This is especially significant around full Moon when the Moon is at its brightest. Another unique quality created by the soft moonlight is the significant dark and sharp shadows that it draws in the environment. John Naylor describes it compared to daylight shadows: "In daylight, scattered and reflected sunlight are bright enough to illuminate all but the deepest recesses. At night, only those nooks and crannies that are directly illuminated by moonlight are visible, hence the inky shadows and dramatic contrasts that characterizes a moonlit scene." [14]. The contrast is created due to the fact, that moonlight is much less bright than sunlight, which therefore only cast soft lighting in the night scene. Besides this, the Moon almost act as a point light source, which has no penumbra shadow but instead only a sharp umbra shadow [14].

Based on this literature review, a schema was created to structure and connect the different moonlight parameters (Fig. 1). The schema divides the Moon's cycle into three overall periods called: Bright Nights, Semi-Bright Nights, and Dark Nights. This division is inspired from the inhabitant's descriptions of the changing darkness levels on Anholt. Each period is defined in relation to the Moon's phase, the illumination level, overall definition of human visual perception, and lux level. Through this schema, it is possible to understand the significance of moonlight in each period. In the further design process, this was used to define the electrical lighting in relation to moonlight.

	Bright Nights	Semi-Bright Nights	Dark Nights
Moon Phase	Waxing- and Waning Gibbous.	First and last quarter.	Waxing- and Waning Crescent
Moon Illumination	95%-100% (Approx. 5 days a month)	75%-95% (Approx. 6 days a month)	0%-75% (Approx. 18 days a month)
Visual Perception	Good visual orientation and significant experience of moonlight colour and shadows.	Vague visual orientation and less significant experience of colours and shadows.	No/very little visual perception.
Lux Level	0.1	0.05	0.0001

Fig. 1: Schema illustrating Moon phase parameters.

4.2. Visualizations of lighting scenarios

In the 3D software Blender, a model of Anholt church and the surrounding area was created in a dark, moonlit scene. This was used as a tool to visualize the effects of electrical lighting on the church and the interaction that it would create with moonlight.

Three criteria defined the intention behind the lighting design: 1) create ambiance and visual orientation in the immediate surroundings of the church, 2) incorporate the Moon as a natural light source, and 3) ensure a light-darkness balance. With these criteria in mind, the lighting design was built up by two lighting elements, which together create a subtle illumination on the church and in the immediate surroundings. The first element is small spotlights in the window arches, which highlight the main architectural qualities of the church. The second element is a wall washer placed under the eaves to create a soft uniform lighting effect on the façade. The church is completely white, and this lighting will softly reflect from the façade, which provide a delicate ambient lighting to the immediate surroundings. It indirectly provides visibility but also a sense of space and direction when it is experienced from the near distance of the town area. To minimize the impact of light emission in the environment the illumination is directed downwards towards the ground. Besides these two elements, small lights are placed along the path to ensure visual orientation on the ground.

After having defined the core elements of the design, four different lighting scenarios were tested to identify the right balance of it in relation to both functional and aesthetical qualities of moonlight (Fig. 2).

	Scenario 1 2700 K. w. downlights	**Scenario 2** 2700 K. w. uplights	**Scenario 3** 3000 K. w. downlights	**Scenario 4** 3000 K. w. uplights
Bright Night	No Façade Light	No Façade Light	No Façade Light	No Façade Light
Semi-Bright Nights	With/without façade light	With/without façade light	With/without façade light	With/without façade light
Dark Nights	With/without façade light	With/without façade light	With/without façade light	With/without façade light

Figure 2: Figure showing the four tested lighting scenarios.

Each scenario explores the combination of CCT level and distribution of window light, and how each of these combinations is experienced, both functionally and aesthetically, with/without façade lighting at 'Semi-Bright' and 'Dark Nights'. Each lighting scenario is furthermore designed based on the above-mentioned light pollution guidelines to avoid any negative impact of lighting in the environment.

In terms of CCT levels, the goal was to determine aesthetical qualities of the design concerning how the color of the electrical lighting and moonlight plays together on the white surface of the church. The color of the light is defined as CCT levels, which, depending on the Kelvin degree, can appear in a cold blueish or warm orange-reddish color tone. As stated earlier, moonlight is perceived as a cold-silvery light with a bluish tinge, and the intention was therefore to test if the electrical lighting should have a warm orange-red color tone of 2700K CCT, which enhances the experience of moonlight by contrasting it. Or if it should be a colder bluish color of 3000K CCT, which enhances the moonlight by supplementing it the same color tone. The reason for not using a higher CCT level, which would match the moonlight color even more precisely, is that this contains a broader part of the blue light spectrum, which is not recommended in a dark sky context [12, 11]. Distribution of window lighting was tested to understand if *uplights* or *downlights* would provide a better distribution of light to highlight the big arched windows on the façade, while at the same time provide a 'cozy' atmosphere to be experienced from the outside. In enhancing the cold-silvery moonlight and balancing the right amount of light and darkness in the space, in terms of human experience, it was essential to furthermore test the effect of façade lighting during the different periods of darkness. This, to determine *if/when* this element should be a part of the design.

Figure 3: Visualization of lighting scenario 1 (left) and lighting scenario 3 (right) with façade lighting at Semi-Bright Nights.

4.3. Online survey

The different lighting scenarios were evaluated through an online survey, which was sent to the inhabitants of Anholt. Through an image comparison survey, the participants had to choose their preferred design solution and comment on their experience of it. A total amount of 24 people participated. The survey questions were designed to prompt answers about the informants' immediate experience of how the church appears in the darkness and how they perceive the balance with and interplay between the electrical light and natural moonlight. The survey was built up as four steps, where the informants had to 1) 'choose the preferred CCT level', 2) 'choose the preferred distribution of window lighting', 3) 'choose whether it is preferred to have façade light at Semi-bright Nights' and 4) 'choose whether it is preferred to have façade light at Dark Nights'. The questions were formulated as following: 1) 'Imagine that you are out on an evening walk, where the full Moon is shining on Anholt church. Which lighting scenario do you think makes the church stand out best in the darkness?' 2) 'Which lighting scenario do you prefer now?' 3) 'The full Moon has decreased; it creates less lighting on the church and in the environment. Which lighting scenario do you prefer?' 4) 'The full Moon is gone, and it is now completely dark. Which lighting scenario do you prefer?'. As a final step, the informants were asked to comment on their choices and experiences of the design. This step was only optional.

5. Design evaluation

The survey results showed that most participants, 77%, preferred the 2700K CCT lighting scenario compared to the 3000K CCT lighting scenario. In terms of distribution of window lighting, the results showed that most people preferred downlights. This in both lighting scenario 2700 K CCT, where 58% preferred it, and in lighting scenario 3000 K CCT, where 100% preferred it. The results regarding façade lighting were more complex. In the 2700 K CCT lighting scenario the results showed that depending on the type of window lighting people preferred a different solution at 'Semi-Bright Nights' and 'Dark Nights. At 'Semi-Bright Nights' with the downlights lighting scenario most participants, 64%, *did not prefer* the effect of façade lighting. But most participants *did prefer* it at 'Dark Nights', 55%. The opposite situation occurred with uplights at 'Semi-Bright Nights'. In this case most participants, 67%,

did prefer it, and at 'Dark Night' most participants *did not prefer* it, 67%. In terms of the 3000 K CCT lighting scenario the results showed that most participants, 60%, *did not prefer* façade lighting with downlights in neither scenario of 'Semi-Bright Nights' or 'Dark Nights'. Since the participants were not asked to explain their choices, it was not possible to draw deep conclusions about these results. But the results do indicate that downlights function better *without* any façade lighting at 'Semi-Bright Nights', while uplights seems to be more balanced when there *is* façade lighting. At 'Dark Nights', uplights on the other hand seems to be more balanced *without* façade lighting than downlights, which in this case seems to be more balanced *with* this effect. In terms of CCT levels, it is only the 2700 K CCT lighting scenario that seems to work best *with* façade lighting compared to the 3000 K CCT lighting scenario, which seems to be better *without* at both 'Semi-Bright Nights' and 'Dark Nights'.
Based on the survey results and evaluation of it, the final design was defined. The 2700K CCT lighting design got more positive response from the participants and was therefore chosen as the final solution in terms of CCT level. One participant described it as "cozy lighting." Another participant furthermore explained: "The warm lighting is more welcoming; the church appears warmer, and it does not disturb the darkness in the same way as the colder lighting." The intention behind the design was to create atmosphere and orientation. With positive comments, such as 'cozy', 'warm' and 'inviting', this was considered achieved with the 2700K CCT lighting scenario compared to the 3000K CCT lighting scenario, which was perceived as more 'sharp' and 'disturbing' in the darkness. In terms of distribution of window lighting and the effect of façade lighting, the final design solution was chosen to be one with downlights, no façade lighting at Semi-Bright Nights and façade lighting at Dark-Nights. Most people preferred this combination. And taken into consideration how delicate moonlight is, especially at 'Semi-Bright Nights', this solution is also the one, which will ensure the most significant experience of moonlight in the environment.
As a final note it is essential to point out, that the survey is not scientifically significant. Instead, the survey was created to engage the user in the design process and provide an overall indication of their perception of a balanced interplay between the components of electrical light and natural moonlight. The results should be viewed as an initial attempt of developing design principles on how to design with moonlight in an architectural context. Developing the lighting design would involve further, more in-depth, examinations of other parameters such as material, lux levels and placements of light sources. These examinations would also involve being situated in the environment at the different moonlight phases to test the effects of light and materials.

6. Final design proposal

The three-step design process has led to the development of a dynamic lighting scheme for Anholt church, where the electrical lighting is designed to turn 'on' and 'off' depending on where the Moon is in its monthly cycle (Fig.4).

	Path Lighting	**Window Lighting**	**Façade Lighting**
Bright Night	OFF	ON	OFF
Semi-Bright Nights	ON	ON	OFF
Dark Nights	ON	ON	ON

Fig.4: Figure illustrating the dynamic lighting design scheme.

This results in a design, where the use of electrical lighting is limited to when there is an actual need for it. At 'Dark Nights', all three lighting elements are turned 'on' to provide atmosphere and visual orientation in the complete darkness. As moonlight becomes more significant during the period of 'Semi-Bright Nights', the façade lighting turns 'off' to let the delicate moonlight shine 'undisturbed' on the façade and let faint shadows draw visible contours in the space. The warm window lighting will contrast the cold moonlight and hereby enhance its significance. Path lighting is still 'on' at this stage to ensure proper visibility on the ground, but as the Moon moves into the last stage of 'Bright Nights', this is also turned 'off'. At this stage, moonlight is at its peak, and by minimizing the effect of electrical lighting to only window lighting, both its functional and aesthetical qualities can unfold naturally in the space.

Fig.5: Visualization illustrating the final lighting design at 'Dark Nights'. View from the vest side (left), view from the north side (middle) and view from the south side (right).

Fig.6: Visualization illustrating the final lighting design at 'Semi-Bright Nights'. View from the vest side (left), view from the north side (middle) and view from the south side (right).

Fig.7: Visualization illustrating the final lighting design at 'Bright Nights'. View from north-east side (left) and view from south-vest side (right).

7. Perspectives

This study stems from a desire to understand why people value living with darkness and shows how a particular sensitivity to the nuances of darkness and thus the power of moonlight can be developed. In this study, Anholt church acts as an example of how this can be done. In future studies, these findings could be developed in relation to the rest of Anholt island in terms of both streetlighting, other public buildings, and lighting in people's private gardens. In future studies it would furthermore be relevant to do field studies with tests of human vision in darkness [19] in relation to the different levels of moonlight.

The case of Anholt church shows an example of how the phases of the Moon can be used as a design element in a night-time environment with almost complete darkness. But can this sensitivity to 'Bright-Nights', 'Semi-Bright Nights' and 'Dark Nights' be transferred to larger, more brightly lit cities? By working with a holistic approach to urban lighting design and by balancing brightness levels between lit zones, brightness levels can be reduced in selected areas [6, 7]. We can create places in cities where it will be possible to regain a lost contact with the naturally created light at night. In cities, we have greater demands for brightness levels due traffic and human feeling of safety, but studies show that by balancing brightness levels and thereby reducing contrasts, a visual contact with spatial and social surroundings can be re-established and thereby a relaxed atmosphere and an increased feeling of safety can be achieved [6, 7].

8. Conclusion

The aim of this project was to investigate how moonlight can be incorporated in an architectural lighting design to balance on the one side natural needs for darkness and on the other side human needs for lighting. The result is a dynamic lighting scheme, where the electrical lighting turns 'on' and 'off' depending on the significance of moonlight.

The findings from this project are not only relevant in the case of Anholt but can act as inspiration for future lighting in bigger cities as well. In a complete dark environment like Anholt, moonlight teaches us, that visual orientation is possible in even the faintest bit of light. If we start taking this into consideration, by simply lowering electrical lighting, we could benefit a lot from this in different ways. Not only would we minimize the issues of light pollution, but it will also let us rediscover the atmosphere of natural darkness and allow us to explore new ways of designing outdoor lighting, that draws attention to and interacts with the unique qualities of moonlight.

The starting point for the design at Anholt church was not lighting but instead darkness. In the beginning of a design process, we should start changing perspective and instead of asking ourselves how much lighting we need, we should focus on darkness and ask ourselves how much we can *minimize* lighting to protect the darkness while still supporting functional and aesthetical needs.

Lighting Designer François Jousse describes it beautifully: "The secrets are very simple. Blend light with the surroundings. Don't annoy the birds, the insects, the neighbors, or the astronomers. If City Hall gave me money to do whatever I want, I'd teach people about the beauty of light." [25].

6. Acknowledgement

The authors of this paper would like to thank Anne Dixgaard, chairman of Dark Sky Anholt, for an inspiring collaboration and the inhabitants of Anholt for sharing their unique insights on natural light and darkness on the island. Also, a thanks to Ellen Kathrine Hansen, Head of Lighting Design at Aalborg University Copenhagen, for important guidance in the process of the project and Lighting Designer Giacomo Lenzi, whom also did a Master Thesis about lighting on Anholt, for a great collaboration and support.

References

[1] *Darksky*. http://www.Darksky.org . Accessed 2. March 2022
[2] Hölker F, Moss T, Griefahn B, Kloas W, Voigt C C, Henckel D, Hänel A, Kappeler P M, Völker S, Schwope A, Franke S, Uhrlandt D, Fischer J, Klenke R, Wolter C and Tockner K 2010 The

Dark Side of Light: A Transdisciplinary Research Agenda for Light Pollution Policy. *Ecology and Society* vol **15**
[3] Falchi F, Cinzano P, Elvidge C D, Keith D M, & Haim A 2011 Limiting the Impact of Light Pollution on Human Health, Environment and Stellar Visibility. *Journal of Environmental Management* vol **92** pp 2714-2722
[4] Lyytimäki J 2012 Nature's Nocturnal Services: Light Pollution as a Non-Recognized Challenge for Ecosystem Services Research and Management. *Ecosystem Services* vol **3** pp 44-48
[5] Longcore T & Rich C 2004 Ecological Light Pollution. *Front Ecol Environ* **2** pp 191-198
[6] Hvass M, Wymelenberg K V D, Boring S & Hansen E K 2021 Intensity and ratios of light affecting perception of space, co-presence and surrounding context, a lab experiment. *Building and Environment* vol **194**, 107680
[7] Hvass M, Waltorp K & Hansen E K 2022 Lights out? Lowering Urban Lighting Levels and Increasing Atmosphere at a Danish Tram Station. In *Lighting Design in Shared Public Spaces* edited by Shanti Sumartojo pp. 151-172 (New York: Routledge)
[8] Crawford D L 2000 Light Pollution, An Environmental Problem for Astronomy and for Mankind. *Memorie della Società Astronomica Italiana* vol **71** pp 11-40
[9] Loncore T, & Rich C 2016 Artificial Night Lighting and Protected Lands: Ecological Effects and Management Approaches. *Natural Ressource Report NPS/NRSS/NSNS/NRR*. pp 1-68.
[10] Dick R 2013 Applied Scotobiology in Luminaire Design. *Lighting Res. Technol* vol **46** pp 50-66
[11] LoNNe 2015 *Statement of the EU-COST Action ES 1204 LoNNe (Loss of the night network)*. http://www.cost-lonne.eu/wp-content/uploads/2015/03/LoNNe-statement.pdf . Accessed 2. March 2022
[12] Longcore T & Rich C 2006 *Protected areas in Europe: Essential for Safeguarding the Nighttime Environment*. http://www.isees.org.il/wp-content/uploads/2016/09/LoNNe-Statement-for-NPAs_2016-%D7%A0%D7%A1%D7%A4%D7%97-8.pdf . Accessed 2. March 2022
[13] 8. Spaceplace 2021 *What are the Moon's Phases?*. https://spaceplace.nasa.gov/moon-phases/en/ . Accessed 2. March 2022
[14] Naylor J 2002 *Out of the Blue: A 24-Hour Skywatcher's Guide* (UK: Cambridge University Press)
[15] Fullmoonphase 2021 *Phases of the Moon*. https://www.fullmoonphase.com/moon-phases.htm . Accessed 2. March 2022
[16] Dickinson D 2017 *How Bright is the Moon...Exactly?*. https://skyandtelescope.org/astronomy-news/how-bright-moon-exactly/ . Accessed 2 March 2022
[17] Zele A J & Cao D 2015 Vision under Mesopic and Scotopic Illumination. *Front Psychol* vol **5** pp 1-15
[18] Illinoislighting 2011 *Bright as the Full Moon: How Much to Light up the Night?*. http://www.illinoislighting.org/moonlight.html . Accessaed 2. March 2022
[19] Boyce P R 2014 Human Factors in Lighting, Third Edition. (USA: CRC Press)
[20] Kyba C, Mohar A, & Posch T 2017 How Bright is Moonlight? *Astronomy & Geophysics* vol **52** pp 31-32
[21] Longcore T 2017 *How Bright the Moon: Correcting a Propagated Figure Error in the Literature*. https://travislongcore.net/2017/08/06/how-bright-the-moon-correcting-a-propagated-figure-error-in-the-literature/ . Accessed 2. March 2022
[22] Ciocca M & Wang J 2013 By the Light of the Silver Moon: Fact and fiction. *Physics Education* vol **48** pp 360-367
[23] Khan S M & Pattanaik S N 2004 Modeling Blue Shift in Moonlit Scenes by Rod Cone Interaction. *Journal of Vision* vol **4** p 316.
[24] Jousse F in Bogard P 2014 *The end of Night. Searching for Natural Darkness in an Age of Artificial Light* (London: Fourth Estate)

IOP Conf. Series: Earth and Environmental Science 1099 (2022) 012047 doi:10.1088/1755-1315/1099/1/012047

Optimizing exterior lighting illuminance and spectrum for human, environmental, and economic factors.

S M Simmons[1], S Baur[1], W Gillis[1], D Burns[2] and H Pickerill[1]

[1]Civil, Architectural, and Environmental Engineering, Missouri University of Science and Technology, Rolla, Missouri, United States
[2]Psychological Science, Missouri University of Science and Technology, Rolla, Missouri, United States

Email: smsgdx@mst.edu

Abstract. With the recent widespread adoption of LED lighting in outdoor areas, numerous concerns have been raised about the potential for harmful effects on humans, animals, plants, and the night sky. These stem from the high blue light content of some LED bulbs and an incentive to increase lighting levels caused by higher efficiency and lower costs. While new lighting installations are often described as environmentally friendly due to their energy efficiency, factors such as light pollution are often neglected or not given enough weight. This research focuses on optimizing the design of exterior lighting for human, environmental, and economic factors using a multi-criteria decision analysis. Based on data in the literature and survey research, illuminance and spectrum alternatives were scored relative to each other using the analytic hierarchy process and multi-attribute utility theory. The findings of this study support the use of artificial illumination at levels similar to a full moon (0.01 fc) and a warm white spectrum (2700K or 2200K), with amber LED becoming a better choice if its energy efficiency and cost effectiveness improve in the future. This methodology can be used in the future as a framework for lighting design optimization in different settings.

1. Introduction

Modern lighting has improved the lives of people around the globe, from allowing society to function more effectively after dark to increasing comfort in the built environment. However, along with its benefits come costs. These include energy usage, which can contribute to climate change, and light pollution, which can negatively impact human health, plants, and animals. In recent years, the rapid transition to LED has revolutionized the lighting industry, but despite presenting new opportunities, this technology also brings the potential for making existing problems worse. While the efficiency and controllability of LEDs allow for the reduction of energy usage and light wastage, these attributes can result in excessive illumination due to decreased operational costs [1]. Additionally, the increased blue light emitted by many LEDs is associated with ill effects on human health [2, 3] and the night sky [4]. The good news is that LED lighting can lead to a reduction in light pollution if the right design choices are made [1, 5, 6]. To benefit both people and the planet, it is imperative that a lighting design considers the factors of human health, the night sky, animals, and plants, in addition to functionality, public perception, energy usage, and cost.

This research optimizes the design of an exterior pedestrian LED lighting system at the Missouri University of Science and Technology (Missouri S&T) campus for human, environmental, and economic factors. Several methods of optimizing lighting designs for multiple criteria already exist in the literature [7, 8, 9]. These all make use of different forms of a multi-criteria decision analysis (MCDA) to select an ideal solution among many possible design alternatives. While an exterior lighting design algorithm has been created that includes light pollution as a principal criterion [7], it does not address factors such as light spectrum. It also does not score alternatives for the illuminance and light pollution criteria, but instead simply eliminates those deemed unacceptable. This research addresses these shortcomings by utilizing the analytic hierarchy process (AHP) and multi-attribute utility theory (MAUT) to compare lighting alternatives in a quantifiable manner, as has previously been done for interior lighting [9], and by incorporating more criteria to cover all major light pollution impacts as well as public perception. The framework used for this case can be used for other exterior lighting design applications and can be modified as necessary to fit different design objectives.

2. Methodology

To create an optimized outdoor lighting design, all significant design criteria were defined and weighted according to their relative importance. After determining the weighted criteria, different alternatives for illuminance and spectrum were scored according to their performance relative to each other and to a baseline. On the condition that they provide an acceptable level of visibility, the alternatives scoring the highest were considered to represent ideal lighting specifications. While not pursued in this study, minimum standards for other criteria could be established as well.

2.1. Defining and weighting design criteria

The following eight design criteria were used in this analysis:

- Functionality—public safety, visibility
- Perception—feelings of safety, comfort and aesthetics
- Human health—melatonin suppression, linked to a multitude of ailments [2, 3]
- Night sky—veiling of stars due to skyglow
- Animals—interference with or harm of various species
- Plants—interference with growth, dormancy period, etc.
- Energy—estimated energy usage of a lighting system
- Cost—capital, operation, maintenance

These criteria were associated with three main objectives: utility (functionality, perception); light pollution reduction (human health, night sky, animals, plants); and economy (energy, cost).

The analytic hierarchy process (AHP) [10], a common method used in multi-criteria decision analyses [9, 11, 12, 13], was used in this study to weight the design criteria. This method involves setting each criterion against all others with pairwise comparisons and allowing decision makers to indicate how favorable one is over the other. The traditional scale of 1/9 to 9 was used, with a score of 1/9 indicating extreme unfavorability of the first criterion over the second, 1 indicating equal preference, and 9 indicating extreme favorability of the first over the second. Intermediate values indicate lesser degrees of favorability or unfavorability.

For this study, a survey containing pairwise comparisons of the design criteria was distributed to students in the university's Civil, Architectural, and Environmental Engineering department. The survey set items representing the criteria against each other—for example, "functionality vs. perception" was represented with the phrases "actually being safe" vs. "feeling safe." For each pair, respondents were asked to indicate which item they believed was more important, and to what degree. To provide simplicity and reduce survey fatigue, a 5-point Likert scale was used and converted to the 1/9 to 9 scale after collecting responses. The choices included the following: "equal importance," corresponding to a 1; "somewhat more important," corresponding to either 1/5 or 5; and "much more important," corresponding to 1/9 or 9. A sample survey question is included in figure 1. The average response among participants was used for the final analysis. It was assumed that if a criterion was ranked "much more important" by all respondents, this would truly indicate extreme importance and merit a 9 on the AHP scale.

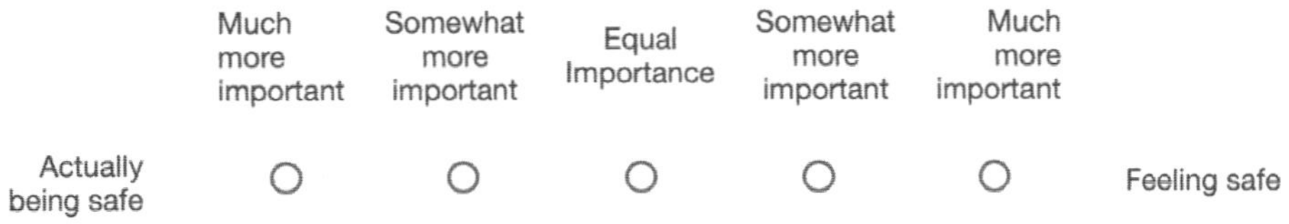

Figure 1. Sample survey question.

Survey responses were initially collected from members of a lighting design course within the department (n = 20, 50% female) as a pilot run, after which slight modifications to the wording were made to provide fairer representation of the criteria. Next, the survey was opened to all students in the department (~ 500 students) and received a response rate of around 10% (n = 53, 53% female). Participation was voluntary and anonymous, and no compensation or incentives were offered in exchange for participation. Responses were screened for credibility with the assistance of an interquartile range outlier analysis, and any spurious responses were removed. Using the average response from the final run of the survey, an AHP Excel template was used [14] to determine the weights of the three objectives, followed by the eight criteria independent of the objectives due to the design of the survey. The criteria were then grouped by objective, and their weights were adjusted in proportion to the objectives' weights. The final weights of the objectives and design criteria are given in table 1. In addition, functionality and perception were divided into subcriteria, the weights of which were approximated as detailed in appendix A.

There was greater than a 95% consensus between the pilot and final survey results. However, there were some noteworthy differences that could not be attributed to any wording changes that were made. Specifically, the weights of human health and functionality declined while those of energy and cost increased. This disparity could be due to differing values among students in the lighting class compared to the department in general. An assumption was made that the results of the final survey are a valid source for designing outdoor lighting for Missouri S&T's campus.

The calculated consistency ratio within the department's collective response was 0.076 for the three objectives and 0.174 for the eight criteria. Typically, a consistency ratio of 0.10 or lower is considered acceptable. However, the threshold is sometimes set at 0.20 [15]. In this case, since the results stem from the mean of more than 50 survey responses, the likelihood of inconsistency due to individual human error or poor judgment is lower. Rather, inconsistencies are more likely caused by the variation of items representing each criterion. For instance, the perception criterion was rated more favorably when represented by feelings of safety as opposed to comfort or aesthetics. Since great care was taken to ensure a balanced overall representation of all criteria, the results are considered acceptable. In addition, between-participant consistency was analyzed for all survey responses except for two that were incomplete (n = 51) using SPSS Statistics software (Version 28). The intraclass correlation coefficient (ICC) was calculated as 0.885 (95% CI: 0.818–0.937) based on a mean-rating, absolute agreement, two-way random model, representing good to excellent reliability [16].

Different methods of determining criteria weights, as well as the employment of this method towards other lighting applications and demographics, can be addressed in future research.

IOP Conf. Series: Earth and Environmental Science 1099 (2022) 012047 doi:10.1088/1755-1315/1099/1/012047

Table 1. Calculated weights of objectives and final design criteria.

Utility		Light Pollution				Economy	
0.221		0.306				0.473	
Function-ality	Percep-tion	Human Health	Night Sky	Animals	Plants	Energy	Cost
0.156	0.065	0.120	0.042	0.073	0.071	0.249	0.224

2.2. *Quantifying relative performance of lighting alternatives*

The following lighting attributes and design alternatives were analyzed in this study:

- Illuminance—alternatives ranging from 0.01 fc (0.1 lx) to 10 fc (~100 lx)
- Spectrum—alternatives ranging from amber to 5000K correlated color temperature

Illuminance and spectrum alternatives are illustrated in figures 2 and 3.

(a) (b) (c) (d)

Figure 2. Illustrations approximating illuminance orders of magnitude: 0.01 fc (0.1 lx) (a), 0.1 fc (1 lx) (b), 1 fc (~10 lx) (c), and 10 fc (~100 lx), (d). Created with assistance of Missouri S&T lighting plans [37] and a LIFX brand light bulb which has a logarithmic dimming profile [48].

(a) (b) (c) (d) (e) (f)

Figure 3. Simulated spectrum alternatives, created using f.lux software as described in Appendix B: 5000K (a), 4000K (b), 3000K (c), 2700K (d), 2200K (e), and amber (f).

For each attribute, an MCDA was performed to compare several alternatives representing a wide range of possible design choices. In addition, significant interaction effects (i.e., variation in one attribute leading to altered performance of a different attribute) were studied. To score the alternatives relative to each other, multi-attribute utility theory (MAUT) [9, 17] was utilized. Utility indices were used to score alternatives with values ranging from 0 to 1, with 0 representing no utility and 1 representing an ideal alternative. Data were obtained from sources in the literature where possible or from other practical methods (e.g., survey research, analysis of market prices and specified product luminous efficiency), and regression models were used to estimate missing data points. Survey research was used to determine the perception utility for light spectrum, and is detailed in appendix B.

To assist in providing fair scoring across criteria and avoiding any bias from arbitrarily assigning utility values, the following scoring system was used:

Positive design criteria (e.g., visibility, feelings of safety, comfort) were scored using equation (1):

$$s_i = \frac{x}{x_{max}} \tag{1}$$

where s_i is the utility score of an alternative for criterion i, x is the value of the metric used to score the alternatives for that criterion, and x_{max} is the maximum value of this metric among all tested alternatives.

Negative design criteria (e.g., crime rate, light pollution impacts, cost) were scored using equation (2):

$$s_i = 1 - \frac{x}{x_{max}} \tag{2}$$

Finally, the total score of each alternative was calculated using equation (3):

$$S = \sum w_i s_i = w_1 s_1 + w_2 s_2 + \cdots + w_n s_n \tag{3}$$

where S is the total score, w_i is the weight of criterion i, s_i is the utility score of the alternative for criterion i, and n is the number of criteria ($n = 8$).

Using this scoring system, for positive criteria a score of 1 was assigned to the best case achievable within the range of alternatives (e.g., best visibility, feelings of safety and comfort, and energy efficiency within alternatives), and a score of 0 represents the theoretical worst case possible (e.g., zero visibility, feeling very unsafe/uncomfortable, zero energy efficiency); for negative criteria a score of 1 represents the theoretical best case possible (e.g., no crime/accidents, no light pollution, no cost) and a score of 0 was assigned to the worst values within the range of alternatives (e.g., highest crime/accident rate, largest light pollution impacts, and highest cost within alternatives). The possible scores for each criterion range from a maximum or minimum theoretically possible value to an extreme value found within the range of alternatives. In some cases, several subcriteria were combined to derive the final score for a criterion, and thus neither the 1 nor 0 point may appear among the alternatives. Due to the nature of this scoring system, the scores of each alternative should primarily be interpreted relative to other alternatives rather than as an absolute measure of utility. The formulation of different possible scoring systems for this analysis could be the topic of future research.

Following scoring by criterion, each alternative was given a total score by taking the sum of scores for each criterion multiplied by the criterion's weight. The results of this analysis are given in tables 2 and 3.

Table 2. Scores for illuminance alternatives.

Criteria	Function-ality	Percep-tion	Human Health	Night Sky	Animals	Plants	Energy	Cost	**Total**
Data Source	[19-23]	[24]	[20, 25-27]	[29-31]	[25, 26]	[28]	[32-37]	[32-39]	
Weight	0.156	0.065	0.120	0.042	0.073	0.071	0.249	0.224	
10 fc (~100 lx)	0.600	0.730	0.000	0.000	0.000	0.000	0.000	0.000	**0.141**
5 fc (~50 lx)	0.587	0.892	0.089	0.038	0.076	0.076	0.564	0.468	**0.418**
2 fc (~20 lx)	0.571	0.876	0.207	0.133	0.176	0.176	0.820	0.681	**0.559**
1 fc (~10 lx)	0.560	0.741	0.297	0.248	0.252	0.252	0.906	0.752	**0.612**
0.5 fc (5 lx)	0.555	0.627	0.386	0.398	0.328	0.328	0.950	0.789	**0.651**
0.2 fc (2 lx)	0.617	0.515	0.491	0.622	0.428	0.428	0.977	0.811	**0.701**
0.1 fc (1 lx)	0.599	0.446	0.556	0.775	0.504	0.504	0.986	0.819	**0.723**
0.01 fc (0.1 lx, approx. full moon)	0.462	0.255	0.772	0.973	0.697	0.697	0.994	0.826	**0.755**

Table 3. Scores for spectrum alternatives.

Criteria	Function-ality	Percep-tion	Human Health	Night Sky	Animals	Plants	Energy	Cost	**Total**
Data Source	[19, 21]		[27]	[27]	[27, 40]	[27]	[33, 41-45]	[38, 39, 46]	
Weight	0.156	0.065	0.120	0.042	0.073	0.071	0.249	0.224	
5000K	0.652	0.914	0.000	0.000	0.000	0.018	1.000	0.291	**0.476**
4000K	0.635	1.000	0.194	0.092	0.063	0.000	0.993	0.290	**0.508**
3000K	0.588	0.965	0.467	0.289	0.128	0.000	0.955	0.285	**0.534**
2700K	0.578	0.931	0.563	0.370	0.198	0.030	0.932	0.282	**0.546**
2200K	0.565	0.850	0.696	0.571	0.243	0.045	0.873	0.247	**0.545**
PC Amber	0.558	0.764	0.924	0.825	0.299	0.111	0.744	0.153	**0.531**
Narrowband Amber	0.558	0.764	0.971	0.921	0.491	0.435	0.461	0.000	**0.474**

3. Results

3.1. Illuminance

The illuminance alternative scoring the highest relative to the others is 0.01 fc (0.1 lx), or approximately the brightness of a full moon [18]. This lighting level is more than two orders of magnitude lower than typical illuminance values recommended for public lighting systems [47]. Levels below 0.01 fc were excluded from this analysis due to the potential for insufficient hazard detection at night [22], which is taken to be a critical factor for good quality lighting. An interaction effect between illuminance and lamp spectrum was found where visibility declined more steeply at low illuminances (< 0.2 fc) when warmer-colored lighting was used [21]. As 2000K HPS lighting was used to compute visibility vs. illuminance in the literature used for this analysis [22], use of a higher color temperature could result in lower illumination levels being required for acceptable visibility.

Despite having the highest score, a light level of 0.01 fc could present a design dilemma, as it provides a minimal level of obstacle detection capability [22] and, despite reduced illumination being associated with less crime [19], is associated with negative safety perception [24]. If a more acceptable level of visibility and perceived safety is deemed to be necessary, higher illumination levels may be called for. An interesting detail to note is that an illuminance of 0.2 fc (2 lx) scores highest for functionality in this analysis. In [21, 22], approximately 0.2 fc is identified as an inflexion point below which visual performance drops steeply, at least for a surface reflectance of 0.20. As certain assumptions were made in the calculation of the utility scores and surface reflectance can vary in real-world scenarios, this finding could be flawed to some extent, and the ideal illuminance could be different for specific cases. An interaction effect was also found in data from [21] where visibility dropped more significantly for elderly subjects than for younger subjects with reduced illuminance; however, this effect was only pronounced below 0.2 fc. A more precise study of how illuminance affects visibility and safety perception for case-specific conditions and demographics, as well as the determination of minimum acceptable illuminance, should be topics for future research.

3.2. Spectrum
The spectrum alternative scoring the highest is 2700K, but this alternative is virtually tied with 2200K (< 0.001 score difference). The interaction effect between illuminance and spectrum discussed in the previous subsection led to 2700K being favored slightly at low illuminances (~ 0.02 fc) and 2200K being favored slightly at higher illuminances (~ 2 fc). However, any slight differences between these two color temperature options are overshadowed by the estimated uncertainty in the criteria weights and from assumptions made in the calculations, as well as from a lack of complete and/or statistically significant data in the literature for some of the metrics. Benefits of choosing 2700K include better utility and economy, whereas 2200K better fulfills the objective of light pollution reduction. Color temperatures higher than 2700K increase light pollution significantly while not providing much benefit to utility or economics, whereas amber light would further reduce light pollution but bring drawbacks to utility and especially economics. If the energy efficiency and cost effectiveness of narrowband amber LEDs increase enough in the future, this alternative will score the highest.

3.3. Final Design Recommendation
Based on the results of this multi-criteria design optimization, the recommended LED lighting design for the exterior pedestrian areas of the Missouri S&T campus is an illuminance on the order of 0.01 fc at a color temperature of 2700K. If a higher illuminance is required for increased visibility and feelings of safety, 2200K may be more appropriate. Due to the interaction effect discussed above, if 2200K is selected at lower illumination levels (< 0.2 fc), a slight increase in illuminance may be appropriate to achieve the same visibility level. A lighting level above 0.2 fc is not recommended, as it is associated with a drop in functionality (due to a paradoxical potential for increased crime) as well as an increase in light pollution impacts.

3.4. Limitations
There are several limitations of this research that could be addressed in future studies. The survey data collected for weighting the design criteria and gauging public perception of different light spectra is limited in extent and represents the collective opinion of a specific group of people. The utility scoring system, while it provides consistency and objectivity through a mathematical model, scores alternatives in a relative manner to a certain degree. The data used to evaluate alternatives were often limited, and assumptions had to be made in several instances. This research also determined optimal lighting attributes independently of each other, potentially not accounting for certain interaction effects. Other lighting attributes such as color rendering index, uniformity, mounting height, distribution, and shielding were not accounted for as well.

4. Conclusion
This research addresses the lack of a comprehensive lighting design that adequately factors in people, the environment, and economics. Through a multi-criteria decision analysis consisting of AHP and MAUT methods, ideal illuminance and spectrum specifications for an exterior pedestrian LED lighting design were determined based on functionality, public perception, human health, environmental impacts, energy use, and cost. The findings of this study support a lower illuminance level than conventional recommendations prescribe, as well as the use of a warm white spectrum. This design is anticipated to be acceptable for visibility while minimizing light pollution, energy consumption, and cost. Future research should study the appropriateness of these findings for different lighting applications. This methodology can also be used as a framework for other design optimization problems.

Appendix A

Due to the limited extent of the survey, subcriteria weights for functionality and perception criteria used in the calculations were estimated using equation (A.1) instead of conducting separate AHP calculations.

$$w_k = w_i * \frac{\overline{a_k} * \overline{v_j}}{\sum(\overline{a_k} * \overline{v_j})} \tag{A.1}$$

where w_k is the weight of subcriterion k of criterion i, w_i is the final weight of criterion i, $\overline{a_k}$ is the average preference (from 1/9 to 9) of subcriterion k, and $\overline{v_j}$ is the average unadjusted weight of opposing criteria j in the pairwise comparisons containing subcriterion k. As an illustration, the feelings of safety subcriterion of the perception criterion was paired against functionality, human health, plants, and energy (average unadjusted weight = 0.152) in the survey and was preferred by a factor of 0.567. The other perception subcriterion, comfort/aesthetics, was compared against functionality, night sky, animals, and cost (avg. unadj. wt. = 0.118) and preferred by a factor of 0.275. The estimated weight of the feelings of safety (FoS) subcriterion is thus given as follows:

$$w_{FoS} = 0.065 * \frac{0.567 * 0.152}{0.567 * 0.152 + 0.275 * 0.118} = 0.047$$

which is about 73% of the weight of the entire perception criterion.

Appendix B

The perception utility scores for light spectrum were determined through a survey given to university students, predominately from the Missouri S&T Civil, Architectural, and Environmental Engineering department. Participation was voluntary, and no compensation or incentives were offered in exchange for participation. The survey consisted of three pictures, shown in figure A1, which were derived from one photograph taken of an area of campus at night (lighting has a known correlated color temperature (CCT) of 5700K). The photos were color corrected to a CCT of approximately 5000K, 2700K, and 1800K (simulating amber) by comparison with the original photo displayed on a screen running f.lux software, which allows for the screen to be set to a specific CCT. Pictures were displayed in a random order to prevent bias.

(a) (b) (c)

Figure A1. Simulated photos for the perception survey: 5000K (a), 2700K (b), and 1800K (c).

Respondents were asked to rate their agreement on a 5-point Likert scale with the following statements:

- "I would feel safe walking around in this area"
- "This lighting is aesthetically pleasing and comfortable to be around"

Next, the three pictures were shown side by side, and respondents were asked to indicate which light color would make them feel the safest and the least safe, as well as which color is most and least aesthetically pleasing and comfortable to be around. An opportunity to explain these preferences or indicate no preference was given. Responses were vetted for consistency, and any instances where the Likert scale ratings were inconsistent with the preferences given in the side-by-side comparison were removed. A total of 56 responses were received, of which six were removed for the feelings of safety (FoS) portion and nine for the comfort and aesthetics portion (FoS: n = 50, 55% female; Comfort: n = 47, 48% female). Between-participant consistency was measured by calculating the ICC for all responses with complete data for both portions (n = 42). The calculated ICC (0.858, 95% CI: 0.642–0.976) represents moderate to excellent reliability between participants.

Results were obtained by averaging the responses for each portion. In addition, a separate analysis was performed after separating data by gender to understand any differences that may be present based on this demographic. Results are assumed to provide a valid data source for designing lighting for Missouri S&T's campus. The overall and gender-separated results are shown in table A1 and figures A2 and A3.

Table A1. Color perception survey data. Scoring ranges from -2 to 2, with a value of 0 representing a neutral response.

	CCT	5000K	2700K	1800K
Feelings of Safety	Composite	1.080	0.980	0.480
	Male	1.182	1.091	0.636
	Female	1.000	0.852	0.296
Comfort/ Aesthetics	Composite	0.149	0.532	0.000
	Male	0.208	0.542	-0.250
	Female	0.091	0.500	0.182

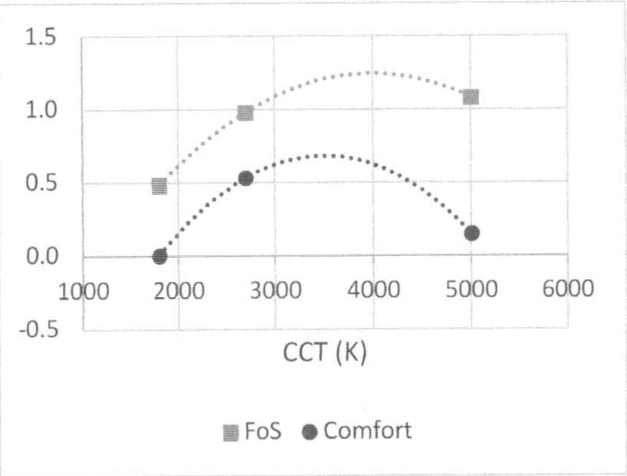

Figure A2. Composite scores.

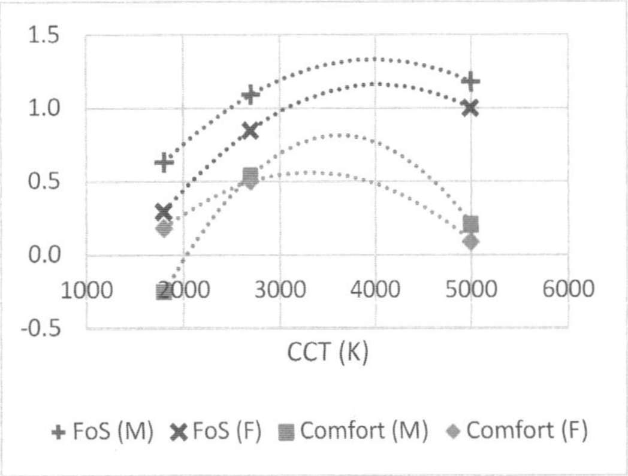

Figure A3. Results by gender.

References

[1] Schulty-Römer N, Meier J, Söding M and Dannemann E 2019 *Sustainability* **11** 6160
[2] Garcia-Saenz A *et al.* 2018 *Environ.l Health Perspect.* **126** 047011
[3] Zielinska-Dabkowska K M 2018 *Nature* **553** 274-6
[4] Hung L, Anderson S J, Pipkin A and Fristrup K 2021 *Environ. Manage.* **292** 112776
[5] Tabaka P 2021 *Energies* **14** 6383
[6] Aubé M, Simoneau A, Wainscoat R and Nelson L 2018 *Mon. Not. R. Astron. Soc.* **478** 1776-83
[7] Pracki P and Skarżyński K 2020 *Sustainability* **12** 1330
[8] Cristea M, Tîrnovan R A, Cristea C and Pica C S 2018 *MATEC Web of Conferences* **184** 04007
[9] Alshamrani O, Alshibani A and Alogaili M 2018 *Buildings* **8** 73
[10] Saaty T L and Vargas L G 2012 *Models, Methods, Concepts & Applications of the Analytic Hierarchy Process* (2nd edn) (New York: Springer)
[11] Madden S, Alles K and Demirel Y 2021 *Biofuels, Bioprod. Bioref.* **15** 1621-37
[12] Sangkakool T, Techato K, Zaman R and Brudermann T 2018 *J. Clean. Prod.* **196** 400-10
[13] Jusakulvijit P, Bezama A and Thrän D 2021 *Energy, Sustainability and Society* **11** 37
[14] Goepel K D 2013 *Proceedings of the International Symposium on the Analytic Heriarchy Process 2013*
[15] Ho D, Newell G and Walker A 2005 *Journal of Property Investment & Finance* **23** 424-44
[16] Koo T K and Li M Y 2016 *J. Chiropr. Med.* **15** 155-63
[17] Jansen S J T, Coolen H and Goetgeluk R 2011 The multi-attribute utility method *The Measurement and Analysis of Housing Preference and Choice* (Dordrecht: Springer)
[18] Kyba C, Mohar A and Posch T 2017 *Astron. Geophys.* **58** 31-2
[19] Steinbach R, Perkins C, Tompson L, Johnson S, Armstrong B, Green J, Grundy C, Wilkinson P and Edwards P 2015 *J. Epidemiol. Community Health* **69** 1118-24
[20] British Standards Institution 2016 *BS EN 13201-2:2015* (London: BSI Standards Limited)
[21] Fotios S and Cheal C 2009 *Lighting Res. Technol.* **41** 321-42
[22] Fotios S and Cheal C 2013 *Lighting Res. Technol.* **46** 362-76
[23] Boyce P R 1985 *Lighting Res. Technol.* **17** 51-71 as cited in Boyce P R 2014 Lighting for pedestrians *Human Factors in Lighting* (3rd edn) (Boca Raton: CRC Press)
[24] Liu M, Zhang B, Luo T, Liu Y, Portnov B A, Trop T, Jiao W, Liu H, Li Y and Liu Q 2022 *Remote Sens.* **14** 826
[25] Grubisic M *et al.* 2019 *Sustainability* **11** 6400
[26] Giménez M C, Stefani O, Cajochen C, Lang D, Deuring G and Schlangen L J M 2022 *J. Pineal Res.* **72** 12786
[27] Aubé M, Roby J and Kocifaj M 2013 *PLOS ONE* **8** 67798
[28] Bennie J, Davies T W, Cruse D and Gaston K J 2016 *J. Ecol.* **104** 611-20
[29] Martínez-Ledesma M, Sanhueza P, Corco C, Astudillo H R, Angeloni R and Damke G (Falchi F and Blanc G) 2019 *Spectral and Zonal Restriction Justification – Review Process of DS043/2012 MMA (Ministry of Environment)*
[30] Kocifaj M and Lamphar H A S 2014 *Mon. Not. R. Astron. Soc.* **439** 3405-13
[31] Falchi F, Cinzano P, Duriscoe D, Kyba C C M, Elvidge C D, Baugh K, Portnov B A, Rybnikova N A and Furgoni R 2016 *Sci. Adv.* **2** 1600377 and https://www.lightpollutionmap.info/ (3 May 2022)
[32] Lithonia Lighting 2021 *D-Series Size 0 LED Area Luminaire*
[33] Lithonia Lighting 2021 *D-Series Size 1 LED Area Luminaire*
[34] Lithonia Lighting 2021 *D-Series Size 2 LED Area Luminaire*
[35] Lithonia Lighting 2021 *D-Series Pole Mount LED Area Luminaire*
[36] Lithonia Lighting 2022 *RADEAN Bollard LED Site Luminaire*
[37] Antella Consulting Engineers Inc 2016 *Campus Lighting – Toomey Hall* construction drawings
[38] Birk M C (Union Electric Company) 2022 *Electric Service* (6th Revised)
[39] Access Fixtures https://www.accessfixtures.com (May 2022)

[40] Longcore T, Rodríguez A, Witherington B, Penniman J, Herf L and Herf M 2018 *J. Exp. Zool.* **329** 511-21
[41] Lithonia Lighting 2021 *D-Series Size 1 Amber Series LED Area Luminaire*
[42] Crossroads LED *Astrophile Series Narrow Band Amber Street Lights* https://crossroadsled.com/lighting-products/narrow-band-amber-street-lights/ (April 2022)
[43] Crossroads LED *Astrophile Series Phosphor Converted Amber Street Lights* https://crossroadsled.com/lighting-products/phosphor-converted-amber-street-lights/ (April 2022)
[44] Crossroads LED *Icon Series LED Acorn Retrofit Luminaires* https://crossroadsled.com/lighting-products/retrofit-luminaires-for-acorn-and-post-top-fixtures/ (April 2022)
[45] Crossroads LED *Prestige Series Luminaires for Decorative and Bell Shaped Fixtures* https://crossroadsled.com/lighting-products/retrofit-luminaires-for-decorative-and-bell-shaped-fixtures/ (April 2022)
[46] Saint Louis Lighting Group (Acuity Brands) price quote (28 April 2022)
[47] Illuminating Engineering Society (IES) as cited in Energy Trust of Oregon and Lighting Design Lab 2013 *Footcandle Light Guide*
[48] Crist R (CNET) 2019 *Smart LED dimming curves: Linear vs. logarithmic* https://www.cnet.com/pictures/smart-bulb-dimming-curves-linear-vs-logarithmic/ (July 2022)

IOP Conf. Series: Earth and Environmental Science 1099 (2022) 012048 doi:10.1088/1755-1315/1099/1/012048

The response of the Bonn Convention on Migratory Species to light pollution

Yana Yakushina[1]

[1] PhD Researcher; University of Ghent, Department of European, Public and International Law, Campus Aula, Universiteitstraat 4, 9000 Gent, Belgium http://www.ugent.be/re/epir

Email: ya.yakushina@gmail.com (alternative: Yana.Yakushina@UGent.be)

Abstract. Light pollution became a major transboundary concern for the environment and biodiversity protection, especially with regard to species migration and spatial orientation. In view of the need for effective nature conservation, the international community has to adapt to the new challenges. The Convention on the Conservation of Migratory Species of Wild Animals, or the Bonn Convention, as the main international framework for migratory species protection, has recently taken initial steps towards mitigation of light pollution adverse impacts. In order to understand the response to the new environmental problem at the international level, this article explores and briefly summarizes the response of the Bonn Convention framework on light pollution as a hazardous threat. The article also discusses the potential role of the Convention in the formation of an international regulatory framework on this matter.

List of Abbreviations

ALAN	Artificial Light at Night
CMS, the Bonn Convention	Conservation of Migratory Species of Wild Animals
COP	Conference of the Parties
MoU	Memorandum of Understanding
EUROBATS	Conservation of Populations of European Bats
EIA	Environmental Impact Assessment
WMBD	World Migratory Bird Days
MoP	Meeting of the Parties

1. Introduction

The constant readjustment of nature to pursue human needs has resulted in adverse impacts on the environment. Something similar has happened with artificial light at night (ALAN). The desire to prolong day activities into nighttime and the widespread of ALAN caused light pollution, a new environmental challenge. Light pollution is defined as every form of artificial light in the wrong place at the wrong time which creates a sky glow, glare, nuisance, and other relevant causes of environmental degradation including some properties of artificial light which emit non-environmentally friendly or inappropriate light [31]. The important role of ALAN in our modern lives cannot be diminished, however, the increased level of nighttime lighting negatively affects various fields, including the environment, biodiversity, and human health. Despite that ALAN has long been a matter of serious concern, only recently it has been recognized as a transboundary ecological phenomenon and has started to receive due attention in both research and regulatory framework.

Acknowledging the significant negative effects of light pollution on biodiversity, the Convention on the Conservation of Migratory Species of Wild Animals (the Bonn Convention, CMS) 1979 [9] became the first international environmental framework that has taken real action against the new challenge. The main purpose of this article is to evaluate the process of adaptation of the Bonn Convention framework to the newly emerged challenge and explore its potential role and contribution to light pollution reduction within international environmental law. In this regard, the article (1) identifies the adverse impacts of ALAN on biodiversity; (2) analyses the Bonn Convention framework; (3) evaluates already adopted legal and policy instruments, and (4) discusses the role of the Convention in the future development of the international light pollution reduction instruments. The article is intended to contribute to filling a gap in international environmental law research relating to the legal protection of biodiversity from ALAN.

2. Light pollution and Biodiversity

During past decades, the scientific literature on the interconnection between ALAN and biodiversity has substantially increased and provided major evidence of the adverse impacts of light pollution. Most of the living species on the planet are nocturnal [1], thus, natural day and night cycles are crucial for their normal functioning. These cycles, inter alia, serve as a source of valuable information on environmental conditions, regulate metabolism, help with the synchronization of circadian clocks, as well as influence behaviour and breeding patterns. Light pollution, on the contrary, intervenes with the environment, causing various hazardous threats. Increased levels of ALAN violate the ability to synchronize diurnal and nocturnal metabolic processes by changing foraging and food patterns [2]. To give another example, light pollution causes a major decline in several insect populations, leading to negative impacts on all species that rely on them for pollination or food [3]. A number of research have also shown that light pollution became a great concern for the spatial orientation of various species since it causes dramatic shifts in the use of space and movement behaviour. The latter is especially relevant for species that navigate themselves according to natural celestial light sources [4]. Nighttime lighting changes predator-prey relations by providing predators with hunting advantages [5]. Moreover, ALAN influences the mobility and migration of various species. If the migration route of eel passes through areas with increased artificial illumination, then the natural timing of their migrations might be disrupted [6]. In addition, the migration behaviour of birds also strongly relies on the natural day and night cycles. ALAN put various migratory bird species at risk by creating disorientation, collisions, entrapments, groundings, and adverse effects on navigation that can result in cases of mortality [7].

Apart from that, light pollution creates significant pressure on habitats, leading to habitat loss and fragmentation. Research has indicated that even species in the protected areas, such as national parks, and Natura 2000, experience a loss of darkness which impacts conservation efforts and can lead to the negative consequences mentioned above [4]. Even though the impacts of ALAN are not homogenous, light pollution affects both aquatic and terrestrial species and habitats. It should be noted, however, that light-induced emissions impacts go beyond the territorial scale and create significant pressure on airspace habitat, which is particularly important for avian species [8]. In this regard, the airspace habitat should be also considered a part of a valuable dimension in terms of conservation the biodiversity from light pollution.

The provided examples highlight the importance of natural environmental conditions for biodiversity and indicate the need to take legal and policy action towards light pollution mitigation in order to prevent further decline of biodiversity.

3. The Bonn Convention: main principles and framework

The environmental challenges are especially crucial for migratory species. The transboundary character of the species' behaviour, consisting in using migratory routes lied through the territories of different countries, as well as the transnational character of the environmental issues, altering species' behaviour and leading to loss of migratory routes, have resulted in the establishment of a legal framework for migratory species conservation.

The main international regime for the protection and sustainable use of migratory species and their habitats is the Convention on the Conservation of Migratory Species of Wild Animals (the Bonn Convention, CMS) which was concluded on 23^{d} June 1979 in Bonn, Germany, and came into force 1st November 1983 [9]. Currently, the Bonn Convention has 133 parties, which have to incorporate legally binding conservation principles into the national law to comply with convention provisions.

To increase the effectiveness of the Convention, a broad institutional structure was established. The main decision-making body is the Conference of the Parties (COP) which meets every two and a half or three years to assess the implementation of the CMS. Another institutional body of the CMS is the Scientific Council which provides recommendations on the adequate protection of migratory species through monitoring of potential threats and research of their impacts. In addition, the Convention has a Secretariat and a Standing Committee. The main function of the Secretariat is to coordinate administrative matters. The Standing Committee provides guidance in policy and administrative matters. Many studies have conducted a more detailed analysis of the institutional structure of the Bonn Convention [12].

The main concept of the protection under the Bonn Convention is the consideration of migratory species a common natural heritage that should be conserved and wisely used in order to avoid any species becoming endangered [9]. For conservation purposes, the Bonn Convention has listed individual species of wild animals into two categories: (1) which are endangered (Appendix I) and (2) which require special international cooperation due to their unfavourable conservation status (Appendix II). In a view of the CMS, endangered species are those that are facing a very high risk of extinction in the wild in the near future [10]. With regard to the second category, the Convention has provided criteria on whether species are having a "favourable conservation status", inter alia, the range of the migratory species is neither currently being reduced, nor is likely to be reduced, on a long-term basis [9], and in case of not meeting one of the criteria, the conservation status should be considered as "unfavourable". The Parties shall "endeavour to provide immediate protection for migratory species included in Appendix I; and endeavour to conclude Agreements covering the conservation and management of migratory species included in Appendix II" [9]. Despite that the conservation measures of the species listed in both Appendices are different, the CMS provides obligations to identify and mitigate hazardous factors and obstacles that may lead to the extinction of all listed migratory species and/or degradation of their habitats.

Apart from Conventional provisions, the CMS framework includes various legal instruments that range from legally binding Agreements to non-binding Memoranda of Understanding (MoU). Both Agreements and MoUs are concluded to ensure the protection of particular species. The special protection attempts have resulted in concluding seven legally binding Agreements and nineteen MoUs [11]. The Agreement on the Conservation of Populations of European Bats (EUROBATS) adopted on 4th December 1991 in London, the UK, for example, aims to protect various species of bats and their habitats.

The international cooperation within the Bonn Convention plays a major role in migratory species protection, facilitating the development of national legislation on this matter and providing a certain level of control to prevent further decline in the number of wildlife species and the loss of their habitats. Caddell [12] rightly notes that the established regime attempts to impose "regulatory control over activities affecting the entire ecosystems of a wide range of disparate species with very different conservation needs". It is important to highlight that the CMS framework always tries to address the newly emerged obstacles to adopt the measures taken towards conservation efforts. In particular, the

CMS framework has adapted to climate change issues and currently contributes to climate change mitigation in terms of migratory species protection.

4. Light Pollution and The Bonn Convention

An increased number of research on ALAN-induced impacts together with actions undertaken by dark-sky protection activists made light pollution a growing concern. The CMS framework became the first international environmental regime that has provided a response to the new growing threat to migratory species and has endeavoured actions for light pollution mitigation [13].

Even though ALAN impacts were considered a disturbance before, the beginning in addressing light pollution can be actually marked by the adoption of Decision 12.17 on Marine Turtles during the CMS COP12 which was held in Manila, the Philippines from 23 to 28 October 2017 [14]. The Decision has entrusted the Scientific Council to prepare an analysis on conservation and threats to marine turtles, including sky glow, and develop new recommendations for their conservation for presentation during the COP13. The increased awareness of ALAN's adverse impacts on marine turtles provided a starting point for consideration of light pollution within the CMS framework.

Resolution 8.6: Bats and Light Pollution adopted by the 8th Session of the Meeting of the Parties (MoP) of EUROBATS held Monte Carlo, Monaco, 8 – 10 October 2018, is the next important milestone in light pollution mitigation within the CMS framework [15]. Resolution 8.6 has acknowledged the value of natural light levels for ecological processes and has noted the growing scientific evidence of the impact of artificial light on bats and the predicted negative consequences for bat populations. To assist the parties in the proper protection of bats species, several recommendations were introduced. The general recommendation was to take into account that artificial light has a negative impact on bats and should be avoided. According to EUROBATS, this can be mainly accomplished by providing that planning of artificial lighting projects is conducted with regard to national legislation on the protection of bats and their habitats and by consideration of artificial light within impact assessment procedures. The Guidelines for consideration of bats in lighting projects became a supplementary document to promote solutions for mitigation of adverse ALAN impacts on bats [16]. EUROBATS Guidelines have established a four-step process for proper protection of bats: (1) to avoid impacts through careful assessment and planning; (2) to mitigate to minimize impacts; (3) to compensate to offset the effects of impacts and (4) to evaluate the effectiveness of mitigation and compensation. Importantly, the adopted recommendations are addressed not only to the parties of EUROBATS but also to the non-party Range States, i.e. any state that exercises jurisdiction over any part of the range of migratory species [9], highlighting the transboundary character of the issue. Although the Resolution and the Guidelines are non-legally binding and are focused specifically on bat species, recognition of the negative impacts of ALAN within EUROBATS has encouraged the CMS framework to pay attention to light pollution as a threat to the other migratory species.

The Bonn Convention itself has considered light pollution for the first time at the CMS COP13 held on 15-22 February 2020 in Gandhinagar, India. The COP13 resulted in the adoption of a number of resolutions and decisions to address light pollution as a threat that is faced by migratory species across the globe.

Despite the inability of the Scientific Council to make any progress related to the implementation of Decision 12.17 on Marine Turtles [23] adopted by the CMS COP12 [14], the Government of Australia has assisted in the protection of marine turtles from sky glow and has proposed the CMS COP13 a draft resolution "National Light Pollution Guidelines for Wildlife including Marine Turtles, Seabirds and Migratory Shorebirds" [17]. The Guidelines aim at raising awareness of the potential impacts of artificial light on wildlife and providing a framework for assessing and managing these impacts. The Guidelines, in particular, set the process of environmental impact assessment (EIA) for the effects of artificial light. The EIA includes a five-step procedure, namely, the description of the project lighting, the description of wildlife, risk assessment, artificial light management plan, and biological and artificial light monitoring and auditing. The main aim of the risk assessment is to consider the importance of the habitat to the identified species, ecological and biological processes of wildlife, the amount and type of artificial light at each phase of project development, and to conclude whether the planned lighting is likely to negatively impact the environment and species. In the event that risk is identified, the project should be modified or additional measures for risk mitigation should be implied. Regular light monitoring and auditing is the final and essential step of the EIA since the collected data will help to adapt and improve lighting management if necessary. The Guidelines also include several technical appendices that provide additional information about best practice lighting design, undertaking EIA, and species-specific information for the management of artificial light for marine turtles, seabirds and migratory shorebirds protection. For the mitigation of the ALAN negative effects, the Guidelines suggest six lighting management principles, for instance, the lights should be kept close to the ground, directed and shielded. In addition, over-illumination of the areas can be avoided by paying special attention to the amount of light produced (lumen) rather than to the amount of energy used (watt).

Another agenda item that addressed light pollution issues during the CMS COP13 was a draft resolution "Light Pollution and Migratory Species" prepared by the European Union and its Member States (the EU) [18]. The EU draft called for the mitigation of the negative effects of ALAN to conserve the darkness of the night needed for the protection of migratory species. The EU also proposed to raise awareness of light pollution through the World Migratory Bird Days (WMBD). According to the EU, WMBD with the main mission to raise awareness of migratory birds and issues related to their conservation will contribute to actions against ALAN-induced impacts.

Both proposed drafts were adopted by the CMS COP13 and were incorporated in Resolution 13.5 "Light pollution Guidelines for Wildlife" and its Annex, which consisted of the Australian Guidelines [19]. The COP13 has recognized artificial light as a threat to the conservation of wildlife, astronomy and human health. In view of the Resolution, light pollution refers to artificial light that alters the natural day and night cycles, causing the night sky brightness. The COP13 has drawn particular attention to the ecological detrimental effects of light pollution. It was stated that ALAN has direct and indirect negative impacts on many species and ecological communities by, for instance, stalling the recovery of threatened species and interfering with a migratory species' ability to undertake long-distance migrations. Acknowledged the need for the right light, in the right place and at the right time and underlined the need to create a balance between public interests and nature conservation, the Resolution recommended the Parties to use the Guidelines and called for the adoption of appropriate measures for mitigation of ALAN adverse impacts.

Given that the adopted Guidelines are focused on limited taxa of species, namely marine turtles, seabirds and shorebirds, the subsequent Decisions 13.138 and 13.139 (Decisions) of the COP13 were aimed at elaboration and improvement of the Guidelines in terms of inclusion of other migratory species protected by the Bonn Convention. The COP13 has requested the Secretary to prepare broader guidelines on how to effectively mitigate light pollution impacts on migratory species under the CMS conservation regime for adoption by the COP14 and to promote World Migratory Bird Days to increase awareness of light pollution [20]. Besides the Secretary, additional tasks were assigned to the Scientific Council. The Council was requested to consider issues related to ALAN impacts on migratory species in its first meeting of the Sessional Committee after the COP13, and advice on proper action against light pollution within the CMS framework. Upon consultation with the Scientific Council, the Secretariat has chosen a step-wise approach, including several stages from the initial research analysis of ALAN effects on different taxa of migratory species to the final development of consolidated guidelines.

Shortly after the CMS COP13, the work on the implementation of the adopted documents has begun. An overview of the progress being achieved to facilitate the implementation of the Decisions was presented at the 5th Meeting of the Sessional Committee of the CMS Scientific Council (ScC-SC5) on 28 June – 9 July 2021[21]. The most important outcome was the Summary of Preliminary Findings of a Review of the Information Available on the Impact of Light Pollution on Different Taxa of Migratory Species of Relevance to CMS [22], which provided comprehensive research findings on this matter. The Summary is divided into seven parts: (1) Introduction, which covers, inter alia, the definition of light pollution, types of lamps, and their light characteristics; (2) Impacts of ALAN on different taxa, including birds, insects, mammals, fish, reptiles and amphibians; (3) Existing guidelines, which identifies a few guidelines and laws relating to lighting standards and/or light pollution; (4) Mitigation, which discloses several approaches to light pollution mitigation; (5) Review of CMS Listed Species, which are subject to protection under the CMS framework but are not covered by the current Guidelines (6) Knowledge gaps, which are crucial for proper measures against ALAN, e.g. no unified approach to light measurement and (7) Suggestions and further steps, which, inter alia, calls for the implementation of a precautionary approach to facilitate efforts towards light pollution reduction. The ScC-SC5 has also recommended the development of three sets of guidelines: (1) generic advice on the reduction of light pollution; (2) concerning migratory birds, and (3) concerning bats. In addition, the provided summary is expected to be used as a basis for the elaboration of amendments to the already adopted CMS Guidelines.

Together with the presentation of the initial research findings, the ScC-SC5 has indicated the start of the process of exploration of possibilities related to the dedication of the WMBD to the light pollution effects. As a result of the subsequent actions, on the 9th of December 2021, it was announced that light pollution is chosen as the main focus of the WMBD 2022 campaign [24]. The campaign is devoted to the adverse impacts of ALAN on migratory birds. For promotion purposes, "Dim the Lights for Birds at Night!" has been selected as the official slogan. The first phase of the campaign will be held in May 2022 and will be focused on raising awareness of light pollution as a threat to migratory birds [25]. The second phase of the campaign will be held in October 2022 and will be aimed at promoting best practices in addressing light pollution [25]. Amy Fraenkel, Executive Secretary of CMS, highlights that WMBD 2022 should not only increase of adverse impacts of light pollution on migratory birds but also should prompt concrete commitments from Parties to reduction of ALAN levels [26].

One of the recent events concerning light pollution consideration within the Bonn Convention framework was a virtual CMS Technical Workshop on Light Pollution held from the 29th of March to the 31st of March 2022 [27]. The main aim was to review and make contributions to the first draft of the new CMS Guidelines which were introduced to the workshop's agenda.

Following the ScC-SC5 recommendation, the new Guidelines have three main dimensions: (1) generic advice on the reduction of light pollution; (2) migratory birds, and (3) bats. The Guidelines consist of eleven general principles and recommendations for adapting lighting design, operation, and planning to conservation purposes. General principles, inter alia, recommend installing artificial lighting only where it is necessary and where it has a specific and defined purpose for safety reasons, minimizing glare, eliminating façade lighting, and making a formal EIA for the ALAN effects on wildlife while planning new lighting installations. The measures recommended for implementation for adapting lighting design and operation are mainly premised on the need for using the specific colour and intensity of the light, setting curfew periods, defining the maximum luminance levels, as well as proper direction and location of the light, general minimization of night-time lighting, and shielding of lights. Interestingly, measures for adaptation of lighting operation include both interior and exterior light. To give an example, while the light spill from interior light sources should be blocked, the need for installed outdoor lighting should be reduced by introducing reflective paints or self-luminous markers. Moreover, the new guidelines also suggest measures in relation to lighting sources for specific locations. Thus, the guidelines provide recommendations for, inter alia, car parks, sports grounds, and security lighting. For the car parks, for instance, the particular measures include limitation of lighting intensity in parking lots and the shutdown of lights after working hours.

In addition to the generic advice, specific measures are established for the protection of migratory birds and bats. The guidelines suggest following general recommendations and implementing the precautionary principle to reduce adverse light pollution effects along with specifically suggested measures, for instance, to use minimum amount of light for tasks, seasonal measures related to migration periods, and compensation measures, including restoration of habitats and dark areas.

Summarizing the analysis of the first draft of the new guidelines, it should be mentioned that the draft is still under consideration and is subject to further revision and amendments. Before the adoption at the CMS COP14 in 2023, the new guidelines are intended to be reviewed by the 6th meeting of the Sessional Committee. Thereby, despite major progress being achieved, the final recommendations on mitigation of impacts on migratory species are still not completely developed.

5. Conclusion and final discussions

Increased levels of light pollution significantly intervene in the normal functioning of biodiversity and habitats. The conducted analysis clearly confirms that the Bonn Convention framework, being aware of the obstacles to proper protection, has provided a response to the new environmental challenge through the recognition of light pollution as a threat to migratory species and the initial work in the development of measures towards mitigation of ALAN-induced adverse impacts. In this regard, several important points could be noted.

The recognition of artificial light as a biodiversity threat within the CMS is a major step towards worldwide light pollution reduction. This recognition might contribute to the call for action to reduce light pollution levels within the other international environmental frameworks. In part, this is due to a strong interconnection between various international environmental protection regimes. Many of the goals and targets defined, for example, in the Biodiversity Convention Framework [28], are reflected in the CMS and vice versa. Moreover, the adoption of measures for reducing ALAN-induced effects at the international level might endeavour the establishment of national legislation and policies on this matter. In addition, the painstaking work in research on light pollution impacts and characteristics of artificial light will benefit the establishment of the best possible solutions for light pollution mitigation, leading to the protection of the environment and well-being.

On the contrary, the Bonn Convention framework has its limitations. The first limitation is related to the scope of species protection. As mentioned before, the CMS regime protects only the species listed in Annex I and Annex II to the CMS and the species that are subject to protection under the separately concluded Agreements or MoUs. In light of this, the species that can also be affected by ALAN-induced impacts will not be conserved. Simultaneously, it should be noted that the Convention provides for the obligation to conserve habitats and migratory routes, which indirectly affects the conservation of species beyond the CMS framework. Another limitation is related to the legal force of instruments adopted within the CMS framework. The CMS COP is the main body entitled to make decisions to improve the effectiveness of the Convention. The adopted instruments, however, are not legally binding and, therefore, cannot create new obligations to the Parties. On the other hand, despite the non-legally binding nature of the COP Resolutions, they still may contribute to the interpretation of the binding Conventional obligations in favour of light pollution reduction. The latter approach is supported by case law and legal studies [29], [30].

To sum up the foregoing, although the legal recognition of light pollution is at the initial stage, the CMS framework has already taken the leading position in terms of action to light pollution reduction. The results of such legal recognition will be visible later, upon the adoption of supplementary documents, indicating specific measures to protect migratory species and their habitats. It is important to note that in order to contribute to the incorporation of such measures into the national jurisdictions of the signatories, the newly adopted documents should provide for binding obligations to light pollution mitigation rather than recommendations due to their legal character and the possibility of being enforced. Since light pollution is rapidly increasing, the example of the Bonn Convention should be followed by the other international environmental instruments to mitigate the potential risks and provide adequate conservation.

References

[1] Hölker F, Wolter C, Perkin EK, Tockner K 2010 Light pollution as a biodiversity threat *Trends Ecol Evol.* D **25(12)** 681-2
[2] Grubisic M, Haim A, Bhusal P, Dominoni DM, Gabriel KM, Jechow A, Hölker F 2019 Light

pollution, circadian photoreception, and melatonin in vertebrates *Sustainability* **11(22)** 6400
[3] Owens ACS, Cochard P, Durrant J, et al. 2020 Light pollution is a driver of insect declines *Biological Conservation* **241** 108259
[4] Sordello R, Busson S, Cornuau J H, Deverchère P., et al. 2022 A plea for a worldwide development of dark infrastructure for biodiversity – Practical examples and ways to go forward *Landscape and Urban Planning* **219** 104332
[5] Ditmer MA, Stoner DC, Francis CD, Barber JR, Forester JD, et al. 2021 Artificial nightlight alters the predator-prey dynamics of an apex carnivore *Ecography* **44** pp 149-161
[6] Vowles AS and Kemp PS 2021 Artificial light at night (ALAN) affects the downstream movement behaviour of the critically endangered European eel, Anguilla anguilla *Environmental Pollution* **274** 116585
[7] Cabrera-Cruz SA, Cohen EB, Smolinsky JA, Buler JJ 2020 Artificial Light at Night is Related to Broad-Scale Stopover Distributions of Nocturnally Migrating Landbirds along the Yucatan Peninsula, Mexico *Remote Sens.* **12** 395
[8] Diehl RH 2013 The airspace is habitat *Trends in Ecology and Evolution* **28** 7 pp 377-379
[9] Convention on the Conservation of Migratory Species of Wild Animals June 13, 1979
[10] UNEP/CMS/Resolution 11.33 Guidelines for Assessing Listing Proposals to Appendices I and II of the Convention *CMS COP11* (Quito, November 4-9, 2014)
[11] Convention of Conservation of Migratory Species of Wild Animals website
[12] Caddell R. 2020 International law and the protection of migratory wildlife: An appraisal of twenty-five years of the Bonn Convention. *Colo. J. Int. Environ. Law Policy* **16** pp 113–156
[13] Efstathia L 2020 CMS COP-13 : International Concern over Biodiversity Decline, Urgent Need for Protection across the Planet *Environmental Policy and Law* **50** pp 207–217
[14] UNEP/CMS/COP12/Decision 12.17 on Marine Turtles *CMS COP12* (Manila, the Philippines, October 23-28, 2017)
[15] UNEP/EUROBATS Resolution 8.6: Bats and Light Pollution *8th Session of the Meeting of the Parties* (Monte Carlo, Monaco, October 8-10, 2018)
[16] Voigt CC, Azam C, Dekker J, Ferguson J, Fritze M, et al. 2018 Guidelines for consideration of bats in lighting projects *EUROBATS Publication Series* **8** UNEP/EUROBATS Secretariat Bonn Germany pp 62
[17] UNEP/CMS/COP13/Doc.26.4.9.1/Rev.1 Light Pollution Guidelines for wildlife including Marine Turtles, Seabirds and migratory Shorebirds, *CMS COP13* Agenda item 26.4 (Gandhinagar, India, January 30, 2020)
[18] UNEP/CMS/COP13/Doc.26.4.9.2/Annex 2 Light Pollution and Migratory Species *CMS COP13* Agenda Item 26.4, (Gandhinagar, India, January 30, 2020)
[19] UNEP/CMS/Resolution 13.5 Light Pollution Guidelines for Wildlife and Annex to Resolution 13.5 *CMS COP13* (Gandhinagar, February 2020).
[20] UNEP/CMS/Decisions 13.138 and 13.139 Light Pollution Guidelines for Wildlife Convention on Migratory Species, *CMS COP13* (Gandhinagar, February 2020).
[21] UNEP/CMS/ScC-SC5/Doc.6.4.3 Light Pollution and Migratory Species 5th Meeting of the Sessional Committee of the CMS Scientific Council *online* 28 June – 9 July 2021
[22] UNEP/CMS/ScC-SC5/Inf.7 Impact of Light Pollution on Different Taxa of Migratory Species of Relevance to CMS 5th Meeting of the Sessional Committee of the CMS Scientific Council (ScC-SC5) *online* 28 June – 9 July 2021
[23] UNEP/CMS/COP13/Doc.26.2.6/Rev.2 Marine Turtles *CMS COP13* Agenda item 26.2 (Gandhinagar, India, February 17-22 2020)
[24] World Migratory Bird Day 2022 Theme to Focus on Light Pollution website
[25] World Migratory Bird Day 2022 Campaign Strategy
[26] "Dim the Lights for Birds at Night!" – The World Migratory Bird Day 2022 Slogan Announced! WBMD website
[27] CMS Technical Workshop on Light Pollution CMS website

[28] United Nations Convention on Biological Diversity June 5, 1992 (1760 U.N.T.S. 69)
[29] Verschuuren J 2008 Ramsar Soft Law is Not Soft at All. Discussion of the 2007 Decision by the Netherlands Crown on the Lac Ramsar Site on the Island of Bonaire (November 25, 2008) *Milieu en Recht* **35** 1 pp 28-34
[30] Trouwborst A 2012 Transboundary Wildlife Conservation in A Changing Climate: Adaptation of the Bonn Convention on Migratory Species and Its Daughter Instruments to Climate Change *Diversity* **4** pp 258-300
[31] Youyuenyong P 2015 Comparative Environmental and Planning Law Relating to Light Pollution Control in England and Other Jurisdictions *Unpublished Ph.D Thesis* (School of Law De Montfort University, Leicester, UK)

IOP Conf. Series: Earth and Environmental Science 1099 (2022) 012049 doi:10.1088/1755-1315/1099/1/012049

Energy-efficient lighting and visually impaired users in homelike environments

P Mattsson[1]

[1]Department of Architecture and Built Environment, Lund University, Box 118, SE-22100, Lund, Sweden
pimkamol.mattsson@abm.lth.se

Abstract. Despite a rapid transition to LED lamps, it is still little knowledge about the effects of these light sources on users with low vision. The study's main objective was to acquire a better understanding about how correlated colour temperatures (CCTs) of non-directional LED affect visual accessibility and perceptions of lighting quality in homelike environments among visually impaired persons. The study took place in a full scale laboratory in which CCTs of 2,700 K and 4,000 K were tested. Participants having normal vision (N=60, 24 males, 36 females, M = 25.50) were divided into three groups of 20 to assess the two lit environments with cataract goggle, severe cataract goggle and normal vision respectively. They were asked to perform everyday activities while the observer recorded the time and conducted observations, and to rate the difficulty of each activity. Thereafter, the participants were asked to rate their perceptions of lighting quality and contrasts in the room. With the same CRI of Ra80, there were no differences in the performance of everyday activities between the two CCTs whereas the differences in the perceived lighting quality and the perceived contrasts were found. The cool white CCT was found to positively affect visual accessibility and perceived contrasts however, its comfort quality remains questionable.

1. Introduction

Lighting is one of the most vital issues for persons with low vision or vision impairment to perform everyday activities [1] [2]. Good quality of lights is very important to improve their vision [2], and make it easier to highlight objects and obtain important information from the environment [3]. It was found that visually impaired persons perceived the lighting quality from the same light sources differently compared to persons with normal vision [4]. Moreover, the lighting from a type of energy-efficient fluorescent light bulbs was perceived as not providing sufficient contrast at home [5].

In recent years, the efficacy of light-emitting diodes or LEDs has significantly improved. The sale of such energy-efficient light sources has continued to increase and shared about 33% of residential lighting sales in 2017 [6]. However, a rejection of this new energy-efficient lighting in a housing area was also found because the lighting quality was considered as not promoting a safe outdoor environment for the residents, especially the elderly [7]. Thus, it is important that LED-lighting can meet the needs of all user groups. As LED lamps rapidly become widely available, quality standards for these light sources (e.g. AS/NZS 5341:2021; IEC 62722-2-1) have been developed and introduced to guide suppliers on how to test important parameters such as light output, colour rendering and energy use that can facilitate users' buying decision. However, it is still unclear whether common standards for lighting quality from LED light sources can be applied to visually impaired users. To this point, illuminance and colour rendering index (CRI) are measurable parameters and often described in lighting design guidelines for users with vision impairment [8] [9]. In the guidelines, most of lighting recommendations were proposed

for task lighting. Studies on human perception of lighting quality provided by LEDs in outdoor environments [7] [10] [13] found that cooler light from LED light sources reduced the perceived comfort quality and pleasantness of lighting particularly, in Northern European countries. While high numerical values of CRI are considered as very important in supporting visually impaired persons, choices of light colours are usually made according to personal preferences [8] [9]. Effects of light colours which are described by correlated colour temperature (CCT) is still unclear. Thus, more studies on how different CCTs of LED lighting are perceived would enable a better understanding of the users' needs.
This paper presents a study in controlled laboratory settings in which effects of CCTs on users with goggles simulating two types of vision impairment and users with normal vision were investigated. The focus was on general or ambient lighting for homelike environments.

1.1. Lighting for homelike environment

Good quality of lighting not only means sufficient amount of light or brightness but also, involves suitable choices of light fittings or luminaires, proper direction of the light, minimal glare, and good contrast [2]. Besides improving vision, good lighting should provide safety and contribute to well-being [14] which, according to Stewart & Ware (in [2]), includes a positive psychological state, and emotional and behavioural control. People's positive mood and well-being could be driven by pleasant ambient lights [15]. Regarding this, cosiness is a significant characteristic of homelike environments that could be supported by the use of warm and soft lights, small spotlights, candles and indirect light creating a nice atmosphere [16]. In Swedish households, ceiling mounted luminaires, and floor standing and up-light luminaires are commonly used to provide general lighting in the room while wall-mounted, window and table luminaires are placed to provide lighting for supporting daily activities and/or decorative purposes [17]. During the recent years, LED lamps have started to replace compact fluorescent and halogen lamps for residential use with existing lamp shades [18]. This may be lead to discomfort glare because LED lights usually consist of small light sources which may give higher light levels compared to fluorescent lamps [9].

1.2. Visual accessibility of the lit environment

Visual accessibility for people with low vision or vision impairment has been described in relation to the ability to perceive important features, orientate and perform intended activities in the environment [19] [20] and also, considered as effortless access to the text in case of reading [20]. The concepts of universal design (UD) which referred to as design of products and environment for people of all abilities and ages [21], has been proposed as a means to facilitate visual accessibility of products and environment (i.e. to make them usable by vision) [20], and considered as guidelines for designing supportive indoor environments [22]. Guidelines for indoor lighting with the emphasis on home environments for visually impaired persons have been developed (e.g. [8] [9] [22]). The guidelines have included recommended illuminance levels, characteristics of light sources, and colour contrasts in the physical environment in which according to Dalke et al. [23], can provide vital cues. Colours of floor, wall, door, doorframes and furniture should be chosen to provide sufficient contrasts. CRI of a light source is recommended as $\geq$ Ra80 for ambient lighting and $\geq$ Ra90 for task lighting to support perceived contrasts or colour discriminations [8] [9]. It is also important to avoid using shiny surface materials and therefore discomfort glare [24]. Considering ambient lighting, it is recommended that indirect and shadow-free soft light should be evenly distributed to give uniform lighting in the room [9] [22]. To achieve this, round shaped ceiling pendants have been suggested as a good choice of lamp shades [22]. Recommended illuminance levels for home lighting for visually impaired persons as well as older adults have been found to range from 100 to 300 lux [25] [26] depending on types of rooms or areas. However, it is unlikely that such relatively high levels would be acceptable in private homes [27].

1.3. Effects of CCTs on the perceptions of the environment

The knowledge about LED indoor lighting for visually impaired persons is still limited with regard to effects of CCTs. For people with normal vision, effects of CCTs were found on cognitive process i.e. 5,

000 K LED lighting could lead to improvements in cognitive task performance compared to 3,500 K fluorescent lighting [28]. It was suggested that with the same illuminance levels, higher CCTs will be perceived as having more brightness than the lower CCTs [29]. It was also found that yellowish lighting increased hesitations in wayfinding behaviour [30]. A study on outdoor lighting for visually impaired pedestrians [31] has found that lighting provided by warm white (3,000 K) LED lamps with CRI of Ra80 better supported accessibility among the pedestrians compared to warm lighting (1,800 K) with considerably lower CRI provided by high pressured sodium (HPS) lamps. In this study, the LED lighting was perceived as having more brightness and more evenly distributed however, the lighting from HPS lamps was perceived as having more comfort quality. This finding is in line with a previous study with pedestrians having normal vision [32].

2. Aim and objectives

The aim was to explore the role of CCTs of non-directional LED lamps in supporting users with vision impairment in homelike environments. The specific objectives were to investigate effects of CCT on:

- performance of everyday activities,
- perceptions of lighting quality, and
- perceptions of contrasts and low contrast in the environment

It was hypothesized that with the same illuminance level, higher CCT ambient lighting would (i) better support users in performing everyday activities, and (ii) be perceived as having more brightness regardless vision conditions. It was further hypothesized that higher CCT ambient lighting would better support contrast perceptions and therefore visual accessibility.

3. Method

3.1. Setting

The study carried out in controlled laboratory settings with no access to daylight. The room (3.50 x 2.50 $m^{2,}$ with a floor-to-ceiling height of 2.75 m) was set up to exemplify a common sitting area in homelike environments in healthcare facilities (Figure 1). Ambient lighting in the room was provided by two non-directional LED lamps (E27); one equipped in a round-shaped ceiling luminaire suspended at 2.20 m above the floor and another one in a floor luminaire providing upward light at 1.70 m above the floor (Figure 2(a),(b)). The lamps were selected in accordance with the recommendations i.e. CRI should be of $\geq$ Ra80 [8] and efficacy of $\geq$ 90 lm/W (requirements for non-directional lamps with light output of $\geq$ 100 lm), and the EU's safety requirements [18]. The lamps with warm white (CCT of 2,700 K) and neutral/cool white (CCT of 4,000 K) available in residential lighting market in Sweden were selected. The former CCT was the most available one in the market. Spectral power distributions (SPDs) of the lamps' two CCTs (Figure 3) were captured by Konica Minolta CL-70F. Average horizontal illuminances of the two lit environment were about 100 lux for the 2,700 K lighting and 110 lux for the 4,000 K lighting (measured at approximately 0.85 m above the floor [33] using a Hagner lux meter). The level of 100 lux was the minimum level suggested by IESNA [26] for home environments for visual impaired people and older adults. Colour and lightness contrasts between the walls, floor, furniture and other smaller elements (e.g. shelves, picture, plants, books, floor lamps), ranging from no lightness contrast to about 0.85. (The measurements were conducted by a NCS®[1] lightness meter providing numeric values from 0.00 (no lightness contrast) to 1.00; according to NCS, the sufficient lightness contrast level for people with low vision or vision impairment is 0.40). The contrasts among furniture, wall and floor ranged from no lightness contrast to about 0.35-0.40.

[1] NCS or Natural Colour System is currently the national colour standard in Sweden; the system is based on how human visually perceive colours (ncscolour.com).

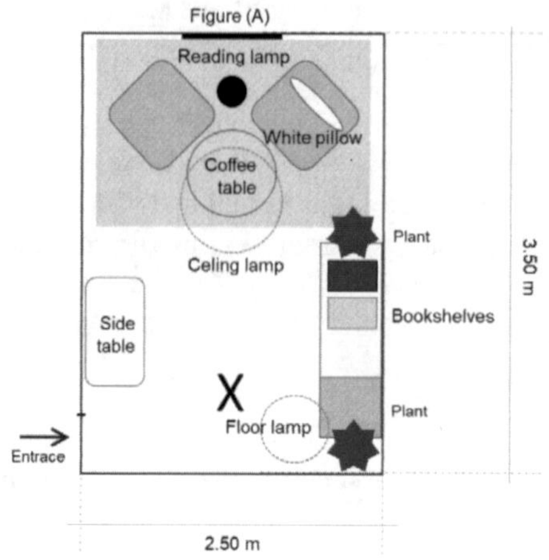

Figure 1. Test room layout

Figure 2(a). Ceiling lamp and floor lamp with the 2,700 K ambient lighting

Figure 2(b). Ceiling lamp and floor lamp with the 4,000 K ambient lighting

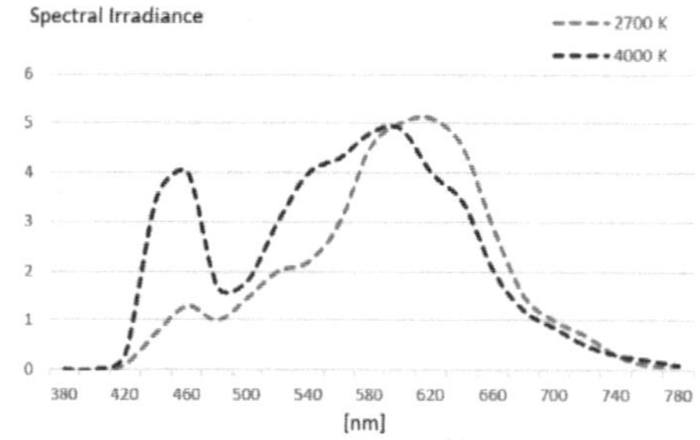

Figure 3. Spectral power distributions of the LED lamps.

3.2. Participants

Persons between 18-37 years having normal eyesight (it could be corrected by wearing glasses or contact lenses) were approached to participate in the study. Participants (24 males, 36 females, mean age = 25.50 years) were informed about the study's aim and procedure, and that the participation is voluntary and would be rewarded with two movie tickets for assessing the two lit environments. The participants' normal visual acuity (20/20) was confirmed by LogMAR chart. Their normal colour vision was confirmed by Ishihara test. They were equally divided into three groups of twenty: two test groups conducted the assessments with goggles with impaired acuity of VA 20/200 (cataract condition) and VA 20/800 (severe cataract condition) respectively, and a reference group conducted the assessments with normal vision. Most participants were Swedish. About 10% of participants in each group came from other countries but had been living in Sweden and/or Northern European countries for more than a year.

3.3. Procedure

Following completion of the eye tests, participants in the test groups had a short training to adapt and feel comfortable to go around with the goggle. Half of the participants started with assessing the lit environment of 2,700 K, and the rest started with that of 4,000 K. After entered the test room, the participants were asked to stand at 3.00 meter from the wall (point X in Figure 1) and looked around. They were given the verbal instruction while adapting to the lighting condition during this time. Under each lighting condition, the participants were first asked to perform seven everyday activities as follows: 1) describing the figure (A) (Figure 4) from the distance of 3 meters, 2) going to sit in the armchair with the white pillow, 3) going to get a glass of water from the side table, 4) going back to sit in the same armchair and putting the glass on the coffee table, 5) going to get the book having blue cover page and orange post-it from the bookshelves, 6) going back to sit in the same armchair and reading the book title on the cover page out loud and 7) turning on the reading lamp beside the chair (Figure 5(a),(b)).

Figure 4. Figure (A) – lightness contrast between the figure and white background = 0.40

Figure 5(a). The lab environment with the 2,700 ambient lighting

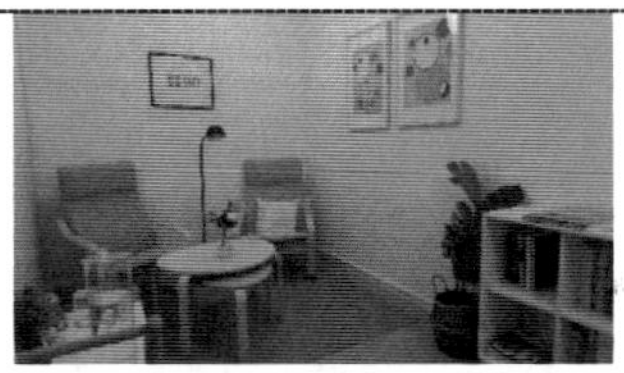

Figure 5(b). The lab environment with the 2,700 ambient lighting

The participants were also encouraged to think aloud while performing the activities to address their specific concerns. The researcher (the author) observed how each activity was performed in accordance with a protocol developed for the study, recorded the time, and asked questions about the difficulty of performing the respective activities. Thereafter, they were asked questions about (i) perceived lighting quality [12] [31] and (ii) perceived difficulty seeing contrasts among walls, floor, furniture and small objects. The answers were given with a 5-point semantic differential scale. After completing the first assessment, the participants took a short break (approx. 3 minutes) outside the test room without taking off the goggle and then, conducted the next assessment. The same sequence of activities was applied to the two assessments and all participants. Each assessment took approx. 15-20 minutes to complete.

3.4. Data analyses

The data were treated by means of statistical analyses using IBM SPSS statistics 27 software. McNemar's tests were used to determine whether there was any association between the different lighting conditions and how the participants performed each of the activities. Non-parametric tests were applied considering small samples ($n = 20$ per group) [34]. Wilcoxon signed rank tests were used to test for differences in recorded time and perceived difficulty of the activities, perceived lighting quality and perceived contrasts. For each lighting condition, Kruskall-Wallis tests were used to test whether there was difference between the groups and further, Mann-Whitney U tests with a stricter significant level of $0.05/3 = 0.017$ were used to determine the differences between pairs of the groups. Think aloud data were used as to supplement the results obtained from the statistical analyses.

4. Results

4.1. Everyday activities

4.1.1. Observation
It was observed that all participants in the reference group with normal vision managed all seven activities under both lighting condition. When asking participants to describe the figure from 3 meters distance (the activity 1)), 95% and 80% of the test group having cataract goggle described the figure correctly under the 2,700 K lighting and the 4,000 K lighting respectively while no one in the test group having severe cataract goggle was able to describe the figure. For the activity 2), i.e. going to sit in the armchair with the white pillow, about 25% of the test group having cataract goggle was hesitant at the beginning and 5% went to the wrong direction under the 2,700 K lighting whereas it was observed that 10% was hesitant at the beginning under the 4,000 K lighting. For the activity 3), i.e. going to get a glass of water from the side table, it was observed that the test group with cataract goggle managed to perform this activity under both lighting conditions without hesitation. About 40% of the test group having severe cataract goggle were hesitant at the beginning and 20% of them went to the wrong direction under the 2,700 K lighting whereas the percentages were 45% and 5% respectively under the 4,000 K lighting. All participants managed the activity 4), i.e. going back and put the glass on the coffee table, under both lighting conditions except about 10% and 15% of the group with cataract goggle that were hesitant at the beginning under the 2,700 K lighting and the 4,000 K lighting respectively. For the activity 5), i.e. going to get the book with blue cover page and orange post-it from the bookshelves, about 5% of the test group with cataract goggle was hesitant at the beginning under the 2,700 lighting whereas the percentage was 10% under the 4,000 K lighting. Of the test group having severe cataract goggle, the percentage of the hesitation was 15% for 2,700 K lighting and 5% under the 4,000 K lighting. After going back to sit, all participants in the test group having cataract goggle managed to read out the book title (the activity 6)) under both lighting conditions. About 30% and 40% of the test group having severe cataract goggle could read out the book title under the 2,700 lighting and the 4,000 lighting respectively. Lastly, it was observed that all participants were able to find the light switch and turn on the reading lamp (the activity 7)). McNemar's tests resulted in no significant associations between the different lighting conditions and how the participants in the respective groups performed each of the activities.

4.1.2. Recorded time
There was only a significant difference in the recorded time between the two lighting conditions for the activity 3), i.e. going to sit in the armchair and putting the glass on the coffee table. A Wilcoxon signed rank test showed that the time for the 2,700 K lighting was significantly longer than that for the 4,000 lighting: $z = 2.40$, $p < 0.05$, $r = 0.31$. Further, a Kruskall-Wallis test resulted in significant differences across the groups for the respective lighting conditions for all seven activities (Table 1).

Table 1. Recorded time (in seconds): mean *(M)*, standard deviation *(SD)*, median *(Mdn)*, and results of between subject differences for the respective lighting conditions ($N = 60$, $df = 2$).

Activity	2700 K				4000 K			
	M	*SD*	*Mdn*	*H*	*M*	*SD*	*Mdn*	*H*
1) Describe the figure				n/a				
2) Go to sit	6.24	3.28	5.80	19.82****	6.10	1.79	6.00	19.72****
3) Get a glass of water	18.98	10.14	15.00	37.77****	18.26	9.72	14.00	29.80****
4) Go back to sit	6.40	1.81	6.00	21.71****	6.00	1.52	6.00	16.72****
5) Get the book	7.25	3.67	6.50	13.07***	8.23	6.99	6.00	13.41***
6) Go back and read	9.45	5.66	8.00	16.14****	10.43	7.19	8.00	22.72****
7) Turn on the lamp	4.84	2.08	5.00	11.42***	4.94	2.26	5.00	9.63***

n/a It is noted that recorded time was not considered for this activity due to variations in verbal expressions of the individuals. ***$p < 0.005$, ****$p < 0.001$

For the 2,700 K lighting, the recorded times of the reference group were significantly shorter than those of the test group having severe cataract goggle for all activities whereas no differences were found when compared to the test group having cataract goggle. There were significant differences between the two test groups. The recorded times of the test group having cataract goggle were significantly shorter than the test group having severe cataract goggle for the activities 2): $U = 76.00$, $z = -3.42$, $p < 0.001$, $r = 0.54$, 3): $U = 12.50$, $z = -5.09$, $p < 0.001$, $r = 0.81$, 4): $U = 73.50$, $z = -3.50$, $p < 0.001$, $r = 0.55$ and 6): $U = 0.00$, $z = -3.79$, $p < 0.001$, $r = 0.60$. Similar results were obtained for the 4,000 K lighting. The recorded times of the reference group were significantly shorter than those of the test group with severe cataract goggle for all activities, and there were no differences between the reference group and the test group with cataract goggle. The significant differences between the two test groups were found for the activities i.e. 2): $U = 66.50$, $z = -3.68$, $p < 0.001$, $r = 0.58$, 3): $U = 43.50$, $z = -4.25$, $p < 0.001$, $r = 0.67$, 4): $U = 74.00$, $z = -3.50$, $p < 0.001$, $r = 0.55$, 6): $U = 1.50$, $z = -4.04$, $p < 0.001$, $r = 0.64$ and also, 7): $U = 110.00$, $z = -2.47$, $p = 0.014$, $r = 0.39$.

4.1.3. *Perceived difficulty*
There were no significant differences in the perceived difficulty in performing the activities between the two lighting conditions. For each lighting condition, the test resulted in significant differences across the groups for all of the activities (Table 2).

Table 2. Perceived difficulty in performing the activities (1= hard, 5 = easy): mean *(M)*, standard deviation *(SD)*, median *(Mdn)*, and results of between subject differences for the respective lighting conditions ($N = 60$, $df = 2$).

Activity	2700 K				4000 K			
	M	*SD*	*Mdn*	*H*	*M*	*SD*	*Mdn*	*H*
1) Describe the figure								
1.1) describe form	3.18	1.74	3.00	53.43****	3.24	1.76	4.00	48.28****
1.2) describe number	3.38	1.80	4.00	51.29****	3.44	1.85	4.75	49.37****
2) Go to sit	4.45	0.87	5.00	23.43****	4.60	0.72	5.00	22.10****
3) Get a glass of water	3.77	1.29	4.00	38.11****	3.82	1.27	4.00	32.01****
4) Go back to sit	4.51	0.85	5.00	9.80**	4.62	0.64	5.00	12.59***
5) Get the book	4.65	0.68	5.00	7.07*	4.53	0.81	5.00	19.94****
6) Go back and read								
6.1) sitting	4.78	0.49	5.00	7.80*	4.75	0.47	5.00	5.64
6.2) reading book title	3.38	1.65	4.00	48.43****	3.42	1.62	4.00	47.40****
7) Turn on the lamp	4.32	0.87	5.00	11.91*	4.43	0.87	5.00	5.04

*$p < 0.05$, **$p < 0.01$, ***$p < 0.005$, ****$p < 0.001$

For the 2,700 K lighting, the reference group rated the perceived difficulty as significantly lower compared to the test group with severe cataract goggle for all of the activities. The differences between the reference group and the test group with cataract goggle were significant for the activities 1.1): $U = 40.00$, $z = -4.96$, $p < 0.001$, $r = 0.78$, 1.2): $U = 80.00$, $z = -4.03$, $p < 0.001$, $r = 0.64$, 3): $U = 76.00$, $z = -3.96$, $p < 0.01$, $r = 0.63$ and 6.2): $U = 66.00$, $z = -4.19$ $p < 0.001$, $r = 0.66$. There were significant differences between the two test groups for the activities 1.1): $U = 0.00$, $z = -5.86$, $p < 0.001$, $r = 0.93$, 1.2): $U = 0.00$, $z = -5.84$, $p < 0.001$, $r = 0.92$, 2): $U = 112.50$, $z = -2.53$, $p = 0.011$, $r = 0.40$, 3): $U = 73.50$, $z = -3.53$, $p < 0.001$, $r = 0.56$ and 6.2): $U = 7.00$, $z = -5.37$, $p < 0.001$, $r = 0.85$. Similar results were obtained for the 4,000 K lighting. The reference group rated the perceived difficulty significantly lower compared to the test group with severe cataract goggle for all activities except the activity 7) i.e. turning on the lamp. The differences between the reference group and the test group with cataract goggle were significant for the activities 1.1): $U = 66.00$, $z = -4.06$, $p < 0.001$, $r = 0.64$, 1.2): $U = 100.00$, $z = -3.57$, $p < 0.01$, $r = 0.56$, 3): $U = 55.00$, $z = -4.39$, $p < 0.001$, $r = 0.69$ and 6.2): $U = 63.50$, $z = -4.20$ $p < 0.001$, $r = 0.66$. Similar to the 2,700 K lighting, the significant differences between the two test groups were found for the activities 1.1): $U = 11.00$, $z = -5.51$, $p < 0.001$, $r = 0.87$, 1.2): $U = 10.00$, $z = -5.63$, p

< 0.001, $r = 0.89$, 2): $U = 104.00$, $z = -2.96$, $p < 0.01$, $r = 0.47$, 3): $U = 83.50$, $z = -3.42$, $p < 0.005$, $r = 0.54$, 5): $U = 106.00$, $z = -2.82$, $p = 0.01$, $r = 0.45$ and 6.2): $U = 10.50$, $z = -5.26$, $p < 0.001$, r =0.83.

4.2. Perceived lighting quality

Both lighting conditions were rated similarly with regard to even distribution, no flicker, and visibility whereas the 2,700 K lighting was perceived as having lower strength quality (PSQ): $z = -4.47$, $p < 0.001$, $r = 0.58$ but higher comfort quality (PCQ): $z = 6.33$, $p < 0.001$, $r = 0.82$ and pleasantness: $z = 4.26$, $p < 0.001$, $r = 0.55$. Table 3 shows significant differences in the perceived lighting quality across the groups. For the 2,700 K lighting, the differences were found for PCQ and perceived visibility. The scores of the reference group rated the perceptions of lighting quality as significantly higher compared to the test group with severe cataract goggle regarding both PCQ: $U = 99.50$, $z = -2.37$, $p = 0.017$, $r = 0.38$ and perceived visibility: $U = 44.50$, $z = -4.52$, $p < 0.001$, $r = 0.72$. In addition, the score of the reference group was significantly higher than that of the test group with cataract goggle for perceived visibility: $U = 84.50$, $z = -3.50$, $p < 0.001$, $r = 0.55$. For the 4,000 K lighting, the significant differences were found for even distribution, pleasantness and visibility of the lighting. The scores of the reference group were significantly higher than the scores of the test group with severe cataract goggle for evenly distribution: $U = 56.00$, $z = -4.18$, $p < 0.001$, $r = 0.66$, pleasantness: $U = 112.00$, $z = -2.47$, $p = 0.17$, $r = 0.39$ and visibility: $U = 8.00$, $z = -5.42$, $p < 0.001$, $r = 0.86$. There were significant differences in perceived visibility between the reference group and the test group with cataract goggle: $U = 76.00$, $z = -3.69$, $p < 0.001$, $r = 0.58$, and between the two test groups: $U = 88.00$, $z = -3.12$, $p < 0.005$, $r = 0.49$.

Table 3. Perceived lighting quality: mean *(M),* standard deviation *(SD),* median *(Mdn),* and results of between subject differences for the respective lighting conditions ($N = 60$, $df = 2$).

	2700 K				4000 K			
	M	*SD*	*Mdn*	*H*	*M*	*SD*	*Mdn*	*H*
PSQ ($N = 57$)	2.91	0.73	2.80	3.62	3.43	0.80	3.40	4.13
PCQ ($N = 56$)	4.04	0.56	4.10	7.26*	2.69	0.73	2.60	0.79
Even distribution	4.10	1.16	4.00	3.68	4.04	1.06	4.00	16.46****
No flicker	4.73	0.69	5.00	3.61	4.63	0.80	5.00	2.42
Pleasantness	4.30	0.92	5.00	1.96	3.47	1.08	3.00	6.54*
Visibility	3.85	1.24	4.00	23.49****	3.73	1.25	4.00	33.90****

*$p < 0.05$, ****$p < 0.001$

4.3. Perceived contrasts

Perceived difficulty in seeing contrasts under the 2,700 lighting were rated as significantly higher than the 4,000 K lighting for seeing lightness contrast at the 0.40 level: $z = -2.20$, $p < 0.05$, $r = 0.28$, furniture: $z = -2.13$, $p < 0.05$, $r = 0.27$ and small items: $z = -2.19$, $p < 0.05$, $r = 0.28$. For the respective lighting conditions, the significant differences were found across the groups (Table 4).

Table 4. Perceived difficulty in seeing contrasts, furniture and small items (1= hard, 5 = easy): mean *(M),* standard deviation *(SD),* median *(Mdn),* and results of between subject differences for the respective lighting conditions ($N = 60$, $df = 2$).

	2700 K				4000 K			
	M	*SD*	*Mdn*	*H*	*M*	*SD*	*Mdn*	*H*
0.40 level	4.48	0.79	5.00	9.55**	4.65	0.55	5.00	17.27****
0.35 level	4.26	0.84	4.50	10.71**	4.43	0.64	4.50	23.56****
0.15 level	4.33	1.02	5.00	12.80***	4.44	0.83	5.00	28.39****
No contrast (0.00 level)	3.44	1.35	3.50	18.85***	3.56	1.38	4.00	30.81****
Furniture (0.00-0.40 level)	4.14	1.03	4.50	16.93****	4.40	0.79	5.00	31.33****
Small items (0.35-0.80 level)	3.30	1.40	3.00	44.12****	3.60	1.43	4.00	44.50****

**$p < 0.01$, ***$p < 0.005$, ****$p < 0.001$

For the 2,700 K lighting, the reference group rated the perceived difficulty as significantly lower compared to the test group with severe cataract goggle for seeing all the contrast levels, furniture and small items. The differences between the reference group and the test group with cataract goggle were significant for seeing furniture: $U = 109.00$, $z = -2.90$, $p = 0.13$, $r = 0.46$ and small items: $U = 29.00$, $z = -4.90$, $p < 0.001$, $r = 0.78$. There were significant differences between the two test groups for seeing no lightness contrast: $U = 65.50$, $z = -3.74$, $p < 0.001$, $r = 0.59$ and small items: $U = 56.50$, $z = -4.10$, $p < 0.001$, $r = 0.65$. Similar results regarding the differences between the reference group and the test group with severe cataract goggle were obtained for the 4,000 K lighting. The reference group rated the perceived difficulty as significantly lower compared to the test group with cataract goggle for seeing lightness contrast at the 0.35 level: $U = 106.00$, $z = -2.89$, $p = 0.01$, $r = 0.46$, no lightness contrast: $U = 105.00$, $z = -2.81$, $p < 0.01$, $r = 0.44$, furniture: $U = 110.00$, $z = -3.35$, $p = 0.14$, $r = 0.53$ and small items: $U = 30.00$, $z = -5.16$, $p < 0.001$, $r = 0.82$. The significant differences between the two test groups were found for seeing the contrast at 0.15 level: $U = 85.00$, $z = -3.31$, $p < 0.001$, $r = 0.52$, no lightness contrast: $U = 75.50$, $z = -3.44$, $p < 0.001$, $r = 0.54$, furniture: $U = 79.50$, $z = -3.44$, $p < 0.001$, $r = 0.54$ and small items: $U = 62.50$, $z = -3.83$, $p < 0.001$, $r = 0.61$.

4.4. Additional information

The information obtained from think aloud data were categorised as perceived difficulty of the respective activities, perceived lighting quality and perceived contrasts. Most of participants in the test groups with the visually impaired goggles expressed difficulties in performing the activities in relation to the vision conditions rather than the lighting conditions. A couple of participants with the goggles expressed that it was easier to perform the activities in particular, going to get a glass of water and reading under the 4,000 K lighting. Regardless the vision conditions, the 2,700 K lighting was often mentioned as giving pleasant or cosy atmosphere in the room. Such warm lighting was also perceived as providing comfort to the eyes. A couple of participants having the goggles perceived that the 4,000 K lighting was sharp and therefore, unpleasant. Most participants in the reference group with normal vision did not detect any differences in the perceived contrasts between the two lighting conditions. Most participants with the visually impaired goggles mentioned that they could see contrasts among furniture, walls and floor, and small items especially clear glasses better in the 4,000 K lighting. Some participants with the visually impaired goggles also mentioned that it was harder to distinguish between white furniture and white wall (no lightness contrast) in such cool lighting and that small items as well as wooden elements stood out and therefore easier to detect them under the 2,700 K lighting. Despite that the cool lighting better supported the perceived contrasts, the participants with cataract goggle expressed that the warm light sufficiently support them to perform the activities and distinguish the contrasts. Regardless the lighting conditions, the participants with severe cataract goggles mentioned that it seemed almost impossible to see lightness contrasts at low levels i.e. 0.10 or no lightness contrast. Furthermore, a couple of them emphasised that high contrasts between small elements (e.g. plants, floor lamp and figures) and their surrounding areas provided environmental cues.

5. Discussion

In this laboratory study, the two CCTs, i.e. 2,700 K and 4,000 K, of non-directional LED lamps available in the Swedish market were investigated whether they provided ambient lighting to support users with vision impaired conditions in homelike environments. With the same CRI of Ra80, the overall results suggested that the different CCTs in this study had effects on performance of everyday activities and perceived contrasts to a very low extent whereas the significant effects were found on perceptions of lighting quality with regard to brightness quality, comfort quality and pleasantness of lighting regardless the vision conditions. The measured SPDs for the two CCTs of the LED lamps were slightly different. The higher CCTs seemed to allow for response to green-blue tints and may thereby suggest a positive effect of high CCT on vision at low illuminance levels. However, both could be considered as having closer alignment with mesopic vision compared to other types of lamps.

The different CCTs in this study did not play a significant role in performance of the activities with regard to how the participants performed the activities, the recorded time and the perceived difficulty despite that the 4,000 K lighting was perceived as having more brightness whereas the 2,700 K lighting resulted in higher comfort quality and pleasantness. The perceived brightness was in line with the measured illuminances levels showing 10 lux higher average illuminance level of the 4,000 K lighting. The mismatch found between perceived strength quality and perceived comfort quality of the 'warm' and 'cool' lights affirmed the findings of previous studies on outdoor lighting considering the pedestrians having different types of vision impairment [31] and normal vision [32]. It was also found that the ambient lighting level of about 100 lux was to some extent, sufficient for supporting the users with moderate vision impairment (VA 20/200) in homelike environments. Both CCTs gave positive results with regard to even distribution which is important to people with vision impairment [22]. Further, the results indicated that the 4,000 K lighting better supported the perceptions of lightness contrasts and therefore, visual accessibility of the environment compared to the 2,700 K lighting. This finding is to some extent consistent with a previous study [31] suggesting the positive effect of cooler CCT on accessibility for visually impaired pedestrians.

Considering the differences across the groups having different vision conditions, the effects of the two CCTs were rather similar. It was obvious that the test groups having visually impaired goggles in particular, the test group with severe cataract goggle, faced more difficulty when performing the activities, and rated the lighting quality and the perceived contrasts as significantly lower than the reference group with normal vision under both lighting conditions. Significant differences in the perceived difficulty among the groups were found for more activities for the 4,000 K lighting compared to the 2,700 K lighting. The differences in the perceived lighting quality among the groups somewhat differed between the lighting conditions in which the differences in perceived comfort quality and visibility were found for the 2,700 K lighting whereas the differences were found for pleasantness, even distribution and visibility for the 4,000 K lighting. Moreover, the difference in visibility was found between the two test groups having visually impaired goggles for the 4,000 K lighting whereas no difference was found for the 2,700 K lighting. These findings emphasised the important role of CCTs in affecting the perceptions of lighting quality and also indicated the positive effect of cool white CCTs on visual accessibility for persons with mild to moderate vision impairment. It was also obvious that the test group with severe cataract goggle rated the difficulty in seeing the different contrast levels, furniture and small items with high scores and that high contrasts between the small elements and their surroundings could provide them environmental cues to perform the activities. Together the results on perceived lighting quality and perceived contrasts indicated the positive effects of cool white CCT however, the results also suggested that the CCT of 2,700 K could provide comfort and pleasant ambient lighting that to some extent supported visual accessibility for visually impaired users with vision impairment. Though the cooler CCT provided more brightness and better supported the perceived contrasts, its comfort quality is still questionable regardless of vision conditions. Some participants mentioned that the 4,000 K lighting was sharp they tended to focus in order to see and perform the activities when having the visually impaired conditions. This may therefore, lead to eye fatigue.

The study had a particular focus on effects of CCTs rather than other parameters such as spectral power distribution, illuminance and flicker and also, small samples with a major percentage (about 90%) of Swedish participants that could limit the ability to generalize the results. It is noted that goggles used to simulate the visually impaired conditions may not accurately imitate actual vision impairment, and that the participants had a limited amount of time to adapt to the goggles. This may not be sufficient to represent ones having vision impairment that occurs gradually or for years [35]. However, the goggles allowed consistent types of vision impairment and therefore, provided a basis for shaping a better understanding about the roles of lights and colours in relation to visual accessibility. Regarding this, it would be desirable to examine and compare the impact of lights on visual perception between simulated and actual vision impairment with larger sample sizes and at the same time improving the assessment methodology.

References

[1] Cornelissen FW, Bootsma A and Kooijman AC 1995 Object Perception by Visually Impaired People at Different Light Levels *Vision Res.* **35**(1) 161-8

[2] Brunnström G, Sörensen S, Alsterstad K and Sjöstrand J 2004. Quality of light and quality of life – the effect of lighting adaptation among people with low vision *Ophthalmic Physiol. Opt.* **24**(4) 274-80

[3] CIE 1997 Lighting needs for partially sighted: technical report 123 *Low Vision* (Vienna: CIE) p109-21

[4] Johansson M, Rosén M and Küller R 2011 Individual factors influencing the assessment of the outdoor lighting of an urban footpath *Lighting Res. Technol.* **43** 31-43.

[5] Raizi A, Boon MY, Bridge C and Dain SJ 2012 Home modification guideline as recommended by visually impaired people *J. Assist. Technol.* **6**(4) 270-84

[6] International Energy Agency - IEA 2019 *Lighting: tracking clean energy progress* Retrieved from http://www.iea.org.tcep/buildings/lighting/ 25 January 2019

[7] Johansson M, Küller M and Pedersen E 2015 Understanding a housing cooperatives' reasons for rejecting energy-efficient outdoor lighting *Lighting Res. Technol.* **47** 876-92

[8] CIBSE – the Society of Light and Lighting 2012 Lighting for people who are visually impaired *Factfile* **8**

[9] CIE 2017 *Lighting for older people and people with visual impairment in buildings: technical report CIE 227:2017* ed Y Akashi, M Cobham, N Itoh, NJ Miller, LJM Schlangen and JHF van den Broek Cools (Vienna: Commission Internationale De L'Eclairage CIE Central Bureau)

[10] Kuhn L, Johansson M, Laike T and Goven T 2013 Residents' perceptions following retrofitting of residential area outdoor lighting with LEDs *Lighting Res. Technol.* **45**(5) 568-84

[11] Pedersen E and Johansson M 2018 Dynamic pedestrian lighting: effects on walking speed, legibility and environmental perception *Lighting Res. Technol.* **50** 522-36

[12] Johansson M, Pedersen E, Maleetipwan-Mattsson P, Kuhn L and Laike T 2014 Perceived outdoor lighting quality (POLQ): a lighting assessment tool. J. Environ. Psychol. **29** 14-21

[13] Juntunen E, Tetri E, Tapaninen O, Yrjänä S, Kondratyev V, Sitomaniemi A and Heikkinen V 2015 A smart LED luminaire for energy savings in pedestrian road lighting *Lighting Res. Technol.* **47** 103-15

[14] Bellia L, Bisegna F and Spada G 2011 Lighting in indoor environments: visual and non-visual effects of light sources with different spectral power distributions *Build. Environ.* **46** 1984-92

[15] Kuijsters A, Redi J, de Ruyter B and Heynderickx 2015 Lighting make you feel better: improving the mood of elderly people with affective ambiences *PLOSONE.* https://doi.org/10.1371/journal.pone.0132732

[16] Bladh M and Krantz H 2008 Towards a bright future? Household use of electric light: a microlevel study *Energy Policy.* **36** 3521-30

[17] Gerdhardsson KM, Laike T and Johansson M 2018 Residents' lamp purchasing behaviour, indoor lighting characteristics and choices in Swedish homes *Indoor Build. Environ.* **0** 1-19

[18] Kofod C and Øbro P 2014 *Availability of non-directional LED replacement lamps* Retrieved from http://www.premiumlightpro.org.uk/fileadmin/uk/Premium_Light_Pro_Indoor_LED_Guidelines.pdf 9 December 2018

[19] Legge GE, Yu D, Kallie CS, Bochsler TM and Gage R 2010 Visual accessibility of ramps and steps *J. Vis.* **10** 1–19

[20] Legge GE 2014 Visual Accessibility: a Challenge for Low-Vision Research *Optom. Vis. Sci.* **91** 696-706

[21] Story MF, Mueller JL and Mace RL 1998 *The universal design file – designing for people of all ages and abilities* (Raleigh, NC: the Center for Universal Design, NC State University)

[22] Thomas Pocklington Trust 2018 *Lighting in and around the home: a guide to better lighting for people with sight loss* ed N Huszarik, P Hodgson, L Watson Retrived from https://www.pocklington-trust.org.uk/wpcontent/uploads/2018/04/Lighting-Guide-2018-PDF-1.pdf 2 February 2019

[23] Dalke H, Conduit GJ, Conduit BD, Cooper RM, Corso A and Wyatt DF 2010 A colour contrast assessment system: design for people with visual impairment *Designing Inclusive Interactions* ed P Langdon, P Clarkson and P Robinson (London: Springer) https://doi.org/10.1007/978-1-84996-166-0_10

[24] Torrington JM and Tregenza PR 2007 Lighting for people with dementia *Lighting Res. Technol.* **39** 81-97

[25] CIBSE 1994 *Code for interior lighting* (London:CIBSE)

[26] IESNA 1998 *Recommended practice for lighting and the visual environment for senior living* (New York, NY: IESNA)

[27] Bright KT, Cook GK, O'Neill LA and Hill SL 2003 Lighting the homes of visually impaired people *Proc. 11th Int. Mobility Conf. IMC11* (Stellenbosch, South Africa) 30 March – 4 April

[28] Hartstein LE, Durniak MT, Karlicek RF Jr and Berthier NE 2018 A comparison of the effects of correlated colour temperature and gender on cognitive task performance *Lighting Res. Technol.* **50** 1050-69

[29] Ju J, Chen D and Lin Y 2012 Effects of correlated colour temperature on spatial brightness perception *Color Res Appl.* **37**(6) 450-4
[30] Kumoglu Suzer O, Olgunturk N and Guvenc D 2018 The effects of correlated colour temperature on wayfinding: a study in a virtual airport environment *Displays.* **51** 9-19
[31] Mattsson P, Johansson M, Almén M, Laike T, Marcheschi E and Ståhl A 2020 Improved usability of pedestrian environment after dark for people with vision impairment: an intervention study. *Sustainability* **12**(3) 1096. https://doi.org/10.3390/su12031096
[32] Rahm J and Johansson M 2018 Assessing the pedestrian response to urban outdoor lighting: a full-scale laboratory study *PLOSONE.* **13**(10) https://doi.org/10.1371/journal.pone.0204638
[33] Küller R 2004 Planning for good indoor lighting *Building Issues* **14**(1) 4-20
[34] Field A 2013 *Discovering statistics using IBM SPSS statistic, 4ed.* (Sage).
[35] Rousek JB and Hallbeck MS 2011 The use of simulated visual impairment to identify hospital design elements that contribute to wayfinding difficulties *Int. J. Ind. Ergon.* **41** 447-58

Acknowledgments
The study was supported by a grant from the Swedish Energy Agency (2018-004710, P46675-1).

Dimming the Lights to Support the Overall Human Wellbeing In An Urban Context

Oktay Akanpinar

48-53 45th St. Unit B3, Queens, New York, USA

oktayakanpinar@gmail.com

Abstract. In modern times we are exposing ourselves to artificial light sources more than our ancestors were and it is conflicting with learned behaviours developed during the evolutionary process over thousands of years. In this review paper, the effect of this rapid change and possible effects on human wellbeing is questioned. Publications detailing, the health consequences of circadian disruptions and the current stage of global artificial light at night publications have been reviewed to see the level of intrusive lighting at night and to understand the possible connection. The current level of light pollution is mainly questioned as it is reported that the human circadian system and melatonin suppression is more sensitive to a lower level of light than has been previously reported [1]. During the time of review "Light at Night and Circadian" terms used in at Google Chrome search engine and, 24,557 publications were found on Science Direct [2], 17,500 publications were found on Google Scholar [3], and 1,240 free publications were found on PubMed [4] sites that are directly related to the circadian rhythm and light at night topics. Among those publications, recently published research papers (2010-2020) have been selected for review to narrow down the reviewed light sources to light emitting diodes (LEDs) as it is noted that the circadian system is susceptible to the blue wavelengths and LEDs are showing a peak in their spectrum at the same sensitivity area. At the end of the review, possible direct connections between the use of artificial light and adverse physiological and psychological effects with more severe health problems have been found. In addition, the current light pollution studies showed that vast majority of the global population is living under light polluted skies and possibly experiencing the negative effects of disrupted circadian system daily and therefore experiencing deteriorating effects of circadian disruptions and personal wellbeing. As the use of artificial light rapidly increases with global population, exposure to artificial light will increase exponentially in the near future and not only human beings but also the rest of the ecosystem will suffer from its negative effects. Because of this rapid change we should start implementing more stringent guidelines written with consideration for the health implications of artificial light while controlling the excessive use of it. Also, as the LED based lighting systems increase their popularity worldwide, we should start researching a possible modification to reduce the impact of the LED lighting on circadian system and its negative effects on human wellbeing.

1. Effects of Circadian Disruptions on Human Wellbeing

Recent research shows that human circadian system is more complicated and delicate than it previously thought. It not only plays an important role in the sleep-wake cycle but also is important for diverse physiological and psychological effects on human wellbeing. As it is widely reported this system gathers its most important cues from environmental inputs and mainly from light-dark patterns.

Aside from rods and cones, which are related to the scotopic and photopic vision, another set of cells have been discovered in the human eye that are directly connected to the suprachiasmatic nuclei (SCN) in the brain and therefore to the circadian system. These relatively small numbers of cells have melanopsin in their cellular structure and they show peak sensitivity against blue-light at around 480nm [5]. Due to their cellular structure and light sensitivity these special cells are called photosensitive retinal ganglion cells (pRGCs). pRGCs contributes to number of non-image forming (NIF) responses to light, such as the pupillary light response, negative masking of locomotor activity, the regulation of sleep-wake patterns, photophobia, and light aversion, as well as effecting mood-related behaviors and cognitive functions [6]. Moreover, it has been discovered that pRGCs have five distinct subtypes (M1 to M5) that show diverse sensitivity against input from environmental changes and providing input to as many as 32 different brain regions [7]. This shows that acute and long term environmental input changes can show direct effects on brain activity, circadian rhythm and its oscillation, and regulation of physiological and psychological effects depending on the regulation of melatonin secretion. These acute or long-term environmental input changes result in phase shift in the healthy circadian system and a resynchronization would be required between the SCN and the peripheral oscillators [8]. These desynchronization effects are widely visible in humans who work during the nightshift, travel transcontinental flight, and change their social patterns constantly. Aside from those high impact changes in the environmental inputs it is shown that smaller impact changes in the light-dark pattern can play a crucial role in circadian disruption, melatonin secretion, and eventually in overall human wellbeing as well.

With the technological advancement and the wide use of Light Emitting Diodes (LEDs) globally, we are encountering those circadian disruptive inputs in daily basis. It is known that the LEDs are show peak spectral distribution in the blue-light portion between 400-500nm depending on their correlated colour temperature (CCT) value. As noted previously, melanopsin shows peak sensitivity at around 480nm and LED lights play a crucial role for entertaining the pRGCs and therefore the oscillation of the melatonin secretion and suppression of melatonin. In their research study, Karatseroes et. al., reported that just four hour of light/dark cycle disruption can lead to modification in normal body temperature rhythms, increase in weight gain regardless of the food intake amount, and results in higher levels of plasma, insulin, and leptin in mice models [8]. Also, the same group reported that the mice models that are in phase shifted light/dark cycles showed slower cognitive behaviors compared to the control group [8]. In addition, Walker II et. al., reported that modified circadian rhythms are frequently observed in persons who are showing major depressive disorder, bipolar disorder, anxiety, and schizophrenia symptoms [9]. Similarly in a recent study Figueiro reported that irregular light–dark patterns or light exposures at the wrong circadian time can lead to circadian disruption and increased risk for disease [10]. Aside from noted psychological symptoms Lax et. al., reported possible connections between circadian disruption and reduction in pRGC numbers in retinal diseases as well as neurodegenerative diseases such as Parkinson and Alzheimer [11].

Additional studies have been conducted to better assess the circadian disruption and the lighting intensity, CCT and the colour of the light, as well as the admission time. Zeitzer et. al., reported that even around 100lx general illumination can shift the circadian rhythm as well as suppress the melatonin production more than eighty percent compared to the control group when administered during the early biological night [12]. Additionally, in their research they noted the lowest melatonin suppression at around 80lux with a steep increase above 200lx [12]. However, it must be noted that during their research Zeitzer et. al., utilized fluorescent based luminaires and a UV restricting lens. Would the same results be achieved if the same experiment were conducted with LEDs? Would we see less melatonin suppression or smaller phase shift in the circadian rhythm, or would the lower light levels create the same effect due to the spectral power distribution curve of the LEDs? Stebelova et. al., conducted a similar study using LED based luminaires at 610nm peak sensitivity and lower ambient illuminance levels (1 and 5lux) during night-time to evaluate the melatonin levels and sleep quality. They reported that whole night exposure to one lux did not affect the measured melatonin concentration, but it did deteriorate the sleep quality and exposure while five lux both reduced the measured melatonin concentration and worsened the sleep quality [13]. A similar experiment was conducted by Arguelles-Prieto, Madrid, Rol & Bonmati-Carrion to understand the effect of different CCTs with similar illuminance levels on the circadian system. Their experiment was conducted using 3000K and 5700K LED light sources at around 550lux ambient light conditions. During their three week experiment, participants are directed to collect their melatonin levels using Dim Light Melatonin Onset (DLMO) protocol during the first week in their home setting under dim light (<10lux) conditions. During the laboratory setting, participants were exposed to dim light for three hours using either 3000K or 5700K light sources and two hours bright light setting under 569lux and 581lux respectively. Following the bright light exposure protocol participants are exposed to dim light again to assess whether melatonin suppression is continuous even after acute bright light exposure. They reported that under both bright light settings melatonin suppression was higher than dim light settings and at the 5700K setting the suppression was slightly higher than 3000K setting [14]. In another study Nagare, Rea & Figueiro questioned whether the position and the colour of the light source would affect the melatonin suppression level. During their experiment they located the light sources at on-axis and off-axis of the eye gaze. They reported that melatonin suppression was significantly higher for the on-axis condition compared to off-axis condition for both blue (451nm) and green (522nm) light conditions [15]. This shows that while the pRGCs are densely located at the centre of the eye they are still collecting light input information even the light source is located at the peripheral location. At another experiment Philips et. al., questioned whether the melatonin suppression against light depends on the light sensitivity of the humans. In their seven week long experiment they exposed their 56 participants to changing light exposures between 10-2,000lux with using an LED light source at 451nm peak spectrum and 4,289K CCT. During their control week they administered less than 1lux using T5 fluorescent lamp source at 545nm peak spectrum and 3,968K CCT. The administration of the light exposure was started 4 hours before the habitual bed time and finalized 1 hour after the habitual bedtime. At the end of their study, they reported that more than half of the melatonin suppression is appeared under 30 lux conditions and the sensitivity of the melatonin suppression was more than fifty-fold per individual [16]. This and other mentioned studies are showing that the human circadian system is highly sensitive to the light and it can be disrupted even with lower level of light compared to typical indoor lighting settings. Some of the mentioned researches are showing that even the current hand held technological devices can produce enough light to disrupt the circadian system.

2. Artificial Light at Night and its Effects on Human Well Being

Changes in our daily behavior have exposed us to artificial light for longer periods of time compared to our ancestors. This exposure is not only happening during the day-time but is also getting more prominent during night-time as well. Night-time natural lighting levels are generally dictated by the Moon, scattered light through the air, the stars and the Milky Way, and zodiacal light, however natural night sky illuminance is increasing due to the artificial light at night being scattered to the atmosphere [17]. The latest reports state that more than four-fifths of the global population and almost all of the United States and European populations, as well as more than twenty percent of the global terrain, are living under light polluted skies that are created by excessive, poorly directed, or noticeable outdoor light [6].

IOP Conf. Series: Earth and Environmental Science 1099 (2022) 012050 doi:10.1088/1755-1315/1099/1/012050

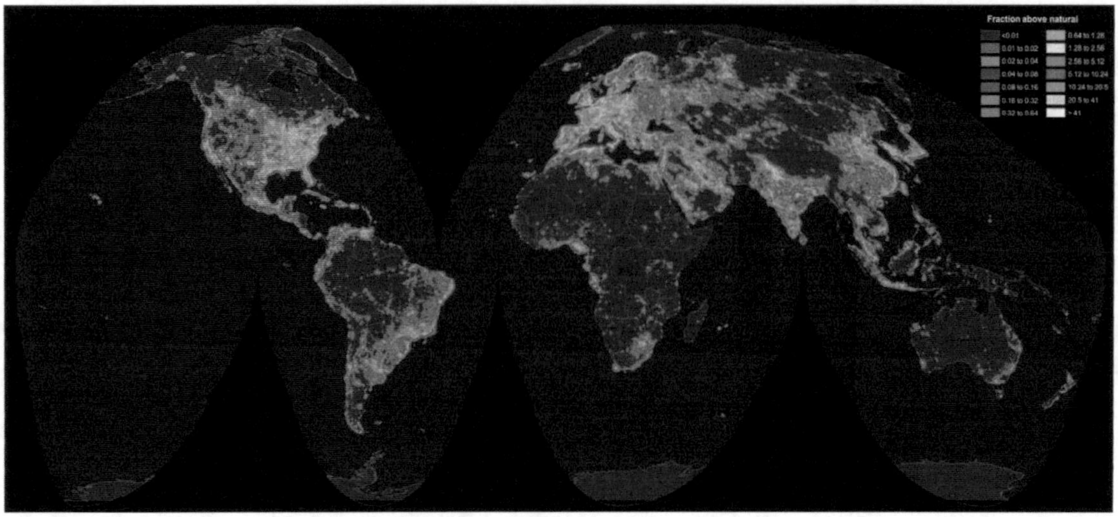

Figure 1. World map of artificial sky brightness [17].

The latest artificial sky brightness world map by Falchi et. al., shows that the vast majority of the densely populated areas in the world are living under constant light polluted skies (Fig. 1). At Figure 1, the black section show the areas that are not affected by light pollution and can be considered as pristine sky, and dark gray sections represent the areas that needs to be protected from future light pollution. Meanwhile, blue sections represent the areas that are considered light polluted, yellow sections, aside from light pollution, show that the general population cannot observe winter Milky Way. Furthermore, in orange sections even the summer Milk Way can't be observed due to the high level of light pollution, and the red indicated sections show the level of light at night that is equal where the transition between scotopic and mesopic vision is taking place [17]. People who are living in the areas that are marked with red and above are not experiencing the mesopic vision during night time due to the excessive light pollution.

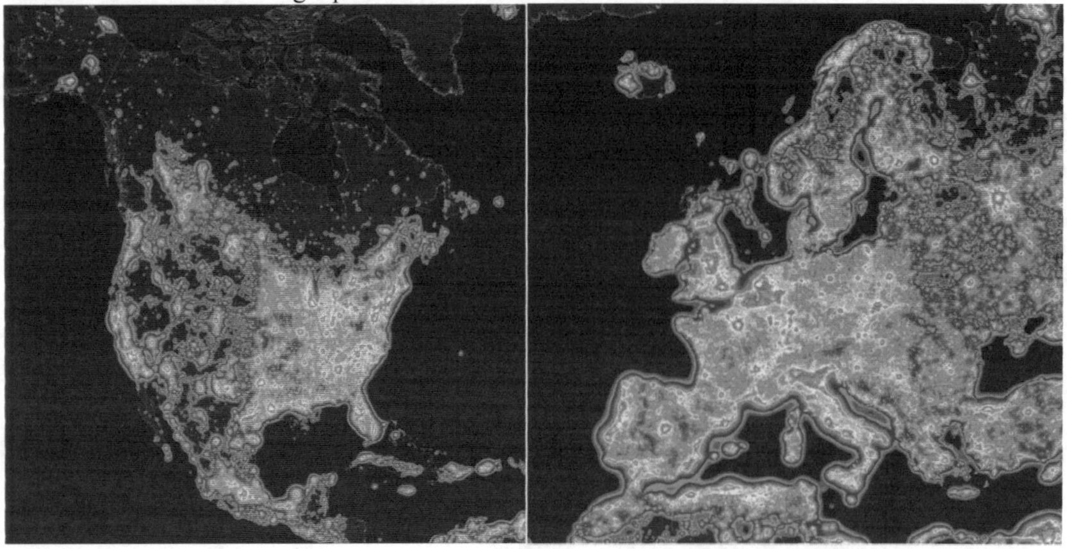

Figure 2. Map of North America's and Europe's artificial sky brightness [17]. Refer to scale at the Figure 1 for more information.

This map indicates that more than thirty percent of the world population cannot observe the Milky Way. Furthermore, more than one-fifth of the world's land surface and more than eighty percent of Europe as well as almost fifty percent of United States experience nights under light pollution [17]. In other words, most of the world's population is affected by light pollution on a nightly basis. As it is reported that the light intensity and exposure time shows disruptive effects on circadian rhythm we must question if the excessive light pollution is playing a role in the sleep-wake cycle and ultimately the overall wellbeing of humanity. In their eight yearlong study Ohayon, M., & Milesi., C., interviewed more than fifteen thousand adults that are living in fifteen different states of varying light pollution levels in the United States to assess the effects of artificial light at night and sleep behaviors. At the end of their study, they reported that the participants who were living in under higher outdoor light night levels more frequently reported delayed bed and wakeup time, reduction in sleep time, and discomfort with their sleep quality, as well as increase in sleepiness during day time [18]. Similar findings are reported by Paksarian et. al., at the end of their three yearlong study, which included over ten thousand adolescents living in the different parts of the United States [19]. Those reported effects are commonly diagnosed in the individual who is suffering from circadian rhythm disorders. In another study, Lee et. al., administered dim light ranging from 5-10lux at the eye level to 30 adult men in two different groups to assess the cognitive function decrease in healthy humans in a two night study. Following the control night and low-level light administration they reported that there was a significant difference for both groups in cognitrone test compared to control night, however there was no significant difference in the vigilance and motor performance test series [20]. On the contrary, in another study conducted by Kang et. al., it was reported that the brain activity of the right and left frontal portion of the brain decreased after nighttime exposure to 10lux illuminance during sleep [21]. That shows an important effect of the nighttime light exposure during sleep as those areas of the brain play an important role in working-memory tasks and the possible decrease in brain activity can create significant reduction in human wellbeing.

Aside from the negative effects of light at night in sleep quality, it is reported that light at night can be a cause for more severe health problems as well. In a workshop held by Lunn et. al., it was reported that maintaining a daily pattern of light exposure is important to preserve the overall health, as the pervasive nature of the light at night can potentially increase the risk factor of cancer, obesity, and heart disease for the general public [22]. In large-scale study conducted among one hundred thousand British women researchers reported a positive relationship between obesity and high level of artificial light at night [23]. Obayashi et. al., conducted a two year study with 528 elderly in Japan to assess the long term effects of the artificial light at night on health and at the end of their study they reported that the group exposed to artificial light at night (average >3 lux) showed significantly higher body weight, waist circumference, glucose, and lipid metabolism [24]. In a supporting study, Fonken et. al., reported that dim artificial light at night (5 lux) alters the food intake time while creating metabolic abnormalities and resulting in obesity in the mice models [25]. In another study that was conducted in South Korea by Kim et. al., they reported a significant connection between breast cancer incidence for women who were living in areas that are affected by higher artificial light at night levels [26]. Similar results were reported by Kloog et. al. as well. In their study, they assessed one hundred forty seven different communities in Israel and breast cancer incidence rate and they reported a positive relationship between the numbers of breast cancer incidences with higher light at night readings [27]. Garcia-Saenz et. al., conducted a five yearlong study in eleven different regions of Spain, which included more than a thousand female controls and around nine hundred male controls that never worked night-shifts to assess the relationship between light at night and prostate and breast cancer incidence rate [6]. Among those controls 88% of females and 70% of the males reported as cancer cases who were living in the areas that are vastly affected by artificial light at night and refurbished with LED streetlights [6]. James et. al., conducted a large scale study between 1989-2013 among one hundred sixteen thousand four hundred thirty registered nurses (with more than forty percent being actively working in night-shift) to assess the relationship between night-shift work and their exposure to artificial light at night in their home setting to the breast cancer incidents among the study group

[28]. At the end of their fourteen year study they reported a more than ten percent increase in breast cancer incident risk among the nurses who were living in areas with higher artificial light at night and worked a night-shift [28]. Another wide scale study conducted by Al-Naggar and Anil to assess the possible relationship between all forms of cancer (lung cancer, breast cancer, colorectal cancer and prostate cancer) and country specific variables (general population, electricity consumption, air pollution and total forest area) that are adjusted according to the protected area light pollution indicator (PALI) and the protected area human influence indicator (PAHI) correlations. According to their findings, there is a significant positive relationship between the increases in numbers of all forms of cancer incidence with higher PALI and PAHI readings [29]. It should be noted that even though their study had a broader impact correlation assessment regarding the total number of all forms of cancer incidence among 158 countries, the higher incidence percentages are reported from countries that are highly affected by the artificial light at night and light pollution. All the reviewed studies reported similar results regarding negative effects of artificial light at night and light pollution on human wellbeing and life threatening diseases.

3. Conclusion

Cities are rapidly increasing their size to accommodate the increasing population and it drives the demand for outdoor and indoor lighting to higher levels. Rapid migration to LED technology and LED based luminaires provides quick solutions to meet this rising demand while increasing the areas getting affected by the light pollution. It is reported that between one tenth to one fifth of global land experiencing some degree of light pollution [30]. Additionally, almost seventy five percent of the seafloor adjacent to cities reported to be affected by artificial light at night to a level that could create negative biological responses for the animals living in that vicinity [31]. It is reported that aside from its negative effects on human wellbeing, artificial light at night shows disruptive effects on different land and water-based organisms while negatively impacting their natural circadian systems [32].

Current global situation shows that the LED based luminaires are being utilized at a faster pace. For instance, just in Switzerland, the number of LED luminaire sales tripled between 2014 and 2019 and in 2019 LED based luminaires were reported to take more than fifty percent of the market share [33]. Aside from the vast use of LED technology, it is reported that the current technology utilized for assessment of light pollution is not capable of capturing the shorter wavelength of the artificial light [17]. This shows a possible problem for future light pollution reports as the LED based luminaires are showing a peak at a shorter wavelength. With the wide acceptance of the LED based luminaires, we might be seeing a possible decline in reported light pollution while experiencing higher light pollution. All the reviewed studies show that the human circadian system is more sensitive even in lower light levels and it can be disrupted by light pollution in many different ways.

It must be noted that, current lifestyle of the global population is contradicting with our bodies circadian system, which been in development for millennia during the human evolution [33]. Therefore, it becomes vital to consider the health related aspects of the artificial lighting in every application while focusing on the possible reductions of the excessive use of artificial light when it is not needed. In addition, new lighting source research should be considered to reduce the negative health effects of artificial light compared to the current LED based luminaires.

References

[1] Vartanian G, Li B, Chervenak A, Walch O, Pack W, Ala-Laurila P et al. Melatonin Suppression by Light in Humans Is More Sensitive Than Previously Reported. Journal of Biological Rhythms. 2015;30(4):351-354.

[2] Light at Night Circadian – Search Results - Science Direct [Internet]. Sciencedirect.com. 2022 [cited 23 March 2022]. Available from: https://www.sciencedirect.com/search?qs=light%20at%20night%20circadian

[3] Light at Night Circadian – Search Results – Google Scholar [Internet]. Google Scholar. 2022 [cited 22 March 2022]. Available from: https://scholar.google.com/scholar?q=circadian+disruption+artificial+light+at+night&hl=en&as_sdt=0%2C33&as_ylo=2010&as_yhi=2022

[4] Light at Night Circadian - Search Results - PubMed [Internet]. PubMed. 2022 [cited 23 March 2022]. Available from: https://pubmed.ncbi.nlm.nih.gov/?term=light+at+night+circadian&filter=simsearch2.ffrft&filter=simsearch3.fft&filter=years.2010-2022

[5] Enezi J, Revell V, Brown T, Wynne J, Schlangen L, Lucas R. A “Melanopic” Spectral Efficiency Function Predicts the Sensitivity of Melanopsin Photoreceptors to Polychromatic Lights. Journal of Biological Rhythms. 2011;26(4):314-323.

[6] Garcia-Saenz A, Sánchez de Miguel A, Espinosa A, Valentin A, Aragonés N, Llorca J et al. Evaluating the Association between Artificial Light-at-Night Exposure and Breast and Prostate Cancer Risk in Spain (MCC-Spain Study). Environmental Health Perspectives. 2018;126(4):047011.

[7] Hughes S, Jagannath A, Rodgers J, Hankins M, Peirson S, Foster R. Signalling by melanopsin (OPN4) expressing photosensitive retinal ganglion cells. Eye. 2016;30(2):247-254.

[8] Karatsoreos I, Bhagat S, Bloss E, Morrison J, McEwen B. Disruption of circadian clocks has ramifications for metabolism, brain, and behavior. Proceedings of the National Academy of Sciences. 2011;108(4):1657-1662.

[9] Walker W, Walton J, DeVries A, Nelson R. Circadian rhythm disruption and mental health. Translational Psychiatry. 2020;10(1).

[10] Figueiro M. Disruption of Circadian Rhythms by Light During Day and Night. Current Sleep Medicine Reports. 2017;3(2):76-84.

[11] Lax P, Ortuño-Lizarán I, Maneu V, Vidal-Sanz M, Cuenca N. Photosensitive Melanopsin-Containing Retinal Ganglion Cells in Health and Disease: Implications for Circadian Rhythms. International Journal of Molecular Sciences. 2019;20(13):3164.

[12] Zeitzer J, Dijk D, Kronauer R, Brown E, Czeisler C. Sensitivity of the human circadian pacemaker to nocturnal light: melatonin phase resetting and suppression. The Journal of Physiology. 2000;526(3):695-702.

[13] Stebelova K, Roska J, Zeman M. Impact of Dim Light at Night on Urinary 6-Sulphatoxymelatonin Concentrations and Sleep in Healthy Humans. International Journal of Molecular Sciences. 2020;21(20):7736.

[14] Arguelles-Prieto R, Madrid J, Rol M, Bonmati-Carrion M. Correlated color temperature and light intensity: Complementary features in non-visual light field. PLOS ONE. 2021;16(7):e0254171.

[15] Nagare R, Rea M, Figueiro M. Spatial sensitivity of human circadian response: Melatonin suppression from on-axis and off-axis light exposures. Neurobiology of Sleep and Circadian Rhythms. 2021;11:100071.

[16] Phillips A, Vidafar P, Burns A, McGlashan E, Anderson C, Rajaratnam S et al. High sensitivity and interindividual variability in the response of the human circadian system to evening light. Proceedings of the National Academy of Sciences. 2019;116(24):12019-12024.

[17] Falchi F, Cinzano P, Duriscoe D, Kyba C, Elvidge C, Baugh K et al. The new world atlas of artificial night sky brightness. Science Advances. 2016;2(6).

[18] Ohayon M, Milesi C. Artificial Outdoor Nighttime Lights Associate with Altered Sleep Behavior in the American General Population. Sleep. 2016;39(6):1311-1320.

[19] Paksarian D, Rudolph K, Stapp E, Dunster G, He J, Mennitt D et al. Association of Outdoor Artificial Light at Night With Mental Disorders and Sleep Patterns Among US Adolescents. JAMA Psychiatry. 2020;77(12):1266.

[20] Lee Y, Cho C, Lee H. Impact of Exposure to Dim Artificial Light at Night during Sleep on Cognitive Function. Chronobiology in Medicine. 2020;2(4):160-165.

[21] Kang S, Yoon H, Cho C, Kwon S, Kang J, Park Y et al. Decrease in fMRI brain activation during working memory performed after sleeping under 10 lux light. Scientific Reports. 2016;6(1).

[22] Lunn R, Blask D, Coogan A, Figueiro M, Gorman M, Hall J et al. Health consequences of electric lighting practices in the modern world: A report on the National Toxicology Program's workshop on shift work at night, artificial light at night, and circadian disruption. Science of The Total Environment. 2017;607-608:1073-1084.

[23] Dominoni D, Borniger J, Nelson R. Light at night, clocks and health: from humans to wild organisms. Biology Letters. 2016;12(2):20160015.

[24] Obayashi K, Saeki K, Iwamoto J, Okamoto N, Tomioka K, Nezu S et al. Exposure to Light at Night, Nocturnal Urinary Melatonin Excretion, and Obesity/Dyslipidemia in the Elderly: A Cross-Sectional Analysis of the HEIJO-KYO Study. The Journal of Clinical Endocrinology & Metabolism. 2013;98(1):337-344.

[25] Fonken L, Workman J, Walton J, Weil Z, Morris J, Haim A et al. Light at night increases body mass by shifting the time of food intake. Proceedings of the National Academy of Sciences. 2010;107(43):18664-18669.

[26] Kim Y, Park M, Lee E, Choi J. High Incidence of Breast Cancer in Light-Polluted Areas with Spatial Effects in Korea. Asian Pacific Journal of Cancer Prevention. 2016;17(1):361-367.

[27] Kloog I, Haim A, Stevens R, Barchana M, Portnov B. Light at Night Co-distributes with Incident Breast but not Lung Cancer in the Female Population of Israel. Chronobiology International. 2008;25(1):65-81.

[28] James P, Bertrand K, Hart J, Schernhammer E, Tamimi R, Laden F. Outdoor Light at Night and Breast Cancer Incidence in the Nurses' Health Study II. Environmental Health Perspectives. 2017;125(8):087010.

[29] Al-Naggar, R., Anil, S. Artificial Light at Night and Cancer: Global Study. Asian Pacific Journal of Cancer Prevention, 2016; 17(10): 4661-4664. doi: 10.22034/APJCP.2016.17.10.4661

[30] Dominoni D, Nelson R. Artificial light at night as an environmental pollutant: An integrative approach across taxa, biological functions, and scientific disciplines. Journal of Experimental Zoology Part A: Ecological and Integrative Physiology. 2018;329(8-9):387-393.

[31] Smyth T, Wright A, McKee D, Tidau S, Tamir R, Dubinsky Z et al. A global atlas of artificial light at night under the sea. Elementa: Science of the Anthropocene. 2021;9(1).

[32] Falcón J, Torriglia A, Attia D, Viénot F, Gronfier C, Behar-Cohen F et al. Exposure to Artificial Light at Night and the Consequences for Flora, Fauna, and Ecosystems. Frontiers in Neuroscience. 2020;14.

[33] Neberich M, Opferkuch F. Standardizing Melanopic Effects of Ocular Light for Ecological Lighting Design of Nonresidential Buildings—An Overview of Current Legislation and Accompanying Scientific Studies. Sustainability. 2021;13(9):5131.

IOP Conf. Series: Earth and Environmental Science 1099 (2022) 012051 doi:10.1088/1755-1315/1099/1/012051

Light and emotions. The importance of context

O Y Vara León[1]

Edificio G-20-4-33 U Lomas de Plateros CP01480 Mexico

orquidea.vara@gmail.com

Abstract. The main objective of this work is to determine if lighting can help triggering emotions in the inhabitants, and if so, if it can be achieved through design. For this, a literature review of several authors, such as Sztulwark, Rose, Chaves, Lozoya and Crilly among others was needed. This was seen through an interdisciplinary approach where several concepts were translated to the lighting design field, built from others, and linked to make them useful for this purpose. The understanding of this concepts will help as a tool to understand the relation between lighting and emotions. Concepts such as context, artifact, use and function will take an important role. Subjects such as communication and the importance of language, as well as the inhabitant's interpretation of the space and its lighting will be touched. Light will be treated as a material and lighting as an artifact with which the inhabitant has a direct contact. The relation of lighting and emotions will be seen differently than that of a direct action-reaction. It was found that the thoroughly understanding of context was the link between light and emotions that should be understood and worked, although the expected result will not always be successful.

1. Introduction

"Light and emotions are deeply linked, being light responsible for triggering emotions and able to change people's mood." I have heard in conferences and informal conversations and read in magazines variations on this statement. This is a statement many people, whether they work with light or not, may agree with, without questioning. In my experience, some people's minds go towards the use of color, others might have a childhood memory where light has an important role, or a magnificent story that proves it true. Nevertheless, the common reaction is not asking why? How? Is this true?

I, myself, have this kind of stories. I remember being moved by light art. I remember being moved by places that lie between poetry and architecture, in which lighting plays an important role. I remember looking for more of those places, for more of that light art and wondering how can they achieve that? And I'm not talking about technical details, but how can the artists or the architects move us? What is it they do that they make us feel something? Do they even plan on making us feel something?

2. Light and emotions

Before wondering how we can trigger emotions through lighting or how are emotions linked to lighting, we should wonder what is lighting and what are emotions. In this first part I will define, for this specific work, the concepts of light, lighting and lighting design, as well as the definition of emotion, for we cannot identify the link between two things we don't understand.

[1] Campo de Conocimiento en Diseño Arquitectónico, Maestría en Arquitectura, UNAM, Cto. de los Posgrados S/N, C.U., Coyoacán, 04510 Ciudad de México, CDMX

2.1 Light, lighting design and lighting

Light, lighting design and lighting are terms we might use equally in any given situation. I find it quite common in everyday conversations to use the word "light" for referring to daylight, electric power, a lamp, a luminary, lighting or lighting design. When the day comes to an end, we say we are running out of light. When we refer to electric power in a certain place, we ask whether there is light or not. When a luminary or a lamp breaks, we say that a light broke. And, finally, when we refer to lighting design in architecture some professionals call it simply "light". Nevertheless, the concepts "light", "lighting design" and "lighting" are to be used to talk about different light related concepts, so they can be differentiated and therefore avoid confusion.

Light is defined by the CIE as the "radiation that is considered from the point of view of its ability to excite the visual system" [1] and as the "radiation within the spectral range of visible radiation" [2] This basically means that light is a radiation that the average human being can see. It is defined in terms of the ability of the average human body to perceive it. Beside these definitions, light has been studied in different fields. It has been given different properties in philosophy and has been a study subject in sciences such as physics.

Nevertheless, to be able to design with light and to manage it properly, we ought to turn it into a construction material such as bricks, mortar, and glass. Lewkowicz [3] says that it is in design and through experimentation with light that you create matter, and it is through its manipulation that it becomes material. With this, it inaugurates the architectural dimension [3] When treating the light as material, with no further meaning, it becomes a construction material with specific properties [3] and can be used as an element of design.

On the other hand, Chaves defines design as a process of choosing and combining elements (material and formal) that will determine the characteristics of a product, the way it is made, distributed, and consumed, before its material production to guarantee the maximum satisfaction of the required need. [4] For Turrillo, design is the way in which man prepares the environment to make it into a world, a possible worls, an inhabitable world, a world that makes sense. [5] On the other hand, Buchanan defines it as the human power to conceive, plan and make products that are useful to human beings. [6] We can say design is a process in which different elements are used in the pursuit of a product with which the inhabitant can interact and make sense of. The process of design looks forward to satisfying the purposes of others, never the sole purposes of the designer. Therefore, lighting design is the selection and combination of different elements, with a purpose or intention, involving the creation of a possible world, so it can guarantee the maximum satisfaction of the inhabitant's needs, whether they are physical or psychological, so he/she is able of making sense out of it.

Knowing what light and lighting design is, we can approach to the meaning of lighting. The CIE defines lighting as the "application of light to a scene, objects, or their surroundings". [7] Lighting, then, is when light stops being an element of design in the mind of the designer, to materialize. It stops being a concept to be part of the material world. We can say that lighting is the representation in the material world of the design process which main material is light. The inhabitant can interact with lighting, whether he/she cannot do so with lighting design.

2.2 Emotions

Throughout history, the term "emotion" has had different meanings depending on the different ways of perceiving the world. The way in which we perceive the world, the stories we tell ourselves, change the way in which people imagine, experience life, and understand themselves. [8] According to Dixon, the word "emotion" has existed in psychology since 1884 as a theoretical word. [8] Before, it was used as an everyday term. There was a time in which this word meant everything understood by feelings, pleasures, pains, passions, affects, sensations, temperaments, fervors, sensibilities, impulses, etc. [8] [9]

Viewed from psychology, "the principal theories of emotion consider this an either physiological, cognitive feeling or a product o the interaction of physical and mental factors." [10] Nevertheless, that definition doesn't take into consideration the importance of culture. On the other hand, "emotions are the relations of the subjective cultural nature, which registry, control and instrumentalization can hardly be reduced to statistical-mathematical patters or to a psicobiological approach". [11] This means that emotions tare way more complex than the body-mind approach presented in the beginning. Emotions depend on the inhabitant and his or he context. When we talk about emotions, we don't talk about a specific reaction related directly to a determined stimulus for all human beings, and any time. Emotions are a product of the received stimulus and the way in which he/she interprets them and reacts to them because of the culture in which he/she is, and the way society validates them.

Emotions have cultural and biological components and the societies in which they are expressed have a deep influence on the way they are expressed, repressed, and the meaning of the emotions when they are named, and values are assigned to some rather than to others. [12] The way emotions are expressed, valued, and recognized change from culture to culture and from language to langue due to the values assigned to them and the words used to describe them. You cannot make sure that an emotion is the same in a different language or valued in the same wat in another culture. When talking about emotions, we rather talk about transfers rather than translations. Najmanovich tells us that: "there is not an exact translation, complete and mechanical from one language to another, but a process of partial translation, metaphorical and creative" [13] and that: "we are translating to the verbal language our corporal experience. This corporal experience is unmeasurable with the language: it belongs to another order." [13] We can say that the same happens with emotions.

Historians agree that there are some basic emotions that exist in everyone regardless of their culture, society or time; nevertheless, the way they are expressed, and their meanings vary dramatically through cultures and centuries [12] The way in which we express emotions and their meaning in different cultures is that which we can see and compare, and, unfortunately, it's not universal. This means that we cannot determine basic universal emotions with measurable and comparable reactions between different cultures and through time. Emotions exist within a determined context and can be expressed as learnt answers to certain situations. They are part of a body experience and a mental process that should happen before we are able to see the physical tells of an emotion if they exist.

In cultural studies of emotions, emotions are analyzed as a cultural construction, (post) affective of the mind-body experience. [11] Emotions depend directly on the context. In another context, the emotion changes. It depends on the culture and its code to be named and valued. The physical shows of the emotion can be learnt since we are kids. This physical tells are validated by the society in which they exist and recognized as emotions. Throughout his/her life, the inhabitant will learn ways to react to certain artifacts, stimulus, and situations. [14]

For being able to explore in a simpler way emotions, several tools have been created. One of such is the concept of "emotional communities". In 2006, Rosenwein defines emotional communities as "groups in which people either identify themselves or adhere to the same rules or emotional expression, based on which emotions are valued or devalued in a similar way." [11] Emotional communities are, therefore, a tool for avoiding mistakes such as generalizing the behavior and the emotions of people due to their geographical or temporary proximity without considering the rest of its context. Two emotional communities can coexist in the same space or time, but their language, values and context are important to tell them apart.

The cultural circulation of emotions is a product of interrelations. Individuals and groups of people can identify with several emotional communities at a time. [11] Even though emotions are personal, these are signified, circulated and act in a collective scale. [11] The social circulation of emotions is the ability emotions have of changing our perception of our surroundings. [11] This is one of the main reasons emotions are so important to us: they can change the way we perceive the world, therefore, the way we react and the way we live. The affective production of emotions tells us that "emotions don't live in subjects or objects but are the effect of the interrelation among things" [11] Therefore, without the relation between things and inhabitants, emotions wouldn't exist.

We can consider emotions as the glue that unites objects and inhabitants and help us perceive our world in different ways. They are a product of the different received stimulus and the culture we live in, and at the same time, they help us relate to the society in which we are at the time and model our perception of the world and therefore, our lives.

2.3 Perception

According to Morris, perception "consists of deciphering meaningful patterns in an entanglement of information." [15] Papalia and Wendkos tell us it is the meaning our brain gives to the sensations that come from our senses. [10] and Niosi resumes it as the way in which we read the world around us for it to make sense. [16] When we talk about perception, we talk about a stimulus, a series of filters that inhabit the memory and a decoding or answer.

The stimulus is the information we get through our senses. This is the reason we can talk about an embodied perception. Our perception of the world depends mainly on our capacity of perceiving those signals as stimulus. If something cannot be felt through our body, it cannot be considered as a stimulus and therefore cannot be perceived. Nevertheless, it is in our brains that this information is deciphered. It is in the brain where this information is compared to patterns and that passes through several filters such as culture, the stories we tell ourselves, who we are, our past experiences, etc. These filters enable us to decode and interpret it. Therefore, we can produce an answer according to who we are. This is how our brain then tells us how to behave and emits judgements.

Depending on the part of the body that receives the stimulus, is the kind of perception we have, being visual perception the most important due to the amount of information that it receives, which is about the 87%. [17] Regardless of this percentage of generated information, sight would have no sense without touch. [18] Visual and haptic perception complement each other so we can have a better understanding of our surroundings. For example, "texture perception depends on the quality of light that touches the building." [19] This is how we can realize that the amount of light, its direction and color give us information that we can relate to haptic perception. We can "touch" a wall without really touching it at all.

2.4 Context

So far, the context has been mentioned in the way we define, and value emotions and we can say it is part of the filters that change our perception of the world. But what is context?

Context is different from the surroundings. While we can talk about the surroundings as the measurable materiality that is near our situation, we can affirm that the context is much more than that. The context of a situation is the "universe of meanings that frame it and affect it in a particular way." [5] It is the "space of meanings that condition the meaning of a point". [20] In other words, we can define the context as a network that holds the situation and makes sense of it, that affects it and modifies it in the eyes of the observer. Therefore, the context depends on an observer and the different kinds of relations between elements he/she makes that build the network and give a meaning to the situation. If the observer changes, the context will change as well.

I find a spiderweb very useful for explaining the context, being the spider the one making the relations among the different elements already given. If the spider changes, or even if the moment changes giving the spider a new approach to that situation, the spiderweb will change. And it is this spiderweb the network of meanings that will give sense to a situation. The situation may be the center of the web, or, in some cases, the prey.

The different links between the existent elements, the links creating the spiderweb, are the context. The elements already exist, but an observer, a reader who can unite certain elements and leave others is needed for there to be a defined, readable context. By creating new links of already existent elements, you create a new context. [5] Sztulwark tells us that "the context of a situation is the field of the significance that that material world represents for he/she who is living, know and seeing it." [21] This means that the context of the situation gives meaning to it, and depends solely on the observer, who does much more than observe. Therefore, we can call him/her the inhabitant.

"To inhabit is to occupy a territory in a material and symbolic way." [22] This means that the inhabitant has an active role when occupying physically a place and having it make sense, through his /her memories, expectations, ideas, and stories. Therefore, the inhabitant is no more an observer. He/she stops being a "passive space consumer" [22] to take decisions about his/her doing in front of indetermination, to build stories and give sense to the material world he/she encounters.

This web involves way more than the physical context. It is a network of connections that can go far physically and temporarily. The context is the "cultural, social, and historical" [5] place in which an action is taking place. This cultural, social, and historical elements help create a meaningful network around the situation. When these connections are modified, the situation becomes meaningless or acquires a different meaning.

So far, we can see that the context permeates on the way we see the world. Culture, society, and the time frame in which it is built are part of the context; therefore, the stories we tell ourselves, and the expectations of society are an important part of context and influence the way in which we perceive the world. These stories, expectations ad time frame define what an emotion is for a certain group of people, give it meaning and value. The memories that we compare our perceptions to so we can act accordingly are part of the context. We need to take that specific context into consideration when designing, as we can be designing for one or several emotional communities that may share geographical location and time frame.

2.5 Project

The human being is always projecting. It is part of our nature to look towards the future. Nevertheless, when designing, the project is an answer to the context. When doing a design project, a different reality is looked for. The project´s aim is to transform reality, to take it from what it is to what is could be [20] while creating new relations and modifying the existing ones. [11] The project looks for a change of reality, for a new possible future. It will not only be something new, but a new element that will create new relations with its surroundings and inhabitants.

The project is made to fulfill a need. This need exists because the world is not perfect, it is unfinished and in constant construction. We call this indetermination, and it makes the project possible. If the world were finished and perfect, there would be no space for projects. Projects exist because we live in an unfinished and constantly changing world. Inhabitants, their needs, and desires are in constant change, and so are the places they look for.

"The project is a relation with space and time, with an unfinished present, with a field of significance that's not finished, with meanings suspended waiting for a new sense." [5] The inhabitant makes sense out of these spaces based on his/her contextualized reality. The inhabitant is always searching, always creating stories, always building illusions. The inhabitant is constantly building the world. The inhabitant projects.

Taking all of these into consideration, we can say that the lighting project is made by an inhabitant for a society that is part of a culture, answering to the temporary conditions of that timeframe, to create a different reality from the one he/she is living in, taking into consideration the technical part about the handling of light as a construction material, and the lighting language of that culture.

2.6 Language and communication

Now that we have mentioned the lighting language, it is important to talk about it. Before we talk about the lighting language a culture might have developed, we need to know what a language is and why it is necessary.

First, we may say that language models our world by telling us what the world is, and what it is not. We cannot live without language. Every institution created by man requires it. [23] By institution, I am referring to everything created by man, such as money, education, or culture. Language models our ideas and the way we see the world. "Language is constituted in and with reality" [24] This is why different cultures, with different languages, give different values to the same objects, situations, and emotions, by expressing them in a different way. If there is no word in a language for a concept, the concept may not be important enough to exist in the culture that uses that language.

"The verbal language is related to reality through meaning; in fact, meaning is the property that converts mere expressions into talk." [23] Similar to verbal language, the lighting project we were previously referring to, can also be full of meanings. Taking into consideration the basic material of lighting design is light or rather the idea of light, it is safe to say that its meaning is changeable depending on the project, the needs it addresses, and the context in which it is developed.

One of the problems of language is its diversification: you can have the same sounds, but each sound, organized in a different way, means something different. We can see the same happening to light, where, handled in a different way, acquired new characteristics that used in certain places can mean something different. You can use the same light elements that compose a lighting in different situations; nevertheless, depending on the way they are arranged, the codes and the context, these elements can have new meanings. Nevertheless, unlike spoken language, there is no translation between situations in lighting. To understand a different lighting language, you must understand their cultural, social, and temporal context. The inhabitant should not try to understand the circumstances surrounding the message to be able to read what the designer wanted to transmit. Understanding the lighting of a place doesn't mean we can understand the lighting of a different place.

Another characteristic of languages is that they have rules. Rules are agreements that are part of a long succession of historical facts. [25] If we talk about lighting as a language, we might notice there are no written rules. If those rules existed, they would have to be understood, even in an unconscious way, by those who interact with the lighting for it to be considered a language. If lighting is thought to be a universal language, it would have to be understood by everyone.

Even though lighting is not a language, it can be a communication phenomenon. Crilly tells us that "Form becomes a temporary communication phenomenon when the inhabitant interacts with the products." [6] The artifacts, on their own, give messages about their function to the inhabitants, who are expected to read the message the closest possible to what the designer envisioned, and act accordingly. [26] The artifact is the message, and the instructions that should be read in the right context, by a reader who is able to understand the cultural codes of the society for which it was designed. If the designer made an artifact with a particular intention and the inhabitant acted accordingly, we can talk about a communication act. The designer needs to use the codes the inhabitants know so they can interpret the message correctly. The interpretation of the artifact determines its use. [27] The designer cannot determine the inhabitants reading as each artifact can be read in different ways, by different inhabitants, in different contexts. [27] The designer is just designing one affordance out of the hundreds that the artifact may have, and hoping the inhabitant chooses that affordance. If the inhabitant does, then we can talk about communication between the designer and the inhabitant.

Therefore, the designer can only expect the artifact to deliver the right message, and the inhabitant to read it accordingly. Nevertheless, "the channel cannot simply deliver the factice function by transmitting the message, but it was to introduce the right cognitive state so the receiver can capture the message's essence." [28] If the inhabitant is not in the right cognitive state, which is impossible to control by the designer, it is possible that the expected interaction is not met. The designer should have in mind that this interaction is just a proposal, that can or cannot be accomplished. The communication can be accomplished or not regardless of the designer's intentions or the artifact.

2.7 Artifact

The term artifact refers to objects, buildings, cities, systems, or services: anything produced by human being to satisfy a need, goal, purpose, or activity. [26] Artifacts are not just objects, but products of design. Artifacts are part of the material culture of a society, such as forks or chopsticks, ut they can also be whole cities, houses, saunas, or the local garbage recollection service. Anything material that is a product of design can be considered an artifact. A children's book of a local legend beautifully illustrated can be considered an artifact, and part of the material culture, but the legend itself is not. Another term for artifact is product of design. Lighting, being a product of lighting design, can be considered an artifact. If we consider lighting as an artifact, it will have all the properties an artifact has.

The function of an artifact is accomplished if it works for whatever it was design to do. [6] Nevertheless, the function of an artifact is different from its use. According to Crilly, "the function is what an artifact does, and the use is what people do with the artifact." [26] The function refers to what the artifact was designed for, and it depends solely on the designer and the artifact, while the use depends on the inhabitant and the way he/she perceives the artifact. For example, if a chair is designed so people can sit in a certain way, and people can sit in that way, then the function is accomplished. But if a kid stands up in the chair to get to the cookie jar, or someone puts clothes in there using it as a storage unit, then we are talking about use and the designer has nothing to do with it.

The artifact can have different affordances, which, according to Gibson, are the action possibilities that the animal and the environment give. [29] Affordances are all the possible uses an artifact can have depending on the user. These are influenced by the inhabitant's context, which include their sociocultural practices and time frame. An inhabitant of a different context can interpret the artifact in a different way. Which means the artifact could be used and understood in a different way by people in a different context.

Therefore, the same lighting design, considering it as an artifact, will not cause the same reactions in different inhabitants, or in the same inhabitant in different moments. Many of these reactions depend on culture and the social group the inhabitant is part of. Even though it has never been taught to the inhabitant to act in certain situations o spaces, the inhabitant acquires the social behavior norms since he/she is a child and acts in the adequate way. Depending on the culture, the inhabitant acts in a certain way in specific social contexts. [30]

An artifact can offer an experience to the inhabitant that has a psychological need to fulfill. [28] Affordances also respond to these psychological needs. This way, an interaction deeper than a physical interaction can be generated. The satisfaction of the psychological need can be seen as an experience. We can refer to it as a communication route. The communication route is between the inhabitant and the designer with an artifact in the middle that is used as a message and a code. It is always possible that the experience is not met, so there will always be the possibility that the interaction is not fulfilled and there is not communication between the inhabitant and the designer. "In the context where the interaction occurs, and exchange occurs between the inhabitant and the system" [28] This exchange affects the inhabitant and the artifact, either physically, psychologically or in its narrative.

The interaction may not be fulfilled because of several reasons. If the code is not shared between the designer and the inhabitant, then the message cannot be understood, and the interaction will not be accomplished. This can be foreseen, and the designer should use a code the inhabitant understands. Nevertheless, there are reasons the designer cannot anticipate, such as the mood of the inhabitant, or whether he/she is not in the right disposition for the interaction to be generated.

On the other hand, when the interaction occurs, it is an exchange that will affect both, the inhabitant, and the artifact, so neither of them remains the same afterwards. The change may be physical or in the stories we tell ourselves about the artifact. For example, the chair may break, changing physically the object and the user may fall, hurting him/herself. Or it may work, giving the user the idea that it is perfect for reaching the cookie jar (story). When talking about lighting, the inhabitant might find a coffee shop so perfectly lit that it becomes his/her favorite spot (story), using it as the designer wanted.

3. Conclusions

Based on the previous concepts built to be applicable to the field of lighting design, we can say that emotions and lighting are linked through context. Context will be understood as the web that affects a situation, and this web will be the connections between different elements in the fields of culture, society, and time. Context will determine how we perceive and value emotions, as well as how we perceive the world. Despite being in the same context, inhabitants perceive the world differently, as we perceive the world through our bodies, and the perceptions are filtered by the different stories we tell ourselves.

If the designer wants to trigger emotions in a certain group of people that resides within a specific context, he/she should study the context, as well as the value of emotions before designing. The identification of the emotional community for which he/she is designing is crucial if a certain emotion is looked for. Otherwise, the objective may not be achieved.

We can say there are more probabilities of triggering an emotion when the designer understands the context of the inhabitant and his or her lighting language. The designer should then research and base the project on these. When designing, if he/she wants to trigger an emotion, questions like what is an emotion in the inhabitant's context? Are important. In case the designer wants to provoke a particular emotion, he/she should ask how is that emotion seen in that specific context? How is that emotions valued? As, if the emotion is not highly valued, the possibilities of provoking it are less. This can happen due to the inhabitant's resistance to that emotion or their lack of interest or ability to recognize it.

In case the designer is looking for physical characteristics that represent such emotion, like a smile, he/she should ask which are those characteristics, how do they look like and what can be expected in that emotional community.

When analyzing the inhabitant's context, it is important to pay attention to the emotional community to which the inhabitant belongs, as well as their likings and collective experiences. We should always have in mind that lighting is part of architecture, and they need each other. Lighting helps create an atmosphere and enables you to perceive architecture visually. If architecture is non-existent, then there is nothing to light up.

Inhabitants interact with lighting, and those interactions are what is important. The architecture, and the lighting, all the materiality the inhabitant can perceive, although important, is not the main thing. The focus should be in the interactions that this materiality can trigger. These interactions make the inhabitants generate emotions thanks to the codes used in a context.

Codes, which also depend on the context of the inhabitant, are extremely important. Designing without knowledge of those codes may result in uses different than the functions designed, resulting in a lack of communication. When no communication is achieved, the probabilities of provoking an emotion diminish.

Lighting is more likely to trigger an emotion in the inhabitant when we know what are the codes, and the stories told in an emotional community. It is easier for us to see if the results are achieved if we know what that emotion looks like. We should design using their lighting communication codes, which are different from the technical codes and respond to the culture and history of that community. And finally, as designers we must understand that the important part of design is the people and their interactions with other inhabitants or objects, rather than the materiality itself.

4. References

[1] CIE, "17-21-012," 12 2020. [Online].

[2] CIE, "17-21-013," 12 2020. [Online]. Available: https://cie.co.at/eilvterm/17-21-012.

[3] I. Lewkowicz, "Sobre el ejercicio de la luz," in *Componerse con el mundo: modos del pensamiento proyectual*, Buenos Aires, Diseño, 2015, pp. 153-156.

[4] N. Chaves, "El diseño: disciplina "vacía"," [Online]. Available: https://www.norbertochaves.com/articulos/texto/el_diseno_disciplina_vacia.

[5] G. Turrillo, "El contexto como invención," in *Seminarios Contexto, compilación de las Cátedras Integradas Mateola, Sztulwark, Turrillo*, Buenos Aires, Ed. Nobuko, 2006, pp. 17-30.

[6] R. Buchanan, "Design Research and the New Learning," *Design Issues,* pp. 3-23, 2001.

[7] CIE, "17-29-001," 12 2020. [Online]. Available: https://cie.co.at/eilvterm/17-29-001.

[8] T. Dixon, ""Emotion": The History of a Keyword in Crisis," *Emotion Rwview,* vol. 4, no. 4, pp. 338-344, Octubre 2012.

[9] U. Frevert, Emotional Lexicons. Continuity and Change in the Vocabulary of Feeling 1700-2000, Oxford: Oxford University Press, 2014.

[10] D. O. S. a. W. O. S. Papalia, Psicología, México: McGrawHill, 1988.

[11] J. Lozoya, "La producción cultural de las emociones: un desafío conceptual para las tácticas," in *El diseño desde la mirada social: un diálogo con las ciencias sociales*, Ciudad Juárez, Universidad Autónoma de Ciudad Juárez-Conacyt, 2021.

[12] S. J. Matt and P. N. Stearns, "Introduction," in *Doing emotions history*, Urbana, University Illinois Press, 2014, pp. 1-13.

[13] D. Najmanovich, "El sujeto encarnado: límites, devenir e incomplitud," in *El juego de los vínculos. Subjetividad y redes:figuras en mutación*, Buenos Aires, Nobuko, 2011, pp. 17-42.

[14] H. H. Rüdiger Ganslandt, Handbook of Lighting Design, Germany: ERCO The Vieweg publishing company, 1992.

[15] C. G. M. A. A. Morris, Psicología, México: Pearson education, 2009.

[16] A. Niosi, "The Perceptual process," 2020. [Online]. Available: https://kpu.pressbooks.pub/introconsumerbehaviour/chapter/the-perceptual-process/#:~:text=Perception%20is%20the%20process%20of,interpreted%20based%20on%20previous%20experiences.. [Accessed 29 Septiembre 2020].

[17] R. Kelly, "Lighting as an integral part of architecture," *College Art Journal,* vol. 12, no. 1, pp. 24-30, 1952.

[18] R. L. Gregory, Eye and Brain: the psychology of seeing, Hong Kong: Oxfor University Press, 1997.

[19] L. M. Roth, Entender la arquitectura sus elementos, historia y significado, Barcelona: Gustavo Gili, 1999.

[20] I. Lewkowicz and P. Sztulwark, "Parte II. Contexto y partido, o (re)pensar el proyecto," in *Aquitectura plus de sentido*, Argentina, Editorial Altamira, 2003, pp. 75-96.

[21] P. Sztulwark, "Fuera de contexto," in *Ficciones de LO habitar*, Buenos Aires, Nobuko, 2009, pp. 101-116.

[22] P. Sztulwark, "Ficciones de LO habitar," in *Ficciones de lo habitar*, Buenos Aires, Nobuko, 2009, pp. 25-45.

[23] C. González Ochoa, "Capítulo 2. Interpretación y diseño," in *El significado del diseño y la construcción del entorno*, Editorial Designio, 2009, pp. 63-86.

[24] P. Sztulwark, "Componerse con la ciudad," in *Componerse con el mundo. Modos del pensamiento proyectual.*, Buenos Aires, Sociedad Central de Arquitectos-Diseño, 2015a, pp. 69-112.

[25] C. Mijares Bracho, "Sobre el lenguaje," in *Tránsitos y demoras. Esbozos sobre el quehacer arquitectónico*, México, Instituto Superior de Arquitectura y Diseo A. C., 2002, pp. 23-31.

[26] T. Vardouli, "Making use: Attitudes to human-artifact engagements," *Design Studies,* vol. 41, pp. 137-161, Noviembre 2015.

[27] N. Crilly, "Design as communication: exploring the validity and utility of relating intention to interpretation," *Design Studies,* pp. 425-547, 2008.

[28] F. Pucillo, N. Becattini and G. Cascini, "A UX Model for the Communication of Experience Affordances," *Design Issues,* vol. 32, no. 2, pp. 3-18, Primavera 2016.

[29] E. Rietveld and J. Kiverstein, "A Rich Landscape of Affordances," *Ecological Psychology,* no. 26, pp. 325-352, 24 Octubre 2014.

[30] M. J. D. Ramstead, S. P. L. Vessiere and L. J. Kirmayer, "Cultural affordances: Scaffolding local worlds through shared intentionality and regimes of attention," *Frontiers in Psychology-Cognitive Science,* vol. 7, pp. 1-21, 26 julio 2016.

Universal lighting design: Towards sustainable lighting solutions that support function, experience and human diversity

SL Nielsen[1] and C Ryhl[2]

[1] Project Developer, Universal Design Hub - Bevica Fonden, Copenhagen, DK
[2] Research Director, Universal Design Hub - Bevica Fonden, Copenhagen, DK
E-mail: stine@bevica.dk

Abstract. This paper introduces the principle of Universal design into the field of lighting design by pointing to and discussing their shared understandings of human diversities, and how functionality as well as experiencing are equally important aspects of designs. This introduction is initialized by the presentation of a thematic analysis of an ethnographic field study investigating how midwives practice and experience lighting in the delivery room during childbirth. Taking on a phenomenological approach, the analysis showcases how lighting both serves a tool to support functional medical practices, and as a tool to affect sensory experiences in the delivery room. As such, the analysis serves to integrate understandings of lighting having both functional effects *and* experiential significance, in addition to pointing out diverse and changing sensitivities to lighting. In this regard, the paper aims to initialize further discussions about how the principle of Universal design may serve as a tool to address and support the quality and sustainability of lighting designs, by embracing human practices, experiences, and differences – leaving no one behind.

1. Introduction: Research into the functional effects and experiential significance of lighting

Flipping through the DS/EN 12464-1 European Standard on lighting for workplaces, one learns how lighting for such environments fundamentally ought to be designed and implemented to “meet the needs for visual comfort and performance of people having normal ophthalmic (visual) capacity” (1).

To inform such lighting design solutions, research within natural science and the engineering fields is carried out on both visual and non-visual effects of light within the visual spectrum. On the visual effects of light, studies have found how higher intensity lighting in the surroundings may increase object visibility, how low luminance ratios between joining spatial areas may be perceived more harmonious and less glaring, and how directional lighting may improve the visual atmosphere of an office space (2,3). Additionally, on the non-visual effects of light, studies have found how short wavelengths of visual light may support alertness, cognitive performance and attention and how long wavelengths of visual light may allow relaxation, heightening sleepiness ratings and decrease reaction time (4–6).

In addition, within the last decade, social scientific studies have begun researching the more experiential significance and use of light beyond its ability to support functional (work) practices. These studies generally focus on the affective qualities of light by studying human bodily practices, experiences and feelings, from an overall phenomenological approach.

This phenomenological attention to the study of light has been named "the atmospheric turn" (7). As claimed by sociologists Joanne Entwistle and Don Slater (8), this attention was in large parts initiated by anthropologist Mikkel Bille and archaeologist Tim Flohr Sorensen in 2007, by them arguing for a general approach and need to examine light as a material and social agent in the everyday lives of people (9). In continuation, anthropologists, sociologists, ethnologists, and geographers have studied the interaction between light and human ethnographically, from general questions of: What does light do? How is light used? What does light mean? to particular groups of people in particular sociocultural settings.

In this regard, social scientific research has pointed to the significance of low light levels to support sensory experiences of secureness and togetherness in domestic environments, and how bright lighting in hospitals may be perceived as "a white killer light" and "like infinite and everywhere" (10–12). As such, the social scientific research links in with the said findings from natural sciences and engineering fields on the soothing effects of low light levels. Nonetheless, current lighting designs for conventional workplaces, such as public office spaces and healthcare environments, shows dominated by even distributions of bright white lighting to serve and support functional and efficient work practices of staff.

2. Scope and aim

This paper unfolds the significance of light within and beyond its ability to support functional (work) practices. This unfolding is carried out by describing the results of a thematic analysis of an ethnographic field study investigating how midwives practice and experience lighting in the delivery room during childbirth. Taking on a phenomenological approach, the analysis showcase how lighting both serves a tool to support functional medical practices, *and* as a tool to affect sensory experiences in the delivery room. As such, the field study serves to integrate understandings of lighting having both functional effects *and* experiential significance, in addition to pointing out diverse and changing sensitivities to lighting.

Equivalent thematic analyses of this field study have previously been published; firstly in 2020, in the scientific journal *Emotion, Space and Society* (13), and secondly in 2022, in the book *Lighting design in shared public spaces* (14). In these publications, the analyses respectively served to unfold how midwives apply a certain bodily felt sensory awareness towards lighting during the process of labour (paper) and to argue for lighting as an aesthetic design element with potential to affect the human body within and beyond current regulatory requirements in lighting design (chapter). However, in this paper, the presentation of the thematic analysis serves to introduce the principle of Universal Design into the field of lighting design by pointing to and discussing their shared understandings of human diversities and how functionality as well as experiencing are equally important aspects of (lighting) designs.

By this, the paper does not suggest any new lighting design method. Rather, it aims to initialize further discussions and investigations into the potentials of Universal design within the field of lighting design – e.g., to inspire and inform a new lighting design methodology and/or conceptual approach to address and support the quality and sustainability of lighting designs, by embracing human practices, experiences, and differences – leaving no one behind.

IOP Conf. Series: Earth and Environmental Science 1099 (2022) 012052 doi:10.1088/1755-1315/1099/1/012052

3. Presentation of study

This paper empirically focuses on a field study carried out at the maternity ward at Rigshospitalet in Copenhagen in 2018, investigating how midwives practice and experience lighting in the delivery room during childbirth (13,14). During a period of four months, 30 semi-structured interviews and countless informal interviews with midwives, and over 500 hours of participant observation during 62 birth processes were carried out across the ward's eight traditional delivery rooms and their one sensory delivery room. By means of these interviews and observations, empirical data was collected on midwives' sensory experiences, bodily practices, and interactions in the delivery rooms, via audio-recordings, fieldnotes, mappings and photos on their physical locations, gestures, postures, rhythms, routes, and routines in relation to the light setting in the delivery room.

The light sources of the sensory delivery rooms included nine LED spots and a dynamic soft screen controlled by a tablet, viewing four preset guiding programs. These programs were designed for midwives to apply and adjust during the different phases of birthing – from 'Arrival' and 'Relaxation' to 'Breathing' and 'Welcome'(ing) of the child (15). As shown in Figure 1, the 'Breathing' program imbued a cold bluish light and the others warmer reddish/orangey hues – in addition to the conventional white 'Work Light' in delivery rooms and healthcare environments in general.

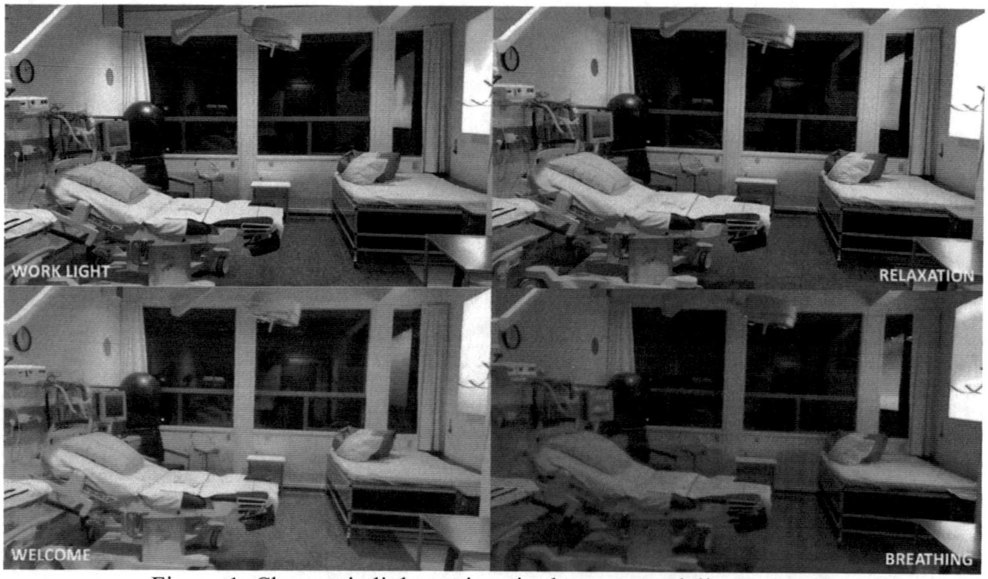

Figure 1: Chromatic light settings in the sensory delivery room.
© Stine Louring Nielsen

A qualitative analysis of the empirical data from the field study was carried following the conventional six-steps of a thematic analysis, as defined by Clarke and colleagues (16). In this regard, data was analysed by an initial process of getting an in-depth understanding of the data datasets by transcribing all semi-structured interviews from audio recordings (10 hours), reading through all field- and descriptive observation notes (over 700 pages), and reviewing all mappings (approx. 450) and photos (approx. 450), to then organizing and coding all data in NVivo in an iterative-inductive process of defining themes by consulting theory and the entire dataset. In this regard, ethnographic data retrieved by informal interviewing, descriptive observations, mappings, and photos essentially served to inform and qualify themes and insights reviled by midwives during their interviews. An in-dept description and discussion of the case study design, as well as the organisation and coding process is laid out in the Ph.D. dissertation *The feel of light: The atmospheric potential of coloured illuminations* (17).

Ultimately, as argued below, the analysis pointed to how lighting in the delivery rooms not only served as a tool for midwives to support their functional practices in the delivery rooms, but also how the lighting served to support sensory experiences and feelings during the process of labour. In addition to pointing out the complexity of midwives' different and changing sensitivities to light. As pointed to in the following discussion, the analytical findings from the field study eventually serves to open for an introduction of the principle of Universal design into the field of lighting design.

4. Analytical findings from field study

4.1 Light supporting functional practices of "emergency situations" and "looking properly at the child"

Analysing the body of empirical data collected during the field study, it became evident how the midwives studied generally applied the conventional white hospital lighting to support their functional work practices in the delivery room.

In this regard, interview transcriptions, fieldnotes, mappings and photos of midwives working in the delivery rooms showed, how midwives would apply the conventional white hospital lighting to support practices of focused and/or speedy physical movements and clear communication. For example, the conventional hospital lighting was applied when midwives needed to take a blood sample from the baby or when they needed to dose analgesics during birthing, or in more serious or alarming situations where midwives had to pull out a baby from a mother's womb or assist doctors in carrying out a c-section. In these cases, midwives would turn up the conventional white ceiling lighting and/or manually direct the operational lamp located in each delivery room above the bedrest, in accordance with where they needed to be physically placed in relation to the birthing mother, possible other staff members and the medical tools needed to carry out their tasks (cf. Figure 2, photo 1).

As such, the thematic analysis pointed to how midwives applied the conventional hospital lighting in the delivery room as a tool to meet their needs for visual comfort and performance as healthcare staff, in line with the DS/EN 12464-1 European Standard on lighting for workplaces (1). However, despite this alignment of lighting standards and actual midwifery practices, midwives described how they generally disliked the conventional bright white lighting in the delivery room. For example, one midwife stated that: *"I rarely use ceiling lights inside these delivery rooms. I find it annoying that it has to be a part of it all. Ceiling lighting is such an institutional thing."*

Along these veins, the thematic analysis of data revealed, how midwives would often hack the conventional hospital lighting in the delivery room by turning off or dimming lights or creating spaces of focal glow. This was for example done by midwives turning on small lights located under the cabinets or pointing operation- and architect lamps to the wall (cf. Figure 2, photo 2). One evening a midwife was turning off the conventional ceiling lighting in the delivery room while explaining why:

> *"I'm trying to get it (the lighting) as muted as possible, to create a sort of cave-like atmosphere. So that there isn't this spotlight shining down unless we need it in an emergency situation or need to look properly at the child. Then we turn it up properly. But otherwise, I try to turn off the top lights as much as possible and only have on the lights above the cabinets. Some of the rooms have a small architect lamp you can just switch on instead. Those I like."*

As pointed to by this midwife and unfolded below, the empirical data from the field study showed how midwives would apply the lighting in the delivery rooms as more than a tool to support their functional work practices.

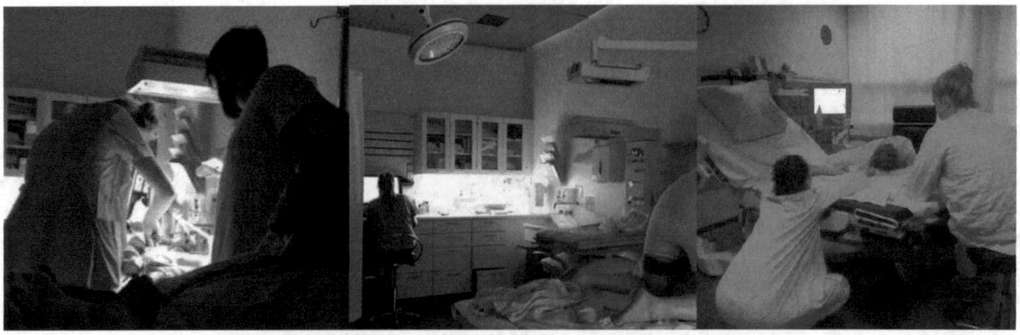

Figure 2: Midwifery practices in delivery room light settings.
© Stine Louring Nielsen

4.2 Light supporting sensory experiences of "safety" and "that womb-like feeling"

Analysing the body of empirical data collected during the field study, it became evident how the midwives to a great extend applied the lighting to support their sensory experiences in the delivery room – hence, beyond its immediate ability to support their functional work practices.

In this regard, the empirical data from the field study showed how midwives generally tuned the lights down to support sensory experiences and bodily movements of intimacy, calmness, and togetherness in the delivery rooms. These practices of turning off and dimming the lighting were carried out by the midwives from their very first entering into the delivery room. Mostly even before the birthing mother and her partner had arrived. In addition, the midwives showed to turn off or mute other sensory stimuli. This was for example done by muting the sound of the alarm system or throw out plastic bags of vomit to remove the stench, cover up visual blood stains from bed rests or floor, and leave drawers and cabinets open to avoid the noise from slamming during childbirth (cf. Figure 2, photo 2).

In relation to muting the sensory stimuli in the delivery room, the midwives mainly showed to apply the warm hues of lighting in the sensory delivery room. As noted by one midwife: *"I especially like the warm tones (...) for me it seems to create safety and cosines. And intimacy."* (cf. Figure 2, photo 3). I similar veins, another midwife explained how the red lighting attuned a combination of calm and enclosing: *"you know that womb-like feeling and a welcoming to the child."*

In this regard, midwives did not only apply the reddish illumination in the delivery room to support certain sensory experiences. They also showed to apply them to support certain physical movements of the body. For example, the descriptive observation notes and mappings collected during the field study uncovered how shoulders would go down when people entered the sensory delivery room, or how doctors' voices would drop, and how body movements would be performed at a slower pace in the warmer illumination. In relation to this, two midwives discussed what was going on in the sensory delivery room, and they stated:

> *"I actually think that I am a bit more careful in there... I think that I am a little slower... (You are calmer?) Yes, I actually think that I move a bit differently in there. Like I think that I would like to support some sort of a sphere of intimacy. That is, anyway, my experience of it."*

On this note, midwives pointed to how the potential of light to support sensory experiences in the delivery room ultimately affected *the ways* midwives performed their functional work practices by means of attuning postures, voices, and paces.

4.3 Light supporting different and changing sensitivities

Ultimately, the midwifery practices and experiences of lighting in the delivery rooms pointed to how midwives generally shared a sensory awareness of how the light affected and supported both functions and experiences during the process of labor.

In this regard, the empirical data on midwives' experiences and orchestrations of lighting in the delivery room revealed, that even though midwives shared a sensory awareness of light, they did not always share a common sensitivity to the lighting, in addition to changing their sensitivity to the lighting throughout the phases of birth.

This ultimately showed whenever midwives would apply the white hospital lighting, even though they did not prefer nor enjoy it, as laid out in section 4.1. More specifically, this was e.g. also the case for one midwife, who addressed her sensitivity to the conventional lighting in the following terms:

> *"Just try and feel it, it's not very enjoyable, this light we are sitting in right now... I just don't like being in it. I would never choose this light if I were to feel safe or relaxed or work."*

Despite her explicitly felt sensitivity to the conventional lighting, said midwife was observed to apply this lighting, whenever an alarming situation occurred in the delivery room and she needed to call and communicate with a doctor or additional staff, or needed to stich up the birthing mother after the delivery of a child. As such, midwives' sensitivities to the lighting showed to change in relation to the particular situation in the delivery room – sometimes foregrounding light as an objectively practical tool to support practical (work) functions and other times as a subjectively felt tool to support sensory experiences.

In addition, midwifes articulated different and divergent sensitivities to the lighting. For even though the thematic analysis of empirical data revealed a commonly shared sensitivity to the lighting amongst midwives – preferring dimmed and reddish lighting over bright white lighting – some midwives expressed how the reddish lighting in the sensory delivery room was *"somnolent"*, *"flickery"* and at times *"an annoyance"*. For one midwife, the reddish lighting even induced headaches and nausea. Thus, even though her bodily reaction was due to a general hypersensitivity to light, causing her to wear sunglasses whenever she went outside in daylight, her sensitivity – and said sensitivities of other midwives – pointed towards a complexity of diversity in midwives' "ophthalmic capacities", sensitivities and needs regarding the lighting in the delivery room.

5. Discussion

5.1 Different bodies practicing and experiencing in lit space

As laid out in the analytical findings section above, the thematic analysis of the ethnographic data on midwives at Rigshospitalet points to how midwives applied the lighting in the delivery rooms as a dual tool to support both functional practices and sensory experiences during the process of childbirth. In this regard, the midwives studied primarily applied the conventional white lighting to support practices of focused and/or speedy physical movements and clear communication. However, when these practices were not needed, midwives commonly showed to hack the conventional lighting by dimming or turning it off, creating spaces of focal glow e.g., by means of architect lamps and applying the reddish lighting made possible in the sensory delivery room – all in all, to support sensory experiences of calmness, intimacy, and togetherness.

In other words, the lighting in the delivery rooms showed its potential to support both *what* midwives were practically doing but also *how* they were doing it, by attuning their sensory experiences –

subsequently pointing to how these two aspects of practicing and experiencing lighting are interlinked.

Ultimately, the midwifery ways of applying lighting in the delivery room reveal a sensory awareness of the body in lit space as both practicing and experiencing, while consequently pointing to how functionality as well as experiencing are equally important aspects of lighting designs. In this regard, the midwives pointed towards a complexity in their sensitivities and needs regarding the lighting in the delivery room, by not always sharing a common sensitivity to the lighting, in addition to changing their sensitivity to the lighting in relation to the phases and situations of birth.

Thus, according to this field study, lighting designers and designs not only need to account for practical and experiential needs, if one wants to generate sustainable lighting designs that doesn't require hacking when implemented into the real world. Lighting designers also need to account for different and changing needs and (hyper)sensitivities to lighting in relation to the particular spatial and situational context.

Fortunately, to most lighting designers, these insights are not new. Since, for long they have strived to stress and integrate their inherent and attained awareness of both human visual and emotional needs into their designs (18,19), not to mention today's trending scope on human biological needs (20). However, having to account for lighting standards such as the DS/EN 12464-1 European Standard on lighting for workplaces in their practice, lighting designers may find themselves limited in their operationalisation and integration of a said insights and awareness of human practical end experiential needs. In this regard, as introduced and discussed below, the principle of Universal design may serve as a future useful tool, in its way of acknowledging the diverse and changing sensitivities and needs of users, and in its general understanding of how, ultimately, functionality as well as experiencing are equally important aspects of design.

5.2 Introducing the principle of Universal design to lighting design

Universal design was originally defined by Architect Ronald Mace to embrace human diversity in the design of our environment (21). The purpose of the principle was to introduce a principle which acknowledges that living with different and changing abilities, permanent or temporary, is a human condition we all share (22,23).

In this regard, Mace ultimately wanted to challenge our understanding of what it is to be a human being and body in spatial design as well as our understanding of what the human body is capable of, if the environment is accommodating to our differences (24–26). Within this framework, Universal design fundamentally acknowledges how, diversity in abilities and user needs includes physical, sensory, cognitive, neuro-diverse, and emotional aspects of being in the world. When recognising the complexity of diversity in our abilities and hence also in our sensitivities and needs, the principle emphasizes both the need to accommodate for diversity in (lighting) design and the diversity in human experiencing and interacting with our designed environment.

In other words, the principle of Universal design offers an inclusive approach to sensory diversity and changeability, which accommodates for sensory experiences being the pre-requisite for functional ability – and thus subsequently emphasising how functionality and experiencing are equally important aspects of design.

Along these veins, Universal design recognises the fact that even though a human being is able to function within a space by means of physical access, physical access in itself does not secure access to what is happening in the space. Accommodating for sensory diversity and sensory experiences of a space or spatial atmosphere, are thus key aspects of Universal design (25,27,28). In this regard, designing lighting for both functional needs of being able to see and navigate in space, *and* experiential needs of feeling socially and emotionally included in space, may serve to promote aspects of both health, well-being and social inclusion, in spatial contexts – including delivery rooms and beyond.

In this regard, integrating the principle of Universal design into the design process of making lighting designs, may support lighting designers in generating sustainable solutions by acknowledging and addressing both functional and experiential aspects of lighting – as well as recognising the fact that, what we have in common as human beings are our different and changing abilities and sensitivities (to lighting). Universal design thus emphasises how diversity in user needs requires diversity in (lighting) solutions – for example by accounting for the needs to turn up, dim and shut off lighting in the contextual and situational context of birthing in a delivery room.

6. Perspectivation: Multidisciplinary as key to generating sustainable lighting solutions

By acknowledging how lighting designs hold potential to support and affect functionality as well as different and changing sensory experiences in space, one subsequently gains an understanding of why it is evident to include both natural scientific and social scientific research in lighting design processes. Since, as pointed to by the midwives at Rigshospitalet, one needs to know how light is experienced and used in order to know what type of lighting is needed in a particular spatial context. As such, the principle of Universal design may serve to promote multidisciplinary perspectives and approaches within lighting design research, linking functional and experiential effect of lighting, as a means to inform our understand of how light supports both *what* we do but also *how* we do it, by attuning our sensory experiences.

In this regard, standards on lighting may benefit of integrating social scientific research in combination with the natural scientific research already informing these standards. Doing so, lighting designers are to a greater extend enabled to design lighting that not only allows people to function in spaces by meeting "the needs for visual comfort and performance of people having normal ophthalmic (visual) capacity" (1), but also allows people to dwell in spaces, by enabling and supporting shared sensory experiences.

A multidisciplinary approach in lighting design research may thus serve as key to generate sustainable lighting solutions by accounting for the body as always practicing and experiencing in space – across (changing) sensitivities, capabilities, and contexts. Understanding the body as a practicing and experiencing entity in constant and complex interaction with a physical space and its stimulations – such as lighting – as well as being unique in its individual and changing needs and sensitivity, requires knowledge and tools to implement in design. In this respect, the principle of Universal design may serve as a design- and planning tool to secure the quality of both functionality and experiences. This is the inherent potential of Universal design and the understanding of human bodies and diversity in functional and experiencing needs in (lighting) design.

7. Conclusion: Towards universal lighting designs

This paper has introduced the principle of Universal design into the field of lighting design by pointing to and discussing their shared understandings of human diversities, and how functionality as well as experiencing are equally important aspects of designs. As such, the paper has indicated how the principle of Universal design may inform sustainable lighting solutions for the future, addressing and acknowledging potential effects of lighting beyond the scope of current standards.

By showcasing how lighting designs affect both functional work practices and sensory experiential effects, lighting designers are encouraged to integrate and generate universal lighting design solutions that embrace people across diverse and changing sensitivities and needs. Acknowledging the diversity and changeability in user needs calls for flexibility and diversity in design solutions, that may not be accounted for by one single design solutions but a catalogue of solutions. Ultimately, integrating the principle of Universal design in lighting design may serve as a useful tool and support for lighting designers in their work of embracing and accounting for diverse human practices, experiences, and differences. In the end, carving the way towards universal lighting designs may serve to support sustainable lighting solutions for the future – by leaving no one behind.

References

1. Dansk Standard. DS/EN 12464: Light and lighting – Lighting of workplaces – Part 1: Indoor work places. 2011.
2. Heintjes F, Fletcher D. Investigations on Improvements to Office Lighting. Architectural Science Review. 2011;36(3).
3. Hvass M, van den Wymelenberg K, Boring S, Hansen EK. Intensity and ratios of light affecting perception of space, co-presence and surrounding context, a lab experiment. Building and Environment. 2021 Mar;194.
4. Lockley S, Evans E, Scheer F, Brainard G, Czeisler C, Aeschlbach D. Short-Wavelength Sensitivity for the Direct Effects of Light on Alertness, Vigilance, and the Waking Electroencephalogram in Humans. Sleep. 2006;29(2):161–8.
5. Katsuura T, Lee S. A review of the studies on nonvisual lighting effects in the field of physiological anthropology. 2019;1–18.
6. Wan SH, Ham J, Lakens D, Weda J, Cuppen R. The Influence of Lighting Color and Dynamics on Atmosphere Perception and Relaxation. 2010;1–4.
7. Bille M, Sørensen TF. Introduktion til materialitetstudier. In: Bille M, Sørensen TF (ed). Materialitet - En indføring i kultur, identitet og teknologi. 1st ed. Frederiksberg: Narayana Press; 2012. p. 13–40.
8. Entwistle J, Slater D. Making space for 'the social': connecting sociology and professional practices in urban lighting design. British Journal of Sociology. 2019;70(5):2020–41.
9. Bille M, Sørensen TF. An Anthropology of Luminosity: The Agency of Light. Journal of Material Culture. 2007;12(3):263–84.
10. Garnert J. From burning flames to the light-emitting diodes of the 2010s. In: Anter KF, Klarén U (ed.). Colour and Light - Spatial Experience. Oxon: Routledge; 2017. p. 47–61.
11. Bille M. Homely atmospheres and energy technologies in Denmark. Living with Light. 1st ed. London: Bloomsbury; 2019.
12. Edensor T. From Light to Dark - Daylight, Illumination and Gloom. 1st ed. Minneapolis: The University of Minesota Press; 2017. 248 p.
13. Nielsen SL, Bille M, Barfoed AB. Illuminating Bodily Presences in Midwifery Practices. Emotion, Space and Society. 2020;37(November):9.
14. Nielsen SL. The Midwifery Feel of Light. In: Sumartojo S (ed.). Lighting Design in Shared Public Spaces. 1st ed. New York: Routledge; 2022.
15. WaveCare. Sensory Delivery Rooms. https://www.wavecare.com/sensory-delivery-rooms1.html. 2022.
16. Clarke V, Braun V, Hayfield N. Thematic Analysis. In: Smith J, editor. Qualitative psychology: A practical guide to research methods. 3rd ed. London: Sage Publications Ltd; 2015. p. 222–48.
17. Nielsen SL. The Feel of Light: The Atmospheric Potential of Coloured Illuminations [PhD Dissertation]. [Copenhagen]: Aalborg University; 2021.

IOP Conf. Series: Earth and Environmental Science 1099 (2022) 012052 doi:10.1088/1755-1315/1099/1/012052

18. Kaçel S, Lau B. Louis I. Kahn and Richard Kelly: collaborative design in creation of the luminous environment. Architectural Engineering and Design Management. 2018;14(4).
19. Brandston H. Learning to See - A Matter of light. Illuminating Engineering Society of North America; 2008. 9–45 p.
20. Boyce P. Human Factors in Lighting. 3rd ed. Boyce P, editor. Boca Raton - London - New York: CRC Press - Taylor & Francis Group; 2014.
21. Mace R. Universal Design in Housing. Assistive Technology. 1998;10(1):21–8.
22. Ostroff E. Universal design: the new paradigm. In: Preiser W, Ostroff E, editors. Universal design handbook. 1st ed. New York: McGraw-Hill; 2001.
23. Ryhl C, Eiriksson M, Sørensen R. Universal design – introduction to a design principle that challenges the idea of being human. Copenhagen; 2021.
24. Story MF. Principles of Universal Design. In: Preisler WFE, Ostroff E, editors. Universal Design Handbook. 1st ed. New York: McGraw-Hill; 2002.
25. Ryhl C, Høyland K. Inkluderende arkitektur. Ryhl C, Høyland K (ed.). Bergen: Fagbokforlaget; 2018.
26. Lid IM. Universell utforming: verdigrunnlag, kunnskap og praksis. Oslo: Cappelen Damm Akademisk; 2013.
27. Ryhl C. Arkitekturen universellt utformet – en ny formidlingsstrategi. Trondheim: Husbanken; 2013.
28. Heylighen A, Herssens J. Designerly Ways of Not Knowing: What Designers Can Learn about Space from People Who are Blind. Journal of Urban Design. 14AD;19:317–22.

Effects of indoor lighting conditions and window views on occupants' well-being and behavior: a systematic review.

Natalia G Vasquez*, Ricardo F Rupp, Rune K Andersen and Jørn Toftum
Department of Environmental and Resource Engineering, Technical University of Denmark, Nils Koppels Allé, Building 402, 2800 Kgs. Lyngby, Denmark

*natgir@dtu.dk

Abstract. This paper reviews 49 studies that addressed how window view, daylighting, and lighting in buildings affect occupants' behavior and well-being. The systematic literature search was performed in November 2021 and focused on office and educational buildings. We quantified the number of papers per study type, study aim, and lighting condition. Predictor categories and methods for data collection were also considered. We analyzed the results according to a structure of records, defined by the number of predictors and type of outcomes from a study. We obtained 106 records. A gap in the number of studies under different lighting conditions and building types was identified. Studies under natural light and studies conducted in learning environments were fewer than studies dealing with artificial lighting in offices. A wide variety of methods for data collection was found. Artificial lighting features and correlated color temperature were the most used predictors. Based on the analysis of records, we found that 61.3% of the associations between predictors and outcomes were statistically significant. The type of effect was not reported in 3.8% of the records-meaning that approximately 35% of the records found no significant associations between predictors and outcomes.

1. Introduction

The built environment plays an essential role for the performance and well-being of occupants [1,2]. Therefore, the indoor environment must be adequate to support people's needs and meet functional demands. Over the last years, researchers have looked at the benefits of proper lighting in schools and offices [3–6], showing that the provision of daylight and the visual connection with the outdoors improve occupants' mood [7–9]. Visual stimuli such as landscapes and views with natural features have been indicated as stress-recovery [10,11] and attention-restoration boosters [12], both of which promote cognitive performance [13,14]. Furthermore, good lighting has been linked to a room's perceived attractiveness, higher satisfaction with the work environment, and work engagement [15], while an unobstructed view and controlled daylighting appeared to improve office employees' decision-making, which was translated into an economic impact based on employee salaries [16].

Regardless of the source, light triggers visual and non-visual responses, which affect human cognition, health and well-being [17]. In buildings, occupant responses to lighting are an important and growing field of research [18]. Available literature reviews have addressed the effects of indoor lighting on task performance [19], examined the evidence on lighting for work beyond visual responses [17,20–22], the effects of windows in the work environment [23], occupants' preferences towards daylighting and control systems [24] and how occupants' behavior can contribute to energy savings [25]. In this paper, we examined the available evidence on how window view, daylighting and lighting in buildings affect occupants' behavior and well-being since, in practice, they are exposed to them at the same time.

2. Method

We conducted a systematic search of studies addressing the effects of window view, daylighting, and lighting on occupants' behavior, perception, performance, and well-being. The systematic search was performed in the Scopus database in November 2021, with the search terms presented in Figure 1. The search was done in the "title", "abstract" and "keywords" fields. Seventy-six studies were selected according to the following criteria: i) papers presenting research results with ii) participants' responses and iii) accounting only for daylighting, lighting, or window view assessments. This paper focused on the current evidence on occupants' health, well-being, and behavior examined in 49 out of 76 papers [26-74].

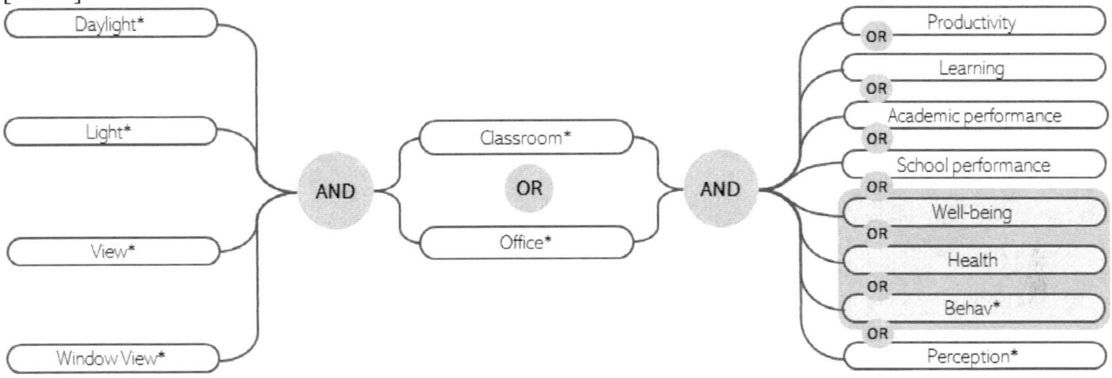

Figure 1. Search terms

We analyzed the results following a structure of records. The number of records was defined mainly by the number of predictors and type of outcomes from a paper. Some studies presented more than one experiment; each was analyzed as separate records.

3. Results

We obtained 106 records from the 49 reviewed papers. The surveyed studies were published between 1978 and 2021 and were performed mostly in offices (n= 31) (Figure 2). Field and laboratory studies were the most frequent type (n= 17 and n= 16, respectively), while interventions were performed in 11 studies. Living-labs (n= 5) and online studies (n= 1) were less frequent -one paper presented results from laboratory and intervention-. Regarding the study aim, 20 studies looked for causality, 23 had an exploratory approach, and eight studies were descriptive. The most investigated lighting source was artificial lighting (n= 27). Combined lighting, i.e., natural and artificial light, was investigated in 18 studies, and only 4 addressed natural light. Regarding the experimental features, 22 studies used fluorescent lamps, and 11 used LED lamps. Nine studies did not report the type of lamp. Four studies addressing the effects of artificial lighting allowed a view of the outside.

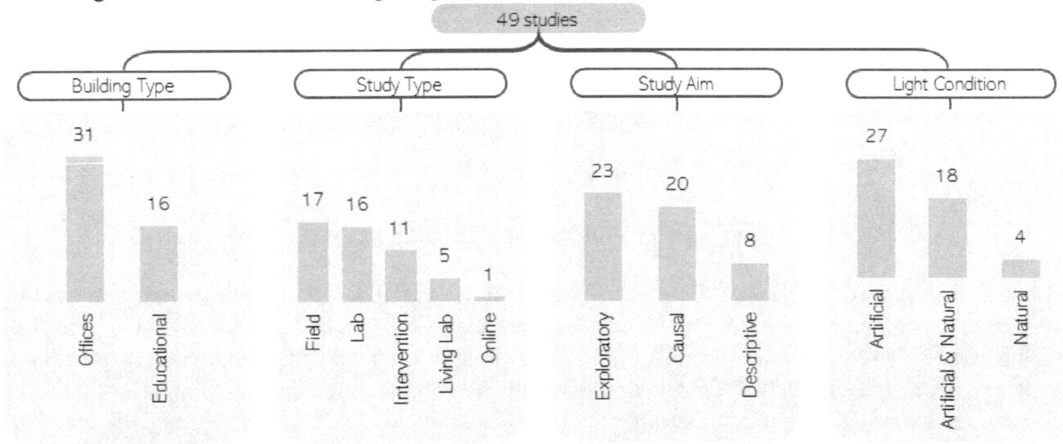

Figure 2. Number of studies per building, study type and aim, and light conditions.

3.1. Predictors, outcome type and method for data collection

We identified 14 categories of predictors, presented in Figure 3. Studies examining artificial lighting and the combination of artificial and natural light used more predictors than studies using natural light. The most used categories were artificial lighting features (type of lamp, type of fixture, lighting program), correlated color temperature (CCT), window features (view features, glazing type, windowed vs. windowless), lighting type (natural vs. artificial, dynamic vs. static), horizontal illuminance (E_h) and flicker. Less used predictors were vertical illuminance, correlated color temperature + horizontal illuminances together, spectral power distribution (SPD), dynamic lighting, light color, light and blinds control (on/off, automatic vs. manual), and daylight and luminous environment (qualitative predictors).

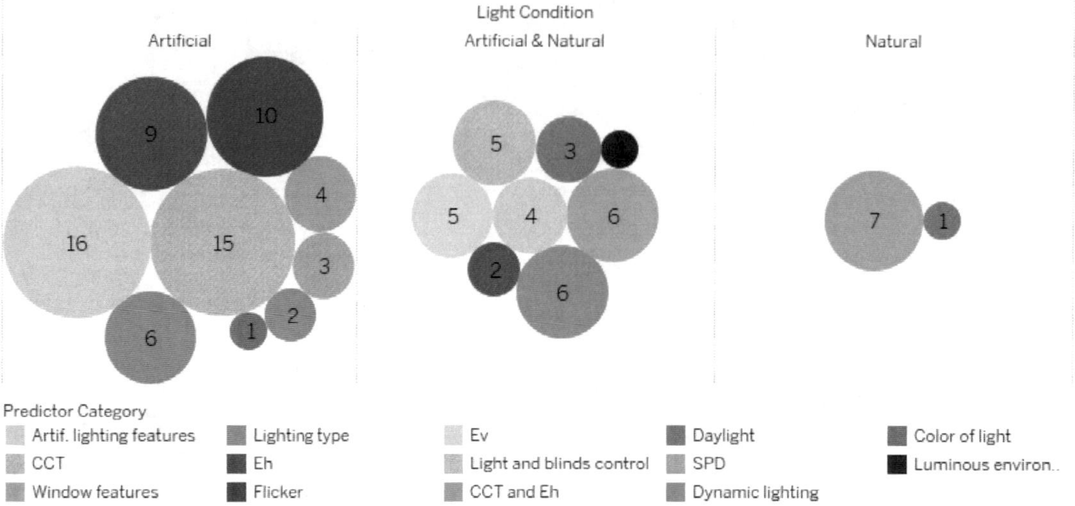

Figure 3. Frequency of records that used each predictor and segregated by lighting condition.

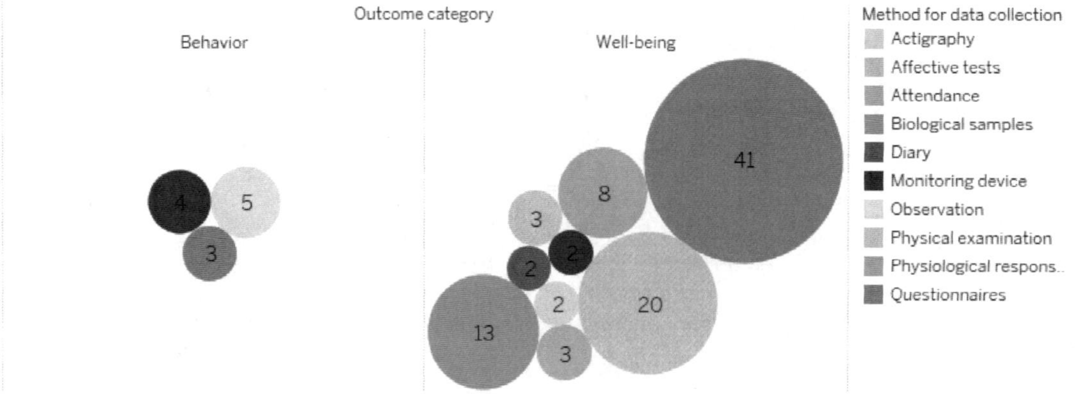

Figure 4. Frequency of records that used each method of data collection.

Most of the reviewed studies (n= 43) addressed the effects of lighting on well-being, whereas only eight studies examined participants' behavior. Two studies [61,67] examined both behavior and well-being outcomes. Three methods were used to collect data about occupants' behavior: observation (5 records), monitoring devices (4 records), and questionnaires (3 records) (Figure 4). Both monitoring devices and observation were used to report occupants' location, i.e., desk occupancy and distance from the window [71,74], and occupants' interactions with lighting and blinds [59,60,62,67]. Questionnaires measured internal responses that affected social conduct and engagement [61]. On the other hand, we identified nine methods to register participants' well-being (Figure 4). Most studies used less invasive methods such as questionnaires (41 records). Karolinska and Epworth Sleepiness Scales-[28,29,31,33,36,42,46,49,52,53] were commonly used to assess alertness/sleepiness. Questionnaires and scales were also employed to self-assess mental effort and workload [40,46,53], mental health and stress [35,56,73], subjective well-being and quality of life [26,35,46,49,56,58,72], some symptoms such as muscular ache, headache, eyestrain and fatigue [36,39,40,44,50,53,56,61,63,67]. Affective tests were also a frequent tool (20 records) to assess subjective mood [27,29–31,33,36–38,41,44,46,55,57,61,65,69,72]. More invasive techniques as biological samples [33,34,36,43,46,50,55,57,66,72] and physiological responses like heart rate [34,40], brain activity [34,61], pupil size and eye fatigue [45,68], and body temperature [70,71] were used in 24 records. Circadian entrainment and motion, mainly during sleep, were tracked using monitoring devices [30,49,51].

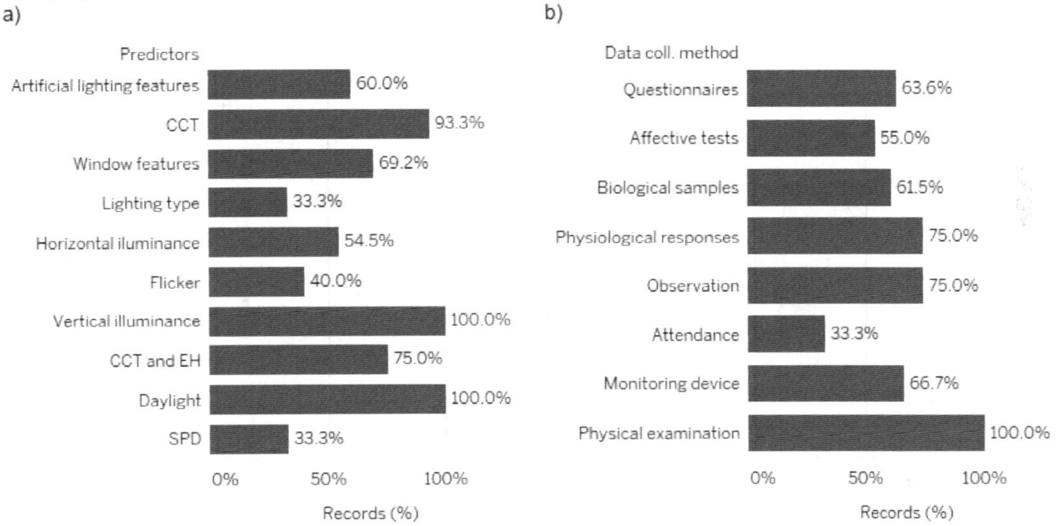

Figure 5. Frequency of records with significant effect according to a) predictor and b) method for data collection.

We analyzed the type of effect, the type of predictor and the method for data collection. Overall, most of the records related to behavior and well-being resulted in a significant association between predictor(s) and outcome (62.5% and 63.8%, respectively). The significance level was not reported in 4 records. Figure 5 includes only data for predictors and data collection methods with more than two records - predictors/methods excluded are: color of light, dynamic lighting, light and blinds control, luminous environment/ actigraphy, and diary. When using as predictors the correlated color temperature (15 records), vertical illuminance (5 records), and daylight (4 records), most of the results showed significant associations. Although research in lighting has moved forward by investigating the effects of the spectral composition of light, only three records used this predictor. Window features were used in 23 records, and 69.2% of these found a significant effect on behavior and well-being. 75% of the eleven records using horizontal illuminance as a predictor found a significant effect on occupants' responses.

Questionnaires, affective tests and biological sampling seem to be the most used methods to collect data. However, 36% and 45% of the records indicated no significant result (Figure 5a). The methods showing a higher frequency of significant results used physiological responses (8 records) and observation (4 records).

3.2. Effects of indoor daylighting and lighting on occupants' behavior and well-being

3.2.1. Behavior: Preschoolers' engagement was higher when the classroom was equipped with LED lamps than fluorescent lamps [32], and pupils were closer to the window mainly because of the amount of light and not necessarily the view [71]. Flicker was found to be correlated with the action and persistent usage of lamps with high-frequency, solid-state ballast [67] but not with social activity or arousal seeking [61]. Furthermore, the need for adjusting illumination levels was linked to controlling lights and blinds[59].

3.2.2. Well-being: Students' attention [28] and mood (affective and emotional responses) [68] were not significantly associated with the type of light bulb [28,72]; however, changes in bulb type led to marginal differences in stress hormones (cortisol) in winter [72], and students in classrooms with full-spectrum fluorescent lamps had better dental health, body growth, development, and attendance than students in classrooms with high-pressure sodium vapor lamps [64]. Workers in offices with parabolic down-lighting (750 lux, 3500 K) had more complaints about eyestrain than those in offices with lensed-indirect lighting (500 lux, 3500 K) [63]; yet, luminaire features were not associated with other physical symptoms such as headache and muscular ache. Participants exposed to an artificial skylight experienced less psychological and physiological stress, higher melatonin secretion, and decreased sympathetic nerve activity (stress-related responses) compared to fluorescent light exposure during the daytime [34]. In a windowless room with an artificial skylight and fluorescent bulbs, participants' mood, tension, and anxiety were improved under the artificial skylight [44]. Job-related stress was lower when the proportion of direct/indirect illumination delivered homogeneous light, but such feature of artificial lighting was not related to self-reported concentration or subjective well-being [56].

Participants' circadian rhythm and sleep [30], subjective alertness [50], and hormone concentration [50,66] were all affected by lighting type. When compared to static lighting, dynamic lighting decreased office workers' sleep time and reported sleep quality [30]. After 30 minutes of exposure to lamps with a blue component, melatonin was suppressed and alertness increased [50]. Variable lighting, however, was not associated with stress and mood [30], stress recovery, subjective alertness, mental health, and physical symptoms [53], or melatonin production [54]. Furthermore, children in windowless classrooms with no daylight and fluorescent lamps had considerably delayed cortisol production when compared to children in windowed classrooms, but no variations in sick leave were associated with the lighting type [66]. Seasonal daylight patterns were found to significantly impact mood, sleep activity, and hormone concentrations [41,43], while dynamic light affected the mood of participants in a simulated office setting but had no effect on melatonin secretion [55]. Under controlled conditions, flicker from fluorescent lighting had no significant influence on mood, emotional state, headache, weariness, and brainwave pattern (EEG) [61]. In contrast to high-frequency lighting, flicker from fluorescent lighting utilizing standard ballasts had a substantial (increasing) effect on self-reported weekly headaches in a controlled field trial [67].

Subjective well-being of children and adults was improved with a higher amount of nature seen from the window and greenery outside [26,35,70], whereas office workers in windowless workplaces had worse self-assessment of well-being than those in windowed offices [49]. Daylighting and greenery outside views also significantly reduced the self-assessed workload and subjective fatigue symptoms [40]. Nevertheless, differences between windowed and windowless rooms and the view content was not associated with students' mood [65], attendance [47], physical activity, or sleep [49]. Depending on the learning task, six qualitative descriptors of the luminous environment could explain students' affective responses (Clear-efficient, Uniform, Cheerful-colorful, Warm-cozy, Surprising-amazing, and Intense-brilliant). Writing-reading tasks, for example, required a light environment that was Clear-efficient, Intense-brilliant, and Uniform [37]. Different SPD (unfiltered bright light at 6300 K and 800 lux, and reduced short-wavelength with 1700 K and 800 lux eye) did not significantly affect mood or neopterin levels, an immune system activation biomarker. On the other hand, short-wavelength light had a significant effect on melatonin secretion [57].

Though cortisol secretion was reduced following exposure to white LED lamps with two CCT (3500K warm light and 6500K blue-enriched light), students exposed to blue-enriched light had lower melatonin levels, and they felt more alert and in a better mood than students exposed to warm white light [33]. Sleep quality and locomotor activity were significantly affected by switching from warm white (CCT 4000K) to blue-enriched light (8000K), which led the participants to office hours entrainment instead of natural light entrainment [51]. Fluorescent and LED lamps with a CCT of 4000K had a positive and significant effect on office workers. They had a better mood, were less sleepy, and had less eye discomfort compared to exposure to lamps with a CCT of 3000K [36]. CCTs between 2700K and 6300K – under controlled air temperature of 22 °C- considerably boosted self-perception of alertness [39]. When compared to lower CCTs (4000K and 2900K), a high CCT (17000K) helped to preserve students' alertness during afternoons in autumn [52] and employees' vitality and mental health [58], though this CCT is not common for indoor lighting. Through different combinations of CCT and E_h, CCT was observed to cause seasonal changes in melatonin and cortisol concentrations, but no effects of illuminance were found [36]. Melatonin levels were higher in summer at 3000K and winter at 4000K (the opposite for cortisol). Gender influenced the effects of CCT and E_h on mood [69]. Positive mood was maintained under cool, dim light (4200K; 300 lux) and warm, bright light (2950K; 1500 lux) for female participants, while negative mood reduced at 3000K and increased at 4000K (reverse for males) [69]. CCT and Eh [48] had little effect on students' motivation. Colored-accent light was not significantly associated with participants' mood [38].

Studies using horizontal illuminance as a predictor resulted in six records with significant effects and five non-significant ones. Significant differences in the assessment of pleasantness were reported only in one of the two conducted experiments when a baseline condition (500 lux) was included in the illuminance pattern [High-low (from 850 lux to 250 lux) versus Low-high (from 250 lux to 850)][29]. Though some studies concluded that changes in horizontal illuminance had no effect on participants' alertness [29,42] or emotional state [31], it was noticed that participants' alertness decreased under dim conditions [31], and fewer children developed myopia in classrooms with higher illuminance [48]. Participants' mood was also lower under dim light (500 lux vertical illuminance); they were tenser and had higher mental effort than in bright (1000 lux) or self-selected illuminance [46].
In Figure 6, we present the type of effect reported in the records of each study.

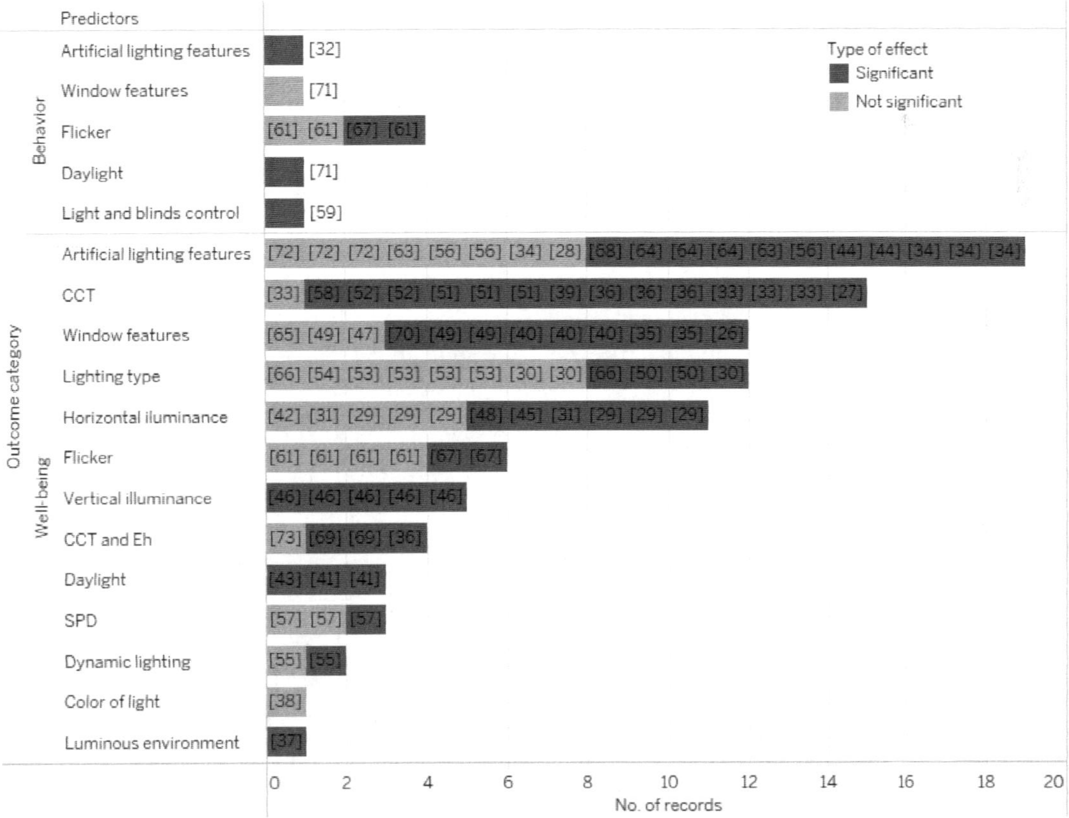

Figure 6. Frequency of records per outcome category and predictor, separated by type of effect-numbers are indicating the reference.

4. Conclusions

We reviewed studies addressing the effects of window views, daylighting, and lighting on occupants' behavior and well-being. The selected papers were found by limited search terms and focused on research conducted in offices and educational buildings. Thus, papers that did not contain any building-related keywords were not caught by the search, and it is a limitation of this review. Nevertheless, we identified a gap in the number of studies under each lighting condition and building type. Studies under natural light and studies conducted in learning environments were fewer than studies dealing with artificial lighting in offices.

Through various light-related predictors and methods for data collection, the studies showed that indoor lighting conditions and window views can affect occupants' behavior and well-being. A wide variety of methods for data collection was identified, mainly in studies on well-being. Based on the analysis of records, we found that 61.3% of the reported results were statistically significant - the type of effect was not reported in 3.8% of the records-meaning that approximately 35% of the results found no significant effect. Changes in the artificial lighting features (e.g., type of lamp, type of fixture, lighting program) and correlated color temperature were commonly used, mainly in studies focusing on artificial sources. Whereas research on the effects of the spectral composition of indoor lighting is growing, in our limited literature search, only three studies used SPD as a predictor. Also, studies under both natural and artificial light should carefully consider and/or control the view to the outside. A combined effect on occupants' responses is particularly expected for windows, which are accountable for both daylight and the outside view unless covered with translucent materials.

5. References

[1] Barrett P and Barrett L 2010 The potential of positive places: Senses, brain and spaces *Intell. Build. Int.* **2** 218–28
[2] Bluyssen P M 2017 Health, comfort and performance of children in classrooms – New directions for research *Indoor Built Environ.* **26** 1040–50
[3] Tenner A D 2003 A Healthy Future for Office Lighting? *J. Light Vis. Environ.* **27** 172–6
[4] Hedge A 2000 Where are we in understanding the effects of where we are? *Ergonomics* **43** 1019–29
[5] Shishegar N and Boubekri M 2016 Natural Light and Productivity: Analyzing the Impacts of Daylighting on Students' and Workers' Health and Alertness *Int. J. Adv. Chem. Eng. Biol. Sci.* **3**
[6] Wu W and Ng E 2003 A review of the development of daylighting in schools *light. Res. Technol.* **35** 111–24
[7] Collins B L 1976 Review of the psychological reaction to windows *Light. Res. Technol.* **8** 80–8
[8] Hopkinson R G 1972 Glare from daylighting in buildings *Appl. Ergon.* **3** 206–15
[9] Collins B L 1975 *Windows and people. A literature survey. Psychological reactions to environments with and without windows* (Washington)
[10] Ulrich R S 1981 Natural Versus Urban Scenes *Environ. Behav.* **13** 523–56
[11] Ulrich R S 1986 Human responses to vegetation and landscapes *Landsc. Urban Plan.* **13** 29–44
[12] Kaplan S 1996 the Restorative Benefits of Nature : Toward an Integrative Framework *J. Environ. Psychol.* **15 VN-r** 169–82
[13] Berman M G, Jonides J and Kaplan S 2008 The cognitive benefits of interacting with nature *Psychol. Sci.* **19** 1207–12
[14] Tennessen C M and Cimprich B 1995 Views to nature: Effects on attention *J. Environ. Psychol.* **15** 77–85
[15] Veitch J A, Stokkermans M G M and Newsham G R 2013 Linking Lighting Appraisals to Work Behaviors *Environ. Behav.* **45** 198–214
[16] Macnaughton P, Satish U, Woo M, Tinianov B and Boubekri M 2021 Economic implications of access to daylight and views in office buildings from improved productivity 1–8
[17] Vetter C, Pattison P M, Houser K, Herf M, Phillips A J K, Wright K P, Skene D J, Brainard G C, Boivin D B and Glickman G 2021 A Review of Human Physiological Responses to Light: Implications for the Development of Integrative Lighting Solutions *LEUKOS - J. Illum. Eng.*

Soc. North Am.
[18] Boyce P R 2010 Review: The impact of light in buildings on human health *Indoor Built Environ.* **19** 8–20
[19] Konstantzos I, Sadeghi S A, Kim M, Xiong J and Tzempelikos A 2020 The effect of lighting environment on task performance in buildings – A review *Energy Build.* **226**
[20] van Bommel W J M and van den Beld G J 2016 Lighting for work: a review of visual and biological effects: *http://dx.doi.org/10.1191/1365782804li122oa* **36** 255–69
[21] Münch M, Wirz-Justice A, Brown S A, Kantermann T, Martiny K, Stefani O, Vetter C, Wright K P, Wulff K, Skene D J and Dk K M 2020 The Role of Daylight for Humans: Gaps in Current Knowledge *Clocks Sleep 2020, Vol. 2, Pages 61-85* **2** 61–85
[22] Veitch J 2005 Light, lighting, and health: Issues for consideration *LEUKOS. J. Illum. Eng. Soc.* **2** 85–96
[23] Veitch J A 2001 A Room with a View: A Review of the Effects of Windows on Work and Well-Being
[24] Galasiu A D and Veitch J A 2006 Occupant preferences and satisfaction with the luminous environment and control systems in daylit offices: a literature review *Energy Build.* **38** 728–42
[25] Gentile N 2022 Improving lighting energy efficiency through user response *Energy Build.* **263** 112022
[26] Lindemann-Matthies P, Benkowitz D and Hellinger F 2021 Associations between the naturalness of window and interior classroom views, subjective well-being of primary school children and their performance in an attention and concentration test *Landsc. Urban Plan.* **214**
[27] Kim D H and Mansfield K 2021 Creating positive atmosphere and emotion in an office-like environment: A methodology for the lit environment *Build. Environ.* **194** 107686
[28] Liu Q, Huang Z, Li Z, Pointer M R, Zhang G, Liu Z, Gong H and Hou Z 2020 A field study of the impact of indoor lighting on visual perception and cognitive performance in classroom *Appl. Sci.* **10** 1–17
[29] Aries M B C, Beute F and Fischl G 2020 Assessment protocol and effects of two dynamic light patterns on human well-being and performance in a simulated and operational office environment *J. Environ. Psychol.* **69** 101409
[30] Zhang R, Campanella C, Aristizabal S, Jamrozik A, Zhao J, Porter P, Ly S and Bauer B A 2020 Impacts of dynamic led lighting on the well-being and experience of office occupants *Int. J. Environ. Res. Public Health* **17** 1–27
[31] de Vries A, Souman J L and de Kort Y A W 2020 Teasing apart office illumination: Isolating the effects of task illuminance on office workers *Light. Res. Technol.* **52** 944–58
[32] Pulay A and Williamson A 2019 A case study comparing the influence of LED and fluorescent lighting on early childhood student engagement in a classroom setting *Learn. Environ. Res.* **22** 13–24
[33] Choi K, Shin C, Kim T, Chung H J and Suk H-J 2019 Awakening effects of blue-enriched morning light exposure on university students' physiological and subjective responses *Sci. Rep.* **9**
[34] Yasukouchi A, Maeda T, Hara K and Furuune H 2019 Non-visual effects of diurnal exposure to an artificial skylight, including nocturnal melatonin suppression *J. Physiol. Anthropol.* **38**
[35] van Esch E, Minjock R, Colarelli S M and Hirsch S 2019 Office window views: View features trump nature in predicting employee well-being *J. Environ. Psychol.* **64** 56–64
[36] Tonello G, Hernández de Borsetti N, Borsetti H, Tereschuk L and López Zigarán S 2019 Perceived well-being and light-reactive hormones: An exploratory study *Light. Res. Technol.* **51** 184–205
[37] Castilla N, Llinares C, Bisegna F and Blanca-Giménez V 2018 Affective evaluation of the luminous environment in university classrooms *J. Environ. Psychol.* **58** 52–62
[38] Kombeiz O and Steidle A 2018 Facilitation of creative performance by using blue and red accent lighting in work and learning areas *Ergonomics* **61** 456–63
[39] Toftum J, Thorseth A, Markvart J and Logadóttir Á 2018 Occupant response to different correlated colour temperatures of white LED lighting *Build. Environ.* **143** 258–68

[40] Ayuso Sanchez J, Ikaga T and Vega Sanchez S 2018 Quantitative improvement in workplace performance through biophilic design: A pilot experiment case study *Energy Build.* **177** 316–28
[41] Adamsson M, Laike T and Morita T 2018 Seasonal variation in bright daylight exposure, mood and behavior among a group of office workers in Sweden *J. Circadian Rhythms* **16**
[42] van Duijnhoven J, Aarts M P J, Rosemann A L P and Kort H S M 2018 Ambiguities regarding the relationship between office lighting and subjective alertness: An exploratory field study in a Dutch office landscape *Build. Environ.* **142** 130–8
[43] Adamsson M, Laike T and Morita T 2016 Annual variation in daily light exposure and circadian change of melatonin and cortisol concentrations at a northern latitude with large seasonal differences in photoperiod length
[44] Canazei M, Pohl W, Bliem H R, Martini M and Weiss E M 2017 Artificial skylight effects in a windowless office environment *Build. Environ.* **124** 69–77
[45] Choi J-H 2017 Investigation of human eye pupil sizes as a measure of visual sensation in the workplace environment with a high lighting colour temperature *Indoor Built Environ.* **26** 488–501
[46] Maierova L, Borisuit A, Scartezzini J L, Jaeggi S M, Schmidt C and Münch M 2016 Diurnal variations of hormonal secretion, alertness and cognition in extreme chronotypes under different lighting conditions *Nat. Sci. Reports* **6** 1–10
[47] Benfield J a., Rainbolt G N, Bell P a. and Donovan G H 2015 Classrooms With Nature Views: Evidence of Differing Student Perceptions and Behaviors *Environ. Behav.* 0013916513499583-
[48] Hua W J, Jin J X, Wu X Y, Yang J W, Jiang X, Gao G P and Tao F B 2015 Elevated light levels in schools have a protective effect on myopia *Ophthalmic Physiol. Opt.* **35** 252–62
[49] Boubekri M, Cheung I N, Reid K J, Wang C-H and Zee P C 2014 Impact of windows and daylight exposure on overall health and sleep quality of office workers: A case-control pilot study *J. Clin. Sleep Med.* **10** 603–11
[50] Wahnschaffe A, Haedel S, Rodenbeck A, Stoll C, Rudolph H, Kozakov R, Schoepp H and Kunz D 2013 Out of the lab and into the bathroom: Evening short-term exposure to conventional light suppresses melatonin and increases alertness perception *Int. J. Mol. Sci.* **14** 2573–89
[51] Vetter C, Juda M, Lang D, Wojtysiak A and Roenneberg T 2011 Blue-enriched office light competes with natural light as a zeitgeber *Scand. J. Work. Environ. Heal.* **37** 437–45
[52] Rautkylä E, Puolakka M, Tetri E and Halonen L 2010 Effects of correlated colour temperature and timing of light exposure on daytime alertness in lecture environments *J. Light Vis. Environ.* **34** 59–68
[53] De Kort Y A W and Smolders K C H J 2010 Effects of dynamic lighting on office workers: First results of a field study with monthly alternating settings *Light. Res. Technol.* **42** 345–60
[54] Hoffmann G, Griesmacher A, Bartenbach C and Schobersberger W 2010 Modulation of lighting intensities and color temperature: Effects on melatonin and cognitive performance *Occup. Ergon.* **9** 27–39
[55] Hoffmann G, Gufler V, Griesmacher A, Bartenbach C, Canazei M, Staggl S and Schobersberger W 2008 Effects of variable lighting intensities and colour temperatures on sulphatoxymelatonin and subjective mood in an experimental office workplace *Appl. Ergon.* **39** 719–28
[56] Fostervold K I I and Nersveen J 2008 Proportions of direct and indirect indoor lighting — The effect on health, well-being and cognitive performance of office workers *Light. Res. Technol.* **40** 175–200
[57] Schobersberger W, Gufler V, Griesmacher A, Bartenbach C, Canazei M, Staggl S and Hoffmann G 2007 Effect of Two Different Lighting Environments on Urinary Neopterin and Sulphatoxymelatonin in an Experimental Night Shift Model *Pteridines* **18** 69–78
[58] Mills P R, Tomkins S C and Schlangen L J M 2007 The effect of high correlated colour temperature office lighting on employee well-being and work performance *J. Circadian Rhythms* **5**
[59] Nicol F, Wilson M and Chiancarella C 2006 Using field measurements of desktop illuminance in European offices to investigate its dependence on outdoor conditions and its effect on occupant satisfaction, and the use of lights and blinds *Energy Build.* **38** 802–13

[60] Moore T, Carter D J and Slater A I 2003 Long-term patterns of use of occupant controlled office lighting *Light. Res. Technol.* **35** 43–57
[61] Küller R and Laike T 1998 The impact of flicker from fluorescent lighting on well-being, performance and physiological arousal *Ergonomics* **41** 433–47
[62] Begemann S H A, Van Den Beld G J and Tenner A D 1997 Daylight, artificial light and people in an office environment, overview of visual and biological responses *Int. J. Ind. Ergon.* **20** 231–9
[63] Hedge A, Sims W R and Becker F D 1995 Effects of lensed-indirect and parabolic lighting on the satisfaction, visual health, and productivity of office workers *Ergonomics* **38** 260–90
[64] Hathaway W E 1995 Effects of School Lighting on Physical Development and School Performance *J. Educ. Res.* **88** 228–42
[65] Stone N J and Irvine J M 1993 Performance, mood, satisfaction, and task type in various work environments: A preliminary study *J. Gen. Psychol.* **120** 489–97
[66] Küller R and Lindsten C 1992 Health and behavior of children in classrooms with and without windows *J. Environ. Psychol.* **12** 305–17
[67] Wilkins A J, Nimmo-Smith I, Slater A I and Bedocs L 1989 Fluorescent lighting, headaches and eyestrain *Light. Res. Technol.* **21** 11–8
[68] O'Leary K D, Rosenbaum A and Hughes P C 1978 Fluorescent lighting: A purported source of hyperactive behavior *J. Abnorm. Child Psychol.* **6** 285–9
[69] Knez I 1995 Effects of indoor lighting on mood and cognition *J. Environ. Psychol.* **15** 39–51
[70] Li D and Sullivan W C 2016 Impact of views to school landscapes on recovery from stress and mental fatigue *Landsc. Urban Plan.* **148** 149–58
[71] Stewart D M 1981 Attitudes of school children to daylight and fenestration *Build. Environ.* **16** 267–77
[72] Gentile N, Goven T, Laike T and Sjoberg K 2016 A field study of fluorescent and LED classroom lighting *Light. Res. Technol.* **50** 631–50
[73] Mott M S, Robinson D H, Walden A, Burnette J and Rutherford A S 2012 Illuminating the Effects of Dynamic Lighting on Student Learning *SAGE Open* **2** 2158244012445585
[74] Boyce P R 1980 Observations of the manual switching of lighting *Light. Res. Technol.* **12** 195–205

Acknowledgments
This paper is based on research undertaken in the EU-funded EERAdata project which has received funding from the European Union's Horizon 2020 research and innovation programme under grant agreement No. 847101. This project has received funding from the European Union's Horizon 2020 research and innovation program under the Marie Skłodowska-Curie grant agreement No 101031380.

Lighting design for diversity: Learning from low-vision rehabilitation

T B Øien[1*] and A K Frandsen[1]

[1]Department of the Built Environment, Aalborg Universitet, A. C. Meyers Vænge 15, 2450 Copenhagen, Denmark
*Corresponding author tuo@build.aau.dk

Abstract. Science understands and handles a phenomenon from different perspectives. Light is handled as daylight, artificial light, task light, and general light, as physics, electrical technology, or safety aids. However, as part of people's everyday lives, the different characteristics of the phenomena fluxes and blend, as people's needs and preferences to lighting differ across a day, in different situations, and not least across a lifetime. Current societal agendas as Leave No One Behind or the sustainable development goals, stress the need for a more holistic approach in design: For the built environment to support wellbeing, comfort, and quality of life, for all.

Disabilities have traditionally been defined in relation to abilities and the normal-bodied, with special aids for special needs, and special spaces for the 'other'. Niches for the other is seen in the field of lighting as special solutions for visually impaired, and correspondingly, lighting is a niche in low-vision services, that has been guided by diagnostics and recommendations of specific lux levels. Thus, more holistic and interdisciplinary initiatives within low-vision rehabilitation that have explored the rehabilitative role of lighting as support of the visually impaired individual in their everyday life, show that individualizing and optimizing the individual's domestic lighting can improve their quality of life. Can the findings from low-vision rehabilitation be translated into the field of lighting design, informing a qualitative approach, and work as an incubator for more diverse user perspectives?

The interactions of the low-vision consultants, visually impaired participants, and their domestic lighting, have been studied in a case study of the role of lighting in low-vision rehabilitation. Participatory observations of 15 consultations and interviews with 17 visually impaired participants form the basis for an analysis of the dynamic human environment interaction at stake, using the theoretical frameworks of affordance and usability. The analysis contributes to a more situated and dynamic understanding of light: The experience of light is context dependent, and lighting needs and preferences are diverging regardless of diagnosis but rather constituted by the dynamic interactions. Consequently, this calls for user involvement and a more diverse user perspective in design processes, which could support more useful designs.

This study implies a promising potential within the intersection of research paradigms and practice: Between quality of life, lighting quality, architectural quality, and user perspectives closer to people's complex and changing needs and preferences.

1. Background

In modernity, the human body has been measured, calculated, and translated into average dimensions, proportion principles, and standards for rational and industrial building construction. Light has similarly been handled as a scientifically measured and standardized object of physics. The Danish Building Regulations require lighting conditions in buildings that ensure persons' comfort, safety, and health, and that the light is sufficient in relation to its intended use. The lighting standards for work environments cover recommendations for the quality of light and stress the importance of the variation of light in design considerations, due to its effect on people's mood, emotions, and mental alertness [1]. Informed by the positivist traditions of lighting research [2] literature on lighting quality often refer to non-visual responses as the effect on the human biology and physics. However, to explore the role and use of lighting in diverse and sometimes more complex issues, the qualitative concern needs to include the social and psychological effects and interactions at stake. Furthermore, even though the situated, contextualized, and perceived light has been explored in academia, through psychology and humanities [3, 4, 5, 6], this understanding of light has only to a limited extent been operationalized in the field of lighting design. More significantly, the diversity of the human body has not been represented in these studies, nor in common lighting principles and standards.

Analytical frameworks can support the investigation of marginals groups, as demonstrated by Lid [7], that explores how heterotopias in situated practices characterize and shape one another. She argues that recognizing the other can enable inclusion, space for human diversity, and participation. The concept of heterotopia is described as a displacement, the 'other' in relation to the 'regular'. "Heterotopias provide an avenue for distance. Isolated yet accessible, heterotopias represent a space that can show us the landmarks of thought by distancing us from those landmarks" [8, p. 66]. As a social space, the heterotopia, engage a continuous dialogue with its 'normal', maintaining powerful positions [9], where the multiple and shifting relational definitions could support social change [10]. As a space of encounter, where the sites are confronted with each other's 'otherness' and potentially changed by the very interaction – a dialogical approach to heterotopia could constitute niches where "difference and strangeness are allowed to remain" [9 p 36]. The clash of difference can render visible new knowledge of both sites, and sometime reorder or reconstitute them, as seen in the dialogic encounters that characterize innovation and interdisciplinary collaboration in our professional practices.

This kind of dialogic encounter was enabled in a pilot project in low-vision rehabilitation, entitled Better Light Better Living (BLBL), from 2017 – 2019. Two low-vision consultants at the Centre of Special Education (CSU) in Slagelse, turned to the field of lighting design, and supported by the new knowledge acquired, they developed and tested, in 60 visually impaired participants, a new methodology for lighting assessments in their services [11]. Traditionally, lighting assessments were practiced based on a scheme of overall recommended lux values for different diagnostics, measures of preferred light levels when seated in a fixed distance of a chart or arrangement of objects in the clinic, and by providing special lighting for special needs. However, compared to the hands-on experience and practice knowledge from working with the visually impaired and with the move to a more user-centered care, a less instrumental and more qualitative approach was needed. The new approach included: a) an individualized lighting assessment in the home, b) a lighting intervention in the lighting lab, and c) a follow up, assessing the changes which had been implemented by the participants themselves.

Exploring and optimizing their own practice by encountering and interacting with the field of lighting, a heterotopia to the field of vision rehabilitation, the low-vision consultants have innovated and moved both their understanding and use of light in their services. Using the parameter of light to recognize and understand the needs and motivation of the clients, and to improve the output of their service, by improving the client's quality of life. The question in this paper is whether we can establish a counterpart of this dialogic encounter within the field of lighting, and whether this can be unfolded as a heterotopia informing or challenging existing design practices? How can the findings from low-vision rehabilitation be translated into the field of lighting design? Can knowledge on lighting from the context of low-vision rehabilitation be seen as a critical case for informing approaches to more holistic lighting design, embracing the diversity of experiences and practices where the light is situated, and work as an incubator and a heterotopia of inclusion?

2. Theoretical framework

To explore and understand the user perspectives embedded in BLBL, between the professionals and the visually impaired, and between the participants and their lighting, we have employed a set of theories that opens to the human-environment interaction at stake. *Usability* has been a key term in product design, describing effectiveness, efficiency, and satisfaction "with which specified users can achieve specified goals in particular environments" [13], as a property of the interaction between the product, user, and a task [14]. Disability is listed as one of the user characteristics associated to usability, along with experience, domain knowledge, cultural background, age, and gender. "…by paying attention to the needs of those with disabilities, it is possible to provide opportunities for the disabled which might otherwise be restricted". These designs could also benefit able-bodied users, demonstrated by handsfree telephones, ball-point pens, and remote controls. The space for maneuver, both literally and metaphorically affects the interaction, where abilities and disabilities can be understood in relation to the concept of *affordances*. The perception of affordances is described as what the properties of the environment offer to the individual, as well as the personal abilities to identify, recognize and act on the properties offered [15]. Affordances are spatially and temporally situated [16], enmeshed in dynamic processes where perception and abilities to act can change, both temporally and during a lifetime. To be actualized, affordances need to be identified and recognized by the individual [17]: Don Norman describes affordances as an important parameter in product design as the way a device presents itself - the bunch of apparently unrelated affordances that the user should be able to understand and learn how to use with a minimum of effort [18]. He argues that technology built from a machine-centered point of view, would benefit from a design-approach that involves the user needs to ensure the proper affordances.

3. Methods

In 2018, BLBL was accompanied by an ethnographic action research study, conducted by one of the authors. BLBL formed an instrumental case study for investigating the current practice of low-vision rehabilitation, and CSU's innovative use of lighting in their service [19]. The empirical material used in this paper includes field notes from the participating researcher's observations of 15 BLBL consultations conducted by the low-vision consultants during the winter season of 2018/19. These consultations were both in the lighting lab of CSU and in the homes of the participants. The sample included both private and rented homes, apartments in multi-storey houses, terraced houses, and detached single-family houses, of 57 – 175 m2, 2 – 7 rooms, and built from 1927 - 1999. Consequently, both typology and orientation of the buildings differed, as well as window composition and distribution, causing different lighting scenarios. Some participants had themselves been involved in decisions regarding the existing artificial lighting in their home, while for others, this was fully or partly decided by previous owners, or included in a lease. Observations focused on the activities and conversations of the intervention, including the participants descriptions of their experience and use of the light in their everyday, the consultant-facilitated assessments and demonstrations conducted in relation to the chosen activities. In 2020, visits and interviews were conducted with 17 participants that had implemented all or some of the recommended changes. These visits were organized without the consultants, and with an objective to map the changes made by the participants as part of the project, or changes caused by the intervention on their mind-set and approach to lighting. Observations also included physical traces of the participants' use of the space, and the results of the intervention seen in adjustments or adaptions of spaces, arrangements, or practices. A semi-structured interview guide supported the visit, covering both the activities and lighting issues processed in the intervention, the overall approach of BLBL, and their personal learning or change process. The interview was partly conducted while moving across the living space, addressing the specific activities and associated lighting scenarios, and partly seated, in a place supporting the conversation, such as at a dining or coffee table. The questions involved their general use and perception of light in their home; the implemented changes and their user-experiences of these, including enabling and disabling aspects; and the role of the assessment and intervention for enabling the changes.

Notes, photos, and sketches were produced in the actual setting, as the photos in Figure 1, while digital field notes were produced immediately after each encounter. Both participatory observations and the visits were documented by rich descriptions, describing both conversations, interactions, and scenarios.

Figure 1. Photo documentations of the intervention: The consultants facilitating the participant's testing of alternative lighting arrangements for a chosen activity.

An analytically focused categorization was conducted in the empirical material, across the individual consultations, to identify and describe a) the visually impaired participants' understanding, perception and use of light in their everyday settings, b) experience of abilities and disabilities regarding their impairment in the specific settings and activities, c) the role of the assessment and intervention on this relation and d) the professional consultants' articulation of and approach to light and the individual. The following analysis was supported by the theoretical frameworks of affordances and usability to unfold and conceptualize the dynamic interactions of the human-environment relation.

Recognizing the relational and interactional aspects of light, focusing on the usability and affordances of the light, we hope to provide insights of the role of light in peoples everyday life of value for lighting design.

4. Findings

While previous analyses have investigated the co-design process within BLBL [20] and the role of lighting in the rehabilitation process [21], the findings here focus on the role of lighting in a more general human environment relation. Based on the personal narratives and settings, yet looking beyond the diagnostics of disability, we aim in this section to get a better understanding of the situatedness of light, intertwined between humans and environment. The case of BLBL show that light is situated in multiple concerns: as part of the physical and social context of the participants, the abilities of the individual, and attuned in the collaborative process around problem shooting and solving at stake in the lighting adjustments and optimization in the specific activities. These three main findings will be unfolded in the following sections.

4.1. Situated as part of the everyday physical and social context

An important aspect of BLBL compared to their previous practices of lighting assessments in low-vision rehabilitation has been to situate the consultation in the physical context of the participant. In the 60 participants, there was cases concerning work environments or school environment, however most participants were elderly, and with issues concerning their everyday life in their home environment. The consultants describe their first encounter with a participant's home environment as a subconscious scanning [19]. This initial inquiry of light, was a patchwork of information and observations, starting with the neighborhood area and close surroundings, and of the typology of housing, before entering the interior space. Any light was situated spatially, for natural light, this spatiality was often related to the light's changing character and movement throughout the space and on the different surfaces, while artificial light was more characterized by its position and orientation. Beyond the interior spaces and surfaces, the different light sources assembled with materials and colors of walls, floors, ceilings, and furniture arrangements. As the consultant was following the participant around the living space, identifying, and discussing issues and possibilities regarding the current lighting, the situatedness was often articulated. In a case where the participant described her problem with insufficient light at her dining table, the consultant examined and described the relation of the light source, the lamp, its size, shape, position and orientation, in order to suggest that the light source was probably good enough, but the many metal shades reflecting the light reduced the amount of light at the table to a degree, which was not optimal for this participant. Another issue stressed by the participant was the transition between the spaces of the apartment, that was determined on the current condition of natural light. For this participant, the sunlight was of great support during the day, while an overcast sky or at night, entering the living room felt like "walking into a wall of light". She had been granted special ceiling light panels but explained that she never used them, as the light was uncomfortable and annoying. The consultant recognized the issue and assessed the specific light, its correlation to the room, and the other luminaires, and summarized based on the descriptions, that due to the scattered and diffuse character of the light, the light did not reach the surfaces, and consequently caused the feeling of entering a 'wall of light'. The suggestion from the consultant was to test different arrangements to get to know the specific issue closer.

The social context of the participants was also impacting the use and perception of light in the home environments. For the participant described above, her social context, was constituted by her eight-year-old son, and friends visiting. Besides her own physical preferences, she also wanted the lighting to fit for social gatherings, meaning that her lighting preferences included both physical and social needs.

4.2. Situated as part of the more-or-less-able bodies' everyday life

The criteria for participating in BLBL included both visual impairment and photosensitivity. One of the hypotheses was that preferences and needs regarding light were not nearly as linked to diagnosis as previously assumed. The results from the BLBL showed no association between the diagnostics and the preferred lighting across the different participants, actual the same individual could choose different types and levels of lighting for different types of activities and in different environmental settings [11]. In fact, for many of the participants, their visual function was not steady, but changing due to their physical condition and conditions in their environment: "I see everything and nothing… you see, it is really hard to describe... it depends, for example, on the light, on the contrasts, and on whether I am tired, or whether I know the place and have been there before…". For some, light was a key to their residual vision, which was illustrated quite clear in a visit to one of the participants. The 50 lux provided by the daylight was all darkness to her, while the 2050 lux measured on the table from the suspended light fixtures she got installed, enabled her to see me and the things on the table.

For most of the visually impaired participants, the vision could not be restored and for many, the process would eventually result in blindness. Conversely, participants suffering from concussion typically experienced the opposite progress, of recovering or regaining a normal condition and a 'normal' everyday life. In this group, the issue was not the visual function, but usually directly related to the experience of lighting: A sensitivity to light and flicker, where luminance differences and contrasts by large were disabling aspects. This could be glare directly from a particular luminaire or from reflections in blank surfaces. Several of these participants preferred the softer twilight hours, and for many it meant different degrees of avoiding light. One participant described her coping as an act of balancing between very little light and darkness, in dialogue with her husband. To get used to the dim morning light in her bedroom, she usually stayed in bed for 45 minutes before she could get up. She described the impact of the light like this: "not as pain, but as a discomfort, like a noise, filling everything in my head". She kept the curtains shut, used a cap and sunglasses indoor, and most artificial lighting were turned off. She tried to manage the everyday with as little light as possible.

Another example from this group of participants showed a more creative approach to the lighting. She used the daylight quite a lot in her everyday life, however with an aim to avoid contrasts. When we moved around the house, she stopped in one of the rooms where a cone of light appeared from behind a blackout curtain: "For me, it has never been about my vision, however I can be in pain, and so it is a physical challenge… just a glimpse of light, like this from behind the blackout curtains, the huge contrast causes problems. From dark to light, can bother me, and from light to dark, I get tired." Transitions between dark and lit spaces was a major disabling issue for many. Illustrated in one participant with reduced flexibility of the only functioning eye, that explained how the adaption from lighter to darker spaces, or the other way around, as a very slow and painful process. A rheumatic disease further increased the sensitivity to light during the periods when the disease was in outbreak. However, when the light was dim, she used much more effort to adjust her eye to see anything at all, which then caused serious migraine. On the one hand, she was happy for and depended on the light, and on the other hand, it could cause great discomfort.

The empirical material shows that the light afforded both disabling and enabling situations. The individual and diverse needs of the participants, from nearly no light at all to far beyond lighting levels for normal vision, were also illustrated in the lighting measures from the follow up, ranging from < 10 -10.000 lux and above [19], which outmatch the normal lighting recommendations (100-1500 lux) for indoor environments [22].

4.3. Situated in complex and dynamic human-environment relations and processes

Where lighting assessments previously had been based on the visual impairment and adjusting illuminance in relation to visual acuity in a fixed position in the clinic, the BLBL embraced the situatedness of the light in the participants everyday lives. Departing from activities and situations of the participants own choice and motivation, engaged the participants beyond their mere visual function, and enhanced the interaction of the individual and the physical and social context.

Facilitated by the lighting intervention in the clinic, where the participants, with guidance and demonstration of the low-vision consultants, tested and compared alternative lighting arrangements for the activities at stake, new knowledge and realization supported the recommendations to be brought back home.

Lighting changes typically involved rearranging existing luminaires and settings or purchasing and installing new, recommended luminaires. As seen in Figure 2, this could be changing and repositioning the light under the cabinets to support activities on the kitchen, such as preparing food or managing the dishes. Or adding indirect lighting by illuminating the wall and ceiling above the cabinets, to optimize the general lighting.

Figure 2. Visuals from one of the homes visited. The photos show: 1) the researcher observing the interaction of the consultant and participant in the lighting assessment, 2) the consultant measuring light and documenting the scenarios, 3) the existing scenario, and 4) the scenario after the recommendations have been implemented.

Where the directed light supported the visually impaired participant in practical activities, the indirect light was primarily used in social occasions. All in all, these were often small adjustments, but they had enabled larger transformations: helping the participants to see other opportunities, recognizing the two-sidedness of affordances where their way of using the light could be adjusted as well. Even participants that had not made any changes to their lighting or physical context reported that their mind-set had changed, resulting them in using a luminaire in a new way, moving the light or moving the activity, or coordinating to the more optimal setting of the natural light if this was the case. In the 60 participants, performance, and satisfaction with the activities, as well as near activity, independency, and self-reliance improved with BLBL [11]. Resuming to activities where they previously had been dis-able due to insufficient lighting and regaining this ability were appreciated.

Actualization of affordances, the processes of complementing environmental opportunities with personal abilities was seen in workarounds and new ways of approaching the everyday practices, and due to the decreased scope of activity caused by the impairment, the activity within this scope became all the more important for well-being and quality of life.

5. Discussion

In BLBL, the low-vision consultants challenged the dominant technical understanding of lighting in their field of rehabilitation and embraced the situatedness and everyday life of the participants by expanding the approach and understanding of light. Aspects of usability, of specific users with specific goals in specific environments [13] seem to be a better fit for qualifying the everyday, than quantifiable measures of averages. The human environment interaction is represented in models of occupational therapy [22] that is somehow embedded in the processes of rehabilitation. The reconfiguration, reintegration, or reinterpretation of abilities in a given context, is the core of the low-vision consultants work, and why approaching the light from different angles seems obvious. They use the light to understand the everyday struggle and motivation of the participant, and to work with this relation to support the participants themselves to engage.

The practice of situating the light in the everyday social and physical context of the participant was informed by knowledge on light concerning our sensory experience of light, but also the use of light. As part of the pilot project, the consultants gained new knowledge on lighting through a series of courses focusing on the use and perception of light, arranged by an architect and lighting designer. This knowledge, of the experience of light, was, opposed to the scientific isolation of a phenomenon in single factors, acknowledging that the factors were perceived simultaneously and interacting. They learned how to observe, experience, and assess the light as part of the context, initially by their own sensory experience, described as an overall evaluation of the light in the room, assessing whether the luminance distribution felt suitable, as a subconscious scan [20], with particular attention on anomalies, things out of place. This knowledge on the situated lighting also included an awareness of the diversity of lighting features needed in different activities and in different spatial configurations.

In the cases of the visual impaired, the experienced and perceived light, was much more differentiated than we know it for normal sighted. On the one hand, glare, flickering, or high luminance contrasts could cause discomfort and constitute huge barriers or disabling conditions. On the other hand, and far beyond the luminance specification per diagnostics, lighting was through the intervention a key for enabling the residual vision for some of the participants. We have used the notion of affordances to help recognize this interdependency between the possibilities afforded by the environment, including the lighting, and the personal abilities to identify, recognize, and act on the properties offered [15]. In the process of BLBL, the participant's understanding and approach to lighting was transformed, in a way enhancing their abilities to appreciate and exploit the potential affordances – for their specific visual function, body, and situation in the given context. The process was further enabled by focusing the lighting assessment on activities and thereby assess the usability. Here effectiveness, efficiency, and satisfaction [13] were asserted in a different scale and pace than the normal-bodied, as endline measures compared to baseline measure of the individual before and after the intervention [11]. In this way the assessment supported an actualization of affordances [17], which is essential in rehabilitation, where the individual needs to create strategies to cope with the abilities and disabilities associated with a functional or cognitive impairment.

The objective of the consultants has been to find lighting appropriate for the individual (optimizing visual function), in the specific tasks, and that could accommodate varying needs by different or flexible/

adjustable lighting. There was an overall need for flexibility regarding lighting, both in relation to the different needs of the visually impaired participant, dependent on the daily or seasonal variations, changes in the impairment, and the needs and preferences of the family members or other in the home. Compared to the previous diagnostic or medical approach to the lighting assessment, the participants were seen as complex individuals with multiple and sometimes contradicting preferences. Lighting preferences could be based on visual performance, social settings, or the level of comfort. The body and the vision were not understood as static diagnosis, but as a dynamic entity that could mobilize change. The notion of shaped affordances helps us illustrate the interrelation of the individual and its context, where either of the parameters can change - in the intervention both the lighting, the use of light, or the understanding of light could help shape new affordances.

5.1. Lessons learned from BLBL and this study that can inform the field of lighting:

1. Light is contextualized, and seen as a qualitative aspect it stretches across architectural quality, to lighting quality and quality of life. Light, both natural and artificial, is particularly important for comfort and wellbeing of visually impaired. The appropriate light is found in the specific relation why mock ups and post-occupancy evaluations would be of great value.
2. The general need of different lighting for different activities, and the different personal preferences, applies to all, not just the visually impaired. Consequently, this diversity calls for user involvement, where focus on usability can support the design process.
3. Beyond usability, the interaction of human and their environment is reciprocal – our understanding and use of an object can affect our understanding of it, as well as our understanding of it can affect our use. This goes both for the designer and the user, and the concept of affordance could potentially scaffold methodological considerations for how to operationalize this relation.

Beyond the mere lighting specifications, all practices working with light employ layers of tacit knowledge. Whether it is the embodied knowledge of end-users, embedded knowledge in lighting technologies, or practice knowledge of lighting designers. As the field of low-vision rehabilitation has recognized the pivotal role of light in their practices and services, we argue that the role of a more diverse user perspective, including low-vision can qualify the field of lighting design. As the before mentioned handsfree telephones, ball-point pens, and remote controls [14], lighting products and technologies have already worked across the heterotopias and several of the special-aid products within lighting are moving into ordinary consumer goods. Everyone may at some point experience challenges with their vision or discomfort from lighting, where a multiple and dynamic approach to lighting can accommodate a more diverse understanding of the user – putting user needs to the point [18]. Incorporating usability studies or the concept of affordance in design practices could ensure qualitative aspects and lighting more sufficient to its use, while linking performance and impact measures to quality of life would be a statement of prioritizing the user perspective. Both could enhance the user perspective in quality of lighting, still missing in the guidelines and standards.

As illustrated in Figure 3, this study has explored the role of lighting in the field of low-vision and low vision within lighting.

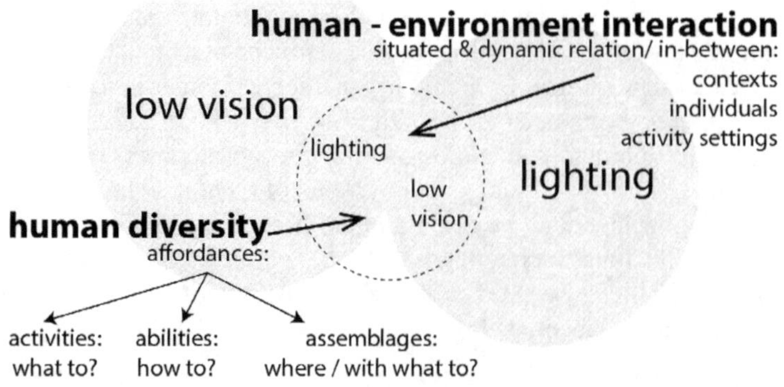

Figure 3. The overlap of lighting and low vision, unfold both diversity and dynamic interaction, of humans and environments. A niche that can be operationalized by the concept of affordances.

Working across these heterotopias, the analysis has identified usability and affordances, as suitable framework for investigating human diversity within the human-environment interaction at stake in BLBL. Investigating the cases regarding the assemblages of lighting technologies and physical and social contexts, the abilities of the individual in the setting of specific activities. This affords to a more holistic, situated, and dynamic understanding of light: The experience of light as multiple and constituted by the dynamic interaction between the individual and the environment, making the abilities of the individual relational and dynamic. This approach requires the time and resources for assessment, exploration, and for including the users in co-design processes as they are the experts in their everyday life, preferences, and perception. Identifying affordances require reflections of the individual on how the surroundings and assemblages enable or disable the specific situation or activity, experienced in use.

There are essential differences between rehabilitation and design. In rehabilitation, the focus is on specific individuals in existing contexts, while in design it is often for an unknown or general user and a yet to come environment. Looking into the experiences of visually impaired and the role of lighting in low-vision rehabilitation, emphasize a much broader user perspective than the 'regular user' of both architectural and lighting design. As stressed by one of the participants: "…working with low vision, it concerns the specific and individual settings. We do not fit a specific lux level, you must test the specific lighting... ".

The other way around, designers will seldom have the opportunity of working as closely with individual users as seen in this inquires, but a more diverse user perspective could accommodate and reach more of us. We have outlined learnings from the field of rehabilitation which we believe could qualify lighting design in meeting the call for social inclusion, in creating inclusive and safe environments, leaving no one behind.

6. Conclusion

In this study, low vision and lighting have been investigated as each other's heterotopia, as counter-sites. The approach to and understanding of lighting in low-vision rehabilitation has been reconstituted to fit the purpose of the rehabilitation of visually impaired in their everyday settings and practices, embracing a situated and contextualized lighting. In this paper we try to turn the tables and direct the dialogical approach of heterotopia to reorder the user perspective in the field of lighting.

Supported by the concepts of usability and affordances we have translated and discussed the findings to the field of lighting design, however, not as lighting characteristics or measures, but as a methodological approach. The examples from the everyday of the visually impaired show light as a) contextualized, in social and physical contexts, b) that lighting needs and preferences diverge independently of visual conditions, why use and activities are important entrance points to optimized lighting, and c) that the relation between users and light is dynamic and affected by the actual interactions between the two and the overall context.

Whether this initiative succeed in enabling a dialogic encounter of diversity in the field of lighting design, and inform or challenge existing design practices, remains to be seen. We hope that it sparks reflections in lighting design practitioners, where a dynamic and relational understanding of the human environment interaction can flourish.

References

[1] Dansk Standard 2021 *Light and lighting - Lighting of work places - Part 1: Indoor work places* DS/EN 12464
[2] Kelly K 2017 A different type of lighting research–A qualitative methodology *Light. Res. Technol* **49** 8 pp 933-942
[3] Böhme G and Thibaud J P (eds.) 2016 *The aesthetics of atmospheres* (New York: Routledge)
[4] Bille M, Bjerregaard P and Sørensen TF 2015 Staging atmospheres: Materiality, culture, and the texture of the in-between' *Emot Space Soc* **15** 1 pp 31-38
[5] Ingold T 2011 *Being alive. Essays on movement, knowledge and description* (Oxon: Routledge).
[6] Pink S and Leder Mackley K 2016 Moving, making and atmosphere: Routines of home as sites for mundane improvisation *Mobilities* **11** 2 pp 171-187
[7] Lid IM 2017 Medborgerskap i spennet mellom ekskludering og inkludering [Citizenship in between exclusion and inclusion] *Rom og etikk: Fortellinger om ambivalens* ed IM Lid and T Wyller (Oslo: Cappelen Damm Akademisk) pp 35-54
[8] Topinka R J 2010 Foucault, Borges, heterotopia: Producing knowledge in other spaces *Foucault Studies* pp 54-70
[9] Meininger H P 2013 Inclusion as heterotopia: Spaces of encounter between people with and without intellectual disability *Journal of Social Inclusion* **4** 1 pp 24-44
[10] Hetherington K 1997 *The Badlands of modernity. Heterotopia and social ordering* (London: Routledge)
[11] Øien TB, Jacobsen AM, Tødten ST, Russotti T, Smaakjær P and Rasmussen RS 2021 Lighting Assessment and Optimization in Low Vision Rehabilitation Improves Participation and Quality of Life in Individuals with Vision Loss *Occup Ther Health Care* pp 1-18
[12] Christiansen AB 1995 Hvilke lystekniske parametre er vigtige for synet? [What lighting parameters are of importance for the vision?] *Fakta om lys* **8** 1 pp 6-13
[13] International Organization for Standardization 2015 *ISO DIS 9241-11: Ergonomics of human-system interaction — Part 11: Usability: Definitions and concepts*
[14] Jordan PW 1998 *An introduction to usability* (London: Taylor Francis)
[15] Gibson JJ 1979 *The Ecological Approach to Visual Perception* (Boston: MA: Houghton-Mifflin)
[16] Chemero A 2003 An outline of a theory of affordances *Ecol. Psychol.* **15** 2 pp 181-195
[17] Moore RC and Cosco NG 2007 What makes a park inclusive and universally designed?: a multi-method approach *Open space: People space* ed CW Thompson and P Travlou (London: Taylor & Francis) pp 105-130
[18] Norman DA 2016 *Living with complexity* (Cambridge: MIT press)
[19] Øien TB 2022 *Belysning i synsrehabiliteringen. Et casestudie* [Lighting in low-vision rehabilitation. A casestudy] (Copenhagen: BUILD)
[20] Øien TB 2022 A Study of Environmental Factors in Low Vision Rehabilitation *Front. Rehabilit. Sci.* **113**
[21] Kunstig belysning [Artificial lighting], Arbejdstilsynet [The Danish working environment authorities], accessed 9 August 2022, https://at.dk/regler/at-vejledninger/kunstig-belysning-a-1-5/
[22] Øien TB 2021 Housing and low-vision rehabilitation – across theories, practices and everyday settings *Proc. from the 4th Conference on Architecture Research Care & Health* **8** 1 ed J van der Zwart, SM Bakken, GK Hansen, E Støa and S Wågø pp 113-119

Acknowledgments
The study has been funded by the Velux Foundation.

IOP Conf. Series: Earth and Environmental Science 1099 (2022) 012055 doi:10.1088/1755-1315/1099/1/012055

Light Sketching for Ecology: A cooperative design tool for balancing human experience and ecological impact

Philip R Ross[1] **and Niek Rutten**[2]

[1] Studio Philip Ross, Eindhoven, the Netherlands
[2] Eindhoven University of Technology, Eindhoven the Netherlands

Abstract. The impact of artificial light at night on ecology is an urgent and continuous consideration in lighting design. People's needs and wishes for light in urban areas often conflict with demands for minimising ecological impact. For example, in city parks people may desire light that offers a pleasant experience of nature and the feeling of safety, while the fauna may thrive best with no light at all. How to overcome this seemingly unresolvable conflict, especially in hybrid environments like city parks? This paper presents Light Sketching for Ecology, a cooperative lighting design tool to help balance lighting demands from the human experience side with demands for minimising ecological impact. Results from a pilot study show the potential of Light Sketching for Ecology to create a win-win situation in which park users are more positive about the park lighting while the lighting's ecological impact is significantly lowered.

1. Background and objective

Artificial light at night impacts ecology, presenting an urgent consideration to lighting design. Light can negatively influence animal biodiversity, sleep-wake cycles, reproduction, population sizes, migration and more [1]. Various aspects of light, such as intensity, colour spectrum, duration and timing play a role in this impact in specific ways [2].

People and animals differ in how sensitive they are to specific wavelengths of light, and animals also differ from each other in this respect. Research on spectral sensitivity of animals and the resulting impact of light on animals is ongoing. In general, avoiding short wavelengths (blueish and UV light) is recommended, since many species are highly sensitive for this part of the spectrum. Longer wavelength (reddish) light is generally considered less impactful, although there are exceptions [3]. From the perspective of ecology, the less artificial light at night, the better.

People's needs and wishes for light in special urban areas, like city parks, however often conflict with demands for minimizing ecological impact. A city park offers social and health benefits for city dwellers, also after sunset. But removing all artificial light, or lighting a park in pure red, would make the park practically inaccessible for people or very unpleasant to use.

How to overcome this seemingly unsolvable conflict, especially in hybrid environments like city parks? We present and discuss Light Sketching for Ecology as a way to balance lighting demands for human experience and demands for reducing ecological impact. This paper first describes the Light Sketching system [4] (section 2), followed by the added ecological functionality and its underlying methodology (section 3). Section 4 describes a pilot study to evaluate the approach, followed by conclusions and discussion in section 5.

2. The Light Sketching system

The goal of the Light Sketching system, developed by the first author for the 'Smart Lighting for Placemaking' project [4], is to enable a nuanced, creative conversation with people about light in their outdoor environment. Talking about light and imagining its effects without seeing it in context is difficult for people without training in light. People need to see and experience light to understand its implications and to creatively engage with it. Light Sketching allows residents and other stakeholders to express their wishes for light in their environment via actual light, live in their environment. They sketch light patterns on a tablet interface, which are transformed into real light patterns in the environment through custom designed street luminaires with advanced lighting capabilities. The following sections describe the two main system elements: The Light Sketching luminaires and the sketching interface. See also this video of the system in action.

2.1. *Light Sketching luminaires*

The system features five custom post top luminaires, designed by the first author, based on the principle of 'pixelated' light. This means that the total light distribution is made up of numerous smaller light spots cast by 38 individually controlled LED modules. These modules have independently controllable RGBWC channels (Red, Green Blue, Warm white and Cold white) and are equipped with a narrow spot optic. The LED modules all together light up a circular area around the luminaire with a diameter of approximately 10m. The luminaires are placed on moveable poles at the height of 4m, which is in the range of much regular park lighting. See Figure 1.

Figure 1. Left: One of the five Light Sketching luminaires, placed on a pole. Right: Several Light Sketching luminaires in a park during a Light Sketching session.

2.2. *Light Sketching interface*

A tablet interface offers an easy and intuitive way to create light patterns with the Light Sketching luminaires, as easy as finger painting. The interface first shows a map of the location with the positions of the five Light Sketching luminaires (Figure 2 left). Selecting a luminaire opens the sketching canvas for that luminaire (Figure 2 right). A colour is selected by touching one of the pads on the right side of the screen: warm white, cold white, black (for less light) and an RGB palette. This colour can then be applied in the sketching area on the left side, showing a top view of the luminaire. Stroking the finger in the area around the luminaire applies a layer of light of the selected colour. This is instantly translated into real light in the corresponding direction by the luminaire. The more a finger touches a particular spot, the more layers of the selected colour are applied, intensifying that colour. This allows people to dose and mix colours carefully and gradually. The luminaires translate the sketching actions directly into real light in the environment, so that the person sketching can experience and adjust the light on the fly. The resulting light scheme on the five luminaires is called a Light Sketch. Figure 3 shows the system in use in a Light Sketching session.

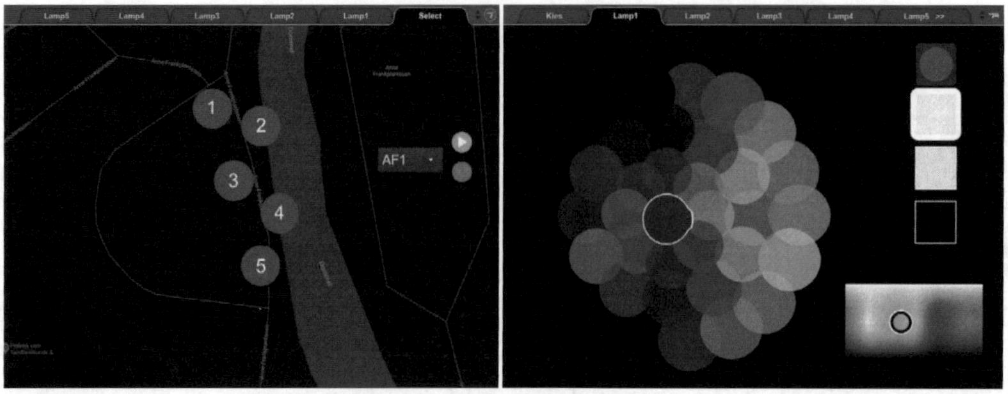

Figure 2. Left: The luminaire selection tab of the interface. Right: Screenshot of the sketching tab, showing the colour selectors on the right and the area for light pattern sketching on the left.

Figure 3. Left: Participant sketching a light-colour pattern on one of the luminaires using the tablet. Right: Overall impression of a Light Sketching session in a city park. See also this video.

3. Ecological functionalities

The previous section describes how the Light Sketching system enables people to sketch and evaluate light from a human perspective. But how can this tool also help consider the ecological effects of light in a chosen environment? To this end, specific animals are given 'a voice' in the Light Sketching process, in two ways: The system incorporates a measure for how impactful the light is to these animals and it offers the possibility to tweak the emitted light to reduce this impact. Both additions to the Light Sketching system are explained in the following sections.

3.1. A measure for ecological impact of light: Effective Irradiance

As mentioned earlier in this paper, species vary in how sensitive they are to different wavelengths of light. Consequently, light of a given Spectral Power Distribution (SPD) impacts different species to different extents. For example, Honeybees are attracted to UV and blue ranges of light, while sensitivity of the Green Turtle spans somewhat longer wavelengths [5, 6]. The 'sensitivity curves', describing the variation of an animal's sensitivity to light across a range of wavelengths, differ between animals. Humans also have specific sensitivity curves. This offers an opportunity to tweak the spectrum of light in such a way that it remains comfortable and pleasant to people, while lowering its impact on particular animals.

To get a grasp on the magnitude of the effect of light with a given SPD on specific animals, we turn to a methodology described by Longcore et al., called 'Rapid assessment of lamp spectrum to quantify ecological effects of light at night' [6]. This calculation method provides a quantitative measure of impact, based on the amount of overlap of the light's Spectral Power Distribution and the sensitivity/response curve of the animal for a range of wavelengths. The resulting measure for impact is called Effective Irradiance. The less overlap, the less the animals are impacted by the light. Refer to [6] for the details of this calculation method. Figure 4 shows a visual illustration. The depicted animal (red curve) is relatively sensitive to blue and green light and less sensitive to red light. Two SPD's of different light sources are drawn in blue. The left SPD has much more overlap with the animal's sensitivity curve, resulting in a higher Effective Irradiance and more impact on this animal.

To allow comparison across different animals, the Effective Irradiance is scaled relative to the animal's response that would be elicited by daylight (D65), see [6]. The measure we use in our calculations is therefore a scaled version of Effective Irradiance.

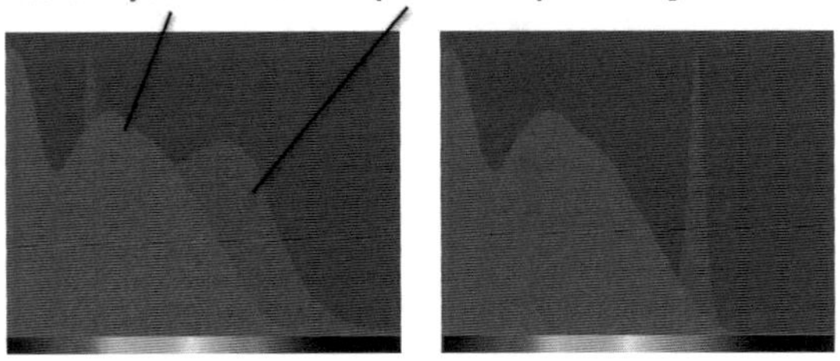

Figure 4. The graphs show two different Spectral Power Distributions of light, indicated by the blue curves, both in combination with a normalised sensitivity curve of an animal (red curves).

3.2. Integrating Effective Irradiance in the Light Sketching system

The next step was to apply the calculation method in the Light Sketching System. We obtained the Spectral Power Distributions of the LED modules' RGBWC channels, and normalised them. Each channel in the system is treated as a separate light source, making a total of 5 channels x 38 RGBWC Modules x 5 luminaires = 950 separate sources. Each channel is dimmable from 0-100% in a linear way. In our calculations, each channel is weighed with its current dimming percentage.

The sensitivity curves of specific animals are also incorporated in the system, so that the (scaled) Effective Irradiance per animal can be calculated. We decided which animals to include based on their presence at the pilot location and their status in terms of protection (e.g. red listed), using the Dutch National Database of Flora and Fauna. We worked at the level of animal groups instead of exact species, because literature did not offer sensitivity curves of the exact species present at the pilot location. We selected the following six animal groups: butterflies, owls, bats, mice, rabbits and songbirds.

One representative sensitivity curve per animal group was extracted from literature, based on availability, to represent the group. The selected sensitivity curves must therefore be seen as approximations of the curves of the animals that we targeted, for pilot purposes. For example, the sensitivity curve of the red listed songbird Swallow, was not available in literature. We took the curve of the Blue Tit as substitute, since it is highly similar according to Dakin et al. [7]. We converted the selected sensitivity curves to so called "action spectra" for the calculation, following the methodology described in Longcore et al. [6]. The six representative action spectra per group were the following:

1. Butterflies: Numata longwing butterfly, *Heliconius numata*, derived from [8].
2. Owls: Tawny Owl, *Strix aluco*, derived from [9].
3. Bats: Pallas's long-tongued bat, *Glossophaga soricina*, derived from [10].
4. Mice: House mouse, *Mus musculus*, derived from [11].
5. Rabbits: European Rabbit, *Oryctolagus cuniculus*, derived from [12].
6. Songbirds: Blue Tit, *Cyanistes caeruleus*, derived from [13].

See Figure 5 for visual representations of the selected action spectra.

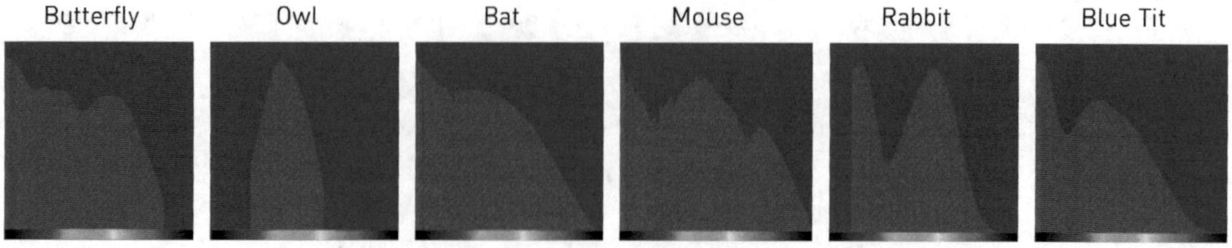

Figure 5. Schematic representation of the six selected action spectra, plotted on a 360-700nm range.

The scaled Effective Irradiance calculation is performed in real-time based on the selected action spectra and the current Spectral Power Distributions of all 950 LED channels. We add all scaled Effective Irradiances of all channels together into what we call the Light Sketch's Total Effective Irradiance. This measure is updated instantaneously when a sketch is altered. This way, we can see for each added stroke of light how it changes the calculated impact on the selected animals.

The Total Effective Irradiance is a number without clear reference to other numbers. To give more reference, we also incorporated the Total Effective Irradiance from the original street lighting at the pilot location in the system. This way, the system is able to report the calculated impact of a Light Sketch as percentage of the normal light situation's impact. The system also reports the energy use in Watts live during sketching.

3.3. Strategies to lower ecological impact

Computing the Total Effective Irradiance of Light Sketches is one thing, but we also aim to reduce the impact. From literature, we derived three strategies to lower ecological impact of a given Spectral Power Distribution:

1. Lowering overall intensity. Very straightforwardly, less light means less ecological impact [3].
2. Shift to red. This means decreasing green and blue components of the light, for which animals are generally most sensitive [14].
3. Metamer colouring. A light spectrum consisting of a small number of narrow peaks likely has less overlap with an animal's sensitivity curve than a continuous light spectrum. White LEDs have peaks, in the blue range for example, but these peaks are bound together in a more-or-less continuous spectrum. Light from white LEDs with a given colour temperature can also be created with only narrow peaks in blue, green and red. See Figure 6. This is a 'metamer' colour. This metamer light has the same colour point, making it hard for people to tell the difference. But it is less impactful than the 'original' white LED light in terms of Effective Irradiance for many animals. Note that the trade-off is that the Colour Rendering Index is lower, and it generally takes somewhat more energy to achieve the same white light intensity in lux with R G and B compared to a white LED.

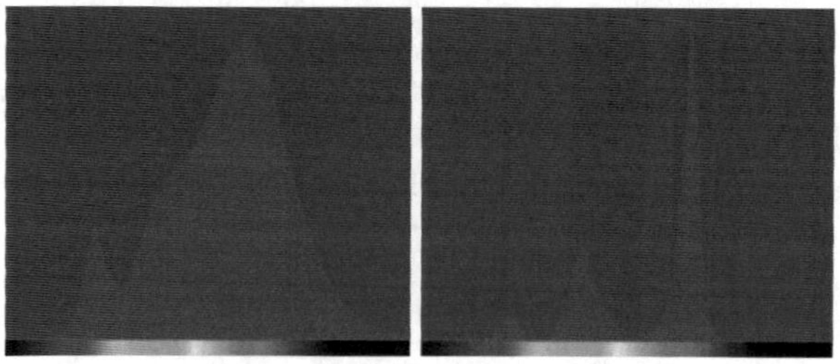

Figure 6. Left: The continuous spectrum of the system's warm white LED. Right: The spectrum of the metamer light, having the same colour temperature, but created with narrow RGB peaks.

3.4. The interface for ecological optimization of light
The Light Sketching tool is able to apply these three strategies on an existing Light Sketch, all in a gradual way from 0-100%. This gradual nature is important, because Light Sketching for Ecology is about finding a balance between human experience and ecological impact. How far can the light be dimmed before people find it too dark? How much of the green and blue component can be removed before people find the environment too red? And to what extent can the light from white LEDs be replaced by metamer (R G B) colours before people object?

The tablet interface has a special ecology tab that allows the session leader to gradually apply the three strategies with sliders. It also gives the possibility to compare the ecological impact of a Light Sketch with a benchmark, such as the impact of the existing public lighting on the location. See Figure 7 for a screenshot and explanation of its elements.

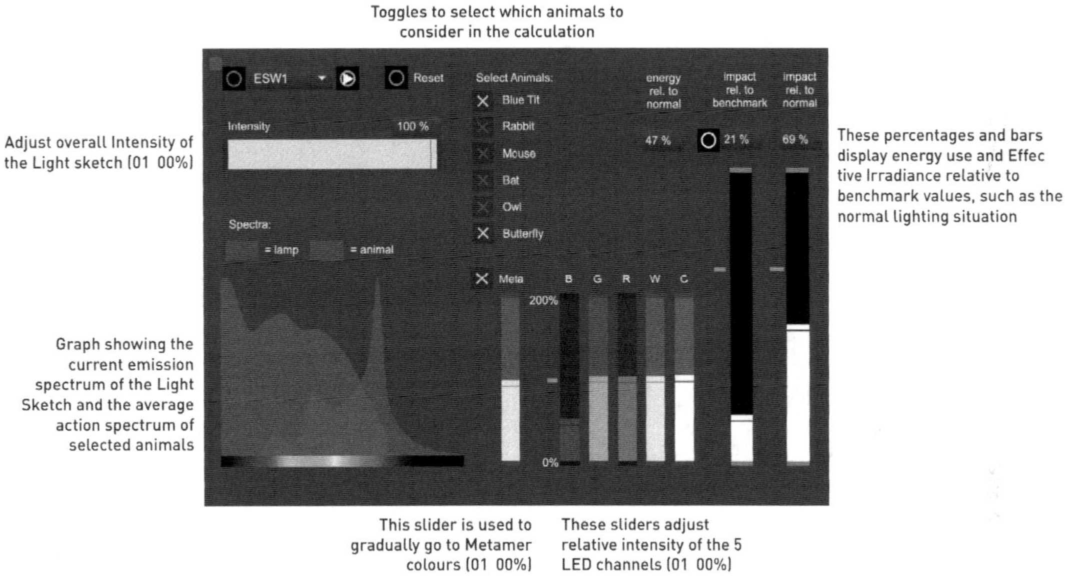

Figure 7. Screenshot of the ecology tab with explanation of the main elements.

4. Pilot study in city park 'Stadswandelpark'
A pilot session was conducted to test the Light Sketching for Ecology system, and to learn if the combination of Light Sketching and the proposed lighting strategies to reduce ecological impact could work in practice.

4.1. The pilot site
The Stadswandelpark in Eindhoven, the Netherlands, is a popular city park during the day, used for walks, sporting, biking and socialising. After sunset, a totally different situation emerges. Many describe it as too dark and eerie, and avoid entering it after sunset. We selected a dark entrance path into the park as the pilot location. This 55m path is normally lit with two 32W Compact Fluorescent lamps placed at 3.5m height. See Figure 8 left. The municipality turned the normal street lighting off for the pilot and we placed our five Light Sketching luminaires in the area around the path. See the map in Figure 8 Right. Luminaires 1 to 4 were placed on 4m portable masts. Luminaire 5 was placed upside-down as uplighter under a tree.

Figure 8. Left: Image of the Stadswandelpark pilot location with regular lighting. Right: Map showing the placement of the five Light Sketching luminaires along the path.

4.2. The pilot protocol

Five participants took part in this pilot in early December. Among them were regular park users and professionals. The professionals, an ecologist and a city public lighting official, did not sketch but reflected on the sketches and the implications of the methodology. The group of park users consisted of three people. One was an elderly woman, living at the edge of the park. She was scared to use the park at night because of the darkness and lack of overview. The second participant was a man in his forties, who visits the park for sport and recreation during the day. The third participant was a woman in her twenties, who normally makes use of the park only at daytime.

After a talk about how they view the park at night, and what they would aspire to do there at night, the participants sketched the light of their preference. They were asked to think aloud and respond to what they saw. They were free to walk around with the mobile sketching interface in their hand, so that they could experience the environment from multiple viewpoints. After they finished sketching, the moderator went through the following steps for ecological improvement:

1. **Dim step A.** The intensity of the sketch is decreased to a minimum of 30%. This step serves two purposes. First, to determine if such dimming is acceptable. Second, the R G and B channels of the used LED modules create whites with less intensity than the W and C channels. To create a metamer light scene with equal impression of brightness, the light sketch needs to be dimmed first.
2. **Metamer colour.** All light created by the W and C channel is gradually replaced by light with the same colour point created by the R G B channels.
3. **Partial shift to Red.** First the B channel is dimmed gradually to 0%. Then the G channel is gradually dimmed to 0%.
4. **Dim step B.** The resulting Light Sketch is dimmed further down gradually to find the minimum acceptable light levels.

4.3. Results

In this section we focus on the results of one of the participants, because it is more instructive to treat one participant deeply than to summarize the results of such a small group into aggerated numbers. The pictures of Figure 9 show the results for participant 1, the elderly woman. All pictures are taken with the same camera settings. The top left picture shows her initial sketch, using a mixture of mostly warm white LEDs, reds and greens, creating a warm atmosphere. She directed light at the bushes and at the trees, away from the path. She explained that she wanted to experience nature, not the path. This also prevented uncomfortable glare from the luminaires, and improved her overview of the area.

The top right picture shows the scene after Dim step A (dim to 30%). The participant was fine with this light reduction. The bottom left picture shows the scene after the Metamer colouring step. She reported she could not tell the difference between the original colours and the metamer colours. The bottom right picture shows the scene after shifting to red. She found dimming Blue to 0% acceptable, but did not want to go lower with dimming Green than 50% of their original intensity. This would make the park look too red for her taste. The final dimming step went down to 11% of the original intensity. She indicated this low intensity was only suitable for a late-night setting, not for the evening.

Figure 9. Pictures of the process of participant 1. The pictures from left to right, top to bottom show the original Light Sketch and the resulting light scenes of steps 1-3.

Table 1 lists the numerical changes of the light sketch after each step in the process of participant 1. The table shows that the ecological impact measure Total Effective Irradiance was reduced to 9% of the original sketch after step 3. Compared to the regular park lighting, the impact was reduced from 515% (original sketch) to 47%. Step 4 (Dim step B) lowers impact even further, but these light levels were considered only suitable for the late night and too dark for evening use.

Process steps	**Tot. Eff. Irradiance compared to original sketch**	**Energy use compared to original sketch**	**Tot. Eff. Irradiance compared to original lighting**	**Energy use compared to original lighting**
0. Original sketch	100%	100%	515%	102%
1. Dim step A	30%	29%	157%	30%
2. Metamer	14%	64%	74%	65%
3. Partial shift to red	9%	52%	47%	53%
4. Dim step B	3%	18%	16%	18%

Table 1: Measures related to ecological impact and energy use per step for participant 1.

Overall, all participants reached calculated impact levels far below those of the original park lighting. The percental decrease of calculated impact per step depended on the SPD's of the Light Sketches. For example, participant 2 and 3 used less Warm White in their original sketch, so Metamer colours had less effect. And shift to red had much more effect for the participant who used the most Blue. Dimming had the strongest effect on Total Effective Irradiance overall.

5. Conclusion and discussion

5.1. Limitations

This pilot study has its limitations, which we briefly treat here. One limitation is that study lasted only one night. A longer-term study needs to confirm if lighting scenes like the ones sketched will remain positively appreciated over time. Light Sketches cannot be seen as full-blown lighting designs, because such designs need to consider more constraints. The participants group was small and cannot represent all pilot site users. The limitations in terms of available sensitivity curves of specific species are already discussed.

The Effective Irradiance-related quantities in this study should be regarded as indications of how impact on animals alters due to specific changes in light. We should not attribute too much meaning to the exact small quantitative differences that emerge from the calculations. Importantly, the Effective Irradiance measure should be placed in perspective: Ecological impact on animals is influenced by more light-related factors than spectrum and intensity. For example, location, direction, timing and duration should be considered as well [2], in interaction with animal behaviour. Therefore, collaboration with an ecologist is certainly advised.

5.2. Conclusions

Despite these limitations, we can draw a number of conclusions from this project. In the pilot, the Metamer colouring contributed to a reduction of Total Effective Irradience up to approximately half of the original white LED based colouring. Strikingly, none of the participants noticed any difference. This points towards benefits of using RGB instead of Warm White LEDs for lowering Effective Irradiance, without compromising human experience. There are trade-offs in terms of CRI and energy use as mentioned earlier, but they can be worthwhile in specific situations.

Shifting to Red was accepted to a significant degree, but not to the extreme of 0% B and G. Still it helped reduce Effective Irradiance significantly, pointing towards the potential of this strategy and of the benefits of using a gliding scale to determine how much of this strategy is acceptable to people, instead of considering only the extremes.

Lowering intensity remains a key factor in reducing ecological impact. Participants accepted quite low light levels. An indication is that the average wattage used for light of all three participants after step 3 (partial shift to red) was approximately 30% of the regular, scarce park lighting (average of 20W LED light compared to 64W CFL regular light). This potential for acceptance of lower light levels also emerged in other Light Sketching sessions [4]. Light distribution seems to be a relevant factor in this. Participants lighted up vertical surfaces (trees and bushes) rather than horizontal surfaces (the path), so that less light made a larger visual impression. Glare was also reduced by sending light away from the path, lowering the contrasts in the scene. This improved overview and enabled the eyes to adapt to lower light levels. Using alternative light distributions that modern LED lighting offers has the potential to make lower light levels acceptable. Note that lighting of vertical surfaces, for example a tree, could introduce specific ecological disadvantages that should be weighed against the advantages.

The gradual application of the impact-lowering strategies, going slowly from 0% to 100% so that participants could indicate which degree of change was still acceptable, helped find a balance between human wishes and ecological impact. The participants were still very satisfied with their Light Sketch after the three impact-lowering strategies were applied to different degrees. And at the same time, the Total Effective Irradiance of the Light Sketches dropped far below the Total Effective Irradiance of the regular, scarce, park lighting. This shows the potential to create a win-win situation where the ecological impact is significantly lowered while the human experience is improved.

Light Sketching for Ecology demonstrates that giving both people and the local ecology a voice in a cooperative design process is possible and worthwhile, especially in hybrid city locations such as parks, where both kinds of considerations need to be balanced. Experience of light in context helps engage people and enables a nuanced and creative conversation about light and ecology. Combining this with the advanced lighting possibilities of LED lighting in terms of SPD, dimming, timing and light distribution opens up new possibilities for the lighting designer to make progress in designing light for people and ecology.

Acknowledgements
We thank the municipality of Eindhoven, the TU/e Intelligent Lighting Institute and GLOW for their support, Travis Longcore and Benjamin Backx for sharing their knowledge and the participants for their contribution.

References

[1] Hölker F, Wolter C, Perkin E K and Tockner K 2010 Light pollution as a biodiversity threat. *Trends Ecol Evol.* **25(12)** 681-2
[2] Schroer S and Hölker F 2017 Light pollution reduction: Methods to reduce the environmental impact of artificial light at night *Handbook of Advanced Lighting Technology* ed Karlicek R, Sun, C C et al pp 991-1010
[3] Spoelstra K 2020 Spectrum dependent effects of artificial light at night. *Biodiverse* **3** 16
[4] Ross P R and Kalinauskaite I 2020 Sketching Smart Light for Placemaking *ILI Magazine* **13** 16-9
[5] Longcore T, Aldern H L, Eggers J F, Flores S, Franco L, Hirshfield-Yamanishi E, Petrinec L N, Yan W A and Barroso A M 2015 Tuning the white light spectrum of light emitting diode lamps to reduce attraction of nocturnal arthropods. *Phil. Trans. R. Soc. B* **370** 2014.0125
[6] Longcore T, Rodríguez A, Witherington B, Penniman J F, Herf L and Herf M 2018 Rapid assessment of lamp spectrum to quantify ecological effects of light at night. *Exp Zool A Ecol Integr Physiol* **329(8-9)** 511-21
[7] Dakin R, Lendvai Á Z, Ouyang J Q, Moore I T and Bonier F 2016 Plumage colour is associated with partner parental care in mutually ornamented tree swallows, *Animal Behav* **111** 111-8
[8] Struwe G 1972 Spectral sensitivity of the compound eye in butterflies (Heliconius) *J. Comp. Physiol.* **79** 191–6
[9] Martin G R 1977 Absolute visual threshold and scotopic spectral sensitivity in the tawny owl Strix aluco. *Nature* **268(5621)** 636-8
[10] Winter Y, López J and Von Helversen O (2003) Ultraviolet vision in a bat *Nature* **425(6958)** 612-4
[11] Jacobs G H 2008 Cone Pigments and Vision in the Mouse *Visual Transduction and Non-Visual Light Perception. Ophthalmology Research* Ed Tombran-Tink J and Barnstable C J pp 353–73
[12] Nuboer J F W, Moed P J 1983 Increment-threshold spectral sensitivity in the rabbit. *J. Comp. Physiol.* **151** 353–8
[13] Hart N S, Partridge J C, Cuthill I C and Bennett A T 2000 Visual pigments, oil droplets, ocular media and cone photoreceptor distribution in two species of passerine bird: the blue tit (Parus caeruleus L.) and the blackbird (Turdus merula L.) *J Comp Physiol A* **186** 375–87
[14] Jägerbrand A K and Bouroussis C 2021 Ecological Impact of Artificial Light at Night: Effective Strategies and Measures to Deal with Protected Species and Habitats. *Sustainability* **13(11)** 5991

AUTHOR INDEX

Akanpinar, Oktay 505
Akuzawa, Y 238
Amoruso, Chiara Ester 442
Andersen, Marilyne 228
Andersen, Rune K 532
Angelaki, S 319
Ashtari, Elmira 130
Avci, Ayse Nihan 340
Barberá, A 432
Baur, S 473
Bay, A 109
Bene, P 159
Besenecker, U C 413
Besenecker, U 319
Besenecker, Ute 390
Bhatia, M 88, 301
Bhusal, Pramod 65
Bjørner, Thomas 267
Bochnia, Anna-Kristin 278
Bordonaro, Elettra 53
Bredal, Kristin Halldis 149
Burns, D 473
Ceponis, Simonas 350
Clausen, Henrik 289
Corell, D D 109, 255
Cosma, L 159
Dam-Hansen, C 109, 255
Danielsson, C B 319
Deodati, A 159
Doulos, L 202
Favero, Federico 390
Flig, Katarzyna Ewa 350
Florio, Pietro 228
Fog, Henrik Schønau 350
Fotios, S 77
França, Ana Judite Galbiatti Limongi 401
Frandsen, A K 192, 542
Fujihira, Ayane 141
Garðarsson, Kristinn Bragi 350
Garrido, B 244
Gentile, N 77
Gillis, W 473
Godley, L 413
Götlind, Peter 309
Grønlund, L 192, 210
Haans, A 97
Hansen, Ellen Kathrine 171, 278, 289, 442
Hansen, Torben 267
Harild, M 192
Harimoto, K 22, 120
Hennig, Vkr 77
Hernández, S 244
Hickcox, K S 32
Huguet-Ferran, A 244
Hvass, Mette 171, 442, 463
Ingi, Dmitrii 65
Ishiwata, Y 22, 120
Jedon, R 97
Johansson, M 77
Juntunen, E 182
Kage, H 22, 120
Kaira, Otori 380
Kántor, D 244
Karaman, Gulsah Dogan 340
Karlsson, Reine 309
Karlsson, Tove 309
Karmann, Caroline 228
Kjellander, Jonas 309
Kofod, C 109
Komazawa, Ken 380
Korakidou, Veroniki 1
Kort, Y De 97
Kostro, André 228
Krause, C 109
Kyriakidou, F 413
Kyttä, Marketta 65
Larsen, Mette Hvam 442
León, O Y Vara 513
Luusua, A 182
Machado, Åsa 309
Maeda, A 373
Martinez, Luciana 53
Martiny, K 255
Mathiasen, N 192
Matos, Jéssica Cristine Da Silva Fonseca 401
Mattsson, P 493
Mavridou, T 202
Miri, Majid 130
Miura, Y 238
Miyake, H 11
Mizokami, Y 373
Mullins, Michael 267
Munk, Sara 350
Nanos, N 202
Nicolas, Houel 361
Nielsen, Sl 522

Nishihara, S ... 11
Oe, Y ... 238, 373
Øien, T B ... 542
Okada, M ... 11
Olaisen, Hilde Noreld ... 149
Pajuste, Mihkel ... 278
Palamas, George ... 41
Parker, Martin ... 65
Pavani, Taísa Dóccosse ... 401
Perednyte, Dovile ... 350
Petersen, Pm ... 255
Petrachi, E ... 159
Pickerill, H ... 473
Pihlajaniemi, H ... 182
Pinho, Paulo ... 65
Polydouli, Effie ... 1
Radetsky, L C ... 454
Rasmussen, Malte Elkær ... 350
Rathsach, Camilla Monberg ... 463
Ross, Philip R ... 424, 552
Ruohonen, Senja Maarit ... 278
Rupp, Ricardo F ... 532
Rutten, Niek ... 552
Ryhl, C ... 522
Sagawa, M ... 373
Saïg, Potard ... 361
Scarazzato, Paulo Sergio ... 401
Scartezzini, Jean-Louis ... 228
Scavée, Anastasiia ... 41
Schledermann, Kathrine ... 267
Schregle, Roland ... 380
Schüler, Andreas ... 228
Schwendinger, Leni ... 424
Sigura, Verda ... 289
Simmons, S M ... 473
Smith, A ... 32
Sternudd, C ... 77
Strandberg, Olle ... 309
Takashima, M ... 373
Teixidó, B ... 432
Thorseth, A ... 109
Thursfield, P ... 330
Toftum, Jørn ... 532
Tomková, Zuzana ... 220
Triantafyllidis, G ... 413
Triantafyllidis, Georgios ... 41, 442
Vasquez, Natalia G ... 532
Ven, R Vd ... 330
Vendramin, G ... 159
Volf, C ... 255
Wienold, Jan ... 228
Yakushina, Yana ... 484
Yamaguchi, H ... 11
Yamaguchi, Hideki ... 141
Yanagawa, Kazunori ... 141
Yoshizawa, N ... 11, 22, 120, 373
Yoshizawa, Nozomu ... 141, 380
Zielinska-Dabkowska, Km ... 413